HANDBOOK OF DAM ENGINEERING

HANDBOOK OF DAM ENGINEERING

Edited by
Alfred R. Golzé
Consulting Civil Engineer

VNR VAN NOSTRAND REINHOLD COMPANY
NEW YORK   CINCINNATI   ATLANTA   DALLAS   SAN FRANCISCO
LONDON   TORONTO   MELBOURNE

Van Nostrand Reinhold Company Regional Offices:
New York Cincinnati Atlanta Dallas San Francisco

Van Nostrand Reinhold Company International Offices:
London Toronto Melbourne

Copyright © 1977 by Litton Educational Publishing, Inc.

Library of Congress Catalog Card Number: 77-8687
ISBN: 0-442-22752-3

All rights reserved. No part of this work covered by the copyright hereon may be reproduced or used in any form or by any means—graphic, electronic, or mechanical, including photocopying, recording, taping, or information storage and retrieval systems—without permission of the publisher.

Manufactured in the United States of America

Published by Van Nostrand Reinhold Company
450 West 33rd Street, New York. N.Y. 10001

Published simultaneously in Canada by Van Nostrand Reinhold Ltd.

15 14 13 12 11 10 9 8 7 6 5 4 3 2 1

Library of Congress Cataloging in Publication Data

Main entry under title:

Handbook of dam engineering.

    Includes bibliographies and index.
    1. Dams—Design and construction—Handbooks, manuals, etc. I. Golze, Alfred R., 1905-
TC540.H28   627'.8   77-8687
ISBN 0-442-22752-3

Preface

Construction of dams in the United States for the storage of water has been slow to develop. In the early years of the nation, pumping of ground water and simple diversion of rivers were adequate to meet most of the country's water needs. By 1900 only about 1600 dams were known to exist.

The rate of dam construction continued slowly until about 1930. The expanding demands of a mid-century growing population saw a substantial increase so that by 1970 approximately 45,000 dams of all types and of all classes of ownership were in service. The rate of construction nationwide at that time fluctuated between 1500 and 2000 dams annually.

The early dams were built to store water to be used for domestic purposes such as those at the Spanish Missions in California. Later the reservoirs performed other functions such as providing flood control, pollution-free hyrdo-power, irrigation storage, and recreation lakes. Hoover Dam on the Colorado River, in service since 1935, is an excellent example of a modern multi-purpose dam and reservoir.

As new sources of water for storage dwindle and recreation and other demands are satisfied, dam construction in the United States may tend to decrease. However, a substantial corps of qualified administrators, civil, electrical and other engineers, contractors and manufacturers will be required to inspect and maintain the existing structures and reservoirs throughout their long service life. Many dams will need to be enlarged; some to be replaced.

This volume contains within its covers the complete story of dam engineering; planning, locations, selection of types, design, and construction. Related design of hydroelectric plants and ship-locks, which impinge on a dam, are given. There is a substantial discussion of environmental influences, and how to prepare for possible earthquake shocks.

As some of the contributing authors of the Handbook are employees or former employees of the U.S. Government, much of their input reflects three quarters of a century of dam building experience. This is worthy of note as is the observation that the entire art of dam engineering has, in fact, been many centuries in development. Authors have made faithful reference in their chapters to published sources of older, but still valid works of others.

The college student in dam engineering will find this Handbook a convenient aid. Its chapter by chapter reference to other texts, expanding each phase of the dam building art, should be especially valuable to a beginner in this field.

To round out the text, there is a summary on construction methods and a legal discussion (in laymen's language) of the phrasing used in plans and specifications for dam construction. For convenient reference, the draft of a Model State Law for Safety of Dams, prepared by the United States Committee on Large Dams, in included.

For engineers and other concerned persons working outside the United States, this Handbook presents a compendium of technical data on dams not found elsewhere. Every effort has been made to have in one convenient volume full information on the best and current American practice in the art of dam engineering.

It is to people everywhere who daily work with dams that this volume is dedicated.

ALFRED R. GOLZÉ

Sacramento, California

Contents

1. PLANNING AND ENVIRONMENTAL STUDIES

Marion Murphy

The Selection Factors	1
The Selection Process	1
The Conventional Approach to Water Resource Development and Management	2
The Planning Approach	3
Plan Formulations	4
The Physical Factors of Plan Formulation	5
The Economic Factors of Plan Formulation	9
Plan Formulation Procedures and Considerations	13
Risk Considerations for Reservoir Sizing with Limited Stream Flow Data	17
Investigations	18
Feasibility-Grade Investigations	20
Flood Control	29
Economic Evaluations	32
Cost Allocations	36
Selection of Dam Sites and Sizes	44
Preconstruction Planning	46
Collection of Basic Data; Quality Considerations and Standards	46
Measurement and Computation of Reservoir Areas and Volumes	56
Basic Data for Project Functions	57
Programming an Investigation	59
Environmental Investigations	61
Beneficial and Adverse Environmental Effects	72
Avoidable and Unavoidable Adverse Environmental Effects	72
Environmental Impact Alternatives	80
The Relationship Between Short-Term Environmental Uses and Long-Term Productivity	84
Irreversible and Irretrievable Commitments of Resources	90
Summarizing of Project Investigations with Respect to Dam Site and Size Selections	90
Appendix 1-A National Environmental Policy Act	91
Appendix 1-B Environmental Quality Improvement Act	94
Bibliography	96

2. HYDROLOGIC STUDIES

Gordon R. Williams

1. General	99
2. Evaluation of Records	99
3. Revision of Existing Records	100
4. Estimating Discharges of Past Floods	102
5. Extending the Period of Record	102
6. Analysis of Runoff Records	104
7. Operation and Routing Studies and Final Storage Selection	113
8. Design Floods	117
9. Analysis of Storms and Floods of Record	120
10. Development of Design Storms	125
11. Development of Design Floods	131
12. Routing of Spillway Design Flood	134
13. Freeboard Against Wave Action	137
14. Sedimentation in Reservoirs	142
References	146

3. MATERIALS SUITABLE FOR CONSTRUCTION

Edward M. Harboe and Richard W. Kramer

3-1.	Investigations	149
3-2.	Surface Exploration	151
3-3.	Subsurface Exploration	155
3-4.	Soil Classification	157
3-5.	Classification of Rocks	166
3-6.	Sampling and Logging of Materials	167
3-7.	Concrete Aggregates	173
3-8.	Materials for Embankments	178
3-9.	Environmental Restraints	185
3-10.	References	186

4. GEOLOGIC AND FOUNDATION INVESTIGATIONS AND EARTHQUAKE HAZARD

Laurence B. James, Cole R. McClure, Arthur B. Arnold, Jerome S. Nelson, Fred Crebbin, IV, F. E. Sainsbury, John W. Marlette, William A. Wahler, George B. Wallace, and George W. Housner

4.0 Geologic Investigation Introduction, *Laurence B. James*	187
4.01 The Reservoir, *Laurence B. James*	189
4.02 Location and Exploration of Sources for Construction Materials, *Cole R. McClure*	195
4.03 The Dam Site	204
4.031 Geologic Mapping, *Arthur B. Arnold*	204
4.032 Geophysical Exploration, *Jerome S. Nelson*	207

	4.033 Soil Exploration, *Fred Crebbin, IV*	214
	4.034 Exploratory Drilling—Bedrock, *F. E. Sainsbury*	222
	4.035 Sites for Appurtenant Structures, *John W. Marlette*	226
4.1	Foundation Investigation	230
	4.11 Earth Dams, *William A. Wahler*	230
	4.12 Concrete Dams, *George B. Wallace*	235
4.2	Earthquake Hazard, *George W. Housner*	248

5. SELECTION OF THE TYPE OF DAM

Dr. Ellis L. Armstrong

I.	Introduction	267
II.	Embankment Dams	270
III.	Concrete Dams	275
IV.	Composite	287

6. DESIGN OF EARTH DAMS

Reginald A. Barron

Purpose	291
General Criteria	291
Selection of Embankment Type	292
General Design Considerations	292
Field and Laboratory Investigations	296
Foundations and Abutments	298
Embankment	302
Appurtenant Structures	313
Construction and Maintenance	314
Instrumentation	316
References	318

7. DESIGN OF ROCKFILL DAMS

Karl V. Taylor

I.	Introduction	319
II.	Early Rockfill Dams with Impervious Facings	320
III.	Modern Concrete-Faced Rockfill Dams	328
IV.	Asphaltic Concrete Deck Dams	332
V.	Timber Deck Dams	336
VI.	Steel Deck Dams	337
VII.	Advantages and Disadvantages of Rockfill Dams with Impervious Facings	338
VIII.	Earth Core Rockfill Dams	339
IX.	Rockfill Dams with Cores of Materials Other than Earth	354
X.	Design and Construction Practices for Modern Rockfill Dams	355
	References	378

8. DESIGN OF CONCRETE DAMS

Merlin D. Copen, Ernest A. Lindholm, and Glenn S. Tarbox

I. General	385
II. Design Criteria	388
III. Design of Arch Dams	393
IV. Design of Gravity Dams	437
V. Design of Buttress Dams	445
VI. Dam Foundations	460
VII. Temperature Control and Joints	469
VIII. Openings in Dams	481
IX. Instrumentation	486
References	497

9. DESIGN OF SPILLWAYS AND OUTLET WORKS

Carl J. Hoffman

I. Spillways

A. General	499
B. Description of Service Spillways	505
C. Structural Design Details	519
D. Bibliography—Spillways	525

II. Outlet Works

A. General	526
B. Outlet Works Components	535
C. Structural Design Details	542
D. Bibliography—Outlet Works	548

10. PART A: DESIGN OF ADJACENT POWER PLANTS

Charles H. Fogg

Relation of Power Plants to Dams	549
Dam and Power Plant Arrangements	553
Relevant Power Plant Design Criteria	560

PART B: DESIGN OF ADJACENT LOCKS

John Mathewson

A. General	562
B. Hydraulic Design	567

C. Lock Walls and Sills	580
D. Lock Gates and Culvert Valves	588
E. Miter Gate and Culvert Valve Machinery	598
F. Electrical Systems	610
G. Lock Appurtenances	614
References	618

11. DESIGN OF RESERVOIRS

Henry L. Hansen

11-1. General Considerations	619
11-2. Structural Integrity and Adequacy	620
11-3. Relocations and Rights-of-Way	628
11-4. Associated Structural Works	636
11-5. Reservoir Clearing	649
11-6. Environmental Considerations	658
Bibliography	664

12. SPECIFICATIONS FOR DAM CONSTRUCTION

James T. Markle

12-1. Introduction	669
12-2. The Technical Provisions	671
12-3. The Standard Provisions	675
12-4. The Special Provisions	680
12-5. Summary	681
Appendix	
Table of Contents of Specifications	685
Notice to Contractors	691
Standard Provisions	694
Special Provisions	730

13. CONSTRUCTION PROCEDURES AND EQUIPMENT

Charles F. Palmetier

I. Introduction	753
II. Construction Scheduling	754
III. Construction Organization	756
IV. Construction Equipment	759
V. Foundation Excavations	762
VI. Construction Plant—Installation	763
VII. Problems In Dam Construction	765

14. PUBLIC SAFETY CONTROLS FOR DAMS AND RESERVOIRS
Alfred R. Golzé

Introduction	771
Federal Controls	773
State Controls	774
Model Law	775

INDEX 787

HANDBOOK OF
DAM ENGINEERING

Planning and Environmental Studies

MARION E. MURPHY
Civil Engineer (retired),
Division of Planning
U.S. Bureau of Reclamation
Sacramento, California

This chapter describes the various considerations, investigations, and analyses necessary to select the site and size of a dam and to specify the operational functions which affect its design. Further described are the procedures necessary or usually followed in arriving at the final site and size selections preparatory to project construction. The factors affecting these selections are usually numerous for the reason that a dam is one part of a water service or water control system which may have several functions.

THE SELECTION FACTORS

Size selection usually follows after site selection, but the two may be interdependent. The following factors, which are discussed below, are pertinent to either site or size selection, or to both:

1. The project function or functions.
2. The physical factors.
3. The economic factors.
4. The environmental considerations.
5. The social considerations.

THE SELECTION PROCESS

The procedures for selection usually involve a series of steps proceeding from a preliminary notion of a development possibility to a specific plan which conforms to all known conditions, and which appears to be the one best suited to the need from all viewpoints.

The five items above are listed in the most usual order of importance in selecting sites and sizes of dams. However, the selection process is not simply one of going through items 1 to 5, one time through, and in that order. Rather, it is a process of exploring ideas, investigating the physical factors, analyzing the economic factors, and weighing the environmental and social considerations in steps of increasing accuracy and refinement until a positive and definite plan is reached. These steps may be considered at first preliminary, then intermediate, and then final plan selections. In this chapter they are designated as reconnaissance-

1

grade, feasibility-grade, and preconstruction investigations. Each step may involve some review and reconsideration of the results of the preceding step because of new and more accurate information, or possibly because of adjustments of the plan objectives. Site locations and sizes of dams are usually established by the feasibility-grade investigation which supports and justifies an authorization for project construction, but some adjustments in sites and sizes are possible in the preconstruction investigation up to the point of making final designs for construction.

Obvious Site Selection Possibilities

There are a few locations in some river basins where the selection of a storage or diversion damsite is so obvious from outstanding physical characteristics that it can be chosen without the usual investigations. Such possibilities are increasingly rare, as most such sites have already been developed. There is also the possibility of deciding, without making extensive investigations, that a storage site should be built to the maximum size afforded by topographic and other structural limitations of the site; this may occur where there is a large water supply susceptible to storage regulation and an absence of other storage sites that could be developed for subsequent project stages. This also is a rare situation.

The Dam Site and Size Decision

Since dams are components of larger water resource development or water regulation projects, the selection of sites and sizes results from the selection and adoption of a complete project plan. The selection is the result of the studies of the selection process as herein described. These studies will identify the most favorable site or sites from among the possible alternatives. They will show recommended sizes resulting from the testing of the project plan with respect to its intended purposes and all the factors which determine the accomplishments, meet the economic and financial requirements, and satisfy the environmental and social considerations.

THE CONVENTIONAL APPROACH TO WATER RESOURCE DEVELOPMENT AND MANAGEMENT

The purpose of all water resource development is to provide improved conditions for the livelihood of man, both at the present time and in the future. The improved conditions are the combination of (1) the satisfaction of social needs, (2) economic benefits, and (3) the improvement of man's environment. The objective is to gain the greatest amount of improvement, with respect to the cost, of the three types of improvements which together make up the total result. The three types of project improvements are considered separately only as a method for arriving at the optimum result possible for the particular management opportunity being considered.

Project effects are found to occur in three categories:

1. Those quantitative in monetary terms.
2. Those quantitative in non-monetary terms.
3. Those non-quantitative, but expressed qualitatively.

The first category includes those effects which are measurable by economic evaluations — comparisons of identifiable benefits and costs. The second and third categories apply both to social and environmental effects, where only some of the effects are quantifiable.

The first part of this chapter deals with the investigations and analyses of the physical factors involved in a water development; these being the hydrologic, geologic, topographic, and biologic circumstances, and the application of engineering-type analyses. The second part describes economic evaluation procedures. These are procedures which measure and compare project accomplishments with cost in monetary terms. The last part of the chapter treats environmental and social considerations. In general, environmental and social effects are not measurable in monetary terms. Their consideration is, nevertheless, extremely important and sometimes overriding of the economic considerations.

There is no simple formula for deciding upon an optimum acceptable plan which consists of the most satisfactory combination of economic, social, and environmental values. The last paragraph of this chapter, summarizing project investigations with respect to dam site and size selections, suggests a procedure for approaching a solution.

THE PLANNING APPROACH

Dams are one of the group of important civil engineering works constructed by man for his physical, economic, and environmental betterment. This list also includes waterways, highways, bridges, pipelines, electrical transmission lines, dikes and levees, railroads, tunnels, jetties, breakwaters, docks, irrigation structures, recreational lakes, and others.

Dams as Key Features of Water Resource Developments

With a few exceptions, dams are water control features — to impound a supply of water, to divert water from a water course, or to raise the elevation of a body of water for power generation. Exceptions are dams to detain or impound a supply of water-borne sediments — the products of natural or man-caused erosion, including wastes from mining operations. Temporary water impoundments may include the control of the damaging peak flows of floods or of water having a damaging chemical quality.

In almost every water project plan or situation one or more dams are key features or important elements of a project plan. However, it is seldom that the dam is the sole or only facility. In a flood control plan, a dam and reservoir may be the only project works, but it is more likely accompanied advantageously with levees and other channel control works. In water supply — irrigation, municipal, industrial, and domestic — and in power generation, dams are one of a combination of project features needed to accomplish the desired project. In such cases, the dam cannot be justified independently of the other project facilities; it must be planned, designed, and constructed to operate efficiently and harmoniously with the combination of facilities to provide the best obtainable project results. To be economically, financially, socially, and environmentally justified it must be evaluated jointly with the group combination of project features and the total plan evaluated and judged as to its merit.

The Planning Process

The planning process for a dam should not be undertaken independently of the planning of other project features and facilities that make up a specific project plan. Planning for a dam is one part of the planning process for the total project objective. The location, size, and design of a dam will be influenced, and often controlled, by the selection of the attainable and best warranted overall project plan.

This is not to say that there should never be any investigations made of potential dams in advance of other project elements. There is an old axiom in planning investigations that it is wise to first look for and examine the possible "weakest link" — the likely most unfavorable element — in a project scheme. Conceivably, this could be the absence of a satisfactory dam site for any reason: geological, environmental, uncertainty of water supply, extreme cost, or otherwise. Such initial information may prevent wasteful or useless expenditure of planning efforts on other parts of a plan. This is only a possible precautionary first step or early step that a good planner will take in starting an investigation.

Starting the Planning Process

Generally the planning will start from a project objective, or group of objectives, or some estimated or assumed project potentialities, and proceed in steps toward devising schemes and evaluating the potentials and costs of alternative plans. In this manner the investigation and planning for dams will proceed, either concurrently or in sequential steps, along with the planning of other project features or aspects. This is discussed further in the pages that follow.

Possible water use or water control project functions using dams include the following:*

  Municipal, industrial, and domestic water supply.
  Irrigation and drainage.
  Hydroelectric power.
  Water quality improvement.
  Salinity repulsion in river deltas.
  Water-oriented recreation.
  Fishery maintenance and improvement.
  Flood control.
  Sediment retention and control.
  Navigation.

Most frequently two or more of these functions are combined as a multipurpose development with some economic or other advantage over separate, single-purpose projects.

PLAN FORMULATIONS

The Fixed and the Variable Conditions. Two generally separate sets of conditions prevail at the time investigations are made and decisions reached on the prospective construction of a new water project, or the reconstruction, enlargement or rehabilitation of an existing water development. These are: (1) the essentially fixed and unchanging physical and engineering conditions, and (2) changing economic conditions and social considerations.

The physical situations of topography and geology remain the same over a long period of time. Likewise, natural water supply conditions do not change, although with the passage of time there is opportunity for obtaining and accumulating greater amounts of reliable statistical information on the amounts and variability of water supplies.

Engineering and construction techniques do change, but at relatively slow rates. Progress occurs and likely will continue to take place in the use of higher strength materials and larger and more efficient construction machinery. Important advances have recently been made in machinery and methods for canal and pipeline construction, suggesting the possibility of further improvements along similar lines. The recent successful development and use of tunnel boring machines in favorable rock conditions gives reason for belief that tunneling costs will be appreciably reduced in the years to come. Yet, such favorable new improvements and advancements each require several years to be "proven out" in construction use, and the rate of improvement is relatively slow. The anticipated improvements of engineering and construction techniques is not a basis for making a decision of either build now or wait a few years.

The second set of considerations which is subject to substantial changes are the economic and social factors. By their nature, most new water control developments are public or partly-public sponsored and owned. This is a change from the past when a great many developments using dams were made by private enterprises for water power generation and a lesser number for water supply for mining and other industrial uses. The change appears to be a permanent one for the reasons that (1) with a few exceptions,* the most advantageous, hence

---

*It should be noted that some functions may be achieved by various means other than dams and reservoirs. This is discussed on p. 000 under "Alternative methods of flood control or protection."

*Hydroelectric pumped-storage power developments are continuing to be built for power system peaking needs.

profitable, water power sites have already been developed throughout the world, (2) most of the world's energy production is now being supplied from petroleum, natural gas, coal, and nuclear fuels from very large known supplies and high incentives for continued explorations for additional supplies, and (3) the direct and indirect values obtainable from flood control and water conservation for irrigation and other uses are so broadly diversified among beneficiaries that there seems to be no financial mechanism by which private investments can collect all fees or tolls for services rendered and thereby be attracted for prospective financial gain. Thus, sponsorship of water resource and control developments is required, with few exceptions, by public agencies which are some part of local, state, or national governments. In many instances the size and cost of a development, the diversity of its benefits, and sometimes the need of subsidy to include a slow-return function bring the project responsibility to the highest available agencies possessing the necessary financial means. These are national governments of the more affluent nations, or international lending institutions for the developing nations.

Economic and Social Influences

As seen from the above, most modern water developments are of vast public concern, and the socio-economic controls for selecting and producing a development are usually exerted by actions and policies of the higher levels of government of the country within which the project is located. A water resource development is only one of many responsibilities of any government, and the economic-social controls are subject to change with the changing goals, management, problems, fortunes, and misfortunes of the sponsoring or assisting government.

The economic and social variables can change significantly in any country within the early life expectancy of a water project. This places much greater uncertainty on a project plan than any ordinary inaccuracies of the evaluations of physical situations. In the absence of possessing the ultimate of wisdom concerning the future, the project planner will stay on firmest ground by planning for existing conditions, modified by estimated growth or changes forecast for some limited period of time. Moreover, if a completed plan evaluation does not show justification for construction, it may be set aside for a few years, then revised and updated with possible justification at the later time.

THE PHYSICAL FACTORS OF PLAN FORMULATION

Water Supply Available for Development. Except for flood control projects water is the essential commodity; in flood control projects its sudden excess is the problem. The occurrence of surface runoff — stream flows — results from weather phenomena which are understood only in general principle. To date, runoff-producing weather conditions are predictable only as seasonal probabilities, with the exception that very short periods (a few hours or days) of weather conditions can now be predicted with some reliability from worldwide weather observations. For water project planning purposes reliance is placed on the premise of recurring stream flows with future quantities and variations similar to those that have occurred in the past. Direct measurements of flows of some streams over several decades, together with measurements of precipitation over periods of a century or more at some locations, support the premise. Acknowledging that a more extreme flood or drought can occur than has actually been measured or observed in a few recent time periods, the historical measurements of stream flows are accepted as the best available forecasts of stream flow supplies for water conservation developments. Theories and principles of probability of variation of stream flow quantities are sometimes applied, but most commonly an actual record of some years of duration is used, unmodified, for calculating the water conservation accomplishment of a plan.

Stream gaging has been and is being performed extensively in the United States. The principal agency performing this work is the U. S. Geological Survey, but many other agencies also measure stream flows in their localities of interest. Records vary in length of time. Where measurements have not been made, or only a very few made, at a specific dam site it is possible to estimate or synthesize streamflow statistics at any dam site by reliable correlation methods. The same conditions and possibilities exist in many other countries.

In the industrially undeveloped countries reliable stream flow measurements and records may be very sparse. A need may exist for a water development without awaiting the time (ten, twenty, or more years) desirable for obtaining a record of stream flow quantities and variations. Here there is a basic question of how much risk is warranted in building a project either too large or too small for its water supply source. A project planner has the additional problem of judging the degree of risk involved and the consequences of taking the risk of proceeding without adequate stream flow information. Provisions for staged developments or other means of adjusting the project size and scope may be important in such a situation.

Flood Flows

Because of the enormous damage or potential damage caused by a flood of the magnitude that occurs once in a hundred years or less, stream gaging records of 10, 20, or 30 years are inadequate, although of some use, in planning flood control projects or for the spillway design of any large dam. In addition to any actual measurements of peak flood flows (usually difficult to obtain even when a stream is being gaged) the project planner uses other techniques of estimating the magnitude of floods (see Chapter 2). These include:

(1) Observations of high water marks evidenced by previous floods and computations of the probable flow from flood channel dimensions.

(2) Records of actual measurements (including duration time) of high rainfall intensities at weather stations in the watershed area above a dam site or nearest comparable location, applied to computations of runoff resulting therefrom. In the application of this method, considerations are given to known principles of precipitation as affected by storm characteristics in the region and the location and topographic shape of the watershed, and to factors affecting runoff rates including watershed vegetative cover, as well as soil and geologic structure affecting rainfall absorption rates. In colder climates the potential of flow from snow melt during a storm is also a factor.

Considering the possibility of the extremely large but extremely infrequent flood, judgment finally rests with the planner and designer as to the size of flood to be controlled; this depending on the damages or losses that would result from a failure. In some instances dams have been built with a secondary or "emergency" spillway, noting that substantial repair expense would be involved if this emergency spillway were ever used.

Locations of Project Sites

Two principal factors determine the location of water conservation project facilities: (1) the areas of water service need, and (2) location or locations of water supply available for development. The connecting link is a water conveyance facility. Where the water conveyance distance is long, or where pumping is required, the cost of conveyance is important in choosing locations for water conservation development. It is desirable, where possible, to locate the source at a higher elevation than the service area to avoid pumping costs. Obviously, there is also economy in having a water source near the place of use. However, in most river basins the most economical developments have already been made, and new project plans call for trans-basin diversions of water or conveyances for long distances, or both. Formulation of a water conservation plan usually requires that investigations and evaluations of water conveyance facilities be made jointly with the studies leading to the selection of the water development site.

The same consideration applies to selection of hydroelectric power generation sites, except that elevation is not a factor in power transmission.

In the case of water storage for flood control, or where water is released downstream from storage for purposes other than water supply, there is, of course, no conveyance cost involved unless diversions are included in combination with storage (such as flood bypasses and power diversions).

Suitability of Available Dam Sites

The saying "you can build a dam anywhere if you spend enough money" only means that some sites are extremely unsatisfactory. Obviously, a site should be in a narrow section of a stream channel and where both abutments have sufficient height for the need. The foundation, including abutments, should be of rock or consolidated materials sufficiently strong to support the structure and they must be watertight or so nearly so that excess leakage can be prevented by sealing any cracks or fissures in the foundation with a grouting material or closing the leakage paths by placing a blanket of impervious material in the reservoir area upstream from the dam site. Unless a concrete dam can be constructed there must be an adequate site for a spillway over or through one of the abutments.

Unless a dam is so small that it can be constructed in a single dry season there must be a means of bypassing stream flows during construction usually through a tunnel in one abutment that can be closed after construction is completed.

Adequate geological inspections of foundation, abutments, spillway, and bypass tunnel sites are necessary to provide assurance of suitability. This requires subsurface explorations and tests, normally performed by diamond-drilling and extraction of core samples of the foundation rock. Such explorations may be supplemented by excavating exploratory tunnels or pits into the abutments or by trenching through loose overlying material. Rock leakage tests are required where rock fissures are found; these tests are performed in the process of drilling core samples. The amount of such exploration depends upon the height and type of dam being considered and the geologic conditions found.

Obviously, a damsite is to be avoided if it is on or very close to a known active earthquake fault.

Chapter 4 discusses in detail the geologic explorations necessary to evaluate a dam site.

Since a dam is a massive structure it is necessary to locate an adequate supply of construction materials within economical hauling distances of the dam site. Depending upon the type of dam, these materials may include aggregates for concrete manufacture, impervious earth materials, pervious materials, and rock for rock fills or riprap.

Suitability of Reservoir Sites

To obtain economical storage capacity a reservoir site should be wide in comparison to the dam site and should be on a stream having a low or gentle gradient to obtain a long reservoir in proportion to the height of the dam. Geological considerations generally require that the site not be on formations that leak excessively, and that there not be any risk of large landslides, rockslides, or rock falls into the reservoir. Consideration should be given to any important mineral deposits in the area which may be of commercial value either at present or at some future time. The mining of some types of mineral deposits above a reservoir site may have accelerated leaching and solution of chemicals which can be concentrated in reservoir storage with objectionable effects.

Many otherwise attractive reservoir sites are on valuable land being used for other purposes: agriculture, forestry, and habitation by people. There may be important roads, railroads, pipelines, and transmission lines through a site. The necessary relocation of such facilities and inhabitants and the loss of the land use may be a major and overriding cost of a specific dam and reservoir plan. Adequate plan formulation requires that consideration be given and comparisons made of all reasonable alternative dam and reservoir sites before selecting the project plan.

Physical Size Limitations

Topographic or geological conditions frequently impose a practical or safe limit on the height of a dam. The quality and quantity of available construction materials may set the limit which a designer will not exceed for safety and stability of the structure. The continuing worldwide experiences in construction and operation of high dams provide guidelines on practical maximum heights. Among the considerations are the problems of constructing and maintaining spillways and outlet control valves operating to dissipate the energy of high-head water discharges.

Practical physical size limitations of reservoirs may be the limiting elevations of the reservoir rim or the risk of land or rock slides if induced by higher storage levels. The other limitations are essentially economic and social ones — increased displacement of people, relocation costs, and loss of land use. There is also the water supply limitation. A reservoir might be constructed so large that it would never be filled while making ordinary and usual water withdrawals for the project functions.

Sedimentation Rates

Every stream carries sediment; every reservoir retains sediment. In time, every on-stream reservoir will fill with sediment and lose its water-storage function. No method has yet been discovered to economically remove and dispose of sediment accumulated in a large on-stream reservoir; the costs considerably exceed those of constructing a replacement storage facility.

Sediment transportation rates vary greatly in different streams, depending on the rates and volumes of stream flows but more importantly on the erosion conditions of the watershed. The erosion conditions vary with the vegetative cover of the watershed and with the type of soil which in turn results from the types of parent rock from which the soils are produced. Except on flat or very gentle land slopes, or areas of high water absorption, erosion takes place with all overland flow, but a great deal occurs in the concentrations of flow — in gullies, tributary channels, and main streams with cutting banks and channels. These are some of the processes which have created rough mountain terrains. Watersheds which have creeping soils or landslides are heavy contributors of sediment as the earth slides continue to feed erosion material into watercourses.

Stream-borne sediments occur in two types — suspended sediments and bed load. Suspended sediments settle slowly in still water and are considerably dispersed over a reservoir bed, tending to concentrate in the deepest and stillest parts of a reservoir, hence near the dam. Some of the lightest sediments may be carried completely through a reservoir and the outlets of the dam, sometimes maintaining persistent conditions of turbidity in project water releases. This happens more frequently where the reservoir capacity is small in proportion to its inflows. The coarser bed load sediments are carried by the higher velocity flows and are trapped in a reservoir near the stream entry points. One concern about such deposits is that, in addition to reducing reservoir volume, they may cause a stream channel to aggrade just above the point where it enters the reservoir, possibly requiring extra clearance to be built on a bridge or other structure at that site.

Suspended sediment is easy to measure by sampling and is now frequently done at various stream gaging stations. Dependable measurements of bed load movement are difficult. The most dependable measurements of sediment are made by periodic re-surveys of a reservoir bed. Provisions for this are sometimes made by establishing permanent range lines across portions of a reservoir. Where existing reservoirs and sediment surveys have been made, it is possible to use the information as a basis of estimating sedimentation rates of a new reservoir if located in a watershed having similar erosion characteristics. Other than this and by use of measurements of the suspended sediment fraction of the total sediment load, a hydrologist must rely on information of watershed characteristics in estimating prospective reservoir sedimentation.

With respect to plan selection and project designs the following guidelines apply:

1. Where there is a choice, avoid constructing storage on a stream having large sediment loads.
2. Provide, as part of the gross reservoir volume, space for the accumulation of sediment in estimated amount at least equal to the intended economic life of the reservoir — usually 100 years or more.
3. Design the reservoir outlet at the dam at least as high as the expected silt accumulation level at the point (the outlet may be higher if the plan calls for maintaining a higher minimum reservoir pool for any purpose).
4. Provide outlets at more than one level if cost is not excessive.
5. Consider, in the project plan, the possibility and the suitability of sites for constructing sediment detention basins short distances above the reservoir.
6. Consider the merits, possible methods, and costs of providing watershed protection and management against accelerated erosion; this is frequently caused by poor land use practices, poor logging practices in timbered areas, or road building without regard to drainage consequences.

THE ECONOMIC FACTORS OF PLAN FORMULATION

Project Benefits. Many enterprises involving capital investments are motivated by financial profit — the expectancy that monetary returns will exceed costs, and that the financial rate of return on investment will exceed that of some alternative use of available investment capital. In the water resources field private enterprise has undertaken a number of single-purpose hydroelectric power developments. More rarely there have been privately owned water service companies. In some cases regulated water from a hydroelectric development has been marketed by private enterprise as a by-product of the power development. Water developments for financial profit have been most attractive where there is least difficulty in collecting payments, as in the case of electric service to consumers.

The values of water resource development for irrigation and stream flow regulation for various purposes, and for flood control are widely dispersed; hence, investments are not altogether justified by the financial return. Thus the principal justification of a water project is most usually based on its benefits. For comparability to investment and operational costs, benefits are stated insofar as possible in monetary values.

There are inherent difficulties and uncertainties in measuring some types of project benefits. Particularly with irrigation, the increased economic benefits of agriculture extend beyond the farmer through a chain of associated activities to the consumer. This involves analyses of direct and indirect benefits; or primary and secondary benefits; or local, regional, and national benefits. Planning decisions may be based on any of the benefit criteria; sometimes, for conservatism, only the direct benefits are used — these being the increased net farm income made possible by irrigation water service.

Simple and generally accepted methods exist and are commonly used for measuring and estimating the benefits of municipal-industrial water service and hydropower generation. In both cases the benefit is considered to be the alternative cost of producing the same thing by the most reasonable available alternative means.* In the case of a municipal-industrial water supply consideration must be given to an alternative supply of equal quality or to adjust the cost of treating an inferior-quality supply to make it equal the intended project supply. In the case of hydropower generation the benefits might be taken as the lesser of: (1) the cost of an alternative equivalent fuel-driven power plant (fossil or nuclear), or (2) the power benefit of supplying a new power-using industry such as the fixation of nitrogen for fertilizer production.

Flood control benefits result principally from the reduction of flood damages. Depending upon the location and types of flood-exposed areas, damages may occur in any or all of the following categories: (1) agricultural, (2) residential, (3) commercial, (4) industrial, (5) utility, (6) public facility losses, and (7) emergency costs. Each category may include different types of losses:

1. Agricultural — Damage to crops, agricultural lands, and agricultural facilities.
2. Residential — Damage to structures, contents, outside improvements, and income losses to residential area inhabitants.
3. Commercial — Damage to structures, fixtures, inventory, loss of business, and to outside improvements, including parking areas.
4. Industrial — Damages to industrial establishments, similar to item 3 above.
5. Utility — Damage to power, telephone, gas, water, and railroad facilities; to oil and gas transportation lines; and to lost revenue due to curtailed or interrupted service as a result of flooding.
6. Public facility — Damage to roads and bridges, levee systems, canals, stream channels, municipal facilities, public schools, and public buildings, all of which are owned or administered by public agencies or non-profit political and semi-political organizations.
7. Emergency costs — Expenditures for flood fighting, repairing flood control works, cleanup operations, caring for evacuated people, and losses to the traveling public resulting from damaged highways and bridges.

In addition to the flood control benefits of eliminating or reducing the damages listed above there may, in some instances, be another flood control project benefit; this being an increase in the values of land protected. An increase in land value may result from the opportunity to make greater use of the land protected. For example, land subject to periodic inundation by floods may be usable only for growing rice or other crops adaptable to occasional flooding; but when protected from floods the same land may be useful for orchards or other high-value cropping. Or it may have still greater value for industrial or urban use when protected against flooding.

In estimating the economic benefits of flood control, projections are made of future conditions of potential flood damage that may be prevented. Projected future potentials are usually somewhat larger than those existing under present circumstances, due to the tendency to make greater use of flood-exposed areas with increase in population. Where future flood damage and flood control benefits are estimated to be greater than for present conditions the future values are discounted to present worth by an appropriate discount factor used in comparing the benefits with costs or alternatives.

---

*In cases where no reasonable alternative source exists, or is in question, benefits may be estimated as the "vendable" (market) value of the product.

Alternatives to flood control by water control methods may consist of flood-proofing the existing structures and/or vacating flood-exposed areas by relocating people together with their possessions to less hazardous areas.

In addition to the economic benefits of flood control there are also social values. The social benefits not possible of economic measurement include protection against loss of life by floods and other consequences.

The usual approach in estimating flood control benefits is to project the magnitude of damage-causing floods and their frequency of occurrence, the damage produced by them, and to compute the damage in monetary terms. Figure 1-1 illustrates the three relationships that are developed. Flood frequency is the probability of occurrence of floods of different magnitudes within a selected long period of analysis. Flood magnitude may be expressed in rates of flow or flood stage levels, or both, according to the data employed in relating flood damages to magnitudes. Extensive collection of data and analyses of the different types of preventable flood damage are necessary for each specific area subject to flooding to estimate flood damages and the benefits of a flood control plan. Since a project adequate to protect against floods may be infeasibly expensive, a flood benefit evaluation is completed by estimating the occurrences, magnitudes, and damages to be expected with a project. The difference between the "with" and "without" damage costs is the project flood control benefit.

The measurement of recreational benefits is, at best, imprecise. A recreational benefit is the amount of money people are willing to pay for admission to a recreation area produced by a project, after adjustment for costs of any amenities that may be provided to enhance the recreational area. This is usually not measurable as people do not accept the commercialization of outdoor, public-project-derived recreation. At the other extreme, the recreational benefits may be claimed as the total money people spend to use the recreational areas. This includes the cost of travel to and from the recreation site and the purchase or rental of all supplies and equipment used in the event (plus admission fee, if any). This has the inherent flaw of assuming that all persons producing the supplies, equipment, and travel facilities used by the recreationist would be otherwise unemployed.

Another approach sometimes used is the alternative cost method — benefits equaling the cost of providing the most inexpensive single-purpose water based recreational opportunity of similar location and equal potential recreational use. But in many instances, this

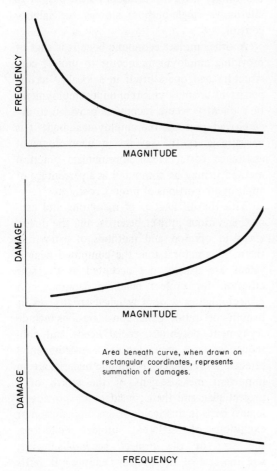

Figure 1-1. Flood frequency-magnitude-damage characteristics.

may produce an unacceptably high figure in comparison to other methods. Recreational benefits, expressed on a per-visitor-day use basis, are usually a consensus of viewpoints derived from the methods described. In multipurpose project planning they are often small in comparison to other benefits and usually do not exert a strong influence on overall plan formulation, primarily because costs for recreation facilities are comparatively large.

Other possible project benefits may include fishery enhancement, water quality improvement, navigation improvement on large rivers, and salinity control where a large river enters the ocean or other body of salt water through delta areas and tidal channels.

Where the fishery is commercialized a project benefit is measurable as the market value of the increased production due to the project, with adjustments, if appropriate, for differences in production expense. Benefits of sport fishery are similar in nature and relative magnitude to recreational benefits. They are measured as a value per fisherman-day of use.

Water quality improvement may occur in downstream flows from storage releases during dry seasons. Existing water quality may be poor due to stream pollution. Although the effort at present is to require polluters to discontinue disposal of pollutants in streams this may not be completely possible. Particularly where a downstream city re-diverts water a benefit of water quality improvement would be the reduced cost of water treatment necessary for use. Other measures of water quality benefit might be increase in net return due to higher crop yield or, if a more suitable measure is not available, alternative costs of water quality control storage.

Navigation (boat or barge transportation) on large rivers may be improved by project releases during low-flow seasons. The benefit may be measured as the savings in cost of transportation over pre-project conditions or by the cost of the least expensive alternative means of providing equal transportation. Recreational (small boat) navigational benefits may be evaluated as savings in cost or in a manner similar to other recreational benefits.

Salinity control may be accomplished by releases of project stored water in low-flow seasons to repulse unwanted intrusion of saline water. The benefit accrues to water users located near the salinity intrusion channels and areas. The benefits can be measured by the increase of crop production and values by improved irrigation water quality, the cost of alternatively constructing and operating water conveyance conduits sufficient to convey water supplies from an upstream source to users in the salinity intrusion areas, or by the cost of an alternative single-purpose storage for salinity control.

Another project economic benefit is that of providing employment income to project constructors and operators if in a locality having persistent unemployment or under employment. In a country where income is provided to unemployed persons, the employment benefit is measurable as the savings in unemployment assistance costs; as an approximate practical method it may be computed as a percentage of appropriate portions of project cost.

With the difficulties of measuring and estimating various project benefits, and the differences in opinion and methods of estimating them, it is natural that the computed benefit values are usually not accepted as the sole criterion for project planning decisions. A project plan is seldom weighed strictly on its benefit-cost ratio. Other considerations include repayment potentials, social needs and consequences, and, most recently, environmental effects. Nevertheless, economic benefits are an important measurement of the worth of a project plan and their consideration provides a logical basis in making decisions on inclusion or exclusion of different project functions, the size of the project accomplishments, and the size of the project facilities. Benefits are also a factor in determining priority of construction.

The following discussion will describe plan formulation procedures based on the use of project benefits.

Project Costs. The costs consist of two components: (1) capital or investment costs necessary to construct the project facilities, and (2) annual or recurring costs of the operation, maintenance, and replacement. Capital costs include interest on the money invested during construction and up to the start of project operation. They also include the costs of planning investigations and project designs. They include the cost of acquiring all rights necessary for project construction such as the cost of acquiring rights to the use of water, water treaties, and litigations.

For project plan formulation, costs are estimated from design of project features and other known cost variables to a degree of accuracy consistent with benefit estimate accuracies for comparisons of plan alternatives. This level of cost estimate accuracy is sometimes described as "reconnaissance-grade (or preliminary) cost estimates." A number of cost estimates of alternative project features and sizes are required for making a dependable plan selection. After a plan has been well identified, cost estimates are refined for improved accuracy to determine financial feasibility and to support a decision on project justification. The latter are sometimes termed "feasibility-grade cost estimates," "contract estimates" or "design quality estimates."

PLAN FORMULATION PROCEDURES AND CONSIDERATIONS

Prime Function of the Development. A plan begins with some definite notion of its principal purpose. This may be flood control, irrigation service, power development, or any other of the possible water project functions. Usually there is also a rough idea of the needed size. Occasionally, there may be two or three functions of equal importance. It is desirable to start a project investigation on a plan to serve the principal function, or, at most, two or three important functions. A first trial size may be chosen arbitrarily or by an early evaluation of the areas or boundaries of the intended service. An evaluation of this first plan, usually on a preliminary or reconnaissance basis, serves to indicate whether there are prospects of feasibility. This initial plan, sometimes termed a "core project," also serves as a base of comparative studies for enlarging or reducing the size, and for adding or deleting other possible project functions.

Multiple-Function Considerations

It is almost always more economical and efficient to combine functions in a single development than to build a separate project for each

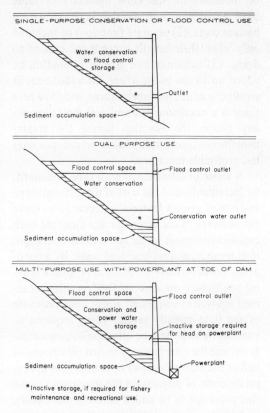

Figure 1-2. Reservoir storage space use – single and multiple functions.

purpose. A single dam and reservoir can be used for any and all functions discussed under "Project benefits," if reservoir storage space is properly allocated and some necessary facilities are included in the design of the dam (see Fig. 1-2). A series of studies is necessary to adequately formulate a multipurpose project plan, considering the different project functions, the most favorable sizes of such functions, and the resultant total project size.

Plan Formulation Analyses

From the economic standpoint the best project plan is attained by obtaining the greatest excess of combined project benefits over the total project cost. Application of this principle of analysis requires that (1) all project benefits be measured on the same monetary or value basis as others, and on the same monetary basis as costs, (2) separate functions be included only when their benefit exceeds the cost of so doing, (3) increments of size of a function be added up to the point where the added benefit equals the added incremental cost, and (4) where there is a maximum size and cost limitation for any reason the function having the greater benefit-minus-cost value takes precedence over less profitable functions.

A series of estimates of each type of benefit, by increments, and the cost of the different type and size of project features is necessary to make the analyses. Cost estimates must include both capital investment costs and annual operation, maintenance, and replacement costs. By amortization principles and use of the appropriate interest rate these costs are combined on either an annual equivalent or present worth basis. Because of the time (usually several years) required to construct a large dam or other major project features, and the time required to first fill a reservoir before the start of project operation, the appropriate costs of interest on the investment up to this point are to be added to investment costs.

For size analyses it is usually desirable to estimate the cost of a dam and the other major

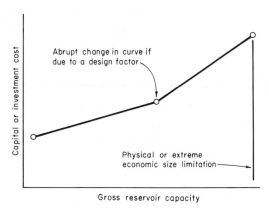

Figure 1-3. Dam and reservoir cost-capacity curve.

project structures for at least three sizes: the smallest size likely to be considered, the largest size (possibly indicated by some physical limitation), and an intermediate size. The intermediate size may be chosen because of an indication of some physical factor that might affect the size-cost trend or relationship. Size-cost curves may then be drawn and the cost of any size read from the curves (see Fig. 1-3).

Chapter 2, "Hydrologic Studies," presents a discussion on the relationship of reservoir size to its water regulation accomplishment. Examples of reservoir yield computations are presented in various publications, including: (1) "Economics of Water Resources Planning" by James and Lee, Chapter 12, "Water Supply," paragraph heading "Developing the Supply," McGraw-Hill, 1971; and (2) United Nations Water Resources Series bulletin No. 41, "Water Resource Project Planning," 1972, Figure IV-4 and Table IV-7 and related discussions.

Size-benefit curves should also be similarly constructed and used because accomplishments also vary with the size of project features, and on a non-linear basis particularly with regard to reservoir storage capacity for both water conservation and flood control (see Fig. 1-4).

Using the above described data a series of analyses will be made to indicate the optimum economic combination of possible different project functions and their sizes and costs. The

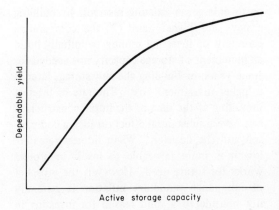

Figure 1-4. Reservoir yield-capacity curve.

analyses may be few in number and simple when only one or two functions are being considered, or they may be numerous when complicated by several functions and by alternative plan possibilities. Some mathematically specialized systems analysis methods have been introduced and used on trial bases for the more complex analyses.

It is rather common that size and function selection analyses will not disclose precise points of optimization (see Fig. 1-5). An optimization curve may be nearly flat over a considerable range of sizes. Here a competent planner makes a series of tests and judgments. First, he will take into consideration the fact that neither the estimates of costs or benefits with which he is working are completely accurate. He will also consider, and possibly test, how much the plan selection would be affected if possible percentages of error exist in any of the estimated data. If one or more estimates appear critical he will request a review and possible refinement of the particular estimate for more reliability. Second, he will consider possible non-economic factors. It may be that the indicated project size is so large that financial constraints would be controlling, or that the amount of funds required for the initial investment would be of doubtful availability. To attract investment it may be desirable to adjust the size toward that of a maxi-

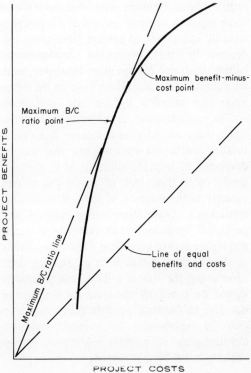

Figure 1-5. Benefit-cost relationships.

mum benefit/cost ratio; usually a smaller size than for the benefits-minus-cost maximum (see Fig. 1-5). Other considerations are the preemption of a dam site at less than its full potential, and the possible merits and opportunities for staged developments.

Pre-empting a Dam Site

Favorable dam sites are frequently limited in number and availability. On a stream system where such is the case there is the possibility of using a site for a dam of some limited size and preventing or restricting any opportunity to make use of the same site for a larger dam that would have been justified under new conditions. This situation presents an argument for initially building a dam to its maximum physical limitation, or of a size at the upper end of the range of possible means of economic and financial justification. Planners who feel that future water resource or flood control development is virtually impossible to predict (sometimes evidenced by past undersized developments) believe that any storage dam in a river basin with limited sites should be built to the maximum size afforded by the site limitations.

The consideration is of least importance when there are additional storage site opportunities within the stream basin that can be developed if needed, although at a probable greater cost than that of additional storage capacity in the reservoir of the dam under consideration. Quite often, topographical and geological conditions exist that enable a new and higher dam to be built at a site immediately or shortly downstream from an existing dam. In selecting the site of a dam, this possibility should be noted by choosing the location of the dam to facilitate possible future construction for replacement by a larger sturcture. Consideration of possible additional storage capacity in the basin is covered below.

Staged Development Possibilities

With few exceptions, the regional needs for water services, power, and flood control continue to grow. This is assured by the growing populations in nearly all parts of the world and their use of water and of land subject to flooding. Long-range projections of such needs are under way or have been made in the United States and elsewhere for planning purposes. Obviously, projects should be built as the need arises; it is uneconomical to build large and costly works far in advance of their need. In order to save costs, planning and construction of projects can be done in sequential steps or with deferred stages of construction, to the extent that physical and design conditions permit.

With respect to water storage functions, there are two principal considerations: (1) the physical opportunities that exist within a basin or water system to construct additional storage capacity by means of new dams and reservoirs, or to enlarge an existing reservoir by building a new and higher dam; (2) the comparative economy of investing money to initially build an increment of storage capacity not needed for some years, or building the new storage later at a higher investment cost. The rate of interest prevailing at the time of the initial construction can have a substantial effect on such a comparison and the decision. When interest rates are low it is more favorable to invest in project works for future need. However, the prospect of long-range monetary inflation is generally not considered a valid reason for investing in advance of need, as benefit values and the ability to repay will tend to escalate proportionally with costs.

Provisions for Future Modifications

Raising an existing dam to increase its reservoir capacity is often considered, but rarely found to be economically justified. Large additional investments are required if the original dam is designed and built with this in mind. The physical conditions at the site have some effect on the cost of raising a dam. A major factor is the economic cost or loss incurred by temporarily stopping or reducing water service from the reservoir during a reconstruction period.

One way of increasing the reservoir capacity of an existing dam may be to design a dam spillway for control gates which add reservoir storage capacity, but to omit the gates until they are needed. The possible savings is small in proportion to the total dam and reservoir cost.

In some instances, provisions for foreseeable project enlargements or modifications are important and worthy of inclusion in the design and construction of a dam. The most usual case is one of providing extra outlet capacity through the dam or its abutment for the possible need of increasing rates of water release. If power generation is an initial or potential function, the extra outlet provision is for pressurized penstock connections. Any such provisions are relatively inexpensive during the initial construction but very expensive as a modification of an existing dam in operation. Examples of the need of larger outlets are: (1) when additional storage capacity will be constructed upstream and the larger quantities of regulated water supplies routed downstream through the existing dam en route to places of water service, and (2) where a hydropower plant is located below the existing dam. With respect to the latter, there are definite inherent factors in power system loads, load growths, and allocations of loads to power plants that cause, with passing of time, peaking power generation to be the more valuable function of a power plant operated from reservoir storage. For instance, if a hydroplant is initially operated to meet a 30 percent plant load factor, but future operation favors a 15 percent factor, the hydraulic capacity of plant and penstocks is doubled.

Environmental Factors in Plan Formulation

In plan formulation it is important to take into account the environmental effects of alternative plans, and occasionally of the size and use of the facilities.

Environmental conditions for people usually are vastly improved as a whole, in the water service and flood protection areas. At the sites of water development and control the effects can be negative. Among the environmental factors usually affected by water control developments are: (1) natural scenic values, (2) wildlife (both land and aquatic species), (3) preservation of endangered species, and (4) opportunities for enjoyment of outdoor solitude.

No one has discovered, or yet agreed on, how to measure these environmental values. A project planner can only be aware of their existence within the affected area and give consideration to alternatives having more favorable or less unfavorable effects. It may even be desirable to produce two alternative project plans — one providing a high degree of environmental protection and the other being the most economically favorable.

Some kind of accommodation is usually sought and reached. This takes such forms as establishing rates of stream flows to be maintained, acquisition of land to be used as wildlife tracts, construction and operation of a fish hatchery, location of new or relocated roads and structures to the advantage of scenic views and viewing sites, vegetative plantings, and providing desirable public access to areas having enjoyment potentials.

Project Beautification

Minor expenditures not specifically fixed are considered justified for the attractiveness of project works. This includes the avoidance or treatment of unsightly scars resulting from construction. This is ordinarily an insignificant factor in plan formulation but, combined with other factors, it has some importance in the selection of alternatives.

RISK CONSIDERATIONS FOR RESERVOIR SIZING WITH LIMITED STREAM FLOW DATA

A planner may encounter the problem of either recommending the size of a storage reservoir based on an inadequate amount of stream flow data, or advising that the size selection and project construction be deferred until a more adequate amount of runoff data can be obtained by a continuation of stream gaging. An important consideration is the amount of risk that is warranted in selecting a reservoir size somewhat too large or small for its stream flow supply, or over or underestimating project accomplishments for this reason. A considerable error in estimating long-term stream flow quantities and variations may result from the use of insufficient amounts of historical data, and as a result, the reservoir may be sized too large or small for the runoff available for project development. In either event the project has less economic merit than could have been attained with optimum reservoir sizing.

The planner is faced with the following questions:

1. Should the reservoir size selection and the sizing of other features of a project plan be deferred long enough to obtain a longer and more representative amount of runoff data? If so, for how long?
2. Should a short period of runoff measurements (supplemented by limited amounts of estimates possible by correlation procedures) be taken as representative of long-term quantities, or should some factor of conservatism be applied to reduce the risk of the project being oversized?

Factors bearing on the answers to these questions are: (1) the urgency or value of immediate construction of the project, or (2) the marginal nature of the project if it so appears from estimated costs and a conservative estimate of the project water supply. In the first instance, a large net value is indicated (either monetary or as benefits) to be realized by the earliest possible construction of the project. The large margin of potential benefits favors taking a risk of either oversizing or undersizing the reservoir and other facilities. There is more to be gained from the early years of project revenues (or benefits) than the cost of continuing the stream gaging program longer and thus possibly achieve a more efficient reservoir size and project design. A further consideration in making a decision on this basis may be whether there is opportunity to construct additional storage if found desirable, or to modify the project plan to gain some use of an oversized reservoir if it is later found that the reservoir is oversized.

In the second instance, where a project plan may be of marginal feasibility even with optimum reservoir sizing, it is desirable to postpone construction until a longer record of stream flows is obtained. Such deferment and collection of additional runoff data reduces the risk of oversizing or undersizing the plan and causing it to be a financial or economic liability. It is almost axiomatic that the more marginal a project plan, the more time and effort necessary to design it efficiently for best possible justification.

A hydrologist can be important and valuable in judging the degree to which a short period of runoff record is representative of long-term average conditions, or the amount and direction by which a short record of observations likely differs from the mean condition. There are data available indicating whether recently observed runoff of a specific stream was above or below normal. However, a hydrologist will point out the degree of error that may occur in attempting to estimate long-term runoff quantities from a short measurement period and the available related factors at his disposal.

INVESTIGATIONS

Reconnaisance-Grade Investigations

It is frequently desirable to make project investigations in two separate and distinct steps: reconnaissance grade and feasibility grade. The principal purpose of the reconnaissance, or preliminary study, is to decide whether or when the considerably more expensive and time-consuming amounts of work are justified to fully assess and test the prospective feasibility of a project plan. Other purposes are to screen out poorer alternatives, of which there may be several, and to identify the types and amounts of basic data that are insufficient or lacking for making feasibility-grade investigations.

The reconnaissance investigations will identify needed data that may be expensive and may require considerable time to obtain. Stream flow records are an example. When lacking, there is the need to start a stream gaging program promptly; then wait a necessary time — often a few years — to accumulate adequate records of the variable flows to support a feasibility determination. Another example may be topographic mapping, which also takes time to obtain; partly because of seasonal climatic and weather conditions limiting the necessary field work, and often because of budgetary limitations or delays.

A reconnaissance investigation will serve to identify the probably scope of a project plan, both as to geographic locations, numbers and types of project functions to be considered, and to show some indication of the magnitudes, and approximate sizes of structures. It should also disclose any major problem areas likely to be encountered. It will facilitate preparing a good estimate of the time and costs of conducting a subsequent feasibility investigation.

Basin Surveys. A basin survey is a specific type of a reconnaissance investigation performed to assess the water resource development potentials and needs of an entire river basin or some geographical area having well-defined boundaries of water supply and potential areas of water use. Usually the basin survey boundary is the watershed boundary of a river basin. However, in flatland areas near the mouths of rivers the watershed boundary is not necessarily a logical boundary of potential service areas, and a basin survey boundary may be adjusted accordingly. The basin survey is intended to insure that all alternative and sequential development opportunities are given adequate attention in the reconnaissance stage of investigations.

Procedures for Reconnaissance Investigations

A complete reconnaissance investigation will be a preliminary version of a feasibility investigation; taking into account all the physical, engineering, economic, social, and environmental factors, as discussed above, but with less accuracy. It is made with available data, supplemented where most important with limited collection of new data, and by preliminary types of surveys. Simple cross sections of a stream at a dam site might be surveyed and used in the absence of detailed dam site topography. Surface inspection of geological conditions at dam sites are made. Subsurface explorations are rarely omitted and only if there are no indications of foundation problems. Ordinarily, some site drilling is made to detect and evaluate adverse conditions. Availability of construction materials is noted as to probable quantities and quality. Where the quality is in serious question, a few samples are taken and tested. Designs and cost estimates are made by "short-cut" methods; these being from curves, tables, and similar information or experience available to designers and cost estimators.

Project accomplishments and their benefits are likewise estimated to the needed accuracy by short-cut methods. For example, the safe yield of a reservoir at various capacities may be approximated by a graphical method rather than by long computations of operation studies which include refinements.

A satisfactory reconnaissance investigation can be made in a rather short time and with a limited amount of work, providing the result clearly shows high prospect of a feasible project plan, or alternatively, shows very poor prospects. Such a result serves the basic objective of being able to reach a decision of whether to proceed with a feasibility investigation. If a reconnaissance study is completed in this brief form and indicates prospective project feasibility, it is then necessary to perform most of the plan formulation work in the beginning phases of the feasibility investigation. From the budgetary standpoint this may be favorable, as adequate planning funds may be better available for the feasibility investigation. However, it is unlikely that a dependable decision can be made from very brief examinations. Few project plans not already constructed have a high ratio of feasibility; more often they prove to be marginal. When this is disclosed the decision for proceeding requires well-founded support.

Examination of Alternatives

In one part of an investigation – either the reconnaissance or feasibility stage – there is need to make extensive examinations of plan alternatives; particularly major alternatives. There is a tendency to initiate and pursue an investigation with ideas somewhat erroneously fixed on the plan. This may come about from a variety of causes; one of which may be the planner's desire to "get on with the job." It is possible to proceed a long way into formulation and evaluation of a specific plan only to later discover that a better alternative had been overlooked. All reasonably possible plan alternatives need to be compared. Unless there is good reason to postpone this to the start of a feasibility investigation it should be done in the reconnaissance investigation.

Along with the examinations and comparison of major plan alternatives the reconnaissance investigation is a favorable time to take note of opportunities for staged developments (see p. 16). Here again, the objective is not to definitely decide on inclusion or exclusion of staged developments, but to indicate the possible merits of so doing and to advise how much further examination appears warranted in a subsequent feasibility investigation.

Summary

A useful reconnaissance planning report will contain:

1. Enumeration of the project functions considered
   a. By service area locations
   b. By approximate magnitudes
2. Description of all plan alternatives considered
   a. Locations and types of project facilities
   b. Approximate sizes
   c. Foreseeable problems; examples:
      (1) Environmental
      (2) Water rights settlements
      (3) Fishery
      (4) Other
3. Evaluation of alternatives examined
   a. Not worthy of further attention
   b. Justifying further investigation for plan formulation
4. Recommendation on further investigation
   a. Proceed promptly, or
   b. Temporarily delay to obtain more basic data (stream gaging or other), or
   c. Suspend indefinitely
5. Inventory and estimate of work required to produce
   a. Feasibility investigation (if recommended
   b. Analyses
      (1) Plan selection and sizing
      (2) Designs and cost estimates
      (3) Economic and financial evaluation
   c. Environmental study and report
   d. Participation and review by concerned agencies
   e. Report preparation
6. Estimate of cost and time schedule for a feasibility investigation

FEASIBILITY-GRADE INVESTIGATIONS

Purpose

The purpose of a feasibility-grade investigation is to determine and demonstrate the soundness and justification of a proposal for construction and operation of a specific project from the standpoint of objectives, accomplishments, benefits, costs, economics, and social and environmental considerations. Alternatively, the investigation should show reasons for lack of justification, if such proves to be the case. The analyses and determinations need to be of such accuracy and authenticity as to support without serious question, the reliability of the findings and the proposal derived from them.

The report of a feasibility investigation provides the results of the findings to all persons responsible or concerned with making the decision to proceed with the development. These persons include members of the sponsoring organization, which may be a governmental agency or otherwise; the agency or organization which will operate and manage the project development; the organization — a governmental body or financial institution — that will finance the construction of the project; and all other agencies and organizations that will be importantly affected by the development. The latter include governmental agencies that have been given jurisdiciton over management and utilization of public resources affected by the development.

The feasibility investigation is not required to be the final end product and the termination of project planning. Relatively minor changes are almost always found to be necessary or advantageous for various reasons during final designs preceding construction, during construction, and during project operation. Thus, the feasibility investigation is not the "last word" in planning but it does provide the basis for making the major decision on commitment of investment and resources for the project development.

Investigation Procedures — Plan Formulation

Unless the feasibility investigation has been preceded by a recently completed, very thorough and high-grade reconnaissance investigation it will be necessary to review and further examine and evaluate the alternative plan possibilities along the lines and for the purposes indicated earlier in the discussion on reconnaissance investigations and plan formulation. Subject to changes brought into consideration from new basic data yet to be acquired, the potential project goals, boundaries, and principal facilities are thereby confirmed or redefined.

Basic Data

Basic data of many categories must be collected or updated to accuracy and dependability levels consistent to the degree of accuracy required for feasibility justification. Desirable or necessary quality standards of various basic data are described under "Basic Data: Quality Considerations and Standards," page 46. For purposes of this handbook it is not necessary to list all the types of data pertinent to service areas and functions, water conveyance and power transmission to service areas. It should be recognized that such data must be collected as well as those pertaining directly to dams.

The basic data for dams include stream flow and design flood information, topographic surveys of dam and reservoir sites, tailwater surveys, construction materials surveys and testing, geological information on foundation conditions of dam sites and reservoir areas, land costs, facilities requiring relocation for the project, reservoir clearing costs, transportation accessibility to construction sites, climatic conditions affecting construction, stream flows and water quality to be maintained during construction, fishery and wildlife values to be preserved, any other important environmental considerations, and the availability of trained manpower.

Plan Formulation — Feasibility-Grade Analysis

Plan formulation procedures and the various factors involved therein were described earlier in this chapter. For the feasibility investigation the project plan should be devised, or confirmed if earlier selected, using the described methods and considerations along with new or updated basic data which may affect plan selection and size.

Design Criteria for Project Functions

Subsequent chapters will deal extensively with the factors that govern the design and resultant cost of any type of dam being considered. These include the physical and engineering factors, and the functional purposes of the dam. At this point; preparatory for making feasibility-grade cost estimates, it is necessary to specify and incorporate into the dam design the facilities and appurtenances necessary for each project function. They are:

1. Flood Control. In the usual flood control operation, water is temporarily stored to reduce a flood peak, then released downstream as rapidly as possible without damage to evacuate the reservoir space for the next flood which possibly may follow in a few days time. In a snow melt flood control operation, releases may be made on a schedule based on predicted runoff, with the objective of filling the reservoir at the end of the snow melt flood period. Releases are made at the "channel capacity" rate; i.e. the maximum rate at which significant flood damage will not be caused or increased by these releases. Large outlets through the dam or abutments are required; these require gates or control valves except in some flood detention structures where the time required for storage evacuation is not considered critical. For a single-purpose flood control reservoir the outlets will be located in elevation near the bottom of the reservoir. For a multi-purpose reservoir the outlets may be partly combined with outlets for other purposes and, for cost economy, may be partly at or near the top of the conservation pool.

2. Water Conservation. The functions may include irrigation, municipal and industrial water services, and possible diversions for other functions. Such diversions are the principal function of a diversion dam. The required outlet capacity of a storage dam and the diversion capacity of a diversion dam are determined by the peak rate of release or diversion required by the seasonal variations of need of the service area supply. Outlets for water release from storage must be made from the minimum conservation pool level or below. These outlets must be controlled by gates or valves, and the gates or valves must be of designs such that rates of discharge can be regulated with sufficient control so as to meet variable service area demands without waste of water due to excess diversion or release rates. The location of outlets depends upon the project plan. Such outlets may be distant and entirely unconnected with the dam if water is taken directly from the reservoir into a trans-mountain diversion tunnel to the use areas. Occasionally a diversion is made directly at the storage dam into a canal or closed conduit. Frequently the water is released downriver from a storage dam for rediversion by a downstream diversion dam. Gates or valves must be designed so that any excess energy resulting from the releasing of water from a storage reservoir with variable water levels higher than the discharged water elevation can be safely dissipated in a stilling pool or an energy dissipation chamber for a closed conduit conveyance system.

Economy of design dictates that where the reservoir is designed for long-period storage carry-over from wet to dry years, and where irrigation service is a function, consideration be given to the rate of release required when the reservoir is drawn down to its minimum conservation storage level. Thus it is unnecessary to design outlets having peak capacity when the least head is available from reservoir water levels for attaining the discharge rate. Figure 1-6 illustrates the situation. It is customary to plan, for overall project economy, for a water supply deficiency to irrigation that may be taken one or two times in a 50-year period during the occurrence of an extreme dry year or a series of subnormal years of runoff.

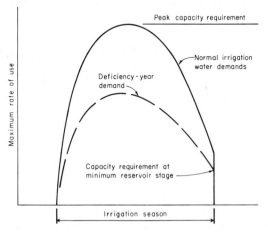

Figure 1-6. Example of variation of irrigation water use.

The amount of such allowable deficiencies depends upon the type of crops being irrigated. Similar deficiencies in water service may be acceptable and planned for downstream releases made for fisheries maintenance, salinity repulsion and river navigation flows, but are not usually planned for municipal-industrial water service and hydroelectric power generation. However, depending upon the circumstances of available standby sources for emergency use, deficiencies may also be planned for hydropower generation and municipal-industrial water supplies.

3. Hydroelectric Power. A hydropower plant is frequently located at or near the toe of a storage dam to operate from the variable head of the reservoir water level. Since dam sites are usually in narrow canyons or stream sections where space is limited; various accommodations are made in layout and design of the dam for best overall project economy. Penstocks are usually embedded in a concrete dam; otherwise they are constructed through an abutment. They may provide all or most of the capacity needed for making downstream releases. Intake levels are located at or below the minimum operational water level of the reservoir. Required flow capacities are equal to the hydraulic capacity of the power plant. This capacity is a complex determination resulting from engineering-economic analyses necessary to determine the optimum size and design of the plant. Much of the complexity is due to the relationship that power is the product of head and rate of flow through the turbine. Lowered reservoir heads require inversely greater flow releases to maintain full plant generating capability, but there is an economic limit to the turbine hydraulic size for the large head variation occurring with a storage reservoir.

If a hydroplant is directly fed from a reservoir, but through a long conveyance conduit to a much lower stream discharge point, the head variation is minor and sizing is simpler. Here the size is essentially determined from the safe yield or dependable volume of water annually available from the reservoir and a selection of plant load factor for peaking power generation.

Hydroplants are sometimes justified at diversion dams where there is a requirement for considerable flows to be released downriver for various needs. The hydraulic size of turbines and penstocks is usually the maximum regulated flow release rate contemplated for the needs of other functions and to supply prior entitlements to downriver flows.

Afterbay Dam and Reservoir

An afterbay dam and reservoir are usually required below the discharge point of any hydroplant operated to produce variable amounts of power at peak periods. The need arises from the objections of causing highly fluctuating downriver flows. When locations are favorable the afterbay reservoir may be made a part of the impoundment created by a diversion dam. The required active capacity of an afterbay reservoir is computed from the schedules of daily and weekly fluctuations of flow resulting from the proposed operation of the upstream peaking power plant.

River Navigation Considerations

When a dam is constructed on a large river used for upstream and downstream navigation it may be desirable and justifiable to construct ship locks to provide passage of watercraft past the dam. Depending upon topographic conditions the locks may be an integral part of the dam or entirely separate structures. The functional design criteria are the dimensions of width and draft of boats and ships to be accommodated, and the estimated numbers to be passed upstream and downstream at peak periods without excessive delay.

Fishery Requirements

Nearly every stream upon which a dam may be constructed is inhabited by one or more species of fish which may be of minor or major importance. The first, and sometimes only requirement, is to continuously release water downstream at appropriate minimum flow rates to sustain the fishery resource. Determination of the appropriate and justifiable rates is a difficult and time-consuming task where downriver releases made for other purposes do not satisfy the fish requirement. It is undertaken by fish biologists who survey the stream habitat for resident species of fish, noting the extent of habitat usable by fish through their life cycle, including spawning and hatching of eggs. These conditions are noted for the range of natural variable stream flows. The numbers of fish involved, hence the value of the resource, are estimated, but with probable low accuracy because of inherent counting or sampling difficulties and the large variations of numbers of fish from year to year. From these surveys the biologist evaluates the usefulness to fish of different rates of flow and thereby recommends the releases to be made from a dam. Usually the result cannot be equated, with any high degree of reliability, to an economic value and it is frequently the policy of a fishery agency not to do so. Recommendations are subject to considerable differences of opinion and frequently are controversial.

A second fishery criterion is the quality of water resulting from a project development; particularly water temperstures as modified by reservoir storage. Biologists can specify water temperature ranges suitable for the involved or most desirable species of fish. Temperatures above and below the desirable range may greatly inhibit propagation and growth, or even be lethal at times of the life cycle. Based on experience and studies of water temperature stratifications in reservoirs it may be necessary to build selective-level outlets at storage dams. This is rather easily accomplished if incorporated in the design of a dam, but difficult and costly as a modification after a dam is built. It is much less expensive for a concrete dam than for a fill-type dam where the outlets would be required in a separate control tower.

A third consideration is the desirability of passing anadromous fish upstream and downstream past a dam, where such fish are an important natural resource. An outstanding example is salmon which migrate from the ocean to upper reaches of rivers and tributaries to spawn, following which the young migrate downstream to inhabit and grow in the ocean. The salmon is a fish which has both commercial and sport fishing value. Upstream migrants can be accommodated past diversion dams and low power dams without great difficulty by fish ladders. Design criteria based on many successful installations are available. A controlling criterion may be the need to pass enough water down the ladder to attract fish to the entrance instead of to other dam outlets. An alternative, usually expensive to maintain because of occasional flood damage, is to construct a screen across the river below the dam to guide fish to the ladder.

Considerable numbers of young downstream migrants can be passed through or past a low diversion dam or fixed-head power dam without providing extensive additional facilities. Due to a tendency of young fish to remain near the surface and near reservoir banks, some find their way to fish ladder and spillway outlets. Others will pass through hydraulic turbines and survive. In the past, these methods have been regarded as "acceptable" by fishery agencies in the sense that methods or reducing the loss were not known. Presently, the view of fishery agencies is that additional measures should be taken, and facilities provided if necessary, to insure a high rate of survival of downstream migrants at these types of dams. Some possibilities are suggested by the accomplishments that have been made at storage dams, as indicated below.

Passage of anadromous fish at storage dams is usually taken to be either impossible or very costly for the attainable results. Because of large reservoir water level fluctuations the conventional type of fish ladder cannot be used. Adult fish can be trapped at the toe of the dam and transported in some form of water container past the dam and released into the reservoir. The passage of downstream migrants is more difficult to accomplish; however, it is being done with considerable success at two installations — Green Peter Dam on South Santiam River, Oregon, and North Fork Dam on Clackamas River, Oregon. A brief description of the situations, the facilities, and the fish passage results at each of these dams follows.

Green Peter Dam Fish Passage. Green Peter Dam and Reservoir is a multipurpose development constructed by the Corp of Engineers in the 1960-decade for flood control, power generation, and other functions. The dam is a concrete structure 360 ft. (111 m) high and 1500 ft. (460 m) long, with a design to fluctuate reservoir water levels through a vertical range of 93 ft. (29 m) in normal seasonal operation. The intake to the power plant penstock is 100 ft. (31 m) below the lowest pool level for normal reservoir operation. The river is the natural habitat of two species of anadromous fish: spring chinook salmon and winter-run steelhead. Coho and sockeye salmon and summer-run steelhead were introduced experimentally during tests.

The dam is constructed with special facilities to pass adult fish upstream to their spawning areas and to pass their progeny, of sizes usually termed "fingerling," downstream through the dam for continued seaward migration.

The unique facility is a fingerling collection device constructed on the upstream face of the dam. There is a fish collection "horn" facing upstream into the reservoir, mounted on supports and a mechanism that permits it to be raised and lowered to follow the 93-ft. fluctuations of reservoir water levels. The horn is 20 ft. high and 6 ft. wide, and designed to operate with the centerline from 15 to 30 ft. deep. The horn structure contracts from the mouth and leads into a fish separator unit (fish screen), a water pump system, and a transport pipe system which delivers the collected fingerling fish to the river below the dam. In operation, water is pumped into the water circulation system in the direction to create flow from the reservoir into the open mount of the horn at a rate to create an approach velocity of 0.1 ft/sec at a distance of 18 ft. from the horn mouth. The flow velocity increases to about 10 ft/sec at the throat of the horn and across a fish separator screen. Most of the pumped water is recirculated to the reservoir; the remainder carries the captured fish into a 24-in. diameter conduit which delivers them to the river below the dam. It was found that the addition of a submerged mercury vapor lamp positioned near the center of the horn entrance increased the attraction of all species of juvenile emigrants except sockeye salmon.

The operation of this facility was tested over a 4-year period, 1967 to 1971, by personnel of the Management and Research Division of the Fish Commission of Oregon. The results are documented in a report dated February 1973.* This agency judged the downstream-migrant transport system successful in passing chinook salmon migrants. This was indicated by test recoveries of marked fish through the transport system approaching or exceeding 80 percent. The success in transporting steelhead smolts remained in doubt at the end of the 1971 test period, with most recoveries of marked fish for the tests being less than 44 percent. Continued surveillance and data compilation on fish runs were recommended following the conclusion of the 4-year tests.

North Fork Dam, Clackamas River Fish Passage. North Fork Dam and Reservoir is the water

---

*"Evaluation of Fish Facilities and Passage at Foster and Green Peter Dams on the South Santiam River Drainage in Oregon," by Emery Wagner and Paul Ingram, Fish Commission of Oregon, Management and Research Division, February 1973. Financed by U.S. Army Corp of Engineers, Portland District. Report published by Fish Commission of Oregon. 17330 S.E. Evelyn Street, Clackamas, Oregon 97015.

storage facility of a single-purpose hydroelectric power development which includes power plants and two smaller downstream dams. North Fork Dam has a hydraulic height of 134 ft. (41 m). The reservoir is fluctuated within the upper 19 ft. (5.8 m) of storage space to provide storage regulation of water released through turbines for generation of power on peaking demand schedules. The remainder of the dam height, of about 86 percent of the total, develops head on the power plant located at the toe of the dam. The large permanent pool of the reservoir provides habitat for fish and space for gradual sediment accumulation in the reservoir.

Water levels are fluctuated in all or a part of the top 19 ft. of the reservoir on a weekly drawdown and refill cycle in meeting the peaking power load. In operation, the average drawdown is 5 ft. The 19-ft. maximum drawdown rarely occurs. The reservoir active storage is not large in comparison to seasonal inflows, and the reservoir spills, on the average, 4 1/2 times per year. This differs from a water conservation and/or flood control reservoir where water storage fluctuation would be greater and reservoir spills would likely be less frequent.

Clackamas River is the habitat of three principal species of anadromous fish: chinook salmon, coho salmon, and "winter-run" steelhead trout. The adults of the three species migrate upstream in three different seasons of the year; their progeny all migrate downstream past North Fork Dam in the same season — principally May and June. There are indications that some of the earlier downstream migrants from tributary streams above North Fork Reservoir remain in the reservoir for a period of growth before continuing downstream past the dam.

Upstream adult migrants ascend past North Fork Dam by a fish ladder. The upper end of the ladder is designed to function with the 19-ft. maximum fluctuations of reservoir water levels; this is accomplished by the ladder extending into the reservoir a sufficient distance so that a series of gated openings connect from the ladder to the reservoir. The openings are constructed at gradually differing elevations so that one or more gates can be opened to provide the desired rates of reservoir outflow into the fish ladder and to provide opportunity for upstream migrants to enter the reservoir from the fish ladder. In operation, gate openings are changed as frequently as necessary to accommodate to the reservoir stage fluctuations. These provisions for upstream migrants also provide one path for downstream migration past the dam by juvenile fish. However, most passage of young fish is by supplemental means.

A downstream-migrant channel was constructed adjacent to the portion of the fish ladder that extends into the reservoir. A downstream flow larger than that required for the fish ladder flow is induced in this channel to collect downstream migrants which have arrived in the vicinity in their instinctive attempt to travel downstream. Most of the flow, of 200 cu ft/sec (5.6 cm), in the channel is created by pumps at the outlet end. The discharge from these pumps is returned by pipe to the reservoir at a point 115 ft. (35 m) beneath the normal surface. Traveling fish screens are operated in the channel a short distance above the pumps to prevent fish from being drawn into the pumps, and to shunt them into an orifice which discharges into the fish ladder. From this point the young fish travel down the dual-purpose fish ladder.

A further unique feature of this installation of fish management facilities is provided at a point farther down the fish ladder. Here, at a separator strucutre, another traveling fish screen is operated to separate the small downstream migrants from upstream migrants. The small fish are conducted into the entrance of a 20-in. (508 cm) diameter pipeline which carries them a distance of some 5 miles (8 km) to a discharge point in Clackamas River downstream from the lower most dam of the development.

Construction of North Fork Dam and the fish passage facilities was completed in 1959. The two low downstream dams had been constructed many years earlier, in 1905 and 1911, each with fish ladders and some associated hatcheries and fish-trapping facilities. Various fishery observations had been made during these earlier years.

Detailed and extensive evaluations were made of the accomplishment of operation of the fish management facilities at and related to North Fork Dam from January 1962 to December 1964. The results are described in a report titled "Evaluation of Fish Passage Facilities at the North Fork Project on the Clackamas River in Oregon" by Robert T. Gunsolus, Fish Commission of Oregon, and George T. Eicher, Portland General Electric Company; report dated September 1970. The Portland General Electric Company is owner and operator of this power development.

Various complications and difficulties were encountered in making the series of field observations, samplings, measurements, and experiments over the three-year test period, and some unanswered questions are the subject of continuing observations. Nevertheless, the evaluations showed successful operation of the fish facilities, and a high degree of success in passing downstream migrants through the reservoir and collection system. The tests which were considered valid indicated passage of all three species of anadromous fish in amounts of from 70 to 90 percent of the total. Success of chinook passage was poorer than for the other species. Mortality loss of the juvenile fish passing through the five-mile pipeline was slight. The mortality rate of fish passing over the spillway of North Fork Dam during occasional reservoir spills was also found to be slight. Special tests indicated that the mortality rate of fish going through the penstocks and turbines of the North Fork Dam power plant was in the range of 25 to 32 percent. The total numbers of fish traveling through the power turbines could not be reliably measured, but were believed to be small since the juvenile fish normally inhabiting the reservoir near the surface would have to dive to the lower most part of the reservoir to be carried into the power penstocks located near the bottom of the dam.

Fishery evaluations are a continuing process, but the results of the three-year evaluation program indicate successful fish management by the methods employed at this installation.

Aeration of Water

In addition to dissolved salts or chemicals from the leaching of watershed soils, and sediments produced by land erosion in the watershed, all stream flows carry some amounts of organic substances from the decay of animal and vegetative matter and the excrements of animals and people. Such organic substances are slowly oxidized by dissolved oxygen in the water, and the gaseous products gradually vented to the atmosphere. When water is impounded in a reservoir there can be a concentration of organic decay substances, and a reduced rate of oxidation and venting of the gaseous products of decay. Algae may also grow in shallow areas of reservoirs when the water is clear, adding to the materials of decomposition. This accumulation of nutrients and other chemical changes due to lack of oxygen sometimes cause the water released from a reservoir to have objectionable taste and odor, to be toxic to some degree, or in extreme cases to have insufficient dissolved oxygen to adequately sustain fish life.

This condition is particularly apt to occur temporarily, for perhaps one or two years, when a new reservoir is first filled. Even though a reservoir site will have been cleared of the larger vegetation much of the small vegetation will remain and decay in the water. For economic reasons, reservoir sites are frequently not cleared below the minimum operating water level, as wood that is always inundated decomposes at extremely slow rates. The tainting of water by decomposition of the original reservoir-site vegetation is usually tolerated because it is temporary.

The other sources of organic pollution are of permanent nature. They are concentrated in the deeper parts of the reservoir where aeration is lacking and it is from these deeper zones that water is released for project supplies in most operating conditions. Analysis and judgment are required as to whether facilities need to be constructed and used for aeration. Som factors are:

1. Nutrient supply to the reservoir will be heavy if there is large and rapid vegetative growth in the watershed, and if there is a large population of animals and people in the watershed.
2. Releases will be considerably re-aerated if project releases flow a considerable distance downstream or through open canals before use, rather than directly into closed conduits.
3. Turbulence from a power plant, outlet valve, or spillway will entrain air in the water and add to the rate of its aeration.
4. Odors from aeration may persist for a distance downstream from the dam, making that portion of a river canyon unattractive for recreational use.
5. Chemical composition of upstream waters and sediment will determine toxic effects due to oxygen deficiency in the reservoir.

A possible aeration problem may be lessened if selective-level intakes are provided for the reservoir outlets, as may be desirable for other water quality controls. In some instances a special aeration facility may need to be provided in connection with the dam.

Cost Estimates

Differing from reconnaissance-grade estimates, the estimates of both construction and operating costs of dams and other major projects facilities are made as accurately as possible. For a dam and other large structures a complete engineering design is made, and the types and quantities of different construction items are computed therefrom. An estimator may then make and apply his estimates of unit costs based on experience records of recent and similar types of construction. In doing so he will take into account the location of the construction site and the transportation and hauling distances from the sources of materials and machinery. In the absence of a fully comparable cost experience record he will estimate costs by the same procedure a construction contractor would use in preparing a bid proposal. Imagining himself as the contractor he will plan a construction program, selecting sizes and numbers of pieces of construction equipment used and the lengths of times of uses. Purchase-minus-salvage costs of equipment are derived,* and hourly operation costs are applied. Manpower requirements, including supervision and administration are estimated and appropriate wage and salary rates for the location applied. The sum of costs makes up the estimate on this basis. To the estimate is added a contingency factor of amount depending upon the probability of encountering unforeseeable construction difficulties, and a factor for cost of engineering and inspection of construction.

The cost of land acquisition is estimated by real estate appraisers, based on experience records of sale value of similar lands and improvements. Costs of relocating roads and public utilities are frequently best made by the agency or organization which owns or constructs such facilities.

Annual operation and maintenance costs are usually estimated by devising an operating plan and noting the numbers and types of personnel required on a full-time or intermittent basis to carry it out. Salaries or wages of the required staff are estimated, as are the costs of materials, supplies, and equipment. Estimated costs of transportation, shop facilities, warehouse, and administrative headquarters are added.

---

*A source of information on equipment cost is a publication "Contractor's Ownership Equipment Expense" by The Associated General Contractors of America.

Replacement costs associated with a dam are usually minor, but there are possibilities of valves and gates requiring replacement or extensive repair within the economic life of the project. Length of life may be considerably affected by the erosive quality of the water discharged and by the amount of use. Sometimes, due to erosion damage, expensive repairs of spillway stilling basins are also required. A cost contingency factor is advisable for these somewhat unpredictable future costs but the amount can be estimated only roughly based on similar experience of old structures.

Length of Construction Period

Considerable time is required to build a dam. Depending upon size and circumstances, periods of from three to ten years are not uncommon. During the construction period money is being expended more or less continuously with the result that there is an accumulation of interest payable on the capital invested before the dam is ready for use. To take this into account an estimate is required of the probable construction schedule and rate of expenditures. This is accomplished jointly by the designer and cost estimator.

FLOOD CONTROL

Alternative Methods of Flood Control or Protection

This being a handbook of dam engineering, it is of importance to note that flood control or protection may be accomplished by a variety of methods of which dam construction is only one. Others include: dikes (or levees, by French and American terminology), flood diversions and bypasses (where physical conditions permit), river channel improvement, flood zoning or other flood plain regulation, flood forecasting and warning, floodproofing of structures, and flood refuge hills or mounds. Two or more methods may be advantageously used in combination; for example, a flood storage reservoir and riverside levees. There are a variety of situations that tend to favor or disfavor each possibility, and the selection of the best method to adopt for a particular flood problem area for initial development or improvement over an existing but inadequate method is often complicated and sometimes controversial.

One set of circumstances tends to prevail in the low delta areas of large rivers near their points of discharge into the ocean. Likely much of the land has been or will be reclaimed from its natural overflow conditions by levees. Flooding is caused both by ocean high tides and by river flood flows. Exceptionally high tides occasionally occur due to storms at sea, unusual differences in barometric pressures, or tsunamis. The water levels due to the ocean flood are further raised if there is a simultaneous occurrence of a river flood.

Because delta areas are at the mouths of large and usually long rivers the volume of flood water produced by the river is great, and the time of the flood flow is of considerable duration. A very large amount of flood water would need to be stored to substantially reduce the flood stage in the delta channels, and such amount of flood storage is usually impossible or economically impractical of attainment. Flood protection is principally obtained by levees, possibly supplemented by the other means that exist or can be provided. Flood control by reservoirs may provide a minor increment toward protection against overtopping of levees, provided such flood storage is justified and also used for protection of areas farther upstream.

In a much different situation, entirely above a river delta and above the area of any backwater influence from ocean tides, different types of flood characteristics prevail. Proceeding upstream the flood volumes are less, as are also the natural capacities of the existing channels. Although river gradients are steeper than in delta channels, very high flood stages can occur due to the limited channel cross sections.

The floods can be damaging both from velocities of flow as well as from inundation of overflow areas. Being nearer the headwaters of a river the time of concentration of flood waters from tributaries is short; the stream rises rapidly in a flood, and the duration of the flood peak is relatively short. With a flood of short duration there is considerable opportunity to reduce the damaging peak flow to slight or non-damaging rates by temporarily storing the most damaging part of the flow in a flood control reservoir. Alternatively or conjunctively levees along the riverbanks might be used, but with one possible disadvantage as compared to delta levees; the higher flood velocities of upstream channels cause risk of levee bank erosion in some localities, requiring that they be constructed erosion-resistant or that considerable maintenance be performed. Lack of such precautions can cause dangerous complacency in assumptions of safety, or critical emergency repair efforts during a flood, or at worst a levee break and large flood disaster. Because of the rapid rise of floods on higher streams, flood warning systems are less effective than for lower valley and delta areas.

It is beyond expectation to encounter two identical flood problem areas; each has its own river and flood plain characteristics including its type of occupancy and use by people. A large amount of data collection of all factors pertaining to flood damage and protection, hydrologic analyses, and studies of alternative plan possibilities are necessary to support a recommendation for a justifiable flood control plan.

Flood Frequencies and Magnitudes

Small floods occur quite often; large floods occur infrequently, and a great flood is a rare event. However, extensive damages occur with infrequent and rare big floods, and their magnitudes and frequencies are of highest importance in planning flood protection.

Estimates of flood frequencies and magnitudes for a specific river location are made from all available historical data of floods of that stream and of comparable streams having similar flood characteristics. Principles of mathematical probability are applied or recognized as a means of estimating the frequency and magnitudes of very large floods for which data of actual experience are lacking or inadequate. The procedures used by a qualified flood hydrologist are described in Chapter 2 entitled "Hydrologic Studies." The product of such studies is usually displayed as a flood frequency-magnitude curve.

Degree of Flood Protection Sought

It is impractical or uneconomical in most instances to attempt to provide full flood control protection to an area for the largest flood that can be imagined; for example, one occurring once in several thousand years. An acceptable degree of flood protection varies with the amount of damage and losses caused by flooding of exposed areas. Damages to rural agricultural areas are much less than to equal areas of intensively developed land occupancy. The degree of protection is therefore designed according to the magnitude of flood damages as well as the cost burden of providing the flood protection. In the United States the Corps of Engineers usually design for control of a "standard project flood." This is defined by the Corps to be that flood which would occur as a result of the most severe combination of meteorological and hydrological conditions of the hydrologic regions, excluding extremely rare combinations. It is a flood of somewhat greater magnitude than has occurred or is expected to occur once in 100 years. It is less than a spillway design flood for the dam of a large reservoir where extreme protection is considered necessary against the failure of a dam due to a flood.

Flood Damage Estimates

These are derived from all the data that can be collected on historical flood damages in the area of study. The current practice in the United States is to promptly collect the field data on flood damages in important areas soon after any large flood and to preserve the information for prospective future flood control analyses. A useful advance service is to publish maps showing the extent of areas flooded, as information of the risk of further developments in flood-exposed areas unless flood control is provided.

A graph can be prepared showing a relationship between flood magnitudes and resultant damages. Information from this graph can then be combined with that of a frequency-magnitude relationship to produce a frequency-damage curve. From this latter curve there can be derived an annual average flood damage estimate quantity which becomes the basis of an economic justification analysis of a potential flood control project investment.

Types of Flood Storage Facilities

Flood Detention Basin. This is the simplest type of a flood control storage impoundment. A dam is provided with an open, ungated outlet of hydraulic capacity such that the discharge at design flood reservoir head equals, or does not greatly exceed, the safe capacity of the downstream river channel, i.e., the capacity without large overflow damage. The height of the impoundment structure is selected to provide a reservoir volume sufficient to temporarily contain the volume of the incoming flood which is in excess of the desired rate of flow to the downstream channel. This flow is determined on the basis of the hydrograph of the maximum sized flood intended to be controlled (design flood). The required reservoir volume is calculated by a routing study of inflow and outflow rates, and storage accumulation. A spillway is provided with the dam to pass flows of larger floods than are managed by the flood control operation.

Flood Control Reservoir. When built as a single-purpose structure the facility is constructed similar to a flood detention reservoir except that the outlet is built with control by a gate or valve. In operation, the control provides for a managed regulation of the outlet flows to achieve a more efficient release of downstream flows during a flood in accordance with downstream flood conditions as affected by tributary inflows below the dam, or otherwise.

When flood control is provided as part of a multipurpose reservoir an upper portion of the reservoir is provided and reserved for the flood control function, with flood releases regulated by gated outlets. Savings in cost of the flood control function can usually be effected in a multipurpose development due to sharing of some of the dam and reservoir costs with other project functions. In addition, it may be possible to interchange use of some reservoir capacity at times between a flood control function and a water conservation function by devising suitable operating criteria.

Reference Information

In the United States the Corps of Engineers, U. S. Army, is the major authority and source of flood control information. This includes both the hydrologic techniques for estimating flood, and the data and procedures for estimating preventable flood damages. The Corps of Engineers has acquired and maintains this capability because of their designated responsibility to advise and recommend flood control projects which are financed or assisted in financing from the investment of Federal funds.

The Soil Conservation Service, Department of Agriculture, is a source of flood control information for dams in "headwaters areas," i.e., small drainage areas, especially rural areas.

Useful reference publications include:

1. U. S. Geological Survey – Water Supply Paper No. 771, "Floods in the United States, Their Magnitude and Frequency" – 1936. Available from U. S. Government Printing Office, Washington.

2. Water Resources Council Hydrology Committee, "A Uniform Technique for Determining Flood Flow Frequencies," Bulletin No. 15, December 1967. Published by Water Resources Council, 1025 Vermont Ave., N. W., Washington, D. C. 20005.
3. Notes on Hydrologic Activities, Bulletin No. 13, "Methods of Flow Frequency Analysis." By Sub-Committee on Hydrology, Interagency Committee on Water Resources, April 1966. Available from Superintendent of Documents, Government Printing Office, Washington, D. C.
4. Economic appendixes of flood control project planning reports of the Sacramento District of the Corp of Engineers, Sacramento, California.

ECONOMIC EVALUATIONS

Build-Up Periods of Project Functions

Preparatory to making an accurate comparison of project costs with benefits, or with repayment requirements, it is necessary to take into account the period of time that will elapse following project construction until the project function is put to the full use made available by the project. The full project costs begin, with slight exception, when construction is completed and the project is ready to start operation. Most project uses grow over a development period of time. Examples are:

1. *Irrigation.* Required farm developments for irrigation take a few years. It is usually estimated that irrigation use will begin with a small amount and grow to full use in 10 years.
2. *Flood Control.* Since a damaging flood may occur any year, the full project flood protection begins when project construction is completed. There is no build-up period involved; however, the value of flood protection is usually estimated to increase with passing years due to more intensified use of and values in flood-risk areas.
3. *Hydroelectric Power.* The build-up time will vary, depending upon the project size with respect to the market need. In an industrially developed country the time will be short; five years or less. In a developing country the time may be many years if the favorable size of the water development is large. Some accommodation to this situation may be made by a plan to add generating units at the hydroplant as the load develops. This may be taken into account, in economic evaluations, as deferred costs.
4. *Municipal-Industrial Service.* Usually the needs are projected to grow continuously over a long period. Circumstances vary, but as a general rule the service from any single project development is not planned to exceed the growth need of this function for more than about 25 years.
5. *Recreation.* Demand must be estimated in each individual case, but it is often considered to reach a maximum potential value in 50 years or less.
6. *Other functions.* These are usually minor in a multipurpose development. Where important, economically, their development periods would be estimated according to the circumstance.

Annual Equivalent Values

In making a financial repayment analysis the estimated quantities of each type of project use or service are listed, year by year, in computing the repayments up to the end of the prospective pay-out period. In the benefit types of analyses the variable amounts of project uses and their values are computed, by appropriate methods, to annual equivalent or present worth values for comparison to costs. Figure 1-7 shows an example of a straight-line build-up followed by 40 years of full use. The diminished importance of benefit value in far-future years, as compared to earlier project years, taking into account interest-on-investment cost, is graphed by use of the non-linear time scale of the second graph. This scale varies with the interest rate involved. If a project function is subsidized by requiring repayment of only investment without interest (U. S. Reclamation policy for irrigation) the first graph is applicable. The second graph illustrates the greater importance of development period length where the cost of interest on investment is considered.

A capital or capitalized value (of cost or benefits) is a summation of a group or series of individual values which make up the total. If an interest rate associated with the passage of time during which the individual values occur is disregarded, the capital value is simply the average of the individual values times the number of occurrences.

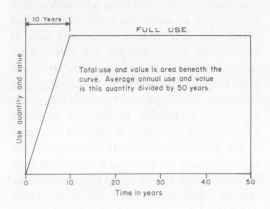

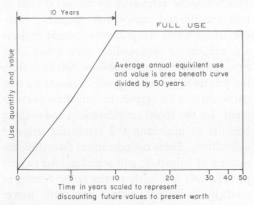

Figure 1-7. Comparative graphs illustrating average and average-equivalent values of a project function with a build-up period of use.

Where interest is considered, each individual value occurring at a future time has an equivalent present worth which is determined by discounting the future value to the present time by use of the appropriate interest discount value. The sum of a series of future annual values, each discounted to its present worth, will be the present worth capitalized value.

One way of evaluating and displaying this graphically is by a diagram in which the time scale is appropriately compressed for future years in comparison to the initial year, to reflect the appropriate interest discount factor and period of analysis. This is illustrated by the second graph of Figure 1-7. Other methods are also applicable.

Types of Evaluations

Three types of economic evaluations are used, either individually or in combination, to assess the economic value of a project. They are:

1. Benefit-cost analysis.
2. Financial, or repayment analysis.
3. Internal rate of return analysis.

There is also a minimum cost method having limited application, subsequently described.

Benefit-Cost Analysis

As indicated previously, this is a simple comparison of estimated project benefits and costs after the usually arduous task of computing and equating them on the same monetary value and time base, using either identical or appropriate interest rates throughout. To appear favorable a plan must show a benefit-cost ratio greater than unity (1/1); the higher the ratio, the greater the favorability. As previously described, some functions such as irrigation have both direct and indirect benefits, and the B/C ratio must be qualified as to which benefit values are used. A decision-making body may hold the view that direct benefits must exceed cost for project authorization, or alternatively, may consider a project favorably when total benefits exceed cost. The prevailing interest rate for the cost of money for long-term investment over the estimated economic life period of the project is normally used for all the comparative cost and benefit computations. This period may be as long as 100 years for a project based on a major dam.

Comparative benefit and cost values are usually reasonably similar for either an analysis period of 50 or 100 years, although the longer period will show increased justification. This increase in justification for the longer period is accentuated when some project function is expected to develop to full use very slowly, for example, a municipal water supply for which the use may not grow to the full project amount for 40 to 50 years. Usually the same interest rate is used for computing all the costs and benefits of a project plan. There are exceptions to this, however, where benefits are computed as the cost of the alternative means of accomplishing the same function. It may be contended that in some instances or circumstances the alternative method of supplying a project function would consist of a development which would be financed, for a sound reason, at a different rate of interest than that applicable to the project plan being evaluated. The use of two different interest rates in evaluating a project plan requires careful thought.

Financial or Repayment Analysis

The purpose of this analysis is to determine whether reimbursement requirements are within the means and willingness of repayment by the project service users, and how much time will be required to recover these repayable costs. As mentioned previously, some project functions are subsidized in various amounts because of government policies and the widespread occurrence of benefits. In the United States, Federal projects are required to repay the full allocated costs, with interest, of hydropower and municipal-industrial water supply, but, in the case of Federal projects irrigation is usually only required to repay its costs without interest, or may be otherwise assisted by requiring repayment only up to the limits of farmers repayment ability for irrigation. For a multipurpose project it is necessary, for repayment purposes, to allocate the project costs between the different functions; the cost allocation methods are described below.

Selection of Repayment Rates

It is obvious that high water and power service rates will repay the investment costs rapidly; low rates will require a long payout period. Since major project structures such as dams and reservoirs have a long physical life there is an opportunity to set repayment rates only as high as necessary to assure the required repayment within this long period. Low rates will, of course, best assure that the full potential accomplishment of the project will be used, and may favor a rapid build-up to full use. The alternative consideration is to obtain repayment as rapidly as possible so that the recovered capital can be re-invested in a project expansion or for some other need. If the project is privately financed the incentive is to set rates at an optimum which should be the highest monetary rate at which the product can be marketed. This would be estimated by market studies of user-demand and competitive supplies. However, most water resource development projects are publicly or semipublicly owned and managed, and governmental policies will vary from the private-investor viewpoint. Reasons for this, particularly with respect to irrigation development, are the social considerations and indirect benefits of increasing and stabilizing irrigated agriculture. These considerations favor the subsidizing of irrigation, and selecting long payout periods with low water service rates. Setting of municipal-industrial water rates and power rates tends to conform to the private-investor viewpoint. But, particularly in the industrially developing countries, the government may wish to set low water and power rates (requiring longer payout periods) as a stimulus to industrial development.

With the best irrigation projects having already been built in the United States, new projects are mainly limited to poorer classes of irrigable land; less productive or more difficult to irrigate.* In these cases the water rate is limited by the farmer's ability to pay and in some cases the repayment period has been extended considerably beyond 50 years. The more usual case is to obtain irrigation repayment in 40 years or less after a development period not exceeding 10 years. Somewhat consistent with this length of time, repayment of other functions in multipurpose Federal reclamation projects of the United States are frequently set at 50 years, and water and power rates devised accordingly.

Repayment Rate Components

There are two separate components that give value to a water or power service: (1) the rate at which the product — water or power — can be supplied upon demand to the customer, and (2) the quantity used. For power, the quantities are (1) the demand and dependable supply of power in kilowatts, and (2) the amount of energy supplied and used in kilowatt-hours or multiples thereof. The cost of the dependable power component is the cost of providing a certain amount of dependable power generating capacity (kilowatts) and maintaining it in readiness to meet a demand. This cost is principally the investment cost, expressed in monthly or annual equivalents, plus the fixed-type operational costs necessary to maintain the power generating capability in readiness for use. In the United States this may be a cost such as $20/kilowatt/year. The cost of the energy component is the cost of generating energy by a power plant ready to do so. In a fuel-driven plant it is principally the cost of fuel consumed to generate a unit quantity of energy, plus operational, maintenance, and replacements costs that vary with the amount of energy generated. In the United States a typical energy cost might be 3 mills ($0.003)/kilowatt-hour. In event there is a fixed relationship between capacity and energy generation and energy use — for example, a plant operating at full capacity continuously to meet a base load — it is possible to express the capacity cost as an amount per unit of energy generated, combine it with the energy cost and express the total as a cost per unit of energy produced.

Similarly, the cost of a water service is made up of two components: the cost of facilities and stand-by operation to deliver water at the peak demand rate (measured as cubic feet per second, cubic meters per second, gallons per minute, or other flow rate as appropriate), and the cost of providing the dependable volume of supply (acre-feet, million cubic meters, million gallons, or other unit) in a time period of use such as a year. Where conveyance facilities (aqueducts and pumping plants) are expensive because of long distances or large pump lifts this cost component is computed and priced separately. The California Water Project is an example. The cost of providing a quantity of water supply for use in a time period such as a year consists essentially of the investment and operational costs of a storage reservoir, or the cost of a diversion dam if there are stream flows available for diversion without storage regulation, or a combination of the two. Once a project is constructed, these costs are usually not variable with the quantity of water used. The repayment rate for irrigation water service is rarely priced in two components because delivery rates are usually closely related to water quantities.

Delivery Points

Project power service is usually measured at a load center where delivery may be made to a

---
*There are exceptions where a project is built to provide dependable water supply for an existing irrigated area having shortages of supply.

power-distributing agency. Water deliveries are usually measured at terminal delivery points at the head of distribution systems for irrigation or municipal-industrial customer service. Power and water transmission losses must be taken into account, as well as the costs of the transmission facilities.

Trial Repayment Analysis

A year-by-year computation is made, listing all factors in columns. Column headings might be: (1) Project year, (2) Irrigation water service, (3) M&I water service, (4) Power service-capacity, (5) Power service-energy, (6) Irrigation revenue, (7) M&I revenue, (8) Power revenue, (9) Total revenue, (10) Investment repayable with interest, (11) Investment repayable without interest, (12) Interest on unpaid investment, (13) OM&R costs, (14) Revenue applied to repayment of investment. These would be modified to fit the specific project case. Project service quantities would be tabulated according to the projected rates of build-up to full use. The computation would be continued through the year of final repayment of investment for all functions. Results will show the prospect of repaying the reimbursable project costs in the desired number of years, or the need to adjust some project rates to accomplish this.

Internal Rate of Return Analysis

This method is an alternative to the benefit-cost analysis previously described. It measures the economic worth of a project plan by determining the rate of interest of money which would be allowable to make costs equal benefits. A project having benefit values that justify a high interest rate of return is more advantageous than one that would only return interest on the investment at a lower rate. Present worth equivalents of annual benefit values and annual OM&R costs are computed for comparison to investment (project construction) costs using the same interest rate as applied to the investment cost. Table 1, "Project analysis by internal rate of return method," shows an example computation illustrating the interest rate determination. The result of this analysis is of interest to the agency or institution considering the financing of the project development.

Limited Cost Analysis

In a rare situation a project may be justified and authorized for construction without regard to its economic worth or repayment ability. The value would be considered to be a social one such as urgent need of increased irrigation to prevent famine, urgent need of water supply suddenly created by a great natural disaster or destruction by war, or an environmental need not measurable in economic terms. The analysis is simply that of comparing any available alternatives and selecting the cheapest one, or the one that is within the available means of financing.

COST ALLOCATIONS

Purpose. In a multipurpose project some facilities serve more than one project function. This is always the case with a multipurpose storage dam and reservoir, and frequently with a diversion dam and conveyance facilities. Different users of the separate project functions are called upon to repay the appropriate costs of the separate project accomplishments. Agreement is necessary on some acceptable sharing of cost of joint-use facilities. This includes operating as well as investment costs.

Cost Allocation Methods

At least nine different methods have been devised and used at one time or another for allocating costs. They are identified by name as follows:*

---

*Reference: "The Allocation of Costs of Federal Water Resource Development Projects", House Committee Print No. 23, 82nd Congress, 2nd Session, Dec. 5, 1952.

TABLE 1-1.

PROJECT ANALYSIS BY INTERNAL RATE OF RETURNED METHOD

Taken from "A Guide to Using Interest Factors in Economic Analysis of Water Projects," June 1970. A publication of the U. S. Bureau of Reclamation.

I. Project Data (assumed for this example)

Construction costs	$160,000,000
Construction period	4 years
Annual O. M. & R. costs	$ 1,000,000

Annual Project benefits

1. Irrigation	$ 6,000,000
2. Power	4,000,000
3. Municipal & Industrial Water	3,000,000
4. Flood control	2,000,000
TOTAL	$ 15,000,000

II. Economic Assumptions

    A. Project life – 100 years, with year 1 being the first year of project benefits

    B. Interest during construction (IDC) – Simple interest during the construction period. Uniform construction expenditures over the 4-year construction period.

Shortcut IDC calculation is therefore:

IDC = construction cost x 4 x interest rate $\div$ 2

III. Solve for Internal Rate of Return using Present-worth

    A. First trial of 7 percent

1. Construction cost	$160,000,000
IDC $160,000,000 x 4 x .07 $\div$ 2	22,400,000
Subtotal investment cost	$182,400,000
2. Present worth of OM & R cost	
1,000,000 x 14.3 #	14,300,000
TOTAL present worth costs	$196,700,000
3. Present worth of benefits	
$15,000,000 x 14.3 #	$214,500,000
4. Benefits minus costs	$ 17,800,000

Benefits are greater than costs. Therefore 7 percent is not the correct rate. The interest rate must be increased to bring values of benefits and costs together.

    B. Second trial at 8 percent

1. Construction cost	$160,000,000
IDC $160,000,000 x 4 x .08 $\div$ 2	25,600,000
Subtotal – investment cost	$185,600,000
2. Present worth of OM & R costs	
$1,000,000 x 12.5 #	12,500,000
TOTAL present worth costs	$198,000,000
3. Present worth of benefits	
$15,000,000 x 12.5 #	$187,500,000
4. Benefits minus costs	$–10,600,000

Benefits are now less than costs. Therefore the internal rate of return has been bracketed between 7 and 8 percent. The exact rate may be found by linear interpolation.

By interpolation, interest rate to equate benefits to cost equals 7.6 percent, equals internal rate of return.

\# Present worth of 1 per period – $\dfrac{(1+i)^n - 1}{i\,(1+i)^n}$ (annuity factor)

Benefit method
Alternative justifiable expenditure (AJE) method
Separable costs – remaining benefits (SCRB) method
Use of facilities method
Separate projects method
Equal apportionment method
Priority of use method
Incremental method
Direct costs method

Each method has certain advantages and disadvantages or shortcomings. Project planners concerned with making a choice of which method to use are referred to the reference document No. 23 listed above which contains a description of the principles involved in each method and commentary on the suitability, differences, and shortcomings of each method.

The Separable Costs – Remaining Benefits (SCRB) Method is the method recommended for general use in allocating costs of Federal multiple-purpose river basin projects.* It has been used extensively since 1952, and is the basic method for use by U. S. Federal agencies;* however, there are occasions where one of the other simpler methods are suitable for a specific cost allocation need.

Some brief observations can be made herein on the factors and principles involved in the different methods. The first three methods listed above make use of computed and estimated project benefits of each project function. Accuracy depends upon the computation of these benefit figures on a monetary basis comparable with each other and comparable with project costs. Some of the difficulties or uncertainties of doing this are indicated in a preceding paragraph titled "Project benefits."

---

*Reference: "Proposed Practices for Economic Analysis of River Basin Projects," Report to the Inter-Agency Committee on Water Resources, May 1958.

*See following paragraphs titled "Payment of Environmental Improvement Costs" and "Environmental Cost Allocations."

Other methods which avoid the use of computed benefits have the appeal of greater simplicity, and ease of understanding and explanation. They employ such principles as proportionate size, proportionate use of facilities, proportionate costs, priority of use, and incremental uses and costs. Depending upon the problem involved, one of these methods may be used in making a sub-allocation of cost of some part of a project plan. For example, the "use of facilities" method has been quite extensively used in allocating costs of a canal between different users. However, it is usually unsatisfactory for allocating costs of a storage reservoir to different project functions.

The SCRB cost allocation method conforms in principle with the method used for selecting and sizing a project plan for optimum economic efficiency as described in the preceding section entitled "Plan Formulation Analyses." Where a series of designs and cost estimates has been made in the plan formulation process for a range of sizes and alternatives of joint-use features (such as dams and reservoirs) much of the required input cost data for the SCRB method will already be available. But if the plan selection has been made in another manner, possibly by a short-cut method, or not controlled by economic efficiency, or otherwise, considerable additional cost estimating will be needed to accomplish an SCRB method cost allocation.

Comparison of AJE and SCRB Allocation Methods

The procedures are similar in most respects, and the results are sometimes nearly the same. The distinction is that the AJE method utilizes figures of "direct" or "specific" costs of physically identifiable features or services which can be extracted from an itemized cost estimate of the multipurpose facility, whereas the SCRB method depends on separate designs and cost estimates made to determine the difference in cost of including or excluding facilities for a single function of the multipurpose plan. The latter are called "separable costs." For example, at a dam providing power generation and flood control the cost of penstocks could be taken as a specific power cost since they essentially serve no function in flood control operation. But if the multipurpose plan also includes irrigation supply the provisions for releasing irrigation water from the reservoir may be principally through the power penstocks, in which event the penstocks are not altogether a specific power cost.

Separable costs for the SCRB method are determined by designing and cost estimating the multipurpose facility with one function removed at a time. That is, in a three-function facility (power, flood and irrigation, for example) one design and cost estimate is made with the power function omitted, another with the flood control omitted, and one omitting the irrigation function.

A number of cost allocations have been made by both the AJE and SCRB methods to test comparative results. In 12 such cases where the plans were essentially dual purpose — power and flood control, with any other function absent or incidental — the difference was never more than 11 percent and averaged about 3 1/2 percent.*

This is not to say, however, that the results would always be as close by the two methods, particularly in plans where more than two project functions involved require cost allocations. Moreover, there is the question of being able to identify all "specific" costs for use in the AJE method. For instance, if a certain amount of inactive reservoir storage capacity is provided to maintain dependable power plant head, what is its cost, and might it not also have a value for fishery maintenance or enhancement?

Priority of Use Method

This method may be applicable in a special case where a definite priority of use of the

---
*House Committee Print No. 23, previously cited.

multipurpose project works has been established by law or other agreement. Its use is based on the premise that if a project is to be operated primarily for one function and secondarily for another, the primary purpose should be charged with a greater portion of the costs.

In computation, the "direct" (or "specific") costs of each function are first identified. The remainder are joint costs to be allocated. Attention is then given to the top priority function. The joint cost to be assigned to it is the lesser of two quantities; either (1) the benefits, less the specific costs assigned to that function, or (2) the cost of the most economical available alternative plan less the specific costs assigned as above. This portion of the joint cost is added to the specific cost to make the total allocated cost assigned to the first priority function.

The process is repeated for the second priority function, using the remainder of joint costs to be allocated, and so on for other functions in descending order of priority. The process results in greater allocation of costs to the higher priority functions than would be done by any other method.

Allocation of Cost to a Deferred-Use Function

Page 16, "Provisions for future modifications" indicates the possible merit of investing some cost in added size of facilities at the time of initial construction for a foreseeable future use. The justification for such design and added investment needs to be carefully established to the satisfaction of the investing agency, which may be the Federal government or otherwise. The basic analysis requires consideration of (1) the benefit value of the deferred use, (2) the time that will elapse between initial construction and the time of beginning use of the deferred function, (3) the incremental project cost of the deferred-use facilities, and (4) the alternative cost of providing the future accomplishment by another means. The decision on making an investment for deferred use is properly made by a benefit-cost evaluation of the of the problem with appropriate data including the interest value of money.

If only a slight investment is made for future-use facilities it may be satisfactory to let the cost be borne by the initial project users. However, if the deferred-use cost is substantial it is desirable, and possibly essential, to identify and segregate such cost for the satisfaction of the project users who are obliged to repay their share of project costs. This requires an allocation of the deferred-use cost, which may be carried as a "bookkeeping" item to be applied when the project is later expanded to provide the deferred service.

In a simple situation the deferred cost may be only that of the specific cost of a construction item included for future use. In other cases it may include an addition to the size or design of a joint-use feature of the project. The cost may be identified as the incremental cost of the project with future-use facilities over that of the project with future-use accommodations. This would be the cost as used for the economic decision to include the deferred-use facilities.

A most analytical allocation, if warranted by the importance of the deferred use allocation, would be to treat deferred use as another separate project function and compute its allocation by the principles and procedures of the AJE or SCRB method. These methods are subsequently described.

Cost of Environmental Protection

Many elements of environmental protection have historically and traditionally been regarded as "essentials" of any project plan, and there has been no need to separately identify the costs of preserving such environmental values. Examples of environmental protection items included in the construction and operation of a dam and reservoir are (1) making suitable downstream releases for fishery maintenance, (2) possibly providing a fish hatchery to help sustain the fishery resource, (3) providing public access to shores of reservoirs, (4) providing replacement areas for wildlife habitat, (5) clearing timber from inundated reservoir areas, and others. In some instances it is desirable, for judgment evaluation purposes, to compute the project costs involved in providing such environmental protection; an example would be the amount of project cost incurred to maintain some dependable rate of flow downstream from a dam for fishery maintenance.

Cost allocations are usually made to some project functions that are environmental in nature; in Federal projects these include fishery improvements in reservoirs, and recreational values created by reservoirs and improved waterways. These costs, which are normally non-reimbursable in Federal projects, are calculated to segregate them from other costs which are repayable by other project beneficiaries.

Cost of Environmental Improvements

If a project plan was first formulated to achieve optimum results without regard to environmental quality (EQ) improvements (Plan "A"), then reformulated to include desired amounts of EQ improvements (Plan "B"), one of three changes would result:

1. Increased costs, to provide the EQ benefits, or
2. Decreased accomplishments of some other project function,* or
3. A combination of the two.

The cost of including EQ improvements in the plan is therefore the sum of (1) increased investment and annual costs of Plan "B" over Plan "A", and (2) the *net* value of reduced non-environmental benefits. The net value being the benefit value attributable to the function accomplished, minus the project cost of providing it.

---

*Example – Power generation might be omitted or reduced because of downstream flow considerations of environment.

Payment of Environmental Improvement Costs

With the newer interest in planning and constructing water resource developments designed to improve the related environmental qualities there is prospective need to make allocations of the cost of providing the environmental improvements. There may be a question as to who pays the cost of the environmental improvement. One proposal, for the case of water quality improvement by Federal projects in the United States is that the investment costs shall be borne equally by the Federal Government and non-Federal entities, and the operation, maintenance, and replacement costs of this function are to be a non-Federal responsibility. It is logical to expect that all or nearly all EQ improvement costs will be non-reimbursable (i.e., paid by Federal or State governments) for the reason that the benefits are widespread, and any practical means of directly collecting revenues for the improvements are lacking. Even so, there is need to segregate the costs of environmental improvements from other costs in a multipurpose developmentt.

Federal Policies of Reimbursement and cost sharing of environmental quality improvement costs were reviewed as of October 1973.

Environmental Cost Allocations

Beginning October 1973, water use and control agencies of the United States government are allocating project costs, including environmental costs, by procedures established by the Water Resources Council (see a following paragraph for full title and reference to the controlling document, published in the Federal Register on Sept. 10, 1973). With respect to a multi-objective project which includes an environmental quality (EQ) improvement function, methods are specified for allocating costs to all functions including the EQ function.

The method is a modification and extension of the separable costs remaining benefits (SCRB) method described in detail in another paragraph of this Handbook chapter. The need for a modified method occurs when an EQ objective not economically justified is included in the plan. In principle, when costs are added to a plan for an EQ objective, or when other benefits of the plan are reduced to provide the EQ benefit, the net economic costs of so doing are allocated to the EQ objective. As indicated in the preceding paragraph, the possible need of such EQ cost allocation arises, or may arise, from the question of who pays for the cost of providing the environmental quality benefit, and how much.

If site and size selection of a dam and reservoir appears to be critically dependent upon the amount of cost allocated to environmental quality improvement, together with other functions, the project planner will wish to peruse the Water Resources Council document, Chapter VII, "Cost Allocation, Reimbursement, and Cost Sharing" for further explanation and illustration of the Water Resource Council's proposed allocations methods.

Principles of an Equitable Cost Allocation*

These may be summarized as follows:

1. In a multipurpose project, the provision of one facility to serve more than one purpose is more economical than providing individual separate facilities.
2. Cost of joint-use of facilities shall be equitably distributed among the purposes served.
3. Savings effected by joint-use of facilities shall be distributed among all purposes served.
4. Costs assigned (maximum allocation) to any one purpose shall neither be greater than benefits it produces nor than the cost of a single purpose alternative producing equivalent benefits.

---

*Reference: United Nations Water Resource Series Bulletins Nos. 26 and 41.

TABLE 1-2. Cost Allocation by Separable Costs-Remaining Benefits Method

To illustrate the separable costs-remaining benefits method of cost allocation, a project has been assumed which serves the purposes of flood control, irrigation, hydro-power, domestic and industrial water supply, fisheries and navigation.

The project facilities consist of a dam and reservoir with appurtenant works, irrigation canals and distribution works, a power plant and transmission facilities and a domestic-industrial water pipe line.

A summary of construction costs is shown in the following table A.

TABLE A. CONSTRUCTION COSTS

Feature		Construction cost (In thousand of dollars)
Dam and reservoir		34,500
Includes power penstocks	$500,000	
Irrigation outlet	$300,000	
Damsite water outlet	$100,000	
Irrigation canal and distribution system		2,500
Power plant and transmission facilities		2,000
Domestic-industrial water pipeline		500
Total construction cost		39,500

The estimated average annual benefits which the project will produce are shown in table B.

For the purposes of this illustration, the project is assumed to have a useful life of 100 years. At the end of this period the salvage value is assumed nil. Some of the facilities, such as power plant machinery, gates, etc., will have a life of less than 100 years.

TABLE B. BENEFITS
(In thousands of dollars)

Annual flood control benefits	500
Annual irrigation benefits	2,500
Annual power benefits	575
Annual domestic-industrial water benefits	750
Annual fisheries benefits	90
Annual navigation benefits	100
Total annual benefits	4,515

In this analysis it is assumed that these short-life items will be replaced at the end of their service lives; therefore a replacement item is added to the annual operating and maintenance costs. The replacement item is the annual sinking fund deposit which will accumulate to the cost of the replaceable facility by the end of its service life. The annual operating, maintenance and replacement costs of the project are shown in table C.

TABLE C. ANNUAL OPERATION, MAINTENANCE AND REPLACEMENT COSTS
(In thousands of dollars)

Dam and reservoir	25
Power plant and transmission facilities	150
Irrigation facilities	110
Domestic and industrial water	15
General expense (administration, overhead, etc.)	35
Total annual costs	335

The separable costs of the various purposes are shown in table D.

TABLE D. SEPARABLE COSTS
(In thousands of dollars)

Cost items	Flood control	Irrigation	Power	Domestic and industrial water	Fisheries	Navigation
Construction costs						
Entire project	39,500	39,500	39,500	39,500	39,500	39,500
Cost with purpose excluded	31,500	32,700	32,800	37,900	39,500	39,500
Separable cost of purpose	8,000	6,800	6,700	1,600	0	0
Annual costs						
Entire project	335	335	335	335	335	335
Cost with purpose excluded	330	215	185	320	335	335
Separable cost of purpose	5	120	150	15	0	0

For use in this illustration of cost allocation, estimates of the cost of single purpose alternates were prepared for the flood control, irrigation, power and domestic and industrial water purposes. These alternatives represent the most likely installations which would be built in the absence of the project and which would produce benefits equivalent to those produced by the project. In the case of flood control, irrigation and municipal and domestic water, the major feature of each alternative is a single purpose reservoir. In the case of power the alternative is a thermal plant with the same capability and production as the project plant. The power benefits produced by the project are also the cost of power produced by the alternative, so in this instance the capitalized benefits and alternative cost are identical. No reasonable alternatives are considered possible for fisheries and navigation purposes. The cost of the single purpose alternatives are shown in table E.

In the example allocation which follows construction costs and annual operation maintenance and replacement costs are allocated concurrently. If other items of annual cost such as taxes, etc., are considered appropriate they can be included in the same manner. In the example, in order to place all items on the same time level, the average annual benefits and annual operation, maintenance and replacement costs have been capitalized (converted to a lump sum by dividing by the "capital recovery factor" for 4 1/2 per cent interest over 100 years) for comparison with the construction costs. The interest rate is arbitrary and the time period represents the assumed useful life of the project. An equally acceptable procedure which would yield the same results would be to use the average annual benefits and annual costs with an annual equivalent of the construction costs.

TABLE E. CONSTRUCTION COST OF SINGLE PURPOSE ALTERNATIVES

(In thousands of dollars)

Flood control	20,000
Irrigation	28,800
Power	12,622
Domestic and industrial water	15,600

TABLE F. SUMMARY OF COST ALLOCATION SEPARABLE COSTS – REMAINING BENEFITS METHOD

(Unit, thousands of dollars)

ITEM	Flood control	Irrigation	Power	Domestic and industrial water	Fisheries	Navigation	Totals
1. Costs to be allocated							46,853
a. Construction costs							(39,500)
b. Operation, maintenance and replacement costs (capitalized)							(7,353)
2. Benefits (capitalized)	10,975	54,875	12,622	16,462	1,975	2,195	99,104
3. Alternative costs	20,000	28,800	12,622	15,600	—	—	—
4. Justifiable expenditure	10,975	28,800	12,622	15,600	1,975	2,195	72,167
5. Separable costs	8,110	9,434	9,992	1,929	—	—	29,465
a. Construction costs	(8,000)	(6,800)	(6,700)	(1,600)	—	—	(23,100)
b. Operation, maintenance and replacement costs (capitalized)	(110)	(2,634)	(3,292)	(329)	—	—	(6,365)
6. Remaining justifiable expenditure	2,865	19,366	2,630	13,671	1,975	2,195	42,702
7. Per cent distribution	6.7	45.4	6.2	32.0	4.6	5.1	100.0
8. Remaining joint costs	1,165	7,894	1,078	5,564	800	887	17,388
a. Construction costs	(1,099)	(7,446)	(1,017)	(5,248)	(754)	(836)	(16,400)
b. Operation, maintenance and replacement costs	(66)	(448)	(61)	(316)	(46)	(51)	(988)
9. Total allocated cost	9,275	17,328	11,070	7,493	800	887	46,853
a. Construction cost	(9,099)	(14,246)	(7,717)	(6,848)	(754)	(836)	(39,500)
b. Operation, maintenance and replacement costs	(176)	(3,082)	(3,353)	(645)	(46)	(51)	(7,353)
10. Annual operation, maintenance and replacement costs	8	141	153	29	2	2	335

Explanation of table F

Item 1. Shows total cost to be allocated composed of:
    a. total construction costs.
    b. total annual costs, capitalized at 4 1/2% over a period of 100 years.
Item 2. Shows the benefits given in table B capitalized at 4 1/2% over 100 years.
Item 3. Shows the cost of single purpose alternates given in table E.
Item 4. Justifiable expenditure is the lesser of lines 2 and 3.
Item 5. The separable costs given in table D composed of:
    a. separate construction costs.
    b. separable annual operation, maintenance and replacement costs capitalized at 4 1/2% over 100 years.
Item 6. Remaining justifiable expenditure is the remainder after subtracting line 5 from line 4.
Item 7. The percentage distribution of line 6, column 8, into its component parts, columns 2 to 7.
Item 8. Remaining joint costs distributed according to percentages shown in line 7. The total joint cost shown in column 8 is the difference between the total separable cost, line 5 - column 8, and the total cost to be allocated (line 1 - column 8).
Item 9. Total allocated cost is the sum of the separable costs (line 5) and the allocated joint costs (line 8).
Item 10. Average annual operation, maintenance and replacement costs as allocated to the various project purposes.

5. The minimum cost allocated to any one purpose shall not be less than the identifiable costs (specific or separable costs) incurred for the said purpose alone.

With reference to the above, specific costs are the costs of individual physical features that serve only one purpose; for example, a power plant. Separable costs are the costs which could be omitted from the total cost of a project if any one purpose was excluded. The separable cost of any one purpose is derived by subtracting the estimated cost of the multipurpose project with said project excluded, from the cost of the project with all purposes included.

Computation of Cost Allocation by SCRB Method

Table 1-2, pages 42 and 43 reproduced from United Nations Water Resource Series Bulletin No. 26, describes and illustrates a cost allocation by the SCRB method. The principal objection to the method is the large amount of input data required. In addition to the estimated benefits and total costs, which are required in other analyses, cost estimates are required for several other sizes of project plans: the construction and annual costs of separate, single-purpose projects to supply the equal, single project accomplishment.

Computation of Cost Allocation by the Alternative Justifiable Expenditure Method

This is a simpler method than the SCRB method. It requires fewer designs and cost estimates to calculate an allocation, but the procedure is less rigorous and hence more susceptible to dispute.

It is based on an arbitrary segregation of facilities into two types: joint facilities (those which serve more than one function), and specific facilities (those which serve only a single function). In this procedure the entire cost of the dam and reservoir is considered as a joint facility even though there may be dead storage which provides only power head, or exclusive storage space which serves only a single function such as flood control or irrigation. Also, in this application, it is assumed that embedded penstocks in the dam or a powerhouse constructed in the dam are specific power facilities even though the elimination of such facilities would not result in savings equal to the cost of the facilities removed. For example, if such facilities were removed, the voids left would have to be filled.

Table 1-3 shows an example computation, with explanatory footnotes. After determining the justifiable expenditure, the specific costs (costs that serve only one function such as the cost of irrigation ditches, power generators, or flood control outlet works) are subtracted from the total justifiable expenditure to secure the remaining justifiable expenditure which is then used as the basis for allocating the costs of the facilities serving more than one function (joint costs).

SELECTION OF DAM SITES AND SIZES

Summary of the Process Steps

Upon completion of the feasibility-grade investigations the sites and sizes of project dams have been provisionally selected. It is probable that these selections will prove to be the final ones, but the possibilities remain of making some further adjustments in the final, pre-construction phases of investigations. An adjustment of capacity of a storage reservoir may prove to be necessary; this, or any new design data may disclose the need of changing the axis location of the dam even though the reservoir site remains the same. Latest criteria on the project functional requirements of a diversion or afterbay dam may require an adjustment of dam height, or in an exceptional case a change of the site.

In the shaping up of the project plan to this point the following steps will have been taken affecting dam site and size selection:

TABLE 1-3. Example of cost allocation by the Alternative Justifiable Expenditure Method
Benefits and costs in thousand dollars or other monetary unit

Items	Project functions				
	Irrigation	Power	Flood Control	Fish and Wildlife	Total
1. Annual costs to be allocated					3,000
2. Annual benefits	2,300	2,500	2,400	50	7,250
3. Alternate annual costs	1,700	2,800	1,800	(a)	
4. Justifiable expenditure	1,700	2,500	1,800	50	6,050
5. Specific costs	0	1,100	0	0	1,100
6. Remaining justifiable expenditure	1,700	1,400	1,800	50	4,950
7. Proportion – percent	34.3	28.3	36.4	1.0	100.0
8. Allocation of joint costs	652	538	691	19	1,900
9. Total allocation	652	1,638	691	19	3,000

Notes:

Item 1. The total annual cost is investment cost (construction cost plus interest during construction), amortized at the appropriate interest rate and estimated useful life period.

Item 2. Annual benefits are adjusted to annual equivalent benefits if there is a growth period for any project function.

Item 3. Alternative annual costs are the costs that would be necessary to provide each function by means of a single-purpose project.

Item 4. The lesser of the annual benefit or the annual cost.

Item 5. A cost incurred only for a specific function – in this example, the cost of power facilities.

Item 6. Item 4 minus Item 5.

Item 7. Calculated by dividing the remaining justifiable expenditure for each function by the total of the remaining justifiable expenditure.

Item 8. Calculated by applying the percentages of Item 7 to the total joint cost, which is Item 1 total minus Item 5 total.

Item 9. The summation of Item 5 and Item 8.

(a) An estimate is not necessary if it is apparent that annual costs would exceed annual benefits.

1. The project functions will have been decided. The project may be single-purpose, or may include two or more functions. This is determined by the plan formulation studies in which evaluations are made to test whether the inclusion of more than one project function is economically justified.

2. The locations of dams and reservoir sites will have been decided. This is done as the result of investigations of the site alternatives, usually in the reconnaissance phases of studies, but subject to possible changes and corrections in later stages. The site selections will have been made by evaluating the physical characteristics of each site, making comparative estimates of costs and potential accomplishments for each site that would be adaptable to a project plan. Consideration will also have been given to whether one site, or set of sites, would have important and possibly overriding advantages or disadvantages to the next-best available alternative from environmental impact standpoints.

3. The gross capacity of any storage reservoir will have been selected. This will have been done by (a) estimating the amount of sediment accumulation that will take place in the reservoir over its economic life period; (b) estimating the amount of

inactive storage capacity needed to maintain a permanent minimum pool for fishery preservation, recreational value, and/or power plant minimum head; and (c) determining the justifiable amounts of active storage capacity for reservoir operation for the one or more project functions. The amounts of active storage capacity will have been selected by the plan formulation procedures whereby costs and accomplishments are estimated for a range of reservoir sizes, benefit values assigned to the different kinds and variable amounts of accomplishments, and comparative economic evaluations made for the range of reservoir sizes. Again, consideration will have been given to whether there would be a serious adverse environmental impact if water was impounded higher than any critical elevation.

The final adoption of dam sites and sizes occurs after a project has been authorized and funded for construction. Some adjustments may be made to either sites or sizes provisionally selected by the feasibility investigation. The adjustments may be warranted by improvements in data collected for making final designs, or by any changes in project scope as required by financial and repayment considerations and limitations.

PRECONSTRUCTION PLANNING

Final designs and specifications for construction are made after a project is authorized and funds provided for that purpose. This normally follows after the feasibility investigation and report have been completed, the report reviewed by all concerned people, and various ideas and suggestions advanced by reviewers to adjust or modify the project plan. Preparatory to making final designs it is appropriate or necessary to consider adjusting the plan to the extent warranted by constructive suggestions, by any possible changes of project objectives and scope, and by any new important basic data that have been obtained during the time interval between the feasibility investigation and the final designs. An example of the latter might be the occurrence during this period of an extreme low runoff season, or a great flood which could alter estimates of critical dry year project water supplies or of flood magnitude-frequency occurrence.

In preconstruction planning, further work is usually necessary or desirable to obtain topographic mapping of greater detail and accuracy than previously obtained, and to make additional geologic investigations. The latter are to reduce uncertainties about foundation conditions at structure sites and about construction materials. Additional sampling and testing of construction materials would almost certainly be done. New and upgraded data on the physical factors can improve the designs and the accuracy of cost estimates, and reduce the contingency allowances for unforeseen or unexpected construction costs.

Plan adjustments may be made for any changes in economic criteria, or for need to give greater or lesser consideration to social and environmental influences than was previously done.

The possible and probable plan adjustments at the preconstruction stage usually require the services of a project planner and associated specialists performing the same categories of work as done in the feasibility-grade investigations. It is advantageous to have a nucleus of the same people in the latter investigations that were in the former.

COLLECTION OF BASIC DATA; QUALITY CONSIDERATIONS AND STANDARDS

Three Levels of Investigation

The required accuracy and reliability of various basic data used in planning depends upon the purpose and intended use of the investigation or study in which the data are used. Three levels of investigation call for progressive degrees of reliability: (1) the reconnaissance investigation, intended to show whether further investigations are warranted; (2) the feasibility investigation, to show whether a project plan should be authorized for construction; and (3) preconstruction investigation, to provide for final designs and specifications for construction of project works, and project financing and repayment. Accuracy of data needed for the three levels is somewhat variable, depending upon its importance in a particular plan. The following remarks on data quality are intended as general guidelines in judging the adequacy of available data and the desired standards of quality of new data to be obtained.

Basic data most usually required for planning of dams and reservoirs are in the following categories:

Hydrology data
Streamflows
Flood flows
Evaporation
Sedimentation
Water quality
Water rights*
Tailwater curves

Geologic data
Reservoir sites
Dam sites
Construction materials

Topographic surveys
Watersheds
Reservoir sites
Dam sites
Borrow areas

Legal data
Water rights
Reservoir site cost data
Land acquisition
Clearing
Relocations
Environmental factors
Fish and wildlife
Recreation and scenic
Historical and
    archeological
Economic data
Economic base for area
    benefitted
Crop data
Land classifications
Market data for various
    purposes

In the following paragraphs the desirable quality of data of each item is briefly indicated for the three levels of investigations.

1. Stream Flow Data

General. Stream gaging varies in accuracy, depending upon physical conditions at the gaging station site and on flow measurement problems. Most published reocrds (example: U. S. Geological Survey) state the probable accuracy of measurements. With some exceptions, recent records tend to be more accurate than older ones. Modern equipment permits continuous recording of water levels and electronic computations of flows therefrom. Some of the early records were obtained by periodic (usually once-a-day) readings of a staff gage, with less accurate results. Because of seasonal and annual variations of runoff, a record of several years is necessary to have good information on wet-year, average, and dry-year stream flows. Longer records are desirable for streams having erratic flows (from desert or arid watersheds) than of streams carrying more sustained flows (from moist watersheds, ground-water drainage, snow melt or ice melt).

Reconnaissance Data

Estimates for various sites can be made by applying a factor to records of a gaging station in the same geographical area. The factor is a ratio of watershed areas and proportional precipitation, to the extent the latter can be estimated from any precipitation records and probable distribution of precipitation as caused by landforms.

Feasibility Data

At least a short stream gaging record at or near a dam site is necessary. Flow estimates can be made therefrom by correlation with concurrent measured flows at a more distant site having a longer record.

Preconstruction Data

A stream gaging station record usable for feasibility data should be continued for additional data.

---

*Hydraulic and hydrologic interpretations from legal water rights data.

2. Flood Flow Data

General. Measurements of high flood flows are very difficult, and often impossible to make, because of adverse weather and travel conditions, debris carried by the flood, shifting streambeds, and flood overflow areas. Maximum elevations (high watermarks) are observed and streamflow rating curves are extrapolated upwards to estimate a peak flood flow.

Reconnaissance Data. Flood flow at a dam site may be estimated from an estimate elsewhere by proportionate watershed areas and storm precipitation patterns, and application of hydrological principles to storm and flood events.

Feasibility Data. Flood magnitudes and frequencies are estimated from all available data of records of precipitation in the watershed and flood runoffs resulting therefrom. Where there is a paucity of such information for the watershed under study, use will be made of additional data for a nearby and similar watershed. Hypothetical storms and floods may be devised by transposing a known storm pattern from one watershed to another.

Preconstruction Data. Additional flood and storm data will be collected for any flood occurring after the feasibility investigation. If not already existing it will be desirable to install two types of observation stations: (1) a peak water stage recorder at or near the dam site; and (2) one or more automatic-type raingages in the watershed — these being of a type that will measure and record rainfall in short time intervals such as 15 min.

3. Evaporation Data

General. Evaporation from a reservoir water surface is usually an appreciable loss of project water supply. Where the reservoir site is desert or arid land the project evaporation water loss nearly equals the gross reservoir evaporation. At the other extreme, where the reservoir site is a swampy area or area of luxuriant vegetation growth the reservoir evaporation loss may be no greater than the pre-project evapotranspiration loss for the same area. Evaporation is measured by evaporation pans designed for the purpose. Evaporation varies with climatic conditions, but records show that annual evaporation does not vary a great deal from year to year and so a record of two or three years is sufficiently representative for project planning purposes.

The simplest, but not entirely accurate method of measuring evapo-transpiration loss is by use of atmometers. Lysimeters are also used for this purpose. It is seldom necessary to measure evapo-transpiration rates of reservoir sites as it can be estimated sufficiently well for the need by noting the predominant types of ground cover and applying known water consumption rates for such vegetation.

Reconnaissance Data. Reservoir evaporation loss is a minor factor and may be ignored or roughly estimated by experiential knowledge of similar sites.

Feasibility Data. If a reservoir area is to be large in proportion to its stream flow supply, an evaporation pan record is collected from an installation in the same climatic region; if none is available an installation should be made and records obtained during the course of the investigation.

Preconstruction Data. If the reservoir will be large it is of some value to obtain an evaporation record at or near the reservoir site for refinement of previously used evaporation data.

4. Sedimentation Data

General. The topic is discussed in a preceding paragraph entitled "Sedimentation rates," page 8.

Reconnaissance Data. First, it is very important to ascertain, at the very start of an investigation, whether the stream under consideration transports large or small quantities of sediment. Existing data on all known factors and evidences of sedimentation should be collected for making preliminary judgments on the magnitude of the problem. Listed somewhat in the order of their value or importance, these data would include:

1. Sedimentation rate taking place in an existing reservoir in the same watershed or a comparable watershed (this information frequently not available).
2. Bed-load sampling of sediment movement of the project stream for a range of high and low flows (this information frequently not available).
3. Suspended load stream flow sampling.
4. Classification of erosion characteristics of watershed land. Such classification surveys and studies have sometimes been made by land use and land management agencies (examples in the United States: the Soil Conservation Service and Forest Service).
5. Present and prospective uses of watershed lands affecting erosion rates (agricultural, mining, or timber harvesting operations and road construction associated therewith). Some data are available from various land use agencies.
6. Rainfall intensity-frequency data. Probable frequency of occurrence of intense storms. Data from weather observation agencies.
7. Air-borne sediment. In a rare situation, a reservoir may be located in a barren region of sandstorms and in the path of sand dune formation and movement. The amount of such sediment trapped by a flowing stream is not detected except as part of bed-load sediment measurements if such are made. If wind-blown sediment movement is intense and thought to be of appreciable importance in reservoir sedimentation rates a sampling program might be necessary. Sampling facilities would consist of sediment traps to detain and measure the coarser grains of material in the wind-blown sediment.

Feasibility Data. Obtain data to fill in the most important deficiencies noted in reconnaissance data collection. Possibly establish sediment sampling stations; make re-surveys of sediment accumulation in existing reservoirs.

Because of possible effects on designs and cost of structures it becomes important at this investigation phase to obtain information on the probable amount of streambed degradation below a dam due to sediment detention in the reservoir. Desilted water released from a reservoir will pick up a new load of sediment from the streambed thus lowering its level to a new hydraulic gradient until or unless limited by bedrock or large stream cobbles and boulders or a downstream control structure. Best data on this phenomenon is the operational experience of an existing dam on the same or similar stream, if available. Other pertinent data are channel depth measurements to bedrock and existence of large cobbles and boulders in the channel materials. If a stream gaging station has been operated at or near the dam site for a considerable number of years the variation of its rating curves, with respect to water surface elevations of low streamflows, provides an indication of how much a channel has scoured during flood flows and a clue of how much it will permanently scour by reservoir detention of sediment.

Preconstruction Data. Data collection for this phase of an investigation would consist of an extension and upgrading of any deficient data of the feasibility investigation.

5. Water Quality Data

General. These data pertain to water quality other than sediment content. Quality constituents of concern may include persistent turbidity, dissolved gasses, dissolved or suspended organic materials, dissolved minerals, and water temperatures. These properties are of concern depending upon the intended uses of the project water supply and sometimes upon the local environmental effects of the project development. All these properties are quite accurately measurable, and in modern stream gaging programs some have been measured or sampled at some gaging stations to the extents thought useful. Additional measurement, sampling, and analysis over a period of a few years during project investigations can usually provide any lacking or insufficient data to the levels of accuracy and reliability needed.

Reconnaissance Data. For any water quality question of particular concern, obtain sufficient data to check magnitude of the problem, if any.

Feasibility Data. Obtain additional data as necessary to evaluate any quality question or problem.

Preconstruction Data. Collect additional data where desirable to improve reliability by longer records.

6. Water Rights Data

General. Project accomplishments, except flood control, are limited by any and all prior entitlements to use of portions of stream flows by others. Entitlements may have been established by jurisdiction of the State of origin and use, or by treaties or similar-type agreements between states or nations. Rights may be in dispute. Further, it may be foreseeable that disputes are likely to arise at a future time when rights and entitlements are not clearly established and agreed upon.

Reconnaissance Data. Where there is any possibility of water shortages among users in future conditions, which is almost always the case, it is necessary to know the magnitudes and times of entitlements of use of the major users before continuing an investigation beyond the reconnaissance stage. Users of very minor amounts of stream flows may be disregarded at this time. Documents of legal entitlements are a starting source of information. Where quantities of water use are not specific in legal records it is necessary to collect records of the quantities of water diverted, and possibly even make flow measurements where records of use are inadequate or in question. Possibilities of excessive and nonproductive diversions of water may be noted.

Feasibility Data. Water right information of all possible concern and question is refined and updated. It may be desirable or necessary to collect data on potential future non-project uses of water that might acquire rights to water use in preference to the project plan consideration (example: upstream or riparian preference uses future entitlements) and the water quantities involved. A project plan may propose some exchange of water service with present users for greater over-all economy, in which event data is collected on the amount, places, and conditions of water use considered for exchange.

Preconstruction Data. All additional data necessary to establish a dependable, non-disputable entitlement to the conditions of use of stream flows for the project functions are collected.

7. Tailwater Curve Data

General. A tailwater curve is used in designing a dam and adjacent structures to show the water surface elevation that will occur downstream from the dam for various rates of flow releases, including spillway discharges of flood flows. The information is needed for the design of stilling basins and of a power plant if one is built at the dam.

Reconnaissance Data. Cost estimates of reconnaissance-grade accuracy usually do not require the use of a tailwater curve.

Feasibility Data. A tailwater curve is developed by hydraulic computations from surveys of river profile gradient and cross sections, and factors of channel roughness for a "control distance" below the dam site. The control distance is sometimes not known until trial computations are made. Approximately, it may be a distance in which the streambed drop is twice the river depth occurring with a large flood. Cross sections of the river channel are surveyed at intervals chosen to be representative of the flow sections over the length of the control distance. Visual observations are made of the channel roughness condition and quantified by reference to standardized coefficients of hydraulic roughness. Data are needed for making an estimate of the amount of channel degradation to be expected below the dam; this is discussed in the preceding section on "Sedimentation Data."

Stream gaging station rating curves, which relate stream flows to gage heights, are useful data if the gaging station is located near enough to the dam site. If the gaging station is within the tailwater control distance below the dam site its rating curve will provide a valuable check on the computations of a tailwater curve. In the absence of a gaging station in the tailwater control zone it is very desirable to install and operate a gaging device for a few years that will record peak high water elevations. The data so obtained is used with data of measured flows at the nearest upstream or downstream gaging station.

Preconstruction Data. In addition to data collected for the feasibility-grade designs, construction and testing of a hydraulic model of the tailwater section of the stream may be warranted in some instances. If so, additional topographic and other surveys may be needed to construct an accurate model.

8. Reservoir Geologic Data

General. The possible concerns are of excessive reservoir leakage, and landslides or rock slides into the reservoir area, as discussed in Chapter 4. Excessive leakage may occur where a part of the reservoir rim is a narrow ridge or where underlying formations are porous types of rock such as limestone or lava. Landslides occur where earth materials are situated on steep hillside slopes. Slides usually take place when the material is wetted from excessive precipitation, or when a part is inundated by reservoir storage. Large rock slides occur from very steep slopes from the same causes. A very large land or rock slide into a reservoir can be a disastrous event, which should never be knowingly risked. Small slides can be tolerated by taking precautions to avoid any use of the slide risk areas for road relocation routes or any other form of improvements. In some situations preventive measures can be taken to reduce the likelihood of a landslide by constructing surface or subsurface drains to prevent excess wetting of a slide area, or even by excavating slide material and disposal elsewhere. These are obviously expensive measures.

Reconnaissance Data. The possibilities of excessive leakage and slides should be given attention early in an investigation. The extent and character of any leaky formations may be noted from regional geological and topographic maps if such are available, supplemented by visual field inspections. Aerial photographs are very useful, and should be procured for this and other purposes. Inspection for slide risk areas should be made by the same methods and by use of the same source information. Occasionally, a land use agency will have mapped the reservoir site area and identified locations and types of soils which are susceptible to sliding when on steep slopes.

Feasibility Data. Where leakage and slide risks are found from reconnaissance inspections there is need to examine the risks in quantitative terms. Detailed topographic and geologic mapping of the potential problem areas may be necessary. Some augering, boring, trenching or other subsurface exploring may be needed to enable a geologist to estimate the thicknesses and adverse characteristics of the suspected formations. Data on experiences resulting from road construction in comparable slide areas are useful.

Preconstruction Data. Further data may be required to definitely specify the preventive and/or corrective measures to be taken during project construction. Such specifications might include: (1) prohibition of use of materials from slide areas for construction materials, road relocations, or access roads; (2) blanketing of leakage areas within the reservoir site with impervious material.

9. Dam Site Geologic Data

General. The data requirements for foundation investigations are discussed in Chapter 4.

Reconnaissance Data. The investigation at this phase is to find out if the site appears to be free of any major adverse qualities. Such qualities might include: (a) badly fractured rock in abutments, (b) landslide or rock slide areas at the dam site, (c) large depth to bedrock below stream channel, or (d) existence of an ancient, refilled stream channel through an abutment. It is usually possible, where there are exposures of bedrock at or near a dam site, to arrive at an opinion of whether the dam site appears promising, based on visual observations. In some cases, however, it may be necessary to do limited amounts of subsurface exploration by augering, digging test pits, or trenching to get any reliable indications of foundation conditions.

Feasibility Data. As indicated in considerable detail in subsequent chapters, programs of core drilling are standard procedures for feasibility investigations. This drilling is supplemented by making water pressure tests of leakage through the rock formations explored, and sometimes by digging exploratory tunnels, auger holes, test pits, or trenches. The quantity of such work varies with each dam site; the total work requirement is enough to enable moderately accurate designs and estimates to be made showing the suitable layout for dam and associated structures, the amounts of excavation required to place the dam on sound foundation, and the amount of corrective work needed, such as to make the foundation watertight by grouting.

Preconstruction Data. The feasibility investigations are supplemented to provide design data for the particular type of dam being designed. The further explorations may include the testing of strength of bedrock, either in place or as large samples.

10. Construction Materials Data

General. The data requirements for construction materials investigations are described in Chapter 3.

Reconnaissance Data. The need is for sufficient exploration to be able to identify where the various types of construction materials — pervious, impervious, rock, and concrete aggregates — of satisfactory qualities would likely be found. Surface inspections with topographic maps or aerial photos may be sufficient. Some sampling and testing may be necessary if there is no experience record of the use of materials from these sources for construction elsewhere.

Feasibility Data. The program will include laboratory tests of the structural properties of the needed types of construction materials, from samples taken by field explorations. The areas and volumes of sampled deposits of materials are surveyed, with auger holes or test pits dug to determine depths.

Preconstruction Data. Depending upon the type of dam and other structures being designed, the explorations and testing are extended to verify the amounts and suitability of available materials.

Topographic Surveys

General. Topographic maps illustrate land forms; they usually show cultures and land ownership boundaries; they show true distances between points, and true elevations. Their contour lines show watershed boundaries, prospective dam and reservoir sites, and water conveyance routes. Used as a base for layout of project plan features, maps provide the means of measuring and computing water storage volumes of reservoirs, volumes of excavation and embankment for dams, canals, dikes, and levees, and related structural features.

The accuracy required of a topographic map depends upon its specific use. Vertical (elevation) accuracy is particularly important for most uses of topographic maps in water resources development because of the importance of water levels and gradients. Except for final designs and calculation of construction payments the accuracy of the horizontal map scale is less critical. A customary standard of accuracy of a topographic map is that no contour line be drawn on the map out of its correct position by more than one-half the contour interval of the map. This sets standards for accuracy requirements of benchmarks, as well as the plotting of contour lines.

Various agencies have devised "standards" of scales and contour intervals of topographic maps for different uses; however, they are not fixed standards. Instead, they are varied between a "large" and a "small" reservoir, or between "gently-sloping" and "steeply sloping" land. These being only relative terms they do not serve to set desirable standards in all cases. In setting a contour interval standard it is always necessary to keep in mind the purpose to which a topographic map is to be used. The original scale of a map, or the scale at which it is furnished for use is important only for legibility of the map contents, since it can be accurately enlarged or reduced in scale according to need by reliable optical projection equipment.

The following contour intervals of maps are suggested, based on the purpose of use and the needed accuracy.

Watershed Areas

General-purpose topographic quadrangle maps are normally satisfactory for delineating watershed boundaries. Greatest difficulty in finding a watershed boundary may occur in a region having several or many lakes. In such an instance, the map information may be best supplemented by stereoscopic aerial photos.

Reservoir Sites

A contour interval that provides five or more contour lines in a reservoir site is sufficient for calculating reservoir capacity for a reconnaissance study. For feasibility-grade studies the reservoir topographic map should contain ten or more contour lines. The imprecise nature of all streamflow measurements and estimates does not warrant extremely detailed reservoir site mapping for a capacity computation. For preconstruction purposes a special survey of the reservoir maximum water line may be warranted for rights-of-way purposes.

Dam Sites

For reconnaissance estimates, mapping which provides ten or more contour lines within the height of the dam under study is satisfactory. This mapping would be available from feasibility-grade reservoir mapping. For feasibility-grade designs and estimates of dams and related structures the topographic mapping should have a contour interval to provide at least 20 contour lines within the height of the dam. This mapping may be supplemented by special surveys for preconstruction needs.

Borrow Areas

Topographic mapping at the same contour interval used for dam sites is usually desirable. Special additional mapping may be necessary for preconstruction sampling and specifications.

Reservoir Site Cost Data

General. The principal items are land acquisition and land clearing costs, and the costs of providing the necessary relocations of existing facilities. The uses of the land may include agriculture, forests, mining, local industries, business establishments, and residences; all normally in private ownerships. There may also be publicly-owned land in parks and forests, school land and facilities, possibly other public institutions, and cemeteries. The principle involved in private property acquisition is to fairly and equitably compensate each landowner and tenant for his loss of being disposed from his property or tenancy. Compensation is usually made by cash purchase of land, and payments to tenants for costs and losses of moving to another location.

In some instances, public land and improvements might be replaced elsewhere, in which case the replacement cost would be a project cost. If there is no replacement of such land, there is no project cost, but there is a negative project benefit.* The moving of a cemetery is a very special problem requiring considerable time to establish a new cemetery, locate the concerned survivors of deceased persons, and arrange permission for relocating graves.

Facilities to be relocated may include roads, highways, railroads, underground pipelines and conduits, power transmission lines, and telephone lines (above ground or underground). Any type of relocation is an engineering and economic problem affected by the terrain available for relocation routes.

---

*Alternatively, this may be considered as an "economic cost."

The principle employed is to make the relocated facility equal in value of use to the original. Where this is not possible a cash payment may be required as an adjustment. Where a relocation is a betterment over the original condition it is sometimes possible to have the improvements funded by the agency or owner of the relocated facility.

Reconnaissance Data. Where it is evident that reservoir costs will be large, considerable attention is necessary at the initial investigation stage. Land acquisition costs can be roughly estimated by reference to any recent property sales in the vicinity. Clearing costs may be approximated from map or aerial photo measurement of amounts of land requiring clearing and cost experience of any similar operations. Possible routes for required relocations would be laid out on topographic maps in sufficient detail to note distances and types of terrain involved. The sizes or capacities of the facilities are ascertained, and costs roughly estimated by short-cut methods. Consultation with the owners or managers of such facilities is highly desirable for information on any planning they may have in progress, and for their views on future use requirements of the facilities.

Feasibility Data. At this stage further data are collected on prices of properties recently sold in the locality, and on the productivity of the land as used for agricultural, timber or minerals production. Data are obtained on the volumes and value of business transacted in the area. The numbers of family units and dwellings are counted, and school enrollment figures obtained. In addition to current-year data, information is obtained on growth trends, as indicated by recent changes and any known plans for new activities.

For required relocations of major utilities, agreements or understandings are obtained with owners or managers of the utilities as to a satisfactory relocation plan. Estimated costs are obtained from the utility agency, or alternatively, the standards of that agency are obtained for use in preparing designs and cost estimates of relocations.

Preconstruction Data. At this stage the previously collected data are updated for any new information on land uses and costs. It is very desirable at this time to accumulate and preserve every item of information pertinent to property acquisition and utilities costs for possible future use in questions or disputes arising after project construction or operation. Such data would include photographs of land and land uses, documented as to locations and dates; data on existing water supplies and systems in use, including wells; sanitation facilities, if any; sources of air and water pollution; and any notable favorable or unfavorable environmental qualities.

Environmental Factors Data

General. Environmental effects in service areas of water resource development or water control projects are discussed separately in subsequent paragraphs on the topic of "Environmental Investigations." In this paragraph attention is given to the local environmental effects of dams and reservoirs in areas that are directly altered or affected by them. These areas are usually the reservoir sites, and downstream sectors of streams from which project water is diverted. Some of the principal effects may be grouped under the following headings: (1) fish, resident and anadromous; (2) wildlife, land animals and waterfowl; (3) recreation, water oriented; (4) scenic; (5) historical; and (6) archeological. Some of these resources are commercialized to degrees; in some localities commercial fishing occurs along with sport fishing. Sport fishing, hunting, boating, and similar activities each have a commercialized aspect in the equipment supplied by manufacturers and vendors. Such commercial value is usually minor in comparison to other project effects and not important in economic terms. These various resources can therefore be principally regarded as environmental values.

In any kind of environmental evaluation the great problem is to obtain a method whereby environmental values can be quantified, or even ranked in order of importance. Extimates can be made in some cases of the number of person-hours or person-days that may be spent enjoying or seeking enjoyment of some environmental experience. That is about all. The methods and instruments to measure the degree of enjoyment of different things by different people have not yet been devised or accepted.

Approaches are made toward evaluating environmental factors in both pre-project and post-project conditions by (1) listing all environmental factors that can be thought of as existing for the location under consideration; (2) attempting to classify them as to relative importance with one another or with respect to a larger region; and (3) using a ranking scale, such as 1 to 10, to assign a quality rating to each element. The degree of success in classifying and ranking environmental factors is indicated by the amount of acceptance or lack of disagreement by all other interested and concerned persons. To have any appreciable quantitative value, a favorable environmental factor must be something that is or can be shared by more than one or a few persons. It is therefore something not destroyed by use, or is a renewable resource. Many of the products of nature, exclusive of those having economic value or needed by man to survive are therefore regarded as having environmental value.

Reconnaissance Data. Observations should be made and information collected of the principal types and amounts of existing environmental resources being used. Observations should also be made of any changes likely to take place in the future, without a project development, which would either increase or decrease the environmental resources and opportunities for their use and enjoyment. Then, by simple comparisons with experiences of similar water resource developments elsewhere, opinions are reached as to the probable magnitude of impact of a project plan on environmental values. This will provide guidance in making tentative or preliminary choices between plan alternatives.

The reconnaissance data collection may likely disclose a considerable lack of reliable information on some of the environmental factors that would be affected by a project plan. If the reconnaissance investigation shows that a feasibility-grade investigation should follow, steps should be taken to begin collection or upgrading of data that will be needed at the subsequent investigation stages.

Feasibility Data. One of the comprehensive data collection and evaluation methods described in a latter part of this chapter may be used as a control list for data collection. Statements should be obtained from the concerned environmental organizations on their views of environmental qualities that would be affected. Information on environmental quantities and qualities should be documented as well as the needs that can be foreseen by photographs and statistics; with particular attention given to those factors indicated to be the most important.

Preconstruction Data. Feasibility-grade data would be upgraded, particularly to support the need or merit of making any accommodations in the design of project features warranted for environmental considerations.

MEASUREMENT AND COMPUTATION OF RESERVOIR AREAS AND VOLUMES

Area Measurements

Because of irregular shapes of reservoir sites, water surface areas at various elevations are almost always surveyed and mapped as a topographic map at some appropriate scale and contour interval. Appropriate contour intervals are indicated in the paragraph headed "Topographic Surveys" and sub-heading "Reservoir Sites." The map is produced or reproduced at a scale which is convenient for accurately measuring the areas shown by contour lines above a dam or dam site by means of a planimeter. Readings of the planimeter are converted to areas in acres or hectares by the appropriate calibration factor of the instrument used, and the scale of the map being measured. A common practice followed to avoid large errors is to trace each contour line area twice without resetting the planimeter dial. The reading obtained by the first operation is noted; the reading obtained by the retracing step will be twice the first if no error has been made. The accuracy depends upon the care taken in tracing the contour lines with the planimeter, as well as the accuracy of the map itself. Planimeters, of which two or more types are manufactured, are reliable and accurate instruments for measuring map areas when properly used.

It is common practice and of convenience in use to plot an area-elevation curve from the map measurement results, with areas as the abscissa, or horizontal scale, and elevation as the ordinate or vertical scale.

Field surveys of land areas defined by a specific contour elevation, and computation of areas therefrom, are not necessary for the purpose of calculating reservoir volumes. In exceptional cases such a survey of a reservoir maximum water line may be needed at the time of project construction to establish the boundaries and areas of land to be used for the project reservoir and other facilities.

Volume Computations. Data on reservoir volumes (also termed reservoir capacity) are needed for various levels of water storage. Reservoir levels are sometimes termed the reservoir water stage. The common and satisfactory method employed is to compute the volume of each horizontal segment of a reservoir between each measured contour elevation area by the simple "average end area" method. That is, the mean, or average of two adjacent measured contour areas is multiplied by the contour elevation difference. The computed

PLANNING AND ENVIRONMENTAL STUDIES 57

TABLE 1-4. Example Computation of Reservoir Volume

Contour elevation (feet or meters)	Area (acres or $m^2 10^4$)	Average area (acres or $m^2 10^4$)	Increment volume (acre-feet or $m^3 10^4$)	Accumulated volume (acre-feet or $m^3 10^4$)
103	0			0
		5	35	
110	10			35
		17	170	
120	24			205
		31	310	
130	38			515
		46	460	
140	54			975
		60.5	605	
150	67			1580

volume quantities are accumulated to produce quantities of the total reservoir volume at each storage elevation. For convenient use the result is usually plotted as a graph of reservoir capacity versus storage elevation, with elevation again plotted as the vertical ordinate. The previously-described area curve and the volume curve are usually plotted on the same graph sheet and given the condensed title of "reservoir area and capacity curves."

An example computation of reservoir volume is shown in Table 1-4.

The simple "average end area" method of computing reservoir volumes is not theoretically precise. A slightly more accurate method, using the same input data, may result from the use of the "prismoidal formula" which is described in surveying and civil engineering textbooks. A difference in results by the two methods occurs only when and to the extent that the area-elevation curve varies from a straight line or simple linear relationship. No difference occurs if the area-elevation curve is a straight line. The slight difference in results does not warrant any objection to the use of the average end area method. Increased accuracy of computed volumes, if in question, can be assured only by a more-detailed reservoir area survey.

BASIC DATA FOR PROJECT FUNCTIONS

Various types of project benefits are briefly discussed in a preceding section (pages 9-13), entitled "The Economic Factors of Plan Formulation." The values that can be assigned to these factors or functions as project benefits are of prime importance in demonstrating project justification. The principal steps necessary in formulating a project plan are listed in the preceding paragraph "Selection of Dam Sites and Sizes – Summary of Process Steps." However, it is beyond the scope of this handbook to describe in detail the many processes employed and steps taken to choose, examine, and formulate a project function plan; then to evaluate its accomplishment for comparison against project cost and decision as to feasibility.

It is necessary for the project planner to be aware of these tasks and of the need of programs and a staff of specialists to perform them. Following are lists of basic data likely to be needed in evaluating the more usual major project accomplishments. Collection and use of these data, where pertinent, either precedes or is done concurrently with collection of basic data of physical factors for plan formulation and designs.

Irrigation function

Prospective service area boundary
Climatic data

Land classification for irrigation adaptability
Soil classification for drainage characteristics
Present agricultural conditions
    Presently-adapted crops
    Crop yields
    Frequency of crop failures, and causes
    Size of farm units – minimum; average; maximum
    Farm investment
        Land
        Machinery
        Motive power – animals or machine
Present irrigation water supply and use (if any)
    Ground water; amount used; quality
    Surface water diversions
    Area irrigated; crop irrigated
    Quantity of water use; rate of application
    Cost of present irrigation supply
    Water rights
    Irrigation drainage system
Crops marketed
Market prices
Net farm income
Land ownership status
    Absentee owners or farmer-owned
    Size of ownerships
Projected future agricultural conditions, without project
(any variations from the above)
Agricultural conditions with project
    Adaptable crops with irrigation
    Fertilizer requirement
    Weed and pest control requirement
    Irrigation water requirement
        Season total water requirement
        Schedules and rates of application
        Available ground water
    Size of economic farm unit
    Type of drainage system required
    Farm investment requirement
        Land
        Water distribution system
        Drainage system
        Farm machinery and motive power
    Credit availability for farm operation
        Interest rates – private credit
        Government-sponsored credit, and interest rates
    Market locations
    Transportation facilities to market
    Market price controls
        Federal policies
        World market prices and trends
    Trends in land ownership sizes
    Disparity between ownership sizes and economic farm unit sizes
        Re-settlement needs and problems.

Municipal-Industrial Service Function

Climatic data
Prospective service area location
Present conditions
    Population
    Existing industries
    Annual and peak-daily water use
    Sources of supply
    Quality of water
    Water treatment for quality
    Reserve supplies
    Status of water rights
Projected future needs
    Population growth projections
    Industrial growth rate and basis of projections
    Types of industry
    Industrial water use requirement
        Quantity
        Quality
    Domestic-use water quality standard
Potential alternative sources available
    Location and type
    Water quality
    Cost factors

Flood Control Function

Area susceptible of flood protection
    Population
    Property value
Topography of overflow areas
History of known floods
    Dates of occurrence
    Flood Stages

Damages
Present flood hazard conditions
  Land occupancy and use in overflow areas
  Damage resulting from small, medium, and large floods
    Private property
    Roads and streets
    Public utilities
    Railroads
  Agricultural land (crop data similar to that for irrigation purpose may be required)
  Experience of direct and indirect flood costs in comparable areas for comparable-size floods
    Emergency relief
    Repair and rehabilitation costs
    Loss of income costs
Projected future damage conditions
  Population growth data
  Industrial growth projections
  Agricultural projections
  Hydrologic-hydraulic change trends, if any
    Land subsidence
    River channel change
    Encroachments on flow channels
    Watershed use changes.
Alternative flood protection measures available
  Levee system
  Levee system in conjunction with flood detention storage reservoirs
  Channel enlargement and improvement
  Diversions and/or bypasses
  Flood plain zoning — removal of residences and major improvements from overflow areas
  Other flood plain regulations such as building code restrictions
  Floodproofing — construction to protect individual buildings from flood damage
  Flood warning systems and evacuation facilities

Hydroelectric Power Service Function

Prospective power system to utilize new power source
  Load center location
  Present conditions
    System peak demand — amount and date
    System load factors — daily, weekly, monthly, and annual
    Daily load curves — typical periods
    Generating plants in service
      Types — steam, hydro, diesel, gas turbine
      Capacities — kw or kva
      Ages — year installed
      Units maintained principally for stand-by use
    Industrial loads served on off-peak or interruptable power service basis
    Main power transmission grid systems
    Interconnection facilities and agreements for power interchange with other systems
  Projected future needs
    Population growth forecast
    Industrial growth forecast
      Type, with respect to power use
      Size
  Alternative sources of new generating capacity
    Base load operation
    Peaking load operation
    Approximate costs
      Investment, per kw
      Operation, including fuel, per kwh

PROGRAMMING AN INVESTIGATION

The many factors to be considered and evaluated in producing a viable plan for a project for which a dam is to be constructed are indicated in the preceding pages. They include the one or more functions to be accomplished by the development, the selection of the sites and sizes of the one or more dams and related structures, the design and cost estimates, the economic justification, a cost allocation and repayment analysis, and environmental considerations. To complete the task one or more reports are required to state the results of the investigation and to present recommendations. Frequently the investigation will extend over several years time. It is unusual when the plan selection is so simple and straightforward that the planning can be done in a year or less. Because of the numerous work items to be performed by various specialists (example: hydrologist, geologist, economist, and others) and because many steps depend upon the completion of some preceding work it is necessary to devise some type of program or work schedule in order to achieve efficiency in the conduct of the investigation.

Planning efficiency is lost or impaired when work steps are done out of order, or not in best sequence. A simple, logical program would be to first select a service area and project function, then turn to the available and alternative sources and means of supplying the service need, and then select the best available plan. This program plan becomes complicated when questions arise such as whether the assumed service area should be larger or smaller; whether other project functions should be included, and if so, how many and to what sizes; and whether stage-development plans would be desirable as the best means of meeting gradually-increasing service area needs. These questions, and others, lead to the need of making preliminary or reconnaissance-type studies whereby alternatives are considered and compared, and a plan is formulated by making analyses of variable accomplishments and benefits, as well as costs.

The making of a program is an attempt to forecast a rather complete course of action for the entire investigation, and to set down a schedule of when each work item should start and end. From such a program a budget of needed funds can be prepared. An ideal program would show:

1. Each work item required for the investigation.
2. Amount of work, time, manpower, and cost to perform each work item.
3. The proper sequence for performing work items.
4. Where major decision points are necessary and are reached in the course of the investigation.
5. The earliest possible starting date of each work item.
6. Necessary completion date of each work item.
7. Total elapsed time to complete the investigation.
8. Total costs of the investigation.

The chances are poor of being able to forecast and devise a program that can be used without alteration from time to time. This is partly because no two investigations are exactly alike. Each possible plan involves explorations into physical conditions and socioeconomic influences that are not well known at the onset. In addition, there are the problems and uncertainties of being able to proceed with work as scheduled for reasons such as unavailable personnel to start when needed, adverse weather conditions for field work, or otherwise. A capable program planner will attempt to foresee and take into account the various constraints and possible delaying factors in an investigation; these may include seasonal periods when field work is stopped or decreased by weather, possible delays in getting work or materials from outside sources, difficulties and delays in getting a work force assembled when desired, and need for time to review and possibly readjust courses of action as may be disclosed by information learned during the course of the studies.

For various reasons the course of an investigation will depart from the preconceived program, and program revisions become necessary. There is a tendency of agencies that operate on annual budgets to revise the investigation program once a year when budget proposals are made, and possibly at a second time in the year when new budgets are established. This possible second revision would be necessary if there should be a financial constraint that would slow the work progress. But for best management of an investigation the program should be revised whenever there is new information from any source that changes the direct or scope of timing of the work. Such updating assists in redirecting work efforts along the most productive lines and to their best advantage in order to complete the investigation.

ENVIRONMENTAL INVESTIGATIONS

Purpose and Objectives

The purpose of making environmental investigations in the planning of water resource developments by dams and other facilities is indicated in the preamble of the U.S. National Environmental Policy Act of 1969[*]: "... a national policy which will encourage productive and enjoyable harmony between man and his environment; to promote efforts which will prevent or eliminate damage to the environment and biosphere and stimulate the health and welfare of man; to enrich the understanding of the ecological systems and natural resources important to the Nation."

This Act further states:

"In order to carry out the policy set forth in this Act, it is the continuing responsibility of the Federal Government to use all practicable means, consistent with other essential considerations of national policy, to improve and coordinate Federal plans, functions, programs, and resources to the end that the nation may—

1. fulfill the responsibilities of each generation as trustee of the environment for succeeding generations;
2. assure for all Americans safe, healthful, productive, and esthetically and culturally pleasing surroundings:
3. attain the widest range of beneficial uses of the environment without degradation, risk to health or safety, or other undesirable and unintended consequences;
4. preserve important historic, cultural, and natural aspects of our national heritage, and maintain, wherever possible, an environment which supports diversity and variety of individual choice;
5. achieve a balance between population and resource use which will permit high standard of living and a wide sharing of life's amenities; and
6. enhance the quality of renewable resources and approach the maximum attainable recycling of depletable resources.

The Congress recognizes that each person should enjoy a healthful environment and that each person has a responsibility to contribute to the preservation and enhancement of the environment."

The above is an expression of U.S. National policy. It is supplemented by laws of several American states indicating similar objectives.

Definitions

The above wording provides considerable insight on the meaning of "environment" and "environmental quality" to be considered in planning a water project. But there are real difficulties in knowing what different people are talking and thinking about:

"Environmental quality is hard to define precisely. Whether the treatment of environmental issues is held to be a science, an art, or, in its extreme, a religion, is a highly subjective matter. Nevertheless, there are good and bad environmental characteristics that can be understood by reasonable men. On the other hand, there are areas of honest differences when what is "good" or "bad" is a function of the observer."[**]

Environment seems to mean any and all external conditions affecting the existence of living things. With regard to people, environment seems to mean the total of all external conditions affecting people individually and collectively. Different living things, including people, affect one another, and so the environments of people are complex webs or networks of both animate and inanimate tangible things that affect people.

---

[*]Public Law 91-190, Jan. 1, 1970.

[**]From "Multiple-Objective Planning in the Development of Water Resources and its Ramifications with Respect to Implementation," United Nations Economic and Social Council, ECAFE Regional Conference on Water Resources Development, September 1972.

With a few exceptions, it is not possible at present to evaluate environmental conditions in economic terms. The exceptions that sometimes exist are cases of serious amounts of pollution—of water, air, and noise. Economic losses or costs do occur with large-scale pollution and are measurable, economically, to some degree; particularly water and air pollution. The cost of water pollution, where it exceeds tolerable levels, may include the loss of some types of irrigated agriculture and fish culture, and the expenditure necessary for water treatment to make it safe for human use and satisfactory for some industrial uses. Economic costs of air pollution have sometimes been estimated as a factor of loss of some types of agricultural production of sensitive crops, the corrosion effects on structures and machinery, the cost of filtering air for some uses, and some health costs identifiable with air pollution. To the extent that such pollution costs can be quantifiable they may be properly considered as economic costs.

Somewhat akin to benefit/cost evaluations there is a concept of "opportunity costs" which has been used to place a value on environmental factors. In application, there may be two or more alternative possibilities which are exclusive of each other. A choice of one method is made at the "cost" of foregoing the value of the alternative.

Beyond this there are the presently unmeasurable environmental effects of pollution of various amounts and types, and the large number of environmental effects, both positive and negative, as itemized and discussed in the following paragraphs, that are important to the well-being of man but not possible of evaluation in economic terms.

It is obvious that the various environmental conditions will range in importance from "minimum" to "maximum" depending upon the specific situation and viewpoint of the affected person or collection of persons. The person situated alongside a mountain stream has a different viewpoint than one living at a lowland site where land is inundated in times of flood, or where the well goes dry in times of drought. Thus, there are varying points of view about the "quality of environment." It might be said that environmental quality, like beauty, is in the eye of the beholder. Those who undertake the roles of judging environmental quality necessarily do so with the realization that their decisions or opinions may not receive unanimous agreement by all interested persons.

Environmental Concerns With Respect to Selection of Dam Sites and Sizes

The environmental effects or impacts of a water resource development or management project extends over the entire project area, or possibly a still larger area of influence. They include the impact effects of dams and reservoirs in the water development and stream flow management areas, and the effects of flood control, water supply, power supply, and/or other services in project service areas.

The local impact effects of dams and reservoirs are principally of concern in the selection of dam sites and sizes. The impacts may be either favorable or unfavorable. More likely they will be unfavorable in the aggregate, and it becomes part of the plan selection process to compare environmental impacts of alternative plans and sizes of dams as one factor of consideration in selecting one plan from among the possible alternatives.

The project environmental impacts in service areas are secondary in importance in selecting dam sites and sizes. They may be of no concern at all if the problem is one of selecting between two alternatives providing equal accomplishments. However, if there is a selection between significantly larger and smaller dam sizes and project accomplishments the service area environmental impact is also important. In any event, the environmental impact effects in service areas are not to be overlooked in evaluating a total project plan.

The Environmental Evaluation Problem

From the foregoing it is evident that: (1) there is need in planning water resource developments, to give adequate consideration to "environment" or "environmental impacts;" (2) the numbers of impacts may be large and of different degrees of importance; and (3) many of the impacts can only be measured by judgments which are, by nature, subject to some degree of disagreement or question. Environmental evaluations being relatively new undertakings, there are no well-established principles, processes, measurements, or methods identified for use in deciding upon the relative weight to be given in considering one environmental value with another, and there are only limited possibilities for considering an environmental value with its economic cost or value.

A sufficient number of project planning evaluations have been made to date to know that (1) in any water resource development there is a large and conflicting number of environmental effects, and (2) some of the environmental effects will be favorable while others will be adverse to greater or lesser degrees. A maximization of favorable environmental effects (or a minimization of adverse effects) in water development has economic cost; either monetary, or in reduced project accomplishments. Although it is possible to quantify the costs of providing environmental protection or improvement in economic terms it is usually not possible to state the environmental values in the same terms.

As shown in the following paragraphs, most environmental values or impacts cannot be expressed in exact quantifiable numbers; the best that have been devised are rating systems. Thus, there is no direct formula that can be applied to arrive at a rational plan selection where compromise is necessary between various economic and environmental considerations. Moreover, planners are faced with the question, since the maximum-possible environmental attainments usually involve additional cost, of who is going to be willing to pay the environmental protection cost; how much, and under what terms?

The remainder of this chapter is devoted to information and discussion on approaches toward the solution or handling of the environmental evaluation problem.

Environmental Categories and Classifications

General. A large number of possible environmental factors have been listed by numerous persons and organizations. In efforts to be systematic and orderly, individuals and organizations have grouped the factors in categories. There is considerable dissimilarity of the grouping systems, but this is probably not very important; the more important thing is to use some system that will be of aid in comparing the importance of different factors with one another for any specific environmental impact analysis. Three different groupings are shown herein for introductory and illustrative purposes.

Water Resource Council Categories. The tabulation entitled "Water Resource Council proposed classification of environmental categories" is listed in the following group. It shows six main classes containing 18 main categories. Most of these categories contain lists of 9 to 11 statistical items or factors for use where applicable in making inventory analyses.

WATER RESOURCE COUNCIL PROPOSED CLASSIFICATION OF ENVIRONMENTAL CATEGORIES

I. Class—Areas of Natural Beauty and Human Categories:
   A. Open and Green Space
   B. Wild and Scenic Rivers
   C. Lakes
   D. Beaches and Shores
   E. Mountain and Wilderness Areas
   F. Estuaries
   G. Other Areas of Natural Beauty

II. Class—Archeological, Historical, and Cultural Elements Categories:
   A. Archeological Resources
   B. Historical Resources

III. Class—Biological, Geological, and Ecological Categories:
   A. Biological Resources
      1. Fauna
      2. Flora
   B. Geological Resources
   C. Ecological Systems

IV. Quality Considerations
   A. Water quality
   B. Air quality
   C. Land quality
V. Class—Irreversibility Considerations
   Category:
   A. Irreversibility
VI. Class—Unique Resources
   Category:
   A. Uniqueness

Reference: Federal Register, December 21, 1971, Vol. 36, No. 245. "Water Resources Council, Proposed Principles and Standards for Planning Water and Related Land Resources, Notice of Public Review and Hearing," pages 24159 to 24162.

Geological Survey Categories. The following tabulation entitled "U.S. Geological Survey—Characteristics and Conditions of the Environment" shows the grouping system devised by that agency for a total of 85 environmental items. This system was devised for use in a matrix, or work sheet, which also lists 100 types of human activities that have possible effect on natural environment in some manner or another. The system is designed to be a very comprehensive one, whereby any one or more of the 100 possible types of impact action could be considered and given a rating of effect on any one or more of the environmental characteristics. In any single proposal or event such as a specific water project plan the number of impacts would be far less than the 8,500 impact capacity of this system.

The 100 possible impact actions are listed under the following headings:

   A. Modification of Regime.
   B. Land Transformation and Construction.
   C. Resource Extraction.
   D. Processing.
   E. Land Alteration.
   F. Resource Renewal.
   G. Changes in Traffic.
   H. Waste Emplacement and Treatment.
   I. Chemical Treatment.
   J. Accidents.

Typical actions in this list that would apply to dams and reservoirs are:

   A. g. River control and flow modification.
   B. m. Dams and impoundments.
   C. b. Surface excavation.

U.S. GEOLOGICAL SURVEY—
CHARACTERISTICS AND CONDITIONS
OF THE ENVIRONMENT*

Categories	Number of items in category
A. Physical and chemical characteristics	
1. Earth	6
2. Water	7
3. Atmosphere	3
4. Processes	9
B. Biological conditions	
1. Flora	9
2. Fauna	9
C. Cultural factors	
1. Land use	9
2. Recreation	7
3. Aesthetics and human interest	10
4. Cultural status	4
5. Man-made facilities and activities	6
D. Ecological relationships	
1. Examples	6
Total	85

*From United States Geological Survey Circular No. 645, published 1971.

Battelle-Columbus Laboratories Categories. By contract with the Bureau of Reclamation, and for the proposed use by this Bureau, Battelle Laboratories produced a classification of "environmental impacts" as summarized on the following tabulation of that heading. As indicated thereon, a total of 79 environmental items were grouped under four main headings and 18 subheadings. The list was designed to be comprehensive for all of the types of activities occurring in water resource and water management development as performed by the Bureau of Reclamation. Of the list of 79 environmental items, about one-half or two-thirds might be affected by the construction and operation of dams and impoundments. The last column of figures on the tabulation headed "Percentage weight of category" is an expression of the importance of each category, obtained by judgment.

Wild and Scenic Rivers Classification

General. This classification is one listed by the Water Resource Council on their "Class I—Areas of Natural Beauty and Human Enjoyment as Category B—Wild and Scenic River." It is of particular significance as it was earlier established by U.S. Public Law 90-542, 90th Congress, S. 119, October 2, 1968 as the Wild and Scenic Rivers Act. Sections 1 to 7 inclusive of this Act are reproduced at the end of this chapter as Appendix C. The Act creates a national

ENVIRONMENTAL IMPACTS[a]

Categories	Number of items in category	Percentage weight of category
A. Ecology		
1. Species and populations	10	14.0
2. Habitats and communities	8	10.0
3. Ecosystems	1[b]	—
B. Esthetics		
1. Land	3	3.2
2. Air	2	0.5
3. Water	5	5.2
4. Biota	4	2.4
5. Man-made objects	1	1.0
6. Composition	2	3.0
C. Environmental Pollution		
1. Water pollution	14	31.8
2. Air pollution	7	5.2
3. Land pollution	2	2.8
4. Noise pollution	1	0.4
D. Human Interest		
1. Education/scientific packages	4	4.8
2. Historical packages	5	5.5
3. Cultures	3	2.8
4. Mood/Atmosphere	4	3.7
5. Life Patterns	3	3.7
Total	78	100%

[a] From Environmental Evaluation System for Water Resources Planning to Bureau of Reclamation, U.S. Dept. of Interior, Jan. 1972 by Batelle, Columbus Laboratories.
[b] Not counted in total since it is a "descriptive-only" item.

wild and scenic rivers system, designates potential inclusions of various named rivers into the system, and prescribes methods and standards by which additional components of rivers may be added to the system from time to time.

Appropriate Federal agencies are required to study and evaluate, in close cooperation with appropriate State and lesser political subdivisions, various rivers, to ascertain if the rivers or parts thereof are potential wild or scenic rivers. In proposing any Federal water development the agencies are required to provide Congress with reports of adequate studies for consideration of designating portions of streams for preservation, alternative to project water resource development or management use.

The Act designates that portions of rivers in the National Wild and Scenic Rivers System be classified in one of three categories:

1. Wild river areas—Those rivers or sections of rivers that are free of impoundments and generally inaccessible except by trail, with watersheds or shorelines essentially primitive and waters unpolluted. These represent vestiges of primitive America.
2. Scenic river areas—Those rivers or sections of rivers that are free of impoundments, with shorelines or watersheds still largely primitive and shorelines largely undeveloped, but accessible in places by roads.
3. Recreational river areas—Those rivers or sections of rivers that are readily accessible by road or railroad, that may have some development along their shorelines, and that may have undergone some impoundment or diversion in the past.

Implementation of Federal Wild and Scenic Rivers Act. The appropriate federal agencies are required to evaluate the wild, scenic, and recreational aspects of the 27 streams named in the act in addition to streams being studied for potential federal water developments. Precise standards are not established beyond those stated by the wording of the act. Questions on the existing and potential future character and use of rivers are being studied at state levels of some states and these studies will influence federal-level views.

Wild and Scenic Rivers Classifications by States. Reportedly in 1971, thirty-eight states of the United States of America were engaged in making some form of inventories or studies of their rivers which have wild, scenic, or recreational potentials; in some cases these studies are authorized or directed by actions of the State Legislature. The studies in California are extensive, having been initiated by act of the State Legislature and designated as the "California Protected Waterways Plan" (initial report, February 1971). A variance by California from the Federal classification is proposed as follows:

"The waterway classification system proposed for California is: *natural; pastoral;* and *developed.* All three categories are designed to conserve, to varying degrees, and in several ways, the extraordinary scenic, fishery, wildlife and outdoor recreational values of our waterways.

A "natural" waterway could have wild, scenic, recreational, and resource attributes.

The "pastoral" waterway would be characterized by rural or farmland uses; support resources and recreation and would contain scenic values.

The "developed" waterway would provide abundant and varied recreational uses; could be very scenic; it might also appear wild in some areas; and would allow controlled developments of the waterway and adjacent lands."

Evaluating Environmental Impacts

General. As noted above, a number of alternative methods of classifying environmental factors have been devised. The principal attention and effort in use (1976) has been given to the classification system of the Water Resource Council (WRC) as indicated in that agency's "Proposed Principles and Standards for Planning Water and Related Land Resources" published in the Federal Register of December 21,

1971.* Guidelines for use in making environmental evaluations in the Water Resource Council classification system were devised in 1972 and are published as Chapter 4 of a Review Draft, "Guidelines for Implementing Principles and Standards for Multiobjective Planning of Water Resources" dated December 1972.* The guidelines were developed by joint efforts of several Federal agencies and were field tested by trial use on several project plans by the Bureau of Reclamation and Soil Conservation Service.

Evaluation Method by Use of the WRC Classification. The recommended procedure for making an environmental evaluation is indicated by the following subject headings:

A. Categories
B. Measurement Standards
  1. Evaluation factors
    (A) Quantity
    (B) Quality
    (C) Human influence
    (D) Uniqueness
    (E) Magnitude
    (F) Significance
  2. Projections
  3. Team approach
  4. Area of consideration
C. Limitations
D. Presentation of material

Categories. These are essentially the same as listed in a preceding tabulation headed "Water Resource Council proposed classification of environmental categories." The categories are further defined as follows:

Open Space and Greenbelts—A landscape used to maximize natural and spatial values in a condition in which nature predominates.

Streams and Stream Systems—Any natural course of waters, whether flowing year-around or on an intermittent basis.

---
*See listing in Bibliography at the end of this chapter.

Lakes and Reservoirs—Both natural and man-made lakes and reservoirs, including water impounded behind a dam.

Beaches and Shores—Land areas adjacent to lakes, reservoirs, and streams that provide access to and from the water.

Wilderness, Primitive, and Natural Areas—Lands included within or having potential for inclusion with the Natural Wilderness Preservation System, or having similar qualities and characteristics.

Estuarine and Wetland Areas—Semienclosed coastal water bodies: lowland areas usually covered with shallow or intermittent water.

Other Areas of Natural Beauty—Sites of nature's visual magnificence and scenic grandeur.

Archeological Resources—Sites and artifacts of ancient human activities.

Historical Resources—Sites and evidences of historical events and social development.

Cultural Resources—Evidences of historic human cultures.

Biological Resources—A broad category covering effects on individuals, species, and populations of living organisms. Two subcategories, flora and fauna, are included.

Geological Resources—(a) Areas of importance as future mineral supplies; (b) areas of interest in studying or displaying the development of the earth.

Ecological Systems—Identifiable communities of organisms and the physical conditions in which they exist.

Water Quality—Chemical, physical and biological qualities of water with respect to its suitability for a particular use.

Air Quality—Chemical, physical and biological aspects of air with respect to conditions that adversely affect man.

Land Quality—Relates to the chemical, physical, and biological aspects of land in relationship to its suitability for particular uses such as farmland, forest land, urban or other.

Sound Quality—Beneficial or adverse effects of sound as it relates to quality of environment (for man).

Visual Quality—Benefits of visually attractive landscapes or adverse effects on pleasant settings.

Uniqueness Considerations—Unusual or extraordinary environmental resources.

Irreversibility Considerations—This is intended to be an evaluation of the environmental effects of any commitment by a project plan or action, of which the effects would be of a permanent nature. It applies only to those effects which would appear to leave no reasonable opportunity to be removed or changed if at some future time the removal or change might be desired.

Measurement Standards

Evaluation Factors—In most cases the environmental qualities or effects, positive or negative, cannot be measured in economic or financial terms. Other scales of measurement or estimation are therefore applied.

Quantity Measurements—Quantities of some environmental features or effects are measurable in numbers such as units of land, water, distances, people, animals, or other specie.

Quality Factor—Judgment or measurement of quality is described and represented by use of a zero-to-ten scale of which zero is the poorest quality known and ten is the best known quality.

Human Influence Factor—The degree to which people use or would use an environmental resource at the site or locality under consideration. Likewise, the degree to which this use would be changed by a project development or activity. The measurement, by judgment, is on a 0-to-10 scale from poorest known to best known.

Uniqueness Factor—The factor is the degree of uniqueness of a specific resource in the plan under consideration with its occurrence elsewhere. The rating, by judgment, is on a scale of 1 to 10.

Magnitude Factor—This is a measure of the amount by which a uniqueness factor is affected by a project action. On a 0-to-10 scale zero indicates complete destruction and 10 indicates no appreciable or measurable effect.

Significance Factor—This factor is applied to "Irreversibility considerations" as elsewhere discussed on a 1-to-10 scale to represent the degree of importance.

Team Approach. Project planning as performed by federal agencies of the United States is undertaken by interagency, interdisciplinary teams organized to include the expertise necessary to evaluate each environmental category. The team includes specialists of each involved category, but the size of the group must not be so large as to make it unwieldy in functioning. Ratings are attempted by the group as a team, but with recognition to the advice of different individuals in their specialty fields. The general project planner would normally serve as team leader. A test of accomplishment is for the team to document and support the team evaluation of the plan or program under consideration.

Area of Consideration. A specific environmental investigation will pertain to a specific development plan and its alternatives; however, the areas of consideration of environmental factors will be variable according to the extent of influence of environmental effects and areas of comparison.

Limitations. Although a measurement or judgment rating is made of most of the environmental factors, usually on a 0-to-10 scale, and degrees of importance indicated, the results are not subject to being combined into a single, simple numerical total. Likewise, there is no established formula or procedure by means of which "tradeoffs" can be made between one environmental factor or quantity and another.

Presentations of Materials. Along with suitable written description and tabulations a graphic method is suggested to display results of an environmental evaluation of alternative project plans or actions. This is illustrated by an example page from the "Guidelines" draft of December 1972 previously cited; the page headed "Environmental Evaluation of Alternative Plans—Quality and Human Influence Summary" (Fig. 1-8). Use of the 0-to-10 scale ratings may be noted.

Evaluation Factors. The first 18 of the above-listed categories can be conveniently examined in three steps:

1. Quantity factors
2. Quality factors
3. Human influence factors

The quantity factors can usually be identified and expressed in numbers—amounts, areas, distances, proportions, percentages. The quality factors are usually judgmental, but supportable in some instances by quality measurements; for example, qualities of water for different uses are indicated by chemical or other analyses.

Human influence factors are judgments of the degree of value of the specific environmental category to people. This factor is therefore the one of prime importance in evaluating environmental impacts. As an example, the quality of water in a specific situation may be of high importance by reason of its types of uses by man; visual quality may be of slight importance if the visual scene is in a remote and isolated area.

Detailed Evaluation Items. The detailed factors for possible consideration of the 18 categories of environmental calssifications are listed in 33 pages of the "Guidelines" (Review Draft of December 1972). A total of 270 factors are listed. Project planners concerned with the selection and sizing of dams and reservoirs for a water control and management development may find this a useful checklist for developing ratings of environmental conditions and impacts by categories. Not all of the detailed factors would be applicable in a single situation. One of the 18 categories entitled "Lakes and Reservoirs" is herein reproduced as an illustration of the number of factors and degree of detail involved in assessing the environmental qualities (Fig. 1-8a). As illustrated by this example summary graph, all pertinent factors would be evaluated and given a rating on a 0-to-10 scale for present conditions, future conditions with no development plan, and for alternative development plans. A summary statement would also be written to describe the most significant items of the environmental inventory, also as illustrated in the "Guidelines" Review Draft publication.

Lakes and Reservoirs

This includes both natural and man-made lakes and reservoirs and other areas of standing water (except those areas classed as wetlands or estuaries) as well as any water impounded behind a dam or other structure where the quantity of water is materially increased. The minimum size of lakes and reservoirs to be included in this analysis should be determined for each specific study. Area of influence should consider recreation potential and satisfaction, plus fisheries and wildlife habitat relative to project area.

Quantity Factors:

1. Total maximum surface areas of natural lakes.
2. Total maximum surface areas of man-made lakes and reservoirs.
3. Total number of natural and man-made lakes.
4. Average surface area of natural and man-made lakes during the prime recreation season (June—September).

Quality Factors:

1. Water quality. Turbidity, debris, chemical components, odor, algae, temperature.
2. Scenic setting. Narrative description.
3. Related land features.

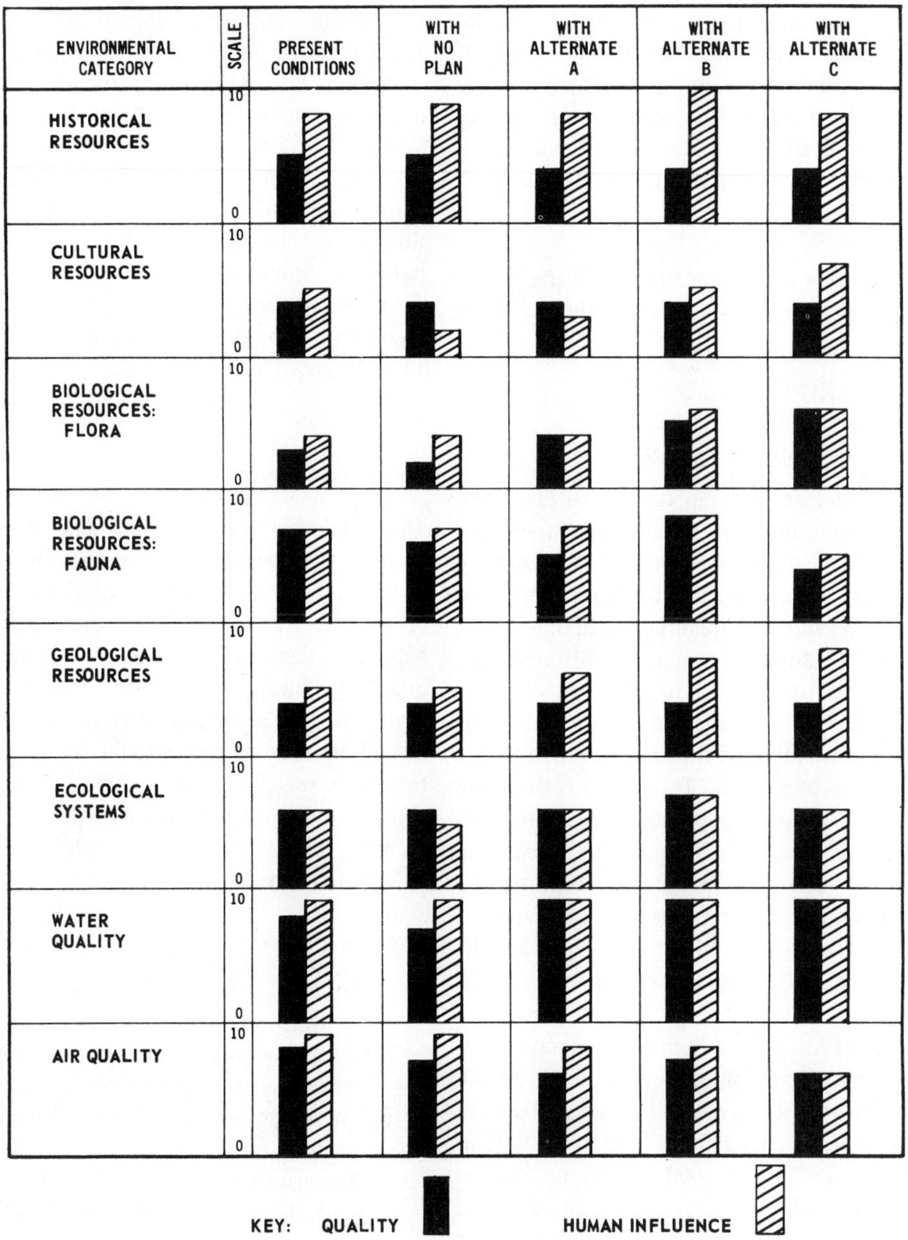

Figure 1-8. Environmental evaluation of alternative plans—quality and human influence—summary (Upper Skinny Fish Basin).

4. Faunal and floral desirability. Presence of insects, nettles, poison oak, algae, aquatic plants (these may be detriments to swimming or boating, but enhance wildlife or fisheries).

5. Productivity. Degree lake or reservoir sustains desirable faunal or floral communities.

6. Fluctuation. Impact on reservoirs and adjacent land.

ACTION PLAN

CATEGORY LAKES AND RESERVOIRS

AREA AFFECTED

EVALUATION FACTORS	ALTERNATIVE CONDITIONS							RATING GUIDE
	1970	WOP	A	B	C			
QUANTITY								
Total max. surface areas nat.								Express quantity factors in acres, miles, volumes, and/or numbers of units. The number that best illustrates the category is to be used as the index.
Total max. sur. areas manmade								
Total natural & manmade lakes								
Av. surface area June-Sept.								
QUANTITY INDEX								
QUALITY								Express quality and human influence factors by comparison to existing situations.
Water quality								
Scenic setting								
Related land features								
Faunal & floral desirability								0—Worst known
Productivity								1-2—Very low
Fluctuation								3-4—Moderately low
Depth								5—Average
								6-7—Moderately high
								8-9—Very high
								10—Best known
SUMMARY RATING								
HUMAN INFLUENCE								Summarize quality and human influence factors separately by inspection.
Relationship to population								
Public access								
Public amenities								
Legal/admin. protection								
legal/admin. restrictions								
Physical protection								
Use without degradation								
Eff. climate on public use								
SUMMARY RATING								

```
Quantity symbols:   + = increase
                    - = decrease
                    0 = no change
```

Figure 1-8a. Example of worksheet—Category: Lakes and Reservoirs.

7. Depth. Adequacy for sustaining year-round fish populations.

Human Influence Factors. Human influence is conditioned in a large measure by the social and economic values of the natural resource to human use or adaptation. The following existing and future uses and demands on the resources should be considered.

1. Relationship to population. The degree that the area can or does receive use.

Consider time and distance factors relative to origin of users and location of resource.
2. Public access. The degree or extent of area open to public use, considering existing or proposed transportation systems.
3. Public amenities. The degree to which public use facilities are developed and maintained in the area. Consider picnicking, camping, fishing, hiking, riding areas, overlooks, scenic viewpoints, golf courses, etc.
4. Legal and/or administrative protection. The degree to which the area is reserved from encroachment by industrial or residential developments.
5. Legal and/or administrative restrictions to orderly public use. Consider both positive and negative effects. (i.e., overuse may be detrimental to the resources; whereas too many restrictions may preclude satisfaction of recreation demands.)
6. Physical protection. The degree of change expected to occur as a result of natural processes, considering the amount of land management practices expected.
7. The ability of the area to accommodate the anticipated use without degradation of natural values. Use in this situation would refer to human impact, including recreation, urban or industrial development, etc.
8. Effect of climate on public use of area.

BENEFICIAL AND ADVERSE ENVIRONMENTAL EFFECTS

The important environmental effects are those that are of most direct concern to the livelihood and well-being of people. The environments of various other living things are of concern to man to whatever degree their existence is important to man's living conditions. As an example, the survival and abundance of domestic animals and fowl are of extreme importance to people. Alternatively, the extinction of the dodo bird is of significance only as an historical example of a biological event of non-survival of a species.

There is a close similarity between important environmental effects and the social needs of man. It is therefore appropriate to consider social values together with environmental effects, and to disassociate them from each other only when taking note of their existence in the specific situation under study.

Among the beneficial environmental-social effects there may be: land use improvements by irrigation, flood protection, improved water supplies for domestic and municipal uses, power supplies without consumption of fuel, water quality improvement, fishery improvement, recreational improvement, and improvement of health. These improvements, partly measurable in economic terms, are among the principal objectives of a project development and water management plan. They are further discussed later on in this chapter.

AVOIDABLE AND UNAVOIDABLE ADVERSE ENVIRONMENTAL EFFECTS

General

Various environmental effects may occur with the construction and operation of dams and reservoirs. These effects may be either large or small, beneficial or adverse, depending upon the pre-project environmental conditions. This portion of the chapter deals with the possible adverse effects. Limiting the consideration to dams, reservoirs, and closely related project works, and excluding consideration of project effects in service areas, the effects may be listed in the following categories:

A. Land converted from present use to reservoir use.
B. Alteration of stream flows.
C. Blockage or interference with fish and wildlife migrations.
D. Landscape appearance.

Within these general categories may be included:

A. Land conversion to reservoir areas.
   1. Loss of fish and aquatic habitat.
   2. Loss of wildlife habitat.
   3. Loss of access to mineral deposits.
   4. Loss of mountain valley areas.
   5. Inundation of historical or archeological sites.
   6. Inundation of exceptional geologic formations.
B. Alterations of downstream flows
   1. Reduction of fish and aquatic habitat.
   2. Reduction of stream flushing flows.
   3. Changes of water quality.
   4. Loss of recreational opportunities.
   5. Loss of scenic values.
   6. Reduction of riparian wildlife habitat.
C. Interference with fish and wildlife migrations
   1. Blocking of anadromous fish runs.
   2. Blocking of animal migration routes.
D. Landscape appearance
   1. Unsightly excavation and waste sites.
   2. Unsightly reservoir banks below maximum waterline.
   3. Abandoned construction facilities.
   4. Erosion scars from construction roads.
   5. Reservoir clearing waste disposal.
   6. Vegetation changes brought about by reduction of downstream flows.

As discussed in the following paragraphs, some are unavoidable, some are unavoidable but can be mitigated to a degree, and some are avoidable when suitable precautions are taken.

A. 1. Loss of Fish and Aquatic Habitat. This refers to habitat in stream channels which are inundated by a reservoir or impoundment. The habitat includes the spawning and rearing areas of fish and the entire food chain of the resident aquatic species. There will likely be a substitution of different types of fish and aquatic species by the reservoir. The net effect requires an analysis for each specific site.

A. 2. Loss of Wildlife Habitat. Trees, brush, grasses, and other vegetation alongside and near streams are part of the food supply of various wild animals. Grazers and browsers may feed in high mountain areas in the summer season but depend upon streamside vegetation for their winter food supplies. Some of this is lost by inundation by a reservoir. Many species of streamside vegetation are adapted to survive short periods of inundation, as during a flood, and so can survive short periods of inundation by a temporary reservoir surcharge during a flood. Mitigation of losses of streamside vegetation for wildlife grazing or browsing may be accomplished by improving other areas for the purpose, but this may be a costly or undependable management effort. Even where lands are adaptable to growth of seasonal browsing/grazing vegetation used by wild animals there is competition for use of the land for agricultural stock grazing, orchards, or commercial forest production.

A. 3. Loss of Future Access to Mineral Deposits. One objective of environmental protection is preservation of resources for future generations. If an existing mining enterprise within a reservoir site is terminated by project construction the loss is measured as an economic cost. But if there are deposits within a reservoir site which lack present commercial value but are of potential future value their inundation may be considered an environmental loss. This is necessarily a speculative consideration, depending both on the possible need of the mineral products at a future time and the extent of their abundance or scarcity elsewhere.

A. 4. Loss of Mountain Valley Areas. A reservoir site in a mountain valley may occupy the only flat land, or most of the flat land in existence within a considerable distance. In such a circumstance such land has future importance for various uses by people due to increasing population growth. The present economic cost may not be a measure of the future value, in which event an environmental loss would be involved if the land was used for a reservoir.

A. 5. Inundation of Historical or Archeological Sites. With some notable exceptions, the most important archeological discoveries to date have been on relatively high land locations, indicating a preference of ancient rulers to construct their monuments of power at the high sites. However, much of ancient civilizations or collections of people were located in valley areas and along waterways, and many artifacts are discovered in such places. Some may occur in reservoir sites. A survey by an archeologist will usually satisfy the question of whether a reservoir site is of archeological importance. Small artifacts are removable to suitable museums. Large and important archeological features may present a strong deterrent to the use of the site for a reservoir. In a most notable case, the construction of Aswan Dam on the Nile River partly inundates great monuments of ancient Egyptian civilizations. The most important of these large monuments were relocated at very high cost to elevations just above the waterline of the new reservoir.

The inundation of a historical site by a reservoir is an environmental loss which may be of slight or major importance. A museum display and record may be sufficient to mitigate the loss of an historical site, but a very important historical site may prevent the construction of a reservoir. An example is the site of gold discovery in California on South Fork American River near Coloma. The site, which is now preserved as Gold Discovery Site State Park is within the site of a major reservoir (Coloma or Salmon Falls Reservoir) which has favorable physical and economic qualifications for water resource development. Because of the gold discovery historical site, alternative water storage developments are being made at greater cost than if the Coloma reservoir site was usable.

A. 6. Inundation of Exceptional Geological Formations. A reservoir site may be situated such that it would inundate, or partly inundate some extraordinary geological or geotopographic feature. Examples would be: waterfalls, large springs, geothermal displays, unique rock outcrops, and caves. Areas which have been incorporated into National Parks or National Monuments for geological attributes would be environmentally-objectionable as reservoir sites. An example is a reservoir site on Colorado River which was removed from consideration because a portion of the reservoir would extend into Grand Canyon National Park.

B. 1. Reduction of Fish and Aquatic Habitat. Diversion of water from a stream channel for service area needs reduces the quantity of water in the channel below the diversion site available to support and maintain aquatic life. The extreme case is where all water is diverted during some time of the year and the stream completely dried up. In consideration of the environmental loss, some flow is almost always left in a stream.

Under natural conditions stream flows occasionally become very low or may even completely cease in periods of drought with the natural consequence of large aquatic habitat losses as natural events. A project may be planned and operated to maintain a live stream, or provide some amount of live stream flow as a minimum by making releases of previously-stored water from a project reservoir for this purpose. Where this is done, the downstream releases of stored water become a compensation or mitigation of the adverse effects of reducing stream flows during the more ordinary periods of water supply.

If large amounts of stored water are released downstream to improve the rates of natural low stream flows the result can be an enhancement of the aquatic habitat over its natural condition. In project developments this usually occurs on the section of a stream between the storage dam and a downstream diversion dam if the two operations—storage and diversion—are not performed at the same point.

The amount of aquatic habitat reduction in a stream below a project diversion dam varies both with the amount of storage regulation and the maintained minimum downstream releases. A large storage reservoir may eliminate all but the highest downstream flood flows; in which event streamside vegetation will encroach on portions of the stream channel which formerly were denuded of vegetation by the force and erosion of the occasional floods. The live stream channel and aquatic habitat is thus reduced in width. In compensation, the maintenance of stream flows stabilizes the aquatic habitat areas and improves its quality.

Opinions vary as to what proportions of total streamflows should be maintained in a stream to avoid significant or serious adverse environmental effects on aquatic habitat. Depending upon various physical conditions the stream may possess the actual or potential capability of supporting large numbers and varied species of fish and other aquatic creatures, or may be very limited in one or more life-supporting capabilities. At different rivers and project sites the recommended amounts of sustained downstream releases for fish and other aquatic habitat have amounted to about 10 percent of the long-term average natural stream flows to one-quarter or one-third the average natural flow.

B. 2. Reduction of Stream Flushing Flows. Streams transport varying amounts of sediments; the amounts depending upon erosion conditions in the watersheds and the rates of stream flow. The seidments are eventually deposited in the ocean or lake into which the stream discharges, or in river delta areas formed near the mouth of the stream. The sediment carrying capacity is in proportion to the rates of stream flow; the large volumes of seidment are carried during floods and other high-flow periods. Project diversions reduce downstream flows, and where the water supply is developed by reservoir storage the higher natural flows are those most frequently reduced.

Tributary streams entering a river below the site of streamflow regulation and diversions debouch sediments in quantities that cannot always be continued in transportation by the main river due to its reduced flows. These excess sediments tend to fill pools and otherwise adversely change the main-river habitat for aquatic species. On rare occasions some of these sediments would be carried downstream by a large flood flow. Theoretically, the situation might be improved by occasionally making large releases from storage as "flushing flows" to rectify an adverse condition of excessive sediment deposition. This has not been sufficiently tested to establish a degree of success.

B. 3. Changes of Water Quality. The water quality parameters of principal importance to environmental conditions affected by dams and reservoirs are: temperature, turbidity, oxygen content, dissolved solids content, and organic matter content.

With respect to environmental values, some biologists tend to believe that the qualities of water occurring in stream systems under natural conditions are the optimum conditions for aquatic environment. The reasoning is that, over very long periods of time, all aquatic species have gradually adapted to the natural conditions; hence, any departure from the natural conditions would be an adverse change to the environment. The complication in the reasoning is that water quality, as well as quantity, does not remain entirely constant because of the occasional year-to-year variable weather and watershed conditions. The variability of storms, the intensity of storm runoff, the occurrence of lightning-caused burns of watersheds, and the proportions of watershed subsurface flow to surface runoff are among the causes of water quality variation. Aquatic species survive because they have some degree of tolerance or adaptability to nonuniform conditions.

It has also been noted that, in a specific locality, only small numbers of some particular specie exist due to environmental conditions that are marginal to their survival. A change of an environmental factor in a favorable direction can permit the specie to flourish. For example, lowering of river water temperature a few degrees may greatly increase numbers of cold water fish species. Conversely, the same change might be adverse to another specie, thus lowering its number or eliminating it from the particular locality.

The usual effects on water quality by reservoir storage are the following:

1. *Temperatures.* Water temperatures in reservoirs are stratified except during infrequent short periods when a "turn-over" of water may take place. Water is warmest near the surface, and colder in the depths. Reservoir releases may be either warmer or colder than natural stream flows, depending upon the location of the reservoir outlets. In the usual designs of dams the water is drawn out from considerable depth and the resultant downstream flows are usually colder than natural flows. This can be adjusted or controlled by constructing selective-level outlets.

2. *Turbidity.* In natural stream flow conditions the high flows are turbid, and low flows tend to become clear. When water is stored, the suspended sediment settles to the bottom; however, the finer the sediment, the slower the rate of settling. Usually water released from a reservoir contains little turbidity, but there are important exceptions. During a large flood a reservoir will completely fill with very turbid water. Because of slow settlement rates the water in the deepest part of the reservoir near the dam will remain turbid for a long period. Releases of turbid water may persist through a dry season, long after natural stream flows have become clear. There are instances where reservoir capacity is relatively small in comparison to inflows, and water in storage remains turbid nearly all the time; again with the highest turbidity tending to persist longest in the deepest water near the dam.

Another phenomenon, termed "density currents," has been noted in reservoirs. Water flows through a reservoir slowly, but not uniformly. Water near the bottom is normally denser due to lower temperature and greater content of suspended sediment, and due to flow behavior factors this water may flow through the reservoir and to the outlets while the overlying warmer and less turbid water tends to remain in place. When this occurs, it adds to the persistence of turbidity in downriver reservoir releases.

Selective-level outlets offer some possibility of regulating turbidity, as well temperatures of water released from a reservoir.

3. *Oxygen Content.* Dissolved oxygen in reservoir water may be depleted in the process of decomposition of organic substances in the water. The depletion can be greatest in deep parts of the reservoir when there is least opportunity for re-aeration of water. The amount of oxygen depletion depends upon the amounts of organic materials brought into the reservoir. Least would be from either arid or arctic-type watersheds; most would be from watersheds of lush vegetation.

4. *Dissolved Solids (TDS) Content.* Dissolved minerals in natural water occurs by the leaching of soils and rock by flowing water. TDS is highest when stream flows are least and all flow is from subsurface sources; it is least during high stream flow conditions with most of the flow coming from overland runoff directly from rainfall or snow melt. The effect of reservoir storage is to mix the higher and lower TDS waters and make resultant releases less variable in this respect.

Shallow reservoirs of large surface area have appreciable evaporation loss, and such evaporation increases the TDS of the water in storage. The effect is greatest when water is held for long periods, from abnormal to subnormal years of water supply. However, it can be computed from a reservoir operational study that the concentration of TDS by reservoir evaporation is only a few percent in a very extreme situation.

5. *Organic Matter Content.* An effect of organic matter content in reservoirs has been mentioned above in the topic of oxygen content. Decomposition of organic matter adds nutrients to the water, with little difference of whether it takes place in a reservoir or in the

stream under natural conditions. Decomposition produces gasses which are sometimes concentrated in reservoir water under pressure and vented at a reservoir outlet creating a localized air pollution condition.

Substantially, all possibly-adverse environmental effects of dams and reservoir on water quality are avoidable when anticipated in advance of project construction and suitable preventive measures taken. The installation of selective-level reservoir outlets provides an opportunity to control any excessive change in temperatures, turbidity, and oxygen content of downstream releases. In a site of extreme concentration and decomposition of organic matter within a reservoir a means of aerating the water upon release will remove gasses produced by the decomposition process and restore dissolved oxygen more rapidly that would take place in a longer stream distance of downstream flow.

C. 1. Blocking of Anadromous Fish runs. Some rivers have important runs of anadromous fish. Notable examples are the Pacific Coast streams of the northern United States and Canada where adult salmon and steelhead trout migrate from the ocean to spawn in gravels of mountaineous streams and from which the young fish migrate back to the sea. The migrations are over long distances to and from spawning areas; frequently approaching the headwaters of rivers and tributaries. Most sites of dams and reservoirs on such streams lie in the paths of migration of these fish.

All dams are obstacles to the free passage of such fish, but low diversion dams and ship locks present the least obstacle, providing adequate fish ladders are installed and properly operated. A storage dam and reservoir creates a barrier that is nearly impossible to overcome. The problem is discussed in "Fishery requirements" (pp. 23-25).

Fish hatcheries are constructed at or near dams that block anadromous fish runs, and are designed to maintain the fish in numbers at least equal to those existing under previous natural conditions. Various problems have arisen in the operation of such hatcheries, but experiences gained from operation of a considerable number of hatcheries on various streams over substantial periods of time show that the hatcheries are generally successful for their purpose. While unavoidable, the adverse environmental effect of dams on anadromous fish can be mitigated with near-certainty by fish hatcheries.

There are also situations in some large rivers where resident fish migrate between upper and lower sectors of the river for their spawning cycle. In these cases, like those of anadromous fish, a storage dam may destroy or greatly reduce the migratory fish resource. For some species of resident fish the distance, or time, of travel of an adult migrant preparing to spawn appears to be important for successful spawning. Where travel distance is blocked by a dam the attempted mitigation of spawning loss by means of a hatchery would need to include ample numbers and sizes of holding ponds containing flowing water of suitable quality for the adult fish until they are ready to be spawned. This type of mitigation has not been as well tested and "proven out" as the successful use of hatcheries for anadromous fish.

C. 2. Blocking of Animal Migration Routes. Deer and other wild animals range from high land areas in summer seasons to low areas in the winter for food. Feeding areas along streams are mentioned in a previous paragraph "Loss of wildlife habitat." In some situations the migration routes cross a stream or tributary which may become part of a reservoir. Although animals will swim across a small lake, a large reservoir is usually considered to be adverse to the best environmental conditions of migrating animals if it lies across a natural migration route. The impact appears to be greatest in a geographical situation where animals attempting to migrate become trapped on a peninsula between two major arms of a reservoir.

D. 1. Unsightly Excavation and Waste Disposal Sites. It is fortunate if all the construction materials needed for a dam can be obtained from within the reservoir area. Usually this is not the case, and there will be exposed borrow areas elsewhere. There will also be piles of waste material from excavation for the dam foundation, spillway excavation, and by-pass tunnels; part of which can be placed within the reservoir area. Borrow pits and spoil piles may present a permanent unsightly appearance on the landscape unless adequately treated. The treatment consists of smoothing and planting suitable cover vegetation. For best results it may be necessary to place a layer of soil over spoil banks to assist growth of vegetation. Borrow areas should be graded such that they will drain rather than impound water. A landscaping plan may also make use of some spoil banks for necessary parking areas for vehicles, equipment storage, or a fish hatchery site. A very small percent of construction cost is usually regarded as a reasonable expenditure to avoid leaving unsightly borrow and disposal areas near a dam.

D. 2. Unsightly Reservoir Banks Below Maximum Water Line. Since the operational function of a reservoir is to alternatively be filled, then drawn down, there will be exposed banks of the reservoir whenever it is less than full. A reservoir used for flood control, or the upper portion of a multipurpose reservoir used for flood control is filled only intermittently (perhaps once in several years) and for short periods of time, and so the exposed lands may grow vegetative cover of small plant varieties between the occasional inundations. Water conservation reservoir pools are usually filled and emptied more frequently and the exposed areas tend to be barren of vegetation. Only where a conservation reservoir is very large in proportion to its streamflow supply will some of the area be exposed for several successive years in sequence and temporary vegetative growth take place.

Being barren of vegetation, most reservoir banks are unsightly when exposed below the high water line. An exception may be a beach recreation area. Another exception is a reservoir in a desert locale where definition of land contour lines by reservoir water lines adds interest to panoramic views. Alternatively, at some sites the impinging of reservoir water-marks on interesting geological formations creates distracting views. Each reservoir site will have its own characteristic impact on environment from the standpoint of bank-line appearance.

D. 3. Abandoned Construction Facilities. Construction camps, storage yards, and workshop areas if abandoned and left in a deteriorating condition after construction can detract from landscape appearance. This is in contrast to a mining "ghost town;" the remains of which are usually regarded as having environmental appeal. The difference may be that the mining ghost town is an historical artifact, but a dam construction camp is not; the historical evidence of dam construction is the dam itself.

It is considered reasonable that some percentage of construction cost should be expended in cleanup after construction. In some instances a construction camp or construction base can be planned, with forethought, for permanent future use as housing, business, and operational use after completion of construction. Such planning can provide for favorable environmental appearance.

D. 4. Erosion Scars From Construction Roads. Temporary construction roads are often built without regard to environmental effect, and upon termination of construction are abandoned in deteriorating condition. One feature of such roads is the excess erosion that may take place due to insufficient provisions and care of drainage, and the resultant formation of new gullies or other watercourses. Another possible situation is that roads constructed on steep slopes near dam abutments increase the likelihood of landslides or rockslides occurring at some time after completion of construction.

These adverse environmental effects are mainly avoidable by precautions taken in the construction of the temporary roads. One

TABLE 1-5. Summary of Avoidable and Unavoidable Adverse Effects of Dams and Reservoirs on Environment.

Reference No.[1]	Potential adverse effect	Mitigation method or effect	Probable degree or importance of adverse effect
A.	*Land use for reservoir*		
A. 1.	Loss of fish and aquatic habitat	Changes of species	New species may be less desirable than original species
A. 2.	Loss of wildlife habitat	Improve other areas for habitat	Full mitigation probably not possible
A. 3.	Loss of future access to mineral deposits	None	Is of importance only if mineral deposits exist
A. 4.	Loss of mountain valley areas	None	Important only in extremely mountainous areas
A. 5.	Inundation of historical or archeological sites	Possibly by a museum	Varies with each individual site
A. 6.	Inundation of exceptional geological formations	Usually not possible	Varies with each individual site
B.	*Alterations of downstream flows*		
B. 1.	Reduction of fish and aquatic habitat	Maintain regulated flows	Full mitigation possible, but frequently not acceptable because of large sacrifice of project accomplishments
B. 2.	Reduction of stream flushing flows	Release occasional flushing flows	Mitigation method not proven to be worthwhile. Degree of environmental effect depends upon specific stream situation
B. 3.	Changes of water quality	Selective level reservoir outlets; water aeration if needed	Somewhat limited experience with selective level outlets indicates good prospects of full mitigation
C.	*Interference with fish and wildlife migrations*		
C. 1.	Blocking anadromous fish runs	Fish hatcheries	Usually capable of full mitigation
C. 2.	Blocking animal migration routes	None practical	Importance depends upon the specific site
D.	*Landscape appearance*		
D. 1.	Excavation and waste disposal sites	Project expenditures required to landscape sites	Satisfactory mitigation usually possible without excessive expenditure
D. 2.	Reservoir banks below maximum waterline	Minor areas may be developed for beaches	Degree of impact depends upon the specific reservoir site
D. 3.	Abandoned construction facilities	Construction clean-up	Full mitigation possible. Important only if not done
D. 4.	Erosion scars from construction roads	Principally by care of drainage	Adverse effects can be reduced but not entirely eliminated within reasonable cost
D. 5.	Reservoir clearing waste disposal	Controlled burning. Marketing max. amounts of wood products	Temporary effect; usually minor, but not entirely avoidable

[1] Refers to paragraph identification in the text.

principal precaution is the installation of adequate drainage structures. These would be permanent or semipermanent types of structures although the road may be used only temporarily. Project construction costs would be increased. In some instances the adverse environmental effects of construction roads are unavoidable.

D. 5. Reservoir Clearing Waste Disposal. In a reservoir site containing man-made improvements such as houses and other buildings the structures are usually disassembled and removed for their salvage value. Most of the necessary reservoir clearing will be the removal of trees and large species of brush. Some trees may have commercial timber value and be removed for such use. Some smaller sizes of trees are occasionally used for fuel or for producing wood chips for manufacture of wood pulp. The large brush and the trees not used for harvesting require cutting and disposal by the best means possible.

The usual disposal is by burning. The smoke of burning adds temporarily to air pollution; an adverse effect, but usually small in comparison to smoke pollution from lightning-caused forest or range fires. Dissolved chemicals leached from the ash residue of burning adds to other dissolved minerals in water first stored in the reservoir. This is a one-time occurrence and apparently not serious with respect to water quality.

With increasing world demand for wood pulp, and with wood scraps being more widely used in new manufacturing processes it appears that more of the trees cleared from a reservoir site will be salvaged for their manufacturing value than at present, and less will be burned merely for disposal. Adverse environmental effect of reservoir clearing is not entirely avoidable, but the effect is temporary and usually minor.

Summary of Avoidable and Unavoidable Environmental Effects of Dams and Reservoirs.

Table 1-5 summarizes the principal adverse effects. Some can be either partially or nearly completely mitigated. The degree of importance of many of the effects depends upon the existing physical and environmental circumstances at specific sites. Some may be so great in importance as to almost certainly prohibit the use of the site for a reservoir. At other sites the effects may be quite minor and the amount of environmental loss easily acceptable when project need is demonstrated.

ENVIRONMENTAL IMPACT ALTERNATIVES

Every well-planned water development or control project produces accomplishments in the areas of benefit that are economically, socially, and environmentally desirable. However, some of the project facilities, notably dams and reservoirs, produce adverse environmental effects to some large or small degree in their areas of direct influence. It is necessary to examine alternative plans of dams and reservoirs to take note of any possibilities of minimizing any adverse environmental effects while accomplishing the major project purposes.

It has been seen, as shown in preceding paragraphs, that the adverse impacts of dams and reservoirs may occur in a number of different ways. It is not possible to rate these impacts in an order of importance that would be correct for all project planning areas. This is because each stream or stream system has its own types and amounts of environmental qualities; some of which are unique, or quite dissimilar from those elsewhere. One stream may possess scenic splendor but be low in aquatic life habitat; another may contain numerous and diverse aquatic species but otherwise be of small environmental importance.

A high dam at one location on a stream may have a high environmental impact but at another location have less impact. A low dam at a particular site may have much less environmental impact than a high dam at the same site. In consideration of environmental impacts it is necessary to devise and evaluate alternative plans giving attention to the specific environmental qualities that would be affected by each plan while also maintaining consideration of the physical factors, costs, accomplishments, and economic aspects of each plan.

The eventual choice of a dam or dams of a project plan is made in consideration of relative accomplishments, cost, social effects, and environmental impacts. Since most environmental values are not quantifiable in monetary terms it is necessary to show the nature and extent of environmental impacts of any and all plan alternatives that appear worthy of consideration in making a plan decision.

Procedures for Devising Environmental Plan Alternatives

The following is one suggested method for selection of alternative plans having different amounts of environmental impact; with each plan accomplishing the water conservation, flood control and/or power generation objectives. Planners will use ingenuity in modifying this method or devising another to fit the circumstances of a specific problem.

A useful first step is to make an examination and inventory of the various important environmental qualities of the stream or stream system under consideration. Environmental factors might be ranked in order according to their importance. This inventory would be of the existing conditions, which may be natural conditions in some instances, but more likely be conditions which have been influenced by human activities up to the present time.

A second step would be to estimate the *future* environmental condition *without the project development*. It may be the same as the existing one, or may be improved or degraded. The future condition might be improved if, as an example, there are definite plans under way to reduce present stream pollution. It may be worsened in various ways. The future environmental condition "without project," to the extent it can be foreseen, becomes the basis of comparison of alternative project development plans.

A next step would be to plan an optimum economic project disregarding all environmental impacts except those that can be positively accepted as being controlling. An example of this is that it would be of no purpose to consider a dam and reservoir within an established National Park or Wilderness Area. The evaluation of this alternative plan will show the maximum possible project accomplishments within the physical conditions and economic limitations. It might be designated as the "optimum economic plan alternative." The adverse economic impacts would be listed and described.

Another alternative plan would be devised wherein an attempt would be made to minimize all important adverse environmental impacts and still produce some portion of the intended project accomplishments. If this "optimum environmental alternative" indicates greatly reduced accomplishments and/or much increased costs from the optimum economic plan alternative it may be set aside in favor of a third alternative.

A third alternative plan would be devised whereby a compromise would be made between the two extreme alternatives, seeking to avoid or minimize the most important adverse environmental impacts while achieving all or most of the potential project accomplishments of the optimum economic plan alternative. Since concessions will be made for environment, this plan will probably cost more per unit of accomplishment than the optimum economic plan.

The results of evaluation of the alternative plans would be summarized to show relative amounts of environmental impacts, project accomplishments, costs, and the costs incurred in providing degrees of maintenance of the "without project" environmental resources. This summary would provide the basis of obtaining agreement and decision on selecting a plan. The indicated environmental cost component may, by policy, be made a reimbursable obligation of the project beneficiaries or a non-reimbursable public (government) cost, or some division between the two.

Example of Environmental Plan Alternatives

There are numerous different situations where environmental values may be affected by dam and reservoir developments. The following is a description of one example which illustrates the idea of examining three alternatives. They are illustrated in Figs. 1-9-1 and 1-9-2.

Fig. 1-9-1 indicates a simplified river system, omitting minor tributaries. This river has various environmental qualities, two of which are of first-order importance: (1) a considerable portion of the river has scenic and/or recreational values; and (2) the main stem of the river is a waterway for anadromous fish migration, and the upper sectors are habitats for fish spawning and nursery areas of newly hatched fish before they migrate downstream. The future environmental condition may be either a little better or a little poorer than at present, depending upon future activities along the river and within the watershed.

The second sketch shows one of several possible alternative developments. This particular alternative includes a large mainstream reservoir which would develop most of the river flow for service area uses, and would provide a large degree of flood control protection for downstream areas. While this alterna-

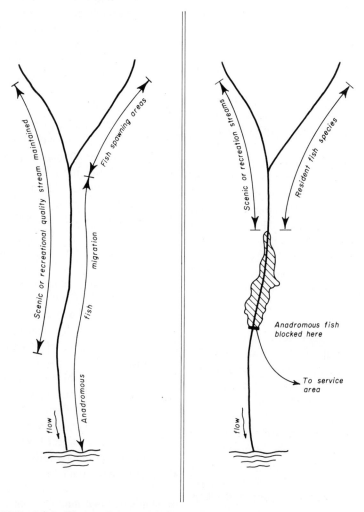

Figure 1-9-1. River environmental alternative plans. (Left) Existing or future conditions without project. (Right) Maximum economic development with large on-stream reservoir.

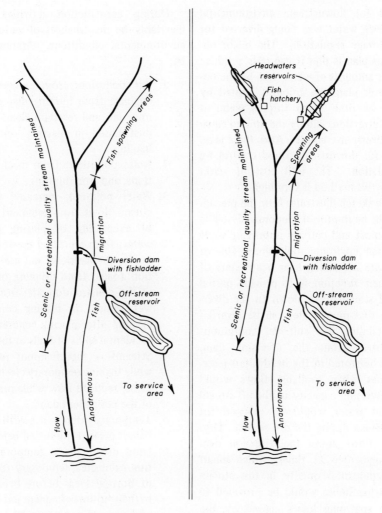

Figure 1-9-2. River environmental alternative plans. (Left) Project plan with maximum environmental protection. (Right) Economic plan with favorable environmental protection.

tive may be the optimum development from an economic viewpoint, there are large adverse environmental effects. The high dam for a storage reservoir blocks all anadromous fish migration, seriously reducing that resource on this stream. Scenic or recreational portions of the stream are reduced in length. The reservoir may offer some recreational use and may assist in supporting some species of resident fish, but these effects might be small in comparison to the losses.

The third sketch shows how water development might be attempted while seeking to maintain the environmental qualities as fully as possible. An off-stream reservoir site has been located where there would be only slight environmental impact resulting from its use. A low diversion structure with fish ladder would not significantly interfere with anadromous fish. A plan of operation would be specified whereby no water would be diverted during low-flow seasons, and substantial flows would be bypassed for downstream environmental needs whenever water was being diverted for off-stream storage regulation. The major objection to this plan is that it can only produce much smaller amounts of water than by the on-stream reservoir plan; the amounts limited by the periods of time when surplus water is available for diversion, and by the limited capacity of the diversion conduit. Also, there is no opportunity for downstream flood control by storage regulation. The unit cost of water developed by this method is very high.

The fourth sketch illustrates how the preceding plan might be improved economically while maintaining most and perhaps improving some of the in-stream environmental values. One or more reservoirs are added near headwaters of streams. Their function is to provide partial regulation of stream flows down to the diversion dam so that more water can be diverted while maintaining the suitable continuous flows downriver from the diversion dam. Water would be stored in the headwaters reservoirs at times when available flows would exceed the diversion capacity to the off-stream reservoir, and water would be released for project diversions during the dry season. Dry-season river flows above the diversion dam would be augmented at times, which might add to environmental quality in this stream sector. Fish hatcheries would be provided to mitigate fish spawning losses caused by the headwaters dams. An overcompensation might also be desirable which would, of course, result in the enhancement of fishery.

THE RELATIONSHIP BETWEEN SHORT-TERM ENVIRONMENTAL USES AND LONG-TERM PRODUCTIVITY

Short-Term Environmental Project Impacts

Some distinctions were made in preceding paragraphs between temporary and permanent project impacts on the environment. A number of possible impacts are of temporary nature and may occur during project construction or during early years of project operations.

During construction activities there will inevitably be disturbances of various kinds to environmental conditions. Among them may be:

1. Air pollution: from exhaust of power engines; from burning disposal of waste materials and reservoir clearing; and from dust raised in blasting, excavating and hauling.
2. Noise pollution: from machinery operations, and from blasting.
3. Water polution: increased turbidity of stream flows from sediment disturbance of excavating or placing materials in water; from increased erosion conditions in site preparations and use of construction roads; from cleaning of equipment at the construction site; from leakage of wet cement in placing concrete; from cleanup after placing concrete; and from accidental spills of fuels or lubricants.
4. Streamflow modification: reduced flows while making temporary closures of coffer dams; reduced flow while making closure of the permanent dam.
5. Temporary interference with anadromous fish, if present (discussed below).
6. Land disturbances: temporary construction campsites; temporary roads; and use of borrow areas before being converted to their future permanent use or condition.

Precautions can be taken to avoid or minimize these types of adverse temporary effects. Some may not be wholly avoidable within possibly-reasonable cost.

If the stream is the habitat of anadromous fish there is a potential and likely adverse impact both during construction and in the early years of project operations. A fish hatchery will almost certainly be required as a permanent installation to mitigate the losses caused by a storage dam and reservoir. To best maintain the continuity and numbers of fish it would be very desirable to have the hatchery in operation before the start of dam construction. Otherwise there may be considerable amounts of fish losses due to temporary stream blocking during construction operations. However, physical circumstances may impose major difficulties in first constructing a permanent-type fish hatchery because: (1) for best accomplishment it is desirable to have the hatchery near the dam, rather than far downstream; (2) the hatchery facilities (including rearing ponds) need to be located near the stream but on a site above the levels of any but an extreme flood; and (3) a hatchery water supply is needed in sufficient amounts and of suitable quality for egg-hatching and rearing of newly-hatched fish. This set of conditions may be difficult to attain prior to and during dam construction. If not practically attainable, the best solution may be to trap adult spawners near the dam being constructed and transport these fish or their eggs to an existing hatchery elsewhere which may have some surplus capacity. Whatever the solution, there will likely be some diminution of the fish resource during the temporary construction period and until a permanent new hatchery is in operation.

There have been frequent experiences that during the first few years of operation difficulties of one sort or another arise which cause partial failures in the hatchery accomplishment. These unanticipated difficulties may be caused by some unfavorable change in quality of water from the new reservoir source, or the type of fish food supplied for rearing pond use, or otherwise. Possibly unfavorable water quality may be temperatures too high or too low for egg hatching and fish rearing, a shortage of dissolved oxygen, or an excess of other dissolved gasses. Corrections, if needed, are possible by making use of selective-level intakes to the discharge conduit of the reservoir supplying the hatchery water and by regulating the amount of aeration of water supplied to the hatchery. A recirculating water system with an aeration facility may possibly be found desirable as an adaptation to the hatchery, depending upon initial operating results and experience. The problems are correctable, but there is a temporary period of perhaps a few years when the fish resource is diminished. Then, because of the life cycle time of the fish species, the full recovery of the quantities of fish may not take place for another few years.

Long-Term Project Accomplishments

Long-term project accomplishments are in three possible categories:

1. Project accomplishments that can be measured in economic terms.
2. Project environmental improvements not subject to economic measurement.
3. Quality-of-life, or social benefits.

In any water development or management plan these three categories of accomplishments occur together to make up the total benefit achieved by the plan.

Project Functions Subject to Economic Measurement. These have been previously discussed. They may include flood control; municipal, industrial, and domestic water supply; irrigation; hydroelectric power generation; salinity repulsion; water quality improvement; recreation; fishery improvement; and sediment control. Although economic values of these functions are measured or estimated, there are additional values not subject to economic measurement. These are discussed below.

Environmental Improvements. A preceding discussion titled "Environmental Impact Alternatives" dealt with impacts closely associated with dams but not within project service areas. The impacts associated with dams were usually negative or neutral; although there is the possibility of environmental improvement from regulated stream flows downriver from a dam. In the service areas, which are the areas where the great majority of people live—the agricultural and urban areas—the environmental effects are both negative and positive. The negative effects are those in which the activities of people reduce the amount, or downgrade the quality of "natural" or pre-civilization conditions. The positive environmental effects are the results whereby people have improved their own well-being in a far-sighted manner that tends to preserve these improvements for future generations, as well as being desirable for the immediate present. The long-range nature of water resource and water control developments, when they are adequately planned, and the long life of major project structures, satisfy the environmental purpose of providing for future generations. Since people, present and future, are the important thing in service areas, the environmental improvements they make for themselves far outweigh the reductions of pre-civilization environmental conditions where they live. It is of importance to consider that people, on the average, spend about eighty percent of their lives in and about their own homes, or nearly all their lives at their homes and working places. Only a small fraction of lifetimes are spent going elsewhere. There is not a clear distinction between peoples' "at-and-near-home" environment and their quality of life or "well-being."

Quality of Life, or Social Benefits. A contemporary writer* provides a list of desirable living conditions for people:

"it is possible to name a number of living conditions that collectively will provide the ingredients for what may be called the 'well-being' of a society. These conditions would include the following:

1. Food, clothes and shelter.
2. Individual and collective security.
3. Luxury and convenience.
4. Good health.
5. Good education.
6. Harmonious family relations.
7. Pleasant working conditions.
8. A clean and stimulating environment.
9. A certain level of culture.
10. A certain level of morality.

There is no doubt that one can think of more items that are important for providing the 'good life.' Also, there is no doubt that every individual will place different emphasis on different items as being meaningful to him. In spite of such diversities, we may safely observe that all of these items, to some extent, have some bearing on what we have called the 'well-being' of society."

It would be unrealistic to claim that long-term water resource project accomplishments are responsible in a large way for improvements of all the ten listed conditions. However, the accomplishments are important to some of the ten, and may indirectly contribute to others. Quite separately from economic benefits, some of the quality-of-life or social benefits can be listed as follows:

Food, clothes and shelter. Irrigated agriculture provides a basis for obtaining a diversity of foods, and some sources of fiber for manufacture of clothes and construction of shelter. These things, in turn, contribute to good health and the benefits that follow therefrom.

Individual and collective security. Three conditions may be recognized under this topic: protection from floods, protection from droughts, and improved dependability of electric power.

   Protection from floods. There is a rather large list of direct and indirect flood control benefits that can be evaluated in economic terms; these have been identified earlier in this chapter. In addition, there are the social well-being effects that are unmeasurable in monetary terms. The most commonly recognized one is the prevention of loss of life.

Anyone who has personally experienced the devastation of a large flood or assisted in the rehabilitation following a great flood can attest to the loss of quality of life. There are the losses of persons drowned, the emergency needs for survival, the risks to health, the loss of personal treasures, the loss of efforts and industry of people to beautify their immediate surroundings, and a loss of the spirit of initiative. Compassionate people come to the aid of flood victims and their laudable aid and intentions help in part to offset the social losses.

---

*Mr. E. Kuiper, "Water Resource Project Economics," Butterworth & Co. Limited, (Publishers) London, England.

Protection from drought. A severe and extended drought has large social effects on the stricken community. Migrations and resettlement of people may result, with the attendant risks to health, severance of community ties and interests, and loss of spirit and accomplishment in seeking individual life improvement. People remaining in the drought-stricken area endure a lowered standard of living from which follows poorer educational opportunities for young people, possibly poorer health, and lessening of the spirit of good will to others less unfortunate.

Improved dependability of electric power. If hydroelectric power generation is a project function it is combined with other units of power generating sources. Thereby, there is some added insurance against disruptions of needed service which might be caused by interruptions of fuel supplies for thermal power plants for any reason. From a long-term outlook there may be noted the social value of producing power from a renewable resource, hydropower, instead of using non-renewable fuels.

Luxury and convenience. Here the social well-being values of municipal and domestic water supplies are of importance. Except for agricultural needs it is possible for people to survive on very small amounts of water, as witnessed by desert nomads, space travellers, mountain climbers, and others. All additional municipal and domestic use is for good health, convenience, and luxury of living. These uses include flushing of toilets, bathing, dishwashing, laundering, housecleaning, cooling, watering of domestic animals, street cleaning, fire fighting, recreational swimming, and growing of lawns and decorative shrubbery. These uses come from water conservation project supplies in the many situations where ground water is unavailable, or of very poor quality, or in short supply because of use for irrigation and industrial purposes. Supplies of ground water are less than existing or projected demands in many areas.

Health. Improved health aspects have been mentioned in the preceding paragraphs. One further factor may be noted if hydroelectric power generation is a project function: the absence of air pollution from burning of fuel for thermal power plants, to the extent that hydropower takes the place of thermal power.

Social Benefits of Income Distribution

A valid economic analysis of a project plan will examine and list the types and amounts of project benefits attributable to the plan. A favorable plan, from the economic viewpoint, will show that total benefits exceed total costs. The degree of favorability is evidenced by the amount these benefits exceed costs.

In the case of water resource development and/or management programs the individual beneficiaries are usually numerous; however, there may be exceptions where a few persons stand to gain large benefits or advantages from water resource development and controls while the remainder of the concerned population gains little or possibly nothing from the development and controls. Such imbalances may be brought about or influenced by the method of project financing, and also by the physical location of project works and their areas of service.

An example of a maldistribution of project benefits can be illustrated by a flood control project which is publicly financed from general tax revenues without repayments required of direct beneficiaries, and where a few property ownerships are so located as to receive a large portion of the benefits. The situation is sometimes described as one yielding vast "windfall" benefits. Agencies such as the U.S. Corps of Engineers identify and evaluate the potentials of land enhancement or "windfall" benefits of a project plan in order that this condition may be taken into account in the decision of whether or not to proceed with the plan.

Maldistribution of benefits is also possible in water use developments. Restrictions against such inequities are usually attempted by regulatory agencies which seek to serve the broad public interest by rejecting a plan having few beneficiaries in favor of one which would serve a larger number more equitably. Such agencies in the U.S. include those of the separate states that grant and control water rights, inter-state Commissions having similar responsibilities and authority on water use of interstate streams, and the Federal Power Commission with respect to development of hydro-electric power.

Still a further example of attempts to avoid inequitable distribution of project benefits in the United States is that of limiting service to any one landowner of an irrigation project constructed with federal funds under reclamation laws; these being projects constructed and administered by the Bureau of Reclamation. The limitation, known as the "acreage limitation," results from provisions of the Federal Reclamation Law. It is intended to distribute irrigation benefits widely by limiting the federally-subsidized irrigation service to individually-owned family-size farms.

The acreage limitation provision of the U.S. Reclamation Law has existed since 1902. The objective of serving the best public interest in federal licensing of hydropower developments has existed since 1922. Examinations of "windfall profit" potentials of flood control developments by the Corps of Engineers are an established procedure. The various authorities granting water rights have, in the adjudication of rights, sought to assure the broadest public benefits from use of the natural resource. These examples serve to show that, although controversial, there is widespread belief in, and considerable acceptance of the idea that society is best served when the benefits (or values) of a water resource development are distributed broadly, rather than being limited or concentrated in the hands of a few. Thus, the degree by which a specific plan favors or permits a broad distribution of newly-created benefits is another one of the several indicators of the value of the plan in comparison to alternative plants.

A further measure of the income distribution value of a plan exists in the generally-accepted belief that income of the poorer

TABLE 1-6. Summary Estimated Income Distribution

Employment Account	Plan B	Recommended Plan	Difference (recommended plan minus Plan B)
Beneficial effects:			
Project construction employment	300 semiskilled jobs for 3 years.	200 semiskilled jobs for 4 years.	−100 semiskilled jobs per year for 3 years, but +200 semiskilled jobs for 1 year.
Project O&M employment	40 permanent semiskilled jobs.	50 permanent semiskilled jobs.	+10 permanent semiskilled jobs.
Employment in service and trade activities induced by and stemming from project operation	850 permanent semiskilled jobs.	900 permanent semiskilled jobs.	+50 permanent semiskilled jobs.
Adverse effects			
Employment in activities induced by and stemming from displaced agricultural operations	15 permanent semiskilled jobs.	50 permanent semiskilled jobs.	+35 permanent semiskilled jobs.
Net beneficial effects			
	300 semiskilled jobs for 3 years.	200 semiskilled jobs for 4 years.	−100 semiskilled jobs per year for 3 years, but +200 semiskilled jobs for 1 year.
	875 permanent semiskilled jobs.	900 permanent semiskilled jobs.	+25 permanent semiskilled jobs.

Source: FEDERAL REGISTER, Vol. 38, No. 174—Monday, September 10, 1973

members of society should be increased to a greater degree than incomes of the more affluent. This idea has support in such well-established institutions such as graduated income taxes and unemployment insurance programs. Aside from individual investment and management activities, incomes derived from water-related projects are in the nature of employment provided to skilled, semiskilled, and unskilled workers. This employment occurs temporarily during construction of new project works, and more permanently in operation and use of the resulting system. When a project plan is clearly visualized it is possible to forecast with some degree of reliability the numbers of new jobs that will be created in the different employment categories. Results can be analyzed statistically and used as one factor in comparing the income distribution value of a specific plan with its alternatives. Such a comparison may be relatively insignificant in selecting between alternative dams of similar locations and sizes, or may be of larger importance in determining whether or not a plan of development is justified and should be undertaken.

Table 1-6 illustrates an example of a summary that might be developed of data pertinent to estimated income distribution attributable to two alternative project plans. The two plans are listed as "Recommended Plan" and "Plan B." The example assumes some loss of agricultural land as required for dam, reservoir or other project works.

Planning Principles and Standards Established by the United States Water Resources Council

The Water Resources Council, an independent Executive Agency of the U.S. Government, established Principles and Standards for Planning Water and Related Land Resources which became effective October 25, 1973. The Principles and Standards were approved by the President of the United States, and thereby became specified for use by all affected agencies of the United States government. Further, they also apply where federal assistance is extended to state and local governments within the United States engaged in sponsoring watershed, water, and land resource programs. These Principles and Standards are published in the Federal Register, Part III, Volume 38, No. 174—Monday, September 10, 1973, Washington, D.C.

The Water Resources Council is composed of the heads of several departments and the Chairman of the Federal Power Commission of the executive branch of the U.S. Government. The Principles and Standards are the result of several years of coordinated efforts to devise best procedures to attain two objectives, economic and environmental in the water resources field:

"A. to enhance national economic development by increasing the value of the Nation's output of goods and services and improving national economic efficiency."

"B. to enhance the quality of the environment by the management, conservation, preservation, creation, restoration, or improvement of the quality of certain natural and cultural resources and ecological systems."

The procedures and detailed methods for applying the Principles and Standards to the economic and environmental aspects of water resource development are rather thoroughly described or outlined in the federal document. They are qualified as being subject to revisions as continued experience, research, and planning conditions may require.

The economic and environmental considerations applicable to problems of site and size selections of dams as described in this chapter are, in general, consistent with the objectives and methods indicated by the Water Resource Council. However, planners employed by agencies of the United States government, or planners whose work product will require review and approval by a concerned federal agency will find it either necessary or highly desirable to be informed of the Water Resource Council's Principles and Standards.

IRREVERSIBLE AND IRRETRIEVABLE COMMITMENTS OF RESOURCES

General

Possible resource commitments for a dam and reservoir development may take several forms. There will be water and land use commitments, and some commitments of construction materials and construction funds. There may also be some commitments of diminished resources such as vegetative, wildlife, archeological, and mineral resources.

Water commitments The commitment of water supply is for the life of the project, which may be 50 years, 100 years, or longer. However, water is a renewable resource, and its use can be altered at some future time according to changes of needs and substitution of possible alternative supplies such as from weather modification or demineralization of saline supplies.

Land use commitments Lands committed for a dam and reservoir are of the nature of a permanent commitment. Eventually, however, a reservoir will fill with sediment and a new valley-type area of land will be formed. But this is not an acceptable replacement of the original land of the reservoir site as it depends on the structural life of the dam.

Land used for forebay, afterbay, and diversion dams is a commitment for an economic project life of more or less than 50 years. Small dams are occasionally demolished after they have served a useful period, and the small water impoundment areas tend to be flushed of sediment accumulation and restored approximately to pre-project conditions.

Construction materials commitments Most of the earth, rock, concrete, and metals used in a dam are a permanent commitment. Some gates and valves have metal salvage value after a dam is taken out of service.

Construction power use commitments This consists of the amounts of fuel or other energy resources used in construction. The energy quantity is frequently expressed in equivalent barrels of oil.

Construction funds commitment The construction cost in monetary units is the commitment of a part of the financial resources of the organization sponsoring the project.

Vegetative commitments The amount of natural and cultivated vegetation growth that is displaced by a dam and reservoir is partly an irretrievable commitment, but can be compensated by increased vegetative growth elsewhere if the project is designed to do so. This is a renewable resource.

Wildlife commitments Loss of wildlife due to a dam and reservoir can usually be mitigated to some degree, or possibly fully compensated elsewhere if it is found to be practical to do so as part of the specific project plan.

Archeological commitment Archeological treasures, where existing in a project area, are subject to loss by dam construction and reservoir inundation, except to the extent they may be moved and preserved prior to construction.

Mineral resources commitment Any mineral resources existing at a dam and reservoir site are subject to permanent loss if they are not extracted prior to dam construction and reservoir filling. In some situations it is possible to combine mineral resource extractions with construction operations.

SUMMARIZING OF PROJECT INVESTIGATIONS WITH RESPECT TO DAM SITE AND SIZE SELECTIONS

The selection process is obviously complicated, or at least complex in nearly all project planning investigations. In simplest terms, the objective is to answer two questions:

1. All factors considered, is this project plan worthy of being constructed?
2. From all standpoints, has the best available alternative plan been chosen or recommended?

To facilitate the deliberations in attempting to arrive at decisions, usually involving a consensus of opinions, it is desirable to list, as concisely as possible, the important attributes of each alternative considered; this in some order or arrangement to aid in noting the comparative factors. The listing might be in the following form:

1. Identification of alternative plan (by a name or symbol).
2. Project accomplishments.
3. Monetary costs (investment; present worth).
4. Economic evaluation
   Benefit-cost.
   Repayment costs.
5. Environmental impacts, principal
   Short-term.
   Permanent or long-term.
6. Social, or quality-of-life benefits.
7. Irreversible/irretrievable commitment of resources.

In such an arrangement the factors would be a comparison against the effects of doing nothing, i.e., not constructing the project. The listing would be made for each plan alternative that was fully considered, omitting the alternatives that were screened out in earlier steps of planning because of having been found to be unsuitable or unwarranted for further consideration.

APPENDIX 1-A

**the national environmental policy act of 1969,
public law 91-190
january 1, 1970
(42 u.s.c. 4321-4347)**

An Act to establish a national policy for the environment, to provide for the establishment of a Council on Environmental Quality, and for other purposes.

Be it enacted by the Senate and House of Representatives of the United States of America in Congress assembled, That this Act may be cited as the "National Environmental Policy Act of 1969."

purpose

SEC. 2. The purposes of this Act are: To declare a national policy which will encourage productive and enjoyable harmony between man and his environment; to promote efforts which will prevent or eliminate damage to the environment and biosphere and stimulate the health and welfare of man; to enrich the understanding of the ecological systems and natural resources important to the Nation; and to establish a Council on Environmental Quality.

**title i
declaration of national environmental policy**

SEC. 101. (a) The Congress, recognizing the profound impact of man's activity on the interrelations of all components of the natural environment, particularly the profound influences of population growth, high-density urbanization, industrial expansion, resource exploitation, and new and expanding technological advances and recognizing further the critical importance of restoring and maintaining environmental quality to the overall welfare and development of man, declares that it is the continuing policy of the Federal Government, in cooperation with State and local governments, and other concerned public and private organizations, to use all practicable means and measures, including financial and technical assistance, in a manner calculated to foster and promote the general welfare, to create and maintain conditions under which man and nature can exist in productive harmony, and fulfill the social, economic, and other requirements of present and future generations of Americans.

(b) In order to carry out the policy set forth in this Act, it is the continuing responsibility of the Federal Government to use all practicable means, consistent with other essential considerations of national policy, to improve and coordinate Federal plans, functions, programs, and resources to the end that the Nation may—

(1) Fulfill the responsibilities of each generation as trustee of the environment for succeeding generations;

(2) Assure for all Americans safe, healthful, productive, and esthetically and culturally pleasing surroundings;

(3) Attain the widest range of beneficial

uses of the environment without degradation, risk to health or safety, or other undesirable and unintended consequences;

(4) Preserve important historic, cultural, and natural aspects of our national heritage, and maintain, wherever possible, an environment which supports diversity, and variety of individual choice;

(5) Achieve a balance between population and resource use which will permit high standards of living and a wide sharing of life's amenities; and

(6) Enhance the quality of renewable resources and approach the maximum attainable recycling of depletable resources.

(c) The Congress recognizes that each person should enjoy a healthful environment and that each person has a responsibility to contribute to the preservation and enhancement of the environment.

SEC. 102. The Congress authorizes and directs that, to the fullest extent possible: (1) the policies, regulations, and public laws of the United States shall be interpreted and administered in accordance with the policies set forth in this Act, and (2) all agencies of the Federal Government shall—

(A) Utilize a systematic, interdisciplinary approach which will insure the integrated use of the natural and social sciences and the environmental design arts in planning and in decision-making which may have an impact on man's environment;

(B) Identify and develop methods and procedures, in consultation with the Council on Environmental Quality established by title II of this Act, which will insure that presently unquantified environmental amenities and values may be given appropriate consideration in decisionmaking along with economic and technical considerations;

(C) Include in every recommendation or report on proposals for legislation and other major Federal actions significantly affecting the quality of the human environment, a detailed statement by the responsible official on—

(i) The environmental impact of the proposed action,
(ii) Any adverse environmental effects which cannot be avoided should the proposal be implemented,
(iii) Alternatives to the proposed action,
(iv) The relationship between local short-term uses of man's environment and the maintenance and enhancement of long-term productivity, and
(v) Any irreversible and irretrievable commitments of resources which would be involved in the proposed action should it be implemented.

Prior to making any detailed statement, the responsible Federal official shall consult with and obtain the comments of any Federal agency which has jurisdiction by law or special expertise with respect to any environmental impact involved. Copies of such statement and the comments and views of the appropriate Federal, State, and local agencies, which are authorized to develop and enforce environmental standards, shall be made available to the President, the Council on Environmental Quality and to the public as provided by section 552 of title 5, United States Code, and shall accompany the proposal through the existing agency review processes;

(D) Study, develop, and describe appropriate alternatives to recommended courses of action in any proposal which involves unresolved conflicts concerning alternative uses of available resources;

(E) Recognize the worldwide and long-range character of environmental problems and, where consistent with the foreign policy of the United States, lend appropriate support to initiatives, resolutions, and programs designed to maximize international cooperation in anticipating and preventing a decline in the quality of mankind's world environment;

(F) Make available to States, counties, municipalities, institutions, and individuals, advice and information useful in restoring, maintaining, and enhancing the quality of the environment;

(G) Initiate and utilize ecological information in the planning and development of resource-oriented projects; and

(H) Assist the Council on Environmental Quality established by title II of this Act.

SEC. 103. All agencies of the Federal Government shall review their present statutory authority, administrative regulations, and current policies and procedures for the purpose of determining whether there are any deficiencies or inconsistencies therein which prohibit full compliance with the purposes and provisions of this Act and shall propose to the President not later than July 1, 1971, such measures as may be necessary to bring their authority and policies into conformity with the intent, purposes, and procedures set forth in this Act.

Sec. 104. Nothing in section 102 or 103 shall in any way affect the specific statutory obligations of any Federal agency (1) to comply with criteria or standards of environmental quality, (2) to coordinate or consult with any other Federal or State agency, or (3) to act, or refrain from acting contingent upon the recommendations or certification of any other Federal or State agency.

Sec. 105. The policies and goals set forth in this Act are supplementary to those set forth in existing authorizations of Federal agencies.

title ii
council on environmental quality

Sec. 201. The President shall transmit to the Congress annually beginning July 1, 1970, an Environmental Quality Report (hereinafter referred to as the "report") which shall set forth (1) the status and condition of the major natural, manmade, or altered environmental classes of the Nation, including, but not limited to, the air, the aquatic, including marine, estuarine, and fresh water, and the terrestrial environment, including, but not limited to, the forest, dryland, wetland, range, urban, suburban and rural environment; (2) current and foreseeable trends in the quality, management and utilization of such environments and the effects of those trends on the social, economic, and other requirements of the Nation; (3) the adequacy of available natural resources for fulfilling human and economic requirements of the Nation in the light of expected population pressures; (4) a review of the programs and activities (including regulatory activities) of the Federal Government, the State and local governments, and nongovernmental entities or individuals with particular reference to their effect on the environment and on the conservation, development and utilization of natural resources; and (5) a program for remedying the deficiencies of existing programs and activities, together with recommendations for legislation.

Sec. 202. There is created in the Executive Office of the President a Council on Environmental Quality (hereinafter referred to as the "Council"). The Council shall be composed of three members who shall be appointed by the President to serve at his pleasure, by and with the advice and consent of the Senate. The President shall designate one of the members of the Council to serve as Chairman. Each member shall be a person who, as a result of his training, experience, and attainments, is exceptionally well qualified to analyze and interpret environmental trends and information of all kinds; to appraise programs and activities of the Federal Government in the light of the policy set forth in title I of this Act; to be conscious of and responsive to the scientific, economic, social, esthetic, and cultural needs and interests of the Nation; and to formulate and recommend national policies to promote the improvement of the quality of the environment.

Sec. 203. The Council may employ such officers and employees as may be necessary to carry out its functions under this Act. In addition, the Council may employ and fix the compensation of such experts and consultants as may be necessary for the carrying out of its functions under this Act, in accordance with section 3109 of title 5, United States Code (but without regard to the last sentence thereof).

Sec. 204. It shall be the duty and function of the Council—

(1) To assist and advise the President in the preparation of the Environmental Quality Report required by section 201;

(2) To gather timely and authoritative information concerning the conditions and trends in the quality of the environment both current and prospective, to analyze and interpret such information for the purpose of determining whether such conditions and trends are interfering, or are likely to interfere, with the achievement of the policy set forth in title I of this Act, and to compile and submit to the President studies relating to such conditions and trends;

(3) To review and appraise the various programs and activities of the Federal Government in the light of the policy set forth in title I of this Act for the purpose of determining the extent to which such programs and activities are contributing to the achievement of such policy, and to make recommendations to the President with respect thereto;

(4) To develop and recommend to the President national policies to foster and promote the improvement of environmental quality to meet the conservation, social, economic, health, and other requirements and goals of the Nation;

(5) To conduct investigations, studies, surveys, research, and analyses relating to ecological systems and environmental quality;

(6) To document and define changes in the natural environment, including the plant and animal systems, and to accumulate necessary data and other information for a continuing analysis of these changes or trends and an interpretation of their underlying causes;

(7) To report at least once each year to the President on the state and condition of the environment; and

(8) To make and furnish such studies, reports thereon, and recommendations with respect to matters of policy and legislation as the President may request.

SEC. 205. In exercising its powers, functions, and duties under this Act, the Council shall—

(1) Consult with the Citizens' Advisory Committee on Environmental Quality established by Executive Order No. 11472, dated May 29, 1969, and with such representatives of science, industry, agriculture, labor, conservation organizations, State and local governments and other groups, as it deems advisable; and

(2) Utilize, to the fullest extent possible, the services, facilities and information (including statistical information) of public and private agencies and organizations, and individuals, in order that duplication of effort and expense may be avoided, thus assuring that the Council's activities will not unnecessarily overlap or conflict with similar activities authorized by law and performed by established agencies.

SEC. 206. Members of the Council shall serve full time and the Chairman of the Council shall be compensated at the rate provided for Level II of the Executive Schedule Pay Rates (5 U.S.C. 5313). The other members of the Council shall be compensated at the rate provided for Level IV of the Executive Schedule Pay Rates (5 U.S.C. 5315).

SEC. 207. There are authorized to be appropriated to carry out the provisions of this Act not to exceed $300,000 for fiscal year 1970, $700,000 for fiscal year 1971, and $1 million for each fiscal year thereafter.

Approved January 1, 1970.

APPENDIX 1-B

the environmental quality improvement act of 1970, public law 91-224, april 3, 1970 (42 u.s.c. 4371-4374)

title ii—environmental quality (of the water quality improvement act of 1970)

short title

SEC. 201. This title may be cited as the "Environmental Quality Improvement Act of 1970."

findings, declarations, and purposes

SEC. 202. (a) The Congress finds—

(1) That man has caused changes in the environment;

(2) That many of these changes may affect the relationship between man and his environment; and

(3) That population increases and urban concentration contribute directly to pollution and the degradation of our environment.

(b) (1) The Congress declares that there is a national policy for the environment which provides for the enhancement of environmental quality. This policy is evidenced by statutes heretofore enacted relating to the prevention, abatement, and control of environmental pollution, water and land resources, transportation, and economic and regional development.

(2) The primary responsibility for implementing this policy rests with State and local governments.

(3) The Federal Government encourages and supports implementation of this policy through appropriate regional organizations established under existing law.

(c) The purposes of this title are—

(1) To assure that each Federal department and agency conducting or supporting public works activities which affect the environment shall implement the policies established under existing law; and

(2) To authorize an Office of Environmental Quality, which, notwithstanding any other provision of law, shall provide the professional and administrative staff for the Council on Environmental Quality established by Public Law 91-190.

office of environmental quality

SEC. 203. (a) There is established in the Executive Office of the President an office to be known as the Office of Environmental Quality (hereafter in this title referred to as the "Office"). The Chairman of the Council on Environmental Quality established by Public Law 91-190 shall be the Director of the Office. There shall be in the Office a Deputy Director who shall be appointed by the President, by and with the advice and consent of the Senate.

(b) The compensation of the Deputy Director shall be fixed by the President at a rate not in excess of the annual rate of compensation payable to the Deputy Director of the Bureau of the Budget.

(c) The Director is authorized to employ such officers and employees (including experts and consultants) as may be necessary to enable the Office to carry out its functions under this title and Public Law 91-190, except that he may employ no more than 10 specialists and other experts without regard to the provisions of title 5, United States Code, governing appointments in the competitive service, and pay such specialists and experts without regard to the provisions of chapter 51 and subchapter III of chapter 53 of such title relating to classification and General Schedule pay rates, but no specialist or expert shall be paid at a rate in excess of the maximum rate for GS-18 of the General Schedule under section 5330 of title 5.

(d) In carrying out his functions the Director shall assist and advise the President on policies and programs of the Federal Government affecting environmental quality by—

(1) Providing the professional and administrative staff and support for the Council on Environmental Quality established by Public Law 91-190;

(2) Assisting the Federal agencies and departments in appraising the effectiveness of existing and proposed facilities, programs, policies, and activities of the Federal Government, and those specific major projects designed by the President which do not require individual project authorization by Congress, which affect environmental quality;

(3) Reviewing the adequacy of existing systems for monitoring and predicting environmental changes in order to achieve effective coverage and efficient use of research facilities and other resources;

(4) Promoting the advancement of scientific knowledge of the effects of actions and technology on the environment and encourage the development of the means to prevent or reduce adverse effects that endanger the health and well-being of man;

(5) Assisting in coordinating among the Federal departments and agencies those programs and activities which affect, protect, and improve environmental quality;

(6) Assisting the Federal departments and agencies in the development and interrelationship of environmental quality criteria and standards established through the Federal Government;

(7) Collecting, collating, analyzing, and interpreting data and information on environmental quality, ecological research, and evaluation.

(e) The Director is authorized to contract with public or private agencies, institutions, and organizations and with individuals without regard to sections 3618 and 3709 of the Revised Statutes (31 U.S.C. 529; 41 U.S.C. 5) in carrying out his functions.

report

SEC. 204. Each Environmental Quality Report required by Public Law 91-190 shall, upon transmittal to Congress, be referred to each standing committee having jurisdiction over any part of the subject matter of the Report.

authorization

SEC. 205. There are hereby authorized to be appropriated not to exceed $500,000 for the fiscal year ending June 30, 1970, not to exceed $750,000 for the fiscal year ending June 30, 1971, not to exceed $1,250,000 for the fiscal year ending June 30, 1972, and not to exceed $1,500,000 for the fiscal year ending June 30, 1973. These authorizations are in addition to those contained in Public Law 91-190.

Approved April 3, 1970.

BIBLIOGRAPHY

Water Resources Planning-General

Economic Commission for Asia and the Far East, "Manual of Standards and Criteria for Planning Water Resource Projects," Water Resource Series No. 26, United Nations, New York, 1964.

Water Resources Council, "Proposed Principles and Standards for Planning Water and Related Land Resources," *Federal Register*, **26**, No. 245, Washington, D.C., December 21, 1971.

Bureau of Reclamation, "Guidelines for Implementing Principles and Standards for Multiobjective Planning of Water Resources," Review Draft, December 1972, United States Department of the Interior, Washington, D.C.

Economic Commission for Asia and the Far East, "Water Resource Project Planning," Water Resource Series No. 41; Sales No. E.73.II.F.7, United Nations, New York 1972.

Anadromous Fishery Protection

Wagner and Ingram, "Evaluation of Fish Facilities and Passage at Foster and Green Peter Dams on the South Santiam River Drainage in Oregon," February 1973, Fish Commission of Oregon, Management and Research Headquarters, Clackamas, Oregon.

Gunsolus and Eicher, "Evaluation of Fish-Passage Facilities at the North Fork Project on the Clackamas River in Oregon," September 1970, Portland General Electric Company, Portland, Oregon.

G. J. Eicher, "Stream Biology and Hydroelectric Power," *Water Power*, **25**, No. 6, June 1973.

Economics

The Institution of Civil Engineers, "An Introduction to Engineering Economics," 1969; The Institution of Civil Engineers, Great George Street, London, S.W. 1, England.

L. Douglas James and Robert R. Lee, *Economics of Water Resources Planning*, McGraw-Hill Book Company, 1971.

Raymond W. Gaines, "Comparison and Evaluation of Economic Criteria and Methods of Analysis Employed by the U.S. Bureau of Reclamation, U.S. Corps of Engineers, and California Department of Water Resources," unpublished thesis, June 1961, California State University at Sacramento.

"The Allocation of Costs of Federal Water Resources Development Projects," 82nd Congress, 2nd Session, House Committee Print No. 23, Report to the Committee on Public Works, December 5, 1952, Government Printing Office.

"Proposed Practices for Economic Analysis of River Basin Projects," Report to the Inter-Agency Committee on Water Resources, May 1958, Superintendent of Documents, U.S. Government Printing Office, Washington, D.C.

Economic Appendix General Design Memoranda, Department of the Army, Sacramento District, Corps of Engineers, Sacramento, California.

Environment

United States Geological Survey, "A Procedure for Evaluating Environmental Impact," Geological Survey Circular 645, published 1971, U.S. Geological Survey, Washington, D.C.

Battelle Columbus Laboratories, "Final Report on Environmental Evaluation System for Water Resource Planning to Bureau of Reclamation, U.S. Department of the Interior," Jan. 31, 1972, Battelle Columbus Laboratories, 505 King Avenue, Columbus, Ohio 43201.

Water Resources Council, "Proposed Principles and Standards for Planning Water and Related Land Resources," *Federal Register*, **26**, No. 245, Washington, D.C., December 21, 1971.

Bureau of Reclamation, "Guidelines for Implementing Principles and Standards for Multiobjective Planning of Water Resources (Chapter 4)," Review Draft, December 1972, U.S. Department of the Interior, Washington, D.C.

"California Protected Waterways Plan (Initial Elements)," February 1971. The Resources Agency, State of California, Sacramento, California.

Council on Environmental Quality, "The Third Annual Report of the Council on Environmental Quality," August 1972, Superintendent of Documents, U.S. Government Printing Office, Washington, D.C. 20402.

Western Systems Coordinating Council Environmental Committee, "Environmental Guidelines," September 1971, Western Systems Coordinating Council, Southern California Edison Company, P.O. Box 351, Los Angeles, California 90053.

Flood Control

Tate Dalrymple, U.S. Geological Survey, "Flood Frequency Analyses—Manual of Hydrology: Part 3, Flood Flow Techniques," U.S. Geological Survey Water Supply Paper 1543-A, 1960. Superintendent of Documents, Government Printing Office, Washington, D.C.

Sub-Committee on Hydrology—Interagency Committee of Water Resources, "Methods of Flow Frequency Analysis," Bulletin No. 13, April 1966, Superintendent of Documents, Government Printing Office, Washington, D.C.

Water Resources Council, Hydrology Committee, "A Uniform Technique for Determining Flood Flow Frequencies," Bulletin No. 15, December 1967, published by Water Resources Council, 1025 Verment Ave., N.W., Washington, D.C. 20005.

U.S. Army, Corps of Engineers, "Engineering and Design, Hydrologic Frequency Analysis," preliminary draft of a manual for Corps of Engineers use of Interagency Committee of Water Resources Bulletin No. 13 (above).

Leo R. Beard, Corps of Engineers, "Statistical Methods in Hydrology," January 1962, Sacramento District, Corps of Engineers.

U.S. Weather Bureau, "Rainfall Frequency Atlas of the United States for Durations from 30 minutes to 24 hours and Return Periods from 1 to 100 Years," U.S. Weather Bureau Technical Paper No. 40, May 1961.

SIGNIFICANT UNITED STATES LAWS RELATING TO ENVIRONMENTAL CONSIDERATIONS OF WATER RESOURCE DEVELOPMENT AND CONTROL

National Environmental Policy Act of 1969, Public Law 91–190 (83 Stat. 852) 42 U.S. Code Section 4321-4347, approved and became effective Jan. 1, 1970.

Executive Order 11,514, March 5, 1970. Protection and Enhancement of Environmental Quality.

Executive Order 11,507, Feb. 5, 1970—Prevention, Control and Abatement of Air and Water Pollution at Federal Facilities.

National Wilderness Preservation System, Public Law 88–577, approved Sept. 3, 1964, U.S. Code 1131 et. seq.

Wild and Scenic Rivers Act, Public Law 90–542, approved Oct. 2, 1968 (82 Stat. 906) 16 U.S. Code 1271 et. seq.

The Federal Water Project Recreation Act, Public Law 89–72, approved July 9, 1965 (79 Stat. 213) 16 U.S. Code 460 1-12 1-21.

Fish and Wildlife Act of 1956, August 8, 1956, ch. 1036, sec 2 (70 Stat. 1119) 16 U.S. Code 742; a through d and 742 e through j.

Fish and Wildlife Coordination Act, as amended, Public Law 85–624, August 12, 1958 (72 Stat 563) 16 U.S. Code 661 through 666.

Acknowledgments. The author wishes to acknowledge and thank the following persons for thier valuable assistance in reviewing portions of the manuscripts of this chapter: Lamoine B. Christiansen, J. Bruce Kimsey, Dee Harper, and Frank Stipak—Bureau of Reclamation, Sacramento, California. Robert A. Williams, The Resources Agency of California, Sacramento, California, William A. Doyle and Rodney Hill, Corps of Engineers, Sacramento, California, Lawrence Korn, Oregon Fish Commission, Clackamas, Oregon, George J. Eicher, Portland General Electric Company, Portland, Oregon.

2
Hydrologic Studies

GORDON R. WILLIAMS
Chief Hydrologist (retired),
Tippetts-Abbett-McCarthy-Stratton
New York, New York

1. GENERAL

This chapter deals with the determination of both the dependable water supply for a dam and reservoir project, and the design floods for the spillway capacity and for diversion waterways during construction. Various aspects of reservoir design such as flooding routing, freeboard against waves, and sedimentation are also included.

The purpose of most dams is to create a storage basin for the regulation of runoff to be used for water supply, irrigation, power development, low-flow augmentation for water quality management, temporary detention of flood flows, or for a combination of these uses. The importance of runoff regulation may be at a minimum in those dams built largely to create a head for power or for navigation or for creation of a water area for recreation, but the availability and dependability of runoff is basic to the functional and economic success of such projects. The variability of runoff within a river basin may be an important factor in determining the location of a dam site, as discussed in Chapter 1. Also, the available water supply and the need for its regulation are key factors in determining the reservoir capacity and the corresponding height of the dam.

2. EVALUATION OF RECORDS

The ideal situation in planning water resources development is to have a relatively long record (30 years or more) of river discharge at or near the dam site being considered. Such records are rarely obtainable near the desired location, even in the United States, where the distribution of hydrologic records is excellent. Often records from adjacent drainage areas must be used in estimating the runoff at the desired site. In underdeveloped countries, records are usually more widely spaced and of shorter duration. Under the latter conditions, judgment must be used in interpreting and extrapolating the available data. It may be necessary to resort to experience in similar climatic zones remote from the area under study.

In evaluating the adequacy of a runoff record, it should be determined if it is representative of the hydrologic experience in the region. Periods of unusually low and high flows must be included. For example, a relatively long record of even 25 years, in the Northeastern United States, which might by chance encompass the years 1937-1961, inclusive, would not be representative because it missed the critical drought conditions experienced earlier and later in the years 1930-32 and 1963-65 and also missed the major flood in 1936. Usually, available precipitation records are longer than runoff records and will give an indication of whether the latter include both extreme wet and dry periods.

Length of record alone is not a reliable indication of the true recurrence interval of extreme events that fall within a record. A search should be made of local histories, and even newspaper files, for general information on past floods and droughts. Such sources may give a general indication of recurrence intervals.

3. REVISION OF EXISTING RECORDS

Records which are current and kept by the government agencies employing modern measurement and reporting techniques[1] may generally be used as printed in official publications. However, early records should be re-evaluated if based on gage heights obtained from nonrecording gages and stage-discharge relationships defined by a small number and range of discharge measurements. The steps in the re-evaluation of a record are given below:

(a) Gage—Height Record.

An ideal record of river stage is one kept by a continuously recording gage. Such a record permits the determination of the average daily gage height at low and intermediate river stages and the complete definition of flood hydrographs. This type of record was not always available in the early years of long records or where funds for data collection were limited. The alternative to a recording gage is a staff gage read once or twice a day, or more often in floods. The average of the gage readings in any day may not give the exact mean but some improvement in the accuracy of the mean may be obtained by making a continuous plot of the readings at the times of observation. Where several readings a day are available in time of flood, a plot will give a better determination of the peaks and volumes of flood rises than an arithmetic average of the readings. Such plotted hydrographs can take account of observed diurnal cycles caused by snow melt or by known daily patterns of runoff resulting from tropical rainfall. Obviously, analyses of this type can be laborious and are only justified when the indicated correction is significant.

(b) Discharge Measurements

In re-evaluating old records, it is usually necessary to assume that the discharge measurements were made by acceptable procedures and can be utilized as computed. If measurements have been made with inadequate equipment or are based on a limited number of observations in the cross section, they still must be accepted as taken. If high-water measurements are based on surface velocities, possibly taken with floats, the relation to the mean velocity in the vertical may be approximated from studies of velocity curves made at lower stages, or by using an average relation, usually taken to be 0.8.

(c) State-Discharge Relation

There are many variables entering into the establishment of the stage-discharge relation at a gaging station. The fact that discharge measurements made in recent years, particularly at low stages, do not define the same relation as those made in earlier years does not mean that the early records should be revised. The channel section, which governs the stage and corresponding discharge may change over the years. The stage-discharge relations in erodible channels may be essentially parallel and may vary between upper and lower limits (see Fig. 2-1). In very flat streams, the slope prevailing at the time of the measurement may be as important as the stage in determining the discharge. Revisions to early records should be considered with caution and rarely should it be assumed that a single relation has applied in the past, except at flood stages (as discussed below), or in a definitely non-erodible channel section.

The most significant revisions to discharge records are often those for periods of high-water, and usually relate more to data deficiency than to hydraulic changes. For example, when a gaging station is first established, evaluation of flood flows must be based on an approximate extrapolation of the stage-discharge relation defined only at low and intermediate flows. Subsequent high-water measurements are used to define more accurately the relation for the high discharges, which is likely to be applicable also to the early records. A consistent flood-flow relation is necessary to establish the correct flood frequency for design. Also, as a substantial portion of the total water supply often occurs during high-flow periods, it is essential that the flood flows be evaluated as accurately as possible by corrections to past records, if justified. A reliable stage-discharge relation at a dam site is also important in establishing an accurate tailwater relation for the design of outlet works and power plants.

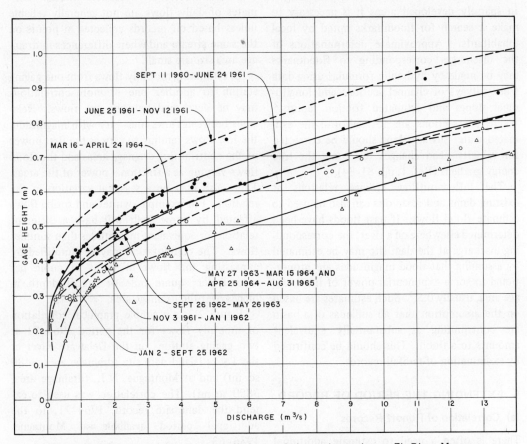

Figure 2-1. Changes in low-water stage-discharge relation gaging station on Ziz River, Morocco.

4. ESTIMATING DISCHARGES OF PAST FLOODS

A flood stage record used in conjunction with a stage-discharge relation is the preferred method for estimating the magnitudes of past floods. However, there are often factors which prevent such a procedure. Some of these are:

- The greatest known flood may have occurred prior to the establishment of a gaging station or after its discontinuance.
- The greatest flood may have destroyed the gaging station.
- Records of past floods are only available at other points in the river basin.

In populated river basins the heights reached by great floods of the past have often been noted on buildings and bridges, and estimates have been made of the corresponding discharges. In sparsely developed areas it is necessary to make a search for floodmarks noted by local inhabitants. Approximate determinations of the discharges corresponding to floodmarks may be made by hydraulic formulas using data from surveys of channel sections and longitudinal slopes and estimated friction factors.[2] Where the sections are not uniform, the estimated water surface slope should be corrected for velocity head changes to obtain the true energy gradient (Ref. 1, pp. 81-84).

The inflow-outflow storage relations at existing dams and reservoirs can also be used to estimate flood flows. If past floods have been determined elsewhere on a river, the corresponding discharge at the dam site may be estimated by assuming that flood magnitudes are proportional to some exponential power of the drainage area, usually 0.5.[3] Such estimates are based on the assumption that all subareas of a basin are contributing at substantially equivalent amounts to a flood. This should be confirmed by examination of rainfall records.

5. EXTENDING THE PERIOD OF RECORD

(a) Correlation of Runoff Records

There is often a need to estimate additional periods in runoff records to increase the length of a short record or to fill the periods where the record was interrupted for various reasons. The procedure is to correlate coincident flows with those at another gaging station in the same or in an adjacent basin which has a longer record. The relation established between the stations can then be used to extend the stream flow record at the dam site. As far as possible, the records to be compared should be for areas about the same size and having similar hydrologic, topographic and geologic characteristics. Records can also be synthesized from statistical data using an electronic computer [see Section 5(c)].

It is usually not necessary to estimate flows on a day-by-day basis, except for studies of flood hydrographs. Average flows for a month are generally adequate for reservoir capacity studies and reservoir operation studies. Estimates of daily flows are not generally reliable unless based on records collected at points on the same stream and where differences in drainage area size are small.

In estimating monthly flows from one gaging station to another one or more correlations may be developed between the flows. Relations between flows may vary with magnitude, the low flows tending to vary as the first power of the contributing drainage areas and the flood flows varying as a fractional power of the areas. There also may be a seasonal variation in the relation. For example, winter and spring flows affected by snow melt will have a different relation for equivalent magnitudes than summer flows. The computation may be simplified by expressing the flows in terms of cubic per second per square mile or runoff depth in inches over the drainage area.

Figure 2-2 is a simple graphical correlation of monthly flows for the period 1964-71 at two gaging stations on the Delaware River at the Tocks Island dam site (drainage area, 3480 sq mi) and at Montague, N.J., (drainage area, 3850 sq mi) The correlation was used to extend the dam site record, 1964-71, to the additional period available at Montague, 1936-63.

If the number of values to be estimated is

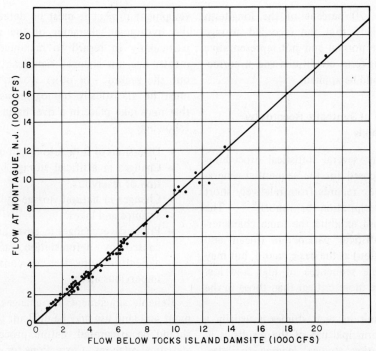

Figure 2-2. Graphical correlation of monthly flows at two gaging stations on Delaware River, period 1964-1971.

not too great, a graphical comparison of the hydrographs at two stations can be made (see Fig. 2-3). The discharge values are plotted on a logarithmic scale as this permits equivalent accuracy for a wide range of magnitudes. The varying ratios between magnitudes are indicated by varying distances between the hydrographs. When a period of record is missing, it may be developed graphically by scaling from the available record using the distances found to be applicable to each magnitude.

(b) Correlation of Runoff and Precipitation Records

If a drainage basin has a sufficient number of precipitation records to provide an index of the average basin precipitation, a runoff-precipitation relation may be developed and used to extend monthly runoff records. Precipitation records are usually available for longer periods than runoff records.

A satisfactory correlation is not always obtained by comparing monthly basin precipitation with the corresponding runoff. Multiple correlation may be required in which runoff is related to various percentages of precipitation in previous months. Separate relations may be required for each calendar month where seasonal effects are pronounced. Numerous trials are sometimes required before a satisfactory correlation is obtained. Use of an electronic computer for this type of analysis can save considerable time.

A long record from a precipitation station, often outside the basin under study, may be useful in indicating sustained periods of high or low precipitation which antedated runoff records in the region. Such precipitation records may be expressed as ratios to their means as a method for readily identifying abnormalities. If, for example, in the period in which runoff records were available, the precipitation record showed an average depth equivalent to 120 percent of the long-term mean, it is likely that the recorded average runoff is above normal and not representative of average long-term conditions and a correction factor should be applied.

(c) Generation of Synthetic Records by Statistical Methods

In recent years, several statistical procedures have been proposed for the sequential generation of synthetic records from relatively short records of precipitation or runoff.[4,5] The generated record exhibits the same characteristics and statistical parameters (mean and standard deviation) as the basic record, but may indicate possible sequences of high and low flows which are more critical than those in the record.

The time unit in such studies is usually a month. The principal use of the results is in reservoir operation studies, leading to determinations of regulated flow and economic reservoir capacity.

6. ANALYSIS OF RUNOFF RECORDS

(a) Homogeneity of Records

Before a record of river runoff can be analyzed for use in the design of a water resources development project it must be determined that the hydrologic character of the river basin, particularly in regard to man-made developments, has not changed significantly throughout the record. In other words, the record must be statistically homogeneous. Changes that may take place in a river basin include the following:

- Diversions of flow into or out of the basin.
- Changes in artificial storage by construction of reservoirs.
- Changes in natural storage by drainage of swamps and lakes.
- Progressive changes in cover and land use caused by deforestation, afforestation, agricultural practices, or construction of impervious areas.

Reliable adjustments for changes in development and land use may be difficult to make but should be attempted, before proceeding with statistical analyses. Corrections for evaporation from the reservoir surface at a proposed dam site are considered separately in Section 7(a).

(b) Hydrograph

The chronological record of flow is termed the hydrograph (Fig. 2-3). It is basic data for all statistical analyses of the runoff of a drainage basin. The hydrograph reveals many aspects

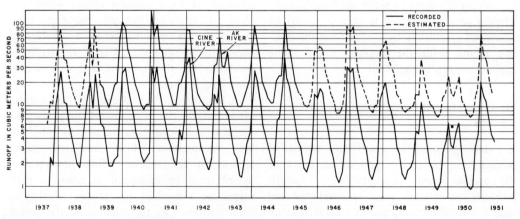

Figure 2-3. Estimating missing periods of runoff records by graphical comparison, Ak and Cine Rivers, Menderes River Basin, Turkey.

of the runoff characteristics of the basin including:

- The seasonal distribution of high and low flows.
- The character of flood flows, whether occurring in isolated flood rises or in a succession of floods rising above a high seasonal base and lasting several months.
- The influence of snow melt.
- The effect of valley storage.
- The contribution of ground-water flow.

The time unit used in compiling and analyzing the hydrograph will vary with its intended use. Annual discharges are usually of interest only in comparing sequences of high and low years. Monthly flows are the most useful unit in reservoir design for municipal water supply, irrigation development, and hydroelectric power. Hydrographs in units of a day or less are needed to develop design floods as described in Sections 8 to 11.

(c) Flow-Duration Curve

If the flows for any unit of time are arranged in descending order of magnitude (without regard to chronological sequence) the percentage of time for which any magnitude is equalled or exceeded may be computed. The resulting array is called a flow-duration curve or cumulative frequency curve (see Fig. 2-4). Such curves are useful in determining the relative variability of flow between two points in a river basin or between two basins. For example, if a stream is highly regulated the curve will approach a horizontal line. The dependable flow is that corresponding to 100 percent of the time.

The relative variability of two flow records may be compared by converting the discharge scale in terms of a ratio to the mean (Fig. 2-5B). Other convenient plotting methods are to use a logarithmic scale for the discharge scale, combined with a probability time scale (Fig. 2-5A).

The duration curve based on a long record is

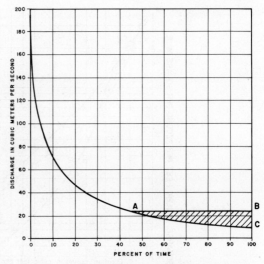

Figure 2-4. Duration curve of monthly discharges.

considered to represent an average year and the percentage scale may be converted to represent the number of months, days or hours in a year. Thus, any subarea under the curve (with arithmetic scales) represents a volume of flow and the total area represents the volume of mean annual runoff.

Flow-duration curves have been used to approximate the amount of storage needed to increase the dependable flow. For example, the horizontal line AB in Fig. 2-4 may represent a new dependable flow, and the required storage needed to obtain this flow is indicated by area ABC. Such a computation is approximate because it assumes the storage volume will be used only once a year.

Power production values may be approximated from the duration curve by converting the discharge scale to kilowatts by multiplying by a selected head, efficiency and a conversion factor. If the time scale is converted to hours in a year, a unit of area represents kilowatt-hours.

The flow-duration curve is particularly useful, in combination with a sediment rating curve (river discharge versus the transported sediment load usually expressed in tons per day), to compute total sediment load to be expected in an average year (see Section 14).

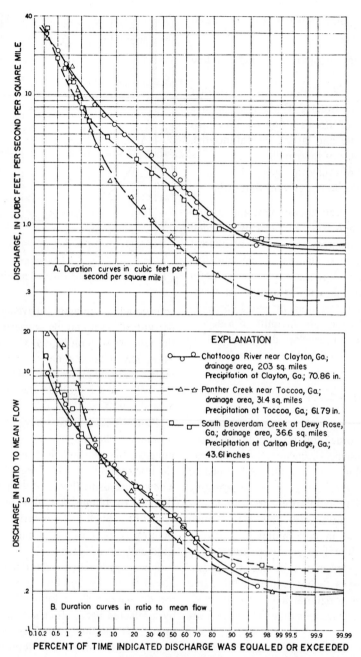

Figure 2-5. Comparison of flow-duration curves for three nearby stations in northeastern Georgia, water year 1952. (U.S. Geological Survey)

(d) Mass Curve

A useful device in the analysis of runoff records is the mass curve which is a plot of the cumulative runoff from the hydrograph against time. The time scale is the same as for the hydrograph and may be in days or months. The volume ordinate may be in cfs-days, cfs-months, acre-feet, million gals per day, etc. The slope of the mass curve is the derivative of the volume with respect to time or the rate of discharge.

The mass curve usually has a wavy configuration (see Fig. 2-6) in which the steeper segments represent high flow periods and flatter segments represent low flows. Uniform rates of withdrawal (draft) may be represented as tangent lines drawn from high points to intersect the curve at the next "wave" The vertical distance between the draft line and the basic curve represents the cumulative difference between regulated outflow and natural inflow, or the required storage. If the draft line does not intersect the mass curve at the end of a year, it means that the reservoir does not refill with that rate of draft and regulation at the proposed draft rate will extend over two years or more. A mass curve for the Ak River in Turkey is shown in Fig. 2-6A.

In estimating storage requirements from the mass curve, it is not necessary to assume a constant rate of regulated flow. For example, if the draft rate to meet a demand for irrigation, water supply, or power varies from month to month, the draft line may be a curved or irregular line (see Fig. 2-6B) and the maximum draft may not occur at the low point in the mass curve. Obviously, it is not necessary to make a plot and the analysis may be made directly on a calculator or with an electronic computer.

An allowance for evaporation should be applied to the mass curve analysis. If the water area does not change significantly during the annual cycle of use, an average correction for each calendar month can be subtracted from the inflow or added to the draft rates. In final design, the best practice is to make allowance

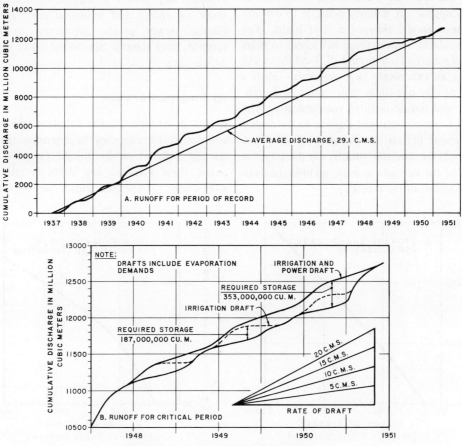

Figure 2-6. Analysis of runoff by mass curve.

for evaporation in the reservoir operation and routing studies so that the effects of variation in water area and seasonal differences in rate of evaporation and rainfall can be evaluated. A detailed discussion of evaporation corrections is given in Section 7(a).

(e) Storage-Draft Curve

The results of a mass-curve analysis can be plotted as a storage-draft curve. This curve gives the storage needed to sustain various draft rates. Examples of storage-draft curves are shown on Fig. 2-7. These curves were computed from the mass curve of the Ak Cayi River shown in Fig. 2-6. Both irrigation requirements and combined irrigation and power requirements are illustrated.

If storage is unlimited, the storage-draft curve will approach the available mean flow as an asymptote. It is rarely possible to develop the mean annual flow of a river basin. For most projects, some spillage will occur in years of high runoff. To impound all flood flows will require an excessively large reservoir. Such a reservoir may not fill in many years, and probably could not be justified economically. The selected rate of regulated flow to be developed will depend (1) on the demands of the water users, (2) the available runoff, (3) the physical limits of the storage capacity, and (4) the overall economics of the project.

The storage-draft curve indicates only the specific storage needed to achieve different rates of regulated flow. The total storage selected for the final development may be considerably greater than that needed for flow regulation as discussed in Section 7(e).

(f) Selection of Design Flow

The hydrologic analyses, combined with economic analyses of costs and benefits for different heights of dam and reservoir capacities will lead to the selection of the reservoir capacity and the corresponding dependable flow that can be justified. The selected design flow may not necessarily be available 100 percent of the time. The proposed water use may permit deficiencies at intervals, for example, a 15 percent shortage once in 10 years. Irrigation water supplies may permit greater deficiencies than those for urban and industrial use. Hydroelectric power plants, connected to large systems, may tolerate substantial water supply deficiencies.

(g) Frequency Analysis

In the hydrologic studies for dam and reservoir design it is necessary to determine the frequency with which hydrologic events (flood peaks, flood volumes, low flows or rainfalls) will occur. Examples shown are for flood

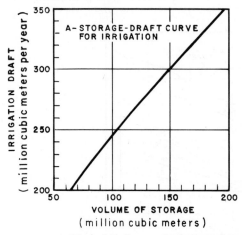
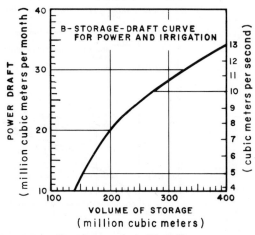

Figure 2-7. Storage-draft curves for multipurpose uses (based on graphs in Fig. 2-6).

frequency but the procedures are applicable to other data.

In approaching the problem of frequency analysis, the engineer is confronted with a multiplicity of technical papers on the subject as it has interested mathematicians and academicians for the past 60 years. Practical experience has shown that no procedure, no matter how well grounded in mathematical and statistical theory, can overcome the fact that hydrological records are not adequate samples (see comment in Section 2). In 1967, the departments and agencies of the U.S. Government through the Water Resources Council, adopted, from several possible procedures, a "uniform technique for determining flood-flow frequencies."[6] Such action on the part of Government was taken admittedly to provide a consistent procedure in carrying out similar activities but its use was qualified by the statement that "the state of the art with respect to flood flow frequency methods ... has not advanced to the point where complete standardization is feasible and appropriate." Other more empirical approaches will be recommended in this section.

Frequency is expressed in either of two ways:

1. *Recurrence interval (or return period)* expressed in years. This is an average interval and does imply that events will be equally spaced in time.
2. *Probability, or percent chance of occurrence* in any year. Percent chance is computed as the reciprocal of the recurrence interval.

The data to be analyzed are the annual maxima, or one flood peak event per year. The year may be the calendar year or the water year (the latter extending from Oct. 1 to Sept. 30, in the United States). Values selected must be from independent events. For example, if a flood occurs at the end of a year, only the peak value should be selected and assigned to the year in which it occurs.

The test for homogeneity of records discussed in Section 6(a) is particularly applicable to data used in frequency analyses.

The independent annual flood maxima should be first tabulated in chronological order and checked to be sure that each year is represented, then the relative magnitudes should be assigned an order number. The theoretical recurrence interval is computed by the formula:

$$T = \frac{n+1}{m} \quad (2\text{-}1)$$

when T is the recurrence interval, in years, n is the length of record or number of events, and m is the order number varying from 1 for the maximum value to n for the minimum value. Other formulas for T have been proposed but Eq. 3-1 has received general acceptance.

The computed points may be plotted on one of several types of graph paper for visual inspection of the array. One type, known as probability paper, has a time scale (abscissa) corresponding to a normal probability distribution (Ref. 5, pp. 8-14) and a logarithmic discharge scale (ordinate) (Fig. 2-8). This paper is also available with an arithmetic scale, but such a scale is usually not suitable for plotting the full range of flood values. A theoretical curve may be drawn through the points using a method similar to that proposed in Ref. 6 or a graphical curve may be fitted to the points (Fig. 2-8). The computed curve may indicate curvature upwards or downwards or an essentially straight line. In any case, extrapolations beyond the limits of the data are not reliable. Computed curves appear to give too much weight to minor floods of frequent occurrence, and in turn overestimate the recurrence intervals of the larger events.

Although perhaps lacking in theoretical justification, flood discharges versus recurrence intervals may be plotted on log-log paper. The indicated trend line will be found to have a curvature towards the time axis or be essentially a straight line through the higher points. As in the case of probability paper, extrapolations beyond the limits of the data must be considered approximate, particularly for records less than about 50 years in length, and for events beyond the 100-year recurrence interval.

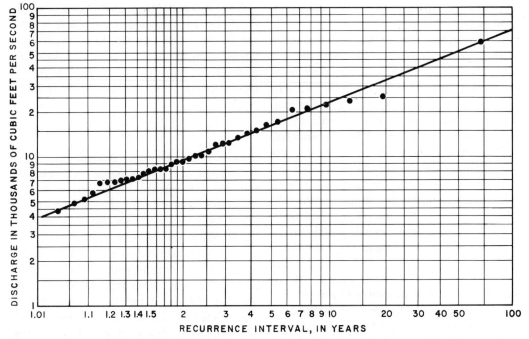

Figure 2-8. Flood-frequency curve on logarithmic-probability scales. Curve fitted graphically. (U.S. Geological Survey)

Because of the admitted fact that flood frequency relations developed for individual stations are subject to considerable error, procedures have been devised to take into account the flood frequency characteristics of other records in a region. Such methods make it possible to approximate flood frequency curves in river basins where gaging station records are not available. A method developed by the U.S. Geological Survey develops two basic relations.[7] For rivers in subregions considered to be hydrologically similar, relations are developed between the mean annual flood and drainage area. A second set of generalized curves indicates the ratios between the mean annual flood and floods of other recurrence intervals. In general, the estimated recurrence intervals have been limited to 50 years. Curves applicable to the Tennessee and Cumberland River Basins are shown in Fig. 2-9. Studies for the entire United States have been completed and are published in a series of Water Supply Papers numbered 1671 to 1689, inclusive.

A method of frequency analysis based on the work of Chow[8] is described below. This method appears to avoid many of the shortcomings of other theoretical methods of curve fitting, and furthermore, permits adjustments based on the characteristics of other records in a region. The basic relation termed by Chow "the general equation for hydrologic frequency analysis" (Ref. 5, pp. 8–23) is:

$$Q = \overline{Q} + \sigma K \qquad (2\text{-}2)$$

in which Q is any annual peak, $\overline{Q}$ is the mean of the annual series and σ is the standard deviation of the series, and K is a frequency factor. The equation may be written in the form

$$Q = \overline{Q}(1 + C_v K) \qquad (2\text{-}3)$$

or

$$K = \frac{Q - \overline{Q}}{\overline{Q} C_v} \qquad (2\text{-}4)$$

where C_v is the coefficient of variation of the series.

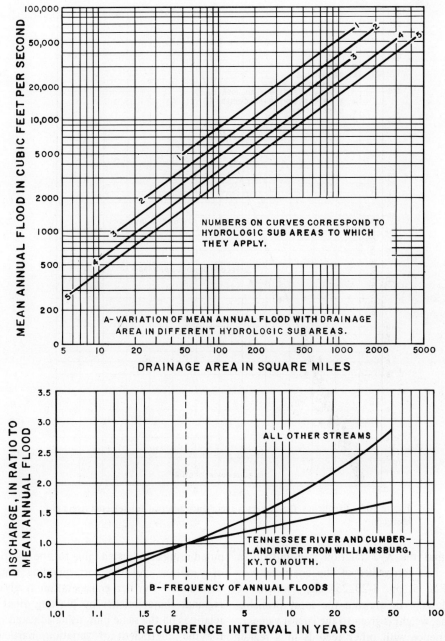

Figure 2-9. Regional flood frequency relations, Tennessee and Cumberland River Basins. (U.S. Geological Survey)

The flood magnitudes in a record are arranged in descending order of magnitude $\overline{Q}$, σ and C_v are computed for the series, and finally K is computed for each flood value from either Eq. 2-2 or Eq. 2-4. The recurrence interval corresponding to each value of K is computed by Eq. 2-1.

A plot is then made of K versus T on semi-log paper (Fig. 2-10) with K on the arithmetic scale and T on the log scale. It has been deter-

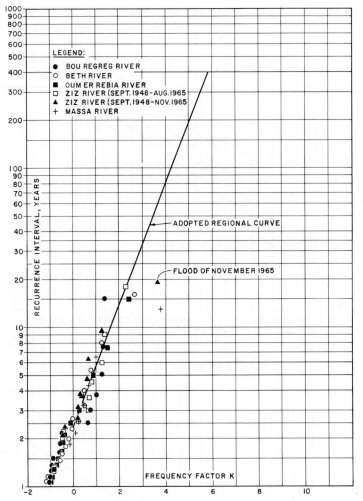

Figure 2-10. Frequency factor versus recurrence interval for selected rivers in Morocco.

mined that the K-T relation is essentially a straight line for values of the coefficient of skew (Ref. 5, pp. 8-23-25) characteristic of flood flows. In drawing the most probable mean line, weight is given to the largest values.

Experience with other records in a region may be taken in two ways. First, coefficients of variation may be computed from the other records, and compared with that for the record at the dam site. The coefficient of variation is affected by length of record, therefore, in determining a suitable regional value, weight should be given to values for the longer records. The K-T relation can then be recomputed using an adjusted value for the coefficient of variation and the corresponding standard deviation. Further consideration is given to regional characteristics by plotting other K-T relations on the same plot, using adjusted values for the coefficient of variation, particularly for the short records. Once a logical K-T relation is developed it can be used to determine a Q-T relation.

Historic floods, for which discharges have been estimated, should be incorporated into a frequency relation even though the indicated recurrence interval is approximate. Local histories may indicate that the largest flood in even a short record is the largest since the early settlement of a region. In such cases the recurrence interval of the largest flood will be arbitrarily modified.

In Section 8(b), the question of a suitable recurrence interval for a selected flood and the effect of an extended construction period on the interval has been discussed. The basic probability theory applicable to this case is discussed below.

A flood with selected design recurrence interval T has a probability of occurrence in any one year of $1/T$. The probability of non-occurrence in any year is

$$1 - \frac{1}{T} = \frac{T-1}{T} \qquad (2\text{-}5)$$

In each year of a proposed construction period of X years ($X < T$) the probabilities of occurrence and non-occurrence of the flood with the recurrence interval T are $1/T$ and $T - 1/T$, respectively, and are independent of what happens in any other year in the period X. The probability of non-occurrence within the period of X years is

$$\left(\frac{T-1}{T}\right)^X \qquad (2\text{-}6)$$

Hence, the probability of at least one occurrence of a flood with a T-year recurrence interval within X years is

$$1 - \left(\frac{T-1}{T}\right)^X \qquad (2\text{-}7)$$

If a value of T of 25 years (probability 0.04) is applicable to a single year and the construction period X is 3 years, the modified probability becomes 0.115 which is equivalent to a recurrence interval of 8.7 years.

If the proposed recurrence interval of T years is to be maintained in X years of construction, a modified return period T^* should be used in which

$$T^* = \frac{1}{1 - \sqrt[X]{\frac{T-1}{T}}} \qquad (2\text{-}8)$$

For example, if a 25-year recurrence interval is desired for maximum flood to be expected in a 3-year construction period, the modified recurrence interval T^* from Eq. 2-8 would be 74 years.

7. OPERATION AND ROUTING STUDIES AND FINAL STORAGE SELECTION

(a) Evaporation Losses

As mentioned in Section 6(d), detailed evaluation of evaporation losses should be postponed until final operation and routing studies, when the actual variation in water area can be considered as well as the seasonal variation in evaporation and coincident precipitation.

Basic data on water surface evaporation in the United States may be obtained from records of pan evaporation.[9,10] Such records overestimate lake evaporation and must be reduced by a coefficient which varies from 0.60 to 0.80 depending on the climate (Fig. 2-11). Figure 2-12 is a map of the United States from Ref. 9, showing estimated annual lake evaporation based on corrected pan evaporation records. The distribution of annual evaporation within the year can be obtained from monthly records similar to those in Ref. 10. The collection of evaporation records at a project site should be initiated in the planning stage.

Evaporation corrections should be made on a monthly basis, using actual past precipitation records at the project site if possible. Use of uniform monthly precipitation corrections is not recommended as it may result in unacceptable approximations. Corrections for monthly evaporation using actual precipitation, particularly in dry climates, may reveal extreme reservoir drawdowns that were not anticipated in preliminary studies.

The importance of evaporation losses from a reservoir depends on two basic factors: (1) the percentage of the drainage area that is to be flooded by the reservoir and (2) the climatic and hydrologic characteristics of the region.

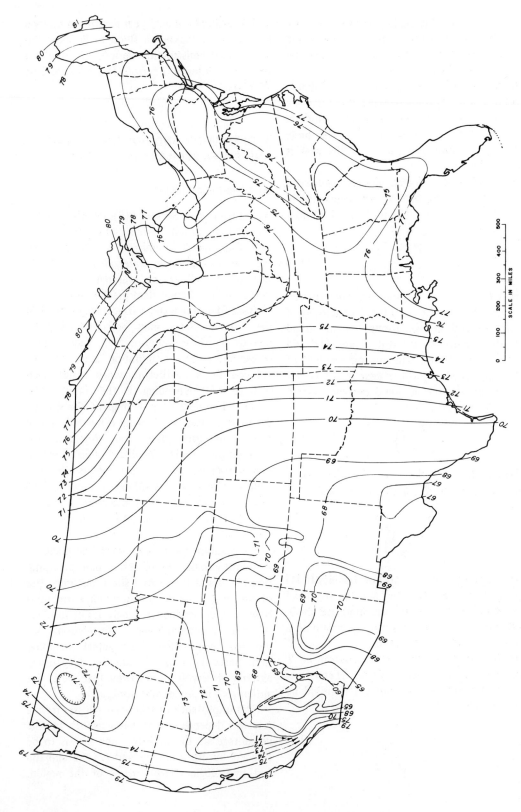

Figure 2-11. Average annual Class A pan coefficient in percent in the continental United States.[9]

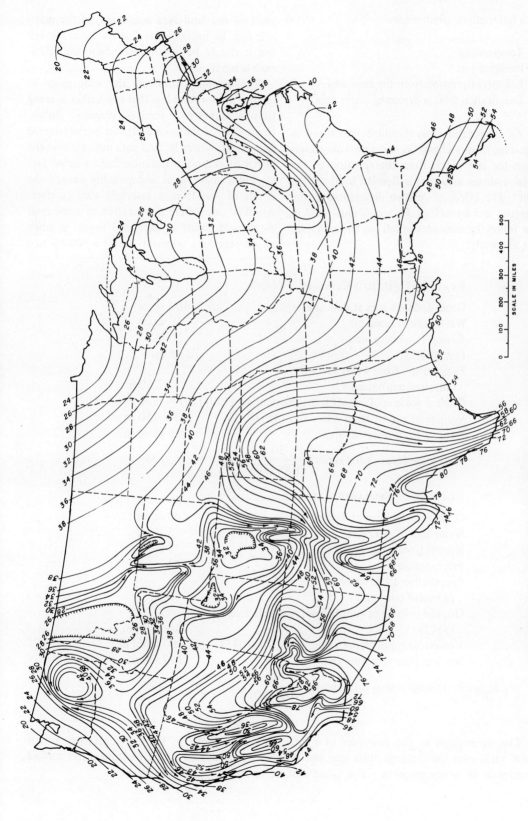

Figure 2-12. Average annual lake evaporation in inches in the continental United States (period 1946-1955).[9]

The latter characteristics are:

Temperature
Precipitation
Evapotranspiration from the land area
Evaporation from a perennial water area

An evaporation loss should be visualized as resulting from a change in the runoff characteristics (or net yield) from that portion of the total drainage area that is flooded by the reservoir. The effect of relative differences in the climatic and hydrologic factors is illustrated in the following examples which use annual values for simplicity.

yield of the land area inundated by the reservoir may be more or less than the basin average and it should be determined where the difference is significant.

The examples illustrate the wide range in relative evaporation loss that is possible in going from a humid to a semiarid climate. An extreme case of the humid climate is the tropical rain forest, where limited data indicate that the depth of evapo-transpiration from a dense jungle cover may equal and possibly exceed the depth of loss from a reservoir water surface. Another case where the creation of a reservoir area could result in increased runoff is when all or part of a reservoir floods a swamp to a

Example 1: Humid Climate

Gross drainage area at dam site	100 sq mi
Water area of reservoir	10 sq mi
Runoff from land area	20 in.
Precipitation	45 in.
Evaporation from water surface (adjusted pan value)	30 in.
Original volume of runoff from reservoir area (20) (10) =	200 sq mi in.
Volume of runoff from water area (45−30) 10 =	150 sq mi in.
Net loss from reservoir area	50 sq mi in.
Percentage loss in total runoff $\frac{50}{(100)(20)}(100)$ =	2.5%

Example 2: Semiarid Climate

Gross drainage area at dam site	100 sq mi
Water area of reservoir	5 sq mi
Runoff from land area	2 in.
Precipitation	24 in.
Evaporation from water surface (adjusted pan value)	60 in.
Original volume of runoff from reservoir area (5) (2)	10 sq mi in.
Volume of runoff from water area (24−60) (5) =	−180 sq mi in.
Net loss from reservoir area	−190 sq mi in.
Percentage loss in total runoff $\frac{190}{(100)(20)}(100)$ =	95%

The assumption in the examples of a uniform yield over the drainage areas may not be applicable to many projects. The prior unit sufficient depth to kill the vegetation, which has been supplementing existing water surface evaporation by transpiration.

(b) Power

Selection of an average flow alone will not permit determination of the benefits from a water resources development project without more detailed studies. Such studies require routing through the reservoir the entire record of flow (corrected for evaporation losses), on a month by month basis, using assumed patterns of use, outlet capacities and, in the case of power, turbine and generator capacities and efficiencies. The reservoir would normally be considered to be full at the start of the operation study, or at least full to normal pool. The latter assumption is reasonable as reservoir storage can be permitted to accumulate as construction progresses.

For power benefits, the energy output will vary in accordance with the inflow, outflow, change in storage and corresponding head, tailwater elevation, turbine capacity and plant efficiency. If the plant is part of a system, the output may be subject to varying demands of the system load curve and whether the plant is to be used as a base load plant or a peaking plant. The routing study will indicate the necessary modifications to the head, storage, and even height of the dam to obtain maximum benefits.

(c) Irrigation

Operations studies for irrigation use should be made using seasonal crop demands and selected outlet capacities. Short-term demands may indicate that the storage needed was greater than that required for a uniform regulated flow (see Fig. 2-6). The proposed annual water use may be greater than that available 100 percent of the time, with the understanding that deficiencies can be tolerated in some years. Studies leading to the determination of maximum benefits might result in some adjustments to the reservoir capacity and height of dam.

(d) Water Supply

Operation studies for projects providing urban water supplies will be similar to those for irrigation projects in that there may be variations in the seasonal demand, especially where more than one source is available, or where there can be transfers to other holding or regulation reservoirs. However, the degree of dependability of flow must be higher for urban water supply than for irrigation projects.

(e) Flood Control

The storage allocated for flood control in single purpose or multipurpose projects is usually based on a definite design flood the control of which is needed for downstream protection. The required storage capacity is based on routing of the design flood inflow coincident with releases not to exceed downstream channel capacities [see Section 8(c)].

(f) Total Storage Requirement

The usable storage needed for single purpose projects can be readily determined as described in Sections (a) to (e). The total usable storage needed for multipurpose projects requires more complex routing studies and numerous trials to obtain the most economic allocations.

In addition to the variable requirement for storage for downstream uses, the total storage may be increased for the following reasons:

- Minimum head on power installations.
- Allowance for the storage of sediments without loss of usable storage.
- Minimum area for recreation use, including seasonal requirements.

The many economic, physical, and environmental factors to be considered in storage allocation and the total storage selected are discussed in Chapter 1.

8. DESIGN FLOODS

(a) General

In the design, construction, and operation of major dams, several floods of different magnitudes and probable frequencies must be determined. These floods are classified as the diversion flood (during construction), the reservoir design flood (for flood control projects), and the spillway design flood. In addition to being based on different hydrologic criteria, each flood is assumed to occur under different conditions with regard to storage and to types and capacities of outlet works.

(b) Diversion Flood

A major problem in the construction of a dam is the passage of flows of all magnitudes, particularly floods, during construction. River flows must be excluded from a part or all of the site during preparation of the foundation and until the dam has reached a safe height. This is accomplished by a cofferdam upstream, and often downstream. The excluded flow may be passed through a portion of the natural valley or excluded entirely from the valley at the dam site by diversion through tunnels in the abutments. The magnitude of the diversion flood determines the capacity of the diversion channels, or the height of the cofferdam and the related capacity of the tunnels when operating under the head created by the cofferdam. Some temporary valley storage may be created by the cofferdam, which in turn may be utilized to reduce the discharge capacity of the tunnels and possibly the height of the cofferdam.

The hydrologic criteria for the selection of the diversion flood cannot be considered independently from the type of dam, the construction program, and the economics of the diversion plan. Obviously, the diversion works cannot be designed to pass the spillway design flood, so a flood with a higher probability of occurrence must be considered. Depending on the magnitude of the project and the economic losses to be caused by overtopping of the cofferdam and damages to the permanent construction, the diversion flood may have a wide range of recurrence intervals, varying between 10 and 100 years, with possibly 25 years being the interval commonly used. If the critical phases of construction are to take place in a certain season of the year, the frequency of the flood may be that pertaining to a particular season, rather than to the entire year.

It should be noted that standard probability procedures [see Section 6(g)] lead to determining the probability of a flood occurring in a given year. If the construction period is extended over a period of several years, the theoretical probability of a selected diversion flood occurring during the construction operation is increased. The modified recurrence interval is approximately equal to the selected interval divided by the length of the construction period. In other words, the recurrence interval of a so-called 25-year design flood would be reduced to about 8 years. Under such conditions, a diversion flood with a greater recurrence interval should be selected. The theory is discussed in more detail in Section 6(g).

The characteristics of a diversion flood may be based on those of a flood of record or may be synthesized in accordance with methods outlined in Section 11. In either case, the frequency of occurrence of the peak and volume should be established in accordance with the methods given in Section 6(g). The complete hydrograph should be treated as an inflow hydrograph and routed through the proposed outlet works, allowing for the effects of any available storage above the cofferdam.

The requirements for the diversion flood in the case of a concrete dam may differ significantly from those for an embankment dam or rockfill dam. Protection against flooding is needed during the preparation of the foundation but overtopping of a cofferdam in this period usually will not cause significant damage except for cleanup and loss of contract time. As the concrete construction work rises above the foundation, high flows may be allowed to pass through openings in alternate monoliths or openings provided for gate controls or power facilities. Sufficient hydrologic studies must be made to insure that high flows will not interfere with the construction program.

(c) Reservoir Design Flood for Flood Control

Storage for flood control may be provided in a single purpose project in which all the capacity, except that for a small minimum pool, is allocated to this purpose; or the storage may be by specific allocation in a multipurpose project. In the latter type of project, the flood control storage is usually the top increment in the reservoir, and such storage is released as soon as practicable after the passage of a flood. Where there is a definite flood season, use of all or part of the flood control storage may be permitted in the dry season.

The volume of flood control storage should be sufficient to impound the runoff from the greatest known flood (less allowable downstream releases) that has occurred from the controlled area. Where the records of runoff are short and experience with known floods is limited, a reservoir design flood based on additional experience in the general region may be developed using methods described in Section 11 for design floods. For a major flood control project, the reservoir should be designed to control a flood having an estimated recurrence interval of at least 100 years.

The benefits from operation of a flood control reservoir are not confined to the immediate reach downstream from the dam. When selecting the capacity of such a reservoir, complete operation and economic studies are needed, particularly if the reservoir is to be one of a system of reservoirs in a basin. Flood detention and later releases involve a change in timing of the natural flood hydrograph. With a system of reservoirs, or even without reservoirs, the changes in timing may not necessarily benefit all downstream points with equivalent effectiveness. This is particularly true if several flood rises can occur in succession, and before the flood storage can be discharged without causing damage. Detailed operation studies are required to work out "rule curves" for system coordination and gate operation at individual dams.

(d) Spillway Design Flood

A spillway for a dam may serve one or more of three principal functions, defined by Cochran[11] as follows:

1. Provides protection against overtopping of non-overflow sections of the dam, acting in conjunction with other outflow facilities, such as regulating outlets or turbines.
2. Limits water surface elevations in the reservoir above the normal full pool elevation to avoid damages upstream from the dam.
3. Supplements regulating outlet for flood control operation when reservoir levels are above the spillway crest.

The spillway design flood is the most important flood to be provided for in the design of a dam. The selected magnitude and probable recurrence interval of this flood is related to the importance of a dam, its functional use, the economic value of the investment, and the potential damages to property and even loss of life that would result from total or partial failure of the structure.

No fixed criteria can be established for the relation between the height of dam, volume of storage, and the factor of safety to be adopted in the spillway design flood. Except possibly in the case of farm ponds, or small recreation ponds in rural areas, failure due to overtopping is unacceptable, even at rare intervals, because of the loss of investment and interruption of functional use. For minor projects, the incremental costs for providing additional spillway capacity may be based on an economic study, providing there is no question of public safety.

Cochran[11] has proposed the following general standards for the hydrologic design of spillways:

Standard 1: Design the dam and spillway large enough to assure that the dam will not be overtopped by floods up to probable maximum categories.

Standard 2: Design the dam and appurtenances so that the structure can be overtopped without failing and, insofar as practicable, without suffering serious damage.

Standard 3: Design the dam and appurtenances in such a manner as to assure that breaching of the structure from overtopping would occur at a relatively gradual rate, such that the rate and magnitude of increases in flood stages downstream would be within acceptable limits.

Standard 4: Keep the dam low enough and storage impoundments small enough that no serious hazard would exist downstream in the event of breaching.

It is common practice in the United States to adhere to the policy of designing and constructing high dams that impound large volumes of water, to conform with security Standard 1. The spillway design flood used in designing projects according to this standard should correspond to the probable maximum flood (see Sections 10 and 11).

Standard 2 has been confined principally to the design of run-of-river hydroelectric power or navigation dams, diversion dams, and similar structures where relatively small differentials between headwater and tailwater elevations prevail during major floods. The standard project flood has been used as the spillway design flood in connection with some major projects where functional Standards 2 or 3 are considered adequate. By definition, the standard project flood corresponds to the most critical flood event that is deemed reasonably characteristic of a particular drainage basin. In watersheds less than a few thousand square miles in size, the magnitude of this flood is usually accepted as being equal to approximately 50 percent of the probable maximum flood hydrograph.

Standard 3 is applicable where Standards 1 and 2 are not practicable of attainment within limits of economics or other practical considerations, and where hazards to life and property downstream in the event the dam fails from overtopping would clearly be within acceptable limits. The occurrence of overtopping floods must be relatively infrequent to make Standard 3 acceptable. Hydrographs based on probability analyses have also been used in deriving spillway design flood hydrographs for Standards 2 and 3.

Standard 4 is applicable to many small recreation lakes and farm ponds. In such cases it is often preferable to keep freeboard allowances comparatively small, in order to assure that the volume of water impounded will never be large enough to release a major flood wave when the dam is ultimately overtopped.

The basic hydrometeorologic and hydrologic studies necessary for the derivation of the spillway design flood (see Sections 10 and 11) first led to the development of the inflow hydrograph at the head of the reservoir together with small tributary hydrographs around the periphery [see Section 12(b)]. The design storm precipitation directly on the water area may also be a significant factor. The ultimate hydraulic capacity of the spillway may be determined by routing the inflow flood through the available surcharge storage above the normal full pool. Many trials, using different spillway lengths and allowable heads, may be necessary before the economic overall project design is obtained. The selected head plus the freeboard above the maximum computed water surface elevation determine the ultimate height of the dam.

Detailed steps in the derivation and routing of the spillway design flood are given in Sections 9 to 12.

9. ANALYSIS OF STORMS AND FLOODS OF RECORD

(a) Basis of Design Floods

The characteristics of design floods should be based as far as possible on an evaluation of actual storms and the corresponding floods of record at or near the dam site, with provision for factors of safety. The characteristics of flood hydrographs are influenced by numerous factors among which are (1) size and shape of the drainage basin, (2) pattern of tributary streams, (3) natural storage in lakes, swamps and river channels, (4) vegetal cover, (5) basic geologic and soils structure, and (6) characteristic rainfall patterns. The collective effects of these factors are revealed in the hydrographs of the floods of record, and there are no adequate substitutes for such records.

(b) Analysis of Storms of Record

Floods of record must be evaluated in relation to the severity of the storms which caused them. Storm analysis, when compared with regional experience, may indicate that much greater floods can be expected.

The major storms over a project basin and its surrounding area are analyzed with regard to orientation, isohyetal patterns, areal coverage, total depth, duration, and short-term intensities. General procedures are summarized in Ref. 12. To obtain sufficient information for this analysis, simultaneous observations should be obtained from all available non-recording rainfall stations (usually daily depths) and from any continuously recording gages.

The first step in the analysis is to plot the isohyetal map of total storm rainfall. Such maps are used to determine depth-area relations for the total storm or for selected time periods (see Fig. 2-13). Isohyetal lines are usually drawn by assuming that a uniform increase or decrease of rainfall depth occurs from one rainfall station to another, along a straight line that may be imagined as connecting the two stations, using the contour map principle. In mountainous topography it may be advisable to draw isohyetal lines considering, in addition to observed rainfall, the topographic and orographic effects where a change in altitude affects rainfall and where rain shadows are known to exist. If sufficient rainfall stations cover the basin, the isohyetal patterns may reveal the existence of one or several storm centers. A persistence in the orientation of the axes of storms may also be indicated. The locations of storm centers and orientation are significant because they reveal predominant meteorological trends and patterns. An analysis of orientation of the major storms of record may indicate that storm centers occur anywhere in a large basin, or they may be consistently located over a specific portion of the basin, or along a definite path. The persistence of particular patterns may justify inclusion of these patterns in a design storm. The average basin weighted rainfall depth is determined by computing the area between isohyetal lines and assigning to each subarea the mean value of the bounding lines.

Another method of determining a basin weighted rainfall depth is called the Thiessen method in which polygons of influence are formed around each rainfall station by straight lines constructed as the perpendicular bisectors of lines connecting adjacent stations (see Fig. 2-14). The observed rainfall at each station is weighted by the ratio of the area of influence at each station to the total area of the basin, and the average basin rainfall is computed by adding the weighted rainfalls. Average basin rainfall computed by the Thiessen method usually differs little from that computed by planimetering isohyetal lines drawn on the assumption of uniform rainfall change along a straight line connecting two stations. However, if the isohyetal lines are varied to take into account orographic or rainshadow effects, the two methods will not give equivalent results. If a difference has been found between the total areal rainfalls as computed by the isohyetal and Thiessen methods, the results found by the latter method may be adjusted by applying a correction factor which is the ratio between the totals found by the two methods. An adjusted Thiessen method is generally used in preference to the isohyetal method where areal rainfall must be determined for many time periods.

The second step in the analysis of storm rainfall over a particular basin is the determination of the depth-time distribution. This is essential in unit hydrograph derivations [see Section 11(c)] and in the determination of loss rates. The time distribution is determined by analysis of observations registered by continuous recording gages. Information obtained

122 HANDBOOK OF DAM ENGINEERING

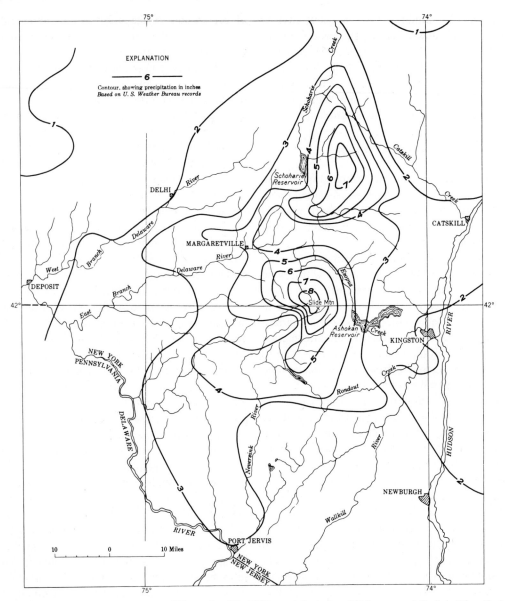

Figure 2-13. Isohyetal map, storm of November 23, 1950, on tributaries of Delaware and Hudson River Basin, New York. (U.S. Geological Survey)

from recording rainfall gages can be found in National Weather Service publications in the United States and is given as clock-hour increments. In many cases, this information is satisfactory as published. Nevertheless, for small or very rapidly draining steep basins, it may be desirable to record rainfall depths in smaller time increments. In such cases, the original rainfall chart of the recording gages will be required. For a given storm, the cumulative rainfalls from various recording gages may be plotted as mass curves on the same graph paper (Fig. 2-15) for a visual appraisal of consistency and a check for obvious errors. By

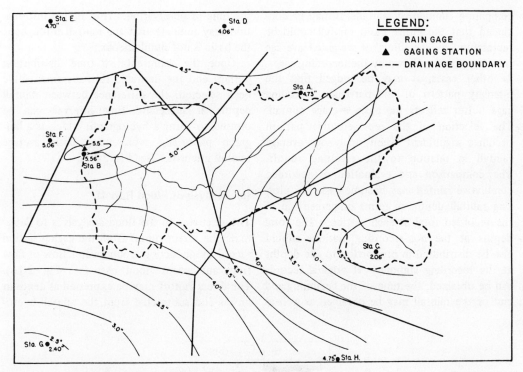

Figure 2-14. Estimating areal rainfall by isohyetal lines and by Thiessen polygons. (U.S. Bureau of Reclamation)

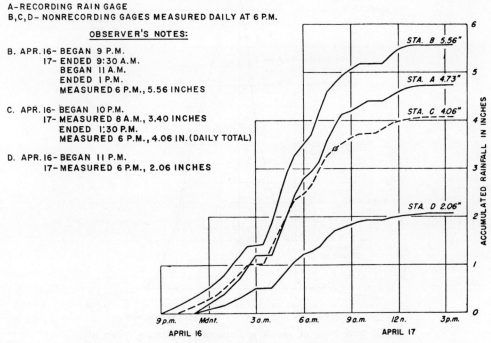

Figure 2-15. Mass curves of storm rainfall. (U.S. Bureau of Reclamation)

comparing cumulative rainfalls, it may be concluded that the entire basin rainfall could be represented by the direct or weighted average time distribution shown by the recording gages. In other cases, it may be judged that the intensity pattern of one particular recording gage better reflects the basin average rainfall. The selection of the representative pattern becomes significant when analyzing average rainfall in relation to the resulting runoff. The comparison and evaluation of recorded cumulative rainfall may become easier by plotting rainfall depths in terms of percentage of the recorded total at each station. The total depths at the non-recording stations should also be distributed in proportion to the depths at the recording stations. If original records can be obtained, the time of the beginning and end of the rainfall may be obtained as well as the time of observation. This limited information may indicate that the rainfall throughout the basin is not simultaneous.

Once the representative time distribution pattern and the desirable time increment have been selected, the difference between rainfall depths at successive time intervals can be plotted to form a hyetograph, which is a bar-graph pictorially representing basin average rainfall intensities (see Fig. 2-16).

(c) Analysis of Flood Runoff

The next step in the flood analysis is to determine the direct runoff from the gaging station record by deducting ground-water flow or flow from antecedent flood rises (see Fig. 2-16). The direct runoff may be expressed as depth in inches and subtracted from the rainfall hyeto-

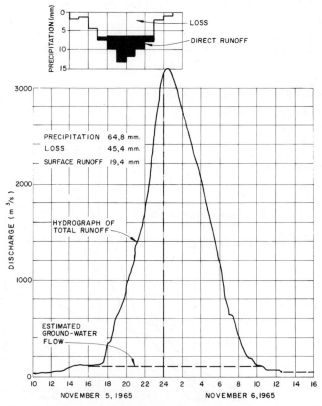

Figure 2-16. Plots of runoff and related precipitation, storm of November 5-6, 1965, Ziz River Basin, Morocco.

graph as shown by the shaded area on Fig. 2-16. The time of the beginning of the runoff is indicated by the rising limb of the hydrograph and determines the point at which the runoff will be subtracted from the hyetograph. Rainfall shown prior to this time can be considered as initial loss. Subsequent loss during the runoff periods is usually considered to take place at a uniform rate because of lack of data to assume otherwise.

10. DEVELOPMENT OF DESIGN STORMS

(a) General Principles

The record of the storm and flood experience in a given river basin often does not give an indication of the flood producing potential of that basin. Therefore, it is necessary to transpose to a basin under study the storm experience from adjacent basins or from basins in the general region that are considered to be subject to the same hydrometeorological influences. The mechanics of storm transposition from place of occurrence to a particular river basin fall into three general categories:

1. Transposition of isohyetal patterns without modification.
2. Transposition of depth-area-duration relations developed from isohyetal patterns.
3. Transposition of maximized depth-area-duration relations.

(b) Transposition of Isohyetal Patterns

The isohyetal pattern, usually that for the total storm precipitation that has occurred in an adjacent area, may be transposed to a basin under study. The area encompassed by the isohyetal lines will usually be greater than the drainage area under study.

This method is not generally used where the data described in Section 10(c) are available, as it has several limitations as follows:

- Unless the storm pattern has already been developed by some government or private organization a large amount of research is required to develop the storm pattern in a region perhaps remote from the project under investigation.
- Transposition of a storm pattern is not justified except in regions with small differences in relief. The area of origin of the storm and the area to which it is applied must both be free of orographic influences on precipitation.
- The assumed orientation of the storm must be in accordance with hydrometeorologic experience in the study area.

Several trial positions of the transposed isohyetal pattern over the drainage area will indicate the maximum total storm precipitation included within the boundaries of the drainage area. The distribution of the storm in time can be obtained from analysis of the hourly precipitation at recording rain gages (usually by mass curves).

(c) Transposition of Depth-Area-Duration Relations

The simplest method of transposing storm data from one area to another in the United States is by use of precipitation data in the form of depth-area-duration relations for individual storms. The data may be available in graphical or tabular form (Table 2-1). Analyses of more than 500 important storms have been made in a cooperative project between the Corps of Engineers, Department of the Army, and the Hydrometeorological Section of the National Weather Service. Printed summaries of these studies may be examined at District and Division Offices of the Corps of Engineers. Similar storm studies have been made by the Bureau of Reclamation for storms in its area of operation and not included in the Corps of Engineers' studies. An earlier study of storms in the Eastern United States[13] was published by the Miami Conservancy, Dayton, Ohio, but the results can be considered largely superseded by the later more detailed studies which gave consideration to the basic meteorology of the storms.

TABLE 2-1. Maximum Rainfall Depth-Duration Data for Storm of August 17-20, 1955, in Northeastern United States.

Area in Sq Mi	DURATION OF RAINFALL IN HOURS								
	6	12	18	24	30	36	48	60	72
Max station	7.9	11.7	14.3	18.2	19.4	19.5	19.8	19.8	19.8
10	7.8	11.1	13.0	16.4	18.5	18.9	19.4	19.6	19.6
100	7.6	10.5	11.6	14.6	17.6	18.1	18.8	19.0	19.0
200	7.4	10.2	11.4	14.2	17.1	17.6	18.2	18.4	18.4
500	6.8	9.7	10.8	13.4	16.3	16.8	17.2	17.3	17.3
1,000	6.2	9.2	10.2	12.4	15.4	15.9	16.2	16.4	16.4
2,000	5.4	8.0	9.4	11.2	14.0	14.5	14.9	15.2	15.2
5,000	4.0	6.3	7.9	9.5	11.7	12.1	12.6	13.0	13.0
10,000	3.1	5.0	6.5	8.0	9.7	10.0	10.6	10.8	10.8
20,000	2.1	3.6	4.9	6.3	7.6	7.9	8.3	8.5	8.5
35,000	1.3	2.5	3.6	4.7	5.6	6.0	6.4	6.5	6.5

If a particular storm is considered applicable to the area under study, the graph or table may be entered at the correct drainage area and the areal depths determined for different durations of time. The data can be used to plot a depth-duration curve for the particular drainage area but this curve does not indicate the true storm sequence. For example, the maximum 6 hours may have occurred at the $\frac{1}{3}$ to $\frac{1}{2}$ point in the total duration. To utilize the storm in computations of corresponding flood flow, the depths at different increments should be arranged in the form of a hyetograph, which will agree with the depth-duration curve but have a sequence that agrees with sequences in actual storms. Several trial sequences may be made to obtain the greatest peak discharge when applied to the unit hydrograph [see Section 11(d)].

Application of depth-area-duration curves based on various isohyetal patterns to drainage areas of different shapes introduces conservatism because it assumes that the isohyetal lines will conform to outlines of the drainage basin under study. For oval-shaped basins the conservatism may be minor but for elongated basins the conservatism may be unreasonable and some adjustment should be considered.

(d) Transposition of Maximized Depth-Area-Relations

Estimates of probable maximum precipitation (*PMP*) in the form of depth-area-duration relations (either as graphs or maps) have been made for the entire United States.[14,15,16] and for specific river basins by the Hydrometeorological Section of the National Weather Service. These studies are published in a series of reports called Hydrometeorological Reports or Technical Papers. These data may be used to develop maximum storms for spillway design in the same way as described above for depth-area-duration relations for a particular storm. Generalized data are available on maps of the United States from which it is possible to interpolate values for any drainage area in the range of 10 to 1000 sq mi.[14,15,16] A typical map is shown in Fig. 2-17. Some of these maps have been reproduced in reference books.*

In locations where determinations of maximum precipitation are not available, precipitation in storms of record may be maximized in accordance with procedures given in Section 10(f).

(e) Adjustments in Storm Transposition

Storm transposition is based on the concept of physical movement of a column of air with a given moisture content from its area of occurrence to the project area. The moisture content (called precipitable water) is defined as the depth of water that would be deposited at the

---
*Ref. 5, pp. 9-63, 9-64; Ref. 12, pp. 597-604

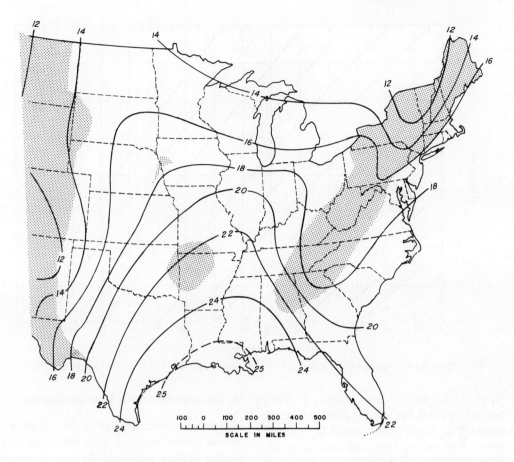

NOTE: AREAS OF LEAST RELIABILITY ARE SHADED.

Figure 2-17. Generalized estimates of maximum possible precipitation in inches for 500 square miles and 12-hour duration over the United States east of the 105th meridian.[14]

base of the column, if all moisture were condensed. The condensation of vapor in the air under a given constant pressure will occur when the air is cooled to a specific temperature, called the dew point. Because dew points are not observed under the same atmospheric pressure, it is the practice to compare them by conversion to a common pressure, taken at sea level, and equal to approximately 1000 mb. Figure 2-18 is used to convert dew points at a given observed pressure to a dew point at sea level using certain simplifying assumptions.

In transposing a column of air from a location at sea level to a basin at another location and elevation the amount of precipitable water available to the second basin will be the water remaining in the air column after moisture losses resulting from passage over any topographic barriers. The computational procedure is to first determine the total amount of precipitable water in a column of air available at sea level by summation from sea level to some upper limit, usually taken as 40,000 ft (or 12.2 km) where the pressure is 200 mb. Charts or tables are available for this computation (see Table 2-2). The amount of water lost in passing over a barrier is the amount of precipitable water between sea level and the elevation of the barrier.

The following is a typical computation for moisture adjustment. An unprecedented 5-day storm, occurred on the Mediterranean Coast of

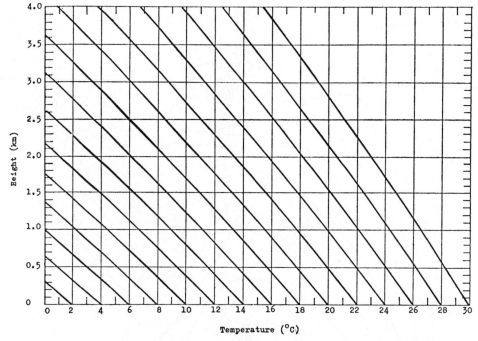

Figure 2-18. Pseudo-adiabatic diagram for dew-point reduction to 1000 mb at height zero.[17]

North Africa in September 1969. It was of interest to know the relative amount of precipitation that could occur inland if the storm passed over an extensive mountain range averaging 3000 m in height. Sea level temperatures during the storm ranged between 20 and 25°C. Because of the long duration of the storm it was assured that the dew point temperature was reached and that it was assumed to be 20°C. The computation is as follows:

Seal level dew point 20°C (68°F)
Inflow barrier, 3000 m (9850 ft)
Precipitable water, sea level
 to 12 km (40,000 ft)   52 mm (2.1 in.)
Precipitable water, sea level
 to 3 km (9,800 ft)   35 mm (1.3 in.)
Precipitable water, 3 km to
 12 km (9,800–40,000 ft) 17 mm (0.8 in.)
Moisture adjustment coefficient: $\frac{17}{52}$ = 0.33

The moisture adjustment coefficient is used to reduce the intensities, total depths, and depth-area-duration relations of the observed storm after it has passed over the inflow barrier to the interior basin.

(f) Maximization of Air Moisture Content

Moisture maximization of storms in place is outside of the scope of most hydrologic engineering studies and must be carried out by meteorologists experienced in such matters. The procedure requires both wind maximization and dew-point maximization.[17] The generalized data given in References 14, 15 and 16 are adequate for most spillway design storm studies in the United States.

(g) Statistical Storm Maximization

In remote areas or in underdeveloped countries sufficient meteorologic data are usually not available for determining probable maximum precipitation through the type of hydrometeorological analyses outlined above. As an alternative, use may be made of a method of statistical maximization derived by D. M. Hershfield,[17,18,19], which is based exclusively on a study of point rainfall records. In this method, use is made of the general equation for hydrologic frequency analysis (Ref. 5, pp. 8-23)

TABLE 2-2. Precipitable water (mm) between 1,000 mb surface and indicated height (m) above that surface in a saturated pseudo-adiabatic atmosphere as a function of the 1,000 mb dew point (°C). (World Meteorological Organization[17]).

Height (m)	1,000 mb TEMPERATURE (°C)														
	16	17	18	19	20	21	22	23	24	25	26	27	28	29	30
200	3	3	3	3	3	4	4	4	4	4	5	5	5	6	6
400	5	5	6	6	6	7	7	8	8	9	9	10	10	11	12
600	7	8	8	9	10	10	11	11	12	13	14	15	15	16	17
800	10	10	11	12	13	13	14	15	16	17	18	19	20	21	22
1,000	12	13	13	14	15	16	17	18	20	21	22	23	25	26	28
1,200	14	15	16	17	18	19	20	21	23	24	26	27	29	31	32
1,400	16	17	18	19	20	22	23	24	26	28	29	31	33	35	37
1,600	17	19	20	21	23	24	25	27	29	31	33	35	37	39	41
1,800	19	20	22	23	25	26	28	30	32	34	36	39	41	43	46
2,000	21	22	24	25	27	29	31	33	35	37	39	42	44	47	50
2,200	22	24	25	27	29	31	33	35	37	40	42	45	48	51	54
2,400	23	25	27	29	31	33	35	37	40	43	45	48	51	54	57
2,600	24	26	28	30	32	35	37	40	42	45	48	51	55	58	61
2,800	26	27	30	32	34	36	39	42	45	48	51	54	58	61	65
3,000	27	29	31	33	35	38	41	44	47	50	53	57	61	64	68
3,200	28	30	32	34	37	40	42	45	49	52	56	59	63	67	71
3,400	29	31	33	36	38	41	44	47	51	54	58	62	66	70	74
3,600	29	32	34	37	39	42	45	49	52	56	60	64	68	73	77
3,800	30	32	35	38	41	44	47	50	54	58	62	66	70	75	80
4,000	31	33	36	39	42	45	48	52	56	60	64	68	73	78	83
4,200	31	34	37	40	43	46	49	53	57	61	66	70	75	80	85
4,400	32	34	37	40	44	47	51	54	58	63	67	72	77	82	87
4,600	32	35	38	41	44	48	52	56	60	64	69	74	79	84	90
4,800	33	36	39	42	45	49	53	57	61	65	70	75	81	86	92
5,000	33	36	39	42	46	50	54	58	62	67	72	77	82	88	94
5,200	34	37	40	43	47	50	54	59	63	68	73	78	84	90	96
5,400	34	37	40	44	47	51	55	60	64	69	74	80	85	92	98
5,600	35	38	41	44	48	52	56	60	65	70	76	81	87	93	100
5,800	35	38	41	45	48	52	57	61	66	71	77	82	88	95	101
6,000	35	38	42	45	49	53	57	62	67	72	78	84	90	96	103
6,200	35	38	42	45	49	54	58	63	68	73	79	85	91	93	104
6,400	35	39	42	46	50	54	58	63	68	74	80	86	92	99	106
6,600	36	39	42	46	50	54	59	64	69	74	80	87	93	100	107
6,800	36	39	42	46	50	55	60	65	70	75	81	87	94	101	108
7,000	36	39	43	46	51	55	60	65	70	76	82	88	95	102	110
7,200	36	39	43	47	51	55	60	65	71	76	82	89	96	103	111
7,400	36	39	43	47	51	56	61	66	71	77	83	90	97	104	112
7,600	36	39	43	47	51	56	61	66	72	77	83	90	98	105	113
7,800	36	39	43	47	51	56	61	66	72	78	84	91	98	106	114
8,000	36	40	43	47	52	56	61	67	72	78	85	92	99	107	115
8,200	36	40	43	47	52	57	62	67	73	78	85	92	100	108	115
8,400	36	40	43	47	52	57	62	67	73	79	85	92	100	108	116
8,600	36	40	43	47	52	57	62	68	73	79	86	93	101	109	117
8,800	36	40	43	47	52	57	62	68	73	79	86	93	101	109	118
9,000	36	40	43	47	52	57	62	68	74	80	86	94	102	110	118
9,200	36	40	43	48	52	57	62	68	74	80	87	94	102	110	119
9,400	36	40	44	48	52	57	62	68	74	80	87	94	102	110	119
9,600	36	40	44	48	52	57	63	68	74	80	87	94	102	111	120
9,800	36	40	44	48	52	57	63	68	74	80	87	95	103	111	120
10,000	37	40	44	48	52	57	63	68	74	80	87	95	103	112	121
11,000	37	40	44	48	52	57	63	68	74	81	88	96	104	113	122
12,000	37	40	44	48	52	57	63	68	74	81	88	96	105	114	123
13,000					52	57	63	68	74	81	88	97	105	114	124
14,000					52	57	63	68	74	81	88	97	105	115	124
15,000										81	88	97	106	115	124
16,000										81	88	97	106	115	124
17,000											89	97	106	115	124

which has the form:

$$X_m = \overline{X}_n + K_m S_n$$

in which X_m is the probable maximum precipitation for a given duration (usually taken as 24 h), $\overline{X}_n$ is the mean and S_n the standard deviation of the maximum annual depths for the selected duration, and K_m is the number of standard deviations that must be added to the mean to obtain X_m. In computing $\overline{X}_n$ and S_n the maximum recorded rainfall is omitted from the computation. The term K_m is an enveloping value determined to be 15 from a study of 2500, 24-hr station values in the United States. K_m also appears to be applicable to estimates of *PMP* for lesser durations. Reference 17, p. 95, states that a value of 15 for K_m is not applicable to all climatic regions and may be less for tropical regions of heavy rainfall and possibly more for arid regions. The reference proposes adjustments depending upon the maximum known rainfall for a given duration.

In applying the method of statistical maximization it has been found desirable to compute not only the standard deviation but the coefficient of variation and compare it with values from other available rainfall records in the general region. A representative value is selected and an adjusted value of S_n is computed and used in the computation of the *PMP*.

When compiling 24-hr precipitation values from non-recording gages it is not sufficient to use calendar-day maximum values unless they are isolated values. If rainfall occurs on a day preceding or following the maximum day, one-half the adjacent values are added to the maximum day. Tests have shown that this procedure gives a reasonable approximation of the true 24-hr total.

The procedure described above serves only to compute the 24-hr *PMP* at a point in a river basin that is assumed to be the center of a design storm. A corresponding depth-area-duration relation applicable to the entire basin, must then be determined.

Trial depth-area relations can be developed from a study of isohyetal maps of the greatest storms in the region. The results may be compared with the generalized relations in Fig. 2-19 to develop the most critical pattern. The rela-

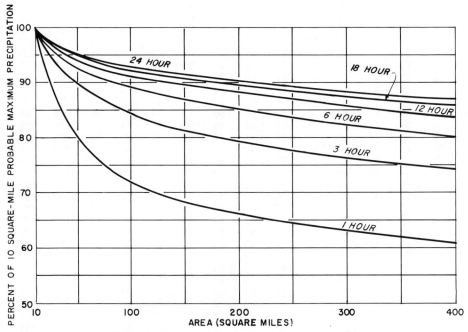

Figure 2-19. Depth-area, or area-reduction, curves.[16]

tions in Fig. 2-19 are not applicable to small-area thunderstorms.

After an area-depth relation has been developed, the next step is to distribute the 24-hr depth into time increments. If recording rain gage records are available, a study should be made of maximum depths that have occurred for intervals of 1, 2, 3, 6, 12, 18, and 24 hr. The depths can be plotted against duration and an enveloping curve drawn through the highest points. As the *PMP* depth will be greater than any recorded depth, it is useful to construct a dimensionless depth-duration curve in which the depths are expressed as percentages of the 24-hr depth as in Fig. 2-20. This figure can be used in judging the degree of conservatism in the derived relation.

The depth-duration curve indicates the maximum depths for different increments of time but not the sequence in which they may occur. Usually, examination of the record does not indicate the critical rainfall sequence. It is, therefore, permissible to arrange the *PMP* incremental rainfall depths (hyetograph) in any sequence that would result in maximized peak runoff rates when applied to the unit hydrograph [see Section 11(c)]. Frequently, the most critical runoff producing rainfall sequence is obtained by placing adjacent to one side of the maximum depth the second, fourth, sixth, etc., largest incremental depths and adjacent to the opposite side the third, fifth, seventh, etc. largest.

11. DEVELOPMENT OF DESIGN FLOODS

(a) Procedures

The development of the design flood from the selected design storm involves four basic steps, as follows:

1. Determination of minimum losses applicable to the design storm and deduction of such losses from the storm rainfall to determine total volume of rainfall excess or runoff.
2. Determination of unit hydrograph for drainage area at dam site.
3. Application of increments of rainfall excess to unit hydrograph to determine hydrograph of surface runoff.
4. Addition of base or ground-water flow to hydrograph determined in Step (3).

(b) Losses

The differences between total storm rainfall and resulting runoff is commonly called the "loss," for want of a better word. It is only a loss in terms of the short-term rainfall-runoff relation, as a portion of the loss may appear later as delayed runoff after the flood has passed. The magnitude of the short-term loss is dependent on many physical features of the basin as well as on antecedent rainfall and the intensity and duration of the storm under study. The magnitude of the loss from a large and heterogeneous basin is largely unpredictable and can only be evaluated from analysis of a number of critical storms and resulting floods. The total loss is usually divided into two parts, namely (1) an initial loss which occurs prior to the beginning of the surface runoff, and (2) a continuing loss which is concurrent with the surface runoff. The concurrent loss is usually taken as occurring at a uniform rate, although

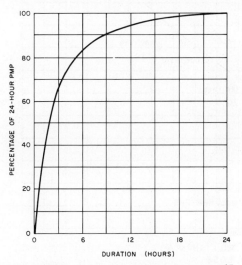

Figure 2-20. Maximum depth-duration curve.[17]

in nature the rate may decrease with storm duration. The loss may be expressed in terms of depth per hour over the basin for comparative purposes. Such rates are only relative and are not indicative of the average rate at which rainfall infiltrates into the soil.

Williams[20] has pointed out "that the determination of loss rates, as residuals between rainfall and runoff results in concentrating all the errors of the basic data in the loss determination. In large basins... where the distribution of rainfall records is inadequate to define storm rainfalls, the errors may be considerable." However, as relative values the computed losses are adequate for the purpose intended.

To determine losses in a given basin under study all major storms and corresponding runoff should be studied in accordance with the procedures outlined in Sections 9(b) and 9(c). The losses assigned to a design storm, particularly the spillway design storm, should be the minimum probable. To aid in the determination of a minimum loss, the results of the rainfall-runoff analyses can be plotted as a rainfall-runoff relation or rainfall-loss relation as shown in Fig. 2-21. The data in this figure are peculiar to a semiarid region or to any region where there has been a non-uniformity of rainfall in storms of record such that the rainfall in certain areas was not sufficient to satisfy the potential loss.

(c) Determination of the Unit Hydrograph

For most river basins it can be determined that surface runoff resulting from equal time periods of rainfall excess (rainfall minus losses) produces similar hydrographs in which the ordinates are proportional to the total volume of rainfall excess and the bases are essentially equal.

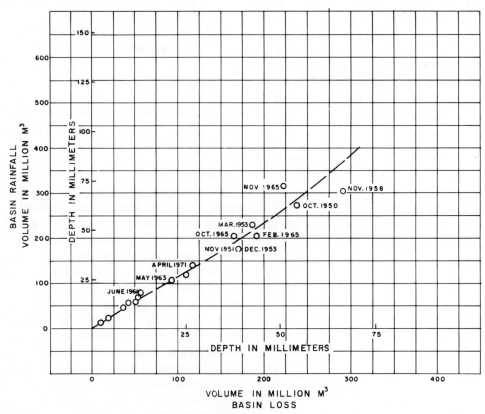

Figure 2-21. Rainfall–loss relation, Ziz River Basin, Morocco.

Such hydrographs are termed unit hydrographs, the term "unit" refers to the equal time intervals. The time interval may vary between 1 and 6 hr, depending on the size of the basin. For the purpose of comparison such hydrographs are reduced to an equivalent volume, taken as 1 inch in the English system and usually 1 or 10 mm in the metric system.

A unit hydrograph which is representative of characteristics of the basin under study, should be derived from analyses of storms and floods of record, following the procedures given in Sections 9(b) and 9(c). As far as practicable the storms should be isolated events in which the rainfall is concentrated in one or two time units, and the resulting floods have a single peak.

The hydrograph is separated into surface runoff and a base flow as described in Section 9(c). The latter may be only ground-water flow or may be delayed runoff from a previous period of high flows.

Subtraction of depth of surface runoff from the rainfall hyetograph (see Fig. 2-16) gives an indication of the duration of rainfall excess and the magnitude and distribution of the average basin losses. If most of the rainfall excess is confined to a single period of time, the hydrograph may be used directly to determine a trial unit hydrograph by dividing the ordinates by the total depth of surface runoff. The unit-time duration assigned to this unit hydrograph will be that corresponding to the duration of rainfall excess (t_r). Several analyses of this type are desirable to obtain a representative unit hydrograph.

The most important characteristics of the unit hydrograph are the peak discharge (q_p) usually expressed in cfs/sq mi and the lag (t_p) which is defined as the interval in hours between the center of mass of rainfall excess and time of peak discharge.

In the utilization of a unit hydrograph to compute a design flood from a design storm, it is important that the time unit adopted for defining the rainfall and the corresponding time unit for the unit hydrograph is small enough to reveal correctly the short term fluctuations in rainfall and resulting runoff. For example, if 40 percent of a 6-hr design storm occurs in 1 hr it is obvious that a 1-hr unit hydrograph must be used, even though the records permitted the derivation of only a 3-hr unit hydrograph. The following general relations have been used to approximate the relation between t_r and t_p where the subscripts R and r indicate longer and shorter time units, respectively:

$$t_{pR} = t_{pr} + \tfrac{1}{4}(t_R - t_r) \qquad (2\text{-}9)$$

$$t_r = \frac{t_p}{5.5} \qquad (2\text{-}10)$$

Detailed procedures for converting unit hydrographs from one time unit to another are contained in several standard texts (see Refs. 3, 12, and 20) and government manuals and will not be repeated here.

(d) Application of Unit Hydrograph

According to the unit hydrograph theory, the hydrograph resulting from a single unit of rainfall excess can be derived by multiplying the ordinates of the unit hydrograph by the depth of rainfall excess, providing both rainfall excess and unit hydrograph correspond to the same time units. If there are several consecutive units of rainfall excess, as in a design storm, a series of component hydrographs may be computed, the origin of each hydrograph being spaced one time unit apart. The final step is to summate the ordinates of the component hydrographs to obtain the total hydrograph. In terms of computer technology, the unit hydrograph is essentially a hydrologic model used for the generation of flood hydrographs resulting from a multi-unit storm. In fact, the procedure is well adapted to computer programming.

(e) Addition of Base Flow

In order to have a total volume equivalent to characteristic basin conditions, a base flow should be added to the design flood computed

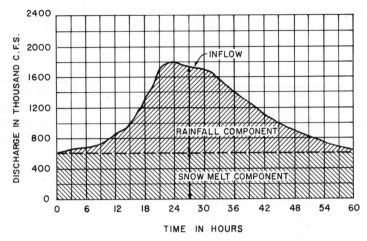

Figure 2-22. Components of Spillway Design Flood Hydrograph, Indus River at Tarbela Dam, Pakistan.

as described above. This base flow would be equivalent to the maximum experienced in a flood of record. Generally the origin would be ground-water flow. However, in regions experiencing snow melt from headwaters, as in the Rocky Mountains in the United States or the Himalayan Mountains in India and Pakistan the base flow may be a substantial part of the total hydrograph. Fig. 2-22 shows an extreme case for the Indus River in Pakistan where the probable maximum precipitation occurs on only about 7 percent of the drainage area and is superimposed on the greatest known snow-melt flood from 93 percent of the drainage area.

12. ROUTING OF SPILLWAY DESIGN FLOOD

(a) General Principles

This section is concerned with the routing of the spillway design flood through the reservoir, including a discussion of possible modifications to the total inflow hydrograph resulting from the creation of the reservoir, and the assumptions to be made regarding reservoir elevation and outlet conditions. The routing is carried out for a number of assumed spillway designs and corresponding hydraulic capacities to determine, along with structural and cost analyses, the most economic design. The important factor in the routing procedure is the evaluation of the effect of storage in the upper levels of the reservoir, termed "surcharge storage," on the required outflow capacity. In computing the available storage, the water surface is generally considered to be level. There will be a sloping water surface at the head of the reservoir due to backwater effect and this condition will create an additional "wedge storage." However, in most large and deep reservoirs this incremental storage can be neglected.

(b) Total Inflow Hydrograph

The hydrograph of the spillway design flood, developed as outlined in Section 11, is derived initially for the drainage area at the head of the reservoir. Additional adjustments are necessary to make it applicable to the total drainage area at the dam site. Such adjustments are necessary, because large reservoirs, many miles in length and covering many square miles, may significantly change the runoff characteristics of the lower reaches of the river, resulting in modifications to the shape and timing of the natural flood hydrograph.

The spillway design flood hydrograph may be made up of three or more components as follows:

1. The hydrograph of the main river at the point where the channel intersects the reservoir surface. There may be two or more principal tributaries for which hydrographs must be developed.
2. The hydrograph of the tributary area surrounding the reservoir. This may be composed of several small streams. Usually a hydrograph is developed for one typical stream and then its characteristics are considered applicable to the total contributing area. The resulting hydrograph will peak sooner than the hydrograph from the main channel.
3. The hydrograph resulting from the design storm rainfall falling directly on the reservoir area. This hydrograph will also peak sooner than the contributions from the main channels, and is generally significant only as a contribution to the total volume of runoff.

The total inflow hydrograph is the sum of the ordinates of the component hydrographs. In combining the hydrographs, no allowance is usually necessary for the time of travel through the reservoir, as the effect of the inflow will be transmitted to the spillway as a pressure wave. With very large reservoirs, the contribution from the peripheral areas and/or the rainfall on the water area may cause an initial peak which is higher than that contributed by the main channel at the head of the reservoir.

(c) Initial Reservoir Elevation

If a reservoir is drawdown at the time of occurrence of the spillway design flood, the initial increments of inflow will be stored, with the corresponding reduction in ultimate peak outflow Therefore, for maximum safety in design it is generally assumed that a reservoir will be full to the spillway crest in the case of an uncontrolled spillway and to the normal operating pool elevation when a gated spillway is used.

There may be exceptions to the above criteria in the case of reservoirs with large reservations for flood control storage. However, even in such cases a substantial part (50 percent or more) of the flood control storage should be considered as filled by runoff from antecedent floods. The effect on the economics and safety of the project should be analyzed before adopting such assumptions.

When the storage is to be used for power, irrigation, or water supply, the reservoir should be assumed to be full to the normal operating pool at the beginning of the spillway design flood.

Any assumption that a reservoir can be significantly drawn down in advance of the spillway design flood (other than one with a definite flood control storage reservation) as the result of a short-term flood warning system, is generally not acceptable for several reasons. The volume that can be withdrawn is the product of the total rate of discharge at the dam times the warning time. Since the warning time is usually short, except on great rivers, the released rate must be the greatest possible without flood damage downstream. Even under the most favorable conditions, it will be found that the volume that can be released will not be significant relative to the volume of the spillway design flood.

(d) Outflow Conditions

The facilities available for discharge of the inflow from the spillway design flood will depend on the type and design of the dam and its proposed use. A single dam installation may have two or more of the following discharge facilities:

- Uncontrolled overflow spillway
- Gated overflow spillway
- Regulating outlet
- Power plant

Uncontrolled and gated spillways may take several forms such as conventional gravity (ogee) section, side-channel overflow, chutes, or shaft (morning glory) spillways.

Uncontrolled Overflow Spillway. With a reservoir full to the spillway crest at the beginning of the design flood, discharge will begin at once and continue at a rate proportional to the three-halves power of the head on the spillway. Surcharge storage is created as the head increases and thus storage will also be proportional to some function of the head, depending on the characteristics of the elevation-capacity curve. The peak outflow will always be less, to some degree, than the inflow.

Gated Overflow Spillway. With a gated spillway, the normal operating level is usually near the top of the gates, although at times it may be drawn below this level by other outlets. The usual purpose in selecting a gated spillway is to make maximum use of available storage and head and at the same time to limit backwater damages by providing a high initial discharge capacity. In routing the spillway design flood, an initial reservoir elevation at the normal full pool operating level is assumed. Operating rules for spillway gates must be based on careful study to avoid releasing discharges that would be greater than would occur under natural conditions before construction of the reservoir. The effect of a large reservoir on the inflow hydrograph has been discussed in Section 12(b). In general, releases should be less than the inflow during the progress of a flood, thereby causing limited surcharge to build up. Furthermore, it is generally the practice in the design of gated spillways for the peak discharge from the spillway design flood to exceed the capacity of the gates at full opening with the result that the maximum water surface will rise above the normal pool operation.

Regulating Outlets. Most dams are designed with low-level outlets, either through the dam or the abutments for normal downstream releases in carrying out their single, or multipurpose functions. The discharge capacity of these outlets is usually small relative to the potential flood flows. These outlets may be assumed to be operating, at least in the initial stages of the design flood, and with due consideration for needs for flood control downstream. As spillway flows increase, it is conservative to assume that these outlets are closed.

Power Plants. When a hydroelectric plant is located at a dam and reservoir, it may be assumed that the turbines are discharging initially, thereby delaying use of the spillway. As in the case of the regulating outlets, the discharge should be limited to non-damaging channel capacity downstream, considering inflow from contributing areas below the dam. As the spillway discharges increase, the elevations of the tailwater below the dam may limit turbine discharge capacity. It is common practice to assume that the turbines are discharging at 75 percent of capacity, unless special conditions may prevent it. An assumption that the plant discharge is completely stopped, as in the case of the regulating outlets, may be unrealistic, as the power production may be required in the region regardless of flood conditions.

(e) Computational Procedures

The relations needed in routing a flood hydrograph through a reservoir are:

- Discharge characteristics (rating curves) of the spillway for various assumed lengths and heads.
- Discharge characteristics (rating curves) of regulating outlets.
- Discharge characteristics of turbines in powerplants, if part of the project.
- Curve of reservoir storage volume versus elevation or head on spillway.

Most textbooks on applied hydraulics and hydrology contain examples of reservoir routing procedures, each of which is based on the same fundamental principle, the storage equation, but contain variations which the individual authors have adopted. The basic principles are repeated here for convenience.

The storage equation may be written in the form:

$$\Delta S = I - O \tag{2-11}$$

where ΔS, I and O represent volumes of storage, inflow, and outflow in an increment of time Δt, which is usually a fraction of a day. All the variables are non-linear, with the inflow hydrograph usually having the greatest variation with respect to time and thus controlling the selection of Δt.

The storage equation may be rewritten in the form:

$$S_2 - S_1 = \frac{(Qi_1 + Qi_2)}{2} \Delta t - \frac{(Qo_1 + Qo_2)}{2} (\Delta t) \tag{2-12}$$

where Qi and Qo represent inflow and outflow rates, respectively, and subscripts 1 and 2 indicate values at the beginning and ending of the time interval Δt.

For purposes of computation, the inflow for the period may be reduced to a single incremental volume where $I = (Qi_1 + Qi_2)/2 \, (\Delta t)$ and Equation (2-12) may be written:

$$I + S_1 - Qo_1 \frac{\Delta t}{2} = S_2 + Qo_2 \frac{\Delta t}{2} \tag{2-13}$$

In Equation (2-13) the storage and outflow at the beginning of the time period and the total inflow, shown on the lefthand side, are known. S_2 and Qo_2 are the only unknowns but storage and outflow are both related to reservoir elevation and therefore a plot between $Qo \, (\Delta t/2)$ versus $S + Qo(\Delta t/2)$ can be made. The units will be in volumes such as day-sec-ft or acre-feet. Some engineers prefer to keep all units in rates by dividing all terms by Δt.

In the routing computation, all the terms on the left side of Equation (2-13) can be determined from the total inflow in time Δt and the storage and outflow relations corresponding to the initial starting elevation in the reservoir. The total volume from the left side of Equation (2-13) gives the value of $S_2 + Qo_2(\Delta t/2)$ at the end of the time interval. From the plot mentioned above $Qo_2(\Delta t/2)$ can be determined, and subtracted from $S_2 + Qo_2 \Delta t/2$ to give the storage (S_2) and the corresponding elevations at the end of the time interval. For the next increment of time, S_2 and Qo_2 become S_1 and Qo_1 in the lefthand side of Equation (2-13). A simple tabulation of data and results avoids confusion. When working in English units Δt should be expressed as fractions of a day. The product of $Q(\Delta t/2)$ then becomes sec-ft-days. As volumes in storage are usually expressed as acre-feet, the product is multiplied by 1.98 (or 2) and the 2 in the denominator cancels out. In metric units Δt will be in seconds, so that all units are in million cubic meters.

(f) Selection of Maximum Water Surface Elevation

Determination of the maximum reservoir level by routing of the spillway design flood hydrograph under various assumed lengths, heads, and possible types of spillway is a basic step in the selection of the elevation of the crest of the dam. The spillway length and corresponding capacity may have an important effect on the overall cost of a project, and the selection of the ultimate spillway characteristics is based on an economic analysis. Among the many economic factors that may be considered are damage due to backwater in the reservoir, cost-height relations for gates, and utilization in the dam of material excavated from the spillway channel.

After the economic water surface elevation is selected, an additional vertical distance to the crest of the dam must be provided for wave action as described in Section 13. Actually this computation may be done concurrently with the reservoir routing in order to have a correct estimate of the total volume in the dam, for use in the economic studies.

13. FREEBOARD AGAINST WAVE ACTION

(a) Basic Considerations

The vertical distance between the crest of a dam or embankment and some specified pool level is termed the "freeboard." It is provided to protect dams and embankments from overflow caused by wind induced tides and waves. The amount of freeboard varies for different structures depending on their importance, need to maintain structural integrity, and the estimated cost of repairing damages resulting from overtopping. Riprap or other types of slope protection are provided within the freeboard to control erosion that may occur even without overtopping.

The reference elevation for setting freeboard is generally the normal operating level, or the maximum flood level. The requirements for both conditions are usually investigated. When the former level is specified, freeboard is generally based on maximum probable wind conditions. A lesser wind condition is used in estimating the freeboard to be used with the probable maximum reservoir level, as it is improbable that maximum wind conditions will occur simultaneously with the maximum flood level. A first step in wave height determinations, is a study of available wind records to determine velocities and related durations and directions.

There are generally three basic considerations in establishing freeboard allowance, as follows:

1. Wind tide
2. Wave characteristics
3. Wave runup

Criteria and procedures for evaluating each of the above have been developed by the Corps of Engineers, particularly its Coastal Engineering Research Center (formerly the Beach Erosion Board); and have been summarized by Saville, McClendon and Cochran.[22] These procedures have received general acceptance for use in estimating freeboard requirements for reservoirs.

(b) Wind Tide

Wind blowing over an enclosed body of water exerts a horizontal force that causes a buildup in level along the leeward shore. There is a similar reduction in the opposite direction. This phenomenon is called wind tide or setup. The following simplified version of the formula originated by the Dutch for the Zuider Zee Project gives reliable results for setup:

$$S = \frac{V^2 F}{1400 D} \quad (2\text{-}14)$$

where S is the setup in feet above the reference level, V is the wind velocity in miles per hour, F is the fetch or water distance in miles over which the wind blows and D is the mean reservoir depth in feet along the fetch. The fetch used to compute wind tide is usually taken as twice the effective fetch [see Section 13(c)]. Figure 2-23 is a graphical solution for this formula.

(c) Wave Characteristics

The characteristics of wind generated waves measured in several large inland reservoirs have been analyzed using the dimensionless parameters, gT/V, gH_s/V^2 and gF/V^2, where g is the gravity constant of 32.2, T is the time between wave crests in seconds, F is the effective fetch in miles and H_s the significant wave height in feet. T is termed wave period.

By definition, H_s is the average of the highest one third of the waves occurring in a particular series. Deep water is defined as depths greater than about one-third to one half the length of the wave. The following relationship was developed for deep-water wave heights in reservoirs:

$$\frac{gH_s}{V^2} = 0.0026 \left(\frac{gF}{V^2}\right)^{0.47} \quad (2\text{-}15)$$

and, for wave period;

$$\frac{gT}{V} = 0.45 \left(\frac{gF}{V^2}\right)^{0.28} \quad (2\text{-}16)$$

Figure 2-24 is the graphical solution for Eq. 2-15 and also relates the wave height to the minimum duration of wind velocity necessary to generate a given height. Figure 2-25 is a graphical solution for Eq. 2-16.

HYDROLOGIC STUDIES 139

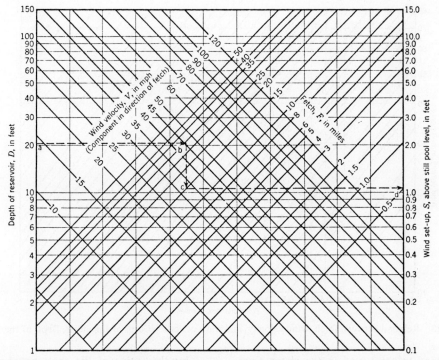

Figure 2-23. Diagram for computations of wind-setup in reservoirs. (Corps of Engineers, Dept. of the Army)

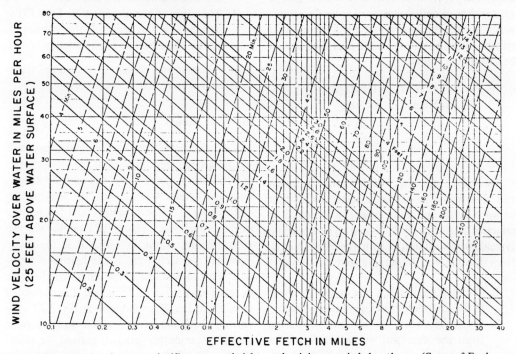

Figure 2-24. Relation between significant wave heights and minimum wind durations. (Corps of Engineers, Dept. of the Army)

140 HANDBOOK OF DAM ENGINEERING

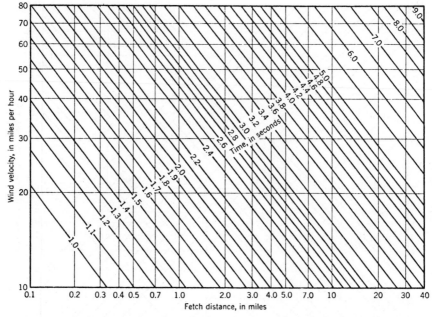

Figure 2-25. Relation between fetch, wind velocity, and wave periods. (Corps of Engineers, Dept. of the Army)

The wind velocity used in the above equations was that measured over water and is greater than velocities measured at land stations, which are the data usually available in reservoir studies. Wind speeds recorded over land should be adjusted as follows:

Fetch in Miles	Percentage Increase
>4	30
2	20
<1	10

To evaluate the effect of differences in reservoir shapes and to obtain a better correlation between fetch and wave height, a weighted or effective fetch has been devised. The steps in a typical computation for estimating the effective fetch in a reservoir are summarized as follows:

1. The maximum fetch line in the reservoir in the direction of the wind is located.
2. Seven secondary fetch lines radiating from the dam and on each side of the maximum fetch are drawn at 6° intervals.
3. The length of each fetch line is multiplied by the cosine of the angle between the line and the maximum fetch line.
4. The sum of the products in Step (3) is divided by the sum of the cosines to obtain the effective fetch distance.

Figure 2-26 is an example of the computation of effective fetch.

(d) Wave Runup

The vertical heights that a deep-water wave will run up a slope are functions of the wave characteristics as measured by the ratio between wave height and wave length, the slope of the embankment, the permeability and the relative surface roughness. Figure 2-27 relates the factors affecting runup, in which R is the runup measured vertically, H_0 is the height of a specific wave in feet, L_0 is the length of the wave in feet (equal to 5.12 T^2 where T is computed from Eq. 2-16. The solid lines in Fig. 2-27 were derived from laboratory results and represent runup on smoothly graded, grassed, or paved slopes. The dashed lines represent wave runup on rubble mounds as in breakwaters. They are

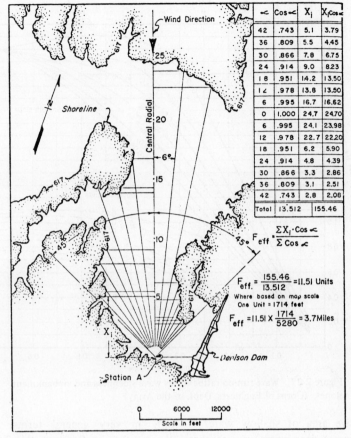

Figure 2-26. Computation of effective fetch for irregular shoreline. (Corps of Engineers, Dept. of the Army)

considered to be representative of the runup that would occur on the roughest of riprapped slopes. Conditions that would be encountered in average riprap protection would be between the two extremes shown.

In freeboard determinations for important structures, waves greater than the significant wave may be considered. Table 2-3 relates statistically the significant wave to other waves occurring in the same series. The runup resulting from wave heights greater than the significant wave height can be estimated from Fig. 2-27 by using the specific wave height in the H_0/L_0 ratio. The length for the significant and greater waves is approximately the same and no adjustment is needed for L_0.

TABLE 2-3. Characteristics of Successive Waves.

Ratio of specific wave height, H, to significant wave height, H_S (H/H_S)	Percent of waves exceeding specific wave height, H
1.67	0.4
1.40	2
1.27	4
1.12	8
1.07	10
1.02	12
1.00	13
0.95	16
0.89	20
0.75	32
0.62	46

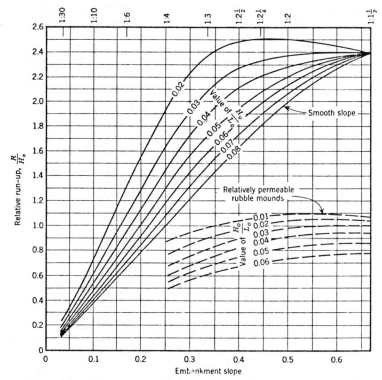

Figure 2-27. Wave run-up ratios versus wave steepness and embankment slopes. (Corps of Engineers, Dept. of the Army).

If an extensive area of shallow water is encountered before a deep-water wave reaches a dam, the wave characteristics and resulting runup will differ from deep-water conditions. References 22 and 23 give relations for shallow water conditions.

(e) Freeboard Allowance

Design freeboard for wave action or the vertical distance between the selected still-water level and the crest of the dam is the computed runup from a selected wave height plus the wind tide. For major embankments a minimum freeboard of 5.0 ft is customary. For other embankments a lesser minimum may be used.

14. SEDIMENTATION IN RESERVOIRS

(a) Factors Affecting Sediment Yield of a Drainage Basin

The volume of sediments carried by a river depends on many factors and can be predicted, only in very general terms, from regional characteristics. Some of the factors affecting sediment production are:

- Climate, particularly annual rainfall and resulting runoff.
- Vegetal cover, which is dependent on rainfall.
- Geology and soil cover
- Land and river slopes
- Land use

A qualitative description of these factors will be of little help in arriving at a quantitative estimate of sediment production per unit of area, and can be no substitute for field data. Of the five factors mentioned above, annual rainfall is perhaps the most important. Langbein[24] made an investigation of the relation between sediment yield of river basins and annual precipitation in regions where the annual temperature is 50°F and concluded the following:

1. The greatest sediment yield per square mile occurs in climates where the annual rainfall is about 12 in. Relation will vary with annual temperature.
2. Sediment yield decreases rapidly when the annual rainfall is less than about 12 in.
3. Sediment yield also decreases as precipitation increases until a minimum is reached in the range of 30 to 40 in.

The Langbein study points out that rainfall has two opposing influences on erosion. In the absence of vegetation, as may be expected in semiarid regions having less than 12 in. of precipitation, erosion increases due to the lack of vegetation, the impact of precipitation, and the erosive effect of overland flow. Above the point of optimum sediment yield, increased rainfall increases vegetative cover which not only breaks the force of the raindrops but helps to hold the soil in place. Unfortunately, no simple correlation is possible between precipitation and sediment because of the widely varying effects of terrain, soil types, temperature and other factors. The generalized relations are, however, an indication as to whether or not sedimentation will be a factor in reservoir planning.

(b) Effect of Reservoir on Sediment Accumulations

Whenever a dam is built across a river, retention of water-borne sediments can be expected. Accumulation of such sediments may in time occupy storage space needed to carry out the basic function of the dam and reservoir. To postpone as long as possible any encroachment on essential active storage, the planned gross storage is increased by an amount necessary for the permanent storage of sediments that may enter during the anticipated useful life of a project.

Some of the factors that determine the rate and total volume of sediment accumulations in a reservoir are:

- Average sediment load carried by a river, usually expressed as tons (dry weight) per square mile of drainage area, or acre-feet per square mile.
- Periods of abnormally high or low runoff, which may result in wide variations from average sediment load.
- Ratio between reservoir capacity and annual volume of inflow, a measure of trap efficiency.
- Grain size of sediments.
- Method of reservoir operation.

(c) Estimating Sediment Accumulations in Reservoirs

There are two basic methods for estimating the sediment accumulations in reservoirs, as follows:

1. *Sampling of suspended sediments*, establishment of sediment-discharge relation, computation of total annual sediment load in the river, and estimation of trap efficiency.
2. *Results of surveys of sediment accumulations* in existing reservoirs, expressed in tons or acre-feet per square mile per year.

The two methods are further discussed in Sections 14(d) and 14(e).

(d) Sampling of Suspended Sediments

In this method, determinations of the sediment load of a stream are made by sampling the concentration throughout a range of discharges. Samples are taken in a specially designed container called a sediment sampler. From each sample the dry weight of sediment is determined and related to the weight of sediment and water together. Knowing the rate of river discharge at the time the sample was taken and the percentage of sediment by weight, the sediment content can be computed as a rate in tons per day. The results of many sediment samples, converted to a daily tonnage, are plotted against the corresponding discharge at the time of the sample to form a sediment rating curve (see Fig. 2-28).

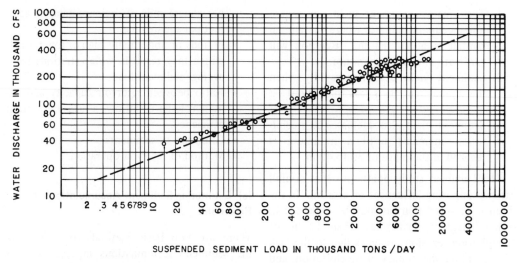

Figure 2-28. Sediment-rating curve, Indus River at Tarbela damsite (prior to 1965).

The points on the sediment rating curve may show considerable scatter, depending in part on drainage area conditions at the time of observation and the rate of rise and fall of the river. It may be advisable to construct several rating curves applicable to different seasons.

The second step in this method is the computation of total sediment load for an average year through use of the flow-duration curve [see Section 6(c)]. From this curve the percent of time, or average number days per year, in which a given discharge occurs can be obtained. The number of days times the tonnage per day corresponding to the discharge gives the annual tonnage. The computation is carried out in increments because of the non-linearity of both the flow-duration curve and the sediment rating curve.

The next step is to determine the percentage of the total sediment inflow that will be retained in the reservoir, or the trap efficiency. Brune[25] has proposed a method, derived from reservoir records, of relating the total sediment trapped (in percent) to the ratio between reservoir capacity and the volume of annual inflow (see Fig. 2-29). It will be seen from Fig. 2-29 that trap efficiency approaches 80 percent (median curve) when the reservoir capacity is equal to about 6 percent of the annual runoff.

Prior to the publication of Ref. 25, trap efficiency was related to the ratio between storage capacity and drainage area expressed as acre-feet per square mile. This ratio is still useful in comparing rates of accumulation in existing reservoirs.[26]

The first two steps described above give an indication of the tonnage of sediment to be trapped in the average year. The total tonnage to be stored will depend on the anticipated useful life of the project, normally considered to be 50 years or more. The final step is to determine the volume that the sediments will occupy during the life of the project, and this is dependent on the estimated unit weight. The unit weight of sediments laid down in a reservoir depends on the grain sizes of the material, the operation of the reservoir, and the period of consolidation. Samples in existing reservoirs have varied from less than 30 lb to more than 100 lb/cu ft. An average figure of 70 lb is often used, where analyses of the incoming sediments are not available.

The most authoritative study on the density of reservoir sediments was made by Lane and Koelzer in 1943[27] and has been widely quoted since then.[28,29,30] Table 2-4 is adopted from Ref. 27 and shows in general terms the effect of the reservoir operation and period of consolidation on the unit weights of specific

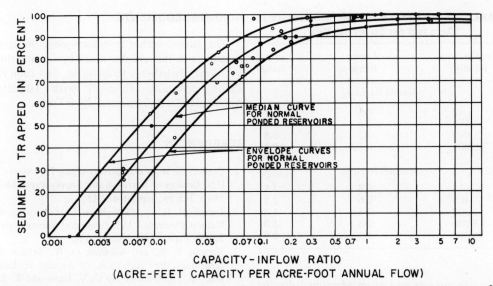

Figure 2-29. Trap efficiency as related to capacity-inflow ratio, type of reservoir, and method of operation.[25]

materials. The terms W_1 and W_{50} refer to unit weights at the end of 1 year and 50 years, respectively. If the percentage of each material is known from samples of the inflow, a weighted density can be computed and used in computing the total volume of sediments to be expected.

(e) Surveys of Existing Reservoirs

The probable rate of sedimentation accumulation in a proposed reservoir may be estimated from records of sediment accumulations in existing reservoirs. The volumes of sediment accumulated over a number of years are measured by topographic and hydrographic surveys. Samples of deposits are taken and the dry weights per cubic foot are determined. The surveys give the volume of deposits, usually expressed in acre-feet per square mile of drainage area per year. The observations of unit weight permit estimates of sediment accumulations in terms of tons per square mile of drainage area per year. The results of about 1000 reservoir surveys by various government agencies have been compiled at intervals of about 8 years since 1949. The latest compilation was published in 1969.[26] Typical values for different humid and semiarid conditions are given in Table 2-5.

In applying the results of a reservoir sediment survey to estimate sediment accumulation in a proposed reservoir, consideration must be given to the following:

1. The observed record must be from a drainage area that is equivalent in size to

TABLE 2-4. Density Values for Use in Design.
(Lb per cu ft)

Reservoir Operation	SAND		SILT		CLAY	
	W_1	W_{50}	W_1	W_{50}	W_1	W_{50}
Sediment always submerged or nearly submerged	93	93	65	72	30	51
Normally a moderate reservoir drawdown	93	93	74	77	46	60
Normally considerable reservoir drawdown	93	93	79	80	60	68
Reservoir normally empty	93	93	82	82	78	78

TABLE 2-5. Rates of Reservoir Sedimentation. (Acre-feet per Sq Mi per Year)

Drainage Area (sq mi)	Lower Range	Upper Range
Humid Regions		
10	0.4	1.5
100	0.3	1.2
1,000	0.2	0.8
5,000	0.18	0.7
Semiarid Regions		
10	0.8	3.5
100	0.6	2.5
1,000	0.4	1.7
5,000	0.3	1.0
10,000	0.2	0.9

the project area or an adjustment for the unit yield must be made.

2. The hydrologic characteristics of the drainage areas being compared should be similar, particularly the annual rainfall.
3. The physical characteristics of the two drainage areas must be essentially the same including topography, geology and soils, vegetal cover, and land use.
4. The trap efficiencies, or the capacity-annual inflow ratios of the existing and proposed reservoirs must be essentially the same.
5. The existing record of sediment accumulations should be at least 10 years long. If only a short record is available for comparison, a regional hydrologic investigation should be made to determine whether the sediment record was obtained in a period of high, low or average runoff.

The unit rate of sediment production, under equivalent climatic and physiographic conditions, varies inversely as an exponential function of the drainage area. Brune[31] has shown that this exponent is 0.15; thus the unit rate of production is proportional to $1/A^{0.15}$. Other studies indicate that the exponent more nearly equals 0.2.

ACKNOWLEDGMENTS

The author wishes to acknowledge the contributions to this chapter made by Constantine Voutyras, Evristhenis Hadjiloukas, and John H. Dixon, hydraulic engineers with Tippetts, Abbett, McCarthy, Stratton, New York.

REFERENCES

1. Corbett, D. M., et al., "Stream-gaging Procedures," U.S. Geological Survey Water Supply Paper 888, 1943, 245 pp., reprinted 1945.
2. Barnes, H. H. Jr., "Roughness Characteristics of Natural Channels," U.S. Geological Survey Water Supply Paper 1849, 1967, 213 pp.
3. Kirpich, P. Z., and Williams, G. R., Section 1, Hydrology, in "Handbook of Applied Hydraulics," p. 1-39, edited by C. V. Davis and K. E. Sorensen, McGraw-Hill Book Co., 3rd Ed., 1969.
4. Fiering, M. B., *Streamflow Synthesis*, Harvard University Press, 1967, 139 pp.
5. Chow, Ven Te, *Handbook of Applied Hydrology*, Section 8-IV, pp. 8-91-8-97, McGraw-Hill Book Co., 1964.
6. Water Resources Council, "A Uniform Technique for Determining Flood Flow Frequencies," Bull. 15, December 1967, 15 pp.
7. Dalrymple, T., "Flood-Frequency Analyses," U.S. Geological Survey Water Supply Paper 888, 1543-A, pp. 25-47, 1960.
8. Chow, Ven Te, "A General Formula for Hydrologic Frequency Analysis," *Trans. Am. Geophysical Union*, 32 (No. 2), 231-237 (April 1951).
9. U.S. Weather Bureau, "Evaporation Maps for the United States," Tech. Paper No. 37, 1959, 13 pp.
10. U.S. Weather Bureau, "Mean Monthly and Annual Evaporation, From Free Water Surface for the United States, Alaska, Hawaii and West Indies," Tech. Paper No. 13, July 1950, 10 pp.
11. Cochran, Albert L., "Spillway Requirements for Large Reservoirs," paper presented before *Texas Sec., AM. Soc. Civil Engrs.*, Fort Worth, Texas, September 1965.
12. Linsley, R. K., Jr., Kohler, M. A., and Paulhus, J. L. H., *Applied Hydrology*, McGraw-Hill Book Co., 1949. pp. 77-90.
13. The Miami Conservancy District, "Storm Rainfall of Eastern United States," Technical Reports, Part V, Dayton, Ohio, 1936.
14. U.S. Weather Bureau, "Generalized Estimates, Maximum Possible Precipitation over the United States East of the 105th Meridian, Areas of 10,

200, and 500 Square Miles," Hydrometeorological Report No. 23, June 1947.
15. U.S. Weather Bureau, "Seasonal Variation of the Probable Maximum Precipitation East of the 105th Meridian for Areas from 10 to 1000 Square Miles and Durations of 6, 12, 24 and 48 Hours," Hydrometeorological Report No. 33, April 1956.
16. U.S. Weather Bureau, "Generalized Estimates of Probable Maximum Precipitation for the United States West of the 105th Meridian for Areas to 400 Square Miles and Durations to 24 Hours," Technical Paper No. 38, 1960.
17. World Meteorological Organization, "Manual for Estimation of Probable Maximum Precipitation," Operational Hydrology Report No. 1, WMO 332, 1973.
18. Hershfield, D. M., "Estimating the Probable Maximum Precipitation," *Proc. Am. Soc. Civil Engrs., Jour. Hydr. Div.*, 87, (No. HY5), 99-116 (Sept. 1961).
19. Hershfield, D. M., "Method for Estimating Probable Maximum Precipitation," *Jour. Am. Waterworks Assoc.*, 57, 965-972 (Aug. 1965).
20. Williams, G. R., Chapter IV, "Hydrology" in *Engineering Hydraulics*, edited by Hunter Rouse, John Wiley & Sons, 1950, p. 309.
21. Hathaway, G. A., and Cochran, A. L., Chapt. 5, Sec. II, "Flood Hydrographs" in *Engineering for Dams*, Vol. I, John Wiley & Sons, 1945.
22. Saville, T. Jr., McClendon, E. K. and Cochran, A. L., "Freeboard Allowances for Waves in Inland Reservoirs," *Proc. Am. Soc. Civil Engrs., Jour. Waterways and Harbors Div.*, WW2, May 1962, pp. 93-124.
23. U.S. Army Coastal Engineering Research Center, "Shore Protection Planning and Design," Tech. Rept. No. 4, 3rd ed., 1966.
24. Langbein, W. B., and Schumm, S. A., "Yield of Sediment in Relation to Mean Annual Precipitation," *Trans. Am. Geophysical Union*, 39, (No. 6), 1076-1084, (Dec. 1958).
25. Brune, G. M., "Trap Efficiency of Reservoirs," *Trans. Am. Geophysical Union*, 34, (No. 3), 407-418 (June 1953).
26. Agricultural Research Service, U.S. Dept. of Agriculture, "Summary of Reservoir Sediment Deposition, Surveys Made in the United States through 1965, Misc. Pub. No. 1143, 64 pp., May 1969.
27. Lane, E. W. and Koelzer, V. A., "Density of Sediments Deposited in Reservoirs," U. S. Interdepartmental Comm. Rept. No. 9, St. Paul District, Corps of Engineers, St. Paul, Minn. Nov. 1943.
28. Koelzer, V. A. Section 4, "Reservoir Hydraulics" in *Handbook of Applied Hydraulics*, edited by C. V. Davis and K. E. Sorensen, McGraw-Hill Book Co., 3rd Ed., 1969, pp. 4-3-4-9.
29. Gottschalk, L. C., Sect. 17-I, "Sedimentation, Part I, Reservoir Sedimentation," in *Handbook of Applied Hydrology*, edited by Ven Te Chow, McGraw-Hill Book Co., 1964, pp. 17-14-17-25.
30. Linsley, R. K., Jr., Kohler, M. A., and Paulhus, J. L. H., *Hydrology for Engineers*, McGraw-Hill Book Co., 1958, pp. 288-291.
31. Brune, G. M., "Rates of Sediment Production in Midwestern United States," U.S. Soil Conservation Service, SCS-TP-65, Milwaukee, Wisc., Aug. 1948, Revised Dec. 1948.

3
Materials Suitable for Construction

EDWARD M. HARBOE
Concrete and Structural Branch
Division of General Research
RICHARD W. KRAMER
Hydraulic Structures Branch
Division of Design
U. S. BUREAU OF RECLAMATION
Denver, Colorado

3-1. INVESTIGATIONS

The investigation for construction materials is directed toward inventorying the available natural materials in the vicinity of the dam site. The inventory should assure that much more than enough material is available. An efficient investigation can be realized only through proper planning of the office, field, and laboratory work. The investigator should be familiar with the published information about the area under consideration, should know how to classify soils and rocks, and understand the geological and engineering characteristics of landforms. He should be knowledgeable in the capabilities and limitations of methods of sub-surface exploration, logging and sampling, and be familiar with the field and laboratory tests used for construction materials. Additionally, an understanding of how the materials are used in the design of an earth dam or the design of a concrete mix and the construction aspect of handling and processing the materials is important.

Investigations normally follow a "learn-as-you-go" procedure and become more detailed as the work progresses and as the project moves from the preliminary stages to the construction stage. Before beginning the field investigation, a thorough search should be made for available information relating to the stream and to the area under consideration including a variety of maps, air photos, and reports.

Topographic maps are indispensable to the investigation. Locations and elevations of exploratory holes, outcrops, and erosional features can be placed on the maps; and the landforms portrayed by the contours indicate to some degree the type of soil and subsurface geologic conditions. Detailed maps will be made later in the investigation of the borrow areas selected for further study and construction use, but until that time, many county, state, and Federal government maps are available. River survey maps may also be available or special maps of selected areas. The availability of topographic river survey and other

149

150   HANDBOOK OF DAM ENGINEERING

special maps and sheets distributed by the U.S. Geological Survey (USGS) is indicated on the index to topographic mapping for the various states. The indices are available, along with a list of agents for the maps, without charge from the USGS.*

Geologic maps can provide considerable engineering information. These maps identify the surface rock and overburden units. Many surface soils are closely related to the type of rock from which they are derived, but soils transported appreciable distances may overlie an entirely different rock type. When the influences of climate, relief, and geology of an area are considered, a very reasonable prediction can be made of the type of soil which will be encountered. Some areas have special map series entitled "Economic Geology," "Surficial Geology," and "Engineering Geology." Often explanatory texts are available with the maps, and full descriptions of many areas are available in the bulletins and professional papers of the USGS. Professional journals should not be overlooked.

The Department of Agriculture has available agricultural soil survey maps and reports for large portions of the United States. These surveys are surficial, extending to depths up to 6 ft. (1.829 m). Although intended for agricultural interests, these reports are of value to the engineer because such items as soil profile descriptions, ground surface conditions, natural vegetation, drainage, meteorological data, flood danger, access, and similar data are included. Some knowledge of the pedological system of soil classification is necessary to interpret the soil descriptions. The textural classification of soils of the U.S. Department of Agriculture charted in Fig. 3-1-1 is used to classify the topsoil or upper horizon of the soil profile. The chart shows the terminology used for various percentages of clay, silt, and sand.[1]

Aerial photography may reveal many sources of material which may easily be overlooked by engineers who depend solely on visual ground investigations. Field surveys are time consuming and expensive. Often possible sources are inaccessible to the investigator due to lack of roads or dense vegetation. From the ground it is sometimes difficult and often impossible to recognize landforms in which materials occur due to sight limitations. Aerial photographic techniques make inaccessible areas accessible to study; they cut the time-consuming and costly job of driving many miles looking, contacting county commissioners, highway personnel, and any other sources of information about possible material locations. Large areas can be searched, prospective sources located, and best entry routes determined and plotted on maps prior to field examination and sampling. An essential part of using aerial photographs for materials investigations is the study of agricultural soils, surficial geologic, topographic, and ground water geology maps; and reports of the area. A knowledge of the elements of geology and soil science will also be of assistance. References 2 through 7 may be consulted for information about aerial photograph interpretation for materials investigations.

Many governmental agencies and private aerial photograph companies sell mosaics and individual aerial photographs. The U.S. Map Information Office compiles a general index of aerial photographic coverage of the United States and its possessions, "Status of Aerial Photography."** An additional reference map is the "Status of Aerial Mosaics," also compiled and distributed by the Map Information Office and published annually. U.S. governmental agencies which hold available photographs and mosaics are the U.S. Geological Survey, Soil Conservation Service, Commodity Stabilization Service, U.S. Forest Service, Corps of Engineers, U.S. Air Force, Bureau of Reclamation, Tennessee Valley

---

*Inquiries may be directed to the U.S. Geological Survey, Denver Federal Center, Denver, Colorado 80225, or U.S. Geological Survey, National Center, Map Information Office, Reston, Virginia 22092.

**The index may be obtained from the U.S. Geological Survey, Map Information Office, National Center, Reston, Virginia 22092.

MATERIALS SUITABLE FOR CONSTRUCTION 151

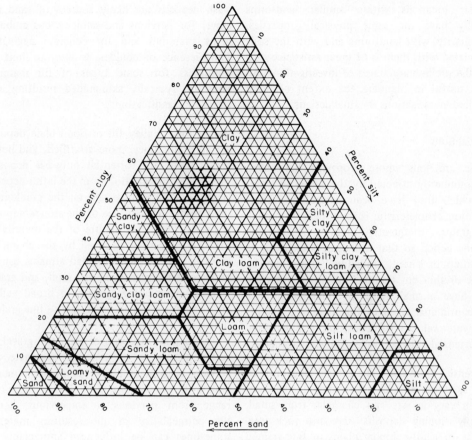

SOIL TRIANGLE
PERCENTAGES OF CLAY (BELOW 0.002 mm.),
SILT (0.002 TO 0.05 mm.), AND SAND (0.05 TO 2.0 mm.)
IN THE BASIC SOIL TEXTURAL CLASSES

Figure 3-1-1. Soil triangle of the basic soil textural classes.

Authority, and the National Oceanographic and Atmospheric Administration.

In Canada, The Map Distribution Office, Department of Mines and Technical Surveys, Ottawa, Ontario, distributes the "Air Photographic Coverage" map. Detailed information on particular areas may be obtained from the National Air Photographic Library at the same address.

3-2. SURFACE EXPLORATION

The properties of soils are often related to the topographic features or landforms in which the soils occur. The ability to recognize terrain features on maps, on air photos, and during field reconnaissance, combined with an elementary understanding of geological processes, can be of great assistance in locating sources of construction materials. The mechanisms which develop soil deposits are water action, ice action, and wind action for transported soils, and the mechanical-chemical action of weathering for residual soils. For the transported soils, each type of action tends to produce a typical landform modified to some extent by the nature of the parent rock. Soils found in similar locations within similar landforms usually have the same physical properties. Familiarity with landforms and with the soils associated with them is of great assistance during the preliminary stages of investigation, and it is useful in planning the extent of more detailed investigations for final designs.

Fluvial Soils

These are soils whose properties have been predominantly affected by the action of water to which they have been subjected. Their common characteristic is roundness of individual grains. Frequently, there is considerable sorting action so that a deposit is likely to be stratified or lensed. Individual strata may be thick or thin but the material in each stratum will have a small range of grain sizes. The three principal types of fluvial soils, reflecting the water velocity of deposition, are identified as torrential outwash, valley fill, and lakebeds.

Torrential Outwash. The typical landforms of this type are alluvial cones and alluvial fans, which vary in size and character from small, steeply sloping deposits of coarse rock fragments to gently sloping plains of fine-grained alluvium several square miles in area. The deposition results from the abrupt flattening of the stream gradient that occurs at the juncture of mountainous terrain and adjacent valleys or plains. The coarser material is deposited first; hence, it is found on the steeper slopes at the head of the fan, while the finer material is carried to the outer edges. In arid climates where mechanical rather than chemical weathering dominates, the cones and fans are composed largely of rock fragments, gravel, and silt. In humid climates, where the landforms have less steep slopes, the material can be expected to contain much more sand, silt, and clay.

Sands and gravels from these deposits are generally subrounded to subangular in shape, reflecting movement over relatively short distances, and the deposits have only poorly developed stratification. The torrential outwash deposits are likely sources of sand and gravel for pervious and semipervious embankment materials and for concrete aggregate. The presence of boulders is likely to limit the usefulness for some types of fill material. Soils are typically skip-graded resulting in a GP or SP classification.*

Valley Fill. Valley fill or flood plain deposits are generally finer, more stratified, and better sorted than are torrential outwash deposits. The degree of variation from the latter depends largely on the volume and on the gradient of the stream. The surface of the stream deposits is nearly flat. The nature of the material in the deposit can be deduced by the characteristics of the stream. Braided streams usually indicate the presence of silt, sand, and gravel, whereas meandering streams in broad valleys are commonly associated with fine-grained soils.

Flood plain deposits of sand and gravel are common sources of concrete aggregate and pervious shell materials for dam embankments. Soils in the various strata of river deposits may range from pervious to impervious; hence, the permeability of the resulting materials sometimes can be influenced appreciably by the choice of the depth of excavation cut. Presence of a high water table is a major difficulty in the use of these deposits, especially as a source for impervious material. Removal of the material from the reservoir floor upstream from the dam may be undesirable when a positive foundation cutoff is not feasible. When considering borrowing from a river deposit downstream from a dam, it should be remembered that such operations may change the tailwater characteristics of the stream channel and that the spillway and outlet works will have to be designed for the modified channel conditions. If tailwater conditions will be affected, borrow operations must produce a predetermined channel, and explorations for final design must accurately define conditions within the channel.

---

*See also Section 3-4., Soil Classification

An important type of stream deposit is the terrace. It represents an earlier stage of valley development into which the river subsequently has become entrenched. Remnants of terrace deposits are recognized by their flat tops and steep faces, usually persistent over an extended reach of the valley. Examination of the eroded faces facilitates the classification and description of the deposits. The extent of the drainage network on the face is helpful in determining relative permeability. Free-draining material has almost no lateral erosion channels, whereas impervious clays are finely gullied laterally. Terraces of sands and gravels are found along streams throughout the United States and are prevalent in the glaciated regions of the northern states. Sands and gravels from terraced deposits usually occur in layers and are well graded. They provide excellent sources of construction materials.

Lake Beds. Lake sediments or lacustrine deposits are the result of sedimentation in still water. Except near the edges of the deposit where alluvial influences are important, the materials are very likely to be fine-grained silt and clay. Stratification is frequently so fine that the materials appear to be massive in structure. Lacustrine deposits are recognizable by their flat surfaces surrounded by high ground. The materials they contain are likely to be impervious, compressible, and low in shear strength. Their principal use is for the impervious core of earthfill dams. Moisture control within these soils is usually a problem since the water content is very difficult to change.

Glacial Deposits

The results of the advances and retreats of the great North American continental ice sheets are represented by recognizable landforms which are important sources of construction materials. Smaller scale evidences of glacial action are found in the high mountain valleys of the Rocky Mountains and the Sierra Nevadas. Glacial deposits are generally heterogeneous and erratic in nature; hence, they are difficult to explore economically. They contain a wide range of particle sizes from clay to silt up to huge boulders. The particle shapes of the coarse grains are typically subrounded to subangular, sometimes with flat faces. Deposits of the glacier proper can be distinguished from deposits formed by the glacier melt water.

Morainal Deposits. Glacial till is that part of the glacial drift deposited directly from the ice with little or no transportation by water. It consists of a heterogeneous mixture of boulders, cobbles, gravel, and sand in an impervious matrix of generally nonplastic fines. Gradation, type of rock minerals, and degree of weathering found in the till vary considerably depending on the type of rocks in the path of the ice and the degree of leaching and chemical weathering. Glacial tills usually produce impervious materials with satisfactory shear strengths, but separation of cobbles and boulders will be necessary in order for the soil to be compacted satisfactorily. Typical landforms containing till are ground moraine which has a flat to slightly undulating surface; end or terminal moraine, a ridge at right angles to the direction of ice movement, which often curves so that its center is farther downstream than its ends; and lateral or medial moraine which occurs as ridges parallel to the direction of ice movement. Low, cigar-shaped hills occurring on a ground moraine, with their long axis parallel to the direction of ice movement, are called drumlins. They commonly contain unstratified, fine-grained soils.

Glacial Outwash. Deposits from glacial melt water are of several types. Glacial outwash plains of continental glaciation and their alpine glaciation counterparts, the valley trains, commonly contain poorly stratified silt, sand, and gravel similar to the alluvial sands of torrential outwash. Eskers are prominent winding ridges of sand and gravel which are the remnants of the beds of glacial streams that flowed under the ice. They generally parallel the direction of ice movement, have an irregular crestline and usually contain sand and gravel with some boulders and silty strata, and are excellent sources of pervious material and concrete aggregate. Kames are low, dome-shaped, partially stratified deposits of silt, sand, and gravel formed by hidden glacial streams. They are round to elliptical in plan with the long axis generally at right angles to the direction of ice movement. They contain material similar to eskers. Glacial lake deposits formed in temporary lakes during the Ice Age are generally similar in character and in engineering uses to fluvial lacustrine deposits. They are normally more coarsely stratified and they may contain fine sand.

Aeolian Deposits

Soils deposited by the wind are known as aeolian deposits. The two principal classes that are readily identifiable are dune sands and loess. Dune sand deposits are recognizable as low, elongated or crescent-shaped hills, with a flat slope windward and a steep slope leeward of the prevailing winds. Usually these deposits have very little vegetative cover. The material is very rich in quartz and its characteristics are limited range of grain size, usually in the fine or medium range sand; no cohesive strength; moderately high permeability; and moderate compressibility. They generally fall in the SP group of the Unified Soils Classification System.

Loessial deposits of windblown dust cover extensive areas in the plains regions. They have a remarkable ability for standing in vertical faces. Loess consists mainly of angular particles of silt or fine sand with a small amount of clay that binds the soil grains together. Although of low in-place density, when remolded loessial soils are impervious, moderately compressible and of low cohesive strength, they usually fall in the ML group or in the boundary ML-CL group of the Unified Soil Classification System.

Residual Soils

As weathering action on rock progresses, the rock fragments are generally reduced in size until the total material assumes all the characteristics of soil. The dividing line between rock and residual soil is not clearly defined, but for engineering purposes it may be considered soil if the material can be removed by commonly accepted excavating methods. A distinguishing feature of many residual soils is that the individual grains are angular but soft. Handling of the material during construction reduces appreciably the grain size, which makes it difficult to predict its performance by laboratory tests. Augering of the residual soils during exploration may grind the soil to a grain size distribution similar to that of the compacted embankment.

It is difficult to recognize and appraise residual soils on the basis of topographic forms. Their occurrence is quite general where none of the previously discussed types of deposits are recognizable and where the material is clearly not rock. Talus and landslides are easily recognizable forms of residual materials moved by gravity. Erosional features may be helpful in evaluating a residual deposit. Since the type of parent rock has a very pronounced influence on the character of the residual soil, the rock type should always be determined when assembling data for the appraisal of a residual deposit. The degree to which alteration has progressed largely governs the strength characteristics.

Rock Sources

The extent of investigations required to locate suitable sources of rock material for riprap, rockfill, or concrete aggregate will be governed by the size and design requirements of the project features and the quantity and quality of material required. An obvious rock source is the rock outcrop which can be located on aerial photographs and on geologic maps by the bunching of contours. Sharp breaks usually indicate hard rock. During field reconnaissance, the countryside should be examined for rock outcrops, cliffs, and road cuts in rock. Slopes below cliffs often have talus deposits. When a bedrock exposure containing satisfactory rock that can be developed into a quarry is not available, consideration should be given to securing rock material by removing boulders from stream deposits and glacial till or from surface boulder deposits.

Preference should be given to locating igneous rock, second choice is massive metamorphics, then well-cemented sandstone or limestone. Desirability of a material decreases with increased weathering and alteration. Sedimentary rocks containing clay should be suspected of weakness. The durability of a rock type can be judged by the examination of the rock in use at a nearby project, in stream channels, or observation of weathering resistance in situ. The spacing of joints, fractures, and bedding planes will control the size of rock fragments obtainable from a deposit. A rock source for riprap should produce fragments which are essentially equidimensional, angular, reasonably well-graded in size from a maximum dimension equal to the blanket thickness to about one-tenth of the blanket thickness. The fragments should be sound (resistant to mechanical abrasion), dense (high specific gravity, greater than 2.5), and durable (resistant to weathering). Elongated thin slabs are undesirable, but fragments having a minimum dimension of about one-third to one-fourth the maximum dimension are not objectionable. Less desirable rock is sometimes used if the cost is low enough, with extra stockpiled for periodic maintenance, rather than securing high-quality rock at great distance and high cost.

3-3. SUBSURFACE EXPLORATION

A systematic program for subsurface exploration started early in the preliminary investigations and carried through to the start of construction is wise. Exploration holes should be laid out on a grid using a coordinate system. The same system should be continued through all the investigations to avoid the confusion of more than one set of coordinates. Holes laid out on a grid system will make plotting subsurface profiles simpler. The spacing of exploration holes will depend upon the variability of the material deposit. During the preliminary stages, only a few widely spaced holes may be necessary to inventory the types of soil available and the approximate quantities. Investigations for final design require information on the quantity and quality of the material, as well as the variations that exist in a deposit. For construction, $1\frac{1}{2}$ to 2 times the design material requirements should be available in outlined borrow areas. Holes spaced at 400-ft (122 m) centers are desirable, but for very uniform deposits the spacing may be increased to as much as 800 ft (244 m). Closer spacing is required for variable deposits than for uniform deposits. The best procedure is to start with widely spaced holes for the preliminary design investigation, then add other holes in between, on the grid system, for the final design investigation. The type of test holes used will depend upon the type of material being investigated and the equipment available.

Test Pits and Trenches

Test pits and trenches are an effective means of exploring and sampling construction materials and may be the only feasible means of obtaining the required information when the material contains cobbles and boulders. Their use affords the most complete means for inspection, sampling, and making in-place density tests. The depth of the test pit or trench is determined by investigational requirements but is usually limited to a few feet below the water table. Pits excavated by a dragline, backhoe, clamshell, or dozer tractor are usually more economical than hand-dug pits. Trenches serve the same purpose as test pits but have the added advantage of providing a continuous exposure showing the continuity or character of soil strata. Test pits or trenches must be cribbed or braced when cave-ins are possible and should be fenced if there is a possibility of people or livestock falling in.

Auger Borings

Auger borings often provide the simplest method of soil investigation and sampling.

Depths of auger investigations are, however, limited by the ground water table and by the amount and maximum size of gravel, cobbles, and boulders as compared to the size of equipment used. Hand-operated post hole augers, 4 to 12 in. (10 to 30 cm) in diameter, can be used for exploration up to about 20 ft (6 m). However, with the aid of a tripod, holes up to 80 ft (24 m) deep have been excavated successfully. Machine-driven augers are of three types: helical augers 3 to 16 in. (7.6 to 40.6 cm) in diameter, disc augers up to 42 in. (107 cm) in diameter, and bucket augers up to 48 in. (122 cm) in diameter. An auger boring is made by turning the auger the desired distance into the soil, withdrawing it, and removing the soil for examination and sampling. The auger is inserted into the hole again and the process is repeated. Pipe casing is required in unstable soil in which the auger hole fails to stay open. The inside diameter of the casing must be slightly larger than the diameter of the auger used. Unless helical casing is used, the casing is driven to a depth not greater than the top of the next sample and is cleaned out by means of the auger. The auger can then be inserted into the borehole and turned below the bottom of the casing to obtain the sample. The auger operates best in somewhat loose, moderately cohesive, moist soil. Holes are usually bored without the addition of water, but in hard, dry soils or in cohesionless sands, the introduction of a small amount of water into the hole will aid the augering and sample extraction. Rock fragments larger than about one-tenth of the diameter of the hole cannot be successfully removed by normal augering methods.

Rock Source Investigations

Subsurface explorations of a rock source are often required to delineate the rock surface when the source is overlain by overburden; to sample the source through core drilling or a blast test; and to test production by development of a quarry face. Core drilling, if appropriate, should be done on a grid system and should include both vertical and angled holes. Observation of the spacing of joints, fractures, bedding planes, and cemented joints which might break apart during production and placing is important as these will control the size of the rock fragments obtainable. The location and distribution of unsound seams or strata will give an indication of the material which must be avoided or wasted during quarrying operations. Unsound seams or strata may be composed of clay materials and shales or other highly weathered material. The attitude of stratified formations and their vertical uniformity must also be observed and reported. A blast test sufficient to remove 10 to 20 cu yd (7.6 to 15 m^3) will give an indication of the production which can be expected. Blast test data should include the shot hole pattern and charge used, photos and the gradation of the material produced. In some instances, it may be desirable to develop a quarry face to determine the production capability of the rock source so that all of the shot materials can be incorporated in the dam design.

Geophysical Exploration

Geophysical exploration differs from excavating or boring and other standard engineering exploratory methods in that the information is always indirect. Geophysical methods include gravity, magnetic, nuclear, electrical, seismic and sonic property measurements of material deposits and geological structures. The two most widely used methods for civil engineering purposes are seismic refraction and electric resistivity. Both methods require special equipment, experienced operators, and trained personnel to interpret the data. These methods are most effective when carried out in conjunction with accurate geological information and direct methods of subsurface exploration. They should always be considered additional to direct subsurface investigation rather than as a substitute. The following discussions are intended to acquaint the reader with these two methods and their potentialities and limitations. Their use is warranted where they can substantially reduce the number of borings.

Seismic Refraction. The seismic refraction method makes use of the fact that different materials transmit elastic waves with different speeds. The speed with which a material will transmit these vibrations is proportional to the square root of its elasticity and inversely proportional to the square root of its density. Hard, well-cemented materials have higher transmission speeds than loose, uncemented ones. Elastic waves generated by an explosion spread outward through the underlying material. When the waves encounter a deeper layer of higher velocity, they are refracted into the formation. Some of the rays, traveling along the upper surface of the high-velocity formation, are further refracted to return to the surface. A number of detectors or geophones are arranged in a line or fan pattern and record the time lapses between the instant of explosion and the arrival of the resulting vibrations. Waves traveling along the surface in soil materials have a low velocity and near the shot point are the first ones received. Waves taking the longer path, through the high-velocity formation, eventually catch up with the slower surface waves, and for more distant pickup points they are the first ones to arrive. This "catching-up" distance from the shot point provides the geophysicist with the data for calculating the depth to refracting horizons and the velocity of wave transmission. Seismic refraction can provide information about the depth of overburden and can distinguish between material types like sand and gravel, and clay, but verification by borings is necessary. The method requires a minimum of interpretation and yields quantitative results.

Electric Resistivity. The electrical resistivities of rock and soil formations may be measured by passing an electric current through the ground between two electrodes inserted into the ground. Two electrodes placed between the outer pair are used to measure the potential drop caused by the resistance of the earth circuit between the two outside electrodes. When the spacing of the four electrodes is equal and in a straight line, the depth to which observations are effective is equal to their spacing. The resistivity of the material is calculated from the measurement of the current flowing (I, amps) into the ground between the outer electrodes, from the potential drop (V, volts) between the two inner electrodes, and from the spacing (d) of the electrodes.

$$R_s = 2\pi d \frac{V}{I}$$

By keeping the electrode spacing setup constant and moving it around the area being investigated, a resistivity map may be constructed which will show how resistivity varies over the area within the fixed depth of the investigation. Such a map may be used to delineate the boundary of the deposit.

The resistivity variation in a vertical column is determined by setting up observations over a location and expanding the electrode separation. When the electrode separation, depth of investigation, equals the depth to a change in material which involves a change in resistivity, the recorded resistivity figures will suddenly increase if the underlying material has a higher resistivity and drop if the stratum has a lower resistivity. Depths to bedrock, water table and material changes may be determined in this manner.

Both of the geophysical methods discussed have two assumptions which are fundamental to their use. The underlying materials are homogeneous horizontally; and each successive layer in depth has a sufficient change in its elasticity or resistivity to provide a contrast at the interface sharp enough to define its position. References 8 and 9 may be consulted for more information about seismic refraction and Ref. 9 for electric resistivity.

3-4. SOIL CLASSIFICATION

Soils are a heterogeneous accumulation of uncemented mineral grains and voids. The

UNIFIED SOIL CLASSIFICATION
INCLUDING IDENTIFICATION AND DESCRIPTION

FIELD IDENTIFICATION PROCEDURES (Excluding particles larger than 3 inches and basing fractions on estimated weights)				GROUP SYMBOLS [1]	TYPICAL NAMES
COARSE GRAINED SOILS More than half of material is larger than No. 200 sieve size [2] (The 1/4" size is about the smallest particle visible to the naked eye)	**GRAVELS** More than half of coarse fraction is larger than No. 4 sieve size. (For visual classifications, the 1/4" size may be used as equivalent to the No. 4 sieve size.)	CLEAN GRAVELS (Little or no fines)	Wide range in grain size and substantial amounts of all intermediate particle sizes.	GW	Well graded gravels, gravel-sand mixtures, little or no fines.
			Predominantly one size or a range of sizes with some intermediate sizes missing.	GP	Poorly graded gravels, gravel-sand mixtures, little or no fines.
		GRAVELS WITH FINES (Appreciable amount of fines)	Non-plastic fines (for identification procedures see ML below).	GM	Silty gravels, poorly graded gravel-sand-silt mixtures.
			Plastic fines (for identification procedures see CL below).	GC	Clayey gravels, poorly graded gravel-sand-clay mixtures.
	SANDS More than half of coarse fraction is smaller than No. 4 sieve size.	CLEAN SANDS (Little or no fines)	Wide range in grain sizes and substantial amounts of all intermediate particle sizes.	SW	Well graded sands, gravelly sands; little or no fines.
			Predominantly one size or a range of sizes with some intermediate sizes missing	SP	Poorly graded sands, gravelly sands; little or no fines.
		SANDS WITH FINES (Appreciable amount of fines)	Non-plastic fines (for identification procedures see ML below).	SM	Silty sands, poorly graded sand-silt mixtures.
			Plastic fines (for identification procedures see CL below).	SC	Clayey sands, poorly graded sand-clay mixtures.

IDENTIFICATION PROCEDURES ON FRACTION SMALLER THAN No. 40 SIEVE SIZE

			DRY STRENGTH (CRUSHING CHARACTERISTICS)	DILATANCY (REACTION TO SHAKING)	TOUGHNESS (CONSISTENCY NEAR PLASTIC LIMIT)		
FINE GRAINED SOILS More than half of material is smaller than No. 200 sieve size. (The No. 200 sieve size is about the smallest particle visible to the naked eye)	**SILTS AND CLAYS** Liquid limit less than 50		None to slight	Quick to slow	None	ML	Inorganic silts and very fine sands, rock flour, silty or clayey fine sands with slight plasticity.
			Medium to high	None to very slow	Medium	CL	Inorganic clays of low to medium plasticity, gravelly clays, sandy clays, silty clays, lean clays
			Slight to medium	Slow	Slight	OL	Organic silts and organic silt-clays of low plasticity.
	SILTS AND CLAYS Liquid limit greater than 50		Slight to medium	Slow to none	Slight to medium	MH	Inorganic silts, micaceous or diatomaceous fine sandy or silty soils, elastic silts.
			High to very high	None	High	CH	Inorganic clays of high plasticity, fat clays.
			Medium to high	None to very slow	Slight to medium	OH	Organic clays of medium to high plasticity.
HIGHLY ORGANIC SOILS			Readily identified by color, odor, spongy feel and frequently by fibrous texture.			Pt	Peat and other highly organic soils.

[1] Boundary classifications:- Soils possessing characteristics of two groups are designated by combinations of group symbols. For example GW-GC, well graded gravel-sand mixture with clay binder.
[2] All sieve sizes on this chart are U.S. standard.

FIELD IDENTIFICATION PROCEDURES FOR FINE GRAINED SOILS OR FRACTIONS

These procedures are to be performed on the minus No. 40 sieve size particles, approximately 1/64 in. For field classification purposes, screening is not intended; simply remove by hand the coarse particles that interfere with the tests.

DILATANCY (Reaction to shaking)

After removing particles larger than No. 40 sieve size, prepare a pat of moist soil with a volume of about one-half cubic inch. Add enough water if necessary to make the soil soft but not sticky.
Place the pat in the open palm of one hand and shake horizontally, striking vigorously against the other hand several times. A positive reaction consists of the appearance of water on the surface of the pat which changes to a livery consistency and becomes glossy. When the sample is squeezed between the fingers, the water and gloss disappear from the surface, the pat stiffens, and finally it cracks or crumbles. The rapidity of appearance of water during shaking and of its disappearance during squeezing assist in identifying the character of the fines in a soil.
Very fine clean sands give the quickest and most distinct reaction whereas a plastic clay has no reaction. Inorganic silts, such as a typical rock flour, show a moderately quick reaction.

DRY STRENGTH (Crushing characteristics)

After removing particles larger than No. 40 sieve size, mold a pat of soil to the consistency of putty, adding water if necessary. Allow the pat to dry completely by oven, sun, or air drying, and then test its strength by breaking and crumbling between the fingers. This strength is a measure of the character and quantity of the colloidal fraction contained in the soil. The dry strength increases with increasing plasticity.
High dry strength is characteristic for clays of the CH group. A typical inorganic silt possesses only very slight dry strength. Silty fine sands and silts have about the same slight dry strength, but can be distinguished by the feel when powdering the dried specimen. Fine sand feels gritty whereas a typical silt has the smooth feel of flour.

Figure 3-4-1. Unified soil classification chart.

INFORMATION REQUIRED FOR DESCRIBING SOILS		LABORATORY CLASSIFICATION CRITERIA	
Give typical name; indicate approximate percentages of sand and gravel; max. size; angularity, surface condition, and hardness of the coarse grains; local or geologic name and other pertinent descriptive information; and symbol in parentheses. For undisturbed soils add information on stratification, degree of compactness, cementation, moisture conditions and drainage characteristics. EXAMPLE:- Silty sand, gravelly; about 20% hard, angular gravel particles ½-in. maximum size; rounded and subangular sand grains coarse to fine; about 15% nonplastic fines with low dry strength; well compacted and moist in place; alluvial sand; (SM)	Determine percentages of gravel and sand from grain size curve. Depending on percentage of fines (fraction smaller than No. 200 sieve size) coarse grained soils are classified as follows:- GW, GP, SW, SP — Less than 5% GM, GC, SM, SC — More than 12% Borderline cases requiring use of dual symbols — 5% to 12%	$C_u = \dfrac{D_{60}}{D_{10}}$ Greater than 4	
		$C_c = \dfrac{(D_{30})^2}{D_{10} \times D_{60}}$ Between one and 3	
		Not meeting all gradation requirements for GW	
		Atterberg limits below "A" line, or PI less than 4	Above "A" line with PI between 4 and 7 are borderline cases requiring use of dual symbols.
		Atterberg limits above "A" line with PI greater than 7	
		$C_u = \dfrac{D_{60}}{D_{10}}$ Greater than 6	
		$C_c = \dfrac{(D_{30})^2}{D_{10} \times D_{60}}$ Between one and 3	
		Not meeting all gradation requirements for SW	
		Atterberg limits below "A" line or PI less than 4	Above "A" line with PI between 4 and 7 are borderline cases requiring use of dual symbols.
		Atterberg limits above "A" line with PI greater than 7	
Give typical name; indicate degree and character of plasticity, amount and maximum size of coarse grains; color in wet condition, odor if any, local or geologic name, and other pertinent descriptive information; and symbol in parentheses. For undisturbed soils add information on structure, stratification, consistency in undisturbed and remolded states, moisture and drainage conditions. EXAMPLE:- Clayey silt, brown; slightly plastic; small percentage of fine sand; numerous vertical root holes; firm and dry in place; loess; (ML)	Use grain size curve in identifying the fractions as given under field identification	PLASTICITY CHART FOR LABORATORY CLASSIFICATION OF FINE GRAINED SOILS COMPARING SOILS AT EQUAL LIQUID LIMIT — Toughness and dry strength increase with increasing plasticity index. (Chart: Plasticity Index vs. Liquid Limit, showing "A" LINE and regions CL, ML, CL-ML, OL, CH, OH or MH)	

TOUGHNESS (Consistency near plastic limit)

After removing particles larger than the No. 40 sieve size, a specimen of soil about one-half inch cube in size is molded to the consistency of putty. If too dry, water must be added and if sticky, the specimen should be spread out in a thin layer and allowed to lose some moisture by evaporation. Then the specimen is rolled out by hand on a smooth surface or between the palms into a thread about one-eighth inch in diameter. The thread is then folded and rerolled repeatedly. During this manipulation the moisture content is gradually reduced and the specimen stiffens, finally loses its plasticity, and crumbles when the plastic limit is reached.

After the thread crumbles, the pieces should be lumped together and a slight kneading action continued until the lump crumbles.

The tougher the thread near the plastic limit and the stiffer the lump when it finally crumbles, the more potent is the colloidal clay fraction in the soil. Weakness of the thread at the plastic limit and quick loss of coherence of the lump below the plastic limit indicate either inorganic clay of low plasticity, or materials such as kaolin-type clays and organic clays which occur below the A-line.

Highly organic clays have a very weak and spongy feel at the plastic limit.

discrete particles and voids of a soil form a particulate system, which distinguishes soil mechanics from solid mechanics. Soil materials have been a part of the earth's structure since the existence of erosional forces which turn the massive mineral accumulations, rock, into soil. Man has always referred to the various soil components with names like clay, sand, pebbles, etc. To the engineer engaged in the design and construction of foundations and earthworks, the physical properties of soils such as their unit weight, permeability, shear strength, compressibility, and interaction with water are of primary importance. In order for engineers both in the field and in the office to be able to communicate with each other, about the soils they are designing and constructing with, a standard for identifying and classifying the soils is needed. A system which describes a soil and categorizes it by distinct engineering properties enables engineers to exchange information and to profit by one another's experiences. Of the several systems for identifying and classifying soils, the Unified Soil Classification System has the most widely accepted engineering usage.

Unified Soil Classification System

The Unified Soil Classification System as presented herein is a modification of Professor A. Casagrande's Airfields Classification. In 1952, the Bureau of Reclamation and the Corps of Engineers, with Professor Casagrande as consultant, adopted the system as their standard. This System takes into account the engineering properties of soils; it is descriptive and easy to associate with actual soils and it has the flexibility of being adaptable both to field and laboratory use. Probably its greatest advantage is that a soil can be classified readily by visual and manual examination without the necessity for laboratory testing. The system is based on:

1. The percentages of various soil components.
2. The shape of the grain size distribution curve.
3. The plasticity and compressibility characteristics of the very fine grains.

Soils are placed into letter groups based upon results of testing for these three items. Knowledge of soil classification including typical engineering properties of soils of the various groups is especially valuable to the engineer engaged in prospecting for earth materials or for investigating foundations for structures. The system has established 15 basic soil groups to define certain distinctive and peculiar engineering properties. The typical names and symbols in each group are given in the classification chart, Fig. 3-4-1. Depending upon its basic properties, a soil is categorized according to these groups and assigned a name and symbol. The groups are broad; therefore, a supplemental detailed word description is also required to point out peculiarities of a particular soil and differentiate it from others in the same group. Two methods for identifying soils have been adopted:

1. Visual method—The visual or field method employs simple manual tests and visual observations to estimate the size and distribution of the coarse-grained soil fraction and to indicate the plasticity characteristics of the fine-grained fractions.
2. Laboratory method—This method in addition to visual observation employs laboratory tests, specifically gradation and moisture (soil consistency), to define the basic soil properties. The laboratory tests are described in Designations E-6 and E-7 of the *Earth Manual*,[10] respectively. These are used when a precise delineation is required and are very useful as a guide or check for the visual classification method when starting to classify a new borrow area.

Only the visual method will be described here.

Visual Method

Special apparatus or equipment is not required; however, the following items will facilitate the work:

a. A rubber syringe or a small oil can having a capacity of about $\frac{1}{2}$ pt.
b. Supply of clean water.
c. Small bottle of dilute hydrochloric acid.
d. Classification chart—Fig. 3-4-1.

The classification of a soil by this method is based on visual observations and estimates of its behavior in the remolded state. The procedure is in effect a process of elimination, beginning at the left side of the classification chart (columns headed Field Identification Procedures) and working to the right until the proper group symbol is obtained. The group symbol must be supplemented by detailed word descriptions including a description of the in-place condition for soils to be used in situ as a foundation. The observations made in the step-by-step procedure given below will provide the information for classifying and describing the soil.

Sample Preparation

1. Select a representative sample of the soil and spread it on a flat surface or in the palm of the hand.
2. Estimate and record the maximum particle size in the sample.
3. Remove all particles larger than 3 in. (76.2 mm) from the sample. Estimate the percentage by volume of cobbles (particles 3 to 12 in. in diameter) (76.2 mm to 30 cm) and boulders (particles over 12 in. in diameter) removed and record as descriptive information. Only that fraction of a sample smaller than 3 in. is classified.
4. Classify the sample as coarse-grained or fine-grained by estimating the percent, by weight, of individual particles which can be seen by the unaided eye. Soils containing more than 50 percent visible particles are coarse grained. Soils containing less than 50 percent visible particles are fine-grained soils.

For classification purposes the No. 200 sieve size (0.074 mm) is the particle size division between fine and coarse-grained particles, Fig. 3-4-2. Particles of this size are about the smallest that can be seen individually by the unaided eye.

Coarse-Grained Soils. If it has been determined that the soil is coarse grained, the soil is further identified by estimating and recording the percentages of:

1. Gravel-size particles, size range 3 in. (76.2 mm) to the No. 4 sieve (about $\frac{1}{4}$ in.) (4.76 mm).
2. (Sand-size particles, size range No. 4 sieve to No. 200 sieve.
3. Silt and clay-size particles, size range smaller than No. 200 sieve, referred to as fines.

If the percentage of gravel is greater than sand, then the soil is a *Gravel* (G). Poorly graded gravel is further described as either coarse (predominantly 3 in. to $\frac{3}{4}$ in.) (76.2 to 19.1 mm) or fine (predominantly $\frac{3}{4}$ in. to No. 4 sieve size) (19.1 to 4.76 mm), Fig. 3-4-2. Gravel is described as being clean if it contains less than 5 percent fines. With more than 12 percent fines, the gravel is described as dirty, but the term "dirty" is usually not used in the description; instead, the properties of the fines that make the gravel dirty are described. These properties are determined from hand tests on the minus No. 40 sieve size fraction. Gravels containing 5 to 12 percent fines are given boundary classifications. An obviously clean soil, with more gravel than sand, will be classified as either:

a. GW, if there is a good representation of all particle sizes in the entire sample,
b. GP, if there is a noticeable excess or

162 HANDBOOK OF DAM ENGINEERING

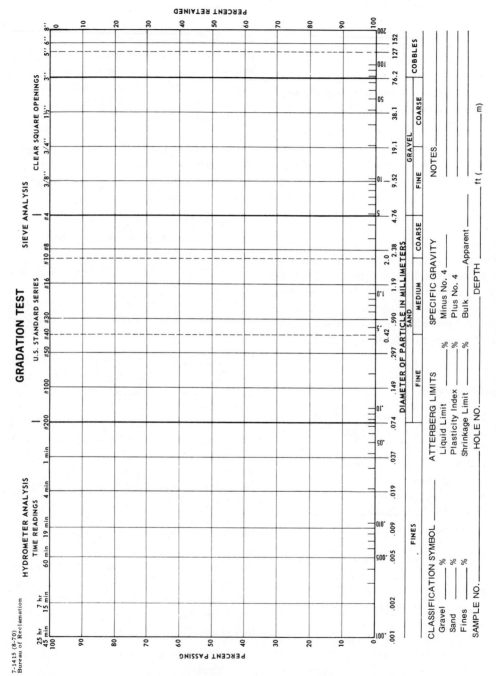

Figure 3-4-2. Form for gradation analysis.

absence of intermediate particle sizes in the entire sample.

If the soil is obviously dirty, the classification will be either:

c. GM, if the fines have little or no plasticity,
d. GC, if the fines are of slight to medium high plasticity.

If the percentage of sand is greater than the gravel, then the soil is a *Sand* (S). Sand is further divided into coarse, medium, and fine, Fig. 3-4-2. These divisions are used to describe the average size of poorly graded sand. Clean and dirty sands are identified as the gravels were above; thus the clean sands will be classified as either:

a. SW, if there is good representation of all particle sizes in the entire sample,
b. SP, if there is a noticeable excess or absence of intermediate particle sizes,

and the dirty sands will be classified as:

c. SM, if the fines have little or no plasticity,
d. SC, if the fines are of slight to medium plasticity.

To complete the description of a coarse-grained soil, the following should be recorded, as applicable:

1. Typical name.
2. Group symbol.
3. Maximum size, distribution and approximate percentage of cobbles and boulders in the total material sample.
4. Moisture and drainage conditions; dry, moist, wet, very wet (near saturation).
5. Approximate percentage of gravel, sand, and fines.
6. If poorly graded, statement of whether sand or gravel is coarse, medium, fine, or skip-graded.
7. Shape of grains; rounded, subrounded, angular, or subangular.
8. Surface coatings, cementation, hardness of grains, and possible breakdown when compacted.
9. Plasticity of fines; none, slight, medium, or high plasticity.
10. Color and organic content.
11. Local or geologic name.

Fine-Grained Soils. A soil that has been determined to be fine grained is further identified by estimating the percentages of gravel, sand, and silt and clay-sized particles. The silt and clay fraction is estimated together and called "fines," because the individual particles cannot be seen with the naked eye. To determine if the fines are silty or clayey, perform the manual identification for dry strength, dilatancy, and toughness, Fig. 3-4-1, on the fraction of soil finer than the No. 40 sieve size (about $\frac{1}{64}$ in., 0.42 mm). The same procedures are used to identify the fine-grained fraction of coarse-grained soils to determine whether they are silty or clayey. The degree of plasticity must always be given.

Various combinations of results of the manual identification tests indicate which classification is proper for the soil being tested.

a. ML has little or no plasticity and may be recognized by slight dry strength, quick dilatancy, and slight toughness.
b. CL has slight to medium plasticity, medium to high dry strength, very slow dilatancy, and medium toughness.
c. OL is less plastic than CL and has a slight to medium dry strength, medium to slow dilatancy, and slight toughness. Organic matter must be present in sufficient amount to influence the soil properties to be placed in this group.
d. MH is slightly plastic, has a slight dry strength, slow dilatancy, and slight toughness.
e. CH is highly plastic, has a high dry strength, no dilatancy, and high toughness.
f. OH is less plastic than the fat clay (CH) and may be recognized by medium to high dry strength, slow dilatancy, and slight to medium toughness. Organic

matter must be present in sufficient amount to influence the soil properties.

The description of fine-grained soils is completed by including the following, as applicable:

1. Typical name.
2. Group symbol.
3. Plasticity characteristics; none, slight, medium, or high plasticity.
4. Moisture and drainage conditions; dry, moist, wet, very wet (near saturation).
5. Maximum particle size. Distribution and approximate percentage of cobbles and boulders in total material.
6. Approximate percentage of gravel, sand, and fines in the fraction of soil smaller than 3 in. (76.2 mm).
7. Hardness of coarse grains, possible breakdown into smaller sizes.
8. Color in moist condition and organic content.
9. Local or geologic name.

Boundary Classification. Soils which are near the borderline between two groups, either in percentages of the various soil sizes or in plasticity characteristics are given dual symbols. The procedure for determining these symbols is to assume the coarser soil, when there is a choice, complete the classification, and assign the appropriate group symbol. Then, assume the finer soil, complete the classification, and assign the second group symbol. Dual symbols have to be used if the percentage of fines is 5 to 12 percent or 45 to 55 percent.

Other Tests and Examples. The acid test using dilute hydrochloric acid is a test for the presence of calcium carbonate. Soils which exhibit high dry strength and which have a strong reaction to the acid may derive their strength from the cementing action of calcium carbonate rather than colloidal clay.

The shine test is a quick procedure for determining the presence of clay. The test is performed by cutting a lump of dry or slightly moist soil with a knife. A highly plastic clay will have a shiny cut surface, while a dull surface indicates silt or clay of slight plasticity. Some typical soil descriptions and classifications are:

Lean Clay, (CL) about 75 percent fines of medium plasticity; about 15 percent predominantly fine to medium sand; about 10 percent angular to subangular gravel, predominantly fine, maximum size $\frac{3}{4}$ in. (19.1 mm); deposits of calcium carbonate; rapid reaction to HCl; light brown, damp.

Clay, (CL-CH) about 90 percent fines of medium plasticity, slight dilatancy, medium to high dry strength; about 10 percent fine sand, maximum size, No. 50; slight reaction to HCl; brown, damp.

Sandy Silt, (ML) about 66 percent slightly plastic to nonplastic fines; about 34 percent fine sand; light brown, dry.

Gravel, (GW) well graded, about 70 percent hard, subround to subangular, coarse to fine gravel, calcium carbonate coated; about 43 percent by weight 3 in. to 6 in. (76.2 to 152 mm), 5 percent +6-in. (152 mm) cobbles, maximum size 8 in. (200 mm); about 25 percent hard, subround to subangular, coarse to fine sand; about 5 percent fines of no plasticity, violent reaction to HCl; brown, damp.

Clayey sand, (SC) about 40 percent fines of medium plasticity; about 40 percent coarse to fine sand; about 20 percent coarse to fine granite type, subround to subangular gravel to 5-in. (127 mm) size; brown, damp.

Silty Sand, (SM-ML) about 54 percent fine sand; about 46 percent nonplastic fines; light brown, dry.

Clayey Gravels, (GC) about 60 percent medium hard to hard, subround to subangular, coarse to fine calcium carbonate coated gravel; about 25 percent hard to medium hard, subround to subangular, coarse to fine sand; about 15 percent 3 in. to 5 in. (76.2 to 127 mm), 10 percent +5-in. (127 mm) cobbles, maximum size 12 in. (305 mm); about 15 percent fines of slight plasticity; rapid reaction to HCl; gray brown, dry.

MATERIALS SUITABLE FOR CONSTRUCTION 165

TABLE 3-5-1. General Classification of Rocks Commonly Encountered

IGNEOUS
(Solidified from a molten state)

COARSE GRAINED CRYSTALLINE	FINE GRAINED CRYSTALLINE (OR CRYSTALS AND GLASS)	FRAGMENTAL (CRYSTALLINE OR GLASSY)
ORIGIN: Deep intrusion slowly cooled	ORIGIN: Quickly cooled volcanic or shallow intrusive	ORIGIN: Explosive volcanic fragments deposited as sediments
Granite Diorite Gabbro Note: Rock names are based on mineral content (see glossary). Color may be used as a rough index as noted above.	↑ Increasing quartz and light minerals RHYOLITE ANDESITE BASALT ↓ Increasing dark minerals Essentially glass (suddenly chilled, few or no crystals) Obsidian, Pitchstone, Etc.	Ash and pumice (Volcanic dust or cinders) Tuff (consolidated ash) Agglomerate (coarse and fine volcanic debris)

SEDIMENTARY
(Sediments transported by water, air, ice, gravity)

MECHANICALLY DEPOSITED	CHEMICALLY OR BIO-CHEMICALLY DEPOSITED

A-UNCONSOLIDATED:
  Clay
  Silt
  Sand   } According to Particle size
  Gravel
  Cobbles
B-CONSOLIDATED:
  Shale   (Consolidated clay)
  Siltstone (Consolidated silt)
  Sandstone (Consolidated sand)
  Conglomerate (Consolidated gravel or cobbles-rounded)
  Breccia   (Angular fragments)
	A-CALCAREOUS:
	  Limestone ($CaCO_3$)
	  Dolomite ($CaCO_3 \cdot MgCO_3$)
	  Marl (Calcareous shale)
	  Caliche (Calcareous soil)
	  Coquina (Shell limestone)
	B-SILICEOUS:
	  Chert
	  Flint
	  Agate } Spring deposit, Vein or cavity filling
	  Opal
	  Chalcedony
	C-OTHERS:
	  Coal, Phosphate, Salines, Etc.

METAMORPHIC
(Igneous or sedimentary rocks changed by heat, pressure)

A-FOLIATED
  Slate: Dense, dark, splits into thin plates (Metamorphosed shale)
  Schist: Predominantly micaceous, semi-parallel lamellae
  Gneiss: Granular, banded, subordinately micaceous

B-MASSIVE
  Marble: Coarsely crystalline, calcareous (Metamorphosed limestone)
  Quartzite: Dense, very hard, quartzose (Metamorphosed sandstone)

REPRINTED FROM U.S. BUREAU OF RECLAMATION
CONCRETE MANUAL 8th EDITION

Clayey Gravel with *Cobbles*, (GP-GC) about 70 percent angular to subangular, flat, poorly-graded gravel, predominantly fine; about 20 percent angular to subangular, flat sand, predominantly coarse; about 10 percent fines of slight to medium plasticity; about 5 percent +5-in. (127 mm) cobbles, maximum size 8 in. (200 mm); violent reaction to HCl; light gray, dry.

3-5. CLASSIFICATION OF ROCKS

In a broad sense, rocks are aggregates of minerals. Some exceptions to this are products of organic decay such as coal and volcanic glasses such as obsidian. Rocks are generally classified into three large groups based on their origin: Igneous, sedimentary, and metamorphic. Table 3-5-1 taken from the Bureau of Reclamation's *Concrete Manual*[11] shows an abbreviated general classification of the most common types generally encountered. Table 3-5-2 from the same *Concrete Manual* shows the principal mineral constituents of common igneous rocks, including some rocks not mentioned in Table 3-5-1.

TABLE 3-5-2. Principal Mineral Constituents of Common Igneous Rocks

Coarsely crystalline rocks	Principal constituent minerals[1]	Finely crystalline or porphyritic rocks
Granite	Q+ O +(P)+ A	Rhyolite
Syenite	O +(P)+ A	Trachyte
Quartz monsonite	Q+ O + P + A	Dellenite
Monzonite	O + P + A	Latite
Quartz diorite	Q+(O)+ P + A or B	Dacite
Diorite	(O)+ P + A or B	Andesite
Gabbro	P + B	Basalt

[1] Mineral symbols:
  Q—quartz (hard, shiny, conchoidal fracture)
  O—orthoclase feldspar (commonly pinkish, unstriated, regular cleavage faces)
  P—plagioclase feldspar (commonly white or nearly so, good cleavage faces which are often striated)
  A—amphibole and/or biotite
  B—pyroxene
  ( )—minerals in parentheses are subordinate in amount

The physical properties of a rock are affected by the properties of the constituent minerals; the degree to which the mineral grains are bound together; the size and arrangement of the grains which produce such structures as bonding and foliation; and the degree of fracturing, jointing, and bedding of the rock mass. Physical properties are least variable in igneous rocks, excluding the effects of fracturing. It is difficult to characterize the physical properties of the highly variable sedimentary rocks because they may range between wide limits. Although average or typical ranges in properties can be established for sound, unweathered specimens of the common rock types, in actual practice it is necessary to evaluate each deposit individually.

In general, the strongest rocks are the most dense, and the weakest rocks the most porous. Porosity of granite and similar igneous rocks and most metamorphic rocks is low, generally less than 1 percent.

Basalt, an igneous rock type, is similarly dense, but in certain areas may contain many small cavities (vesicles) and in other areas may be extremely vesicular. Porosity of limestone, which is a sedimentary-type rock, ranges from 0.5 to 15 percent, and in unusual types such as coquina, up to 25 percent.

Among the strongest and hardest rocks are quartzites, igneous rocks like granite and basalt, sound gneiss, and some schist. Compressive strengths of 15,000 to 30,000 lb. per sq in. (1,055 to 2,110 kg/cm^2) or more can be obtained. Some of the hardest, densest sandstones and siliceous limestones approach these strengths. Most limestones, marbles, dolomites, and sandstones, however, are intermediate in strength and hardness, with compressive strengths of about 2,500 to 15,000 lb. per sq. in. (176 to 1,055 kg/cm^2).

The most durable rocks are igneous rocks and massive quartzite and gneiss. Of these rocks, the fine-grained varieties such as basalt are generally tougher and wear better under abrasion than coarse-grained varieties. Foliated and laminated metamorphic rocks, such as schist and slate, are hard but split readily and fall apart under abrasion. In general, limestones and sandstones are moderately tough under abrasion. Shale is weak, tends to soften when wet, and disintegrates when exposed to weather.

3-6. SAMPLING AND LOGGING OF MATERIALS

Samples

Samples are required to identify and classify the soil and rock properly, and they are essential for laboratory tests on earth materials, concrete aggregate, and riprap. To a large degree, samples are used to judge the results of explorations for construction materials. Obviously, erroneous conclusions will be drawn if the samples are not truly representative of the explorations. The importance of obtaining representative samples cannot be overemphasized. This requires considerable care because of the variations in natural deposits of earth materials. Representative samples are comparatively easy to secure in trenches, test pits, and cut banks, because the various undisturbed strata can be inspected visually. Auger holes do not permit a visual inspection of the profile, and it is more difficult to secure representative samples.

Samples may be either individual or composite. An individual sample is a single sample representing one stratum or type of soil. A group of individual samples representing all the strata from a single test pit may be combined in the same ratio as the thicknesses of the strata from which the samples were taken to form a composite sample. Also, a sample may be composited to any desired ratio of the soil types involved. It is usually preferable to obtain individual samples from each stratum of a soil deposit rather than to composite the sample in the field, since the most desirable depth of excavation may be determined by trying several composite samples in the laboratory. Each stratum over 12 in. (30 cm) in thickness should be sampled. If samples are omitted, notes on the logs should give reasons why they were not taken. Representative samples of stockpiles are essentially composite samples in that parts are taken at random over the pile. If the sample is taken from one location, the selected area should be excavated deep enough to avoid segregated materials.

The sizes of samples required depend on the nature of the laboratory test which may be required. Table 3-6-1 gives suggested sizes of samples. Samples of 75 lb. (34 kg) or more should be placed in bags or other suitable containers which will prevent loss of the fine fraction of the soil and moisture. Samples of silty and clayey soils proposed for borrow which are obtained for engineering properties tests should be protected against drying and should be shipped in waterproof bags or other suitable containers to retain the natural moisture. Samples of sands and gravels may be shipped in closely woven textile bags and should be air-dried before placing in the bag. When the sack samples are shipped by public carrier, they should be double sacked to decrease the probability of bag breakage. Small samples for moisture determination are placed in moistureproof containers with airtight covers and completely filled. All samples must be clearly and completely identified.

Undisturbed samples are not often obtained for materials investigations. An exception might be when all the explorations are made by augering and a relatively undisturbed sample of cohesive material is taken by thin-wall, double tube, or other such samplers for in-place density and moisture testing. Undisturbed sampling is discussed in Chapter 4. In-place moisture and density information is required for the enbankment design and construction. Borrow irrigation requirements or the necessity to design the embankment for wet soil placement will be established by the in-place moisture content. The amount that borrow materials will shrink or swell (residual soils) is determined from the in-place density determinations. In-place densities are usually taken in test pits by field density methods. Samples not tested should be stored, for possible future examination and testing, until

TABLE 3-6-1. Samples of Construction Materials

Purpose of material	Minimum sample size	Remarks
Individual and composite samples for classification and engineering properties tests	Sufficient material passing the 3-inch (76.2-mm) sieve, to yield 75 pounds (34 kg) passing the No. 4 sieve size	Include information on percentages by volume larger than 3- and 5-inch (76.2- and 127-mm) size including maximum size. Cohesive soils—natural water content; ship in moistureproof bags. Noncohesive soils—air dried; ship in closely woven cloth bags
Permeability—gravelly soils	300 pounds (136 kg) passing a 3-inch (76.2-mm) sieve	See above
Relative density—gravelly soils	150 pounds (68 kg) passing a 3-inch (76.2-mm) sieve	Air-dried
Moisture samples, identification samples of soil, soil samples for sulfate determination (reaction with concrete)	1 pound (0.45 kg)—fill container completely	Samples should represent range of moisture and type of materials. Sealed in pint or quart ($\frac{1}{2}$ or 1 l) size containers
Concrete aggregate	600 pounds (272 kg) of pit-run sand and gravel. If screened, 200 pounds (91 kg) of sand, 200 pounds (91 kg) of No. 4 to $\frac{3}{4}$-inch (19.1 mm) size, and 100 pounds (45 kg) of each of the other sizes produced. 400 pounds of quarry rock proposed for crushed aggregate	For commercial sources, include data on ownership, plant, and service history of concrete made from aggregates
Riprap	600 pounds (272 kg) representative of quality	Method of excavation used, location of pit and quarry
In-place density and water content of fine-grained soils above water table	8- to 12-inch (20- to 30-cm) cubes or cylinders	Sealed in suitable container

[As identification on sample tags, give project name, feature, area designation, hole number, and depth of sample]

the dam is completed and has shown satisfactory operation.

Sampling Open Excavations

Open excavations are sampled by trimming an area of the side wall of the test pit or exposed surface to remove all weathered or mixed soil. The trimmed surface is examined to determine the sequence, thickness, classification, and description of each stratum of material. This information is recorded on a log form as illustrated in Fig. 3-6-1. The number of samples taken and the depth each represents are noted on the form. The sample is obtained by trenching the trimmed surface with a cut of uniform cross section and collecting the soil on a canvas spread below the trench. The minimum dimension of the sampling trench should be at least four times the diameter of the largest gravel particle in the soil. When sampling an individual soil stratum, special care is required to prevent inclusion of material from other strata. Where a composite sample is desired, the cross section of the sampling trench is kept constant through all the strata to be represented by the sample. If the soil being sampled is a gravelly soil containing 25 percent or more of plus 3-in. (76.2 mm) material, it is usually advantageous to take representative portions of the total

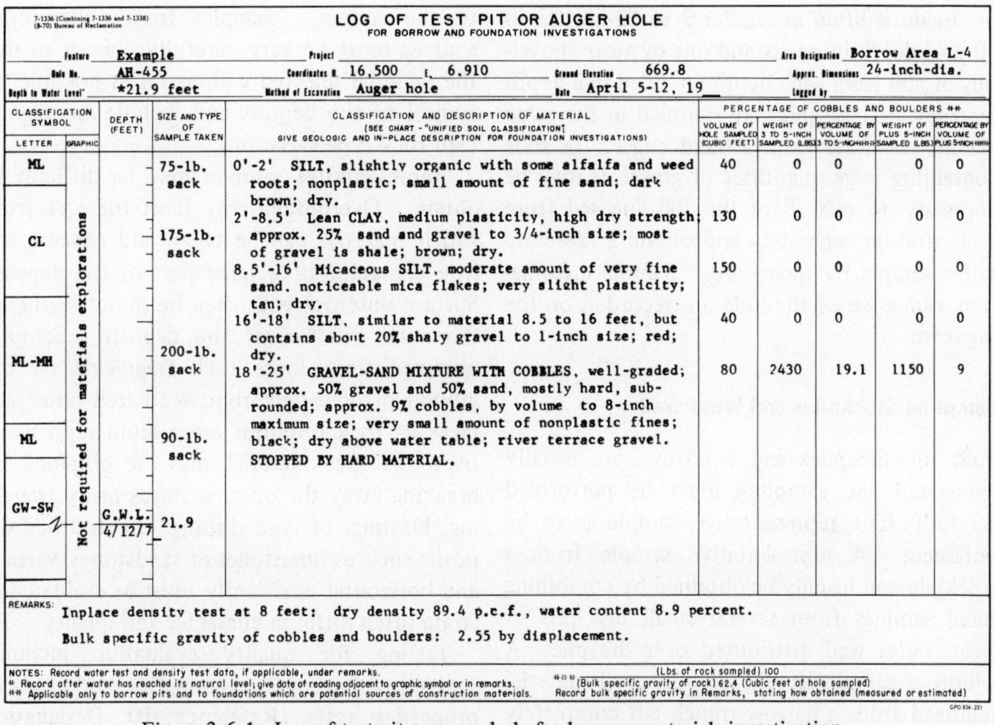

MATERIALS SUITABLE FOR CONSTRUCTION 169

LOG OF TEST PIT OR AUGER HOLE
FOR BORROW AND FOUNDATION INVESTIGATIONS

7-1336 (Combining 7-1336 and 7-1338) (8-70) Bureau of Reclamation

Feature	Example	Project		Area Designation	Borrow Area L-4
Hole No.	AH-455	Coordinates N 16,500 E 6,910	Ground Elevation 669.8	Approx. Dimensions	24-inch-dia.
Depth to Water Level*	*21.9 feet	Method of Excavation Auger hole	Date April 5-12, 19__	Logged by	

CLASSIFICATION SYMBOL		DEPTH (FEET)	SIZE AND TYPE OF SAMPLE TAKEN	CLASSIFICATION AND DESCRIPTION OF MATERIAL (SEE CHART - "UNIFIED SOIL CLASSIFICATION") GIVE GEOLOGIC AND IN-PLACE DESCRIPTION FOR FOUNDATION INVESTIGATIONS	PERCENTAGE OF COBBLES AND BOULDERS ##				
LETTER	GRAPHIC				VOLUME OF HOLE SAMPLED (CUBIC FEET)	WEIGHT OF 3 TO 5-INCH SAMPLED (LBS)	PERCENTAGE BY VOLUME OF 3 TO 5-INCH	WEIGHT OF PLUS 5-INCH SAMPLED (LBS)	PERCENTAGE BY VOLUME OF PLUS 5-INCH
ML			75-1b. sack	0'-2' SILT, slightly organic with some alfalfa and weed roots; nonplastic; small amount of fine sand; dark brown; dry.	40	0	0	0	0
CL			175-1b. sack	2'-8.5' Lean CLAY, medium plasticity, high dry strength; approx. 25% sand and gravel to 3/4-inch size; most of gravel is shale; brown; dry.	130	0	0	0	0
				8.5'-16' Micaceous SILT, moderate amount of very fine sand, noticeable mica flakes; very slight plasticity; tan; dry.	150	0	0	0	0
ML-MH			200-1b. sack	16'-18' SILT, similar to material 8.5 to 16 feet, but contains about 20% shaly gravel to 1-inch size; red; dry.	40	0	0	0	0
				18'-25' GRAVEL-SAND MIXTURE WITH COBBLES, well-graded; approx. 50% gravel and 50% sand, mostly hard, sub-rounded; approx. 9% cobbles, by volume to 8-inch maximum size; very small amount of nonplastic fines; black; dry above water table; river terrace gravel.	80	2430	19.1	1150	9
ML			90-1b. sack	STOPPED BY HARD MATERIAL.					
GW-SW		G.W.L. 4/12/7 21.9							

(Not required for materials explorations)

REMARKS: Inplace density test at 8 feet: dry density 89.4 p.c.f., water content 8.9 percent. Bulk specific gravity of cobbles and boulders: 2.55 by displacement.

NOTES: Record water test and density test data, if applicable, under remarks.
* Record after water has reached its natural level; give date of reading adjacent to graphic symbol or in remarks.
Applicable only to borrow pits and to foundations which are potential sources of construction materials.
(Lbs. of rock sampled) 100 / (Bulk specific gravity of rock) 62.4 (Cubic feet of hole sampled)
Record bulk specific gravity in Remarks, stating how obtained (measured or estimated)

Figure 3-6-1. Log of test pit or auger hole for borrow materials investigation.

excavated material, such as every fifth or tenth bucketful, for the depth sampled rather than to sample the side wall of the excavation.

When the collected sample is larger than needed for testing, the size may be reduced by quartering on a canvas. The material to be quartered is placed in a uniform conical pile. The pile is then flattened to a uniform thickness, by spreading the pile uniformly in all directions, and marked into quarters. The two diagonally opposite quarters are removed and the remaining material is mixed, quartered, and reduced until the desired sample size is obtained.

Sampling Auger Holes

Small auger holes cannot be sampled and logged as accurately as an open trench or test pit because the entire profile cannot be inspected and representative strata cannot be selected for sampling. Hand augers 4 ins. (10 cm) in diameter or larger, depending upon the type of soil and in-place conditions, should be used so that samples will be large enough to provide sufficient material for laboratory testing. As the hole is advanced the soil removed from the hole is deposited in individual piles in an orderly depth sequence. New piles should always be started when significantly different materials are encountered. In preparing an individual sample from an auger hole, consecutive piles of similar material may be grouped to represent each stratum, by combining all materials from each pile within the stratum or depth being represented, to form the desired size of representative sample. For a composite sample, all materials from each pile representing the strata to be composited are combined. The combined samples are then reduced by quartering, as described above, to obtain the desired sample size.

When a large diameter (8 in. (20 cm) or larger) power auger is used in fine-grained soils, the blade is lifted at regular 9 to 12-in. (23 to 30 cm) depth intervals and one or more shovelfuls of soil from each depth are piled in a depth sequence. These piles are sampled in the same manner as those from a hand auger. In soils containing large quantities of gravel, it may be necessary to pile all of the soil augered from each stratum separately and obtain a representative sample by quartering. Sample information and a log of the hole are recorded on the log form.

Sampling Stockpiles and Windrows

Soils in stockpiles and windrows are usually segregated and sampling must be performed carefully if a representative sample is to be obtained. A representative sample from a stockpile can usually be obtained by combining small samples from several small test pits or auger holes well distributed over the pile. A windrow sample should comprise all material removed from a narrow trench cut completely across the windrow and from top to bottom. Representative soil samples from both stockpiles and windrows should initially be quite large, and after thorough mixing should be reduced to desired size by quartering.

Riprap Samples

Riprap sampling must be carefully performed because riprap has an important function in the operation of hydraulic structures and riprap production and placement constitutes a relatively costly construction operation. The sample size should be at least 600 lb. (272 kg) and represent proportionally the quality range from poor to medium to best as found at the source. If the material quality is quite variable, it may be preferable to obtain three samples which represent respectively the poorest, median, and best quality material available. The minimum size of individual fragments selected should be ½ cu. ft. in volume (14 dm^3), if possible. An estimate of the relative percentages of each material quality should be made and included as information relating to the source. Samples from undeveloped sources must be very carefully chosen so that the material selected will, as far as possible, be typical of the deposit and include any significant rock-type variations.

Representative samples may be difficult to obtain. Overburden may limit the area from which material can be taken and obscure the true character of a large part of the deposit. Surface outcrops will often be more weathered than the interior of the deposit. Samples obtained from loose rock fragments on the ground or collected from weathered outer surfaces of rock outcrops are seldom representative. Fresher material may be obtained by breaking away the outer surfaces or by trenching, blasting, or core drilling. In stratified deposits such as limestones or sandstones, vertical and horizontal uniformity must be evaluated as strata often differ in character and quality.

Testing for quality evaluation includes detailed petrographic examination and physical properties tests (Reference 10, Designation E-39). Physical properties tests include specific gravity and absorption; and sodium sulfate soundness, and Los Angeles abrasion, which give an indication of the relative hardness or toughness and handling breakdown of rock fragments.

Logs Of Explorations

The logs of exploration for test holes and excavations are written records of the data on materials and conditions encountered in the exploration. They provide fundamental facts upon which many subsequent conclusions are based, such as the need for additional exploration or testing, structure feasibility, embankment zoning, construction cost and methods, and instrumentation for structure performance. The information on a log may be used over a period of many years, not only for the engineering aspects of dam design but also the log may form a part of contract documents or be required as basic evidence in court in case of dispute. Each log must therefore be factual, accurate, clear, and complete. It should not be misleading. Examples of log forms are shown in Fig. 3-6-1 for the log of a test pit or auger hole, Fig. 3-6-2 for the geologic log of a drill hole. The geologic log would be encountered when using required foundation excavation for construction material, in quarry explorations, or when geologic investigations fall in a material source. Logs of trenches are best presented on a drawing and would contain the same information as the log form, except that the material strata variations can be shown.

The headings on the forms provide spaces for supplying identifying information as to project, feature, hole number, location, elevation, dates started and completed, and the name of the person responsible. Summary data such as depth to bedrock and to water table are useful. All of this information is important; any omissions should be justified. The body of the log form is divided into a series of columns covering the various kinds of information required according to the type of exploratory hole.

A log should always contain information on the size of the hole and on the type of equipment used for boring or excavating the hole. This should include the kind of drilling bit used in drill holes and a description of the excavating equipment or type of auger used. The location from which samples are collected should be indicated on the logs, and the amount of material recovered as core should be expressed as a percentage of each length of penetration of the barrel. The logs should also show the extent and the method of support used as the hole is deepened, such as size and depth of casing, location and extent of grouting used, type of drilling mud, or type of cribbing in test pits.

Information on the presence or absence of water levels and comments on the reliability of these data should be given on all logs. The date measurements made should be recorded, since water levels fluctuate seasonally. Water levels should be recorded periodically from the time water is first encountered and as the test hole is deepened. Perched water tables and water under artesian pressure are important to note. Since it may be desirable to obtain periodic records of water level fluctuations in drilled holes, it should be determined whether this is required before abandoning and plugging the exploratory hole. Soils that are potential sources of borrow material should have their natural moisture condition reported. Very dry borrow materials require the addition of large amounts of moisture for compaction control, and very wet soils containing appreciable fines may require extensive processing in order to be usable. For simplicity the natural water content of borrow soils should be reported as either dry, moist, or wet. A soil that is reported dry should be one that will definitely require the addition of moisture in order to compact it properly in a fill. A soil reported as being moist is reasonably close to the Proctor optimum water content. Soils that are reported as wet should be obviously well beyond the optimum water content. Borrow pit holes are logged to indicate divisions between soils of different classification groups. However, within the same soil group significant changes in moisture should be logged.

It is important to determine the percentage by volume of cobbles and boulders encountered in explorations for sources of embankment materials. The log form for test pit or auger hole includes a method for obtaining the percentage by volume of 3 to 5-in. (76.2 to 127-mm) rock and rock over 5 in. (127 mm) in size. The method involves weighing the rock, converting weight to solid volume of rock, and measuring the volume of hole containing the rock. This determination can be made either on the total volume of stratum excavated or on a representative portion of the stratum by means of a sampling trench.

Explorations for borrow material should include a statement under "Remarks" on the log giving the reason for stopping the hole. For all other types of holes, a statement should be made at the end of the log that the work was completed as required, or a statement explaining why the hole was abandoned. Material should not be described as bedrock, ledge rock, slide material, or similar interpretative terminology unless the exploration actually

172 HANDBOOK OF DAM ENGINEERING

Figure 3-6-2. Geologic log of drill hole.

penetrated such conditions and samples were collected to substantiate these conclusions. Terminating statements similar to the following would be considered satisfactory: Hole eliminated due to lack of funds, hole caved in, depth limited by capacity of equipment, encountered water, unable to penetrate hard material in bottom of hole.

Soil and rock descriptions should follow the classification systems discussed previously. The use of a general type name such as loess, caliche, adobe, in addition to the soil classification name, may be helpful in identifying in-place conditions.

Potential quarry sites for concrete aggregate, riprap, or rockfill should have the following items reported:

a. Ownership.
b. Location, indicated by map, with reference to section, township, and range.
c. General description.
d. Geologic type and classification.
e. Joint spacing and fracture systems.
f. Bedding and planes of stratification.
g. Manner and sizes in which rock may break on blasting as affected by jointing, bedding, or internal stresses.
h. Shape and angularity of rock fragments.
i. Hardness and density of rock.
j. Degree of weathering.
k. Any abnormal properties or conditions not covered above.
l. Thickness, extent, estimated volume, and average depth of deposit—type, extent, and thickness of overburden.
m. Accessibility (roads affording access to highways or railroad, giving distance, load limitations, required maintenance, privately owned, and other pertinent information).
n. Photographs and any other information which may be useful or necessary.

If commercial quarry deposits are considered, the following additional information should be obtained:

a. Name and address of plant operator—if quarry is not in operation, a statement relative to ownership or control.
b. Location of plant and quarry.
c. Age of plant (if inactive, approximate date when operations ceased).
d. Transportation facilities and difficulties.
e. Deposit extent, plant, and stockpile capacity.
f. Plant description (type and condition of equipment for excavation, transporting, crushing, classifying, loading, and restrictions, if any).
g. Approximate percentages of various sizes of material produced by the plant.
h. Location of scales for weighing shipments.
i. Approximate prices of materials at the plant.
j. Principal users of plant output.
k. Service history of material produced.
l. Any other pertinent information.

When rock deposits other than quarries are considered for riprap use, the rock properties and deposit should be described in the same manner as for quarry rock where applicable and, in addition, the deposit description should indicate shape, average size, and variation in the size of the rock.

3-7. CONCRETE AGGREGATES

A durable concrete is one which will withstand the effects of service conditions to which it is subjected such as weathering, chemical action, and wear. The durability is influenced by the physical nature of the component parts, and although performance is largely influenced by mix proportions and degree of compaction the aggregates constitute nearly 85 percent of the constituents in mass concrete and good aggregates are essential for durable concrete.

Economy of construction usually dictates that the aggregate sources for concrete dam construction be within reasonable haul distance from the project. Therefore, the highest quality aggregate may not always be the source selected. Laboratory tests of the physical properties of the materials permit adapting design and materials handling and construction procedures to make the most advantageous use of the properties of the materials. However, the aggregate that can be delivered to the jobsite for the least cost may not be the most economical material. Cost of additional processing and additional cement to produce concrete of suitable quality may exceed the cost of obtaining aggregate from a more distant source.

Strength

Compressive strength is one measure of the quality of concrete. To give resistance to failure through the aggregate by crushing, the aggregate should be hard and tough, which usually implies a high specific gravity. The Los Angeles abrasion test and the sodium or magnesium sulfate soundness tests are used to give indications of the relative hardness or toughness of the aggregates.[12] They also give an indication of the breakdown that may occur during stockpiling, handling, and mixing.

The sulfate soundness test is sensitive to variations in temperature, concentration of the sulfate solutions, temperature of the oven while drying to a constant weight, and techniques of the operator. Also the performance of aggregates immersed in a solution such as in the sulfate soundness test may not strictly correlate with the performance of these same aggregates when surrounded by cement paste. However, aggregates with more than 10 percent weight loss after 5 cycles when sodium sulfate is used or 15 percent when magnesium sulfate is used should be considered of marginal quality. In a similar manner the Los Angeles abrasion test, although not definitive in itself, does show a relationship between the quality of the aggregate and the compressive strength of the concrete. Aggregates showing more than 10 percent weight loss after 100 revolutions or more than 40 percent weight loss after 500 revolutions are generally considered to be of questionable quality.

Workability

Workability of concrete is dependent upon the shape and size of the aggregate particles as well as the proportions of the size fractions. A highly angular, coarse aggregate will usually require more sand, cement, and water to produce a workable, easily placed concrete than will aggregates that are smooth and rounded. Flat and elongated particles tend to tear the matrix during finishing operations.

The sand characteristics affect the workability to an even greater degree than do those of the coarse aggregate. Uniform grading from coarse to fine is important to good workability. Coarse sands in concrete result in harshness, bleeding and segregation and tend to require more cement to compensate for deficiencies in fine sand sizes, while an excess in such fines results in high water requirements. Very coarse sands may be blended with fine sand from windblown deposits to improve workability. For good concrete workability the fineness modulus of the sand should not be less than 2.3 nor more than 3.1.

With the importance of uniformity and proper proportioning on production of workable and placeable concrete, it is necessary that the aggregate be delivered to the batch plant in the desired condition. Final screening at the batch plant will usually be profitable in avoiding problems in poor workability that can result from concentrations of undersize and oversize materials.

To utilize the greatest practicable yield of suitable materials in a deposit, crushing of oversize material and excesses in any size fraction are usually permitted to make up deficiencies in the grading. Crusher fines that will pass a $\frac{3}{16}$-in. (5 mm) screen should be wasted. Crushed sand, is used to make up deficiencies in the natural sand grading, should be produced by a suitable ball or rod mill, disk, or cone crusher or other suitable equipment that will manufacture sand predominantly cubical in shape and free from objectionable quantities of flat and elongated particles. The crushed sand and coarse aggregate should be blended uniformly with the natural sand and coarse aggregate in order to maintain uniformity to the concrete.

Contaminating Substances

Aggregate is commonly contaminated by silt, clay, shale, mica, coal, wood fragments, other organic matter, chemical salts, and surface coatings and encrustations. Such contaminating substances in concrete combine in a variety of ways to cause unsoundness, decreased strength and durability, and unsightly appearance as well as making mixing and placing more difficult. Fine materials such as silt and clay increase the mixing water requirement and tend to work to the surface of the concrete, resulting in a higher incidence of cracking and checking. Organic materials and coatings on particles may inhibit development of maximum bond between hydrated cement and aggregate, may affect the normal hydration of the cement, or react chemically with the cement constituents. Coal and fossilized wood can cause weak zones and local stains and surface blemishes.

Clays are subject to swelling and shrinkage when subjected to cycles of wetting and drying. When clay seams occur in rocks used for coarse aggregate, poor performance frequently occurs because of the swelling pressures. The disruptive forces are further aggravated by freezing and thawing. Clays occurring as part of sand particles have been known to contribute to severe slump loss during placement and generally poor performance in service.

The particular application of the concrete will determine to what extent contaminating substances can be tolerated. Higher percentages of deleterious and contaminating substances can be tolerated in interior concrete of massive dams than can be permitted in surface concrete subjected to freezing and thawing, face concrete in zones of fluctuating water levels, or in high-strength concrete required in thin arch structures. Where it is necessary to prevent adversely affecting the strength, durability, or appearance of the concrete from the effects of contaminants, the following will serve as a guide:

Contaminating substances	MAXIMUM CONTENT, PERCENT BY WEIGHT	
	Fine aggregate	Coarse aggregate
Coal	1.0	1.0
Clay lumps	1.0	0.25
Shale	1.0	1.0
Material passing 200 screen	3.0	1.0
Total of other substances such as alkali, mica, coated grains, soft flaky particles and loam	2.0	1.0
Sum of all contaminating substances	5.0	3.0

Soundness

Soundness is usually defined as the ability of an aggregate to resist large or permanent changes of volume when subjected to freezing and thawing, heating and cooling, or wetting and drying. Aggregates that can absorb water so as to become critically saturated are potentially vulnerable to freezing. Rock particles that are physically weak, easily cleavable, or which swell when saturated are susceptible to breakdown through exposure to natural weathering processes. The use of such materials in concrete increases water requirement, reduces strength, or leads to premature deterioration by weakening the bond between cement paste and aggregate or by inducing cracking, spalling, or popouts. Shales, friable sandstones, some micaceous rocks, and various cherts are frequently physically unsound materials.

The freezing and thawing test on concrete requires considerable equipment and time to conduct, but there is evidence that it provides the best indicator of an aggregate to perform.

Since there are many variables such as the amount of air in the concrete, the moisture condition at time of test, and the age of the concrete at time of test, it has not been possible to establish specification limits on the soundness of the aggregates based on the results of the freezing and thawing test. However, air entrained concrete that will not withstand 500 cycles of freezing and thawing in water usually shows poor performance under freezing exposures.

Chemical Suitability

Some aggregate materials undergo chemical changes which are injurious to concrete. Such reactions may be of various kinds, including reaction between aggregate material and the constituents of cement, solution of soluble materials, oxidation by weathering, and complicated processes which impede the normal hydration of cement.

Reaction between certain aggregate materials and the alkalies in cement is associated with expansion, cracking, and deterioration of concrete. The expansion is caused by osmotic swelling of alkali silica gels, which are produced by chemical interaction of alkalies released by hydration of the cement and the siliceous rocks and minerals of the aggregates. Rocks and minerals known to react deleteriously with cement alkalies are volcanic rocks of medium to high silica content; silicate glasses (artificial or natural, excluding the basic type such as basaltic glass); opaline and chalcedonic rocks (including most cherts and flints); some phyllites, tridymite, and certain zeolites. In general, aggregates which are petrographically similar to known reactive types, or which on the basis of service history or laboratory experiment are suspected of reactive tendencies, should be used only with cement which is low in alkalies or with the addition of pozzolanic materials that have shown ability to prevent harmful expansion due to alkali-aggregate reaction.

BENEFICIATION

Where suitable aggregates are not readily available the quality can sometimes be improved by one of the various means of beneficiation which can reduce excessive amounts of porous, lightweight, or deleterious particles. Three systems that have been developed for improving aggregate quality are hydraulic jigging, heavy media separation, and elastic fractionation.

The hydraulic jig system employs vertical pulsations of water through a layer of aggregate in a nearly horizontal flat bed. Deleterious particles that are of low specific gravity migrate to the surface during upward pulsations of water through the aggregate and can be scalped over a weir and discarded.

Heavy media separation also works on the principle of a difference in specific gravity between the sound and the unsound particles. In this process the aggregate is processed through a heavy liquid bath of water, finely ground ferrosilicon and magnetite mixed to a specific gravity of about 2.5. The heavy aggregate particles sink in the fluid and are removed from the bottom of the chamber by augers or other lifters. The lightweight particles float in the medium and are removed over a weir. Proper operation of the separation usually restricts the operation to separation of a single size fraction of aggregate at one time.

Elastic fractionation operates on the principle that hard, dense aggregates usually have a higher modulus of elasticity than soft porous particles and will therefore rebound farther when bounced off an inclined hard steel plate. The aggregates to be processed are elevated to an overhead hopper. The hopper discharges into a vibratory feeder which controls the feed and drops the aggregate onto the inclined hardened steel plate. The dense elastic particles rebound a greater distance than the soft, porous particles and are collected in bins located at distances from the plate to collect the aggregate of the desired quality. A variation of this method uses a revolving steel drum in place of an inclined plate. The particles are dropped on the revolving drum a distance in advance of the top centerline of the drum. Hard elastic aggregate particles bounce opposite to the direction of rotation while soft particles receive a velocity in the direction of the rotation of the drum.

All of the above methods of beneficiation have limited rates of operation which tend to restrict their use in aggregate processing for dam construction.

Aggregate Gradation

Aggregate gradation has a definite effect on the workability of the concrete. However, suitable concrete for dams can usually be made with aggregate having a wide range of gradations. Even gap-graded aggregates, where certain size fractions of the aggregate are left out completely, can produce high-quality concretes. Gap gradings can be used economically where deficiencies in certain sizes occur naturally in a deposit but there appears to be no advantage to purposefully gap grading an aggregate where a source contains continuously graded material.

It is more important to maintain a constancy in the grading once a grading has been established than in striving for the optimum grading. As far as workability of the concrete permits, it is economically desirable to deviate from an ideal gradation to a gradation that permits maximum utilization of a deposit with minimum wasting and minimum crushing to make up deficiencies.

Theoretically, the larger the maximum aggregate size the less the cement that will be required in a given volume of concrete to achieve a desired quality. Plant considerations for processing, batching, mixing, transporting, placing and consolidating generally restrict the maximum practical size of coarse aggregate to 6 in. (150 mm) in size. Deficiencies in the larger size fractions in some deposits frequently require reducing the amount of 6-in. (150 mm) size fraction to less than the theoretical preferred amount. Forms and reinforcing steel usually do not restrict the maximum size of coarse aggregate in mass concrete structures, but in the areas of the structure where they do form a constraint the maximum size of aggregate should not be larger than one-fourth the dimension between sides of forms nor larger than two-thirds of the minimum clear spacing between reinforcing bars.

It has been demonstrated[13] that to achieve the maximum cement efficiency there is an optimum maximum size of aggregate for each compressive strength level to be obtained with a given aggregate and cement. Because of this, thin arch dams with relatively high design strength requirements tend toward a smaller maximum size of coarse aggregate.

The American Concrete Institute Committee of Mass Concrete has published [14] the following tables as generally practiced grading limits for fine and coarse aggregates. Mix design investigations may indicate that deviations outside of these limits could be economically advantageous in individual instances.

Production Control Testing

Periodic tests of aggregate materials should be made during the progress of construction to determine specific gravity, moisture content of the aggregate, and to determine the relative percentages of the different size fractions. Specific gravity is usually a quick indicator of

TABLE 3-7-1. Grading Limits for Fine Aggregates

	PERCENTAGE RETAINED	
Sieve designation	Individual—by weight	Cumulative—by weight
$\frac{3}{8}$ in. (9.53 mm)	0	0
No. 4 (4.76 mm)	0–8	0–8
No. 8 (2.38 mm)	5–20	10–25
No. 16 (1.19 mm)	10–25	30–50
No. 30 (0.60 mm)	10–30	50–65
No. 50 (0.30 mm)	15–30	70–83
No. 100 (0.15 mm)	12–20	90–97
Pan fraction	3–10	100

TABLE 3-7-2. Grading Limits for Coarse Aggregates

Maximum size aggregate in concrete, in.	PERCENTAGE OF COARSE AGGRETAGE FRACTIONS (CLEAN SEPARATION)					
	Cobbles 3"–6"	Coarse gravel $1\frac{1}{2}"$ to 3"	Medium gravel $\frac{3}{4}"$ to $1\frac{1}{2}"$	FINE GRAVEL		
				$\frac{3}{16}"$ (#4) to $\frac{3}{4}"$	$*$ $\frac{3}{8}"$ to $\frac{3}{4}"$	$\frac{3}{16}"$ $*$ (#4) to $\frac{3}{8}"$
mm	(76–152 mm)	(38–76 mm)	(19–38 mm)	(5–19 mm)	(10–19 mm)	(5–10 mm)
$\frac{3}{4}$ (19 mm)	0	0	0	100	55–73	27–45
$1\frac{1}{2}$ (38 mm)	0	0	40–55	45–60	30–35	15–25
3 (76 mm)	0	20–40	20–40	25–40	15–25	10–15
6 (152 mm)	20–35	20–32	20–30	20–35	12–20	8–15

*These columns used only when fine gravel is separated into two sizes.

change in aggregate quality. Low specific gravity frequently indicates porous, weak, or absorptive material. Aggregates having a specific gravity less than 2.5 are generally considered not suitable for use in mass concrete. Deviations in aggregate gradation, that are not quickly corrected, may affect the workability of the mix and require an adjustment in the water-to-cement ratio or the aggregate-to-cement ratio to provide concrete of adequate quality.

Fluctuations of the moisture content, particulary in the sand, create serious problems in producing uniform concrete. A change of 1 percent in the moisture content of the sand that is not corrected at the mixer can change the water-cement ratio by two hundredths or cause a 1-in. (2.5 cm) variation in slump. Multiple stockpiles that permit 24 hr of free drainage will usually stabilize moisture content to a degree that adequate control can be maintained.

3-8 MATERIALS FOR EMBANKMENTS

Most soils are suitable for use in embankments. Some soils are more suitable than others, but the term "unsuitable" is limited to those materials which will deteriorate in the fill such as soils containing large quantities of soluble salts or organic matter. The successful utilization of soils in rolled earth dams depends upon careful inventory of all materials available, so that the designer may select the most efficient material configuration for his design. To this end the inventory of materials for embankment construction involves:

1. Locating the deposits.
2. Identifying the deposits as to type of material.
3. Determining the quantity available.
4. Showing the significant features of occurrence.
5. Sampling and testing to determine engineering properties.

The investigation may be summarized by the equation: *Inventory* carefully all construction materials to be considered as borrow + *Report* clearly and fully the significant features of occurrence and properties = (*Adequate* + *Economical*) design. Chapters 6 and 7 describe the design of earth and rockfill dams. However, if the investigator keeps in mind the elements which make up a design, his exploration for materials will be greatly aided.

Design Considerations

The main design factors are:

1. Permeability
2. Stability
3. Compression and shrinkage

4. Piping and washing out of fines
5. Economical utilization of materials

Permeability. Materials can be grouped as being impervious or pervious according to the degree of watertightness they will attain when compacted in the embankment. The terms "pervious" and "impervious" are relative. Soils having a coefficient of permeability of less than a foot per year (10^{-6} cm per sec) are considered impervious, while soils with a coefficient greater than 100 ft per year (10^{-4} cm per sec) are pervious. The combination of these materials used in an embankment must keep seepage within tolerable limits. Both types in varying amounts are always needed.

An important point to bear in mind is that the properties of the soil may change when the soil is compacted in the embankment. A seemingly pervious in situ gravelly or sandy soil with some fines can under compactive effort become very impervious, especially when the coarse grains are well-graded. Weathered formation which excavates in large chunks may, under tamping rollers, break down to form an excellent water barrier. To get imperviousness, adequate compaction and proper gradation (minimum limit on the amount of fines and a maximum limit on the amount of coarse grains) are required. Coarser grained soils when devoid of fines are pervious, easy to compact, and little affected by moisture. Natural impervious blankets upstream of the dam should not be removed if a positive cutoff under the dam is not obtainable.

Stability. The ability of enbankment materials when compacted in an embankment to resist shear forces is a measure of the stability of an earth dam. The shear strength of embankment soils depends upon many factors. Some of them are grain size, gradation, grain shape, and drainability when loaded. In general, the larger the grains and the more well graded the grains are, the better the material's shear resistance will be. Soils high in clay and silt content usually develop high pore pressures when loaded. The pore pressure acts to reduce the embankment's resistance to shear forces. A coarse-grained material containing sufficient amounts of clay and silt to assure reasonable imperviousness is preferred over soils high in percentage of fines.

Compression and Shrinkage. Soils high in clay content will compress under the load of subsequently placed layers of embankment and tend to develop high pore pressures. These soils will also be of lower density than well-graded soils. The combination of lower density and pore pressure buildup in soils is not desirable, but such materials can be used in an appropriate design. Materials high in clay content will shrink upon drying and crack. When such materials are used in an embankment they will dry out and crack during extended periods of low reservoir levels if they are not provided with cover material. During construction shutdowns, such portions may also experience this cracking. Specifications should provide for key trenches and reworking of the upper layers when construction resumes.

Piping and Washing of Fines. Normal wave action will create velocities high enough to carry away particles of gravel size. In order to prevent erosion of the water barrier or core, a filter arrangement consisting of a layer of sand and gravel (bedding) against the earthfill and a layer of riprap over the bedding is used. The blanket of riprap for slope protection is generally of large, durable, well-graded angular rock. The bedding material should have gravel-size particles which approximate the size of the voids in the riprap blanket and should also contain a sufficient amount of finer particles to prevent passage of earthfill through the bedding and riprap. Such a material arrangement grades the materials toward the reservoir, from the finest to the coarsest. In areas where rock is plentiful an upstream graded rockfill zone will also serve as erosion protection. The downstream slope will require material to protect it from erosion due to surface runoff.

Where rock fragments are used to provide downstream slope protection, almost any rock that does not break down on exposure to air or water is acceptable. Size is not critical except the fragments should be at least gravel size, and the maximum size is controlled by the blanket thickness. Rounded gravel is acceptable, and special investigations for securing blanket material from appreciable distances are justified only when rock is absent or sod blankets are not feasible. In some situations where a positive cutoff in the foundation is not provided, a filter material arrangement similar to the upstream slope treatment may be required at the downstream toe to prevent piping from foundation seepage.

Riprap is also used below stilling basins and other energy-dissipating structures where high velocity turbulent flow is reduced to non-eroding velocities to prevent damage to downstream channels. The quantity required is usually small, but the quality requirements may exceed those of dam riprap. In some instances, it will be necessary to secure high-quality riprap for protection of the structures while lower quality material may be used on the dam.

The quality and size of fragments required will depend on:

1. Freezing and thawing.
2. Wetting and drying.
3. Severity of wave action.

Economical Utilization of Available Material. It so often happens that the ideal material for each zone of an embankment is seldom found close to the site. However, most any large deposit of a soil can usually be utilized in some portion of the embankment. Areas of anticipated required excavation for a cutoff trench, appurtenant structures, and channels must not be overlooked as potential material sources. Required rock excavation should be considered as a source of riprap or rockfill. Most important is that all types of apparently suitable material in deposits large enough to be considered as borrow areas should be carefully examined and fully reported.

Construction Considerations

The more important construction factors are:

1. Accessibility
2. Natural moisture
3. Workability

Accessibility. Accessibility and type of transportation facilities available have an important bearing on the desirability of a material source. Locating suitable sources of materials within a practicable hauling distance is most important and cannot be overemphasized. A haul distance of less than 2 mi (3.218 km) is desirable, but sometimes longer hauls may be required when circumstances warrant the extra cost. The accessibility of a deposit must be considered in conjunction with haul distance. Sometimes a more distant borrow area that is close to a road or at a location where a road to the dam site can easily be built may be selected in preference to a less accessible one closer to the dam site. Elevation of borrow area with respect to the dam is important. Hauling material up or down great distances is costly. Some seemingly inaccessible areas may be adaptable to conveyor or train systems, but bear in mind that these systems are costly and are generally only economical where large quantities of materials are to be handled.

Often it is desirable to locate borrow areas in the reservoir area. In this way an acquired portion of land does double duty, i.e., serves as a borrow area and later as a reservoir area. However, it is not desirable to locate all areas at low elevations where high water due to stream diversion or natural flood may inundate the area and stop work. Such areas should be designated for pre-diversion or pre-flood season use.

Natural Moisture. The natural moisture content of a borrow area may be a factor in its choice or rejection. If an area is extremely dry, its adaptability to irrigation and its location with respect to a source of water must be considered. Depth of water table and its seasonal fluctuation is another factor. The possibility of drainage should be considered, however, drainage is often costly and not always successful. In fine-grained soils you may not be able to excavate closer to the water table than 5 to 6 ft (1.5 to 1.8 m). There have been instances where heavy hauling equipment traveling the borrow area pumped water up from the water table and turned the area into mud. Likewise it has also happened that the contractor could take a shallow cut, allow the base to dry, and then take another shallow cut. Both instances are dependent upon the local conditions.

Workability. The workability of a material is often the deciding factor in the choice between otherwise suitable materials. By workability we mean the ease, from an economic standpoint, with which satisfactory engineering soil properties can be achieved of a particular soil. The cost of procuring a unit volume of soil and placing it in a structure varies widely not only according to the soil type, but also according to the size of the structure, the kind of equipment available, and the labor market. A soil can be made workable by irrigating, draining, mixing with other soil, and separation. Each one of these methods increases the unit cost over that unit cost of merely excavating and placing an ideal soil. A very wet clay of high plasticity, for example, cannot be drained practicably, is impossible to compact properly, and should be avoided. A material containing a large percentage of plus 5-in. rock could be separated, but if sufficient rock can be obtained more cheaply elsewhere and earthfill is also available, this source would be excluded in the final design. Pre-irrigation may make excavation easier and be cheaper than excavating dry and adding water later.

There are no test procedures by which the property of workability may be given a quantitative value. Rather, it is necessary that all pertinent information concerning a soil or a borrow pit be tabulated so that the various design possibilities can be evaluated in the light of the current economic conditions when the final design is prepared. A complete inventory of potential borrow areas is required in order to make a proper selection of borrow areas.

Materials With Special Properties

There are frequently situations where explorations for special materials with specific properties are required. The requirement for such materials may be because of design features dictated by geological or topographical conditions, or deficiencies in locally occurring materials. The quantity of special materials is usually small compared with the total embankment volume. Often these materials are obtained from sources at considerable distances from the site. Special materials include fine-grained soils for impervious linings or blankets; sand and gravel for filter blankets, soil-cement, drains, road surfacing, and erosion protection; and rock fragments for riprap, rockfill, filter blankets, and drains.

It is usually not economically feasible to secure material with the desired ideal characteristics. The extent to which the desirable characteristics are necessary varies with the purpose for the material. For example, in drain blankets quantity may be substituted for quality to some extent. Most often, economic considerations play a large role in the selection of material sources. Special processing of close sources may be more economical than long hauls of satisfactory pit run material. There must be a definite improvement in the quality of a material with increasing distance from the site of utilization to justify long hauls.

Impervious Materials. Impervious soils may be required for blanket and lining material when small storage dams and diversion dams are constructed on pervious foundations, or when terminal equalizing or regulating reservoirs have extensive permeable beds in their foundations. Such material should, when compacted, have a permeability that is low compared to that of the foundation. Highly plastic clay materials are seldom necessary or desirable. A large reduction of water loss, such as for canal linings, is generally not the intent for the blanket, but rather to reduce seepage gradients to prevent piping at the toe of the dam. When hydraulic gradients are high, gradation requirements will be necessary to prevent piping of the blanket or lining into the foundation. Blending of different soils may be practicable and can provide a superior blanketing material. Material exposed to alternately wet and dry conditions and in some cases to freezing and thawing should be free of shrinkage and swelling characteristics.

Pervious Materials. Pervious materials are required for filters and drains associated with the construction of concrete structures, for blankets under riprap, for use in drainage zones within an embankment, for drainage blankets under the downstream toes of earth dams, for transitional materials to prevent piping, and for road surfacing. These applications require that a substantial amount of gravel-size particles be included in the material.

While the quantity requirements of pervious material for filters and drains are usually small, the quality must be high. The principal utilization of this material is for the prevention of hydraulic uplift. Therefore, the materials must be very free-draining, and, at the same time, comparatively high heads must be dissipated without movement of either the filter material or foundation soil. A single layer of material may often be inadequate to meet the filtering requirements and a two-layer filter blanket will be required. Fine sand, silt, or clay is objectionable and processing by washing, or screening, is required to produce acceptable material from most natural deposits. Although gradation requirements will be different, materials for filters are commonly secured economically from sources acceptable for concrete aggregate. Neither particle shape nor nature of the minerals contained in the pervious material is very critical, and processed concrete aggregates rejected on these accounts can be used for the construction of drainage blankets and drains if suitable gradation is obtained.

For sand and gravel blanket material under riprap, the primary requirement is coarseness. Because of this requirement, blanket material for this purpose is often secured from the rock fines developed in quarrying operations. However, if a deposit of coarse gravel can be found within a reasonable distance from the dam site it will usually prove to be economical to develop such a source. Quantity requirements will be quite large and special processing by screening or other means will be costly. The principal purpose of this type of blanket is to prevent waves that penetrate the riprap from eroding the underlying embankment. A limited amount of fine material is not objectionable even though some of it will no doubt be lost through wave action. The material should be durable. Most gravel deposits are adequate; however, some deposits have been found which contain quantities of unsuitable material. Such deposits include ancient gravel beds that have deteriorated by weathering and talus or slope wash deposits where water action has been insufficient to remove the soft rock.

The materials used for drainage blankets under the downstream toes of earth dams should be pervious with respect to both the embankment and the foundation soils. The coefficient of permeability of such materials may therefore be quite low. Particle size is relatively unimportant and the material may contain large quantities of coarse to fine sand. It should not, however, contain appreciable quantities of silt or clay. Volume of blanket material is to a large extent a substitute for quality and if an ample supply can be found close to the point of use, a search for better materials at greater distances is seldom warranted.

Materials for road and access area surfacing or base courses are sought primarily for strength and durability. The preferred material for surfacing will consist primarily of medium to fine gravel with enough clay to bind the material together and relatively small amounts of silt and fine sand. Similar material without silt or clay is preferred for base courses.

Riprap and Rockfill. Rock fragments are required in connection with earthwork structures for the protection of earth embankments or exposed excavations from the action of water either as waves, turbulent flow, or heavy rainfall. Rock fragments associated with protection from wave action or flowing water are designated as riprap. The term "rockfill" is commonly applied to the more massive bodies of fill in the dam embankments consisting of rock fragments, which are primarily to provide structural stability. Rockfill is also used in drainage blankets, blankets under riprap, and blankets for the protection of embankments from the erosive action of rainfall. If the rockfill material used for the shell is adequate for slope protection, a separate need for riprap may be eliminated.

The size of rock fragments for use as riprap is very important. Fragments up to 1 cu yd (0.76 m^3) in volume are required in dams associated with large reservoirs and in even the smallest earth dams with small reservoirs, fragment sizes up to one-fourth cu yd are desirable. Riprap blankets are also used below spillway and outlet works stilling basins and other energy-dissipating structures where high-velocity turbulent flow is reduced to non-eroding velocities. The quantity required for the basins is small but quality requirements may exceed those of reservoir riprap blankets.

Table 3-8-1 may be used as a guide for selecting a rock source for riprap. The riprap should be reasonably well graded in size within the limits shown.

While joint-bedding systems do have much influence on the maximum sizes which can be obtained from a quarry, it should also be recognized that quality takes precedence over quantity and control must be exercised over the quarry blasting operations. In some instances, even though marginal material may be used on the dam, it will be necessary to secure a high-quality riprap blanket for the protection of the appurtenant structures. Where investigations disclose only moderate quality rock in the vicinity of the site, investigations should be extended to establish the location of a high-quality source.

When rockfill is used to provide surface protection from rainfall almost any rock fragments that do not break down on exposure to air or water are acceptable as rockfill. Shale and some siltstones are about the only types of rocks considered unacceptable. Size is not a critical criterion, except that fragments should be at least gravel size and upper limit is controlled by the specified thickness of the blanket. Rounded gravel if readily available is frequently used for this purpose. Substitutes include sod blankets, which can be produced cheaply. The sod blankets will require locating topsoil which is generally available close to the dam site if climatic conditions are satisfactory for growing sod.

TABLE 3-8-1. Riprap Gradation

Nominal thickness of riprap, ins.	(cm)	Maximum size, lb	(kg)	GRADATION, PERCENT BY WEIGHT SIZE OF ROCK FRAGMENTS			
				$\frac{1}{2}$ cu yd to 1 cu yd (0.4 to 0.8 m)	$\frac{1}{2}$ cu ft to $\frac{1}{2}$ cu yd (0.014 to 0.4 m)	Less than $\frac{1}{2}$ cu ft (<0.014 m)	Sand and rock dust
18	(46)	1,000	(450)	–	More than 75	Only enough to fill voids between larger rock	Less than 5
24	(61)	1,500	(680)	–	More than 75		Less than 5
30	(76)	2,500	(1,130)	–	More than 75		Less than 5
36	(91)	5,000	(2,270)	40–50	50–60		Less than 5

Soil Cement. In areas where rock suitable for riprap is not available, or available only at high cost, soil cement can provide an alternate for slope protection. Soil cement may be competitive with rock when the haul distance to a suitable rock source exceeds about 20 mi and soil suitable for soil cement is available at a short haul distance. Although almost any material can be used to produce soil cement, certain materials are more desirable from the standpoint of the amount of cement required for durability and compressive strength, and efficiency in processing and placing the materials. Any soil proposed for use in soil cement is tested by a standard series of laboratory tests to determine cement and processing requirements, and expected in-service performance.

A few guidelines can be established for the more desirable gradations of naturally occurring or processed material, based on actual slope protection applications.[15,16] For maximum economy and most efficient construction, the recommended soil contains no plus 2-in. (50.8 mm) material, at least 55 percent of the material passes the No. 4 sieve, and between

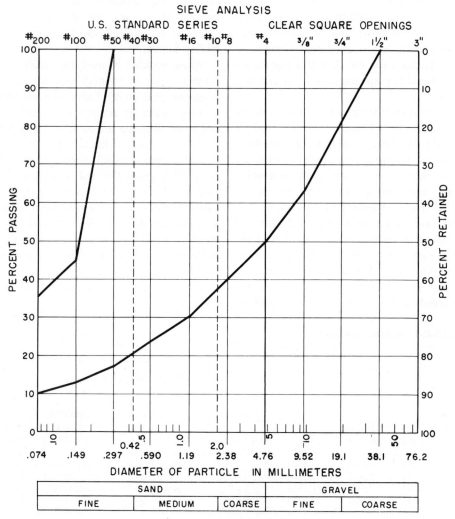

Figure 3-8-1. Gradation range of soils for soil-cement.

5 and 35 percent passes the No. 200 sieve. The gradation range of soils used in soil cement for embankment slope protection is shown on Fig. 3-8-1. Intermediate gradations should maintain a good distribution of particle sizes from the smallest to the largest. The cement volume requires decreases as the soil gradation becomes coarser; however, mixing and placing become more difficult. An additional consideration when selecting soil for soil cement is the presence of "clay balls" (rounded balls of fines and sand which do not break down during ordinary processing). Alluvial sand deposits which might otherwise be acceptable often contain layers of silt and clay. These layers are caused by low flows in the depositing stream and all alluvial deposits have this characteristic to a certain extent. "Clay balls" in the material tend to go through the processing intact and do not disperse through the sand. A small amount of minus 1-in. (25 mm) clay balls is not considered sufficient grounds to reject otherwise suitable material. The material can be screened to remove larger clay balls before the material is introduced into the mixing plant. During the investigations the material should be screened at its natural moisture to obtain an estimate of the amount of "clay balls" present in the deposit. The "clay balls" in an auger hole sample will normally be of rather small size due to the action of the auger. However, the presence of over about 10 percent "clay balls" even of the smaller sizes will probably indicate problems during construction. Excavation procedures will not usually break down the clay lenses and the result is large clods of clay within the soil stockpile.

3-9. ENVIRONMENTAL RESTRAINTS

In the United States, there are Federal, State, and local laws relating to environmental protection, which not only restrict the selection of concrete aggregate and embankment materials, but also the methods the contractor uses in his excavation, processing, and hauling operations. Final selection of material sources may be strongly influenced by the ability to conform economically to the restraints imposed by these regulations. Prevention of scarring or defacing of the natural surroundings, or the preservation of trees, natural shrubbery, and vegetation can eliminate some sources from utilization completely and permit only selective borrowing from other available sources. Many requirements include that upon completion of the work, the area must be smoothed and graded in a manner to conform to the natural appearance of the landscape and reseeded. In some areas, trees must be preserved and the location of haul roads is subject to approval. Borrow pits and quarry sites may be required to be brought to stable slopes with slope intersections rounded and shaped to present a natural appearance. Restoration of quarry sites to meet environmental restraints may be exceptionally difficult and costly. Blasting operations should be controlled to prevent scattering of rocks, stumps, or other debris outside the work area.

Turbidity control usually requires settling ponds or tanks and the use of flocculating chemicals for the effluent of all aggregate wash water. Turbidity restrictions can also make aggregate deposits in beds of flowing streams difficult to excavate. The restrictions vary with the stream and should be determined for each site.

Excavating, processing, and hauling equipment must have such devices as are reasonably available and utilize such practicable methods that will minimize atmospheric emissions or discharges of air contaminants. Equipment and vehicles that show excessive emissions of exhaust gases due to poor engine adjustments or other inefficient operating conditions may be denied permission to operate until corrective repairs or adjustments are made.

The emission of dust into the atmosphere is not permitted during the manufacture, handling, and transporting of concrete aggregates or embankment materials, and it is necessary to use such methods and equipment as are necessary for the collection and disposal or prevention of dust during these operations. All roads must be maintained dust-free by sprinkling or by applying oil and/or bituminous surfacing to the roads.

REFERENCES

1. United States Department of Agriculture, *Soil Survey Manual*, Handbook No. 18, 1951.
2. Neely, Doss and Abdel-Hady, Mohamed A., "Aerial Photos to Locate Granular Fill Material," *Journal of the Construction Division*, A.S.C.E., Paper No. 6166, October 1968.
3. Kiefer, Ralph W., "Airphoto Interpretation of Flood Plain Soils," *Journal of Surveying and Mapping Division*, A.S.C.E., Paper No. 6819, October 1969.
4. Lueder, Donald R., *Aerial Photographic Interpretation*, McGraw-Hill Book Co., Inc., New York, 1959.
5. Miller, Victor C., *Photogeology*, McGraw-Hill Book Co., Inc., New York, 1961.
6. American Society of Photogrammetry, *Manual of Photographic Interpretation*, 1960.
7. American Society of Photogrammetry, *Manual of Color Aerial Photography*, 1968.
8. Henbest, O. J., Erinaks, D. C., and Hixson, D. H., "Seismic and Resistivity Methods of Geological Exploration," Technical Release No. 44, Soil Conservation Service, Department of Agriculture, 1969.
9. Society of Exploration Geophysicists, *Seismic Refraction Prospecting*, Edited by A. W. Musgrave, 1967.
10. Bureau of Reclamation, *Earth Manual*, 2nd Edition, Department of the Interior, 1974.
11. Bureau of Reclamation, *Concrete Manual*, 8th Edition, Department of the Interior, 1974.
12. Annual Book of ASTM Standards, Part 10, "Concrete and Mineral Aggregates," 1972.
13. American Concrete Institute, "Symposium on Mass Concrete," Special Publication No. 6, 1963.
14. ACI Committee 207, "Mass Concrete for Dams and Other Massive Structures," *ACI Journal*, Proceedings, **67**, No. 4, April 1970.
15. DeGroot, Glenn, "Soil-Cement Slope Protection on Bureau of Reclamation Features," Report No. REC-ERC-71-20, Bureau of Reclamation, Department of the Interior, 1971.
16. Portland Cement Association, "Soil-Cement Slope Protection for Earth Dams," Skokie, Illinois.

4 Geologic, Foundation, and Seismicity Investigations

LAURENCE B. JAMES
Chief Engineering Geologist (retired)
California Department of Water Resources
Sacramento, California

JEROME S. NELSON
Chief Geophysicist
Harding-Lawson Associates
San Rafael, California

F. E. SAINSBURY
President, Boyles Bros. Drilling Co.
Salt Lake City, Utah

COLE R. McCLURE
Manager of Engineering Geology
Bechtel Incorporated
San Francisco, California

W. A. WAHLER
President, W. A. Wahler & Associates
Palo Alto, California

GEORGE B. WALLACE
Research Director
Spokane Mining Research Center
U. S. Bureau of Mines
Spokane, Washington

DR. GEORGE W. HOUSNER
C F Braun Professor of Civil Engineering
Division of Engineering and Applied Science
California Institute of Technology
Pasadena, California

FRED CREBBIN, IV
Manager of Mining Engineering
CH2M-Hill
Corvallis, Oregon

ARTHUR B. ARNOLD
Chief Geologist
Bechtel Incorporated
San Francisco, California

JOHN W. MARLETTE
Supervising Engineering Geologist
California Department of Water Resources
Sacramento, California

4.0 GEOLOGIC INVESTIGATION— INTRODUCTION BY LAURENCE B. JAMES

Whereas in the past the nature and extent of the stresses in the dams themselves were of primary importance, the intense study in recent years which has been made into these structures by mathematical analysis, detailed investigation by electronic computers and, not least, by the precise and brilliant studies by models, have reduced the problem so far as the structures are concerned to almost a precise science.

On the other hand, all this knowledge is of no avail if the foundations on which the dam is to rest are in doubt. The recent failures are all concerned with foundation movements or geological slips .... It is from the Geologist and the combined science of soil mechanics and rock mechanics that the engineer must obtain the detailed and scientific information he requires to confirm his own experience and judgment of the foundation on which rests the safety of his dam ....[1]

The foregoing statement by Mr. Guthrie Brown is as true now as when he included it in his presidential address to the 8th International Congress on Large Dams at Edinburgh, Scotland, in 1964. Furthermore, more recent events have emphasized the need to extend investigations for dams to include the reservoirs they impound and even to areas beyond. For example, in 1963 a landslide at Vaiont Reservoir, Italy, generated an immense water wave which overtopped the world's highest concrete arch dam.[2] This wave left its marks 780 ft (238 m) above reservoir level. Remarkably, the dam withstood the overpressures to which it was subjected; however, the wave continued downstream into the town of Longarrone leaving an estimated 2500 fatalities in its wake. Within 60 sec Vaiont Reservoir was choked with more than 312 million cu yd (239 million cu m) of earth and rock slide which extended from the toe of the dam upstream for a distance of about 1 mi (1.61 km).

The failure in 1963 of Baldwin Hills Reservoir located in Los Angeles County, California, provides an illustration wherein geologic factors and man's activites outside the reservoir area combined to destroy a conservatively designed dam.[3] Baldwin Hills dam was a homogeneous earthfill embankment 155 ft (47.2 m) high, constructed in conformance with standards acceptable for structures of this type when built on reliable foundations.[4] However, the dam was located in a seismically active and unstable region. The Inglewood fault, source of historic earthquakes, lay 500 ft (152 m) to the west; and branches of this fault passed through the reservoir floor and dam abutment. The entire project was situated at the edge of Inglewood Oil Field, and extractions of oil and gas from deep-seated formations had created a bowl of subsidence, the rim of which enclosed the dam and reservoir. Contemporaneously localized areas within this bowl were experiencing uplift as a result of secondary oil recovery operations which entailed high pressure reinjection of fluids into the oil-bearing horizons.[5] The long-term effect of these conditions was to depress, tilt, and stretch the reservoir. A slight offset and separation developed along one of the faults that passed through the reservoir creating an avenue of leakage beneath the dam. Because the foundation was erodible, this opening was rapidly enlarged leading to complete failure of the dam, the loss of 5 lives, and approximately $15 million in downstream property damage. Thus, the failure of Baldwin Hills reservoir resulted from a combination of factors relating to regional geology, dam foundation, and local tectonic or seismic activity. It demonstrates the importance of the investigations which are the subject of this chapter. Furthermore, it showed that man's activities at some distance from a dam may contribute to the causes of its failure and that the builders of dams must be alert to both contemporary and possible future activities which may endanger their structures.

Today the planning, design, construction, and operation of dams frequently involves the services of geophysicists, seismologists, earthquake engineers, and legal experts in addition to geologists and engineers experienced in soils and rock mechanics. This chapter has been prepared by authorities in these several fields. Its intent is to focus attention on the aims and objectives of geology, foundation, and seismic investigations; to discuss procedures of investigation commonly employed; and to provide useful criteria.

During the author's experience with the California Department of Water Resources over the past three decades, 263 engineering geologists were employed for various periods of time. The author had the privilege of serving with many of these professionals on a variety of projects, which presented problems involving geology and other disciplines. It was observed that, for the most successful programs, the concerned disciplines worked together as well-integrated teams. This relationship usually prevailed where the geologist members had acquired knowledge of the principles and practices of civil engineering and some familiarity with the problems confronted by other team members. With this background, they were able to recognize and appreciate the needs of the planners, designers, and operators of the project and to interpret geology in terms that were meaningful to these people.

In practice, every dam requires geologic investigation tailored to detect and evaluate conditions that will affect its design, construction, and operations. Since these conditions are generally revealed as site examination progresses, the investigation should be a concatenation of activities wherein each succeeding event is programmed on the basis of the findings from the foregoing events.

Taken in sequence, the main parts of a geologic investigation generally include:

1. A canvass of the literature for reports, maps, and photos.
2. Reconnaissance of the project area to detect geologic conditions which could affect project feasibility.
3. Geologic mapping of the dam and reservoir site and location of sources of construction materials.
4. Subsurface exploration, which may include geophysical surveys, drilling, and exploratory pits or adits.

5. Foundation evaluation entailing soils and or rock mechanics investigations.

Earthquake-related hazards should be identified early in the investigation. They are evaluated in increasing detail in the subsequent stages of dam design.

It is the responsibility of the engineering geologist to identify geologic conditions which could endanger life or property if a dam is constructed. The most significant geologic hazards include landslides, earthquakes, land subsidence, leakage through points or erodible formations and the presence in critical locations of liquefiable sediments. The case histories of dam and reservoir failures include reviews of catastrophies attributable to each of these hazards.

It is incumbent on those responsible for building dams that the risks associated with geologic hazards be considered in the earliest stages of site investigation. In the final analysis, the evaluation of these risks and the decisions based on these evaluations are matters of judgment involving earth science, soil mechanics, earthquake engineering, and law. The prudent project manager therefore will confer with the geologist, seismologist, geophysicist, foundation and structural engineer, and lawyer as necessary to assure that he is fully informed on these aspects of the investigation. The current practice for large dams is to convene a board of appropriate specialists from these fields to proffer concerted judgment. Generally, it can be demonstrated that the benefits in terms of increased safety, improved design, and reduction in construction costs justify the fees for such technical services.

Although safety is the paramount consideration, geology or seismology-related elements often determine the feasibility of the project and strongly influence planning, design, construction, and operational activities. For example, an appraisal of the mineral resources to be inundated by a reservoir is important to the legal process of acquiring a reservoir site; the availability of construction materials often dictates the location and type of dam to be constructed; seismicity and the presence of faults, active or inactive, influences the design of the dam and its major components; the occurrence and movements of ground water may affect planning and scheduling of construction operations including foundation excavation for the dam and its appurtenant structures and for the driving of associated tunnels; the stability of the slopes surrounding the reservoir may determine the maximum reservoir level that can be maintained without danger of overtopping the dam by a landslide-generated water wave; in some regions the rate of slope movement into the reservoir and consequent loss of storage space will limit the useful life of the project; severe leakage through the walls or floor of reservoirs can lead to the abandonment of a project; and seepage causing swamping in areas downstream or adjacent to reservoirs may precipitate legal actions after the project becomes operational. These, then, are elements which should be considered in the investigation of the dam site and reservoir area.

4.01 The Reservoir—by Laurence B. James

The geologic investigation of every reservoir site should include consideration of the following possible problems:

1. Leakage
2. Peripheral landslides
3. Active faults, seismicity, and contemporary tectonic deformation
4. Land subsidence

The engineering geology report should always direct a statement to each of these problems indicating that it has been investigated and elaborating if the problem proves to be significant. If the investigation discloses no problem of possible consequence, this observation should be stated in the report and supported with the evidence therefor.

The investigation should also include an appraisal of the mineral resources located within the reservoir area and which would be rendered unobtainable by construction of the project.

The location and appraisal of sources of materials for constructing the dam are generally included as part of the reservoir study; although this search may extend to areas beyond.

This section of Chapter 4 deals with these elements of the reservoir investigation.

4.011 Reservoir Leakage. Serious leakage most commonly occurs in regions underlain by limestone or volcanic rocks; although significant problems have been experienced at reservoirs located on aeolian (wind blown) sediments or connected with buried alluvial channels. A number of case histories of reservoirs within these several rock categories were considered at the National Meeting of the Association of Engineering Geologists at Seattle, Washington, in 1968. The proceedings of that meeting are recommended reading for geologists and engineers concerned with the investigation of reservoir sites.[6]

Leakage can be hazardous when it occurs through erodible formations. Where geologic mapping has established the presence of such formations in the floor or walls of reservoirs, potential avenues of leakage should be investigated for the possibility of piping. Piping may be defined as erosion due to underflow. It commences at the point where a seepage path emerges and works backward (upstream). Piping through the foundations of dams and walls of reservoirs has been the cause of serious failures.[7] For piping to occur through erodible formations, there must be a sufficient seepage force to dislodge and transport the soil particles through which the underflow is moving. This condition arises when the hydraulic gradient exceeds the buoyant weight of the soil particles.[8] It is expressed in soil mechanics terminology as follows:

$$\frac{h}{L} > \frac{G-1}{1+e}$$

Hydraulic gradient
> Buoyant weight of particles

h is the difference in elevation between the reservoir stage and the point of seepage emergence, L is the length of the shortest path of percolation, G is the specific gravity of the soil particles and e is the void ratio of the formation through which the leakage occurs.

When geologic examination discloses the existence of a possible seepage path wherein the left-hand side of this equation may be greater than the right, piping is possible. Thin ridges of poorly consolidated sediment in the walls of a reservoir should be examined carefully since under this condition the comparatively short length of L and large value of e combine critically. Similarly, reservoir walls comprised of deposits of low density sediments such as particles of diatomite shales, porous volcanic cinders, and the like should receive special attention because of their low buoyant weights.

In some instances, leakage from reservoirs may be governed by the position of the natural water table in the formations that comprise the walls of the reservoir. Where that water table continuously stands higher than the reservoir stage, it is obvious that leakage cannot occur. Thus, where portions of a reservoir perimeter consist of permeable formations, it is advisable to conduct an investigation to determine the presence (or absence) of ground water, the characteristics of the aquifers, and direction of underflow.[9]

4.012 Reservoir Landslides. The perimeters of reservoirs should be examined to identify potential slides which could damage the dam, spillway, appurtenant structures or which could create dangerous or damaging water waves. Preliminary investigation for slide areas in large reservoirs may be accomplished by aerial photo interpretation or by visual examination of the terrain from airplanes or helicopters. Unstable slopes so identified may be shown on areal geology maps of suitable scale. Where slides are potentially dangerous due to precarious location or because they appear capable of generating large water waves, subsurface exploration may be warranted to determine their volume and to estimate the potential velocity of the unstable land mass. If the geology and topography appear conducive to high velocity sliding, the construction and study of a hydraulic model may be desirable. Such modeling involves highly specialized hydrodynamic techniques whereby reasonable similitude between reservoir model and prototype can be assured.

Landslides and soil creep may substantially decrease the economic life of reservoirs by diminishing their usable storage capacities. A study by the California Department of Water Resources for a proposed major reservoir of 536,000 acre-ft (6.61×10^8 cubic meters) capacity located in the State's northern coast ranges indicated that approximately 21 percent of the total storage could be lost over the 100-year estimated life of the project.[10]

The first step in estimating the volume of unstable mantle material in a reservoir perimeter is the preparation of a detailed geologic map delimiting areas of creep and of potential sliding. These areas are identified by the geologist from geomorphic evidence.[11] Recent slope movements may be manifested by vegetation, i.e., uprooted and tilted trees or by the display of sheared roots in slide scarps. Most estimates of the average depth of slides on soil mantles rely heavily on judgment and are generally at best rough approximations. Where contacts with underlying bedrock are exposed as in the walls of gulleys, the depth may be measured directly. Geophysical surveys can be helpful where the unstable mass consists of loose (low velocity) material resting on a firm basement. Exploratory shafts or drilling provide more positive measurements but are costly. Down hole surveying techniques are becoming popular. These surveys rely on a recording inclinometer, which is lowered in the hole to detect points of warping and displacement. The determination of the base of a slide is not always possible from such records and experience is required for their interpretation.

Where active creep is suspected, traverses across the questionable areas may be surveyed periodically to measure the rate of displacement of stations established on the soil mantle. On large slides, geodetic techniques are sometimes employed for this purpose, the positions of survey monuments being determined by triangulation from stable outcrops located on the opposite wall of the reservoir. Such surveys provide the data required to estimate the rate of creep. They should be repeated for a sufficient number of years to assure that the effects of variable factors (such as precipitation) are taken into account.

Obviously, judgment is required in using the results of slide surveys. Secular phenomena such as strong earthquakes or unprecedented floods can trigger large mass movements which may invalidate estimates of the rate of soil movement which are based on surveys conducted over a short period of years.

Reservoir Seismicity. Earthquake hazards are considered in greater detail under Section 4.2 of this chapter. It will therefore suffice here to review the geologists's role in evaluating these hazards. The geologic investigation should include the mapping and examination of faults within and in the vicinity of the reservoir area and the mapping and description of geologic and soil conditions which may affect the characteristics of earthquake ground motions Active faults should be identified and their lengths, interrelationships, and habits studied. The geologist must also be alert for conditions conducive to such earthquake-related secondary hazards as landslides, soil liquefaction, subsidence, and tectonic deformation.

In the simplest of terms, there are two categories of earthquake hazards to dams and reservoirs—the danger of structural rupture due to a seismic displacement or deformation and the possibilities of damage related to severe earth shaking. Both of these dangers are influenced by geologic factors, and it falls on the geologist to identify conditions that may influence the selection of the site and the design of the facilities and assist the engineering staff in their evaluations.

Damage from displacement may occur if an offset occurs along an active earthquake fault on which a structure, tunnel, or conveyance facility has been inadvertently located. Facilities sensitive to tilting, such as canals and certain types of pumps or generators, may be affected by tectonic deformations which happen during earthquakes or which are part of the slow process of crustal warping that takes place in regions of active mountain building.

An active fault is one on which an earthquake or displacement has occurred in comparatively recent geologic time. For the purpose of dam building, the definition generally has been interpreted to include faults evidencing activity during the Recent geologic time period—i.e., roughly within the past 10,000 years. More conservative criteria have been recommended for classifying as active, faults in the vicinity of proposed nuclear power plants.[12]

Regardless of the time frame adopted, any of the following evidence will serve to classify a fault as acitve:

1. Ground disruption along the fault trace or offsetting of facilities constructed by man or of geologically recent physiographic features (such as alluvial fans or stream channels);
2. Records such as news accounts and diaries or accounts of reliable observers, from which it can be reasonably inferred that the fault was the probable source of a historic earthquake; and
3. An alignment of earthquake epicenters along the fault trace or its subsurface projection as determined by analyses of records from seismographs.

Unfortunately, the lack of all the above criteria does not prove that a disruption will not occur along a fault within the life of a man-made structure.

In some instances, it can be established by careful geologic mapping that a fault plane is overlain by an unbroken formation of considerable age thereby establishing the antiquity of the faulting. For most engineering purposes, it is assumed that the likelihood of revived movements along a fault which has been quiet during Recent geologic time is remote.

In certain regions the earth's crust is undergoing deformation, which may or may not be accompanied by faulting and earthquakes. This tectonic activity, as it is called, is usually slow and its effect on man-made facilities is imperceptible during their life spans. However, in the case of certain machinery and gravity flow systems of low gradient such as canals and conveyance tunnels, tectonic movements can cause serious problems. Therefore, where large pumps, generators, or lengthy aqueducts are proposed for construction in tectonically active regions, an effort should be made to measure rates of crustal strain accumulation. Such investigation may entail geodetic surveying, installation and monitoring of tiltmeters, and periodic levelling.[13,14]

The San Francisco earthquake of 1906 provided some of the earliest observations of the affects of geologic factors on the severity of earthquake ground motion.[15] Generally speaking, those observations disclosed that ground motions are amplified in loose sedimentary deposits. The geologist's map should therefore delimit such formations and in particular it should show saturated deposits since these materials may exhibit the so-called "jelly effect" or may undergo liquefaction with potentially serious consequences.[16,17] The logs of subsurface explorations in conjunction with measured or estimated soil properties can be used in models to compute the response of the soils to strong earthquakes.[18]

4.014 Land Subsidence in the Reservoir Area.
The following six kinds of environment may be susceptible to subsidence:

1. Oil fields
2. Ground water basins
3. Aeolian, mudflow, or low density soils
4. Tectonically active regions
5. Limestone terrains

6. Regions of active or historic subsurface mining operations

Land subsidence is known to have occurred at more than 40 oil fields in the United States[19] and has been identified in a number of foreign countries. It has been noted previously that the failure of Baldwin Hills dam and reservoir with consequent loss of 5 lives and severe property damage has been attributed at least in part to subsidence caused by oil field operations. At Wilmington field, California, subsidence exceeded 29 ft (8.84 m) and an area of 24 sq mi (62.2 sq km) receded at least 2 ft (0.61 m). At this field the consolidation of deep strata was halted through a program of field repressurization wherein sea water was reinjected through a network of deep wells.[20] The possibility of such subsidence should be considered where reservoirs will occupy areas from which petroleum products (1) are being extracted or (2) which are geologically promising from the standpoint of a future oil or gas discovery.

Subsidence due to ground water extraction was first reported in 1933. Since that date, 10 areas of significant water-related subsidence have been identified throughout the world.[21] In the San Joaquin Valley of California, the magnitude of settlement has ranged from a few inches to a maximum of 23 ft (7.01 m) and an estimated 3,000 sq mi (7769 sq km) has been affected. This phenomenon occurs only in areas underlain by aquifers. Because of the accelerated increase in utilization of ground water, it is becoming more widespread. No failures of dams have been attributed to this kind of subsidence to date. It has caused serious changes in the gradients of canals—particularly in California.

Low density soils found in some localities experience a restructuring and compaction when they are saturated. The result is a collapse of the ground surface and the appearance of peripheral cracks. This kind of subsidence, sometimes called hydrocompaction, has been experienced in regions of aeolian (wind deposited) sediments and of moisture deficient alluvial debris.[22] Such soils are sometimes quite sensitive. In San Joaquin Valley, California, slight subsidence due to seepage through concrete irrigation ditches is sufficient to initiate cracking of the linings. The result is further leakage leading to a progressive failure. The end result has been the abandonment of ditch irrigation in some areas and the substitution therefor of sprinklers. A substantial section of the California Aqueduct was constructed through materials of this type.[23] In this instance, the problem was surmounted by an intensive program of presubsidence wherein water was ponded along the aqueduct alignment and the soil allowed to consolidate prior to excavation of the canal prism.[24]

Areas susceptible to hydrocompaction are probably best detected by a coordinated program of geologic investigation and soils testing. The behavior of suspect soils can be appraised by constructing test ponds on suspect formations at specific sites and monitoring the elevations of a network of benchmarks in and surrounding them.[25]

Investigations for subsidence related to tectonic activity are conducted by geologists and seismologists in the course of the evaluation of site seismicity. This subject is discussed elsewhere in this chapter.

Areas underlain by carbonate rocks such as limestone or dolomite may be prone to subside. These rocks are soluble in slightly acidic ground waters leading to the creation of underground chambers. The roofs of such natural caves ultimately collapse to form crater-like sinkholes. In some regions where solution processes are active, the terrain is pocked by a number of these features forming a distinctive physiographic surface known as karst topography. On December 2, 1972, the largest recent sinkhole in the United States was created in Alabama from the collapse of a dolomite chamber. A crater measuring approximately 459 ft (140 m) long, 377 ft (115 m) wide, and 160 ft (49 m) deep resulted. Aerial reconnaissance of this region suggested that perhaps as many as 1000 sinkholes have been formed in an area of about 6.2 sq mi (16 sq. km).[26] Thus, the identification of soluble carbonate rocks during the geologic mapping of the reservoir is cause for concern. The dissolution of such rocks may result in reduced bearing strength as well as severe subsidence through roof collapse. Serious reservoir leakage may also occur through solution channels.[27] Furthermore, the cost of remedial treatment of carbonate formations by grouting or by backfilling large chambers through open cuts is generally high and may be prohibitive.

The investigation of reservoir sites situated in regions of contemporary or historic mining activity should include the locating of any underground workings, the collapse of which would endanger proposed facilities. The possibility of reservoir leakage through such workings should also be considered.

REFERENCES

1. Brown, J. Guthrie, President of the International Commission on Large Dams, 1964-67, "International Commission on Large Dams Congress, 8th International," Transactions, Vol. V., p. 48.
2. Kiersch, George, A., "A Case Study of the Vaiont Reservoir Disaster," Proceedings of the State of California, The Resources Agency, Landslide and Subsidence Geologic Hazards Conference, Los Angeles, Calif., May 26-27, 1965.
3. James, L. B., "Failure of Baldwin Hills Reservoir, Los Angeles, California," Geological Society of America, Division on Engineering Geology, Engineering Geology Case Histories Number 6, 1968.
4. California Department of Water Resources, "Investigation of Failure Baldwin Hills Reservoir," April 1964.
5. Hamilton, D. H. and Meehan, R. L., "Ground Rupture in the Baldwin Hills," *Science*, 172 (No. 3981), (April 23, 1971).
6. Association of Engineering Geologists, "*Reservoir Leakage and Groundwater Control*," Bulletin of the Association of Engineering Geologists, Vol. VI, No. 1, Spring, 1969.
7. Sherard, J. L., Woodward, R. J., Gizienski, S. F., Clevenger, W. A., *Earth and Earth-Rock Dams*, John Wiley & Sons, 1967.
8. Taylor, D. W., *Soil Mechanics*, John Wiley & Sons, Eighth Printing, 1955.
9. Herrera, Carlos Garcia, "Geological and Geohydrological Studies for Angostura Dam, Chiapas, Mexico," Geological Society of America Bulletin, Vol. 84, May 1973, p. 1733.
10. California Department of Water Resources, "Middle Fork Eel River Landslides Investigation," Memorandum Report, September 1970.
11. National Academy of Sciences, Highway Research Board, "Landslides and Engineering Practice," Special Report 29, 1958.
12. United States Atomic Energy Commission, "Nuclear Power Plants—Seismic and Geologic Siting Criteria," Code of Federal Regulations (Proposed), *Federal Register*, 36 (No. 228), Nov. 1971.
13. California Department of Water Resources, "Earthquake Engineering Programs," Bulletin No. 116-4, May 1968.
14. California Department of Water Resources, "Geodimeter Fault Movements Investigations in California," Bulletin No. 116-6, May 1968.
15. California State Earthquake Investigation Commission, "California Earthquake of April 18, 1906," 1908.
16. Peacock, W. H., and Seed, H. B., "Liquefaction of Saturated Sand Under Cyclic Loading Simple Shear Conditions," Report No. TE-67-1, University of California, Berkeley, Calif., July 1967.
17. Seed, H. B and Lee, K. L., "Studies of the Liquefaction of Sands Under Cyclic Loading Conditions," Report No. TE-65-5, University of California, Berkeley, Calif., December 1965.
18. Seed, H. Bolton and Idriss, I. M., "Influence of Soil Conditions on Ground Motions During Earthquakes," *Journal of the Soil Mechanics and Foundations Division*, American Society of Civil Engineers, 95 (No. SM1), pp. 99-137 (1969).
19. Yerkes, R. F. and Castle, R. O., "Surface Deformation Associated With Oil and Gas Field Operations in the United States," Land Subsidence Vol. I, Pub. No. 88, AISH International Association of Scientific Hydrology and UNESCO, 1969.
20. Allen, D. R., "*The Mechanics of Compaction and Rebound*, Wilmington Oil Field, Long Beach, Calif.," Department of Oil Properties, City of Long Beach, May 1969.
21. Poland, J. F. and Davis, G. H., "Land Subsidence Due to Withdrawal of Fluids," Geological Society of America, Reviews in Engineering-Geology II, 1969.
22. Lofgren, Ben E., "Land Subsidence Due to the Applications of Water," Reviews in Engineering Geology II, Geological Society of America, 1969.
23. Golze, Alfred R., "Land Subsidence—Why the State is Concerned," Published in the Proceedings for Landslides and Subsidence Geologic Hazards Conference, Los Angeles, California, May 1965.

24. California Department of Water Resources, "Design and Construction Studies of Shallow Land Subsidence for the California Aqueduct in the San Joaquin Valley," Interim Report, December 1964.
25. Shelton, M. J. and James, L. B., "Engineer-Geologist Team Investigates Subsidence," *Journal of the Pipeline Division*, American Society of Civil Engineers, May 1959.
26. LaMoreaux, Philip, E., and Warren, William M. (Geological Survey of Alabama University, Alabama) "Sinkhole," *Geotimes*, p. 15, March 1973.
27. Moneymaker, Berlin C., "Reservoir Leakage in Limestone Terrains," Paper presented at the Symposium on Reservoir Leakage and Ground Water Control, Association Engineering Geologists National Meeting, Seattle, Washington, 1968.

4.02 Location and Exploration of Sources for Construction Materials—by Cole R. McClure

Investigations of construction materials are usually concerned with one of two questions; namely, is there a sufficient quantity of a specified construction material available within a reasonable haul distance, or the more comprehensive question of what construction materials are available in the area. In considering the construction of a dam, hopefully answering the latter question is the objective of the investigation. The proper selection of the type of dam to be built at a site must include a careful evaluation of the construction materials available. Many times the availability of construction materials determines the type of dam which can economically be considered at a given site. In this section of the book, the more comprehenisve investigation will be described, although the same approaches will apply to a study of the availability of a specific construction material. See also the discussion in Chapter 3.

The importance of a thorough construction materials evaluation to the planning and design of a dam is apparent, but it is also critical to the actual construction phase. One of the chief bases for "change orders," disputes between the owner and the contractor, is the availability of a sufficient quantity of suitable construction materials. The economic impact of the availability of construction materials, as described in the specifications, must be considered with great care by both the experienced engineer and contractor. Not only must the material be available where, when, and as specified, but the physical properties of the material must be described. Also, the borrow area must be as described to the bidders, i.e., access, depth to ground water, ease and depth of excavation and available waste disposal area.

It has always been desirable to obtain the construction materials for a dam upstream of the proposed axis, within the reservoir area, because of economics and right-of-way consideration. But professional responsibility, and in some cases, legal requirements, demands protection of the environment of the area by avoiding unsightly excavations or rehabilitating them. Therefore, these are factors which must be given serious consideration in locating borrow areas.

4.021 The Preliminary Phase

General. One of the most valuable sources of information is the actual production records of construction materials. They are usually available through governmental geologic agencies. These records generally include the total annual production of the various construction materials produced in the area and some description of the various sources. Local highway departments generally have test results and performance records on aggregate, road base materials, and frequently on riprap. It would be a simple solution to the materials problem if an established, proven source had the capability to produce the materials required for the proposed dam construction; however, this is almost never the situation encountered in practice. The chief benefit which can be derived from the study of performance of existing sources is the geologic correlation of the existing borrow areas with possible new borrow sites.

In the remote areas of the world detailed geologic maps, data on existing quarries and borrow areas, and good topographic maps are usually lacking. In these areas aerial photographs and other remote sensing tools are most helpful. However, reconnaissance of the area is the only method of obtaining reliable data. Helicopter reconnaissance is often the least expensive and certainly a most effective means.

The preliminary phase of a construction material investigation is generally conducted by an engineering geologist. The effectiveness of this study depends on his experience and how well he is briefed by the project engineer on the requirements.

The published geologic maps of an area are most useful in locating possible sources of construction materials. These maps show the distribution of various geologic units and the legend generally provides a brief description of each of these units. Potential sources of construction materials may thus be identified and located and further studies planned. There are many types of geologic reports which may provide some useful information to the construction materials investigator. Reports on the general geology of the area usually contain a description of the various construction materials found there. But even reports on the ground water geology, various mining aspects, petroleum geology, and other geologic data seemingly unrelated to construction materials may provide much useful data.

The preliminary stage of the investigation should include reviewing the geologic data on the area such as the geologic maps and all types of geologic reports, descriptions of existing quarries and borrow areas, their production records, and materials performance histories. Existing quarries and borrow areas are a great source of data. A review of the production records from each operation, and its geologic description as noted on geologic maps and reports, provides information on the characteristics to be expected at other sites. Many governmental agencies, both local and federal, keep records of the quantity of each type of construction material produced, frequently including an estimate of the reserves of each area. Performance histories are also frequently available from these local agencies. Interviews with the various suppliers of construction materials often develop data on problems others have experienced.

Evaluation of the Preliminary Investigation. A preliminary evaluation of the construction materials potential of an area should be directed toward providing estimates of the quantities of all the various types of construction materials available within a specified area. Actual exploration, such as drilling, sampling, and testing, is not performed during this preliminary phase. The evaluation is based on the collection of existing data and reconnaissance of the area. The discussion of each of the types of construction materials follows:

Concrete Aggregate. This is the only material described in this section which has rather well-defined specifications. Most of the concrete aggregate throughout the world comes from stream-channel deposits. Such deposits are subjected to abrasion caused by stream transportation which generally produces well-rounded, hard, and frequently clean aggregate. The presence of deleterious materials and poor grading characteristics is an aspect which often requires special attention in evaluating stream deposits.

The physical properties which are most important in the evaluation of potential aggregate sources are:

1. Rock types
2. Degree of weathering, soundness
3. Hardness
4. Strength
5. Size and grading
6. Particle shape
7. Coating
8. Organic impurities

The more acceptable and economical deposits should contain ample material, usually 3 or 4 times the estimated requirements for preliminary evaluation. Ideal deposits should not have large overburden thickness, poor grading characteristics, e.g., gap-grading or too great a percent of fines, high ground water conditions, difficult access or great haul distance.

As noted earlier, performance histories and test results of concrete aggregate sources are frequently available from highway departments. If not, similar information can be obtained by examining various concrete structures, e.g., bridges, highways, etc., where the aggregate sources are known. Potential aggregate reactivity must be considered at this stage and some evaluation of this problem can be made by obtaining data on the type of cement to be used and the admixtures, such as pozzolans, to be used with the aggregates. For example, if low alkali cement was used with the aggregate being considered, the performance records are valid only with the same type of cement. These same aggregates could be reactive when used with other types of cement.

Concrete aggregates can be produced by quarrying and crushing rocks. It is expensive but required if natural aggregates are not available within reasonable haul distances. Crushed aggregates produce harsher mixes. To achieve the desired workability, more water is needed; to then maintain a required water-cement ratio, more cement is needed. Thus, greater costs are involved. It is especially desirable to try to locate natural deposits of the fine aggregate.

Riprap. Armor stone, natural slope protection, or riprap, is the construction material which may be the most difficult to locate. Riprap is usually natural rock placed on an excavated or constructed slope to act as protection against wave action. Fresh, sound, and massive rock, such as granite, basalt, limestone or quartzite, is frequently used. Specifications for gradation of riprap are usually worded to conform with expected severity of wave action. Usually, riprap should have an unconfined compressive strength greater than 7000 psi, a specific gravity not less than 2.5, and not have a wet-shot test loss greater than 40 percent.

Suitable quantities of a dense, sound riprap in sufficiently large blocks to meet design requirements are frequently not available within a reasonable haul distance of a dam site. Haul distances for riprap have exceeded 150 mi. There are, however, examples of investigation which failed to find suitable riprap sources that were found in the same area by later investigation.

A preliminary study of possible riprap sources should begin with examination of the geologic maps and reports. The geologic description of the materials occurring in the area usually provides the investigator with some suggestion of the rock types which most likely would yield suitable riprap. Field reconnaissance provides more reliable data for preliminary evaluation of the potential riprap sources. For example, rocks which may be likely sources of riprap commonly have bold outcrops and talus slopes which can provide a geologist with some indication of the suitability of the material for riprap. The field examination and the review of the geologic literature should provide adequate data for a preliminary estimate of the volume of rock present, the percent of waste material anticipated, size, distribution and fragment shape, all based on assumed quarrying operations. Each possible source area should contain at least 3 or 4 times the estimated required volume of riprap if it is to be considered worthy of further consideration.

Many rocks are not suitable for riprap. Certain geologic and physical properties are necessary, such as suitable rock type, strength, hardness, and durability, and they must be moderately dense. Stratification, jointing, and fracturing must be considered in estimating the size distribution and fragment shapes developed by quarrying operations. Usually, flat, platy fragments are undesirable and specifications require that no one dimension can be more than 3 times any other. These fragments should be quite angular to achieve an interlocking action. Angular fragments will resist displacement better than rounded boulders of comparable size and density. Gradation limits are usually strict and must be considered when estimating the percent of waste which will be produced at each quarry.

Most igneous rock types can be suitable for riprap. Only the more massive metamorphic rocks, such as meta-volcanic rocks, gneiss, and massive schists, are generally acceptable. The platy slates, phyllites, and fissile schists may be acceptable. Many sedimentary rocks are not suitable; however, massive or thickbedded limestone and well-cemented sandstone may be acceptable.

Rockfill. Requirements for materials for rockfill dams are similar to those for riprap, but perhaps not quite so demanding. The rocks must still be strong, hard, durable, and relatively dense, but the gradation limitations are much less restrictive. Hence, the percent of waste is much less for rockfill. The need for angular shaped fragments is usually much less important, although flat or platy fragments are undesirable.

Suitable rockfill materials should be sufficiently hard and durable to resist major changes in gradation during hauling, stock piling, and placement. The materials used especially in the outer portions of the upstream and downstream slopes must be capable of resisting weathering. Another important aspect of the study of rockfill sources which differs from the evaluation of riprap sources is the haul distance. A riprap source may be economical, although it is many miles from the axis; a rockfill source must be within a few miles of the site. The large quantities usually required for a rockfill dam indicate the economic need for a short haul distance. Environmental considerations suggest that a borrow area ideally should be in the area to be inundated by the reservoir.

Earthfill Materials. The preliminary evaluation of the earthfill materials which may be suitable for dam construction should be directed to providing an accurate estimate of the quantities and description of materials available. The designer has a wide range of embankment designs which he can consider. The concept of having the earthfill materials investigation directed toward developing specific quantities of pervious and impervious materials is generally not valid in the preliminary phase of the project. Many capable earth-dam engineers feel that a safe dam can be built with almost any earth materials. If good description of the materials available in an area is provided to the designer early in the planning stage, he can begin to evaluate the most desirable combinations to use for his design. During the second phase of materials investigation, the exploration can be directed toward evaluating the potential for specific materials.

In evaluating potential embankment sources, the haul distance, access, estimated stripping and waste, location of ground water in the potential borrow area, and the moisture content of the materials must be considered in addition to the physical properties of the materials. A desirable borrow area would contain: relatively uniform materials, free of organic matter; a reasonably thick deposit of borrow material overlain with a minimum thickness of overburden; ground water deep enough not to interfere with excavations.

Materials for the stability section or pervious zone usually are sandy materials with good shear strength and are free-draining, relative to the core or impervious section materials. Good core material contains sufficient fines to be relatively impermeable when compacted in the embankment. If the material contains sufficient granular material, it should provide some shear strength and make it more workable.

A random-fill or homogeneous section design is frequently considered when the more readily available materials are neither clearly pervious or impervious. The material for drains usually involves limited volumes and hence can be processed materials.

As stated earlier, the objective of the construction materials investigation is to locate and describe as accurately as possible the depth, areal extent and physical properties of the various types of materials occurring within reasonable haul distance of the site.

4.022. The Secondary Phase

After the preliminary phase of the exploration for construction materials is completed, and the evaluation is presented in a report, new requirements for the construction materials are established. These new requirements are guides for the secondary phase of the construction materials investigation. The objective of this phase of the materials investigation is to locate suitable materials to meet the probable requirements for the types of dams being considered for final planning.

This stage includes preliminary drilling, sampling, and testing of the most promising sources in order to establish that adequate quantities of suitable material are available. It is usual to locate approximately 2 or 3 times the actual volume of material needed for the dam construction.

A more complete and thorough investigation of existing geologic data, existing quarries and borrow areas, their test results, performance records and production data is required at this time. An additional reconnaissance of the area should also be completed at this stage of the investigation.

Remote-sensing methods ranging from black and white photography to line scanning devices, such as thermal infrared or side-looking radar (SLAR), and satellite imagery may be employed to good advantage at this stage, especially in remote areas. These methods should provide a basis for further study of the more promising sources. Aerial photographs have been used in the search for engineering construction materials over the last few decades, using black and white stereo photographs. Within the present decade, technology has advanced to include other complementary remote sensing tools; namely, SLAR, thermal infrared scanning, color infrared photography, and satellite multispectral imagery.

U.S. Army Corps of Engineers' studies in the Mekong Delta and other studies in the Mississippi Delta show color infrared photography to be superior to black and white photography for identification of beaches, abandoned channels, and point bars. The tonal contrast between land/vegetation/water boundaries accounts for this superiority. These land forms are all good sources of engineering materials in the deltaic environment.

Whereas color infrared photography records reflected solar radiation, the tones on thermal infrared imagery are a result of temperature differences caused by varied emissivities of materials on the surface. Thermal infrared can be useful in mapping areas of shallow ground water. These maps may also provide suggested source areas for construction materials where shallow aquifers are buried river channels or thick glacial deposits.

For regional construction materials inventories, satellite and radar imagery are useful because of their synoptic coverage. Regional structural and landform features can be seen more readily on small scale imagery than on larger scale air photos, which allows for prediction of the types of materials that will be encountered.

Radar provides its own illumination and is not dependent on weather conditions. For example, in Darien Province, Panama, where photography had been unsuccessful because of ever-present cloud and impenetrable vegetation, radar studies showed that it is possible to differentiate alluvial deposits of sand and gravel from bedrock.

LANDSAT imagery has worldwide coverage and is often the only readily available source of imagery for remote areas. The standard data products include black and white prints in four wavelength bands or color composites, all of which can be interpreted in much the same way as aerial black and white or color infrared photos, but allowing for the small scale involved. The regional viewpoint makes it possible to predict possible materials locations.

At this stage of investigation, field studies, including geologic mapping or the checking of existing geologic maps, should be completed for the more promising areas. Limited sampling of selected areas and preliminary testing will provide more reliable data on the area. New estimates of the volume of materials available in the area should be based on the additional data collected since the initial reconnaissance.

A review with planners, designers, estimators, and construction specialists is appropriate at this time. These reviews bring out questions, new approaches, and suggestions which should be incorporated in the final exploration for the construction materials. Frequently, this review provides the basis for narrowing the scope of the study and permits the investigator to concentrate his efforts on the more probable types of materials to be used for the project.

4.023 Final Exploration. The objective of the final exploration program is to confirm the location of adequate quantities of suitable materials for an evaluation of the specific sources of construction materials.

Large diameter, bucket auger drilling is probably the most efficient method of exploring for unconsolidated construction material. Trenching and test pits excavated by backhoe are also very efficient methods of exploring unconsolidated deposits. All of these methods permit accurate, detailed logging. Logging of these exploratory holes should be accomplished by a capable specialist. They should record the usual geologic and soils descriptions, and their observations are of great importance during drilling and sampling; for example, notes on organic material, moisture content, water levels, degree of cementation and uniformity of the materials. Samples of the materials should be obtained in sufficient quantities for testing. These methods of exploration also provide data on the methods of excavation most suitable for material, depth of overburden, thickness and uniformity of the material and depth to ground water. Geophysical methods such as seismic and electrical resistivity are not usually needed in exploring these materials as the bucket auger is economical and the drilling is completed so quickly; however, in many areas the geophysical methods may be used effectively to provide some data on the borrow areas, e.g., water table, depth to consolidated rock, etc.

Aggregate exploration is usually accomplished with dozer, backhoe or clamshell or where equipment is scarce and labor is inexpensive, by hand excavated test pits using caissons or timber and lagging. The methods of exploration may vary locally, but the location and distribution of the test excavations must be designed to determine accurately the uniformity, quality and quantity of the deposit. Sampling is quite critical at this stage and it is necessary that representative grading be obtained. Instruction such as found in ASTM standards must be followed in detail. Special consideration must be given to zones occurring below the water table and sampling techniques developed to assure that fines are not lost during the sampling.

Potential aggregate reactivity must be carefully evaluated during this final phase of aggregate exploration. If significant quantities of potentially reactive materials may be present, the planning and design engineers must be alerted to the problem. There are several possible methods of solution, e.g., low alkali cement, pozzolans or finding another aggregate source which contains non-reactive aggregates. The following is a partial list of materials which are frequently reactive:

1. Siliceous minerals such as cherts, opal, chalcedony.
2. Certain rhyolites, andesites.
3. Tuffaceous rocks.
4. Zeolites.
5. Some phyllites.

Explorations for riprap and rockfill is usually based on careful geologic mapping, diamond core drilling, geophysical logging, geophysical surveys and test blasting. The geologic mapping should stress rock type, jointing, fracturing, weathering, overburden thickness and detailed descriptions of representative outcrops. Core drilling should be used to provide representative coverage of the potential quarry site. The drilling should enable the geologist to "prove" a volume of rock equal to twice that required for the dam. This estimate should not include waste rock in this volume. Geologic logs should have detailed lithologic descriptions and should describe fracturing, jointing, core lengths and degree of weathering.

Electrical resistivity, sonic, density and other types of geophysical logging can be helpful in estimating the degree of fracturing and the density and modulus of elasticity can be calculated. While these geophysically determined values may not be precise values, they are very accurate for a relative determination and are reliable in a qualitative sense.

Geophysical surveys, especially seismic refraction surveys can provide good information on the bulk properties of the rock. For example, if the P-wave velocities are less than 10,000 ft/sec, the rock is suspect and requires considerable further investigation. If the rock velocities exceed 12,000 ft/sec a more optimistic view of the quarry potential is indicated. Small test blasts can provide considerable data on the quarry site. In addition to information on the

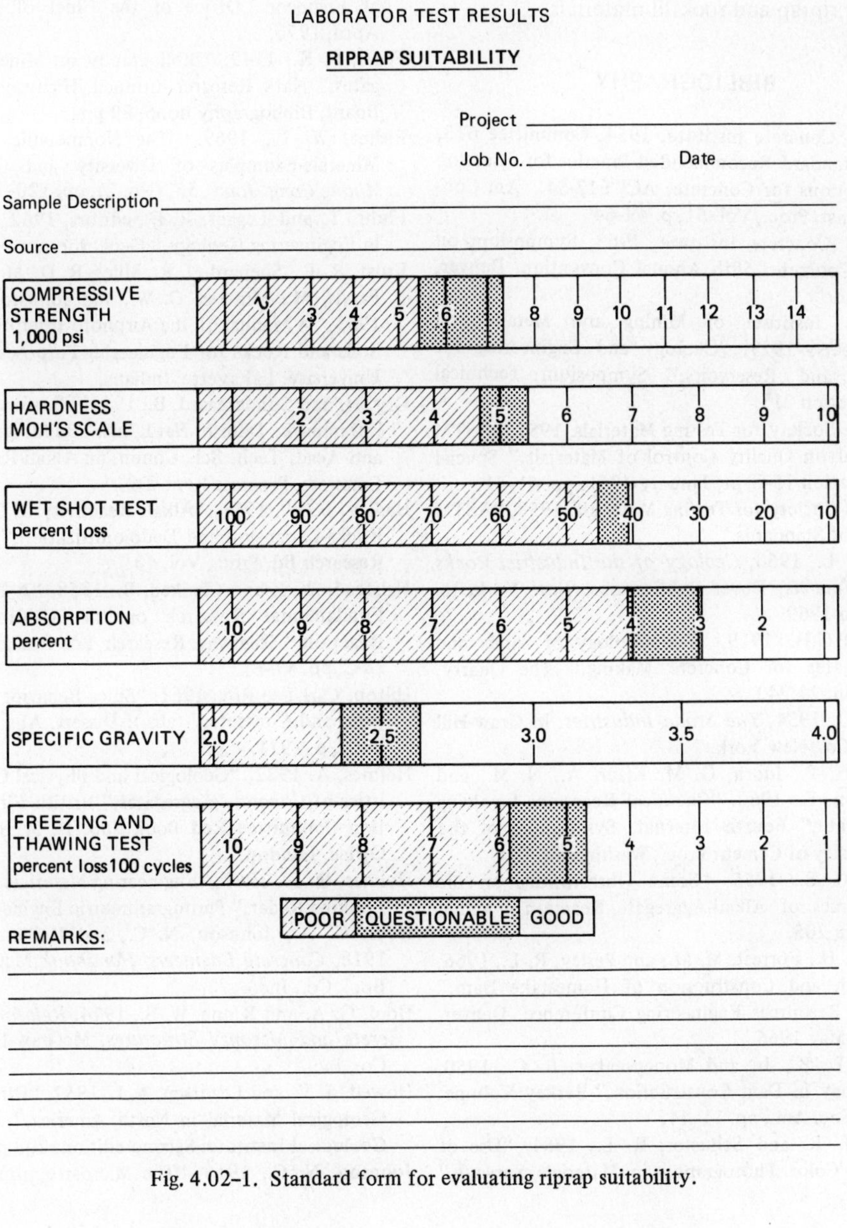

Fig. 4.02-1. Standard form for evaluating riprap suitability.

quality of the rock, the test blast should provide data on estimated drilling and blasting costs.

An evaluation of the physical properties of rock samples may be obtained by laboratory tests. Unconfined compression tests, hardness, absorption, specific gravity, freezing and thawing determinations, and the wet-shot tests are used to make such evaluations. The following chart may be used in evaluating the suitability of rock for riprap and rockfill materials.

BIBLIOGRAPHY

American Concrete Institute, 1954, Committee 613, ACI Standard Recommended Practice for Selecting Proportions for Concrete, ACI 613-54: Am Concrete Inst. Proc., Vol. 51, p. 49–64.

American Concrete Institute, 1963, Symposium on Mass Concrete: 58th Annual Convention, Denver, Colorado.

American Institute of Mining and Metallurgical Engineers, 1929, "Geology and Engineering for Dams and Reservoirs," Symposium Technical Publication 215.

American Society for Testing Materials, 1951, "ASTM Manual on Quality Control of Materials," Special Publication 15-C, pt, January 1951, pp. 41–51.

American Society for Testing Materials, 1958, "ASTM Book of Standards."

Bates, R. L., 1960, *Geology of the Industrial Rocks and Minerals*, Dover Publications, New York, reprinted 1969.

Bowell, P.G.H., 1919, "The Geology of Sands and Aggregates for Concrete Making," The Quarry, London, 24:242.

Bowles, O. 1934, *The Stone Industries*, McGraw-Hill Book Co., New York.

Bredsdorff, P., Idorn, G. M. Kjaer, A., N. M., and Poulson, E., 1960, "Chemical Reactions Involving Aggregate," Fourth Internat. Symposium of the Chemistry of Cement Proc., Washington, D. C.

Brown, L. S., 1955, "Some Observations on the Mechanics of Alkali-Aggregate Reaction," ASTM Bulletin 205.

Burke, H. H., Forrest, M. M., and Perley, R. L., 1966, "Design and Construction of Homestake Dam," Water Resources Engineering Conference, Denver, Colo., May 1966.

Burwell, E. B., Jr. and Moneymaker, B. C., 1950, "Geology in Dam Construction," Berkey Volume, Geol. Soc. Am., pp. 11–44.

Chaves, J. R. and Schuster, R. L., 1964, "Use of Aerial Color Photography in Materials Surveys," Highway Research Rec., No. 63 (Nat'l. Acad. Sci-Nat'l. Research Council Pub. 1247).

Creager, W. P., Justin, J. D., and Hinds, J., 1945, *Engineering for Dams*, John Wiley and Sons, Inc., Chapman and Hall, Ltd., New York.

Crosby, I. B., 1939, "Geology for Dam Sites," presented to Power Division, Am. Soc. Civil Engrs., San Francisco meeting, July 1939.

Davis, A. C., 1934, *Portland Cement*, Concrete Publications, Ltd., London.

Dept. of the Army, 1970, "Engineering and Design Stability of Earth and Rock-fill Dams," Corps of Engineers, Office of the Chief of Engineers, April, 1970.

Fears, F. K., 1949, "Bibliography on Mineral Aggregates," Nat'l Research Council, Highway Research Board, Bibliography no. 6, 89 pp.

Fisher, W. L., 1969, "The Nonmetallic Industrial Minerals-Examples of Diversity and Quantity," *Mining Cong. Jour.*, 55, (No. 2), pp. 120–126.

Fluhr, T. and Legget, R. F., editors, 1962, "Reviews in Engineering Geology," Geol. Soc Am., v. 1.

Frost, R. E., Shepard, J. R., Miles, R. D., Montaro, P., Parvis, M., Mintzer, O. W., and Johnstone, J. G., 1953, "A Manual on the Airphoto Interpretation of Soils and Rocks for Engineering Purposes," Purdue University, LaFayette, Indiana.

Gry, H. and Sondergaard, B., 1958, Flintfore Komster i Danmark, Danish Nat'l. Inst. Building Research and Acad. Tech. Sci., Comm. on Alkali Reaction in Concrete, Progress Rept. D2.

Hadley, D. W., 1961, "Alkali Reactivity of Carbonate Rocks-Expansion and Dedolomitization," Highway Research Bd. Proc., Vol. 40.

Halstead, W. J. and Chaiken, B., 1958, "A Review of Fundamental Research on the Alkali-Aggregate Reaction," Highway Research Bd. Research Rept. 18-C, pp. 47–51.

Hiltop, Carl Lee Roy, 1961, "Silica Behavior in Aggregates and Concrete," (abs.): Dissert. Abs., Vol. 21, no. 7, p. 1911.

Holmes, A. 1922, "Geological and Physical Characteristics of Concrete Aggregates," British Fire Prevention Committee Red Book 256, H. M. Stationery Office, London.

Holmes, R. F., 1967, "Engineering Materials and Side-Looking Radar," Photogrammetric Engineering.

Hool, G. A., Johnson, N. C., and Hollister, S. C., 1918, *Concrete Engineers' Handbook* McGraw-Hill Book Co., Inc.

Hool, G. A. and Kinne, W. S., 1944, *Reinforced Concrete and Masonry Structures*, McGraw-Hill Book Co., Inc.

Howell, J. V. and Levorsen, A. I., 1957, "Directory of Geological Material in North America," American Geological Institute, Second edition, 208 pp.

Johnson, N. C., 1915, "The Microstructure of Concrete," Amer. Soc. for Testing Materials Proc., Vol. 15, pt. 11.

Kiersch, G. A., 1955, "Engineering Geology-Historical Development Scope, and Utilization," Quarterly-Colo. School of Mines Vol. 50, no. 3.

Kirn, F. D., 1953, "Design Criteria for Concrete Gravity and Arch Dams," U. S. Dept. of the Interior, Bureau of Reclamation, Engineering Monographs, No. 19.

Knight, B. H. and Knight, R. G., 1948, *Builder's Materials*, Edward Arnold, London, 394 pp.

Kynine, D. P. and Judd, W. R., 1957, *Principles of Engineering Geology And Geotechnics*: McGraw-Hill Book Co., Inc.

Lea, F. M. and Desch, C. H. 1956, *The Chemistry of Cement and Concrete*, St. Martin's Press, New York, 637 pp.

Leggett, R. F., 1939, *Geology and Engineering*, McGraw-Hill Book, Inc., New York.

Leighour, R. B., 1942, Chemistry of Engineering Materials: McGraw-Hill Book Co., Inc., New York.

Lemish, J., et al., 1958, "Behavior of Carbonate Rocks as Concrete Aggregates (abs.)," Geol. Soc. Amer. Bull. Vol. 69, no. 12, pt. 2.

Lerch, W. 1959, "A. Cement-Aggregate Reaction That Occurs With Certain Sand-Gravel Aggregates," Portland Cement Assoc. Jour. Research and Development Laboratories, Vol. 1, pp. 42-50.

Loughlin, G. F., 1927, "Quality of Different Kinds of Natural Rock for Concrete Aggregate," *Jour., Concrete Inst.*, 23:319.

Marsal, R. J., 1969, "Mechanical Properties of Rockfill and Gravel Materials (Proprietes mecaniques des encrochements et graviers)," Internat. Conf. Soil Mechanics and Found. Eng., 7th, Mexico Proc., Vol. 3, Suelos, pp. 499-506.

Mather, B., 1969, "Pozzolan: Reviews in Engineering Geology," Vol. 2, (D. J. Varnes and G. Kiersch, editors), Boulder, Colo., Geol. Soc. Am., pp. 105-118.

Mather, K. K., 1953, "Crushed Limestone Aggregates for Concrete," Min. Eng., Vol. 5, No. 10, pp. 1022-1028: AIME Trans. 1953, Vol. 196, 1954.

Mather, K. K., 1958, "Cement-Aggregate Reaction - What is the Problem?" Geol. Soc. America Eng. Geology Case Histories, no. 2, pp. 17-19.

Mather, K. and Mather, B., 1950, "Method of Petrographic Examination of Aggregates for Concrete," Amer. Soc. for Testing Materials Proc., Vol. 50, pp. 1288-1312.

Mather, L. B., Jr., 1948, "Aggregates for Concrete (abs.)," Geol. Soc. Am. Bull., Vol. 59, No. 12, pt. 2, p. 1340.

McConnell, D., et al., 1950, "Petrology of Concrete Affected by Cement Aggregate Reaction in Paige S., chm., Application of Geology to Engineering Practice," Geol. Soc. Amer., Berkey Volume, pp. 225-250.

McConnell, D., Mielenz, R. C., Holland, W. Y., and Greene, K. T., 1948, "Cement-Aggregate Reaction in Concrete," Am. Concrete Inst. Proc., Vol. 44, pp. 93-128.

Mead, W. J., 1936, "Engineering Geology for Dam Sites," Second Congress on Large Dams, Wash. D. C.

Mead, W. J., 1937, "Geology of Dam Sites," Civil Eng., Vol 7.

Melville, P. L., 1949, "Remarks on the Geology of Concrete Aggregates (abs.)," Va. Acad. Sci. Proc. 1948-1949, p. 1939.

Mielenz, R. C., Greene, K. T., and Schieltz, N. C., 1951, "Natural Pozzolans for Concrete," Econ. Geology, Vol. 46, pp. 311-328.

Mielenz, R. C., 1958, "Petrographic Examination of Concrete Aggregate to Determine Potential Alkali Reactivity," Highway Research Bd. Research Rept. 18-C, pp. 29-35.

National Resources Committee, 1938, "Low Dams - A Manual of Design for Small Water Storage Projects," Government Printing Office.

Orr, D. G. and Quick, J. R. 1971, "Construction Materials in Delta Areas," Photogrammetric Engineering.

Plum, N. M., Poulsen, E., and Idorn, G. M., 1958, "Preliminary Survey of Alkali Reactions in Concrete," Ingenioren, Internat. Ed., Vol. 2, No. 1, pp. 26-40.

Portland Cement Association, 1952, "Design and Control of Concrete Mixtures," Tenth Edition.

Rexford, E. P., 1950, "Some Factors in the Selection and Testing of Concrete Aggregates for Large Structures," Min. Eng., v. 187 (2), no. 3, pp. 395-402; AIME Trans. 1950, Vol. 187, 1951.

Rhoades, R. F. and Mielenz, R. C., 1948, "Petrographic and Mineralogic Characteristics of Aggregates," Am. Soc. for Testing Materials Spec. Tech. Pub. No. 38, pp. 20-48.

Rhoades, R. F., 1950, "Influence of Sedimentation on Concrete Aggregate," in Trask, P.D., ed., Applied Sedimentation, pp. 437-463.

Ries, H., 1942, "What Use the Engineer Makes of Geology (abs.)," Royal Canadian Inst. Proc., Ser. 3A, Vol. 7, 1941-1942, p. 25.

Sorbe, V. K., 1969, "Aggregate Serviceability Study (abs.)," Geol. Soc. Amer. Abs with Programs, 1969, pt. 5, Rocky Mts. Sec., p. 76.

Swenson, E. G. and Chaly, V., 1956, "Basis for Classifying Deleterious Characteristics of Concrete Aggregate Materials," *Am. Concrete Inst. Jour.*, 27, (No. 9), pp. 987-1002; discussion, Vol. 28, (No. 6), pt. 2, pp. 1447-1450.

Trefether, J. M., 1949, *Geology for Engineers*, D. Van Nostrand Co., Inc. Princeton, New Jersey.

United States Army, Corps of Engineers, 1953, "Annotated Bibliography Concerning Aggregates,"

Ohio River Division Laboratories, 1940–1950, iii, 212 pp., Mariemont, Ohio.
United States Dept. of the Interior, Bureau of Reclamation, 1946, "Appendix of Geology of Dam and Reservoir Sites," Region II Branch of Project Planning, Sacramento River Tributary Plan.
United States Dept. of the Interior, Bureau of Reclamation, 1955, "Concrete Manual," Sixth Edition.
United States Dept. of the Interior, Bureau of Reclamation, 1960, "Design of Small Dams."
United States Dept. of the Interior, Bureau of Reclamation, 1963, "Earth Manual," First Edition.
Viens, E., 1934, "Does Cement Protect Poor Quality Aggregates? Yes and No," Jour. Am. Concrete Inst., 5.
Weaver, K. N., 1965, "Geologist's Role in America's Cement Industry," Mining Eng., Vol. 17, No. 1, pp. 51–54.
Woolf, D. O., 1958, "Field Experience with Alkali-Aggregate Reaction in Concrete," Highway Research Bd. Res. Dept. 18-C, 8–11.

4.03 The Dam Site

4.031 Geologic Mapping—by Arthur B. Arnold

Introduction. "No phase of dam site investigation pays larger dividends in proportion to the amount of time and money invested."

The above quotation was taken from the Geological Society of America, Berkey Volume, 1950, and indicated the importance attached to surface geologic work at that time.

Geologic mapping for dams today requires even more detailed investigations than in the past since most of the more desirable sites have been utilized and the ones remaining are usually very complicated.

Most of the early design concepts for large hydro projects are based upon interpretation of the geologic map. The main difference between this type of geologic mapping and other types is that this mapping is done for a particular purpose and the map should convey as much information as possible to the designers of the project so they will be aware of the geology of the site and utilize this knowledge in selecting the location and determining the design of the structures.

Geologic features uncovered during construction should be mapped as they will have an influence on any redesign that becomes necessary due to different foundation conditions than expected. The most frequent use of the map during construction is to assist in establishing new cutslopes that become necessary due to removal of landslides or extending the foundation deeper than contemplated. The map is also helpful in planning and performing foundation grouting.

The geologic map is frequently the only "as built" drawing of the foundation conditions and is useful in evaluating any stability, settlement or seepage problems that may occur during the operation of the reservoir.

The final geologic map may also prove valuable in settling future claims and may prove to be an important document to consulting boards of inquiry in the event of a future serious structural failure.

Observe, Interpret and Record. Observation, interpretation and recordation are the three elements of the geologic mapping study. These must be performed with extreme care to produce a useful map. This mapping plays such an important part in planning, construction and operation of the project that it should be performed under the direction of and close supervision of an experienced engineering geologist skilled in recognizing and interpreting those geologic features that are important from the standpoint of project safety and design.

Observation. The quality of the geologic map will largely depend upon the acuteness of the observations made in the field. The geologist should make it a point to examine as many outcrops and structural features as he can, and these examinations should include identifying the rock type, a determination of the strike, dip, joints and structural trends and photographs of the significant geologic features.

Preparing a geologic map can be as important as preparing a legal document and, in fact, the map may be used in claims resulting from a construction contract. Therefore, when preparing the map the features delineated should be clearly marked as to whether they were observed or inferred.

Gullies, canyons and streambeds are the windows to the subsurface and should be "walked out" and examined. These features may prove as meaningful as an exploration trench.

Rodent holes and any other animal, natural or man-made excavations or openings should be observed as they provide knowledge of the subsurface; also, these features can lead to leakage through abutments or, as in the case of some limestone areas, reservoirs that will not hold water. Mine shafts and connecting tunnels have also caused serious leakage problems in reservoirs. Oil wells in the reservoir area can provide useful lithologic data and should be accurately located on the map, as they may require special treatment before construction of the dam.

Water wells that are within the reservoir area or in the area affected by the reservoir should be investigated and pertinent information recorded. The logs of such wells may contain useful clues concerning the thickness and characteristics of formations. They may also provide an opportunity to obtain hydrologic information such as ground water level measurements, formation permeability and water samples for chemical analysis. These data may provide guidelines for subsequent studies for the evaluation of dam foundations or reservoir leakage problems. The information may help in interpreting the geology and could indicate possible seepage and drainage problems that must be considered in the design of the dam.

In many instances, faults act as ground water barriers and cause a high phreatic zone along their trend. This wet area can be recognized by springs or strange vegetation characteristics. Aerial photos, and particularly infra-red, will usually show moist areas quite well, and observation of this feature may also be helpful in delineating landslides.

Vegetation can be a useful guide in mapping in that certain types will prefer soils derived from a particular geologic formation or unit.

Interpretation. In preparing the geologic map, it is necessary to interpret the geologic structure as the field work is progressing so the areas that may hold the key to understanding the geologic feature are not overlooked. This interpretation can be aided by preparing geologic cross sections as the mapping is in progress.

A review of published geologic work in the general area may also provide clues to unraveling the "geologic mysteries" at the site. Discussions with geologists who have worked in the area are also helpful in providing other opinions as to the significance of certain geologic features.

Interpretations made from the preliminary geologic map can aid in selecting locations for explorations that will provide the most information at the least cost.

Recordation. The record of the field work is the geologic map and cross sections, and the neatness and clarity of the maps will increase their usefulness. Geologic descriptions should be in uncomplicated terms and relative to the engineering purpose for which the mapping was done. A text describing the significance of the geologic features should accompany all stages of geologic mapping.

Preliminary Geologic Mapping. Topographic maps may not always be available, or may be of such a scale that they are useless for geologic mapping purposes. In such instances, it may be necessary to conduct surveys to provide a surface topographic base. Good vertical aerial photographs can be utilized as a base for preliminary work.

Geologic mapping during the preliminary investigation should be of sufficient detail and encompass a large enough area so that it can be used for making comparative studies for dams at different locations and for different heights and types. It is assumed that before the geologist begins mapping for a dam site he has reviewed all geologic literature, photographs and maps that pertain to the area and has obtained the best topography for mapping that is available.

Before starting to map, it is helpful to climb a high point that affords a good view of the area. This vantage point assists in getting oriented and in getting a feel for the geologic structure. A helicopter or plane ride is a great help in getting familiar with the area, and in photographing and locating access to difficult locations.

In preliminary studies where good topographic maps are not available, an altimeter is a useful tool, and knowing your approximate elevation can assist with locating yourself in the field. Elevation points observed on low ridges may restrict the height of the dam and reservoir and are important features to observe during the preliminary mapping.

In most preliminary and planning studies, alternatives and changes can lead to considering a dam higher or at a slightly different location than the one first considered in the geologic mapping assignment. Therefore, the geologic mapping for this stage of an investigation should cover a broad enough area so that some geologic knowledge is available for alternative studies.

In a preliminary geologic mapping assignment for a dam, very often the type of dam, concrete, rockfill or earthfill, has not been determined and the geologic mapping should include information on availability and haul distances for construction materials for various types of dams.

The geologic map should be prepared in such a manner that it can be clearly understood by design engineers, as well as identifying rock types, location of faults and other pertinent geologic features. Important engineering decisions will be based upon an interpretation of the geologic map and the geologist should participate in making these important early decisions, as he can provide considerable insight concerning the significance of the geologic features.

Design Geologic Mapping. The design stage of geologic mapping is supplemented by subsurface exploration and is, therefore, more accurate and detailed than that performed for the preliminary work.

The design geologic map should be at the same scale as that being used by the design engineers working on the project ($1'' = 100'$ or $1'' = 50'$).

At this period of geologic work, it is essential that the geologist and the engineer communicat frequently and that geologic features revealed during exploration are transmitted rapidly to the engineers. Geologic sections should be kept up-to-date, as they will assist in evaluating the geology and help in explaining the foundation conditions to the designers.

Usually, the exploration program is reviewed and revised every few months. During these review periods, up-to-date geologic maps and sections will help in indicating what has been learned thus far and what changes or additions, if any, should be made in the investigation.

In the final design of the project, the geologist should participate in the preparation of the plans and specifications of the dam so that the geologic problems will be considered in the design and construction of the project.

Construction Geologic Mapping. On most dam projects, after locating construction materials, geologists are assigned full-time to plan and prepare a geologic map of the foundation of structures and of tunnels, perform additional exploration that may be required, and supervise the grouting program.

It is important during construction to keep up-to-date drawings that reflect the foundation conditions as they are actually encountered. If the foundation is much different than anticipated, then a redesign may be necessary, such as resloping an excavation. Up-to-date geologic maps and cross sections are necessary to evaluate what these changes mean to the overall stability and will help in establishing the angle and configuration of a new cutslope.

The geologic maps prepared during construction are often used in claims arising out of the contract and should, therefore, be as carefully prepared and filed as the other contract documents.

Operation. During the operation of a large hydro project, especially during the first filling of the reservoir, problems of seepage, landslides and settlement sometimes occur, and it becomes necessary to evaluate the significance of the problem and how to correct it. The geologic map and cross sections prepared during design and construction are the base from which the analysis of a foundation problem begins, and therefore it is important that the geologic maps be available for use during the operation and maintenance of the project.

Special Problems. Faults discovered in dam foundations have necessitated lowering the height of the dam, redesign of the embankment and in some cases complete abandonment of the project. Therefore, every effort should be made to determine location, activity, extent and history of any fault located in or near the foundation of a dam during the design investigation.

When a fault is discovered during construction, it can necessitate serious and costly redesign problems, such as moving a tower or the abutment of a dam. The rock in and along a fault is usually broken and contains gouge or soft clay material. In order to improve the foundation, additional grouting or overexcavation becomes necessary.

Landslides during construction are a frequent and difficult problem that usually occur when the foundation is being stripped and shaped for the dam. Large slides can affect the integrity of an abutment, increase the embankment quantities and delay the contractor. Geologic work performed during the design stage of the investigation should investigate slope stability in the general area as well as at the dam site. If the area is one that has a history of landslides, then the investigation of slope stability at the dam site should be more comprehensive than usual.

Temporary cut slopes and cuts made for the contractor's convenience (such as haul roads) should be designated as well as the permanent cut slopes. Most slides occur when the dip of the bedding is undercut or when some geologic feature, such as a fault or joint, has an adverse attitude that affects the cut slope stability.

REFERENCES

Publications that describe methods and application of geologic maps are listed below. Additional information concerning geologic mapping can also be obtained by referring to the bibliographies listed in these references.

1. Application of Geology to Engineering Practice, the Geological Society of America, Berkey Volume, 1950, "Geology in Dam Construction," Burwell and Monemaker.
2. "Geology and Engineering," Robert F. Legget, 1962, pp. 182-190.
3. "Design of Small Dams," U.S. Department of Interior, Board of Reclamation, 1960, pp. 110-117.
4. "Principles of Engineering Geology and Geotechnics," Krynine and Judd, 1957, pp. 270-301.
5. "Geology in Engineering," Schultz and Cleaves, 1955, pp. 357-369.
6. "Field Geology," F. H. Lahee, 1941.

4.032 Geophysical Exploration—
by Jerome S. Nelson

Concept of Engineering Geophysics. Geophysics is the application of principles of physics to the study of the earth. However, applied engineering geophysics is the measurement and interpretation of physical properties of small scale, shallow features that exist within the earth. Some of these features are geologic faults, bedrock configuration, type of rock and geologic structure. It is fairly common knowledge that these features very often have bearing on the location of construction and construction of civil engineering projects.

Defining the nature and concepts of engineering geophysics starts with the basic meaning of geophysics. Physics is best understood by remembering that such subjects as electricity, heat, light, density, magnetism, velocity and other physical properties form the basis of this science. When we look for these same parameters in soil, rock or water, we are specializing in the science and art called geophysics.

In differentiating from geology where the science is observational and largely visual, geophysics is essentially a series of remote measurements of physical properties which are analyzed and interpreted. It is often required practice that most geophysical surveys need other forms of control just as the geologist needs drilling and sampling to shape his geologic thinking and interpretation.

The exploration manager who decides on the use of geophysics in some phase of his study should be sure that his geophysical consultant is not only a reputable one but that he also possesses experience in the particular area in which geophysical consulting is called for.

Geophysics is most commonly employed as a surface mapping method for the delineation of the subsurface. Classification based on the surface, airborne, or bore hole method refers only to operational procedure.

Physical Basis of Geophysics

Seismic.

Basic Physics. Seismic method utilizes the fact that elastic waves travel with different velocities in different rocks.

The principle involves starting the elastic wave (or seismic wave) at a point (explosives or weight drop) and determine at a number of other points the time of arrival of the energy wave that is refracted or reflected through different rock layers.

Refraction

The propagation of elastic waves for seismic refraction occurs when the critical angle is such that the angle of refraction (r) is 90° (see Fig. 4.032.1). The ray travelling along the interface produces oscillatory stress at every point that is refracted to the surface. The wave will arrive at the detector along the minimum time path whether direct (surface) or refracted (subsurface layers). Ultimately, we compute the velocity and depth of these layers from the travel time data. These layers can be geologic material and/or density change.

For angles of incidence, other than the critical, ($r = 90$), elastic waves are not only refracted but reflected.

Quantity Measured. We are measuring travel time in milliseconds of refracted and reflected seismic waves generated by artificial sources.

Method of Measurement. The ground motion generated by an explosion or other sources is detected by a series of geophones as a function of time. The geophone converts the physical motion (wave) to an electrical impulse, which is fed to a seismograph, where it is amplified, filtered, and then recorded on photographic paper.

Special Requirements. In seismic field work, access is needed to lay out cables and geophones. Our usual geophone array in shallow refraction are from 220 to 1100 ft long and should be confined to areas that are either relatively flat or have an even slope. The depth of investigation is increased, by varying the distance between shot point and geophones.

Data Presentation. The refraction data are usually presented in a table of velocities and in cross section of seismic layers. See Table 1.

Interpretation. Using the velocities and the critical distance obtained from the plots of energy travel times, depths are computed according to various formulas for each layer. A multilayer problem has a correspondingly more complex geometry and equations. As in actual practice, problems of insufficient energy in the ground, blind zones, hidden layers, and complex geology will require special computing procedures.

Electrical.

Basic Physics. Electrical methods are actually a field application of Ohms Law in which

$$R = \frac{V}{I} \text{ Resistance} = \frac{\text{Voltage}}{\text{Current}}$$

A direct or low frequency alternating current (I) as introduced into the ground via two electrodes and potential (V) measured at two other electrodes. In relatively porous rock formations, the resistivity is controlled more by the water content and quality within the formation than by the rock resistivity.

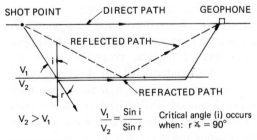

Figure 4.032-1. Propagation of elastic waves for seismic refraction.

TABLE 1. Seismic Velocity Ranges California Materials, California Department of Highways

MATERIAL			RANGE
Name	Type	State	Velocity Ft/Sec
Soil	Clayey		700–1,300
Mud	Bay		1,000–1,400
Gravel	Clayey		800–1,800
Sand	Beach	Wet	3,300–5,000
Sand	Beach	Dry	1,500–3,000
Shale	Cretaceous	Weathered	2,150–4,600
Shale	Cretaceous	Fresh	7,000–12,000
Sandstone	Franciscan	Weathered	2,200–6,000
Sandstone	Franciscan	Fresh	12,500–15,000
Sandstone	Franciscan	Jointed	4,000–7,000
Quartz-Diorite		Weathered	3,000–6,000
Meta-sediments		Fresh	10,000–14,000
Slate	Mariposa	Weathered	2,600–4,000
Slate	Mariposa	Fresh	5,400–14,000

Quantity Measured. The unit of resistivity is the ohm-meter, or ohm-foot, and is an extremely variable physical property in earth materials. The range is illustrated by graphite 10^{-6} ohm-meters and dry quartzitic rock 10^{12} ohm-meters. Common resistivity values are

Materials	in ohm-meters
Fresh water sand	20–100
Brackish water sand	4–20
Clay shale	2–10
Dense fm	1000–10000
Drinking water	10–100

Method of Measurement. One well known method of measurement is the Wenner technique of four electrodes in a straight line at equal spacing. Current is induced through the outer electrodes and the potential drop is measured between the inner electrodes. The basic equation from this configuration is $\rho = 2\pi a(V/I)$ ("a" is separation).

Special Requirements. Acess is needed to lay out cables—the length depends on depth of investigation. Electrical disturbances may cause measurement troubles in developed urban areas. Buried pipe or metal and surface electrical conditions may limit usefulness.

Data Reduction. Resistivity values can be presented in the form of a contour map, a depth curve, or a profile.

Interpretation (Refer Meidav). Although computation is simple, interpretation can be complex.

1. Resistivity contour map—trends, gradients, resistivity distributions
2. Resistivity profile—locations of faults, buried highs
3. Resistivity depth curve—matching field and model curves to determine true resistivities.

Magnetic.

Basic Physics. Many rocks possess a magnetic susceptibility so that when placed in the earth's field, they acquire appreciable magnetization of their own. These will show up in the main field as anomalies. The object of this geophysical technique is to determine shape, depth and magnetic properties of the source of the anomaly and to relate magnetic properties to the rock type. Most magnetic anomalies are produced by the mineral magnetite, a trace constituent in various rock types (ref. Parasnis).

Quantity Measures. The main magnetic field of the earth is roughly equal to that of a magnet close to the center of the earth and inclined at 11°. The maximum intesity is 0.65 oersted and 1 oersted = 1 line force/sq cm.

(1 practical unit is the gamma 1 gamma = .00001 oersted = 10^{-5}) Earth magnetic field

is 30,000–60,000 gammas with local anomalies ranging from 10's of gammas to a few thousand gammas.

Method of Measurement. Field instruments called magnetometers can be either sensitive magnetic balances that measure a component of the earth field or those that measure the total field like the proton precession.

Special Requirements. Ground magnetic stations should be accurately located. Airborne procedures require photographic coverage for flight line control. Topography and geologic structure control the direction of flight.

Data Presentation. Before magnetic readings can be used, corrections for temperature, diurnal (daily), a normal and sometimes a terrain must be applied.

Interpretation. Much of the interpretation of magnetic readings involves only qualitative examination of contour maps observing regional trends, dominant anomalies, gradients and gross correlation with the geology. The next step would be to examine a profile across some anomaly. These anomalies take the form of contoured highs over rises in bedrock, lows over sedimentary basins, and abrupt changes in magnetic gradient at contacts, faults, or dykes.

Gravity.

Basic Physics. Gravitational methods are analogous to magnetic methods in that measurements are made of a natural field of force. Variations in gravity are caused by differences in density of the materials of the earth. Factors which affect density are: composition, depth of burial, texture, mineralization, porosity and metamorphism. (Grant and West).

Quantity Measured. A special name (gal) has been given to gravity units (after Galileo).

1 gal = 1 cm/sec/sec and the more common milligal or 1/1000 of a gal = .001 gal.

Fluctuations in gravity due to geologic features vary from less than 1 milligal to 10's of milligals. Gravimeters are actually sensitive spring balances that can measure gravity variations on the order of 0.01 mgal. Thus, the weight of constant mass in the gravimeter is affected by gravity variations (g) due to density changes of various subsurface materials.

Method of Measurement. Gravity data is usually taken over a previously surveyed grid of stations. Readings can be accomplished by one man with a gravimeter at an average rate of 30 to 50 stations/day.

Special Requirements. Topographic surveys for station elevation accuracy should be to the nearest foot vertically. Station grids can be from 25 ft to $\frac{1}{2}$ mi depending on detail required. Density values can be determined from either surrounding rock outcrops or from drill cores.

Data Presentation. Gravity data is presented in the form of contour maps or profiles which have been corrected for elevation, latitude, topography and free air. Data reduction is now a routine computer operation.

Interpretation. Such anomolous features on contour maps as highs and lows, regional trends, gradients, and amplitude are evaluated. Individual field profiles across significant anomalies are then compared with model curves to ascertain the most probable geologic picture.

Geophysical Bore Hole Logging.

Electrical Logging. Electrical well logging has come into widespread use during recent years. Generally, in bore hole geophysics we are applying the techniques of surface geophysical methods, and by use of downhole probes, measure and record the same physical properties continuously with depth.

(1) Resistivity—With various downhole electrode arrangements, the resistivity in ohm meters is measured for different penetration intervals around the bore hole.

(2) Spontaneous Potential—In drilling of any well, liquid junction voltages arise from chemical differences between bore hole and formation fluids.

Radioactive.

(1) Natural Gamma Rays—This probe basically *measures the clay content of a formation.* Most rocks and soils contain

small but measurable quantities of radioactive elements, mostly potassium, uranium, and thorium which release gamma rays.
(2) Gamma-gamma—This probe is used to measure the density of a formation. *The higher the bulk density* of the formation, *the smaller the number of gamma-gamma rays that reach the detector.*
(3) Neutron—The neutron method is a measurement of the *hydrogen content of a formation.* If the material is rich in hydrogen, there will be a high density of slow neutrons around the source.

Sonic. Velocity logging is a bore hole seismic refraction problem similar to surface techniques. This log will give you an indication of rock type fracturing and porosity.

Fluid Logs.

(1) Temperature—continuous or gradient
(2) Resistivity—resistivity of fluid between two closely spaced electrodes (water quality)
(3) Flow Meter—records number of impeller revolutions against time (fluid movement from one aquifer to another)

Caliper. Simply a measure of diameter of hole with depth.
Other.

(1) Directional—by means of a camera, pendulum and compass all housed in a probe, the inclination and direction of drill hole deviation can be determined
(2) Television—a 3-in. diameter probe that gives a 360° scan of the bore hole wall

Engineering Problems and Geophysics. The exploration manager using geophysics for dam site investigation must decide which method can be used for his specific engineering problem. The geophysical method selected will be the one that, in the judgment of the geophysicist, can delineate the physical anomalies or contrast of the earth materials at the site. The site may be any structure associated with the main dam such as spillways, penstocks, power plants, reservoirs and others. In some situations the best approach may require more than one method.

Table 2 lists the engineering-geologic problems than can arise during a dam site investigation. The column headings designate the type of data often required. Listed under each column heading are the geophysical methods that can be applied to the specific engineering problem. In addition, several of these methods can be adapted for airborne, bore hole or marine measurement.

Geophysical Costs

Cost Classification. Charges for geophysical services are quite variable on the North American continent. Charges for all kinds of services offered by geophysical service companies may be classified into three categories:

Charges per line mile (in aerial surveys) or per lineal foot (in some types of seismic work and bore hole logging). This is commonly used in those areas where data acquisition and equipment used are standardized. Mobilization costs are extra.

Charges per unit time of the geophysicist, technicians, helpers, vehicles, equipment in the field and in the office. This scheme is used where the project is run under field conditions which are not amenable to precise pre-evaluation. A variant on this scheme is the method of charging per crew day, including interpretation and final report. Mobilization costs are extra.

Charges based on a lump sum contract, which assure the client of getting a report on a particular project but leave the consulting organization free to select the proportion and type of various techniques to be employed. This method is favored in complex projects, or when an experienced and reputable consulting firm is employed and the client must know in advance the exact cost of the project.

Ground Geophysical Surveys. A consulting firm will often submit a list of costs, based on the unit charges for its personnel. These are widely variable (by as much as a factor of two),

TABLE 2. Engineering Problem and Geophysical Method

Engineering-Geologic Problem Dams, (Channel, Abutments Reservoir, Spillway Power Plants)	EXPLORATION DATA REQUIRED AND GEOPHYSICAL METHOD APPLICABLE						
	Depth, Thickness Profile	Mapping & Contours	Location	Quantities	Quality Classification	Monitoring	Velocities Elastic Moduli Strength
Materials							
Alluvium-sand gravels	S,E,G,M	S,E,G,M	S,E,G,M	S,E	E		
Weathered Rock	S,E	S,E	S,E	S,E	S		S
Bedrock	S,E,G,M	S,E,G,M	S,E,G,M	S,E	S	S(Micro)	S
Quarry	S	S	S	S	S		
Ground Water							
Water Table	E	E	E				
Seepage		E	E			E	
Porosity		E			E		
Geologic Boundries	S,E,G,M	S,E,G,M	S,E,G,M				
Cavities-Caverns	S,E,G,M	S,E,G,M	S,E,G,M				
Buried Channels	S,E	S,E	S,E				
Cut & Fill	S,E						
Landslides	S,E	S,E		S			
Blasting-Ripping-Excavation			S	S			
Seismic Activity						S(Micro)	
Dynamic Response						S(Micro)	S
Embankments						S,S(Micro)	S
Foundations						S,S(Micro)	S
Reservoir Loading						S(Micro)	
Abutment Loading						S(Micro)	
Corrosion Potential	E	E	E		E	E	
Electrical Grounding Mats			E		E		
Grouting Effectiveness	S	S				S	S
Concrete Deterioration	S		S				S
Fracture Evaluation						S,E	S
Stress Relief Zone	S,E		S,E			S(Micro)	S
Reservoir Seepage			E	E	E		
Reservoir Silting	S(M)	S(M)	S(M)	S(M)		S(M)	

Geophysic Code and Types of Measurement
S   Seismic Refraction—surface, cross hole, bore hole measurements
S(M)   Marine—Seismic Subbottom, Sonar
S(Micro) Microearthquake, Ground Noise, Surface
E   Electrical Resistivity—surface, borehole, marine
M   Magnetic—Surface, airborne
G   Gravity—Surface

but the average figures shown in Table 3 probably represent a middle value between the extremes. The highs and lows will be about ±40 percent of these figures. The charges shown in Table 3 normally include the overhead charges of the home office of the consulting organization. Large organizations and well-established consultants tend to charge the higher figures.

Seismic Refraction. Hammer seismic exploration required one operator plus one or two helpers. In small-scale projects where decisions must be made on the spot as to the extent of the survey, depth to be reached, changes in alignment of profiles, etc., a geophysicist should be present on the project. Approximately one office day per field day must be allowed for interpretation and final reporting in small-

TABLE 3. Average Unit Charge per Man for Ground Geophysical Surveys (1973)

Manpower	Cost per day in dollars
Senior Geophysicist	200–250
Geophysicist	160–180
Senior Operator (party chief)	125–150
Operator/Draftsman	90–100
Helper	50–55
Instrument and Equipment Charge	0.3–0.6% original cost/day

scale surveys. In longer surveys, the ratio of office time to field time decreases. In the exploration of complex geologic conditions or very short surveys, the interpretation and report writing effort may be as much as twice that of the data collection.

Charges for single-channel seismic field operation, interpretation, and final report vary from $250 to $550/day, depending on size of project, manpower, and equipment. A single-channel seismic refraction crew will survey anywhere between 5 and 25 stations/day, varying in length from 100 ft (31 m) to 350 ft (92 m). The productivity of the crew depends upon terrain, distance between stations, noise level in the area, and type of equipment used. All seismic refraction traverses are assumed to be of the forward-and-reverse type.

Twelve-channel seismic refraction requires a larger crew and normally requires explosives. Operation and interpretation charges vary between $500 and $800/field day. Up to 10 spreads (a layout of geophones) are normally accomplished per day. The length of each spread will vary between 200 ft (60 m) and 2000 ft (600 m) normally, although much longer spreads can be carried out wherever required. The speed and efficiency of the operation are greatly increased if dynamite shotholes have been predrilled.

Electrical Resistivity, Gravity and Magnetics. The charges for shallow and deep resistivity exploration are fairly similar to the charges for single and multichannel seismics, respectively. Production rates vary from 15 soundings/day for a shallow exploration depth of 50 ft (15 m) to 100 ft (31 m), to about 3 soundings/day for a depth of 1000 ft (300 m), to one sounding/day if the depth of exploration is 3000 ft (900 m) to 10,000 ft (3000).

Gravity and magnetic survey charges for these services are usually a combination of the manpower costs, Table 3, for a geophysicist, helper and an instrument charge plus mobilization. Add to this, charges for a topographic survey for location of station grid and profiles if required.

Bore Hole Geophysics. The cost of individual logs varies from $0.06/ft ($0.2/m) to $0.20/ft ($0.6/m). The price is considerably lower per log if several logging functions are combined into a single run; for example, a gamma ray log will cost aboue $0.12/ft ($0.4/m) by itself, but a combination of gamma ray and neutron logs will cost only $0.14/ft ($0.5/m).

All logging services are subject to minimum charges per toll, per hole, and per operation. In dam site exploration, it is often these minimum charges which apply, rather than the per foot charge, because of the shallowness of such holes.

Aeromagnetic and Other Aerial Surveys. The charge for aeromagnetic surveys, including final maps and preliminary interpretation, varies between $5.00 and $35.00 per line mile. The variation in charges is primarily due to the variation in mobilization costs.

Infrared radar imaging costs vary between $5.00 and $10.00/line mile. The lower figure refers to an aerial survey of fairly large extent (on the order of 1000 lineal miles) and for little or no processing of the data. It is doubtful because of the cost of mobilization that an airborne survey will be run for less than about $1000, even at a location fairly close to the home base of the service company.

It should be added as a final note on charges for geophysical services that because of the gradual but constant effect of inflation, it would be advisable to add five percent to seven percent per year after 1973 for the charges listed for geophysical services. This hopefully might provide a reasonably accurate estimate of prices for a number of years to come.

REFERENCES

Collett, L. S., "Remote Sensing Geophysical Applications to Hydrology," Proceedings of the 7th Hydrology Symposium, National Research Council of Canada, 1969, pp. 237-260.

Cook, K. L., and Van Nostrand, R. G., "Interpretation of Resistivity Data," USGS Professional Paper 499, United States Government Printing Office, Washington, D.C., 1966.

Dobrin, M. D., *Introduction to Geophysical Prospecting*, McGraw-Hill Book Co., Inc., New York, 1976.

Eaton, G. P. and Watkins, J. S., "The Use of Seismic Refraction and Gravity Methods in Hydrogeological Investigations," Proceedings of the Canadian Centennial Conference on Mining and Ground Water Geophysics, Ottawa, Canada, 1967.

Flathe, H., "Possibilities and Limitations in Applying Geoelectrical Methods to Hydrogeological Problems in the Coastal Areas of Northwest Germany," *Geophysical Prospecting*, 3, 1955, pp. 95-110.

Griffiths, D. H. and King, R. F., *Applied Geophysics for Engineers and Geologists*, Pergamon Press, 1969.

Hall, D. H. and Hajnal, Z., "The Gravimeter in Studies of Buried Bedrock Topography and Density Contrasts in Overburden," *Geophysics*, 27, 1962, pp. 939-951.

Johnson, A. I., "An Outline of Geophysical Logging Methods and Their Uses in Hydrologic Studies," USGS Water Supply Paper 1892, United States Government Printing Office, Washington, D.C., 1968.

Keller, G. V. and Frischknecht, T. C., *Electrical Methods in Geophysical Prospecting*, Pergamon Press, 1966.

Keys, W. Scott, "Borehole Geophysics as Applied to Ground Water," Economic Geology Report No. 26, Mining and Ground Water Geophysics, Geological Survey of Canada, 1967.

Lennox, D. H. and Carlson, V., "Geophysical Exploration for Buried Valleys in an Area North of Two Hills, Alberta," *Geophysics*, 32, 1967, pp. 331-362.

McGinnis, L. D. and Kempton, J. P., "Integrated Seismic, Resistivity, and Geologic Studies of Glacial Deposits," Circular 323, Illinois Geological Survey, 1961.

Meidav, T., "Hammer Reflection Seismics in Engineering Geophysics," *Geophysics*, 34, 1969, pp. 383-395.

Meidav, T., et al., "Application of Non-Explosive Reflection Seismics in Urban Areas," (abstract only) Proceedings of the 1969 National Meeting, Association of Engineering Geologists, San Francisco, Calif., 1969.

Musgrave, A. W., ed., "Seismic Refraction Prospecting," Society of Exploration Geophysicists, Tulsa, Okla., 1967.

Ogilvi, A. A., Ayed, M. A., and Bogoslavsky, V. A., "Geophysical Studies of Water Leakage from Reservoirs," *Geophysical Prospecting*, 17, 1969, pp. 36-62.

Orellana, F. and Mooney, H. M., "Master Tables and Curves for Vertical Electrical Soundings Over Layered Structures," Interciencia, Madrid, Spain, 1966.

Stam, J. C., "Modern Developments in Shallow Seismic Refraction Techniques," *Geophysics*, 27, 1962, pp. 198-212.

Wyllie, M. R. J., *Fundamentals of Well Log Interpretation*, Academic Press, New York, 1963.

American Society of Civil Engineers, Ground Water Management, ASCE Manuals and Reports on Engineering Practice No. 40, 1972.

4.033 Soil Exploration—by Fred Crebbin, IV.

The primary objective for a properly constructed drilling exploratory program is to obtain information by sample extraction to determine the distribution, types, and physical properties of the subsurface soils and rock formations at the site. The state of the art of soils engineering has progressed to a point where information obtained solely from shallow bore holes and from test data derived from such holes can no longer provide all of the data required to properly interpret soil characteristics at a particular project site.

Soil exploration methods can be broadly classified as indirect or direct. The indirect method involves aerial photographic interpretation, geologic reconnaissance, seismic and geophysical techniques including many electronic "down the hole" devices used to correlate physical drill hole data. The direct method involves drilling holes into the ground and extracting samples for visual identification or laboratory analysis as required. Although there are many approaches to the problem of identifying soil types and structural characteristics, no singular method provides all of the information required for all of the types of soil associated with any particular dam site location. Depending upon the magnitude of a particular project, a combination of disciplines involving indirect methods in the preliminary phase followed by direct methods for correlation are generally utilized. The direct method of exploration will be discussed in further detail. Under the broad category of direct soil exploration there are several types of equipment in common use for drilling and advancing the drill hole and likewise a variety of sampling devices available for obtaining samples. Since the method of drilling does not necessarily dictate the method of sampling it is proposed to discuss these separately.

Sampling

General. Sampling is the activity of advancing a bore hole and extracting samples on a continuous or intermittent basis. The basic standard procedures for advancing drill holes are similar for distrubed and undisturbed sampling alike, and it is highly probable that a combination of these two sample types may be required in any one exploratory hole.

Sampling Interval. Initially in the primary exploration phase, exploratory borings should provide continuous samples for classification and testing. The economics and scope of a particular project may dictate that only a few of the proposed boring locations be sampled continuously in an attempt to correlate strata conditions. In general practice, the "Bracket Technique" is used whereby alternate primary holes, offset along the centerline, are sampled continuously until general subsurface data appear to show strata continuity or correlation in respect to soil type and depth. When reasonable correlation is found to exist, intermediate borings can then be established in particular areas to substantiate the primary boring data. The intermediate borings should be sampled at not more than 5 foot intervals, and at each strata change as it is recognized in the advancement of the boring. Where particular problem soils are found or anomalous features encountered that require additional exploration, continuous sampling should be resumed with adjacent primary borings until sufficient information is gained to fully understand the scope and physical properties of the soil and rock formation at the site.

Sample intervals related to rock coring are usually recognized to be continuous and can be accomplished with 5 ft. incremental core barrels up to 20 ft in length. The total depth of conventional exploration borings usually extends to the bedrock contact zone where a core barrel is adapted for continued drilling into rock to a minimum predetermined distance. Rock coring depths will vary depending upon the size of dam structures and obviously the type of rock formations encountered. As rock structures are visibly identified it may be advisable to change from vertical borings to batter borings in an attempt to drill at normal angles to the dip of the formation resulting in larger cross sectional areas of core for improved observation of shear planes, weathered zones and fault identity.

Undisturbed Sampling. A completely undisturbed sample is impossible to obtain; however, careful application of recommended techniques will minimize disturbance and provide a suitable sample for testing. It is recommended that a small portion of each undisturbed sample be taken from the bottom of the sampler and used for visual classification. The sample along with a complete log description is helpful in assigning laboratory tests. The minimum size and amount of undistrubed sample to be obtained should be defined and coordinated with the testing facilities available. In general, the trend toward larger diameter samples results in less disturbance of the center core and improves identification of thin lenses which otherwise may be overlooked in a smaller sample.

The minimum sample sizes of undisturbed samples for various standard laboratory tests on material containing no particles retained on a No. 4 sieve are recommended as follows:

Test	Min. Sample Diameter (In.)
Unit Weight	3.0
Permeability	3.0
Consolidation	4.75
Triaxial Compression	4.75
Uniconfined compression	3.0
Direct shear	4.75

Disturbed Sampling. A disturbed sample is one which contains the same components as the undisturbed sample, however the in-situ soil structure has been altered. Appropriate minimum sizes for disturbed samples must be determined and should be the largest size practical utilizing commercially available sample devices.

The minimum weights of disturbed samples on material passing a No. 4 sieve are as follows:

Test	Minimum Sample Required Lb. (Dry Weight)
Water Content	0.5
Atterberg Limits	0.5
Shrinkage Limits	0.5
Specific Gravity	0.2
Grain size analyses	0.5
Standard Compaction	30.0
Permeability	2.0
Direct shear	2.0
Consolidation	2.0
Triaxial Compression	2.0

Sampling Devices and Application for Use in Undisturbed Soil Sampling.

Sampler Types. A variety of samplers are commercially available for undisturbed sampling. Although samplers differ somewhat in design and method of use, they basically are all variations of "Push-tube" type samplers or rotary core barrels. The Push-tube samplers must be capable of penetrating soil by a smooth continuous pushing action at a rate in the realm of 1.5 ft/min maximum. The tube is pushed preferably by hydraulic drive from the drilling rig proper.

The various types of sampling devices are explained in "Subsurface Exploration and Sampling of Soils for Civil Engineering Purposes," by M. J. Hvorslev, 1949. Several of the most frequently used sampling devices will be discussed in the following paragraphs.

Thin Wall Tube Sampler. These samplers are used to obtain relatively undisturbed samples of cohesive soils. They shall be 2-in., 2½-in., or 3-in. O.D. seamless tubes 30 in. long and shall be constructed of 16 to 20 gauge steel or brass. The tubes shall have a uniform, sharp cutting edge machined to a small angle of bevel and turned in slightly to provide a positive inside clearance. The tube shall be clean and free from rust, both inside and outside and may be reused as long as they remain in such prescribed condition. The Shelby Tube head shall have a clean, operating ball-check valve.

The Shelby Tube should be seated at the stratum to be sampled, at least 1 ft. below the bottom of any casing in the boring, and if more than 6 in. of cuttings or wash are apparent at the sampling elevation, the boring shall be recleaned prior to sampling. The tube shall be pushed into undisturbed soil a distance of 20 in. in a continuous hydraulic or jacking method. No rotation of the sampling device shall take place and no driving of the sampler will be permitted. The resistance to penetration and the rate of penetration shall be recorded on the boring log. Following the stroke, there should be a waiting period of from 1 to 10 min, depending on the sensitivity of the soil. The sampler then should be rotated one turn and brought to the surface. The boring shall be kept full of drilling fluid as the sampler is brought to the ground surface. In very stiff or hard soils, a 12 gauge tube may be used.

The thin wall sampler or "Shelby tube" is used primarily in a cased hole. This sampler performs well in soft to medium clays or silts and in some silty sands of medium density. Loose sands or gravels particularly when saturated cannot be satisfactorily extracted with this sampler.

Clay penetration rates are from 2 to 4 in./min. Drilling fluid pressures range from 35 to 40 psi. Recommended rotation for sand is 75 to 100 rpm. Penetration 4 to 6 in./min. Drilling fluid pressures 15 to 25 psi.

Double tube core barrels are often designed to accommodate leaf spring core catchers to retain soil core. The use of core catchers should be discouraged as they frequently disturb the sample.

One method to improve sample recovery in dense material is to shut off the drilling fluid flow when the sample has been cored to within 3 in. of the desired depth. This will wedge cuttings between the inner and outer barrel shoes, causing the inner barrel to rotate and shear the soil core. The plug formed assists in core recovery and is discarded.

Sampling Devices and Application for Use in Disturbed Soil Sampling

Sampler Types. The selection of samples in this category predominately depends upon the nature of the material to be sampled and the purpose for which the sample is intended.

Augers. Augers range in all sizes from light weight post hole hand augers to large diameter bucket augers. Augers are best suited for drilling above the water table. Hollow core, helical augers provide an open center section which acts as a casing and allows access for other forms of samplers to extract material above or below water table. These continuous flight hollow core augers do not have to be withdrawn between sample runs and allow visual identification of soils as they are rotated. Limitations are generally depth in excess of 100 ft and the presence of gravel, running sands or extremely dense strata.

Split-barrel Sampling. (Also called a Split Spoon). In disturbed soil sampling it is usually required that a representative sample be recovered of the soil stratum using a standard split-barrel sampler. In order to facilitate comparison of the penetration resistance of the various strata encountered and in order to furnish satisfactory data for design purposes, all split-barrel sampling is generally accomplished under the following specifications. The sampling device, a standard split-barrel sampler, 18 in. of split length with an inside diameter of $1\frac{3}{8}$ in. and an outside diameter of 2 in., shall be in good working order with an unflattened and sharp shoe, unbent and unsprung tube section, and a ball-check valve kept clean. The sampler and drillrods should be so assembled that no movement or loss of energy is caused due to loose threads. No unnecessary energy should be consumed in using a hammer slide or steel cable driver.

Core catchers shall be used with the split-barrel sampler when necessary to insure adequate sample recovery in cohesive soils. A split-barrel sampler equipped with a flap-valve or sand trap shall be used to recover samples of "running" cohesionless soils if the sampler fails to recover a satisfactory sample after two (2) attempts at a given elevation. The use of such retaining devices shall be so noted on the boring log.

The sampler shall be seated on the stratum to be sampled at least one foot below the bottom of any casing in the boring with one (1) blow of the drive hammer. The sampler shall be then driven at a constant rate with no interruptions by a drive hammer weighing 140 lb and falling 30 in. The distance the sampler shall be driven shall be 18 in., and the blow count for each six (6) in. of penetration shall be recorded on the boring log. At least 10 in. of sample should be recovered. If less than 10 in. of sample is recovered, the boring should be cleaned to two (2) ft below the top of the preceding sample and another attempt made for a valid sample. If four (4) or more in. of cutting or wash are present in the sampler, the length of cuttings should be recorded on the boring log. The samples obtained by split-barrel samplers shall be representative of the stratum sampled and shall not consist predominantly of "wash." "Wash samples" are not acceptable unless, in the opinion of the Site Engineer, failure to obtain a "dry sample" is not the fault of the methods or equipment of the contractor. For any one sampling interval, should the number of blows required to drive the sampler 12 in. exceed 100, the sampler need only be driven until the number of blows reaches 100, and the amount of penetration obtained shall be recorded on the boring log.

In soil containing coarse gravel, the $1\frac{3}{8}$-in. I.D. split-barrel sampler may be replaced by a 2-in. I.D. split-barrel sampler, at the discretion and direction of the engineer. If any sampler is driven on coarse gravel that becomes lodged in the shoe of the sampler, this fact should be noted on the boring log.

Displacement Samplers. Displacement samplers are often used to advance bore holes for disturbed sampling. The sampler contains a retractable cone point which is driven to a desired sample depth, the plug or point is retracted and the sampler is once again driven to obtain the sample. The Memphis and Porter Samplers are examples of displacement type samplers and have a $1\frac{1}{4}$ in. O.D. by 1 in. I.D. barrel and drive shoes. The Porter sampler is fitted with brass segmental liners consisting of seven segments, 1 in. O.D. by $^{15}/_{16}$ in. I.D. by 6 in. long. The samples are sealed and transported in the segments. Samplers of this type are limited to relatively shallow depths of 30 to 40 ft. and are difficult to drive in dense granular material.

Fragmentary Samplers. Fragmented samples are often very useful in obtaining correlation data on depth of bedrock and rock type. The drill rig is equipped with a "down the hole hammer" device that advances the hole by percussion fracturing of the strata. The drilling medium is generally compressed air which drives the cuttings to the surface and deposits them in a cyclone separator for examination. The samples are entirely mixed allowing visual examination, however, strata delineation is extremely difficult. This type of sampling can be accomplished rapidly in the range of 60 ft/hr.

Stationary Piston Samplers. The thin wall stationary piston samplers are used to obtain samples of very soft to stiff clays, silts and sands both above and below the water table. This sampler is utilized in a cased hole and has two principle advantages: (1) It is fully sealed at the bottom and can be lowered through fluid and cuttings without fear of sample contamination. (2) By pushing the sampler downward while holding the piston stationary, the top of the sample is protected from distorting pressure at the top. This type of vacuum seal is generally more effective than the ball check valve in the Thin Wall Tube Sampler previously described. The tube in the Stationary Piston Sampler is assembled with the piston locked in the down position flush with the bottom of the tube or cutting shoe. This sampler requires that actuating rod sections are added within the drill rods as the hole is advanced. The actuating rods should extend 6 to 8 in. above the mating drill rod. These extensions remain connected to the piston and are clamped at the ground surface, preventing movement of the piston during the sampler advance. After completion of the drive, additional clockwise rotation releases the activating rods for withdrawal. The split cone clamp is adjusted by clockwise rotation of the drill rods and prevents downward movement of the piston during withdrawal of the sampler. The piston moves relative to the sampler during the push only and helps pull the sample into the tube and retains it. Some versions of the stationary piston samplers utilize a vacuum release rod within the lower section of actuator rods to aid in removing the sampler head.

Double Tube Samplers. Two examples of double-tube samplers are the Denison Sampler and the Denver Sampler. Both are similar in design and can be utilized in uncased holes to recover samples from dense or highly cemented strata where the Thin Wall Tube Samplers are unable to penetrate or extract satisfactory undisturbed samples. The Double Tube Samplers are basically a double tube core barrel similar to that used in rock drilling and utilizes a thin wall stationary inner tube which accepts the sample cut by the rotational penetration of the outer barrel. The inner barrel can be adjusted in length relative to the outer barrel cutting surface and extended beyond the outer barrel when drilling soft formations to prevent sample erosion, likewise, retracted for drilling in dense formations. Recommended rotation for clay is up to 100 rpm.

Drilling. The following are minimum specifications that should be acceptable guidelines for any basic soil drilling program. The objective here is to advance the boring between samples so that a clean and stable boring is made available for the sampling operation; and, of equal importance, so that the soil at the sampling level is not disturbed or contaminated during the drilling operation.

When a predominance of granular soils is expected on a project, it is, therefore, suggested that the requirements be met as set forth herein regarding the use of casing or drilling mud. In no case should a boring made by a drilling rig be acceptable where no drilling fluid and casing, air, or drilling mud is used to remove cuttings from the boring.

In general, casing should be used throughout the project as a means of maintaining an open, stable hole. Drilling mud may be used in lieu of casing to maintain the sides of a drill hole if the use of such material is approved by the project. The soils engineer shall recommend the type of mud to be used and the methods of mixing and circulating in the field prior to use by the contractor.

The driller should not attempt to use drilling mud in lieu of casing without the approval of the site engineer. If the drilling mud prevents the driller from obtaining satisfactory or non-contaminated samples, then the use of mud should be abandoned and casing shall be used.

Casing should be driven vertically through the soils or other materials by means of a 300-lb. hammer falling not more than 24 in. The blow count of the hammer shall be recorded on the boring log for each foot of drive. The bottom of the casing shall be kept at least 1 ft. above the location of the next sample. After the casing has been driven to the desired depth, the soil in the casing shall be removed by washing, jetting, chopping, or some other suitable means.

The drilling bit should contain side discharge ports in lieu of face discharge type. Drag bits in which the discharge is directed against the blades are suitable and not subject to jetting out the hole. To insure a clean hole after the desired sample depth is reached, the drill bit and fluid circulation are to be continued until drill cuttings are in suspension or removed from the hole. Bit rotation in sand should not exceed 100 rpm at 4 in. to 6 in./min penetration. Drilling mud pressures of 15 to 25 psi are generally adequate to stabilize holes in sand with heavy drilling mud. Bit rotation in clayey soil should not exceed 300 rpm at penetration rates of 1 to 2 in. per minute. Drilling mud pressures for medium weight consistency should range from 25 to 40 psi. It is generally recognized that the thinnest mud that will satisfactorily stabilize the hole is the most efficient.

It is necessary that the cuttings be inspected continuously during the drilling to detect changes in material. When mud is used or when drilling fluid becomes mucky, a screen and a pail of fresh water shall be used to help detect stratum changes. A slow rate of drilling is a necessity so that stratum changes may be detected before the stratum is completely penetrated.

The water level in the drill hole should be maintained at a higher level than the free ground water level at all times during the drilling and sampling operations.

Procedures for Handling Soil Samples. The sampler should inspect and package soil samples on a clean working surface so as to prevent contamination of the soil sample and to avoid errors in his work. The following procedures are related as examples for specific types of samples and represent general requirements for disturbed and undisturbed sample handling.

Split-Barrel Samples. Split-Barrel Samples are disturbed during the sampling operation, therefore, it is not necessary to retain the structure of the sample during packaging. There are instances, however, when soil structure is evident in a Split-Barrel Sample; this fact should be noted on the boring log and the structure preserved, if possible, during packaging.

Disturbed samples are tested in the laboratory for: visual identification, moisture content, and for other classification tests. It is important that a natural moisture content determination be performed upon soil at its natural moisture content. Drilling mud or free water should be scraped or wiped from the surface of the sample, and the sample should be placed in a pint jar as quickly as possible. The jar should be air tight. The jar should be labeled and the lid should be marked with: job number, boring number, sample number, and depth of sample.

Shelby Tube Samples. A Shelby Tube Sample is relatively undisturbed as it is brought up to the ground surface. Great care should be used so that the sample remains undisturbed until it arrives at the laboratory.

When the Shelby Tube Samples are to be extruded in the field, the cutting edge of the tube must not have been bent out of shape prior to extruding. If a bent edge cannot be straightened by pliers, a hacksaw should be used to remove the damaged end of the tube. The sample should be extruded at a constant rate and no unnecessary stress should be applied to the sample. A wire saw should be used to remove the best 6 in. section of the sample. This section should be wrapped in wax paper and the edges coated with paraffin applied with a small brush. About 1/4 in. of paraffin should be allowed to stiffen in the bottom of a quart carton prior to placing the 6 in. sample. Paraffin should be used to fill the remaining space in the carton around and above the sample. The paraffin should be kept at the lowest temperature possible, and still remain liquid, because paraffin shrinks upon cooling and will leave large air spaces in the seal.

The carton should be labeled immediately after placing the sample in the carton, top end up, so that cartons do not become mixed. In addition to the 6 in. section of soil placed in the carton, a 3 in. section of the same material should be placed in a pint jar. This jar sample will assist the engineer during assignment of laboratory testing. A hand penetrometer reading should be made upon the jar sample in the field.

Should it be necessary to preserve the Shelby Tube Sample inside of the tube, a 3/4 in. length of the soil from the cutting edge should be removed and placed in a pint jar. Paraffin should be placed in both ends of the tube, followed by caps and tape. The tube should be labeled and also marked with ink or paint.

Preparation of Field Drilling Log. The information set down on the field log must be correct, and the more complete it is, the better and should contain the following:

a. Job name, boring number, boring location surface elevation, weather observations.
b. Water level observations, taken at 0 and 24 hours after the completion of the boring and at the completion of the job, depending upon project requirements.
c. A complete record by depth and number of each sample taken or attempted.
d. A record of sample type attempted and amount of sample recovered.
e. A record of the blow count required to drive the sampling device for each 6 in. of penetration and the recovery.
f. A complete classification of the material encountered and the limits of each soil or rock stratum. The following sequence will be followed in classifying a soil or rock.
 (1) Condition
 (2) Color
 (3) Texture
 (4) Modification

All samples must be correctly and completely marked, using an ink marking pen and gummed labels, immediately after they are taken. All samples must be tightly capped or sealed. The following tables outline typical standard descriptions for soil classification.

Soil Classification

Condition. The condition of a soil mass refers to its relative compactness (granular soil), or to its relative consistency (cohesive soil). The condition of a stratum should be noted where major changes occur. For example, if very soft clay is encountered at 10 ft. and it becomes soft at 20 ft., this fact should be noted on the field log.

Relative Compactness of Granular Soils.

Description	D_r-%	Blows/foot
Very loose	0 to 15	0 to 4
Loose	15 to 35	4 to 10
Medium Dense	35 to 65	10 to 30
Dense	65 to 85	30 to 50
Very Dense	85+	50+

Relative Consistency of Cohesive Soils.

Description	Qu–tsf	Penetrometer	Blows/foot
Very soft	0.0 to 0.25	0.0 to 0.25	0 to 2
Soft	0.25 to 0.5	0.25 to 0.5	2 to 4
Firm	0.5 to 1.0	0.5 to 1.0	4 to 8
Stiff	1.0 to 2.0	1.0 to 2.0	8 to 15
Very stiff	2.0 to 4.0	2.0 to 4.0	15 to 30
Hard	4.0+	4.0+	30+

The condition of Cohesive Soil can also be determined by the following method:

Very Soft   Tall core with slump under its own weight.

Soft   Core can be pinched into two between fingers.

Firm   Core can be imprinted easily with fingers.

Stiff   Core can be imprinted only with considerable pressure of fingers.

Very Stiff   Core can be imprinted very slightly with fingers.

Hard   Core cannot be imprinted with fingers.

Color. The color of the soil should be described in clear and concise terms. Terms such as "brownish," "blackish," "dove gray," and "sky blue," should not be used.

If color appears to be a cross between gray and brown, describe the color as gray-brown or brown-gray using the modifying color preceding the major color. Some soils are mottled and the color description should read as follows, depending upon the degree of mottling: "mottled, brown and gray," or "brown with gray mottling."

Texture. The following soil components and sizes shall be used on all logs; (Unified Soil Classification System–T.M.–3-357, Waterways Experiment Station, 1953):

Component	Size
Boulders	Larger than 8"
Cobbles	3" to 8"
Gravel	
Coarse	3/4" to 3"
Fine	No. 4 to 3/4"
Sand	
Coarse	No. 10 to No. 4
Medium	No. 40 to No. 10
Fine	No. 200 to No. 40
Silt and Clay	Below No. 200

The typical soil names are:

Sandy Gravel
Gravel
Gravelly Sand
Sand
Silty Sand
Silt
Clayey Silt
Silty Clay
Clay
Organic Clay
Organic Silt

Organic materials should be classified as follows:

    Sedimentary Peat
    Fibrous Peat
    Woody Peat

Most soils are composed of a major component, as described previously, and a minor component or components.

Minor components should be noted on the field log as follows:

Term	Range
Trace	0-10%
Little	10-20%
Some	20-35%
And	35-50%

It should be noted that gravel is rounded and rock fragments are noticeably angular.

Modifications. Modifications, preceded by the word "with," include all added information concerning the soil condition, texture, color, and all inclusions which may be present. Well graded, poorly graded, organic odor, maximum size—one inch in diameter, contains wood chips, and with numerous fissures are typical modifications.

The following are two sample soil descriptions:

> Loose to medium-dense dark gray fine to medium sand, trace fine gravel, with wood chips.
>
> Stiff to very stiff brown-gray silty clay, little fine to coarse sand, trace fine gravel, with gray clay streaks.

4.034 Exploratory Drilling in Bedrock by F. E. Sainsbury. The U.S. Department of the Interior[1] states "The purpose of drilling and logging is to secure evidence of the 'in place' conditions of the rock." The intent of this section is to acquaint the reader with the various methods of drilling that can be utilized to enable him to secure that evidence.

In selecting a drilling method, consideration must be given to the complexity of the formation structure, accessibility to the area, and the magnitude and extent of faulting, folding or fracturing of the formation. Initial drilling may also indicate the need for a second or a third method.

The most commonly used methods of exploratory drilling in bedrock are:

> Diamond core drilling
> Percussion drilling
> Rotary non-core drilling
> Churn or cable tool drilling
> Shaft drilling
> Calyx drilling

Because more dam site feasibility footage is drilled using diamond drill methods than all the other methods combined, we will concentrate on this area.

Diamond Core Drilling. James D. Cumming[2] states "The principle of core drilling comes to us from remote antiquity." However, the first machine to utilize diamonds set in an annular ring was built in 1862-63 and was operated by manpower. From this humble beginning we now have rigs with the capability of drilling to 15,000 ft. in depth.

The diamond drilling equipment manufacturers (DCDMA)[4] have established standards for the industry that enable an engineer to write specifications with which all drilling contractors can comply. These same "standards" also indicate nominal hole and core sizes which allow the engineer to calculate hole volume and core diameter.loss through washing.

Diamond drilling always requires the following:

A. *A source of power converted to rotation.* Large surface rigs are usually powered with diesel or gasoline motors; small surface rigs, underground or gallery drills with air or electricity. A recent newcomer to the industry utilizes hydraulic pressure to power a down-the-hole hydraulic motor.

B. *A thrust exerted upon the bit sufficient to force the diamonds or tungsten carbide inserts to cut into the rock at the hole bottom.* Studies by engineers of a major diamond bit manufacturer[3] suggest the following bit rotation, weight, and coolant.

TABLE 1. Suggested Operating Parameters for 25/Carat Stones

Rock	Bit Description	RPM Min.	RPM Max.	Weight on bit	GPM Approx.
Soft	EX	900	1600	1500	3
Hard		700	1500	2500	
Soft	AX	600	1100	2000	6
Hard		400	900	3500	
Soft	BX	350	700	3000	9
Hard		250	600	5000	
Soft	NX	200	600	4500	14
Hard		100	500	7500	
Soft	NC	100	400	5500	22
Hard		100	250	9500	

They also suggest this thrust may be developed by:
1. Weight of the drill string only.
2. Use of a hydraulic head or pull down to impose downward thrust upon the drill string and the bit.

3. A "screw feed head" which develops drill string thrust through a gear differential arrangement.
C. *Diamonds or tungsten carbide inserts set into a bit which is attached to the core barrel.* The selection of the correct bit style is a technical skill obtained through years of experience and no attempt will be made to acquaint the reader with these skills. It is normally best to leave bit selection to the specialists.
D. *A coolant and flushing media introduced through the drill string at the surface.* Most frequently a liquid (clear water, drilling mud, water and soluble oil mixture) is used to cool the bit cutting surface, remove the rock cuttings from around the diamonds and carry these cuttings to the surface. (The cuttings may then be collected for correlation with the core samples.) A secondary and less frequently used coolant is compressed air.
E. *A method of collecting, protecting and extracting the core.* In most formations a double tube, swivel-type core barrel will obtain acceptable core recovery. In soluble, highly fractured or loosely cemented formations a triple tube core barrel with a split or rubber sleeve inner tube will usually increase the recovery percentage.
F. *Extraction of the core from the hole bottom.* This is accomplished by:
1. "Conventional" methods whereby the entire core barrel with the core sample is retrieved by extracting the entire string of rods.
2. Extraction of the inner tube containing the core sample by use of a release assembly and a cable. This is known as "wireline drilling" and is the most common in use today because it does not require extracting the entire string of rods. Not only does this method save time and reduce work, but hole deterioration is virtually eliminated. Since "cave" is not permitted to enter the hole, core recovery is usually improved and bit life is increased. All of these factors combined enable deeper holes to be drilled at less cost per foot. In the attached schematic illustration, Figure 4.034-1(a) shows the inner tube in a locked position inside the core barrel full of core. In (b) the "overshot" has been lowered into the hole, and has depressed the fingers of the locking head. The inner tube and core are shown being extracted by use of a cable and winch. (c) displays the empty inner tube free falling toward a locking position in the core barrel. In (d) the inner tube has advanced to the locking position and is latched ready to receive core.
3. A double tube drill string which permits the core to be pumped to the surface through the inner tube. This method, introduced in recent years, is very effective in formations which resist coring because the entire sample of the formation is recovered either as core, chips or sludge. The use of the double tube rod string and the sample extraction through the inner rod eliminates contamination from the hole wall and assumes an accurate sample. A disadvantage in this method is that the core is subjected to prolonged action of the coolant. Also, soluble zones may go into solution while enroute to the surface, requiring that both core and sludge be correlated for formation evaluation.

Percussion Drilling. A secondary and increasingly important method of exploration is percussion non-core drilling done either with a wagon (airtrack) drill or with a down-the-hole hammer. Either method is less expensive and faster than core drilling and makes an ideal hole for use with down-the-hole photographic and television cameras or electronic and geophysical

224 HANDBOOK OF DAM ENGINEERING

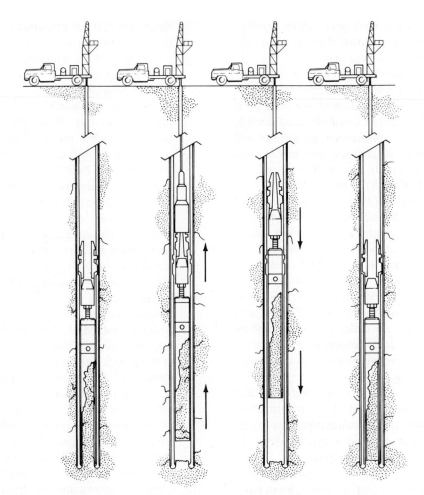

Figure 4.034-1. Extraction of core procedure.

logging equipment. The development of these small cameras now enable engineers to study formations "in place" and should greatly reduce the need for large diameter (24" to 36") drilling.

Both the airtrack and down-the-hole hammer method utilize compressed air to exert impact upon the bit. The airtrack applies a hammering effect upon the top of a string of steel rods. Its effectiveness is limited as the rods eventually absorb all the impact.

The down-the-hole hammer method has greater capability because the rods conduct the compressed air to the hammer which in turn impacts directly upon the bit at the hole bottom.

From both methods a sample is obtainable by capturing the chips and dust as they are exhausted from the collar of the hole. Effectiveness and sampling are both adversely affected by moisture in the hole and drilling limitations are frequently determined by the inflow of water. Accuracy of the samples is also in direct ratio to the competence of the hole wall. Obviously, consideration must be given to the accuracy when the volume and/or weight of the captured sample differs greatly from that which should be available from a given sized hole. Alternate hard and soft bedding or zones are especially difficult to accurately sample. This is because the abrasive action of the chips in the exhausted air erodes into the soft zones causing contamination of the sample.

Rotary Non-Core Drilling. Rotary non-core drilling is frequently used when rock formation type or change information is desired. For this purpose it is a rapid, usually inexpensive method of drilling. Generally a large, heavy rig is used which enables a tremendous weight to be put on the bit through the use of "collars," a heavy string of rods and the rig itself. Bits are a roller "tricone" type with either hardened chisel point design or with tungsten carbide inserts. Compressed air, which permits the collection of chip dust samples, is usually the coolant. If too much water is encountered, a drilling mud is substituted but this greatly complicates the sample collection and evaluation.

Rotary methods are of little or no value in formation competency, and fracturing studies. The rotary holes are, by nature of their size, ideal for use with in hole industrial TV cameras or for installation of geophysical or electronic instruments.

Churn, Keystone or Cable Tool Drilling. Primitive man must be given the credit for inventing this method of drilling after he learned that a hard object repeatedly dropped upon the same spot eventually chipped or wore away the rock and in time made a hole in the formation. When steel came into use, this method enabled the drilling of larger and deeper holes but required devising a method of extracting the crushed rock from the bottom of the hole. This problem was solved by adding a small quantity of water to the cuttings, making a slurry, and then bailing this slurry from the hole.

Today a churn drill is a large rig usually trailer or truck-mounted and powered with a diesel or gasoline engine. The bit is of forged or tungsten carbide tipped steel and weighs from 100 to 1000 lb, depending on its diameter. This bit is attached to a stem which is normally solid steel slightly less than the bit size in diameter. The stem measures from 15 ft. to 30 ft. in length and weighs between one and five tons. To this is attached a cable leading to the drill. Impact is achieved when an arm on the drill alternately depresses and releases a span in the cable which alternately raises and drops the stem and bit to the hole bottom.

When crushed rock accumulates in the hole bottom, the force of the blow is cushioned and these cuttings must be removed. For this procedure water is added into the hole, the stem and bit are withdrawn, and a bailer is lowered. The slurry enters the bailer and is entrapped with a flapper valve on the bottom. The bailer is then extracted, the slurry is screened and cleaned and an evaluation of the rock type is made from the chips. Exactness of information is questionable because cave or loose rock along the hole walls could fall into the hole and contaminate the sample.

Use of churn drilling for dam site investigation has decreased markedly during the past 30 years and is now used only when other methods of drilling cannot be utilized.

Shaft Drilling. While this drilling method was developed for and is primarily used in the mining industry, it will undoubtedly be utilized in the construction industry in the future. Equipment currently in use is capable of drilling holes from 36 in. to 96 in. in diameter. A major contractor has indicated his intent to construct a drill capable of drilling 12 ft. diameter shafts in the near future.

Shaft drilling could replace Calyx drilling on projects where a hole of sufficient diameter to accommodate a geologist must be drilled. An advantage of this method is the reduction in time required to complete a hole. A disadvantage is the high cost of mobilization and site preparation because of the massive rigs.

Shaft drilling would not normally provide a core sample that is obtained from Calyx drilling. However, chip samples are recovered from shaft drilling through reverse circulation methods. This normally suffices because of later visual inspection.

Calyx Drilling. Calyx or shot drilling is frequently employed when a large diameter hole and core sample is required. Generally, the need for this method is determined by previous drilling methods. Calyx drilling is a rotary coring method wherein a large diameter, blunt-nosed steel barrel is rotated in the hole. Cutting is accomplished by pouring or injecting steel shot into the hole. The shot is trapped beneath the blunt bottom of the barrel and the subsequent rolling and tumbling action wears away the formation. Normally, the core is removed every two or three feet. Various methods are employed to break off this large core. The most common is to lower a man into the hole and to then drive a wedge between the core and the hole wall. An alternate method is to place prima cord or another explosive alongside the core at the hole bottom and then shoot it off. Usually a hole is jackhammered into the top of the core, a pin with an eye is inserted, and the core is lifted out of the hole with a hoist. Calyx cores of 3, 5, 8, and 10 ft. in diameter have been extracted from some dam sites. Generally, a geologist is lowered into the completed hole to log the hole walls and to take pictures.

An engineer, when involved in a formation evaluation study should recognize that each method of drilling is most desirable in a particular situation. Normally, however, it is not practical to employ all the aforementioned methods. One or two methods can usually accomplish all that is required and other techniques should be reserved for special or unusual situations.

REFERENCES

Material for this portion of Chapter 4 is based on the following publications:

1. U.S. Department of the Interior, Bureau of Reclamation. "Handbook on Small Dam Construction."
2. "Diamond Drill Handbook," by James D. Cumming, B. A. Sc., P. Eng., Mining Engineer.
3. Technical Manual Mining and Construction Volume 1, Christensen Diamond Products Company.
4. Standards, Bulletin No. 3, Copyright 1970, Diamond Core Drill Manufacturers Association.
5. Schematic prepared by Don Harward, Boyles Bros. Drilling Company, Engineer, and depicts Boyles' patented wireline core barrel.

4.035 Sites for Appurtenant Structures—by John W. Marlette

Spillways. Geologic exploration for the spillway should begin with preparation of a geologic map of the spillway and adjacent areas. Subsurface exploration, usually exploratory drilling or trenching along the spillway alignment is then done.

Insofar as spillway foundations are concerned, bearing capacity can be important, but usually the weight of material excavated is greater than the weight of the spillway, so the foundation generally is able to support the structure proposed. Soft materials should be identified by the geologic investigation so they can be removed. Care must be exercised to ensure there are no bedding planes, joint sets, fracture systems or other rock structures with unstable orientation that could cause landslides in the foundation.

Stability of both cutslopes proposed and natural slopes adjacent to the spillway should be studied with great care. Potentially unstable slopes are most apt to fail after saturation by heavy rain or during earthquakes—times when the proper functioning of the spillway is important to prevent over-topping of the dam. Stability of rock slopes requires thorough geotechnical analysis, taking into account bedding planes, joint and fracture patterns, ground water conditions and the physical condition of the rock. Stability of slopes in earth materials should be analyzed with appropriate soils mechanics techniques. Stability may be verified after construction by installing slope indicators to monitor performance of the slopes.

Where deep cuts are made for the spillway, particularly in shales and siltstones, considerable rebound may occur. The best way of dealing with rebound is to allow an interval of, say 3 or 4 months, after excavation to allow the rock to finish rebounding before placement of concrete. Rebound can, and should be, monitored by installing rebound gages at invert elevation depth prior to excavation. Such gages should be monitored until rebound has diminished.

Excavations made in finer-grained sedimentary rocks such as clay shales and siltstones frequently slake, primarily because of changes in moisture content caused by exposure to the atmosphere. After final cleanup and the foundation is ready for concrete placement, slaking can cause the foundation to deteriorate to an unsuitable condition unless promptly covered with concrete in a matter of hours. If this happens over-excavation is required to prepare again a foundation suitable for concrete placement. The geologist should detect slaking in drill core and fresh cuts, and if slaking is a problem, the spillway excavation should be stopped short 2 or 3 ft. above the final grade. The final few feet should be removed just before concrete placement so the foundation is covered before slaking can start.

Water should be prevented from leaking through the foundation in the spillway area. Leakage generally is checked at the weir, or control area, of the spillway—and if the spillway is adjacent to the dam, certainly no further downstream than the grout curtain of the dam. Generally, watertightness is developed by extending the grout curtain from the dam on across the spillway structure, when the spillway is adjacent to the dam, but where foundation rock cannot be grouted, watertightness may be achieved by impervious blanketing. The geologic investigation should give the designer the information needed to develop a system compatible with the geologic conditions.

One of the first questions in preliminary design is whether, or not, the spillway has to be lined. The need for lining is determined by geologic conditions and velocity of the spillway discharge. Softer, sedimentary rocks that erode easily unquestionably need to be lined. The decision on the need for lining is more difficult to determine in harder sedimentary rocks and in crystalline rocks where rock structures are the important consideration. If the rock has platy or flaggy structure, such as volcanic flows or thinly bedded sandstones, it will be susceptible to plucking by the rush of water, particularly if unfavorably oriented.

Plucking can also occur in closely jointed or fractured rock where the blocks of rock between joints are small. Massive, hard crystalline rock generally does not have to be lined, but, unfortunately, this kind of rock seldom occurs in spillways.

In stilling basin, energy dissipator, or flip bucket areas at the end of the spillway, the design should be compatible with foundation conditions, so erosion will not damage either the structures, or the surrounding area.

Invert slabs in the spillway chute frequently are anchored or keyed to the underlined foundation. Selection of which method to use depends somewhat on geologic conditions. Keyways are difficult to cut satisfactorily in hard crystalline rocks, and therefore anchor bars are preferable. In softer sedimentary rocks where keyways can be cut, keyways may be more desirable.

Tunnels. Regardless of the type of tunnel under consideration, initial exploration procedures are about the same. The initial step should be the preparation of a detailed geologic map along the tunnel alignment. From this map detailed geologic cross sections should be prepared showing geologic conditions anticipated along the tunnel alignment. Either concurrent with, or after this phase of work, exploratory drilling should be done along the tunnel alignment to determine physical conditions of rock at tunnel grade. Information on rock condition from exploratory drill holes may be supplemented by determining seismic velocities along tunnel grade by seismic refraction surveys. Continuous diamond coring should be done in the vicinity of the tunnel grade and these cores should be retained for future testing and prospective bidders.

During exploration drilling, water pressure tests should be made so permeability of the rock can be determined, and also it is important to note the levels of the ground water encountered. Temporary casing should be installed in drill holes so water levels can be monitored. Springs in the area also should be canvassed and measured. These ground water data are needed to determine how much water will be encountered during tunneling and also how tunnel construction affects ground water conditions.

Basic questions with answers dependent on geologic conditions are: can the tunnel be built? Will it have to be lined? Will the portal areas be stable? Is there a possibility for tunnel collapses? Will raveling or running ground be encountered during the tunnel excavation? Will ground water be a problem? What kind of support? Many of these questions have to be answered, in a subjective way, primarily from evaluation of core obtained from drill holes along the tunnel alignment, and the geologic mapping.

Information on the modulus of deformation frequently is needed by the designer. A rough approximation of the modulus of deformation can and should be obtained by laboratory testing of core samples, but it should be recognized laboratory tests are made of a sound piece of core. The fracturing of tunnel wall rock by tunnel construction lowers the deformation moduli, so laboratory tests of drill core usually yield values for deformation moduli about 4 to 10 times greater than moduli determined by in-situ tests in tunnel walls. For this reason, if deformation modulus is critical, test adits should be driven along the tunnel alignment and test chambers made for jacking tests. It is important to orient test chambers so jacking stresses will have the same orientation as design stresses.

For tunnels whose upstream end will be underneath the reservoir, precautions must be taken to ensure the tunnel does not become a conduit or a pathway for leakage, particularly along the outside wall of the tunnel lining. In hard fractured crystalline rocks this can be handled best by making an impermeable curtain of grout around the tunnel in several places. The geologic investigation, should indicate the best geological location for permeability barriers along the tunnel. In soft, fine-grained sedimentary rocks that are relatively unfractured, grout penetration is difficult to achieve, therefore, a permeability barrier cannot be developed by grouting. Under these conditions, the permeability barrier can be achieved by building cutoff collars and the geology along the tunnel should be studied carefully to select the best spots for cutoff collars.

Powerhouses. Exploration programs for underground plants should start with a detailed surface geology map and preliminary exploratory drill holes to develop a preliminary evaluation of geologic conditions at the underground site. During this phase of investigation, geologic structures such as joint sets, faults or fracture patterns may become evident which would make reorienting or relocating the underground chamber desirable. To more accurately evaluate underground conditions exploratory adits are necessary to observe behavior of the rock during excavation, to perform additional rock mechanics tests such as determination of the deformation modulus, to determine residual stress, and to support properties and other underground information needed by the designers.

One of the fundamental questions to be answered by the geologic exploration program is—are there geology-related conditions which would prevent the construction of a safe and reliable facility at an acceptable cost? The answer depends upon both the rock conditions at the proposed site and the dimensions of the machine hall. In the United States, an underground machine hall is usually about 70 ft wide, about 150 ft high, and the length will depend on the number of units. At sites where ground is soft and closely fractured, it is extremely unlikely that a machine hall of such size can be built underground, and the geologist should immediately inform the designer of this so surface plant sites can be located. Crystalline rocks, such as granitic and metamorphic rocks, are more favorable for underground plant sites, than sedimentary rocks. Few underground plants in the United States and Canada have been built in sedimentary rocks.

Another important question to be answered by the geological exploration is the kind of support to be used. The tendency generally is to use rock bolt reinforcement of the roof and walls of the plant, because it is the cheapest and easiest type of support system to install. However, if rock is closely fractured or too soft to be adequately reinforced by rock bolts, then other types of support systems must be used. Concrete arches commonly are used to support the roof of the plant. Arches may either be supported on posts or on a rock haunch. If a haunch support is contemplated, the geologist should determine that the rock is suitable for excavation of a haunch and that there are no rock structures which would cause the haunch to fail. In general, the determination of the kind of support system to be used and the amount of support to be used is a subjective decision made after careful evaluation of geologic and rock mechanic data.

The deformation modulus of the rock, other elastic properties and orientation of the stress field in the rock are data that the designer of the underground plant usually desires. Test adits should be excavated through the machine hall area and a test chamber constructed, if necessary for the rock mechanics test desired. Tests should be oriented with the geologic structures so as to obtain data applicable to design conditions.

Because underground plants are usually in the dam abutments, water ponded in the reservoir changes the ground-water conditions and can cause ground water to envelop the plant. Geologic investigations should evaluate rock permeability around the plant, so the magnitude of the water problem can be evaluated. One method of protecting the plant from ground water is to use a grout curtain as an impermeable envelope around the power plant, and provide a drainage system to pick up any water that does leak through the curtain. Successful installation of such a system requires thorough knowledge of geologic conditions around the plant.

Geologic problems with surface plants are considerably different from those of underground plants. The main geologic problems that generally arise are stability of both construction and permanent slopes surrounding the plant, stability of the penstock foundation, adequacy of the foundation and problems that might arise from rebound. Also ground water conditions that may affect either construction of the plant, or its operation are important considerations.

Once detailed geologic mapping of the plant site is completed a subsurface exploration program for a surface plant site can be developed, usually exploratory drill holes, accompanied by trenching. If there is some concern about the bearing strength of the material, for example if the plant is an alluvial foundation, an attempt should be made to dig trenches, or make some other sort of excavation down to the invert elevation of the proposed foundation and perform plate bearing tests to evaluate bearing capacity.

Stability, particularly of rock slopes surrounding the plant and penstock area, should be determined by preparing very detailed geologic maps, being careful to gather all information on rock structures, for they have the largest influences on stability of rock slopes. Great attention should be paid to small defects such as small shears with clay that could become weak when saturated. Where the cut slopes are in earth materials, stability should be evaluated by soil mechanics techniques.

Rebound is difficult to evaluate precisely beforehand. It can be expected that rebound will be worse in sedimentary rocks, particularly shales and siltstones, and will be virtually negligible in crystalline rocks such as granite. Where significant rebound is anticipated, it is desirable to allow several months for completion of rebound between completion of site excavation and beginning of concrete placement. Rebound can be monitored by installing rebound gages in drill holes prior to construction and measuring as excavation progresses. Sometimes rebound causes a problem in the side slopes of construction cuts adjacent to a structure. If the walls around the structure are placed before the rebound is completed—rebound can crack the concrete in the wall. If this type of problem is anticipated, a buffer material can be placed around the walls.

Where ground water is above foundation level, particularly in alluvial materials, pump tests should be made to evaluate the ground water conditions and to provide information for designing dewatering systems for the foundation. When a dewatering system is required, geologic investigations should also be made to ensure dewatering will not impair local water supplies—a situation that can lead to expensive litigation. Geologic studies should be made to determine how the water table fluctuates through the year and to estimate what the water conditions will be at the time of construction and on completion of the project.

4.1 FOUNDATION INVESTIGATION

4.11 Earth Dams, by W. A. Wahler & Associates, Palo Alto, California

Foundation properties such as permeability, density, consolidation and shear strength must be explored, tested and analyzed in order to design a dam that is safe, feasible, and appropriate for the site conditions.

There are many books and publications that describe the field and laboratory procedures for conducting tests to evaluate the physical properties of foundations and embankments. A selected list of these references follows sub-section 4-116. The bibliographies of these references should be referred to for additional sources of information on the subject.

4.111 Permeability.
Permeability is one of the most important foundation properties to assess, as it will have a direct bearing on the design of the dam and can represent a significant cost of the structure.

An accurate evaluation of the permeability can also be the most difficult foundation property to determine due to the heterogeneity of most foundations and the limitations of test procedures.

Before selecting the type of permeability investigations to be performed, a geologic study of the site should be made. The geologic investigation will assist in determining the location and type of permeability data necessary for the design of the dam.

Permeability studies conducted for a dam investigation consist of field tests, such as those performed in drill holes, permeameter tests, and laboratory tests.

4.112 Field Tests.
One of the most common field permeability methods is performed in exploration holes. These tests are relatively inexpensive and can be performed during the exploration drilling of the site. Samples from the holes should be recovered for identification and laboratory testing. Permeability tests performed in drill holes are outlined below and described in more detail in Ref. No. 1.

Constant Head Test. This test is a measurement of the quantity of water required to keep the hole full.

Falling Head Test (Gravity Test). The falling head test is a measurement of the rate of drop in the water level in the hole.

Pressure Test. In a pressure water test, sections of the hole are sealed off with a packer and then subjected to water under pressure (to simulate reservoir head). The pressure, time and quantity of water loss are then recorded.

Permeameter Test. The permeameter test is a relatively inexpensive and simple field test and is recommended for use on soil foundations where knowledge of the permeability is required on specific permeable layers of alluvium. The test is usually performed in 6-in. or larger diameter holes, 5 to 50 ft deep. The test measures the rate at which water flows outward from a hole under constant head. This test is also described in Refs. 1 and 3.

Pump Tests. Pump tests are the best way to evaluate the overall permeability of a deep alluvial foundation when the materials are saturated. However, they are expensive and require considerable time to perform and interpret. A pump test should only be performed after the geologic features of the site have been investigated.

A considerable amount of experience and judgment is involved in selecting the location and depth of the pump and observation wells, the conduct of the pump test and interpretation of the results. Therefore, it is recommended that the test be supervised by a person with a good knowledge of ground water geology and hydraulics.

A pump test consists of pumping a well with a constant rate of flow and measuring the drop in water level in the pumped well (drawdown) and in one or more observation wells. The test can last from 8 hr to several weeks depending upon the geohydrologic properties of the alluvium. When the pumping is stopped, the water levels in all the wells are measured (recovery phase of the pump test) until a static water level is determined.

The information on drawdown and time is plotted and matched with known curves and the coefficient of transmissibility and permeability determined.

There are many good reference books that are helpful in planning, conducting and analyzing pump tests. Refs. 6 and 7 contain descriptions and case histories of permeability determinations based upon pump tests.

Laboratory Permeability. The laboratory permeability test consists of measuring the rate of flow of water through a sample. There are many variations to this test in order to simulate the construction and operation phases of the reservoir, as they apply to stability studies of the embankment.

The Bureau of Reclamation Earth Manual (Ref. 1) contains a description of test procedures and Refs. 4 and 5 discuss the application to embankment design and seepage studies.

4.113 Density

General. One of the important engineering properties of a soil is its density (weight per unit volume) and by controlling this property, the strength of a soil in bearing and shear resistance is controlled.

Foundations. The density of a foundation becomes very important when sections of a dam will be placed on alluvium that contains soft or loose soils.

Low density soils in the foundation of a dam can settle when they become loaded and saturated and cause cracking in the embankment; and should these soils be subjected to an earthquake, a failure could occur by liquefaction.

The identification of low density soils can be made by normal foundation exploration methods such as undisturbed sampling, penetration tests, and trenching, but actual measurement of in-place density must be made in the field.

For granular soils it is frequently necessary to determine how dense the material is in the field as compared to how dense or how loose it can exist (relative density). If this value falls below 70 percent, it could indicate that the material would be unsuitable for the foundation of a dam without modification.

The procedure for an in-place density test is well illustrated in Ref. 1 and practical applications are described in Ref. 5.

Embankments

Pervious Soils. The density of granular, free draining soils, such as those used in filters and drains in dams, is usually controlled by the relative density test. The maximum and minimum density is determined in the laboratory (Ref. 1, Desig. E-11) and a density between the maximum and minimum is then specified as the placement density desired for the embankment.

When the rock content of a soil is greater

than 25 percent and the maximum size approaches 12 in., the sampling and testing to control the density of the fill becomes more difficult and in order to obtain representative samples large volumes of material must be tested. During construction of a dam where thousands of cubic yards of rocky soil is placed daily, it would be difficult to obtain more than a few in-place densities per day; therefore, the primary density control must be visual inspection. Ref. 5 contains examples of in-place density tests.

Impervious (field). The density test used in the field on fine-grained or impervious soils is described in Ref. 1, Desig. E-24. For quality control of the embankment the field values must be compared with laboratory maximum dry density and the optimum water content. During construction of a dam a field density test is usually made for about every 2000 cu yd of material placed.

A rapid compaction control that yields the maximum dry density and an approximation of the water content can be performed within 1 hr. The following day the exact moisture content is available and corrections of the values or field control changes can be made. This method is also described in Ref. 1.

Nuclear Moisture-Density Determination. The nuclear moisture-density determination is a fast method that is becoming widely used for compaction control during construction of embankments. It consists of directing gamma-rays from a radioactive source into the soil and measuring the reflected radiation. The measurement is then referred to a chart and the wet density determined. The moisture content is determined in a similar manner. This method of testing can greatly reduce the number of in-place and laboratory density tests required for quality control of the embankment. Ref. 13 describes the development, application and construction of the nuclear probe and the advantages and disadvantages of its use.

Impervious (laboratory). The usual laboratory method of performing a maximum dry density test (Proctor Compaction Test) is described in Ref. 1, Desig. E-11.

The test determines the relationship between the water content of the portion of a soil passing the No. 4 sieve and the resulting densities obtained when the soil is mechanically compacted in a mold. The Proctor maximum dry density is the greatest dry weight obtained by this procedure.

A modified AASHO Density is used on earth dams when it is desirable to obtain greater densities. In this test, the applied energy is greatly increased. The soil is compacted in 5 equal layers with each layer receiving 25 blows of a tamper which weighs 10 lb and falls 18 in. The maximum density thus indicated is called the Modified AASHO Density (or Modified Proctor Density).

4.114 Consolidation. Consolidation is the gradual reduction of a soil mass resulting from an increase in compressive stress.

Consolidation is related to the density of a foundation and is significant in designing the embankment and predicting how much settlement will occur due to the load of the dam and what stresses will develop in the foundation during and after construction.

Consolidation tests are performed in the laboratory on undisturbed samples of the foundation. The test consists of loading a laterally confined sample compressed between porous plates. The loads placed on the sample should approximate the stresses that will be imposed by the embankment.

The test results are plotted to show the degree of consolidation after the application of the load. Consolidation tests are also performed on samples of the embankment material to determine the settlement characteristics of the fill.

The details of consolidation test procedures are described in Refs. 1 and 3.

4.115 Shear Strength. The shear strength of most rock foundations is stronger than the overburden soils and the embankment and usually does not represent a foundation problem. However, weak sedimentary rocks such as shales and siltstones can have a low density and high water content and should be investigated to determine their shear strength.

If the dam is to be constructed on alluvium that contains deposits of silts and clays then shear failures in the foundation should be investigated.

Field Tests

Block Shear. The block field shear test consists of excavating around a 2 ft. to 3 ft. block of rock, applying a vertical confining load and by using a horizontal jack shearing the rock on the predetermined horizontal or inclined surface. The movements of the block are monitored with dial gauges to determine the strains and the moment of failure. The pressure required to shear the block per unit area of the sheared surface determines the shear strength.

Standard Penetration Test. The standard penetration test is the most widely used field method to determine some of the strength characteristics of soils. The shearing resistance of sandy soils can be estimated from correlations developed between penetration resistance (blow count) and shear strength.

The standard penetration test consists of driving a 2-in. diameter sampler 12 in. into the soil with a 140-lb hammer falling 30 in. Descriptions of this test are given in Ref. 1. Ref. 11 contains a review of all the various types of penetrometers. This test must be used with caution, as it is not applicable for many types of soils.

Vane Shear Test. The vane shear test is inexpensive and relatively simple to perform but is limited to low-strength cohesive soils (silts and clays).

The test consists of placing a four bladed vane in the undisturbed soil at the bottom of a drill hole and rotating it from the surface to determine the torsional force required for failing a cylindrical surface sheared by the vane which is converted to the shearing resistance of the cylindrical surface. The test is described in the Bureau of Reclamation Earth Manual Designation E-20, Ref. 1 and Ref. 3.

Laboratory Tests. The selection of the type of shear test to perform depends upon what stress and drainage conditions are being analyzed in the dam. There are two main types of laboratory shear tests, the direct shear and the triaxial shear.

Triaxial Shear. Triaxial tests are used in analyses of slope stability and bearing capacity and to compute earth pressures where shear strength is mobilized.

The triaxial test samples should be undisturbed and typical of the major strata. Three to six samples are necessary to define the strength characteristics of the particular soil. The 3 main types of triaxial tests are the unconsolidated-undrained (*UU*), consolidated-undrained (*CU*), and consolidated-drained (*CD*). Ref. 12 is a detailed study of the triaxial shear test.

Direct Shear. The direct shear test is probably the most common shear test used in the United States. It consists of placing the sample in a cylindrical or rectangular box and applying a normal stress. The shear stress is then applied at the side forcing the failure to occur on a predetermined plane. The problem of unequal distribution of stress along the failure plane, the predetermined failure plane which may not be the weakest part of the sample, and the changing moisture condition in the soil during the test require considerable judgment as to applying the results of the test. In fissured clays and soils that are disturbed by landsliding, the residual shear strength may be the controlling property in the design of cut slopes and the stability of embankments placed on disturbed materials. Reference 12 describes shear tests in more detail and contains a diagram of the triaxial test.

A field shear test can take several weeks of preparation and several days to perform. Variations of this test have been used on soils.

Reference 8 describes a block shear test performed at Auburn Dam in California by the Bureau of Reclamation.

4.116 Special Tests. Dynamic triaxial shear testing is done primarily to determine if pulsating loads, such as those developed during an earthquake, can cause the soil to liquefy. These tests are extremely complicated to perform and require considerable experience and judgment to interpret and apply the results. References 9, 10, and 15 contain additional information on dynamic soils tests and the application of the test results.

Torsion Shear Test. The torsion shear test is sometimes used on weak rock foundations to investigate the shear strength along the bedding or other weak structural features such as the joint, fracture or slide plane. The test requires considerable field preparation and equipment. Torsion shear tests were performed by the State of California, Department of Water Resources at Pyramid Dam and on the Corps of Engineers' Hannibal Dam and the Philadelphia Electric Co. Muddy Run Dam in Pennsylvania. Reference 14 describes the method and contains diagrams of the equipment.

Jacking Tests. Jacking tests in alluvium and some sedimentary rocks to determine the elastic properties of a foundation can be performed on the surface and in large diameter auger holes. The test consists of jacking a plate approximately 1 sq ft in area against the surface of the ground or the wall of the hole and measuring the loads and deflections. From this data a modulus of deformation can be determined. The jacking test methods for a dam site on hard rock foundations are described in 4.035–Tunnels and 4.121.

Test Fills. Test fills will usually provide valuable information on the excavation and placement characteristics of soils and an opportunity to evaluate variations of water content and compactive effort on the densities obtained.

Due to the cost of constructing and testing a "Test Fill" they are used primarily during the design of a large dam. Reference 5 describes some of the advantages of constructing a test fill during the design investigation.

REFERENCES

1. "Earth Manual," First Edition–Revised (1963), U.S. Bureau of Reclamation, Denver, Colorado.
2. "Design of Small Dams," First Edition (1960), U.S. Bureau of Reclamation, Denver, Colorado.
3. U.S. Navy, Bureau of Yards and Docks (1961), "Design Manual, Soil Mechanics, Foundations, and Earth Structures," NAVDOCKS DM-7, Washington, D.C.
4. Cedergren, H. R., (1967), *Seepage, Drainage and Flow Nets*, John Wiley & Sons, New York.
5. Sherard, J. L., Woodward, R. J., Gizienski, S. F., and Clevenger, W. A. (1963), *Earth and Earth-Rock Dams*, John Wiley & Sons, New York.
6. Walton, W. C. (1970), *Groundwater Resources Evaluation*, McGraw-Hill Book Co., New York.
7. Todd, D. K. (1959), *Groundwater Hydrology*, John Wiley & Sons Inc., New York.
8. Wallace, G. B., Slebir, J., and Anderson, F. A., (1969), "Foundation Testing for Auburn Dam," 11th Symposium on Rock Mechanics, University of California, Berkeley.
9. Seed, H. B., "A Method for Earthquake Resistant Design of Earth Dams," *Journal of Soil Mechanics and Foundation Division*, ASCE, 92, (No. SM1), January 1966.
10. Lee, K. L. and Seed, H. B., "Strength of Anisotropically Consolidated Samples of Saturated Sand Under Pulsating Loading Conditions," Report No. T.E.-66-3, University of California, Berkeley.
11. Sanglerat, G. (1972), *The Penetrometer and Soil Exploration*, Elsevier Pub. Co., New York.
12. Bishop, A. W. and Henkel, D. J., Second Edition 1962, *The Measurements of Soil Properties in the Triaxial Test*, Edward Arnold (Pub.) London.
13. Roy, S. E. and Winkerkorn, H. F. (1957) Scintillation Method for the Determination of Density and Moisture Content of Soils and Similar Granular Systems. Soil Density Control Methods, Highway Research Board Bulletin 59, Washington, D.C
14. Hartman, B. E. (1967), "Rock Mechanics Instrumentation for Tunnel Construction," Terrametrics, Inc., Golden, Colorado.
15. Wahler, W. A. (1975), *Seismic Design of Embankment Dams*, Transactions of the Fifth

Panamerican Conference on Soil Mechanics and Foundation Engineering, Buenos Aires, Argentina.
16. Wulff, J. G. and Perry, C. W. (1976), "Efficient Methods of Site Appraisal and Determination of Type of Dam: A Discussion of Basic Philosophy and Procedures in Project Planning and Site Selection," 12th International Congress on Large Dams," Mexico City, Mexico.
17. Wahler, W. A., "Dynamic Analysis Builds Safer Dam for Less," *Engineering News-Record*, July 5, 1973.
18. Wahler, W. A. and Schlick, D. P. (1976), "Mine Refuse Impoundments in the United States," 12th International Congress on Large Dams, Mexico City, Mexico.

4.12 Concrete Dams—by George B. Wallace

As the reservoir rises behind a concrete dam, it presses the dam into the canyon walls and valley floor. To compute the stresses and deflection of the dam and the abutment reactions, it is necessary to know how much the rock foundation deforms under the combined loads of gravity, temperature, and water. Under sustained loading the rock continues to deform or "creep" causing changes in the stresses and shape of the dam. Under reduced loading as the reservoir is lowered the dam tends to resume its unstrained position and it is important to know whether the rock will follow it and maintain tight contact with the abutments or whether it will "set" and leave a void.

The deformation characteristics of foundations for concrete dams are significantly influenced by the density, orientation and width of joints and cracks near the loaded rock surfaces and the compressibility of gouge found within faults and shear zones. To evaluate the resistance of the foundation to forces tending to push the dam downstream it is necessary to know the shear strength along potential planes of weakness. Since the strength of discontinuities as well as the shearing forces are influenced by the magnitude and direction of the in situ stresses these must be determined. Measurements of in situ stresses are also needed to determine the foundation deformation and degree of slabbing that may occur as a result of their relief as the excavation of overburden and rock for the keyway is accomplished.

4.121 Deformation Modulus.
Four basic types of equipment are used to measure in situ deformation moduli:

1. Bore hole jacks and dilatometers.
2. Flatjacks installed in slots cut into rock mass.
3. Plate bearing tests.
4. Radial jacking tests.

Jacks designed to fit inside exploratory (NX) drill holes and associated equipment for measuring dilation as the jacks expand against the walls of the bore hole can provide qualitative information on deformability of rock masses at depths of several hundred feet.[1] Although they are quite useful in extending results of larger jacking tests described below they are not sufficiently accurate in hard rock to provide quantitative design data for large structures.

Thin flat jacks[1] may be installed in a slot cut in a rock mass by a diamond disc saw and inflated to load the sidewall of the slot. The jacks should closely fit the slot and be wide and deep enough to develop a plane stress field and reduce the effects of the restraint provided at edges of the slot. Slots approximately $1/4$ in. (7 mm) thick, 6 ft (150 cm) deep and 12 ft (300 cm) long have been used by the National Civil Engineering Laboratories, Portugal, and by the U.S. Corps of Engineers.[2] Deformeters located at several points within the flatjacks measure the expansion of the slot (δ) as the jacks are inflated. For any given intensity of pressure (q) the deformation modulus (E_d) may be computed by the equation:

$$E_d = \frac{Kq}{\delta}$$

Where K is a constant the value of which depends on the position of the deformeters and the shape and size of the loaded area.

The most widely used test methods for obtaining deformation modulus for design purposes is the uniaxial plate bearing test. In past years most of these tests have been performed by placing a circular rigid bearing plate against a carefully prepared rock surface and loading it with a hydraulic jack. The average deflection of the plate (δ) the area of the plate (A) and the total applied load (Q) is used to compute the deformation modulus (E_d):[3]

$$E_d = \frac{KQ(1-\nu^2)}{\delta\sqrt{A}}$$

Where K is approximately 0.89 for a rigid die and 1.00 for a uniformly loaded surface. Poisson's ratio (ν) is generally determined from laboratory tests of core samples.

In recent years circular flatjacks with a small hole in the center have been preferred over rigid solid circular plates because they distribute the load more uniformly, facilitate measurements of rock deformation at and below the loaded surface and because much higher loads can be applied over larger test areas at lower costs. The flatjacks and associated equipment are light compared to heavy hydraulic jacks of equivalent capacity and consequently provide mobility and are easier and quicker to install.

A schematic of the test configuration[4] used by the Bureau of Reclamation at Auburn Dam site, California, is shown in Fig. 4.12-1. The site preparation involves carefully excavating parallel test surfaces directly opposite each other on the walls of a test adit. Closely spaced holes are percussion drilled so that they bottom on the same plane. The rock between the holes is removed by air tools or light explosive charges. The equipment consists of four aluminum columns bearing on one large baseplate at each column end. A concrete mortar pad containing an embedded circular flatjack approximately 3 ft (0.91 m) in diameter is placed between the baseplate and rock surface to be tested. Underneath the center of both pads, NX holes are drilled into the rock mass, for installation of gages to measure rock movements.

Deformations are measured at several different depths within each hole by a multiple position extensometer. The extensometer (REX-$7P$) illustrated in Fig. 4.12-1 measures deformations between the collar of the bore hole and 7 anchors located at various depths within the hole. Each extensometer consists of a sensor head containing seven Linear Vari-

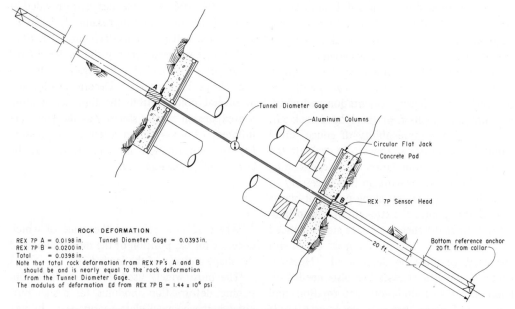

Figure 4.12-1. Schematic of uniaxial test with deformation values at 1000 psi load.

able Differential Transformers (LVDT's) each connected by a metering rod to one of the 7 anchors. Anchor locations are determined by means of a borescope or television camera to enable measurement of joint closure as well as elastic strain. The load is applied to the rock surface through the mortar pad by pressurizing the flatjacks in increments of 200 psi (14.06 kg/cm^2) up to 1000 psi (70.31 kg/cm^2). In general, incremental loads are maintained for 24 to 48 hr. Rebound and set are measured after unloading the jacks in between each incremental load.

In addition to the extensometers, the rock movement due to the test loads is measured by a conventional tunnel diameter gage. The measuring points for the tunnel diameter gage are fixed to the extensometer sensor heads so that the two measuring systems can be compared. To facilitate this, it is important that the opposing instrument holes be drilled exactly on the same axis. Two in. (5.08 cm) holes located in the center of each flatjack enable the tunnel diameter gage to be inserted through the concrete pads and contact the gage points on the extensometer sensor heads.

The deformation modulus (E_d) may be found using the deflections (δ) measured at the loaded surface or at any depth (Z) below the surface when the flatjacks apply a pressure of intensity (q) to the rock surface by the following equation:

$$E_d = \frac{Z^2}{\delta} q(1+\nu) \left[\frac{1}{(a_1^2 + Z^2)^{1/2}} - \frac{1}{(a_2^2 + Z^2)^{1/2}} \right]$$
$$+ 2(1-\nu^2) \frac{q}{\delta} [(a_2^2 + Z^2)^{1/2} - (a_1^2 + Z^2)^{1/2}]$$

Where:

ν = Poisson's ratio
a_1 = radius of hole in flatjack
a_2 = outer radius of flatjack

For $Z = 0$, the equation reduces to:

$$E_d = \frac{q}{\delta} 2(1-\nu^2)(a_2 - a_1)$$

The above equation may be used with the tunnel diameter gage measurements if it is assumed that the rock on both ends of the test "set-up" deforms the same and therefore δ equals the gage reading divided by 2.

Radial Tests. Radial jacking tests[4] are performed to evaluate loads on larger volumes of rock and to bring into play some of the structural features within the rock mass. Loads up to 30 million lb. were applied in tests by the Bureau of Reclamation to a carefully prepared test bore about 8 ft. (2.44 m) in diameter and length. The radial test equipment illustrated in Fig. 4.12-2 consists of structural ring sets with flatjacks fastened to their outer circumference. Sixteen flatjacks 8 ft. long by 16 in. wide are attached to the ring sets. Between each jack and the ring sets, a chipped wood composition board is inserted to serve as a cushion, distribute the load to the sets, and prevent damage to the jacks. Concrete is placed between the flatjacks and the rock to transmit desired loads to the rock surface. Extensometers, similar to those described above, are used to measure deformation of the rock at many locations beyond the loaded surfaces. Loads are applied in the same increments as for the uniaxial tests. Figure 4.12-3 illustrates a depth deformation curve for a

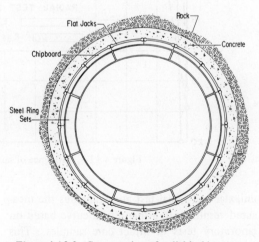

Figure 4.12-2. Cross section of radial jacking test.

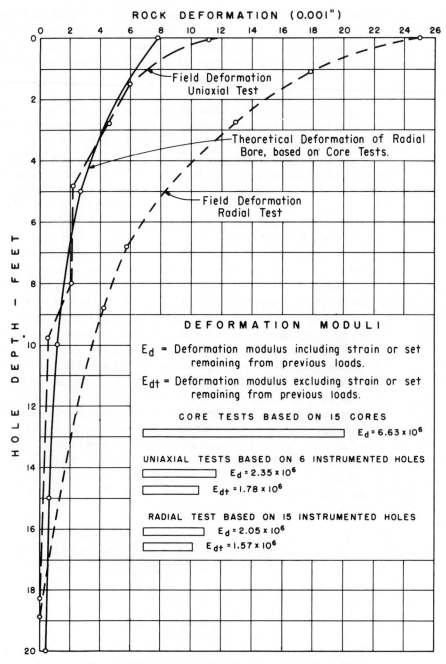

Figure 4.12-3. Influence of sample size on rock deformation.

uniaxial and radial test and compares the measured results with a theoretical curve based on laboratory tests of intact core samples. This figure shows a significant difference between the deformation modulus calculated from the laboratory and field tests. However, the difference between the results of the uniaxial and radial field tests are not so pronounced.

GEOLOGIC, FOUNDATION, AND SEISMICITY INVESTIGATIONS

4.122 Compressibility of Gouge. Equipment used to determine the compressibility of gouge in fault zones is illustrated in Fig. 4.12-4. It consists of a load bearing plate in contact with a cement grout pad, spherical bearing block, pipe column, hydraulic ram, shoe plate, and staging required to block the setup on the rock face opposite the test surface. Three dial

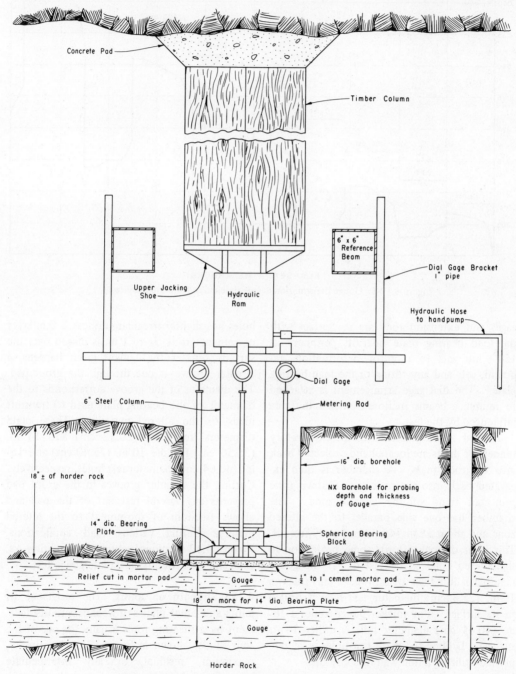

Figure 4.12-4. Schematic of typical plate gouge test setup.

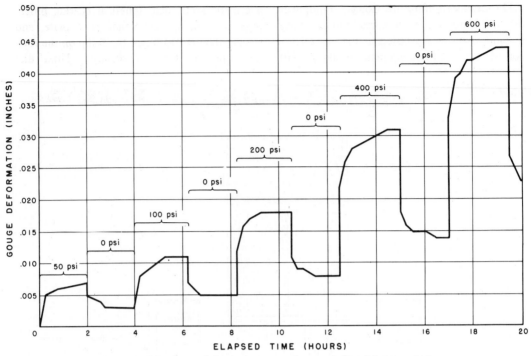

Figure 4.12-5. Gouge deformation versus time curve for plate gouge test.

gages and extension rods are in contact with the load bearing plate near its periphery at 120° intervals to monitor deformation, rebound, set, and any tilting of the load bearing plate. The dial gage arrangement is attached to reference beams anchored outside the area influenced by the test.

Altered zones which intersect exploratory tunnels or drifts are located and projected back into the rock mass. A chamber is then excavated back into the rock mass following the projected trend of the altered zone. The chamber has one side parallel to the altered zone and about 18 in. (45.72 cm) away from it. This layer of rock affords some protection to the altered zone to be tested. If the gouge material is less than 18 in. (45.72 cm) thick, a 10-in. (25.4 cm) diameter test hole is drilled to its surface. For thicker gouge zones a 16-in. (40.64 cm) diameter hole is drilled to its surface. Drilling for these test holes is accomplished utilizing a thin wall masonry bit.

Since the material at the bottom of test holes has slightly irregular surfaces, a thin layer of cement grout ½ to 1 in. is placed over the surface. After the cement grout hardens, a circular groove is cut through the grout pad. The diameter of the groove corresponds to the diameter of the bearing plate used to transmit loads to the altered material. These plate diameters are 8 in. (20.32 cm) and 14 in. (35.56 cm) for the 10 in. (25.40 cm) and 16-in. (40.64 cm) diameter test holes, respectively. Cutting the circular grooves in the grout pad prevents any lateral restraint of the pad and allows freedom of movement to the altered material during various load applications. Figure 4.12-5 shows gouge deformation vs. time and load for a typical plate gouge test.

4.123 Strength of Joints. The resistance of joints and other planes of weakness to displacement by a shearing force usually fits one of the four modes shown in Fig. 4.12-6. The initial, maximum, residual, or any intermediate strength of the joint may be described by the

GEOLOGIC, FOUNDATION, AND SEISMICITY INVESTIGATIONS 241

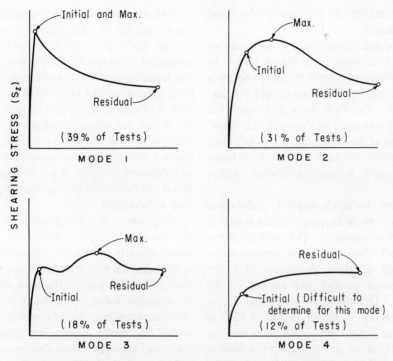

Figure 4.12-6. Four typical modes of joint strength and displacement (percentages based on 121 direct shear tests of Nx cores from Auburn and Grand Coulee projects).

Coulomb-Navier criterion which is developed in terms of force but is frequently converted to an average stress by dividing the force on the shear zone by its gross area.

$$S = C + N \tan \phi$$

Where

S = shearing stress in direction of joint movement
C = apparent cohesion
N = stress normal to direction of joint movement
ϕ = angle of friction

Mode 1 is typical of joints having high cohesion. The initial strength is defined as the point where the change in shear stress versus displacement relationship is maximum prior to reaching the first peak strength. When undulations along the joint are sheared off as they frequently are at higher normal loads, the strength of intact rock may influence the maximum joint strength as shown by Mode 2. In this case the joint surfaces move under increasing shear load upward along a series of controlling undulations until their combined cross sections are reduced enough to allow the undulations to shear off. Occasionally, a secondary set of controlling undulations, whose adjacent surfaces are not initially in contact, may be brought together as displacement proceeds and the maximum strength of the joint occurs as shown in Mode 3. Mode 4 is typical of repositioned samples or fractured surfaces which have experienced previous relative movements. In effect, an increasing number of higher-strength surfaces are "seated" or matched as displacement occurs and the resistance increased until the residual-maximum strength of the newly fractured surface is reached.

Both field and laboratory tests of joints are schematically illustrated in Fig. 4.12-7. In both cases a normal stress, N_z, and a shearing stress, S_z, are applied to a zone of rock containing the joint. The rock above and below the shear zone is restrained as shown in the figure. The thickness of the shear zone is just sufficient to fully contain the joint. Joints with larger undulations will require somewhat thicker shear zones.

As the test loads are applied, undulations along the joint surfaces cause vertical as well as horizontal displacements. The resultant direction of block displacement is computed by measurements of the components of movements measured parallel and normal to the shear zone. This direction referenced to the applied stress S_z is shown in Fig. 4.12-7 as Angle I. The figure also shows equations for computing the shear stress, (S), on the joint in the direction of movement, stress on the joint normal to the direction of movement (N), and a value of shear stress (S_c) under a constant normal load.

Many undisturbed joints have a primary strength much greater than their strength following movement. By restraining a joint as it is being prepared for field test, its integrity can be substantially preserved and its undisturbed primary strength determined. The initial normal load for the test should be as close to the anticipated normal load as possible. Subsequent normal loads usually range from low to high so that surface alterations produced by high normal loads are not present during the test at lower normal loads.

Tan ϕ and values of cohesion for the Coulomb-Navier equation may be obtained from results of field or laboratory tests by statistically computing the best straight line through *appropriate* values of N and S. The choice of whether to select values of N and S from initial, maximum, or residual points along the shear stress-displacement diagrams should be based on the intended application.

Primary strength is not generally used in computing tan ϕ. Subsequent tests at various normal loads are performed, and the tan ϕ computed from the N and S values taken from the initial points on the shear stress-displacement diagrams. An estimate of primary cohesion is obtained by substituting this value of tan ϕ and the primary strength obtained from the first test into the Coulomb-Navier equation. Primary strength at other normal loads can then be computed; however, it would be prudent not to extend the results beyond the range in which they are obtained.

At present, it is not possible to determine the strength of joints supporting large rock masses directly from tests. Even relatively large in-situ samples may have different values of I than the actual joints. However, it is possible to estimate values of I from joint profiles established using comprehensive geological studies and make appropriate adjustments to the shear strength values determined by field and laboratory tests.

Foliation planes, joints, and shear zones found in exploratory tunnels in the dam foundation may be tested as illustrated in Figure 4.12-8 to determine shear stress-displacement relationships under a variety of normal loads. Blocks of rock are cut with a diamond saw so that the base of each block is on the joint, shear zone, or particular foliation plane to be tested. The blocks shown in the figure are 15 in. (38.1 cm) square and 8 in. (20.32 cm) high. A metal frame is placed around the test block with a mortar filling to ensure loading the block uniformly. Normal loads are applied utilizing a spherical bearing block, large hydraulic jack, and staging to the tunnel wall opposite the test block. Shear loads are applied to the test block by pivot shoe, jack, and spherical bearing block at an angle of 10° from the shear zone to reduce the effects of overturning moments. Photoelastic studies indicate that this load configuration produces optimum distribution of stress along the shear zone.

In a test of foliated amphibolite at Auburn Dam site in California, normal loads were ap-

GEOLOGIC, FOUNDATION, AND SEISMICITY INVESTIGATIONS 243

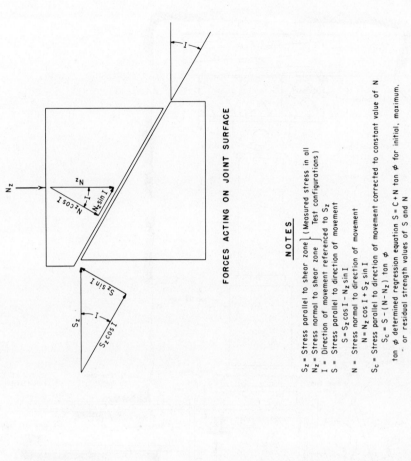

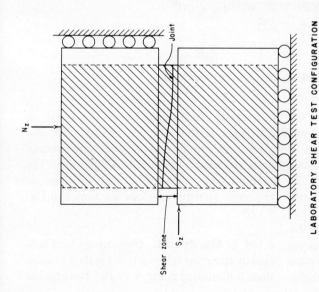

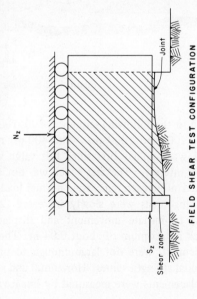

Figure 4.12-7. Schematic diagram and nomenclature for shear strength tests.

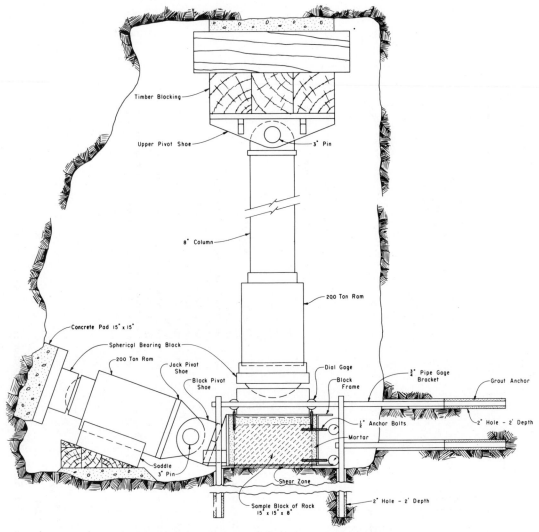

Figure 4.12-8. Shear and sliding friction testing equipment.

plied in the following sequence: 400, 100, 300, 500, 700, and 1000 psi. The 400-psi normal load was applied first since this value was of more interest. The test block was not repositioned after each successive normal load. The shearing loads were slowly increased (25 psi/min) to initiate movement and displace the block a maximum of about 0.04 in. These displacements were not large enough to produce residual strength values. Horizontal and vertical displacements were measured by linear variable differential transformers ($LVDT$'s) as well as dial gages as shown in Fig. 4.12-8. The displacements and the ram pressures were continuously recorded on electronic XY plotters. The resulting shear stress-displacement relationships and shear strength envelope are shown in Fig. 4.12-9.

4.124 In Situ Stresses. Overcoring a bore hole gage to determine in situ stresses in dam foundations is illustrated in Fig. 4.12-10. The gage has

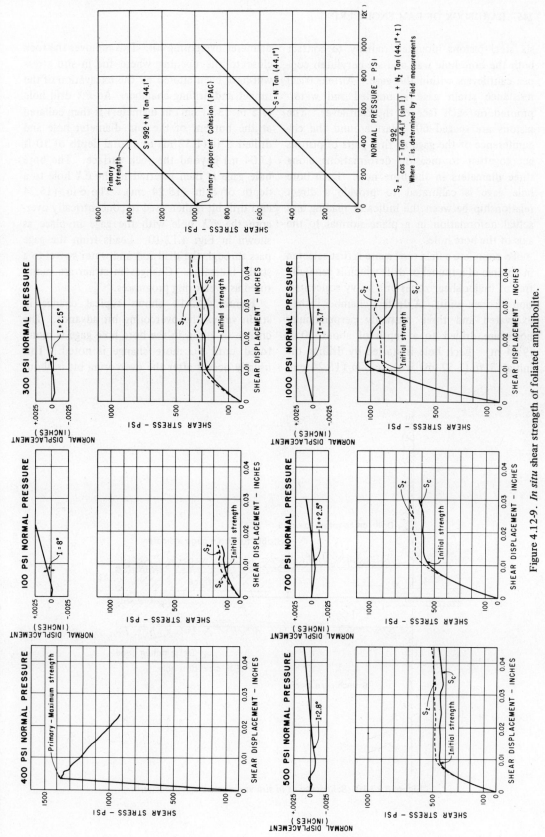

Figure 4.12-9. *In situ* shear strength of foliated amphibolite.

six steel pistons mounted radially to contact both the bore hole wall and six beryllium copper cantilevers within the gage. An electrical resistance strain gage is bonded and waterproofed on each face of the cantilever. The pistons are spaced 60° apart around the circumference of the gage so that pairs of pistons act together to measure deformations along three diameters in the bore hole. Each bore hole gage is calibrated to provide a direct relationship between the indicator reading and actual deformation in a plane normal to the axis of the bore hole.

To complete a field test, a site is first selected in an area of a tunnel or drift which is generally free of major shear zones or a heavy concentration of open joints. A drilling chamber is then excavated and three mutually perpendicular holes are drilled to a depth of about 30 ft (9.14 m). Each hole is started by drilling the initial 5 ft (1.52 m) with a 6-in (15.24 cm) diameter overcoring bit. This removes the rock close to the opening where the in situ stress will have been altered by the excavation of the tunnel and drilling chamber. An *EX* drill hole (1.5 in. (3.81 cm) in diameter) is then collared at the bottom of the 6-in.-diameter hole and drilled 5 ft (1.52 m) to a total depth of 10 ft (3.04 m) beyond the rock surface. The bore hole gage is then inserted in the *EX* hole to a depth of 6 in. (15.24 cm). The 6-in. (15.24 cm) drill bit is then used to concentrically overcore the *EX* hole with the gage in place as shown in Fig. 4.12-10. Leads from the gage pass through the drill rod and water swivel to a strain indicator. Changes in strain are monitored as overcoring progresses.

Figure 4.12-11 shows typical change in strain versus the overcoring bit advance. The change in strain in the bore hole gage is monitored until no more change is noted. This usually occurs after the overcoring bit is about

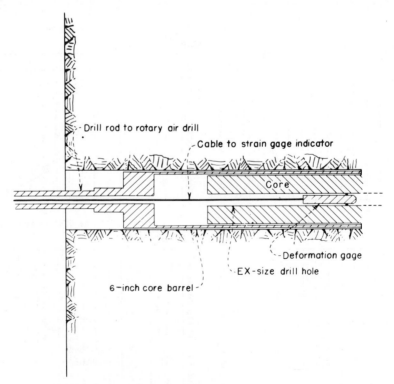

Figure 4.12-10. Schematic of *in situ* stress relief test procedure.

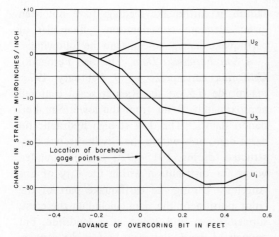

Figure 4.12-11. Stress relief test change in strain as overcoring advances.

6 in. (15.24 cm) beyond the location of the bore hole gage. At this point the relief of all the stress in the annular core is considered complete and the bore hole gage is then advanced to another station about 12 in. (30.48 cm) down the EX hole and the overcoring process is repeated. At the end of the EX hole, the gage is removed and the annular core is broken and retrieved. The EX hole is extended 5 more ft, (1.52 m), the gage is reinserted, and overcoring is resumed. The retrieved annular core is loaded biaxially in the laboratory to determine the elastic modulus of the rock. The data from each drill hole may be utilized to determine the two-dimensional stress field or data from three orthogonal holes may be combined to determine the three-dimensional stress field. The two-dimensional solution is from Merrill and Peterson.[5] The three-dimensional solution of stress is from Panek.[6]

4.125 Results and Evaluation. The results of the in situ jacking tests can be used in conjunction with laboratory tests of cores to determine the deformation moduli of the entire foundation. Knowing the properties of each geological feature, the foundation can be analyzed as a composite continuum using the finite element method.[7]

The ratio of deformation modulus determined from an in situ test (E_d) and that based on laboratory tests of cores (E) extracted from instrument holes for the field test is dependent on the density, orientation and width of cracks near the loaded surface. These may be empirically expressed as a Joint Shear Index.

$$\frac{E_d}{E} = \text{Joint Shear Index}^8$$

By observing the geological features in the instrument holes at the test site with a borescope or a television camera and from the drill core Joint Shear Indexes can be established for measured values of E_d/E. Joint Shear Indexes and values of E can then be determined from observations and laboratory tests of core taken from exploratory drill holes blanketing the entire foundation. With this information values of E_d may be computed from the above equation for each zone in the foundation. These along with the measured and estimated in situ stresses may be used in an appropriate finite element model to determine the foundation behavior and its influence on the dam.

Major joints and shear zones can also be included in the finite element model and their strength determined from the in situ and laboratory tests of safety against sliding of rock masses. Also the shear-displacement properties of the joints can be used in rigid block studies[9] to determine the stability of blocks of rock near the valley walls.

REFERENCES

1. "New Techniques in Deformability Testing of In Situ Rock Masses," by M. Rocha, ASTM STP 477, American Society for Testing and Materials, 1970.
2. "Development of Equipment for Determining Deformation Modulus and In Situ Stress by Means of Large Flatjacks," by E. J. Deklotz and B. P. Boisen, ASTM STP 477, American Society for Testing and Materials, 1970.
3. "Foundation Modulus Tests for Karadj Arch Dam," by Messrs. Waldorf, Veltrop and Curtis. *Journal of Soil Mechanics and Foundation Division*, Proceedings of the American Society of Civil Engineers, July 1963.
4. "Foundation Testing for Auburn Dam," by G.

Wallace, E. Slebir, and F. Anderson, II Symposium on Rock Mechanics, U.S. National Committee for Rock Mechanics, 1969.
5. "Deformation of a Borehole in Rock," by R. H. Merrill and J. R. Peterson, Report of Investigation No. RI 5881, U.S. Bureau of Mines, 1961.
6. "Calculation of the Average Ground Stress Components from Measurements of the Diametral Deformation of a Drill Hole," Bureau of Mines Report of Investigations No. RI 6732, 1966.
7. "Applications of the Finite Element Method in Geotechnical Engineering," Edited by C. S. Desai, U.S. Army Waterways Experiment Station, 1972.
8. "Deformation Moduli Determined by Joint Shear Index and Shear Catalog," by J. L. Von Thon and G. S. Tarbox, Rock Fracture Symposium, Nancy, France, 1971.
9. "Method of Stability Calculations for Rigid Elements Resting on Plane Surfaces and Application of the Results to Stability Calculations for Rock Slopes," by Walter Wittke, Institute of Soil Mechanics and Foundation Engineering, Karlsruhe, Germany, 1965.

4.2 EARTHQUAKE HAZARD—BY GEORGE W. HOUSNER

4.21 Seismic Hazards

Many regions of the world are subject to potentially destructive earthquakes.[1,2] The border of the Pacific Ocean is particularly prone to seismic activity; approximately 80 percent of the world's earthquakes occur along this ocean border. A second seismic belt runs across northern India into Turkey, Greece and Italy. In addition destructive earthquakes also occur in other parts of the world, although much less frequently. In highly seismic regions, such as California, where earthquakes occur frequently enough so that they are actively studied, the seismic hazard is well understood. On the contrary, in regions of low seismicity, where destructive earthquakes occur very infrequently, the seismic hazard is often not given the attention it deserves and is not well understood, and the actual danger to inhabitants may be greater than in California or Japan where appropriate preventative measures are taken.

The occurrence of an earthquake in the vicinity of a dam can cause damage, or even failure, if earthquake loadings have not been given adequate consideration. Examples of earthquake damage to dams are the following:

a. *San Fernando Dam.* The San Fernando, California earthquake of 9 February 1971, magnitude 6.5 on the Richter scale, centered approximately 5 mi from the lower San Fernando dam.[3] This 144 ft high hydraulic fill earth dam, now within the city limits, was constructed during 1915-1920 and impounded water for distribution within the City of Los Angeles. During the earthquake part of the front portion of the dam, including the crest, slipped down and failed as a landslide, but enough of the downstream half of the dam remained in place to contain the water. Approximately 80,000 persons living below the dam were evacuated for 2½ days until the water level was reduced to a safe elevation.

b. *Eklutna Dam.* The Alaska earthquake of March 27, 1964 damaged the Eklutna earth dam which was located approximately 60 mi northwest of the epicenter of this magnitude 8.4 earthquake.[4] The dam was built in 1939 and strengthened in 1952. The height of the dam was approximately 50 ft. Substantial damage was sustained by the dam and its appurtenances, and it was decided to replace it with a new dam rather than to repair the old dam.

c. *Sheffield Dam.* This small earth dam failed completely during the magnitude 6.3 Santa Barbara earthquake of 1925. There was approximately 20 ft of water behind the dam at the time of the earthquake and the failure released 45 million gallons of water.[5,6]

d. *Koyna Dam.* This 300 ft high concrete, gravity dam was damaged by the south India earthquake of December 10, 1968.[7,8] This magnitude 6.5 earthquake occurred close to the dam and an accelerograph within the dam recorded a peak acceleration of approximately 50 percent g. The dam sustained a horizontal crack near the upper third point and many of the appurtenances of the dam were damaged.

e. *Hsinfengkiang Dam.* This 300 ft. high diamond head buttress dam in China was damaged by the magnitude 6.2 earthquake of

March 19, 1962. The dam sustained a horizontal crack approximately 50 ft. below the crest.[9]

The main hazards to a dam from an earthquake are the following: surface faulting through the dam, strong ground shaking, water waves in the reservoir produced either by earthquake ground motions or landslides and rock falls. Overall ground deformations associated with nearby faulting may also be hazardous to a dam. The potential hazard to a dam from an earthquake depends on how large the earthquake is and how near the dam site it is. It is not possible to predict when and where earthquakes will occur and how large they will be, therefore considerable judgment is involved in assessing the seismic risk at a dam site. Most seismic countries have prepared seismic zoning maps to assist planners and builders. These maps divide the country into a number of zones of different degrees of seismic hazard. An example of such a seismic hazard map is shown in Fig. 4.2-1. This map of the United States was zoned by a committee of seismologists and it indicates, in a not precisely defined way, the degrees of seismic hazard in various parts of the country. Such crude zoning maps are the first level of seismic hazard assessment. They are used, for example, in building codes to regulate the earthquake design of ordinary structures. For especially important structures, or facilities, it is customary to make a second level of seismic hazard assessment. This more detailed and thorough investigation includes the following items:

a. *Seismological investigation.* Study is made of the past occurrence of earthquakes in the general region of the site and on this basis estimates are made of the probability of occurrence of future earthquakes on the supposition that the future will be similar to the past. In order for this approach to be valid, a sufficiently long seismic history must be available. For example, it would be inappropriate to suppose that in California the next 10 years will be similar in earthquake occurrence to the past 10 years but it would be reasonable to suppose that the next 100-200 years will be similar to the past 100-200 years. In less seismic regions even longer intervals of time would be needed. The seismic history of the United States is very short compared to countries like Italy and China where the history extends back several thousand years.

b. *Geological investigations.* An evaluation is made of the tectonic processes in the general site region; faults in the general region are identified, and the degree of activity of the faults is estimated. This geologic information supplements the seismological data. It provides information pertinent to seismic activity over intervals of time measured in many thousands of years.

c. *Soils and local geology investigations.* Investigations are made of geologic formations and soil deposits on the site area to assess their possible behavior during earthquake shaking and how they might affect the ability of a structure to resist earthquakes.

The foregoing three investigations provide information on the seismic hazard at a proposed project site. They give information as to the magnitudes and frequencies of occurrence of earthquakes that can be expected in the general site area; likely active faults are identified and the nature of faulting that might be expected is discussed, etc. Judgment based on all this available information must then be used to establish appropriate earthquake design criteria for the project.

4.22 Earthquakes and Reservoir Loading

The earthquakes that damaged Koyna Dam in India and the Hsinfengkiang Dam in China are thought to have been related to the filling of the reservoir. In both cases as the reservoir was being filled numerous small earthquakes in the general vicinity of the reservoir were observed; and later the damaging earthquake occurred. Similarly, earthquakes were observed in the vicinity of the 350 ft high Kariba Dam on the Zambesi River in Africa. The reservoir was full for the first time in August 1963 with 4700 ×

250 HANDBOOK OF DAM ENGINEERING

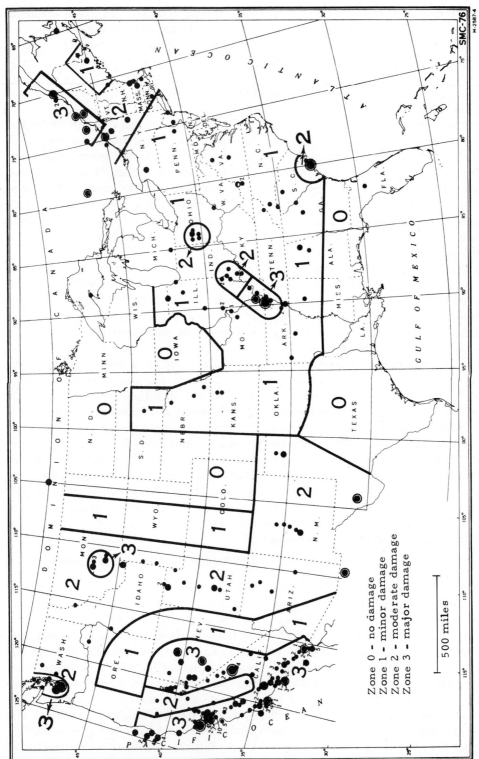

Figure 4.2-1. Seismic zoning map of the United States. This is an example of a seismic hazard map which divides the United States into four zones of different degrees of estimated hazard. The map was originally compiled by the U.S. Coast and Geodetic Survey; but as shown here has been modified in the northeastern part of the country. The boundaries between zones were given distinctively arbitrary shapes to make clear that the precise location of the boundary line is just an estimate.

10^9 cu ft of storage. From August 14 to November 8, 1963 nine earthquakes occurred in the vicinity having magnitudes ranging from 5.1 to 6.1. The filling of the reservoir was started in December 1958 and the first tremors were recorded in 1961. A total of 155 tremors were recorded in 1962 and 135 were recorded from January 1 to July 31, 1963. Some 750 small aftershocks were recorded between August 14 and September 30, 1963. Other cases of earthquakes associated with reservoir loadings have also been reported.[10]

On August 1, 1975 a Magnitude 5.7 earthquake occurred near the town of Oroville, California. The earthquake originated about 10 kilometers south of the 770-foot-high Oroville earth dam and about 10 kilometers beneath the ground surface. The peak acceleration recorded on the crest of the dam was approximately 10 percent g; and there was no evidence of damage. The earthquake was preceded and followed by numerous smaller shocks during 1975; but no seismic events were observed in the vicinity at the time of initial reservoir filling some 5 years earlier. Stress analysis indicates that the reservoir weight (4 million acre feet of water) caused a small decrease in shear stress on the fault. The consensus is that the dam and reservoir probably had no connection with the earthquake.

How the reservoir loading is related to earthquake generation is not well understood at present and is still the subject of research. Presumably, it is related to the extra weight on the earth's crust and possibly also to additional water and water pressure diffused to considerable depth. It is thought that the reservoir loading is just the triggering mechanism to release an existing state of stress in the earth's crust. Both Koyna Dam and Kariba Dam were in locations that were not highly seismic. Potentially damaging earthquakes associated with reservoir loading have not been observed in California, which is highly seismic. It is now customary to install one or more seismographs in the vicinity of a new major dam to record shocks having magnitudes in the range of 1.5 to 3.5, to monitor any change in frequency of occurrence during and following the filling of the reservoir.

4.23 Earthquake Generation

The shaking of the surface of the ground during an earthquake is a consequence of the passage of seismic waves (stress waves). These seismic waves are generated by slip on an existing fault which causes the sudden release of a state of stress in the earth's crust.[1,2] Because of certain tectonic processes (seafloor spreading) stresses are building up in the earth's crust in highly seismic areas such as California and Japan and periodically the stress is released by the occurrence of earthquakes; possibly, a stress buildup at a much lower rate is also taking place in less seismic areas.

Most earthquakes are observed to originate from displacements on four types of fault, as shown in Fig. 4.2-2. Each of these four examples represent a shear stress failure on an existing fault. The diagrams in Fig. 4.2-2 are idealized; for example, some strike-slip displacement usually accompanies the main dip-slip displacement on a thrust fault, and vice versa. Also, during an earthquake, displacement may be observed on splinter faults that are associated with the main fault. The ground motion produced by these 4 different types of faulting are not strongly influenced, at some distance away from the fault, by the details of the faulting process. Close to the fault, however, there can be quite significant differences in ground shaking and in ground deformations depending on the details of the faulting mechanism.

When the area of fault that slips is small, a small earthquake is generated; when the area that slips is large, a large earthquake is generated. A much greater amount of strain energy is released by the slip during a large earthquake than during a small earthquake. The size of the slipped area is determined by the length of slip on the fault and by the depth of slip on the fault. Earthquakes whose Richter magnitude

252 HANDBOOK OF DAM ENGINEERING

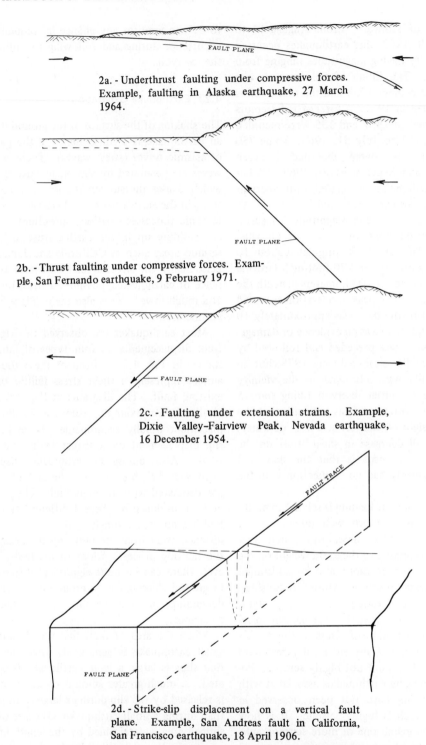

2a. - Underthrust faulting under compressive forces. Example, faulting in Alaska earthquake, 27 March 1964.

2b. - Thrust faulting under compressive forces. Example, San Fernando earthquake, 9 February 1971.

2c. - Faulting under extensional strains. Example, Dixie Valley–Fairview Peak, Nevada earthquake, 16 December 1954.

2d. - Strike-slip displacement on a vertical fault plane. Example, San Andreas fault in California, San Francisco earthquake, 18 April 1906.

Figure 4.2-2. Faulting mechanisms. These four examples illustrate how earthquakes are generated by shear stress failures along existing planes of weakness (faults). The sudden release of stress by the fault slip generates seismic waves.

exceed 6.5 are often accompanied by evidences of surface displacement on a fault; that is, for these earthquakes the slipped area is so large that it reaches the surface of the ground as indicated in Fig. 4.2-2. Smaller earthquakes are usually not accompanied by evidences of surface fault displacement; that is, the slipped areas of these earthquakes are so small that the displacement does not extend to the surface of the ground.

Earthquakes that produce destructive ground shaking usually are generated by fault slip in the upper 20 mi of the earth's crust; most California earthquakes originate at a depth of 5 to 10 mi. The San Fernando, California earthquake of 9 February 1971, for example, was generated by displacement on a thrust fault such as shown in Fig. 4.2-2b. The slipping started at the lowest point, approximately 8 mi beneath the surface of the earth, and the slip propagated upward to the ground surface. The maximum fault displacement was estimated to have been approximately 6 ft; the strike-slip component of displacement was estimated to have been a bit less than the dip-slip component. The 1906 San Francisco earthquake (magnitude 8.2) originated from slip on about 250 mi of the San Andreas fault as shown in Fig. 4.2-2d. The maximum observed strike-slip displacement on the surface of the ground was 21 ft.

4.24 Magnitude and Intensity of Earthquakes

The size of an earthquake, that is, the size of the slipped fault area, is described in a quantitative way by the so-called Richter Magnitude of the earthquake. C. F. Richter of the California Institute of Technology defined the magnitude of an earthquake for shallow shocks as

$$M = \log_{10}(CA) = \log_{10}\frac{A}{A_0} \quad (1)$$

where M is the magnitude of the earthquake, A is the maximum amplitude of horizontal ground motion recorded by a standard seismograph at a distance of 100 km from the center of the disturbance, C is an arbitrary constant and A_0 is an amplitude of 0.001 mm. The standard instrument has a natural period of 0.8 sec, has almost critical damping, and a nominal static magnification of 2800. It is seen that, in effect, the magnitude of an earthquake is a measure of the ratio of the peak recorded amplitude, A, to a standard peak amplitude, A_0. In practice, the recordings must be made at distances that are large compared to the dimensions of the slipped fault area. The recordings are then extrapolated to a distance of 100 km from the center of the shock. For best results an average value of M is determined from a number of recordings from different seismological stations.[1]

Earthquakes of magnitude 5.0 or greater can generate ground motions sufficiently severe to be potentially damaging to structures; though for the smaller magnitude shocks the depth of slip beneath the ground surface is an important factor in the intensity of shaking. For magnitudes less than approximately 5.0 the ground motion is unlikely to be damaging because of its very short duration, even though the peak acceleration may be large in the case of very shallow slip. Table 1 gives an idealized relation between magnitude and length of slipped fault.[2] The great Chile earthquake of 1960 had an assigned magnitude of 8.5 and slipped over a length of fault approximately 600 mi. The 1964 Alaska earthquake had an assigned magnitude of 8.4 and slipped over a length approximately 450 mi. The 1906 San

TABLE 1. Idealized Relation between Magnitude and Length of Slipped Fault

Magnitude	Length (Miles)
8.8	1000
8.5	530
8.0	190
7.5	70
7.0	25
6.5	9
6.0	5
5.5	3.4
5.0	2.1
4.5	1.3

Francisco earthquake had an assigned magnitude of 8.2 and slipped over a length approximately 250 mi. The El Centro, California earthquake of 1940 had an assigned magnitude of 7.1 and slipped over a length of approximately 40 mi. The Baja, California earthquake of 1956 had an assigned magnitude of 6.8 and slipped over a length of approximately 15 mi. These facts illustrate that the energy released in the form of seismic waves does not originate at a point source, as is the case of an underground nuclear detonation, but originates in a volume of rock that is greater for large earthquakes than for small shocks. If, as appears to be the case, the average shear stress released by slip on a fault is approximately the same for earthquakes of larger magnitudes it follows that slipped area is directly related to the magnitude. It is this fact that makes the magnitude meaningful to engineers. The magnitude indicates approximately the size of the earthquake source and, hence, the approximate area affected by ground shaking.

The epicenter of an earthquake, as recorded by seismologists, is not the center of the earthquake. In the early days of seismology it was thought that seismic waves originated at a point, the hypo-center, and that the point on the surface of the ground directly above was the epicenter. Now that it is known that seismic waves originate over an area of fault, the term epicenter is used by seismologists to indicate the point on the surface of the ground directly above the point where fault slip began. The first compression waves and the first shear waves that reach the seismograph originated at the point of initial slip and it is these waves that are used by seismologists to locate the so-called epicenter. For example, the epicenter of the 1960 Chile quake was near the city of Concepcion and the length of slipped fault extended southward a distance of approximately 600 mi,

TABLE 2. Modified Mercalli Intensity Scale of 1931 (Abridged)

I. Not felt. Marginal and long-period effects of large earthquakes.

II. Felt by persons at rest, on upper floors, or favorably placed.

III. Felt indoors. Hanging objects swing. Vibration like passing of light trucks. Duration estimated. May not be recognized as an earthquake.

IV. Hanging objects swing. Vibration like passing of heavy trucks; or sensation of a jolt like a heavy ball striking the walls. Standing motor cars rock. Windows, dishes, doors rattle. Glasses clink. Crockery clashes. In the upper range of IV wooden walls and frame creak.

V. Felt outdoors; direction estimated. Sleepers wakened. Liquids disturbed, some spilled. Small unstable objects displaced or upset. Doors swing, close, open. Shutters, pictures move. Pendulum clocks stop, start, change rate.

VI. Felt by all. Many frightened and run outdoors. Persons walk unsteadily. Windows, dishes, glassware broken. Knickknacks, books, etc., off shelves. Pictures off walls. Furniture moved or overturned. Weak plaster and masonry cracked. Small bells ring (church, school). Trees, bushes shaken visibly, or heard to rustle.

VII. Difficult to stand. Noticed by drivers of motor cars. Hanging objects quiver. Furniture broken. Damage to poor masonry, including cracks. Weak chimneys broken at roof line. Fall of plaster, loose bricks, stones, tiles, cornices, unbraced parapets and architectural ornaments. Some cracks in fair masonry. Waves on ponds; water turbid with mud. Small slides and caving in along sand or gravel banks. Large bells ring. Concrete irrigation ditches damaged.

VIII. Steering of motor cars affected. Damage to fair masonry; partial collapse, no damage to good masonry. Fall of stucco and some masonry walls. Twisting, fall of chimneys, factory stacks, monuments, towers, elevated tanks. Frame houses moved on foundations if not bolted down; loose panel walls thrown out. Decayed piling broken off. Branches broken from trees. Changes in flow or temperature of springs and wells. Cracks in wet ground and on steep slopes.

IX. General panic, poor masonry destroyed; fair masonry heavily damaged, sometimes with complete collapse. Frame structures, if not bolted shifted off foundations. Frames racked. Serious damage to reservoirs. Underground pipes broken. Conspicuous cracks in ground. In alluviated areas sand and mud ejected, earthquake fountains, sand craters.

X. Most masonry and frame structures destroyed with their foundations. Some well-built wooden structures and bridges destroyed. Serious damage to dams, dikes, embankments. Large landslides. Water thrown on banks of canals, rivers, lakes, etc. Sand and mud shifted horizontally on beaches and flat land. Rails bent slightly.

XI. Rails bent greatly. Underground pipelines completely out of service.

XII. Damage nearly total. Large rock masses displaced. Lines of sight and level distorted. Objects thrown into the air.

therefore strong ground shaking was experienced 600 mi from the epicenter.

The Modified-Mercalli intensity number is a short-hand way of describing the effect of the ground shaking. The intensity scale, as shown in Table 2, is correlated with certain classes of observed effects of ground shaking.[1,2] The intensity is therefore an indication of the severity of the effects of ground shaking at a point, whereas the magnitude is a measure of the size of the earthquake. For any single earthquake the Modified-Mercalli contour map, if correctly constructed, gives an approximate description of the relative intensity of ground shaking. However, a Modified-Mercalli VII assigned to one earthquake is not necessarily indicative of the same intensity of ground shaking as a VII assigned to another earthquake by another person. In many parts of the world there are no recordings of destructive ground shaking and the only indication of the severity of ground shaking is the assigned Modified-Mercalli number; this is particularly the case for earlier historical earthquakes. The Modified-Mercalli intensity numbers assigned in one country are very difficult to correlate with intensities assigned in another country.

When slip occurs on a fault, seismic waves are generated which travel outward and affect a greater or lesser surface area depending on how large the earthquake was. Points close to the causative fault will be shaken harder than points some distance away. This attenuation of shaking with distance from the causative fault is illustrated by the diagram in Fig. 4.2-3. The contour lines (smoothed idealizations) in Fig. 4.2-3 may be thought of as representing how the peak accelerations in the ground motion attenuate with distance from the fault. Actually, in most earthquakes the ground shaking is recorded only at a small number of points and the total area of ground shaking is not well defined, however, the areas covered by strong shaking are so important in engineering that it is valuable to have even an approximate understanding of them.

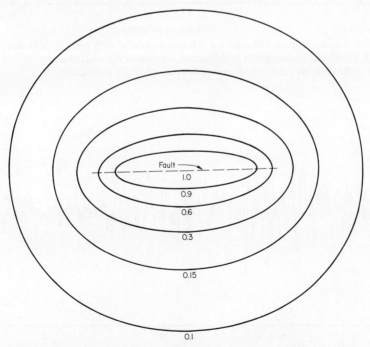

Figure 4.2-3. Idealized diagram of intensity of shaking in the vicinity of the causative fault. This illustrates that the ground shaking is most intense close to the fault and attenuates with distance from the fault.

4.25 Earthquake Ground Motions

The potentially destructive ground shaking during an earthquake is measured by accelerographs. These commercially available instruments remain at rest until activated by the initial shaking of the ground. This turns on the instrument and the three components of ground acceleration are recorded; the instrument then shuts itself off and is ready for the next earthquake. These instruments are often located on or near the abutments of dams to record the ground shaking and are also placed on the dams themselves to record how they vibrated during the earthquake (Fig. 4.2-12). Figures 4.2-4, 5 and 6 are examples of recorded

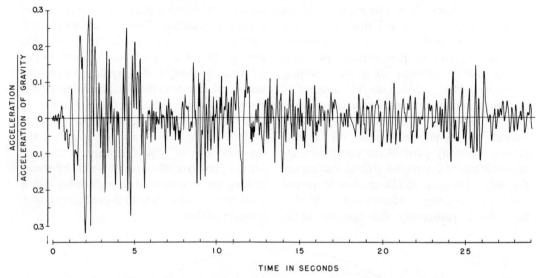

Figure 4.2-4. North-South component of ground acceleration recorded at El Centro, California during the earthquake of May 18, 1940. The accelerograph that recorded the motion was located approximately five miles from the surface trace of the causative fault that generated this magnitude 7.1 earthquake.

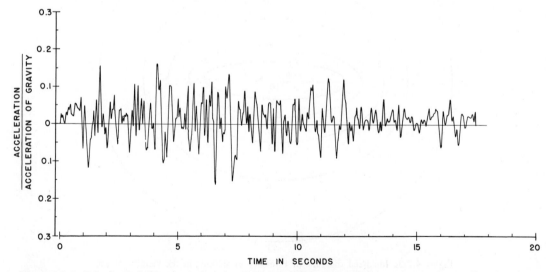

Figure 4.2-5. Tehachapi, California earthquake of 21 July 1952. This N21E component of ground acceleration was recorded approximately 25 miles from the causative fault of this magnitude 7.7 earthquake.

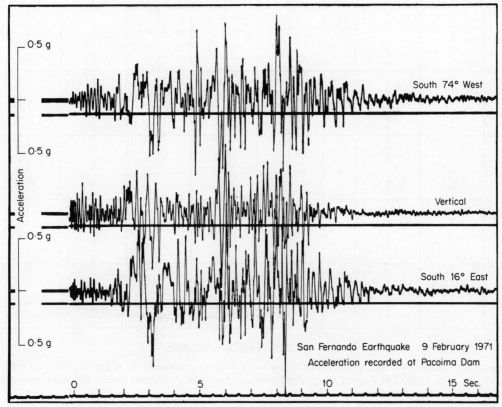

Figure 4.2-6. Acceleration recorded at Pacoima Dam during the San Fernando earthquake. The accelerogram that recorded this motion was located on a narrow mountain ridge above the abutment of Pacoima Dam and was directly above the causative fault that generated the earthquake (9 February 1971). This is the strongest ground shaking ever recorded. Pacoima Dam, a 372 ft high arch dam, was not damaged by the ground shaking but the abutment of the dam experienced cracking and dislocation.

ground acceleration. Figures 4.2-4 and 5 are representative of horizontal components of ground shaking in California. The most important characteristics of ground acceleration are described by the following three quantities; (a) the amplitude of the accelerations, (b) the duration of the strong shaking, and (c) the frequency with which the accelerogram crosses the axis. The last characteristic describes the frequency content of the ground motion. In California earthquakes the accelerogram crosses the axis approximately 7 to 10 times/sec, whereas in Peru the accelerograms have 15 to 20 crossings/sec. Because of this, Peru ground motion has somewhat different effects on structures than does California ground motion.

Figure 4.2-6 is the ground acceleration recorded on a steep ridge above the abutment of Pacoima Dam during the San Fernando, California earthquake of 9 February 1971. This is the strongest earthquake ground motion ever recorded. The dam was located near the edge of the San Gabriel mountains directly above the causative fault, (see Fig. 4.2-2b). This recorded motion, presumably, is not the same as that at the base of the dam (which was not recorded) and is not typical of ground motions to be expected near dams unless the dam and abutment bear the same relation to the causative faults as was the case during the San Fernando earthquake (magnitude 6.5). The 372 ft high Pacoima arch dam was not damaged by the

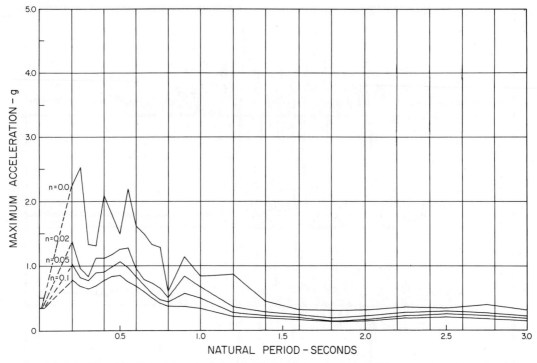

Figure 4.2-7. Acceleration response spectrum for the North-South component of the El Centro, California earthquake of May 18, 1940. The shape of the spectrum curves is characteristic of strong ground shakings recorded in California during large earthquakes.

earthquake but the abutment above which the accelerograph was located was cracked and deformed.

At points very close to the causative fault the peak acceleration can, under certain circumstances, be very high, perhaps exceeding 50 percent g, however in other circumstances the peak acceleration very close to the fault may be lower than it is at some distance from the fault. This peak acceleration seems to depend mainly on the proximity to the slipped fault, the details of the faulting mechanism, and the nature of the local geology, and does not appear to be strongly dependent on the magnitude of the earthquake. At distances greater than a few miles from the causative fault, the maximum acceleration of the ground seems to depend primarily upon the magnitude of the earthquake and secondarily upon the influence of the travel path of the seismic waves, and less strongly on the details of the faulting mechanism. Table III gives an idealized relation between the maximum acceleration and the magnitude of the earthquake for points on firm, relatively deep alluvium not especially close to the causative fault.

The duration of the strong phase of ground shaking in the accelerogram is very significant so far as the destructive effect is concerned. In the case of earth slides and similar ground effects the duration of even the weak shaking following the strong phase can have a significant effect. For example, at Anchorage, Alaska during the 1964 earthquake the strong phase of ground shaking was followed by an additional 3 min of attenuated shaking that was perceptible.[4] The latter shaking was sufficiently strong to keep large earth slides moving that had been initiated by the stronger shaking. Idealized values of duration of strong shaking are given in Table 3. The table gives the duration of the strong phase in the general vicinity of the fault; however, at greater distances from the fault the duration of shaking is longer and the intensity of shaking is less. Also, the higher frequencies of ground shaking attenuate with distance more quickly than do the long-period components. If two consecutive earthquakes occur, the effective duration of strong shaking can be greater than indicated by Table 4.

The vertical accelerations recorded during earthquakes are usually less intense than the horizontal accelerations and have higher fre-

TABLE 3. Maximum Ground Accelerations and Durations of Strong Phase of Shaking on Deep Alluvium

Magnitude	Maximum acceleration (% g)	Duration (sec)
5.0	9	2
5.5	15	5
6.0	22	8
6.5	29	12
7.0	37	17
7.5	45	25
8.0	50	32
8.5	50	37

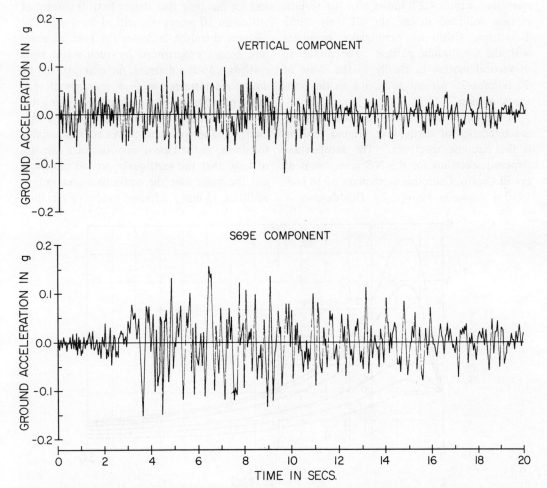

Figure 4.2-8. Vertical ground accelerations. As compared to the horizontal ground acceleration, the vertical acceleration has smaller amplitudes and somewhat higher frequencies. The accelerograms shown were recorded at Taft, California during the Tehachapi earthquake of 21 July 1952.

TABLE 4. Area in 1000 mi² Covered by Ground Acceleration (% g)

Acceleration	M						
	5.0	5.5	6.0	6.5	7.0	7.5	8.0
≥ 5	0.4	1.6	3.6	6.8	13	28	56
≥ 10		0.6	1.6	3.6	7.6	14	32
≥ 15			0.6	2.0	4.4	9.6	21
≥ 20				0.9	2.5	6.0	14
≥ 25					1.3	4.0	10
≥ 30					0.25	2.0	6.4
≥ 35						0.6	4.0
≥ 40							1.2

quencies. Figure 4.2-8 shows how the vertical motion recorded during the 21 July 1952 Tehachapi, California earthquake compares with the longitudinal motion. The amplitude of vertical motion is usually in the range of 50 percent-75 percent as intense as the horizontal motion.

A commonly used method of exhibiting the characteristics of earthquake ground shaking is the response spectrum. The acceleration response spectrum for the N-S component of the El Centro, California earthquake of 18 May 1940 is shown in Fig. 4.2-7. This diagram is based on the oscillatory response of a single-mass oscillator that is excited into vibrations by the ground motion.[2] The quantity that is plotted in Fig. 4.2-7 is the peak acceleration experienced by the one degree-of-freedom oscillator (structure) when its base is subjected to the N-S component of recorded ground acceleration. The linearly elastic oscillator will experience different accelerations depending upon the natural period of vibration and upon the amount of viscous damping. Figure 4.2-7 plots the peak acceleration for oscillators ranging from 0.2 sec period to 3.0 sec period, and for damping that ranges from 0 percent of critical to 10 percent of critical damping. The diagram therefore indicates the peak acceleration forces experienced by such single mass systems. Such a diagram, for example, can be used to determine the peak response of each mode of vibration of a building, supposing it is a linearly elastic system with viscous damping.[2,11] The spectrum curves are not suitable for dams, as the base dimensions of a dam are so large that the earthquake ground motion is not the same over the entire base area and, in addition, so many different modes of vibration

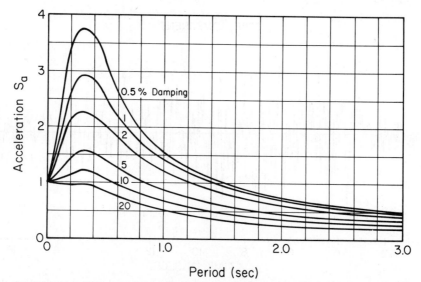

Figure 4.2-9. Average spectrum curves. These average spectrum curves are based on the response spectra of recorded strong ground motions of large western United States earthquakes. Such smooth spectrum curves are often specified as design spectra for earthquake criteria.

participate strongly in the motion of the dam that the spectrum approach is not practical. The spectrum is suitable, however, for buildings and equipment that may be associated with the dam project.

Figure 4.2-9 shows a so-called design spectrum.[2] This is a set of smooth curves whose shapes are similar to the average shapes of response spectrum curves calculated for strong shaking generated by large western U.S. earthquakes. Such smooth design spectrum curves are usually specified for design criteria. For specially important structures, it is customary to calculate the time-history response to design accelerograms rather than to use spectrum curves.

If, on the basis of geological and seismological investigations, it has been determined what magnitude of earthquake at what distance might be anticipated at the proposed dam site, it is often desired to investigate the possible response of structures and dam to the ground shaking likely to be experienced. One way of estimating the nature of ground shaking likely to be experienced is to locate an earthquake recording made under circumstances similar to those determined for the site; that is, approximately the same magnitude of earthquake at approximately the same distance and recorded on geological formations that are not too different from those at the site. Because destructive earthquakes occur relatively infrequently, and recording accelerographs are limited in number, earthquakes of all magnitudes at various possible distances have not been recorded and, therefore, some other approach must be taken to estimate the ground shaking likely to occur at the site. An approach that is often used is to construct a simulated earthquake accelerogram that corresponds to the magnitude of earthquake and the distance of causative fault that have been specified. This can be done by means of a digital computer calculation of a random function that is filtered so as to have frequency characteristics similar to those of actual recorded earthquakes. The filtered function is then multiplied by an amplitude shaping function which gives the desired amplitude, duration and attenuation. Figure 4.2-10 shows such a simulated accelerogram that corresponds approximately to the ground motion recorded 5 to 10 mi from the causative fault of a magnitude 7.0-7.5 earthquake. When the average response spectrum shape is specified (as in Fig. 4.2-9) and the amplitude shaping function is specified, the computer can generate as many simulated accelerograms as desired. These are random statistical samples of ground accelerations whose average response spectrum curves are shaped as those in Fig. 4.2-9, and which have the same duration of shaking. An accelerogram such as that shown in Fig. 4.2-10 can be used as a so-called *design earthquake*. When making response calculations it is customary to repeat the calculation for several different ground motions so as to iron out statistical fluctuations that might be involved. The two horizontal components of the design earthquake would be similar, and the vertical component would have smaller amplitudes and have higher frequency components. Such a design earthquake specifies the time-history of ground motion at a point, and the design of the project is based on this specification.

4.26 Design and Analysis

During an earthquake a dam is a vibrating structure excited into motion by the ground shaking. An earth dam, for example, has many modes of vibration with different natural frequencies. This is shown in Fig. 4.2-11. These resonance curves were measured on the Bouquet Canyon Dam, a 200 ft high earth dam belonging to the Los Angeles Department of Water and Power. The dam was excited into vibrations by a shaking machine which could exert an oscillatory back-and-forth force at any desired frequency. The machine was mounted on a concrete pad on the crest of the dam and was operated for several minutes at each of a variety of frequencies of vibration; each dot on a curve in Fig. 4.2-11 represents a particular frequency of vibration at which the machine was operated.

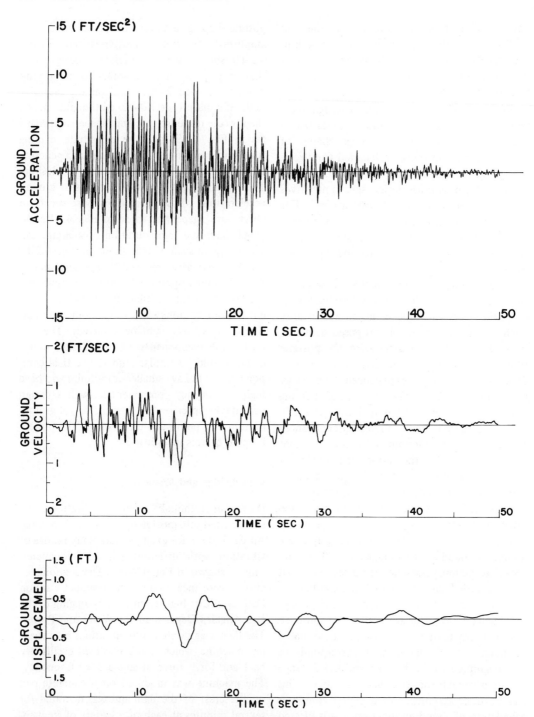

Figure 4.2-10. Simulated earthquake ground motions. A portion of a stationary random function was generated by the digital computer, was then filtered so as to have frequency characteristics similar to real earthquakes, and was shaped to correspond approximately to the amplitude and duration of ground shaking generated by an earthquake of magnitude 7.0–7.5 at a distance of approximately five to ten miles from the causative fault.

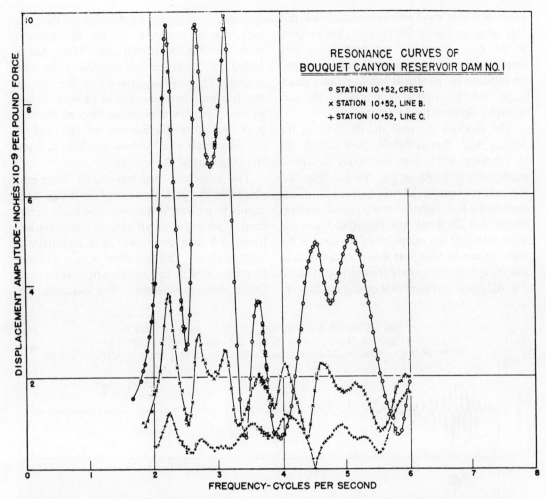

Figure 4.2-11. Experimental response curve for Bouquet Canyon Dam. A shaking machine was installed on the crest of the 200 ft high Bouquet Canyon earth dam and the dam was subjected to an alternating horizontal force in the direction perpendicular to the axis of the dam. By running the machine at different frequencies and measuring the motion of a point on the crest of the dam, a response curve was constructed shown as Curve A. Curve B was constructed from measurements made approximately half way down the face of the dam and Curve C was constructed from measurements made approximately two thirds of the way down. Each peak on the response curve indicates resonance with the natural frequency of one of the modes of vibration of the dam.

At each speed of the machine the frequency was noted and the amplitude of motion at the crest of the dam at midpoint was recorded. The machine was then operated at a slightly higher frequency and the amplitude of motion recorded. This step was repeated and the curve marked A was thus determined. Curves B and C are for measured amplitudes made at the mid-height and at the quarter-height of the dam respectively. Each peak on these curves represents resonance with a natural frequency of vibration of one of the modes of vibration of the dam. The shape of the mode of vibration can be determined by measuring the amplitudes at various points over the surface of the dam. The amplitude of the alternating force exerted by the shaking machine was in the range of 2000 lb to 10,000 lb; and the amplitudes were very small and were measured with a very sensitive seismic instrument. The material in the dam, therefore, was undergoing very small strains during the test, whereas during an earthquake the strains would be very much larger and the physical properties of the dam might be significantly different.

The motions recorded on the crest of the 200 ft high Santa Felicia Dam during the 9 February 1971 San Fernando, California earthquake is shown in Fig. 4.2-12. The dam was approximately 20 mi northwest of the earthquake in a region of strong ground shaking. Figure 4.2-12 shows the recorded transverse acceleration of the midpoint of the crest of the dam. It can be seen that this recorded motion has an appearance quite different from that of Fig. 4.2-5 and it is clear that during the first ten seconds of motion the accelerations of the crest were mainly produced by the fundamental mode of vibration of the dam, which had a period of vibration of approximately one second. The peak acceleration on the crest of the dam was approximately 18 percent. It can be seen from the displacement record that the peak oscillatory displacement of the crest of the dam relative to the base was 2 in. at 5 sec from the start of the motion.

The procedure that has usually been employed in the past for earthquake design of a dam is to include in the prescribed loads a constant 5 percent g or 10 percent g acceleration force, and then to proceed with essentially a static analysis. This procedure is very difficult to correlate with the actual earthquake forces that a dam experiences. For example, the

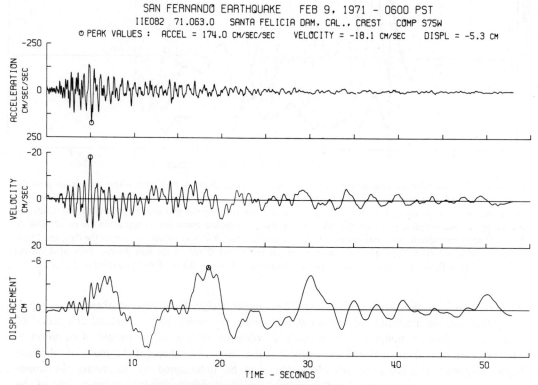

Figure 4.2-12. Earthquake motion of crest of Santa Felicia Dam. The dam was located approximately 20 miles northwest of the center of the 9 February 1971 San Fernando, California earthquake. The motion shown is that perpendicular to the axis of the dam, and it can be seen that the strong shaking is predominately the vibration of the fundamental mode of the dam; this is clearly shown in the earlier portion of the velocity record.

Santa Felicia Dam whose motion is shown in Fig. 4.2-12, experienced 18 percent g peak acceleration at the crest, which was undoubtedly more than the design acceleration forces, but the dam did not experience any significant damage during the earthquake. Similarly, the concrete gravity Koyna Dam was designed for a stated 5 percent g acceleration force (equivalent static) but during the earthquake an accelerograph within the dam recorded accelerations as high as 50 percent g. The difficulty in correlating the equivalent static horizontal force method of design with what actually happens during an earthquake, and the difficulty in knowing what the true factor of safety is when the equivalent static method of design is used, makes it desirable to develop a method of design that is based on dynamic analysis and that relates more closely to the actual behavior of a dam during an earthquake. Some dynamic analyses have been made for earth dams, arch dams and concrete gravity dams by representing the dam and surrounding ground by an equivalent mathematical model composed of finite elements, and then by means of a digital computer determining the vibrations of the finite elements when the sub-base of the dam is shaken like some recorded earthquake motion.[2] The difficulties with this type of approach is that all the physical parameters are not known accurately and hence have to be estimated when making the dynamic analysis; and, furthermore, with the present state of the art the sub-base of the dam is given an acceleration which is the same over the entire base, which is not realistic. It appears, however, that it is only by developing the dynamic method of analysis that earthquake designs of dams will be put on a rational basis. More recordings of the earthquake motions of dams and their abutments are needed and more research is needed on how to model a real dam by means of finite elements that take into account the real dynamic properties of the material of which the dam is constructed.

When an earthquake occurs with a full reservoir behind the dam the vibrations of the dam will induce dynamic water pressures and these must be taken into account when making the design. Also, the earthquake ground shaking may induce water waves in the reservoir; and if there is fault displacement through the reservoir there may be generated relatively large water waves and these must be taken into account in the design.

When establishing the earthquake design criteria it is important to take into consideration both the allowable design stresses and the design earthquake. These should not be established separately for they both have an influence on the safety factor.

It is necessary not only to improve the earthquake design of new dams but improved methods must be developed for dealing with the earthquake hazard posed by existing dams. In California the State Division of Dam Safety has a continuing program requiring investigations of the earthquake safety of existing dams in the state. The analysis requires an assessment of the safety of the dam when subjected to realistic earthquake ground shaking. Failure analyses of dams are also required in which the path of the water, the depth, etc., is analyzed for a hypothetical dam failure.

REFERENCES

1. Richter, C. F., (1958), *Elementary Seismology*, W. H. Freeman and Co., San Francisco.
2. Wiegel, R. L., Editor, (1970), *Earthquake Engineering*, Prentice-Hall, Inc.
3. Jennings, P. C., Editor, (1971), "Engineering Features of the San Fernando Earthquake," California Institute of Technology, Pasadena, California.
4. "The Great Alaska Earthquake—Engineering," (1973), National Academy of Sciences, Washington, D.C.
5. Ambraseys, N. N., "On the Seismic Behavior of Earth Dams, Proceedings of Second World Conference on Earthquake Engineering," (1960), International Association for Earthquake Engineering.
6. Seed, H. B., "Soil Problems and Soil Behavior," in Reference (2).
7. Berg, G. V., Das, Y. C., Gokhale, K. V., and Settur, A. V., "The Koyna, India Earthquakes,

Proceedings Fourth World Conference on Earthquake Engineering," (1969), International Association for Earthquake Engineering.

8. Housner, G. W., "Seismic Events at Koyna Dam, Rock Mechanics-Theory and Practice," Proceedings of Eleventh Symposium on Rock Mechanics (1970), American Institute of Mining, Metallurgical and Petroleum Engineers, New York.

9. Sheng, Chung-Kang, et al., "Earthquakes Induced by Reservoir Impounding and Their Effects on the Hsinfengkiang Dam," Proceedings 1970 ICOLD Conference, Madrid, Spain.

10. Rothé, J. P., "Earthquake and Reservoir Loadings," Proceedings of Fourth World Conference on Earthquake Engineering (1969), International Association for Earthquake Engineering.

11. Newmark, N. M. and Rosenblueth, E., (1971), *Fundamentals of Earthquake Engineering*, Prentice-Hall, Inc.

12. Jennings, P. C., Housner, G. W., and Tsai, N. C., "Simulated Earthquake Motions for Design Purposes," Proceedings of Fourth World Conference on Earthquake Engineering (1969), International Association for Earthquake Engineering.

13. Okamoto, S., "Introduction to Earthquake Engineering," University of Tokyo Press (1973).

14. Seed, H. B., et al., "Analysis of the Slides in the San Fernando Dams during the Earthquake of February 9, 1971," Earthquake Engineering Research Center, University of California, Berkeley, California, June, 1973.

15. Bolt, B. A. and D. E. Hudson, "Seismic Instrumentation of Dams," *Journal of Geotechnical Engineering Division*, American Society of Civil Engineers, JP11, November 1975.

5

Selection of the Type of Dam

ELLIS L. ARMSTRONG
President, Armstrong Associates,
Engineers & Consultants
Salt Lake City, Utah

I. INTRODUCTION

A. General

It is only in exceptional circumstances that an experienced designer can say that only one type of dam is suitable or most economical for a given dam site. Except in cases where the election of type is obvious, it will be found that preliminary designs and estimates will be required for several types of dams before one can be shown to the best solution from the standpoint of direct costs and all other factors. It is, therefore, important to emphasize that the project is likely to be unduly expensive unless decisions regarding selection of type are based upon adequate study after consultation with competent engineers.

In the selection of type for important structures, it is also usually advisable to secure the advice of an experienced engineering geologist in connection with the relative applicability of possible types to the foundation at the site.

In numerous cases, costs of providing protection from spillway discharges, limitations of outlet works, and the problem of diverting the stream during construction have an important bearing on the selection of type. In certain cases, type selection may also depend upon the availability of labor and equipment. These may be particularly important considerations when the element of time is involved. Inaccessibility of the site may also have an important bearing on the selection.

The selection of the best type of dam for a particular site calls for thorough consideration of the characteristics of each type, as related to the physical features of the site and the adaptation to the purposes the dam is supposed to serve, as well as economy, safety, and other pertinent limitations. The final choice of type of dam will generally be made after consideration of these factors. Usually, the greatest single factor determining the final choice of type of dam will be the cost of construction. The following paragraphs discuss important physical factors in the choice of the type of dam.

B. Topography

Topography, in large measure, dictates the first choice of type of the dam. A narrow stream flowing between high, rocky walls would naturally suggest a concrete overflow dam. The low, rolling plains country would, with equal fitness, suggest an earthfill dam with a separate spillway. For intermediate conditions, other considerations take on more importance, but the general principle of satisfactory conformity to natural conditions is a safe primary guide. The location of the spillway is an important item that will be governed very largely by the local topography and will, in turn, have a material bearing on the final selection of the type of dam.

C. Geology and Foundation Conditions

Foundation conditions depend upon the geological character and thickness of the strata which are to carry the weight of the dam, their inclination, permeability, and relation to underlying strata, existing faults, and fissures. The foundation will limit the choice of type to a certain extent, although such limitations can frequently be modified, considering the height of the proposed dam. The different foundations commonly encountered are discussed below:

1. Solid rock foundations, because of relatively high bearing capacity and resistance to erosion and percolation, offer few restrictions as to the type of dam that can be built upon them. Economy of materials or overall cost will be the ruling factor. The removal of disintegrated rock together with the sealing of seams and fractures by grouting will generally be necessary.
2. Gravel foundations, if well compacted, are suitable for earthfill, rockfill, and low concrete gravity dams. As gravel foundations are frequently subject to water percolation at high rates, special precautions must be taken to provide effective water cutoffs or seals.
3. Silt or fine sand foundations can be used for the support of low concrete gravity dams and earthfill dams if properly designed, but they are not suitable for rockfill dams. The main problems are settlement, the prevention of piping, excessive percolation losses, and protection of the foundation at the downstream toe from erosion.
4. Clay foundations can be used for the support of earthfill dams but require special treatment. Since there may be considerable settlement of the dam if the clay is unconsolidated and the moisture content is high, clay foundations ordinarily are not suitable for the construction of concrete gravity dams, and should not be used for rockfill dams.

   Tests of the foundation material in its natural state are usually required to determine the consolidation characteristics of the material and its ability to support the superimposed loads.
5. Nonuniform foundations. Occasionally, situations may occur where reasonably uniform foundations of any of the foregoing descriptions cannot be found and where a nonuniform foundation of rock and soft material must be used if the dam is to be built. Such unsatisfactory conditions can often be overcome by special design features. Each site, however, presents a problem for appropriate treatment by experienced engineers, and no attempt is made in this text to treat such unusual problems. The details of the foundation treatments mentioned above are given in the appropriate chapters on the design of earthfill, rockfill, and concrete gravity dams.

D. Materials Available

Materials for dams of various types, which may sometimes be available at or near the site, are:

1. Soils for embankments.
2. Rock for embankments and riprap.

3. Concrete aggregate (sand, gravel, crushed stone).

Elimination or reduction of transportation expense for construction materials, particularly those which are used in great quantity, will effect a considerable reduction in the total cost of the project. The most economical type of dam will often be the one for which materials are to be found in sufficient quantity within a reasonable distance from the site.

The availability of suitable sand and gravel for concrete at a reasonable cost locally and perhaps even on property which is to be acquired for the project is a factor favorable to the use of a concrete structure. On the other hand, if suitable soils for an earthfull dam can be found in nearby borrow pits, an earthfill dam may prove to be the most economical. Advantage should be taken of every local resource to reduce the cost of the project without sacrificing the efficiency and quality of the final structure.

E. Spillway Size and Location

The spillway is a vital appurtenance of a dam. Frequently its size and type and the natural restrictions in its location will be the controlling factors in the choice of the type of dam. Spillway requirements are dictated primarily by the runoff and streamflow characteristics, independent of site conditions or type or size of the dam. The selection of specific spillway types will be influenced by the magnitudes of the floods to be bypassed. On streams with large flood potential, the spillway will become the dominant structure, and the selection of the type of dam could become a secondary consideration.

The cost of constructing a large spillway is frequently a considerable portion of the total cost of the development. In such cases, combining the spillway and dam into one structure may be desirable, indicating the adoption of a concrete overflow dam. In certain instances, where excavated material from separate spillway channels can be utilized in dam embankment, an earthfill dam may prove to be advantageous. Small spillway requirements often favor the selection of earthfill or rockfill dams, even in narrow dam sites.

The advisability or practice of building overflow concrete spillways on earth or rock embankments has generally been discouraged because of the more conservative design assumptions and added care needed to forestall failures. Inherent problems associated with such designs are: Unequal settlements of the structure due to differential consolidations of the embankment and foundation after the reservoir loads are applied; the need for special provisions to prevent cracking of the concrete or opening of joints which could permit leakage from the channel into the fill, with consequent piping or washing away of the surrounding material; and the construction delays necessitated by the requirement for having a fully completed and seasoned dam before spillway construction can be started. Consideration of the above factors, coupled with increased costs brought about by more conservative construction details such as increased lining thickness, increased reinforcement steel, cutoffs, joint treatment, drainage, and preloading, have generally led to selection of alternative solutions for the spillway design such as placing the structure over or through the natural material of the abutment or under the dam as a conduit.

One of the common spillway arrangements is the utilization of a channel excavated through one or both of the abutments outside the limits of the dam, or at some point removed from the dam. Where such location is adopted, the dam can be of the non-overflow type which extends the choice to include earthfill and rockfill structures. Conversely, failure to locate a site for a spillway away from the dam requires the selection of a type of dam which can include an overflow spillway. The spillway overflow can then be placed so as to occupy only a portion of the main river channel, in which case the remainder of the dam could be either of earth, rock, or concrete.

F. Environmental Considerations

In the past few years, environmental considerations have become very important in the design of dams. These factors can have a major influence on the type of dam constructed. They are discussed more fully in the following sections.

The principal influence of the environmental laws and regulations on selection of a specific dam type is the need to consider maximum protection for the environment which can affect the type of dam, its dimensions, location of spillway and appurtenant facilities. The cost of providing environmental protection may raise the total cost of the dam substantially above that of the structure alone. Environmental regulations usually require detailed considerations and presentations concerning one or more alternates to the dam favored by the owners. This does not mean the alternates must necessarily be the same type as the favored dam. A more detailed discussion of the environmental requirements for dams and reservoirs appears in Chapter 1.

II. EMBANKMENT DAMS

A. Introduction

Embankment dams are those constructed of the natural materials of the earth; namely, soil and rock in their many and varied forms. There are two types of dams as follows:

>   Earthfill dams
>   Rockfill dams

These main types of dams will be subdivided further in the appropriate following sections. Types of earthfill and rockfill embankments will be discussed and the conditions which make one type preferable to another will be reviewed. With proper design, the major influencing factor is that of cost to meet all conditions of the design.

To begin the type selection for an embankment dam, the designer must gather all available design data. From the topography, he makes an initial selection of the embankment type that appears to be economical. Appurtenant designs are then fitted to the dam site and quantities of materials determined. Diversion requirements are determined and incorporated into the designs. The cost of each type is estimated, and the most economical one is selected.

B. Earthfill Dams

1. Types of Earthfill Dams. The types of earthfill dams are as follows:

>   Homogeneous.
>   Modified homogeneous.
>   Zoned.
>   Hydraulic fill dams.

These types are shown graphically in Fig. 5-1.

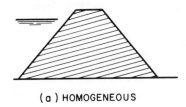

(a) HOMOGENEOUS

(b) MODIFIED HOMOGENEOUS

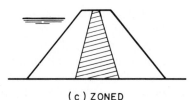

(c) ZONED

Figure 5-1. Types of earth dams.

SELECTION OF THE TYPE OF DAM

2. Site Topography. Earthfill dams have been constructed in almost every conceivable type of topography but are generally not used where deep gorges with extremely steep abutment walls are present. This is due mainly to a lack of adequate materials, river diversion requirements for flood control, construction sequencing, and the greater suitability of many types of concrete dams. Broad valleys usually lead to central core dams. Between these two topographic extremes, many types of earthfill dams are possible. Broad valleys or plains locations from long dams may lead to the use of several different types of earthfill sections, depending upon the types of material available.

Narrow valleys can dictate the type selection through the influence of construction working space. Topography which is mountainous may dictate the type selection by limiting the materials which may be economically hauled to the site. Topography dictates the axis selection for the dam, and this can influence the earthfill type selected. The topography shown in Fig. 5-2 (a), along with the type of material available, resulted in the proposal of a zoned earthfill dam as shown in Fig. 5-2 (b).

3. Geology and Foundation Conditions. The dam site geology, reservoir geology, and the site foundation conditions influence the selection of the type of earthfill dam. It is essential that the dam geology be carefully studied by a competent engineering geologist. Large shear zones may cause relocation of the core and thus a type change and fault location can force the change from a homogeneous earthfill to a zoned dam. If the reservoir slopes are unstable it may also force reconsideration of the initial type of dam selected.

The given set of foundation conditions will influence greatly the type of earthfill dam selected. Few foundations are homogeneous, and the extent of exploration available at the time of design will affect the type selection. If a type is selected on inadequate field exploration, it should be one that can easily be modified.

Alluvial foundations mean seepage problems will be present and their permeability and depth can mean changes in the earthfill type used. If special cutoffs are used, a definite type can be suggested depending on the cutoff location. For example, upstream cutoffs point to sloping core embankments.

Foundations of glacial material often present special design problems and can easily cause changes in the dam type.

Also to be considered is the potential earthquake to which a structure may be subjected. If the magnitude of the earthquake is large it can cause a change in the proportions of the dam or a complete change of embankment type or a major type change.

4. Availability of Materials. This is an important item in selecting the major type of dam (concrete or embankment) and also the earthfill which will be used. The available materials and their excavation costs dictate the type of earthfill structures to be considered. An abundant supply of a single material points to the use of a homogeneous dam. Sufficient quantities of both previous and impervious materials lead to the use of zoned dams. The amount of each type of material available from borrow areas influences the type of earthfill dam used. If materials are to be used with stockpiling, they must be used early in the construction period. Materials from a borrow area may also be available only at certain times of the year and only to certain depths due to groundwater or climatic conditions. Available excavation machinery could, in some instances, force a change in the type of dam. Haul distance from the borrow area influences cost and therefore the type of dam.

The potential of hydraulic excavation and its possible economy should not be overlooked. Although this type of dam is subject to liquefaction there may be cases where it can be safely and economically used.

5. Influence of Spillway and Outlet Works Selection. The spillway and outlet works are sometimes a large proportion of the cost and

272 HANDBOOK OF DAM ENGINEERING

Figure 5-2a. Topography, Dallas Creek Project, Colorado.

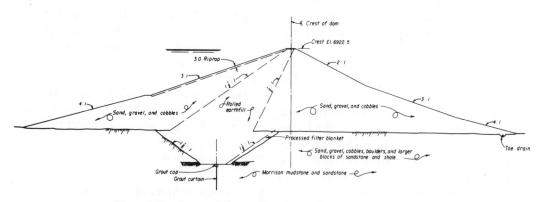

Figure 5-2b. Proposed dam section, Dallas Creek Project, Colorado.

thus are important in determining the selection of a type. In some cases the type of outlet works and spillway used for an earthfill dam will influence the type of dam selected. There are two types of spillways used for embankment dams: (1) tunnel type (2) chute type. Chute type spillways are used the most and are constructed in cuts in one of the abutments. Side channel inlets are used for narrow valleys which require large excavations for the approach channel.

The type of outlet works generally has little effect on the type of embankment dam selected. It does not affect the overall cost of the project and thus whether an embankment dam will be constructed.

6. Climatic Conditions. Climatic conditions can often play a large part in the selection of the type of dam. Dams with thinner cores have often been used to offset rainy weather conditions which would slow or stop the rolled fill compaction process. The wet compaction methods used in Finland and Sweden are also used to compensate for short rainy construction periods.

Climatic conditions often influence how deep borrow area excavations can be and the magnitude and timing of the floods expected at the site. These factors can determine the type of dam that is selected for a specific site.

7. Reservoir Operations. The reservoir operation does not often affect the type of dam selected. If the reservoir is for flood control, it may affect the type of dam used but it is more likely that the design details and not the type will be more greatly affected. If the reservoir fluctuations are to be rapid a zoned dam must be used to maintain stability of the upstream slopes.

C. Rockfill Dams

1. Types of Rockfill Dams. The rockfill dam types considered herein are as follows:

Central core
Sloping core
Diaphragm

These types are shown in Fig. 5-3.

Diaphragm rockfill dams can have numerous advantages. They provide greater safety against downstream sliding than the central core type. Another strong advantage of the diaphragm dam is that grouting may be done along with rock placement. If the use of the reservoir permits, it can be also be drawn down at the end of each year to check the diaphragm. Uplift pressures present no problems. Benefits also arise from the fact that the diaphragm can be placed after the embankment is constructed. The diaphragm is available for inspection and repair. If

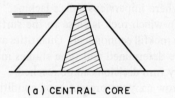
(a) CENTRAL CORE

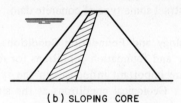
(b) SLOPING CORE

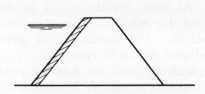
(C) DIAPHRAGM

Figure 5-3. Types of rockfill dams.

the abutments of the dam contain deep overburden deposits, the excavation required can become excessive and the diaphragm dam may be more advantageous. Central and sloping core rockfill dams allow the use of an economical source of earth material. Sloping core dams are thought by some to provide better resistance to seismic forces during an earthquake. Finn[4] has produced some evidence that central core dams may be less advisable in earthquake areas.

One disadvantage of diaphragm dams which should be considered is that they are subject to weathering. This can exclude the use of diaphragms for use with pumped storage projects which require rapid drawdown.

2. Site Topography. Site topography for rockfill dams is mountainous terrain or steep valley areas where impervious soil is lacking. Topography in which rock is close to the surface can make a rockfill economical. Quantities available must be determined and type studies made before any firm decision can be reached.

Narrow canyons present good conditions for central core rockfills since high core pressures will be effected on the foundation. Steep canyon walls at the site can mean settlement problems for core type dams and it might be better to construct some type of concrete dam.

3. Geology and Foundation Conditions. The geology and foundation conditions for rockfill dams are important influences in the type selection. Geological conditions at the site can cause relocation of the core and thus a change in the rockfill type.

Foundations for rockfill dams must be capable of resisting settlement and deformation. Sand and gravel foundations can usually be used but they must have a cutoff constructed to bedrock. Clay foundations are generally not adequate for rockfill dams.

Foundations which are deep and pervious can sometimes be suitable for a rockfill dam but extensive grouting may be necessary. Generally a different type of dam is indicated.

4. Availability of Materials. Materials used in rockfill dams must be studied carefully. In most cases the spillway and outlet works and the appurtenant features supply the greater portion of the rockfill. Care must be taken when examining a quarry since rock just back from the face may contain excessive fines. If a material shortage is envisioned a diaphragm dam may be economical. Test fills can often be used to advantage if it is undecided whether or not a given rockfill material can be used in the embankment. Shear strength tests can also be used to decide on the use of available material.

5. Spillway and Outlet Works Selection. Choice of the type of spillway and outlet works play an important function in the selection of a rockfill dam. In general, chutetype concrete spillways are used. As noted for the earthfill-type dam, spillways and outlet works have a major effect on the cost of the structure as compared to a concrete dam. Much of what was said about spillways under Section II-B for earthfill dams also applies to rockfill dams. The amount and type of excavation material is an important factor in determining whether the dam will be an earthfill or rockfill type, both from the cost and environmental viewpoints.

6. Climatic Conditions. Rockfill dams are used when harsh climatic conditions, either cold or wet are encountered. Placement can continue in cold weather where earthfill construction would have to stop. Rockfills with a diaphragm can be adapted to adverse weather conditions and have been used extensively. Rockfill dams can also be designed to pass floods during construction over or through the partially complete structure. In dry climates where lack of water may prevent the use of earth core material a diaphragm type rockfill dam can be used.

7. Reservoir Operation. Reservoir fluctuations with rapid drawdown present special problems and rocktill types with cores which are not adversly affected by pore pressures should be used. Sloping core dams can be more greatly affected by pore pressure effects. Reservoir operations may require that the structure be built higher a few years after initial construction and rockfills are especially adaptable to this type of program. If the reservoir is to be brought into use as the structure is built, a diaphragm type can be economical.

III. CONCRETE DAMS

A. Concrete Arch Dams

1. Site Topography

(a) Width-to-height ratio. One of the primary factors in determining the suitability of a site for an arch dam is the width-to-height ratio. Ideally, the ratio should be relatively small (3.0 or less) for the use of a thin arch dam. Figure 5-4 shows an arch dam constructed on such a site. However, arch dams can be designed to give good economy for the wider sites (Fig. 5-5). Some designers would consider the use of arch dams for sites with width-to-height ratios as high as 10:1. The reduced load carrying capacity of the longer arches in a wider site will require that a greater percentage of the load be transmitted vertically to the foundation by cantilever action. This results in greater thicknesses approaching those required for a gravity dam as the width-to-height ratios of sites being considered approach the upper bound.

Use of polycentered or elliptically shaped

Figure 5-4. Buffalo Bill Dam in Wyoming, an arch dam in a narrow canyon.

Figure 5-5. Gibson Dam in Montana, an arch dam in a wide canyon.

arches to add rigidity and increased load-carrying capacity to the long arches permit a reduction in thickness and increase the range of width-to-height ratios which are favorable to the use of an arch dam.

(b) Canyon shape. Canyon shape is an important consideration in the design of an arch dam which affects the thickness requirements and, consequently, the economy. Thus, when the site is marginal for other reasons, canyon shape may be the factor which determines the acceptability of an arch dam. The V-shaped sites are, generally, more favorable to the design of an acceptable arch dam than are the U-shaped sites. Other considerations being equal, the arch dam designed for the V-shaped site will require less volume of concrete than the one designed for the U-shaped site. The stress distributions in dams designed for the V-shaped sites are usually more acceptable because a larger percentage of the load is resisted by arch action in the lower part of the dam. An arch dam constructed in a V-shaped site is shown in Fig. 5-6 and one in a U-shaped site in Fig. 5-7.

Although not often a controlling factor in the selection of the type of dam, a symmetrical or nearby symmetrical site is desirable for an arch dam. Another type of dam may prove to be a more economical structure if the site is extremely asymmetrical.

2. Foundation Conditions

(a) Foundation material. The material of which the foundation is composed may be the

Figure 5-6. Monticello Dam in California, an arch dam in a V-shaped site.

Figure 5-7. Glen Canyon Dam in Arizona, an arch dam in a U-shaped site.

controlling factor in determining which types of dams can be considered for a site. A foundation of competent rock is essential for any arch dam. The rock should be sound and durable to provide a satisfactory foundation for the life of the structure. Rocks which decompose or deteriorate upon exposure to water, atmosphere or pressure are unacceptable as foundation material. Even though some deposits of these undesirable materials are present, the foundation may still be acceptable for an arch dam if these deposits are not large or continuous through the contact area so that treatment is feasible. Often such foundations can be satisfactorily improved by removing the poorer material and replacing it with concrete.

(b) Foundation deformability. Foundation deformations have a significant effect on stress distributions within an arch dam. An overall foundation modulus of elasticity somewhat less than the concrete modulus of elasticity is the most favorable condition for an arch dam. However, foundations with deformation moduli as low as 500,000 psi are satisfactory for an economical design of this type. The increased deformability of such a foundation may require added thickness at the concrete-to-rock contact and, consequently, to lead to a greater volume requirement for an arch dam. Zones of foundation rock with a high modulus of elasticity can cause stress concentrations in an arch dam near the contact.

(c) Foundation strength. If a high dam is being considered, compressive strength of the foundation is an important consideration in the selection of the type of dam. A foundation with relatively low strength material may require an increase in thickness of the dam near the concrete-to-rock contact to distribute forces from the structure over al larger area of foundation. The resulting increase in concrete volume may show another type of dam is more suitable.

(d) Foundation discontinuities. Faults, shears, jointing, foliation, and nonhomogeneities exist, to some extent, in most foundation rock formations. Their frequency, size, continuity, and orientation are important factors in determining the suitability of a foundation for any type of dam. The existance of a large or active fault zone at the dam site may preclude the selection of an arch dam. However an inactive fault, which can be economically excavated and backfilled with concrete or consolidated in some other manner can be satisfactory for an arch dam. Shears and joints in the foundation do not make the site unsuitable for an arch dam if their orientation and shearing resistance are such that satisfactory safety against sliding can be assured. The presence of zones of more readily deformable or weaker material in the foundation should have little effect on an arch dam because of the inherent ability of such a structure to bridge over these zones if they are of limited size. If the zones of weaker materials are extensive, an increase in thickness of the dam over its contact with the less rigid materials may be necessary with a resulting increase in volume.

(e) Permeability. Permeability of a foundation may influence the decision as to the acceptability of a site for construction of any type of dam. Generally, grouting curtains and drainage systems are provided beneath a dam to prevent excessive leakage and the build-up of hydrostatic pressures within the foundation. Because an arch dam does not depend on its weight to resist loading, the development of hydrostatic pressures at the concrete-to-rock contact are not as critical to the stability as are some other types.

3. Influence of Spillway Type. The spillway is one of the more critical elements in the design of project facilities. Frequently, site restrictions on its location will be one of the controlling factors in the selection of the type of dam. Spillways discussed are the orifice and the overflow types, which are constructed integrally with the dam; and the channel or tunnel type excavated through one or both abutments and

SELECTION OF THE TYPE OF DAM 279

Figure 5-8. Hungry Horse Dam in Montana, showing intake for tunnel spillway near right abutment.

separated from the structure. Figure 5-8 shows the intake structure for a tunnel spillway at Hungry Horse Dam. If site conditions are such that the spillway must be located in or near the main river channel, the types of dams which can accommodate an orifice or overflow spillway should be considered. An arch dam can be used under these circumstances if required spillway capacity can be accommodated.

Both overflow spillways at the crest, as shown in Fig. 5-9, and orifices through the upper part of the dam (see Fig. 5-10) have been used satisfactorily with arch dams. Although spillway tunnels can be used with most types of dams, the relatively short length of tunnel required to transport the water under a thin arch dam will be more economical than the longer spillway tunnels needed for some of the other types. Spillways located in areas away from the dam usually are not deciding factors in the selection of the type of dam.

4. Climatic Conditions. Temperature fluctuations are an important consideration in the design of arch dams. If large seasonal temperature changes occur at the dam site, stress changes induced in a thin concrete arch dam may be unacceptably large. The added expense of insulation on the faces of the dam, special grouting procedures, or greater thickness of concrete may preclude the use of an arch dam under these circumstances.

5. Reservoir Operation. The reservoir operating schedule controls, to some extent, the load-

Figure 5-9. North Fork Dam in California, showing an overflow spillway in an arch dam.

ing combinations used in the design. If the reservoir operation scheduled combined with the seasonal temperature fluctuation creates critical loading combinations, undesirable stresses will result in an arch dam with an accompanying increase in thickness and required volume of concrete. If the volume increase is large, another type of dam may be more economical.

6. Availability of Construction Material. Construction materials for dams of various types may be available at or near the site. Sometimes the most economical type of dam is the one for which acceptable construction materials exist within a relatively short distance from the site. Adequate deposits of suitable aggregate for concrete at economical cost locally or on the property to be acquired for the dam site are favorable to the use of a concrete dam. A limited supply of aggregate may still be sufficient to economically construct an arch dam.

B. Concrete Gravity Dams

1. Site Topography. Although gravity dams are suitable for use with most canyon widths and shapes, some topographic features will favor the use of a gravity type structure. One of these is the wide site with a relatively flat canyon floor. The nearly uniform height of the blocks reduces the tendency for twisting transmitted by keyed joints between adjacent blocks. This twisting is caused by the differential deflections of adjacent blocks due to differences in height. As discussed in Section III-A-1, a site with a width-to-height ratio of 10:1 or less can be used for an arch dam. However, a gravity dam may prove to be more economical when the ratio is higher than 6:1. Figure 5-11 shows a gravity dam in a wide site.

2. Foundation Conditions

(a) Foundation materials. Most foundation materials are acceptable for consideration of a

Figure 5-10. Morrow Point Dam in Colorado, showing spillways through the dam.

gravity dam under 50 ft in height and with a difference between headwater and tailwater less than 20 ft. Even silt, clay, fine sand, or gravel foundations can be used. However, tests should be made of bearing values as well as of consolidation characteristics. Generally, sound rock foundations are required for gravity dams having a headwater-tailwater differential of more than 20 ft or greater that 50 ft in height. Materials which decompose or deteriorate upon exposure to water, atmosphere, or pressure are questionable foundation components and should be carefully evaluated before a gravity dam is considered.

(b) Foundation Deformability. The deformability of foundation rock is not a major consideration in the design of gravity dams. If contraction joints are to be grouted in the gravity dam, highly deformable foundation and abutment material may produce undesirable stress concentrations, however. For foundations other than rock, settlement and deformability characteristics should be investigated before consideration of a low gravity dam.

(c) Foundation strength. Strength or bearing capacity and shearing resistance of the foundation materials can be very important factors in determining the suitability of a site for a gravity dam. For the higher gravity dams to be founded on rock, shearing resistance is the more critical factor. If shearing resistance of the foundation material is low, an increase in volume of con-

Figure 5-11. Friant Dam in California, a gravity dam in a wide site.

crete may be required. The greater cost of the larger volume of concrete or of possible foundation treatment may preclude the use of a gravity dam under such circumstances. Bearing capacity is of much greater significance for the low gravity dams being considered where the foundation is composed of silt, sand, clay, or gravel. The acceptability of the bearing capacity and shearing resistance of such materials should be determined in each case by the appropriate tests.

(d) Foundation discontinuities. Often situations may occur where a nonuniform foundation exists at the site. This can be the result of either variations in foundation materials or the presence of faults, shears, or jointing. Such conditions can be improved by foundation treatment so that the effect on choice of type of dam may not be significant. If differential settlement can result, however, special studies should be made to determine possible effects on a gravity dam.

(e) Permeability. Generally, grouting curtains and drainage systems are provided for concrete dams on a rock foundation to reduce seepage and the buildup of hydrostatic pressures within the foundation. Such pressures have a significant effect on the stability of gravity dams and can be the cause of increases in base thickness and volume of concrete required. Permeability of foundation materials such as silt, sand, clay, or gravel depends on the type of material, stratification, and homogeneity. The control of erosion, seepage, and hydrostatic pressures under a concrete gravity dam constructed on such materials often requires the use of combinations of upstream aprons, downstream aprons, drains, and cutoffs. The increased cost of these special

treatments may indicate that another type of dam is more suitable for the site.

3. Influence of Spillway Type. The topography at some sites will require that the spillway be located within the dam. Where the required spillway capacity is small, this may not be one of the major factors in determining the type of dam to be used. However, if the required capacity is large and a major part of the dam must be devoted to the spillway, consideration should be given to the use of a gravity dam such as is shown in Fig. 5-12. In many cases where an overflow diversion dam is required in the river channel, a small overflow gravity dam may prove to be the most acceptable.

4. Climatic Conditions. As has been discussed, in locations where large seasonal temperature fluctuations occur, an arch dam may be unacceptable because of high stresses due to temperature. If the contraction joints in a gravity dam are ungrouted, these temperature fluctuations should have little effect on the stability of the gravity blocks.

5. Reservoir Operation. Generally, the operating schedule for the reservoir has little effect on a gravity dam. Thus, reservoir operation which may control the design loading combinations of water level and temperature for an arch dam or the size and cost of the spillway for an embankment dam may not have an appreciable effect on the gravity dam stresses or overflow spillway costs.

6. Availability of Construction Material. As noted in Section I-D, local availability of adequate deposits of suitable aggregate for concrete may be a major concern in determining the economy of constructing a concrete dam. A gravity dam will require a relatively large amount of both aggregate and portland cement. If the local supply of aggregate is not adequate or the distance cement must be transported from the source is long, another type of dam may prove to be more suitable.

Figure 5-12. Grand Coulee Dam in Washington, showing large capacity overflow spillway.

C. Combination Arch Dams with Gravity Tangents

1. Site Topography. Combination arch dams with gravity tangents or thrust blocks should be considered for those locations where the upper portions of otherwise satisfactory arch dam sites are nonsymmetrical or have much flatter slopes. A thrust block can be used on the long side of a nonsymmetrical site to provide more symmetry for the dam and a better stress distribution (Fig. 5-13). Thrust blocks can be included on both sides to produce shorter, less flexible arches when the upper abutments are flatter than the rest of the profile. Another situation favorable to the use of the arch dam and thrust block combination exists when the orientation of the topography does not provide an abutment for the upper arch elements. In addition to the thrust block, a small dam or dike connecting it with the nearest relatively impervious formation will usually be necessary to provide a water barrier.

2. Foundation Conditions. The presence of unsatisfactory foundation materials, faults, joints, or shears in the upper part of the dam site can be unfavorable to the selection of an arch dam. However, the unsatisfactory foundation can often be replaced by gravity tangents or thrust blocks. These thrust blocks are designed to adequately resist the forces transmitted by the arches which abut against them.

3. Influence of Spillway Type. A gravity tangent can provide an economical location for an overflow spillway (see Fig. 5-14) when it is to be located within the dam. Such a spillway can be used for larger flows than one over the crest of the dam in which the free falling water must be stilled in a relatively deep pool to prevent damage at the toe of the structure.

D. Concrete Buttress Dams

1. Site Topography. Buttress dams are suitable for a variety of canyon shapes and widths.

Figure 5-13. East Canyon Dam in Utah, an arch dam, gravity thrust block combination.

Figure 5-14. Diablo Dam in Washington, showing overflow spillways located in gravity sections.

The most desirable canyon is one having gently sloping walls, however. Such conditions permit more desirable connnections between the canyon wall and the dam. More stable buttresses can be obtained near the canyon walls with less difficulty than when precipitous walls are encountered.

For the wider sites, a buttress dam, if not eliminated from consideration by other conditions, should require less materials than a solid gravity dam. Several types which may be used are multiple-arch, slab and butress, and massive-head buttress dams.

2. Foundation Conditions. Generally, buttress dams should be considered only when rock foundations are available. However, low buttress dams can be constructed on poor rock or earth foundations if footing slabs are used to keep bearing pressures within allowable limits. Multiple arch dams require relatively small volumes of concrete (Fig. 5-15) when compared with gravity dams or massive-head buttress dams. However, the thin arch barrel must be reinforced and the cost of formwork is high. Possible differential movements of adjacent buttresses should be investigated and the effect on the arch barrel determined.

Slab and buttress dams (see Fig. 5-16) and massive-head buttress dams (fig. 5-17) are less susceptible to detrimental effects of differential movement between adjacent buttresses. If foundation studies indicate possible differential movement from one part of the site to the next, these types of buttress dams should be considered.

Buttress dams have an advantage over solid gravity dams where uplift pressures are concerned. The hydrostatic pressure or uplift at the base of the dam is relieved between the buttresses and the small remaining uplift pressure on the buttresses does not materially affect the

Figure 5-15. Grandval Dam in France, a multiple arch dam.

Figure 5-16. Stony Gorge Dam in California, an example of slab and buttress construction.

stability of the structure. However, relatively impervious foundations are desirable because excessive seepage losses may occur over the comparatively short path of water travel under the narrow dam.

3. Influence of Spillway Type. If topographic conditions or spillway capacity requirements are such that the spillway cannot be located away from the dam, an overflow spillway can be provided over the buttresses.

Figure 5-17. Malga Bissima Dam in Italy, showing massive head buttresses.

4. Availability of Construction Material. Necessary construction material may not be available at or near the dam site in sufficient quantities for most types of dams. If the site is remote, transportation of materials and equipment can also be costly. The smaller quantities of materials needed for buttress dams can add to their desirability for such sites.

IV. COMPOSITE

A. Introduction

Composite dams use the advantages of both the embankment dam and the concrete dam. They generally consist of concrete gravity or buttress sections in combination with earthfill or rockfill sections. The concrete dam portion incorporates the ability to pass flood flows over or through the section during construction and act as the spillway after construction, and the earth or rockfill sections take advantage of low cost construction and local materials.

B. Site Topography

Except for the hydraulic considerations mentioned above topography is the main reason for the selection of a composite dam. The topography for combination dams is usually associated with plains locations or the zone of land where plains and mountains meet, but seldom with mountainous terrain itself.

Wide valleys often provide favorable sites for the use of combination dams. Wide valleys and a location adjacent to mountainous terrain often must provide for large flood flows. These flows present stream diversion problems if only an embankment dam is considered. The tunnel capacity required to pass floods during construction and the large spillways required often make the project uneconomical. If, however, a concrete gravity or buttress section is used as the spillway portion of the dam, the project becomes economical.

An artist's concept of Pueblo Dam, a composite dam, on the Arkansas River at Pueblo,

Figure 5-18. Pueblo Dam, a combination dam on the Arkansas River in Colorado.

Figure 5-19. Angostura Dam in South Dakota, showing combination of a concrete gravity dam with an earthfill embankment.

Colorado, now under construction (1973), is shown in Fig. 5-18. The spillway section consists of a buttress type and the embankment section is an earthfill dam. Figure 5-19 shows a combination dam using a gravity dam spillway section.

C. Foundation Conditions

Foundation conditions are the second most important factor in selecting a combination type dam.

In cases where the river section of a wide valley provides a competent rock foundation close to the surface and the overburden increases toward either or both abutments, composite dams can be economical. If a combination concrete-embankment structure is being considered, the concrete portion must have a foundation of acceptable quality. If the concrete part of the dam has a maximum net head (headwater to tailwater) of more than 20 ft, it should have a rock foundation. Concrete portions of combination dams which have net heads less than 20 ft can be constructed on the more pervious materials such as sand and gravel, provided the allowable bearing pressures are not exceeded.

REFERENCES

1. U.S. Department of the Interior, Bureau of Reclamation, "*Design of Small Dams*," 2nd Edition, Government Printing Office, Washington, D. C., 1973.
2. Sherard, J. L., et al., *Earth and Earth-Rock Dams*, John Wiley and Sons, Inc., New York, 1963.
3. Visnetini, G., "Varying Conditions Affecting Dam Design-Some Experiences," *World Dams Today*, Japan Dam Association.
4. Finn, N. D., Liom and Khanna, J., "Dynamic Response of Earth Dams," *Third Symposium on Earthquake Engineering*, Sah Cement Company, New Delhi (India), November 1967.

6

The Design of Earth Dams

REGINALD A. BARRON
Consulting Engineer
Guilford, Connecticut

PURPOSE

The purpose of this chapter* is to discuss the basic concepts relative to the safe and economical design, construction and operation of earth dams. The following criteria must be complied with to provide a safe structure:

GENERAL CRITERIA

1. The foundations, abutments and the embankment must be stable for all conditions of construction and operations. While of a serious nature, some embankment and foundation distress can be tolerated during construction provided corrective measures are promptly taken.
2. Seepage through the embankment, foundation and abutment must not exert excessive forces on the structure nor must piping of material be permitted. Excessive seepage forces will result in instability of the dam or abutments while piping, if not controlled, will eventually result in the release of the pool.
3. The top of the dam must be high enough to allow for settlement of the dam and foundation and also to provide sufficient freeboard to prevent waves from a maximum pool from overtopping the dam.
4. The spillway and outlet capacity must be such as to prevent overtopping of the dam. It is especially important that the spillway be capable of this alone because of the possibilities of the outlet being rendered inoperable.
5. The slopes of the spillway and the outlet works must be stable under all operational conditions. It is especially important that the slopes adjacent to the inlet be stable under drawdown of the pool and that the

---

*This chapter is a revision of the Engineer Manual EM1110-2-2300 (Ref. 1) which was written by J. R. Compton and S. J. Johnson of the Waterways Experiment Station, Vicksburg, Mississippi, under the general supervision of the writer while Chief, Soil Mechanics Branch, Civil Works Directorate, Office, Chief of Engineers U.S. Army, Washington D.C. The original version of this manual was written by G. E. Bertram and the writer.

291

slopes of the spillway cut be stable even under earthquake conditions.

The foundation, abutments and potential sources of borrow materials for construction of the dam must be studied in detail. This requires carefully made explorations as discussed in Chapter 4. These studies will assess the strengths and seepage characteristics of the foundations: the stability of the abutments, especially during pool drawdown and the seepage potentials, and the dangers of piping through the joints, bedding planes, and solution cavities. Studies of the various borrow sources and required excavations must be made. This will include types of soils, the thickness and distribution, variations of soil water contents and the water table with the seasons, thickness of required stripping and the various practical methods of borrow excavations to best use the available soils. Of vital importance to borrow usage is the variation of the climate and possibility of flooding on the use of borrow areas.

The influence must be considered of seismic action on the stability of the dam, the abutments, the cut slopes of the spillway and inlet and outlet works, and especially the susceptibility of the foundation to seismically induced liquefaction. This is of prime importance if the dam is to retain a pool of significant volume.

SELECTION OF EMBANKMENT TYPE

In general, there are two types of embankment dams: earth and rockfill. The selection is dependent upon the usable materials from the required excavation and available borrow. It should be noted that rockfills can shade into soil fills depending upon the physical character of the rock and that no hard and fast system of classification can be made. Rocks which are soft and will easily break down under the action of excavation and placement can be classified with earthfills. Rocks which are hard and will not break down significantly are treated in Chapter 3.

The selection and the design of an earth embankment are based upon the judgment and experience of the designer and is to a large extent of an empirical nature. The various methods of stability and seepage analyses are used mainly to confirm the engineer's judgment.[2,3]

Conditions which favor the selection of an earth dam for a site are:

1. Significant thickness of soil deposits overlying bedrock.
2. Weak or soft bedrock which would make the resistance to high stresses from a concrete dam difficult to obtain.
3. Abutments which are either of deep soil deposits or of weak rock.
4. Availability of a satisfactory location for a spillway.
5. Availability of sufficient and suitable soils from required excavation or nearby borrow areas.

Environmental Conditions

Federal and state laws now exist which require prevention of damage to the environment. These laws apply to design, construction and operation of the project. Some problems which may require solutions are contamination of rivers by hydraulic fill operations, erosion from borrow pits, construction roads and quarries.

GENERAL DESIGN CONSIDERATIONS

Freeboard

All earth dams must have sufficient extra height known as freeboard to prevent overtopping by the pool. The freeboard must be of such height that wave action, wind setup, and earthquake effects will not result in overtopping of the dam. In addition to freeboard, an allowance must be made for settlement of the embankment and the foundation which will occur upon completion of the embankment. The hydraulic freeboard can be determined using the procedures given in Ref. 4 and 5.

An extra allowance for freeboard to counter earthquake induced settlement is required by the U.S. Army Corps of Engineers, either (1) the maximum pool plus normal freeboard or (2) the elevation of the flood control pool plus 3 percent of the height of the dam above stream bed. It should be noted that this allowance is for settlement of the embankment under seismic action. If a compressible foundation exists, an extra allowance in the freeboard must be made. If there are dangers of a massive slide in the reservoir or a block tilting induced by earthquakes sending a wave over the top of the dam, then additional freeboard allowances and design provisions must be made.

Top Width

The width of the earth dam top is generally controlled by the required width of fill for ease of construction using conventional equipment. To facilitate construction near the top of the dam the internal zonation should not be excessively complicated. If a highway is to cross the dam then this will control the top width. In general, the top width should not be less than 30 ft. If a danger exists of an overtopping wave caused either by massive landslides in the pool or by seismic block tipping, then extra top width of erosion resistive fill will be required.

Alignment

The alignment of an earthfill dam should be such as to minimize construction costs but such alignment should not be such as to encourage sliding or cracking of the embankment. Normally the shortest straight line across the valley will be satisfactory, but local topographic and foundation conditions may dictate otherwise. Dams located in narrow valleys often are given an alignment which is arched upstream so that deflections of the embankment under pool load will put the embankment in radial compression thus minimizing transverse cracking.

Abutments

Three problems are generally associated with the abutments of earth dams: (1) seepage, (2) instability, and (3) transverse cracking of the embankment. If the abutment consists of deposits of previous soils it may be necessary to construct an upstream impervious blanket and downstream drainage measures to minimize and control abutment seepage. The upstream impervious blanket should be designed with a transverse slope to permit it to withstand the seepage forces out of the abutment during rapid pool drawdown. In many cases the abutments may contain badly jointed rocks which will require treatment to minimize seepage. This can be accomplished by use of cement grout using more than one line of grout holes (see Chapter 3). Also, it may be necessary to flare the dam at the abutments to provide larger impervious and drainage contacts. If the rock has been subject to solution activity, then treatment may be either the use of surface slabs of concrete or internal mining and construction of concrete anti-seepage walls in the adits.

If the abutments are weak then the embankment fill may be widened to provide a support to prevent sliding under the action of downstream seepage and upstream pool drawdown. This treatment may be of special significance if the outlet works is a conduit located at the toe of a weak abutment; blockage of the inlet by abutment slides must be prevented.

Where steep abutments exist, especially with sudden changes of slopes or with steep bluff, there exists a danger of transverse cracking of the embankment fills. This can be treated by excavation of the abutment to reduce the slope, especially in the impervious and transition zones. The transition zones, especially the upstream, should be constructed of fills which have little or no cohesion and a well-distributed gradation of soils which will promote self healing should transverse cracking occur. To minimize cracking the impervious fill may be compacted at slightly higher water content than normal to render the fill more plastic.

Surface drainage should be provided at the junction of the fill and abutments to avoid rain and snow melt run-off erosion. The best treatment is a flexible rock pavement resting up a filter bed; this will permit the protection to adjust to minor settlements and frost heaving.

Stage Construction

It is often possible, and in some cases necessary, to construct the dam embankment in stages. Factors dictating such a procedure are (1) a wide valley permitting the construction of the diversion or outlet works and part of the embankment at the same time; (2) a weak foundation requiring that the embankment not be built too rapidly to prevent overstressing the foundation soils; (3) a wet borrow area which requires a slow construction to permit an increase in shear strength through consolidation of the fill. In some cases it may be necessary to provide additional drainage of the foundation or fill by means of sand drain wells or by means of horizontal pervious drainage blankets.[6]

It is sometimes not possible to use stage construction either because the valley is too narrow for a closure section of stage construction. For such cases the rate of fill placement and the internal zonation of the fill and the embankment placement water content should be such as to permit completion of the fill without danger of excessive straining of the fill or foundation as a result of low shear strengths. It may be necessary in such cases to use flat outer slopes or stabilizing berms. Instrumentation of the fill and foundation may be required. It is often necessary to protect the lower toe of the transverse slope of the first stage embankment with riprap to protect it against high velocity erosion by the river. Prior to closure, the riprap and its bedding must be removed to prevent seepage piping through the embankment.

River Diversion

The diversion of the river is a critical operation in the erection of an earth dam and the timing and method are significant parts of design. The factors affecting the method of river diversion are hydrology, site topography, geology and construction programing. The diversion works, either a tunnel or a conduit, must be constructed first. If the valley is wide enough, a part of the embankment may also be constructed. It may be necessary to relocate the river and to construct a low cofferdam around the first stage embankment. At the proper time the river is diverted using up and downstream cofferdams; these may be of a temporary nature sufficient to affect diversion during a period of low water. Once diversion has been made, the permanent cofferdams are constructed; the most important of which is the upstream. This is often included as part of the permanent part of the embankment. If possible the major part of the upstream cofferdam should be constructed of free draining soils which compact readily under water. If an upstream inclined impervious blanket is required for the cofferdam, consideration should be given to either its removal at the end of the need for the cofferdam or else it should be designed so that it can fail upon drawdown without damage to the main embankment. If an impervious blanket is used, the downstream pervious zones should be designed and constructed so that the impervious fill is not piped into the downstream pervious zone by seepage during a temporary pool during construction. The design of the cofferdam should be made by the designer to insure it against operational failure during the construction of the embankment.

A major cause for concern of the upstream cofferdam is the possibility of its being overtopped by a flood during construction. This may be minimized in some cases by leaving low gaps in a part of the spillway if such is located in the stream bed. If the upstream cofferdam top is wide, or if a significant part of the closure has been constructed and a danger of overtopping exists, then consideration should be given to using the part of the downstream fill as a temporary source of borrow to quickly raise the cofferdam. This may be of significance if the borrow haul is long or is in the upstream flood plain.

Sometimes it may be worthwhile to consider a cofferdam design which will permit overtopping to a minor extent. This was done at Blakely Mountain Dam[1] and at other projects.

If the free drop of water is significant then this method may be questionable.

Embankment Closure

Two main problems exist with the design and construction of earth dam closures. First, because of the rapid rate of construction, a slow gain of soil shear strength and consequent high pore pressures may occur in foundation and embankment; and secondly, if the transverse slopes of the previously constructed embankments are too steep, there exists a danger of transverse cracking. A slow gain in shear strength can be minimized in the foundation by means of not too rapid construction or through use of foundation sand drain wells. A slow gain in shear strength in the closure embankment may be controlled by the rate of construction, the minimization of the impervious core and design of a major part of the closure embankment to use pervious fill. If none of these can be done and a slow gain in shear strength is a problem, then sand drain wells or horizontal pervious drains may be used to accelerate drainage and thus the reduction of high pore pressures and a gain in shear stength. Consideration should also be given to flattening of the outer slopes of the closure section to reduce the shear stresses induced by the dead weight of the embankment.

If it appears that a slow gain of shear strength with associated high pore pressures may be a problem for the closure, then the foundation and embankment should be instrumented with piezometers and horizontal and vertical displacement monuments. If adverse conditions develop, corrective measures can be taken before a construction mishap occurs. Under no condition should outlet conduits be built in or near closure sections.

Seismic Problems

An earth dam should never be located on or near an active fault. The determination of a fault activity is a geological problem and is discussed in Chapter 4. It is often necessary, however, to construct dams in seismically active areas and for these cases defensive design measures should be taken. These consist of (1) not building upon loose foundation sands which may be subject to liquefaction under seismic shocks. A relative density of at least 70 percent should exist, if not, then the sand should be densified or removed. (2) The impervious zone should be made plastic by compacting at a somewhat higher water content; (3) enlarging the impervious core; (4) flattening the outer slopes; (5) increasing the height of the dam to allow for seismically induced settlement either by compaction or sliding; (6) increasing the width of the crown; (7) increasing the widths of the filter zones and constructing the upstream zones of cohesionless soils which will readily move into any downstream cracks; (8) increasing the width of the zones at the abutments. The outlet works should not be located within the embankment, but rather a tunnel should be used with a control shaft having a minimum of a free standing portion. The spillway should not be in the embankment and any cut slopes should be conservatively designed so that the discharge capacity will not be reduced by landslides. In some cases, where block tilting or large landslides in the reservoir may occur which may cause the pool to overtop the dam, consideration should be given to making the downstream zone of the embankment of clays to resist the erosion and to increase the top width of the embankment.

Embankment Cracking

The design and construction of an earth dam should be such as to prevent cracking, especially transverse cracks which may lead to failure by piping. Types of cracking are (1) transverse, (2) longitudinal, and (3) horizontal. Care should be taken in the post-construction installation of peizometer or boring explorations that such action does not split the dam by means of excessive hydraulic pressures developed during the boring operation. If possible, installations should be made by means of augers. Transverse cracking develops as a result of tension in the fill. This can be caused by excessive differential settlements, or by abrupt changes in the slopes of the abutment. These cracks may be prevented by proper sequence of construction which allows a compressible foundation to consolidate prior to placing fills adjacent to more rigid foundations and by trimming the rock abutment to provide slopes with less abrupt changes. Steep river slopes of soil should also be degraded for the same reasons. When stage construction is used the transverse slopes should not be excessive, especially when the foundation contains compressible soils, an end slope not steeper than 1 vertical or 3 or 4 horizontal is suggested. When compressible soils exist within the foundation and a cutoff trench is used then the compressibility of the trench backfill should be about the same magnitude as the foundation soils. Longitudinal cracking often occurs when the core is less compressible than the outer shells. Such cracking often occurs as the shells adjust to their dead weight or to the first filling of the pool. Such cracking is not dangerous but should not be confused with cracking caused by an incipient embankment slide. Horizontal cracking of the core may result if the core is compacted too far on the dry side of the optimum water control. Saturation by seepage may cause subsequent settlement. If the core is narrow and contained by relatively incompressible shells, then core settlement with concomitant arching may result in horizontal cracking. Hydraulic fracturing may also occur on the first pool filling if the horizontal earth pressure is less than that of the water pressures. Prevention of such cracking can be obtained by proper design and construction through avoidance of very narrow, central, impervious cores and the use of interior, overly compressible fills immediately adjacent to much more rigid outer fills.

FIELD AND LABORATORY INVESTIGATIONS

Geological Explorations

The geological studies of the structure sites and potential borrow areas must be such as to permit a determination of the suitability of the foundations including abutments, the extent of foundation treatment, the excavation slopes for the outlet works, cutoff trench (if needed) abutments and spillway. Also the characteristics of the materials from the required excavations and from potential borrow areas. It is very important that the aid of an engineering geologist be used in this study to develop the details of the regional and local geology. This includes a detailed development of the surface and the subsurface geology as the explorations proceed.

The extent of the exploration program will be determined by the stage of the studies and the size of the project and the complexities of the site geology. The borrow areas should be explored at least in a preliminary manner, early in the studies so that the character and quantities of soils can be estimated for the later embankment design studies. A detailed discussion of geological engineering is given in Chapter 4. (see also Refs. 8, 9, 10 and 11).

The foundation of the dam should be explored to develop the rock and soil profile, to determine the character of the soil foundations, especially if pervious soils are present that will require seepage control, loose sands which may liquefy under seismic actions, weak sensitive clays, organic soils, or soils subject to collapse upon wetting such as loess and clay shales which may either expand and lose shear strength upon unloading or be subject to strain softening. The rock should be explored to determine the presence of fault gouge seams, deposits of soluble gypsum, worked out mines or filled or partly filled solution channels or open joints. In addition, the groundwater table should be located and its variation with the seasons recorded. Artesian pressures in the soils and rocks should also be detected if such exists. The abutments must be explored to determine seepage conditions in the soils and rocks during pool storage and reservoir drawdown; especially adjacent to the intake area of the outlet works.

The valley walls should be explored to permit a determination of the possibilities of excessive seepage losses and also the possibility of rim instability. Locations of saddle dams and reservoir crossings should be explored to ascertain foundation conditions and potential problems. The explorations of outlet works and spillway sites are of special importance, especially for problems of cut instability and as sources of construction materials. All water wells should be located and water level fluctuations with seasons noted. Oil and gas wells should also be noted and a study made of old records to determine future treatment needs.

Borrow areas must be explored to determine the extent of stripping, depth and type of soils, the natural moisture contents and the seasonal variations of the water contents and water table. Sufficient samples must be obtained to determine permeability and compaction characteristics, shear strengths of the compacted soils and the swell or loss factors between the pits and compacted fills. The flooding potential will be of significance to valley borrow areas, and the feasibility of using such soils throughout the construction schedule. It is highly desirable if valley borrow area use is planned, that additional upland areas be available during periods of flooding or wet seasons.

The means of exploration are visual inspection, air photos, test pits and trenches, calx holes, adits in the abutment rock, auger and machine borings. Especial care is needed in sands, firm shales and soft, weathered rock to recover samples. Where extensive pervious soil deposits occur it may be necessary to perform pumping tests to obtain data for use in seepage and foundation dewatering studies.

Test Fills

While, in general, sufficient experience exists as to the use of various soils and their reaction to various types of compaction equipment, in some cases where new equipment is proposed or where difficulties with soils are contemplated, it may be desirable to construct test fills. This is true for silty soils and for soft rocks which will degrade under excavation and compaction. The desired information is the lift thickness, number of roller passes, placement water content, maximum permissible particle size, degradation of soft rock under placement and rolling equipment, and the engineering properties of the fill, especially compacted density, shear strength and permeability. The test fills may often be located on access roads, or on fill areas of the dam where the height is low. In general, the tests fills should be made in the study stage of the project to permit use of the data in the design and to be made available to the bidders.

Laboratory Testing

The discussion of the laboratory testing of soils is given in Refs. 2 and 12. Soil tests are of two general types: identification or classification and engineering properties. It is essential that classification tests (which may often be visual) be made on a wide range of the exploratory samples before an extensive program of testing is undertaken to determine the engineering properties. Detailed tests for engineering properties should be made using carefully selected representative samples. The mixing of soils from different strata should be done only when such will be done by shovels, etc. in the borrow pits, otherwise the mixed samples will not be representative and may be highly misleading. Consolidation tests should be done only in devices which are wide enough to minimize sidewall friction, a minimum diameter of 3½ to 4 in. is suggested. Special care must be exercised in preparing and testing samples to determine the shear strengths of soils. Cylindrical specimens should be of a uniform nature with no built-in weaknesses such as compaction planes. It is highly desirable that the specimens be molded using a grain size approximating the natural material. In some cases this may require specimens having a diameter of 6 to 12 in. The maximum particle size should in all cases be not more than $1/6$ of the specimen diameter.

Representative samples of the site foundation and borrow areas should be retained for inspection by the bidder and also until all contract claims are settled. Test results, but not engineering interpretation of the data by the designers, should also be available to the bidders for study.

FOUNDATIONS AND ABUTMENTS

Seepage Control

The control of seepage through the foundation and abutments is very important and requires a thorough design study. The source of the seepage is mainly from the reservoir, but often the reservoir seepage may divert seepage which prior to construction seeped down to the valley floor. The source of the seepage is not too significant, its potential damage to the structure by piping and rendering soil masses unstable are the items which need control. Control may be obtained by cutoffs, upstream impervious soil blankets to reduce the amount of seepage. Control is also accomplished by downstream pervious drainage blankets, pipe drains, relief wells, and drainage adits into the abutments.

Cutoffs

Significant reduction of seepage can be obtained through use of compacted backfill trenches which penetrate through the upper pervious deposits to impervious soil layers or to bedrock. The cutoff trench is an extension of the impervious zone. The base width of the trench should be equal to at least a quarter of the net difference between the maximum pool elevation and minimum tailwater. In all cases the base must be wide enough to permit the operation of construction equipment and the use of temporary dewatering devices. All rock joints (especially transverse joints) should be cleaned of soil debris and backfilled with lean concrete. If the cutoff trench extends into bedrock, then the joints on the downstream face should be cleaned out and backfilled with lean concrete. If the rock is badly fractured a satisfactory treatment is to spray on a premixed sand-cement. The impervious backfill must also be protected against being piped into open pervious, foundation soils; where necessary filter layers must be used to prevent such damage. If the bedrock is soft or weak, then damage from the treads of the placement and rolling equipment must be prevented. This can be done by limiting the trackage upon the finished rock surface to rubber tired vehicles including rollers for the first few lists. Ground water seepage into the open trench must be prevented so that placement of the backfill can be done in the dry. Control of the ground water can be done by means of sump pumps, well points and deep wells. The latter are often required to prevent blow-ups of thin slabby bedrocks. This prevention of blow-ups is important until after completion of bedrock grouting. Temporary dewatering pipes should generally be removed prior to completion of backfilling. In no cases should such pipes be left in places where such traverse the cutoff.

Slurry Trench Cutoffs

When the depth of excavation is great or when the control of ground water is difficult, a cutoff can be achieved through the use of a slurry trench. Excavations through pervious foundations by backhoe, dragline or other means have extended downwards to 80 ft. The width of the trench is dependent upon excavating equipment. The sides of the trench are supported by means of a bentonite slurry. The top of the slurry should be at least 5 ft above adjacent ground water. If this head difference becomes too small side caving may occur. Sand suspended in the slurry will settle out and must be removed to prevent the formation of a soft, compressible layer under the backfill. The sand can best be removed by air lift pumps. The backfill should be a well graded mixture of gravel, sand, silt or clay with some of the slurry used to give plasticity. The use of coarse gravels and sands will render the backfill less compressible. Placement of the backfill is started with a clamshell bucket and when a sufficiently flat slope has been obtained further placement is done by a bulldozer. Great care must be used to be sure that the base of the trench is upon bedrock and not on large boulders. The location of the trench is based upon judgment. While trenches under impervious cores have been used, it is the writer's opinion that the best location for the trench is beyond the upstream toe of the dam so that repairs may be made if necessary. A detailed discussion of a slurry trench used at West Point Dam is given in Ref. 13.

Concrete Cutoffs

Concrete cutoff walls have been used where the depth of cutoff is very large. At Allegheny Dam[14] the maximum depth was over 180 ft. Much deeper cutoffs have been constructed for cofferdams (over 400 ft. deep). These walls are 2½ to 3 ft thick and the excavations are drilled with special equipment. The walls are supported by bentonite slurry. Concrete is placed by tremies and it is important that the discharge end be low enough so as not to lose the charge; otherwise discontinuities and segregation in the concrete wall will occur. Because of the thin nature of the walls it is considered best to locate the wall beyond the upstream toe of the dam, as at Allegheny Dam, to minimize cracking induced by the embankment weight.

Sheet Pile Cutoffs

Steel sheet pile cutoffs have been used in the past but not often in recent years. Seepage occurs through the interlocks, butt splices and spaces where the sheets are driven out of the interlocks. A minor imperfection in the steel sheet piling can render the cutoff ineffective where the average path of seepage is long.

Upstream Impervious Blankets

Upstream impervious blankets are often used to minimize foundation seepage. It is important that downstream seepage control measures be used with these blankets to collect seepage which will always occur when the cutoff is not complete. In some cases it is necessary to use upstream impervious blankets to reinforce thin spots in natural blankets. The impervious blankets often must be carried up the adjacent abutment slopes. In these cases care must be used to insure that the blankets will be stable under drawdown of the pool and that the blanket soil will not be piped into openings in the abutment rock. The upstream impervious blankets are dependent upon the permeability of the soil and the thickness and lengths. A discussion of design procedure and analysis is given in Ref. 3.

Foundation Grouting

The grouting of foundation rocks is discussed in Chapter 3; however, a few ideas will be presented here. The purpose of grouting is to reduce seepage through the rock and also to prevent or minimize the piping out of the soil filled rock joint, seams and solution cavities. It should never be forgotten that the soil left in such openings is always subject to future piping and if the deposits are signigicant, as in solution cavities, then surveillance must be maintained during pool operations. Grouting is also done to prevent piping at the soil bedrock interface.

A grout curtain is placed beneath the impervious fill zone by drilling holes into the rock and then forcing a grout mixture into the openings. In the past, one line of grout holes was often used; however, present practice leans towards the use of multiple lines of holes, especially for the rock area just below the soil. This creates a broader seepage barrier zone. The grouting pressures must be carefully controlled to prevent splitting and lifting the rock; thin bedded rocks are especially susceptible to this damage. All joints and seams exposed at the base of the cutoff trench should be cleaned and backfilled with lean concrete. This also applies to faults and solution cavities; this surface work should be done after completion of nearby grouting so that grouting will not force out the new filling. In all cases all rock blasting should be done before grouting. If the rock is badly cavitated by solution it may be necessary to mine out the soil filling and to erect formed concrete walls. The sealing of mines is also necessary. This may be done by adits which are filled with concrete and then grouted. In some cases, especially with coal mines, it may be necessary to build up a concrete barrier using closely spaced holes. The concrete is allowed to flow in under gravity followed by pressure grouting. Since such mine workings often contain underclay and debris deposits, careful surveillance is necessary after the pool operation is initialed.

Downstream Drains

The use of upstream measures to reduce seepage should be supplemented with downstream seepage control devices. A very common device is a pervious drainage blanket which serves to collect seepage from both the foundation and embankment. Such drains should also be placed over the downstream abutment to collect seepage emerging from the abutments. The extent, thickness and perviousness of the drain should receive careful design attention, especially the need for filter requirements.

Pressure relief wells are often used along the downstream toe of the dam to collect seepage. These wells perform well if the pervious foundation has a natural impervious cover. The wells consist of a screen surrounded by a filter layer in the pervious foundation. These connect to a tight riser which discharges into a collector pipe or open ditch. The top details are important and are needed to prevent backflow of high tailwater which could mud up the filter. Prevention of damage by vandals is also necessary.

Installation of the wells is often done using the reverse rotary method in which the sides of the hole are supported through the use of a drilling mud. After installation of the screen and filter the mud is removed by pumping and the well developed by surging which removes some of the fine sand from the filter immediately adjacent to the screen. Bentonite will gel and its removal is difficult or impossible, hence it should never be used for relief well installation. The relief well system should be designed to collect the expected seepage. A peizometer should be placed between wells to measure the seepage uplift and if necessary, additional wells may be installed later. Relief wells should be inspected periodically by sounding for sand and pumped to check their discharge capacity under varying heads. Wells with reduced capacity should be redeveloped and, if necessary, replaced (see Refs. 3 and 15).

The seepage from drainage blankets should be collected and discharged at low spots. This can be done using pipe drains. In no case should drain pipes be allowed to extend under the earth dam. If broken by the weight of the dam or destroyed by corrosion, a drain pipe could become an uncontrolled seepage channel. All drain pipes should be readily accessible for repairs and maintenance. Relief wells cannot collect all foundation seepage, especially when the upper impervious layer is thin or not too impervious. Therefore, such wells need to be supplemented by perforated collector drains. For all cases the drainage material must meet the filter gradation criteria discussed below.

In some cases drainage tunnels or galleries are used to collect seepage from under the dam or from the abutments. Drain wells are drilled from the tunnel into the bedrock or underlying soil. Such tunnels or galleries should be large enough to permit ready inspection and the use of drills to extend or add new drains. Tunnels have been used by the Corps of Engineers, U.S. Army, at Chief Joseph Dam, Oahe Dam and Howard A. Hanson Dam. At the latter project the wells were installed from the abutment surface and then connected to the tunnel as it was mined.

Stability Berms

Stability berms are sometimes necessary to protect weak abutments against the seepage forces caused by a sudden drawdown of the pool. In general, these potentially unstable abutments are soil, but in a few cases they may be of rock either badly faulted or with a joint system dipping towards the pool at a nearly unstable angle.

Foundation Preparation

The design of embankment dams resting upon soil foundation is controlled by the in situ shear strength of the foundation and the embankment. If the foundation soils are weak, then there are a number of choices open to the designer. The outer slopes may be flattened, stability berms may be used, the weak foundation soils may be excavated if the deposit is not too thick, stage construction may be used to allow a gain in the foundation strength or sand drains may be used to accelerate the consolidation of the soils and thus an increase in shear strength.

The foundation is prepared to receive the dam by clearing all vegetation, grubbing out stumps and large roots, stripping off of the sod, top soil, boulders, organic soils, spoil fills, etc. Organically stained soils may be left in place. Pockets or thin near-surface layers of soft, compressible soils should be removed. Stump holes and other excavations are filled and compacted by power hand tampers. The foundation is then plowed and rolled to compact the upper surface. The compaction roller should be as heavy as permissible and used not only to compact the upper foundation, but also to locate deposits of near-surface soft soils. If a cutoff trench is not required, then an inspection trench having a mimimum depth of 6 ft should be excavated and backfilled with compacted soil. The purpose of this trench is to locate farm tile drains, pervious zones, abandoned pipes and other unsuitable features not detected by foundation explorations. If the abutments are soil with steep slopes, then the slopes should be flattened to minimize transverse cracking of the dam. The amount of flattening is dependent upon the designer's judgment aided by finite element strain studies.

Loose fine sand foundations may be liquefied by earthquake vibration. The minimum permissible relative density used by the Corps of Engineers, U.S. Army is 70 percent, but this may be subject to change depending upon judgment and future research. Densification may be obtained by vibration or driving sand piles, but this is costly. An alternative is to reduce the average shear stress in the foundation by using very flat slopes. If the loose sand deposits are shallow then excavation of the sand should be considered.

Rock foundations should be cleaned of all loose rock. The uneven surface should be trimmed to permit the operation of spreading and compaction. Open joints and seams should be cleaned as deep as feasible and backfilled with lean concrete. Open joints in the downstream face of the cutoff trench should also be cleaned out and concreted over. Badly fractured rock, especially shale and coal seams existing between firmer sandstones and limestones should be cut back to permit the construction of a concrete seal. Solution joints should be cleaned out and backfilled with concrete. In many cases it will be necessary to trim the rock prior to concreting. When the downstream face of the cutoff trench is badly fractured then a cover coat of sand cement may be used. The use of batch mix rather than a nozzle mix is considered to be best. The rock surface should be moistened prior to spraying. Drummy cover should not be accepted as it will crack under the action of compaction and the dead weight of fill.

Shale foundations should have the final excavation made just prior to placement of backfill so that drying out of the shale will be minimized. If an excavation is made into the shale for a cutoff then the faces should be protected by a sealer to prevent excessive drying and cracking.

Key walls are no longer used. The preparation of the rock trench to receive the base damages the rock and the projecting key renders compaction of the adjacent soil fill difficult. Open joints under the upstream part of the dam should be filled with lean concrete. For the downstream area such joints should be filled with drainage soil so that seepage emergence may occur without piping. The soil packed joints should contact the downstream horizontal drainage blanket. Dead faults crossing the dam site should receive careful treatment. They should be cleaned out as deeply as possible and then backfilled with concrete. Afterwards the contact between the walls of the fault and the concrete should be carefully grouted. Drainage of the fault should be provided in the downstream area of the dam.

The design of earth dams having permafrost in the foundation is in the developing stage and much remains to be learned. A few dams have been built on permafrost by the Canadian and Russian engineers. If possible, the permafrost should be removed; this can be done either by excavation if the deposit is thin and near the surface; it may be melted by steam jets or, if time is available, by removal of the top humus layer so that the permafrost may be degraded by an average above-freezing temperature. If the permafrost deposit is too thick and deep for any of these methods, then consideration may be given to an embankment designed to rest directly on the permafrost.

Abutment Preparation

Rock abutments should be prepared in the same manner as the rock surface of the valley. It is very important that all overhangs be removed, especially in the impervious and transition zones. This trimming should be done prior to the abutment grouting. In some cases it may be that rock overhangs can be eliminated through use of concrete using forms as necessary. The maximum vertical rock face should not exceed 5 ft. Very steep high abutment should be cut back to a flatter slope in the impervious and transition zones. While no exact value can be given slopes of 2 vertical on 1 horizontal have been used. Mines and exploration adits must be backfilled with concrete or sealed with a thick concrete wall. Treatment of solution cavities is most important. They should be cleaned out and backfilled with concrete. If the solution activity is extensive then mining may have to be used so that heavy concrete wall extending into the abutment may be constructed. For such cases adits should be built downstream of the walls to permit drainage, inspection and further treatment if necessary.

A weak rock foundation requires a careful study; this includes such cases as compaction shales, erodable sandstones, coal seams and damage resulting from rock rebound caused by the valley formation. Treatment is generally limited to adjusting the outer embankment provision of downstream drainage.

EMBANKMENT

Embankment Soils

Most soils are suitable for use for embankment construction, however, there are physical and chemical limitations. Soils which contain excessive salts or other soluble materials should not be used. Substantial organic content should not exist in soils. The action of organic material is to increase the soil's compressibility and lower its shear strength, thus such soils are limited to low structures. Lignite, sufficiently scattered through the fill to prevent the danger of spontaneous combustion, is not objectionable. Fat clays with high liquid limits may prove difficult to work and should be avoided. Soft rocks which readily break down through the action of excavation, spreading and compacting such as soft sandstones and compaction shales have been used successfully on a number of projects. Fine uniform silts and rock flours are very sensitive to excess water and are very difficult to compact. It should be noted that silts and silty soils have low optimum compaction water contents and rapidly loose undrained, unconsolidated shear strength as the compaction water content increases, thus their use requires careful control of the water content. The practical use of soils depends upon its natural water content and the weather during the construction season. Soils with excessive water content require drying either at the borrow pit or on the fill. Rain will seriously interfere with this operation and may require use of drier soils from upland locations. The use of soils having an excess of water will result in compressible low strength fills which will limit the maximum height of fill. On the other hand, in dry areas or during a local drought, the in situ water content of soils may be too low. These soils will become objectionably compressive upon saturation and water must be added. If the amount of additional water is small, this can be added on the fill and mixed in by disking. In other cases it may be more economical to add the water at the borrow pit by sprinkling and then to allow the water content to stabilize before using the soil in the fill.

The design of the embankment should be based on a careful study of the available soils, their in situ water contents, the need for drying or wetting the soils to a workable range, the expected weather and length of haul to produce a safe, economical embankment. Large cobbles and boulders should be removed prior to rolling because they will render rolling difficult if not impractical. Various devices exist to do this. If the amount of oversize is minor, then the removal can be done on the fill, otherwise, the use of a screening or separation plant is necessary. Soils with high shrinkage properties should be protected by an outer layer of low shrinkage soils to prevent shrinkage cracks impairing the safety of the dam against piping while storing high pools.

For projects where storage of water has high economical value, the impervious zone should consist of fine grained soils; however, for flood control projects or where a certain minimum water release must be maintained, then the use of more pervious soils should be considered. If such soils are used, then an adequate seepage collection system must be provided.

The embankment should be zoned to obtain a safe, economical structure using materials from required excavation and nearby borrow. The zoning should be such as to provide a suitable impervious zone, transition zones between the impervious zone and the outer shells, seepage collecting zones and adequate stability.

Embankment Dams

There is no typical earth dam. All are designed and constructed to meet the condition at each particular site. However, there are certain general similarities which permit presentation of types of designs. Schematic sketches are given in Fig. 6-1. Seepage is prevented or minimized by an impervious zone located in the central or upstream part of the dam. The exact extent of the zone is dependent upon the amount of impervious soils available from required excavation and borrow areas. If the embankment rests upon a pervious foundation, then a cutoff may be used; or if the foundation is too thick, then a slurry trench or an upstream impervious blanket is used. The impervious zone is supported by up and downstream shells. The shell may be either pervious or random depending upon available soils. The outer slopes of the shells are controlled by the shear strength of the embankment and the foundation. The conditions controlling the stability are strength during and at the end of the construction, rapid drawdown of the reservoir, earthquake, and slow progressive loss of shear strength under constant stresses after completion of the dam. Seepage is collected by either a pervious downstream shell or by a combined inclined and horizontal drainage blanket. For cases where a cutoff is not used, then the seepage collection system is augmented through use of a toe drain and relief wells.

The width of the impervious zone is dependent upon the amount of impervious soils available. While a very narrow core may appear to be feasible, the width must be such as not to result in high seepage gradients. The core width must also be sufficiently wide to prevent its arching onto the shells and developing horizontal cracks. The bottom width should not be less than a quarter of the net head between maximum pool and minimum tailwater. If there is a cutoff trench this applies to its base width. The top of the impervious zone is controlled by practical placement and should not be less than 10 to 12 ft. To obviate narrow zones which may be difficult and expensive to place, the zoning of the top part of the dam may be to omit the shells and the drainage zones above top of spillway or spillway gates. If a paved roadway crosses on top of the dam, then the soil used in the upper part of the impervious zone should be such as to serve as an adequate road sub-base and not subject to frost heave where cold winters prevail.[16]

304   HANDBOOK OF DAM ENGINEERING

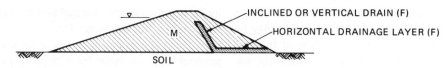

a. Homogeneous dam with internal drainage on impervious foundation

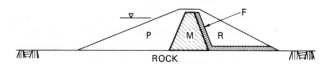

b. Central core dam on impervious foundation

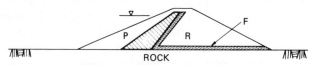

c. Inclined core dam on impervious foundation

LEGEND
M = IMPERVIOUS
P = PERVIOUS
R = RANDOM
F = SELECT PERVIOUS MATERIAL
US = UPSTREAM

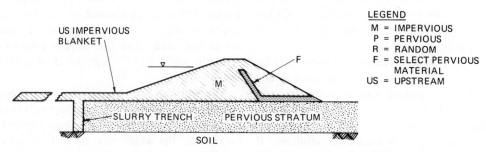

d. Homogeneous dam with internal drainage on pervious foundation

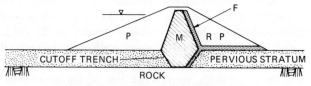

e. Central core dam on pervious foundation

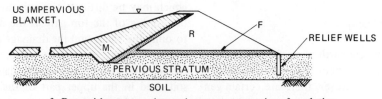

f. Dam with upstream impervious zone on pervious foundation

Figure 6-1. Types of earth dam sections.

Soils which may range from impervious to pervious are commonly classified as random. Maximum use of such materials may be made as shown in Fig. 6-1. Such material is used as it comes from excavation provided its moisture content is suitable. Use of zones for random material often permits the minimization of stock piling. The strength of random fill used in design studies should be that of its weaker and generally finer soils. When random zones are used, suitable transition zones must be used to prevent seepage piping soil from a finer into a coarser zone. Also, zones of sufficient drainage capacity must be provided to prevent seepage breaking out on the downstream slope and rendering it unstable. Homogeneous embankment which may be entirely of random fill may be constructed when the borrow areas are such as not to permit selection of soils for specific zones; however, in such cases the specifications should require, to the extent possible, that the more pervious soils be placed towards the outer slopes and the more impervious towards the center. It is important that the very outer zone be as pervious as possible to permit ready drainage of rain infiltration so that, if the downstream slope is grassed, mowing will not be impeded by a soft, soggy condition.

Examples of cross sections of embankment dams recently constructed by the Corps of Engineers, U.S. Army, are shown on Fig. 6-2 and 6-3. Prompton Dam is an example of a homogeneous random fill with internal drainage. The reservoir is intended for flood control with a small pool for recreation. Alamo Dam is a zoned embankment where a plentiful supply of sands and gravels for the outer shells were available. The project is for flood control. Borrow selection for these two dams was rather simple. The borrow selection for Milford Dam and W. Kerr Scott Dam were more complex. For Milford Dam only a token inspection trench was used. Seepage was minimized by a central impervious zone and a long upstream impervious blanket. Seepage is collected by an internal inclined and horizontal downstream drain and by a system of downstream foundation relief wells. A significant amount of the embankment fill was obtained from required excavation of the spillway and the outlet work channel. The materials for the W. Kerr Scott Dam were from highly weathered residual soils and rocks and from alluvial borrow derived from the erosion of residual soils. The soils were mainly of low plasticity and of a silty nature. The more impervious soils were placed in an impervious zone located upstream of the center line. The inclination of the core provides a greater distance over which to dissipate the downstream thrust of the pool forces. Random soil was used in both the upstream and downstream shells and the seepage is collected by an internal drainage system. A dumped rock zone provides protection against wave erosion. The rock was obtained from the less weathered rock of the spillway excavation.

The design of earth dams, in permafrost areas, is in the development stages. If the permafrost can be removed or degraded then a normal design may be used. If the permafrost is left in the foundation then it may degrade slowly because of the warm seepage waters from the pool. This degradation will cause settlements of the foundation and possible piping and cracking of the foundation and dam. To minimize this danger, the embankment should be constructed of cohesionless soils so that it will readily deform and prevent embankment cracking. The slopes should be flat because of the lack of an impervious section and the dam should be overbuilt to allow for future settlement. Constant post-construction inspection. must be had and prompt correction must be made to control any excess settlement or excessive downstream seepage.

Seepage through the embankment is minimized by the impervious zone, a cutoff, if used, or by an upstream impervious blanket if a cutoff is not used. Seepage through the embankment is collected by either a pervious downstream shell or by a combined inclined and horizontal drainage blanket. The soils of the drain should be compacted to at least 70 percent relative density to prevent seismically in-

306 HANDBOOK OF DAM ENGINEERING

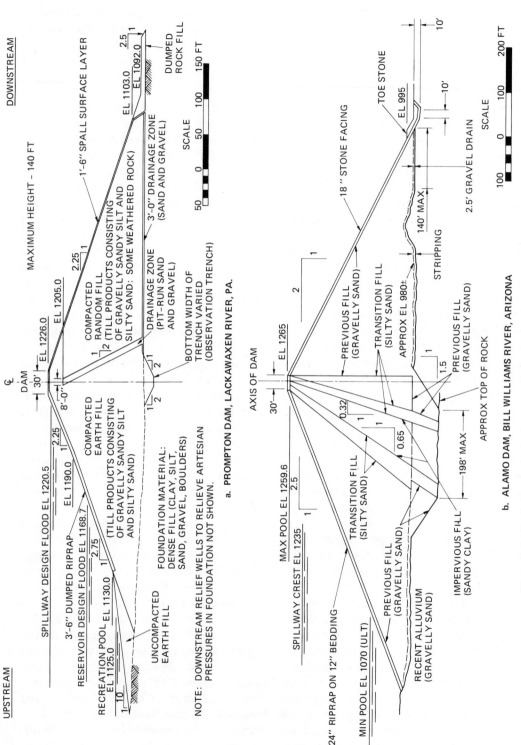

Figure 6-2. Typical embankment sections, earth dams (Prompton and Alamo Dams).

THE DESIGN OF EARTH DAMS 307

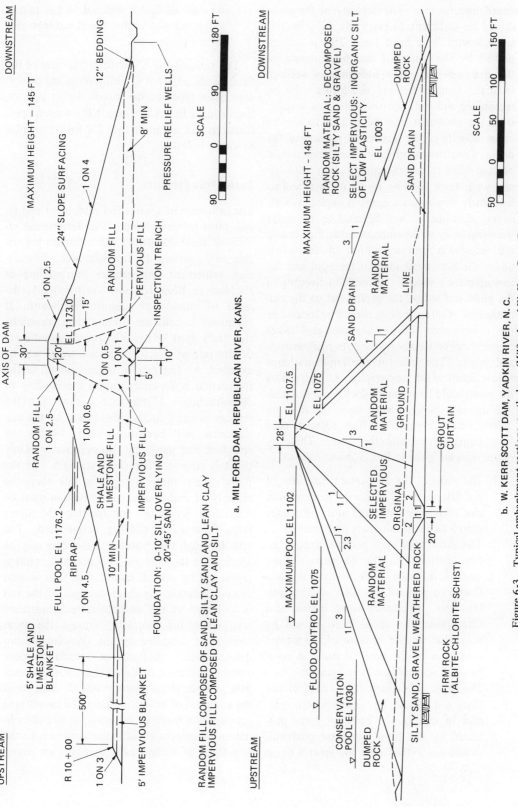

Figure 6-3. Typical embankment sections, earth dams (Milford and W. Kerr Scott Dams).

duced liquefaction. The capacity of the drain should be sufficient to carry off the collected seepage with little head loss. This is accomplished by the thickness and perviousness of the drainage soil. The influence of seepage from the foundations and abutments must be recognized and for a narrow valley the seepage from the abutment may be such as to require extra capacity of the horizontal portion of the drain. Construction of the inclined drain should be controlled to prevent construction traffic or rain wash from forming layers of finer soil so that ready downward seepage is impeded. Collector pipes should not be used in or under the main body of the embankment. Pipe breakage, corrosion, and separation of the joints under the horizontal load of the pool and the embankment's weight can result in clogging of the pipes and worse, the piping out of the embankment. Collector pipes should be located at the downstream toe of the drainage blanket where it can be readily reached for maintenance and repairs. The collector pipe should discharge at low points of the topography. At high points the pipe should be segmented with a sealed end. The drainage material must meet the filter requirements to prevent the finer adjacent soils from being carried into the drain. These requirements are given in Ref. 3 and as follows:

1. The ratio of the 15 percent finer size of the filter material to the 15 percent finer size of the base material must be greater than 5 to insure adequate perviousness.
2. The ratio of the 15 percent size of the filter material to the 85 percent finer size of the base material must be less than 5 to prevent migration of finer into the filter. If crushed stone is used as a filter, then the above ratio should be somewhat smaller because of the greater void sizes of the crushed stone as compared to screened sands and gravels.
3. The grain size distribution curve of the filter soil should be approximately parallel to that of the base soil being protected by the filter. A filter gradation should not extend over too great a range of sizes so that it will not be too impervious and also so that it will not segregate during placement.

The ratio of the 85 percent finer size of the filter or soil adjacent to well screens and perforated drain pipes to the screen size or perforations should be greater than 1.0 for round perforations and not more than 1.2 for rectangular or square holes.

Excess Pore Pressures

The influence of foundation settlement and its rate must be considered. The detrimental effects of these conditions will influence the design and possibly the site selection. Foundation settlement will require overbuilding of the dam to allow for post construction settlement to provide an adequate freeboard. If compressible soils are present then consolidation tests must be made to permit an estimate of the rate at which the excess water will be expelled.[12] The gain in shear strength of the foundation is dependent upon the rapidity of consolidation. If the rate is slow then the stresses induced in the foundation by the embankment will be carried partly by the soil structure and partly by the pore water. Only the soil structure has shear strength. If the foundation's shear strength is low then the rapidity of embankment construction must be controlled by stage construction—the outer slope flattened or stability berms used. The rate at which the consolidation occurs may be accelerated through use of horizontal drainage layers and by vertical sand drains. For no gain in consolidation the shear strength of the soil is indicated by the unconsolidated, undrained test, while for complete drainage the shear strength is indicated by the consolidated undrained test. Analyses of foundation pore pressure are given in Ref. 17 and stability analyses procedures are given in Ref. 2. Because of the variation of natural soils and the simplifying assumptions made in the shearing and consolidation theories, a conservative approach to the selections of design shear strengths are necessary. Pervious foundation soils will generally consolidate as rapidly as the stresses are increased. Exceptions may occur in varved soils where escape of the excess water is mainly horizontal to the unloaded toe area and for seismically induced liquefaction of loose sands and silts.

Fine Grain Soils

Embankment soils are subject to consolidation under the overlying load. Factors influencing the rate are grain size, excess water content, rate of construction, and the length of drainage paths. The latter is influenced by such features as pervious shells and drainage layers. Overly wet soils will consolidate slowly with a consequent build-up of excess pore water pressure and slow or minor gain in shear strength. If excessively wet soil is used in the embankment, then the construction rate must be adjusted to the slow gain in shear strength or the outer slopes flattened or stabilized with berms. Vertical sand drains have been used within a partially constructed fill to accelerate drainage. These drains must be connected to an upper drainage layer and also to any pervious layers within the foundation. Piezometers are necessary to monitor the rate of excess pore water dissipation.

Embankment Slopes

The stability of an embankment and its foundation is dependent upon the types of soils present in the foundation and used in the embankment and upon the zonation of the embankment. Stability procedures are not discussed in this chapter but are presented in Ref. 2. It should be noted that all methods of stability analyses are approximate and the final selection of the outer slopes must be tempered with judgment and experience. If the foundations or embankment soils tend to creep or to strain soften, then the safety factors must be much higher than normal.

In some cases horizontal stability berms are used in lieu of flat outer slopes. If these berms are exposed above the pool or on the downstream side then consideration must be given to surface drainage collection to prevent objectionable erosion.

For cases where the reservoir will have a permanent minimum pool, the upstream riprap is extended only a short distance below such low water. A narrow berm should be provided at the toe of the riprap. This is a sacrificial berm to be expended by wave wash of the raising pool should the required time be excessive. Often this berm material can be obtained from nearby borrow waste which is not needed in the embankment construction. If excessive waste is available for a short haul, it may be placed as a waste berm upstream of the embankment toe. Here it may act to increase the foundation seepage path or to increase the stability. If a waste berm is used on the downstream area, then care must be used not to block the drainage from the foundation or the embankment.

Compaction Requirements

The strength of the impervious and semi-impervious soils depends upon the compacted densities. These depend in turn upon the water content and the weight of the compacting equipment. The design of the embankment is thus influenced by the water content of the borrow soils and by the practicable alterations to the water content either prior to placement of the fill or after placement but prior to rolling. If the natural water content is too high, then it may be reduced in the borrow area by drainage, or by harrowing. If the soil is not excessively wet then it may be possible to dry it on the fill by harrowing. This, however, will impede placement operations. If the soil cannot be dried sufficiently, then the slopes must be flattened to adjust for the weaker shear strength of the wet soil or else a drier source of soil borrow must be obtained. If the soil is too dry it should be moistened in the borrow area either by sprinkling or by ponding and then permitted to stabilize the moisture content before use.

Wetting on the fill should be confined to adjustment of moisture losses occuring during hauling and placing operations and losses in the exposed layer by drying prior to placement of the next layer. The range of placement water content is generally between 2 percent to 2 or 3 percent wet of the standard Proctor optimum water content.[12]

The natural water content of borrow soils and the ease of adjusting the moisture is significant in the design and construction of an earth embankment. Soils compacted wet of optimum are weaker and more plastic than soils compacted dry of optimum. The drier soils are more rigid and brittle and are subject to settlement and cracking upon saturation by the pool. Cracking can be to a significant depth in the dryer soils because of the cohesion induced by capillarity. Such cracking will occur not only upon saturation of the fill but by plastic or consolidation distortions caused by a weak foundation.

The stability of a fill during and just after construction is determined mainly by the undrained shear strength of the soils. At water contents greater than optimum the compaction of the soil mass is prevented by the presence of the excess water. Thus the undrained shear strength is lower than that at optimum. Any added stresses are imposed upon the pore water with a consequent build-up in excess pore pressures. The main danger of excess pore pressures is that the undrained shear strength cannot develop in proportion to the added stresses. Thus the height of an embankment and its outer slopes are controlled by the undrained shear strength which the soil can develop for a given water content.

The stability of the embankment after a period of time sufficient to allow expulsion of the excess water is dependent upon the consolidated undrained shear strength. In general this strength is a maximum at or near the optimum water content. Soils compacted at significantly wetter or dryer water contents will have lower shear strengths. Also, the soils compacted on the dry side of optimum will be more compressible and subject to cracking when saturation by seepage.

Embankments constructed on weak, compressible foundations will have outer slopes dictated by the foundation conditions. To minimize cracking of the embankment, the soil should be compacted on the wet side of the optimum water content, thus making it weaker and more plastic.

Large differential settlement and cracking may occur to the embankment if the abutment slopes are too steep or have benches and near vertical faces. To minimize embankment cracking, the soil should be compacted somewhat on the wet side of optimum. This fact and the reduced undrained shear strength of the wetter soil must be recognized in the design.

The soil densities obtained by conventional rolling equipment is about that obtained by the standard Proctor test. Higher densities are possible through use of heavier rollers or thinner lifts and more passes of the roller. The design densities and thus the shear strengths and the soil strain characteristics will be dependent upon the water content, the lift thickness and the type of roller selected. This is dependent upon the judgment and experience of the soil designer. Usually it is assumed that the field soil densities will range between 95 to 100 percent of the Standard Proctor maximum.

As already indicated, the compaction water content is significant in regards to the undrained shear strength which can be developed in soils. The selection of the design water content is dependent upon the judgment and experience of the designer. Factors which must be considered are:

1. The natural water contents of the borrow soils and the practicable adjustments which can be made.
2. The undrained-unconsolidated and also the consolidated-undrained shear strength during construction, and the operations of the pool.
3. Weather conditions.
4. The significance of the design of the

foundation and the embankment shear strength, consolidation and deformation.
5. The possibilities of embankment cracking and preventative measures which can be taken.
6. Settlement of the embankment and foundation.
7. The height and type of dam.
8. Trafficability of the embankment soils subject to construction equipment.

The shear strength of the fill materials should be determined for the range of probable placement content and densities.[2]

It is desirable that the laboratory and field compaction be similar. In many cases this may not be so and can lead to construction difficulties. The designer should be aware of these differences and if necessary test fills should be constructed with variations in placement water content, lift thickness, number of roller passes, and types of rollers. It is essential that the embankment adjacent to the abutments be properly compacted. For cases of steep abutment, this may be done by sloping up of the fill so that the draw bars of the rollers clear the rocks. If this cannot be done then the fill must be placed in thin lifts and compacted by mechanical hand tampers. All overhangs should either be removed or filled with lean concrete prior to fill placement. Placement of a new fill against an existing fill is important. The dried and eroded face should be cut back to acceptable soil to permit a good bond. Requirements for roller weights, dimensions and speeds are given in Ref. 18.

Pervious soils should be compacted to at least 80 percent of relative density. This applies to drainage layers and zones. The standard Proctor compaction test may be better suited than the relative density test for pervious soils having a few percent of fines passing the No. 200 sieve. For such cases the density should be at least 98 percent of standard.

Pervious, free drainage soils can be placed in 3 to 4 ft lift, in shallow water and compacted by heavy crawler tractors. The fines have an important effect on such compaction and sands should have less than 10 percent fines and well graded gravels even less. This property of placement under water is important when making diversions, cofferdams or starting pervious shell fills or wet foundations.

Soft rocks will break down under excavation, placement and compaction operations. The rock fill should be considered to be the same as that of a well graded, dirty pervious fill.

Slope Protections

The downstream slope and berms must be protected against rain and snow melt runoff. In general, grass turf is the most desirable means of protection. This is obtained by seeding a layer of fertilized top soil at the proper time of year. The grass should be kept short by mowing and a slope of 1 vertical or 2.5 horizontal is about the practical maximum upon which mechanical equipment can be operated. Vegetation which conceals the outer surface from ready visual inspection should not be used. Trees and shrubs which may interfere with inspection should not be permitted. In arid areas or where material is readily available, it is practical to protect the outer slope with a layer of gravel or rock spalls. The part of the downstream toe subject to high tailwater should be protected by riprap design either for erosive water currents, waves or both.

The upstream slope must be protected against rain and snow melt runoff and also against wave wash. The most common form of protection is dumped riprap. This is a well graded mixture of rock which requires some rehandling to prevent excessive segregation and pockets of fines spalls. The bedding layer beneath the riprap should be designed as a filter to keep wave wash from leaching out the soil fill. For thick riprap layers it is necessary to use a layer of rock spalls between the riprap and the bedding layer. While the use of quarry run rock may be cheaper, it is often necessary to process the rock on a grizzly screen to eliminate excessive fines and to obtain spalls when necessary.

Riprap sizes and thickness required are based upon the estimated wave height which in turn depends upon the maximum wind velocity, duration of high winds, fetch and configuration of the reservoir. Maximum wave height may be estimated using Fig. 6-4 and the average stone size from the following expression:

$$W_{50} = \frac{62.4\, SH^3}{2.63(S-1)^3} \cot \alpha$$

where

W_{50} = average of 50% finer rock size in pounds
S = the bulk specific gravity of the riprap
$\cot \alpha$ = the inclination of the outer slope
H = maximum wave height in ft

The above expression is the result of a wave tank research by the Corps of Engineers, U.S. Army.[19] The value 2.63 is empirical and is dependent upon the gradation of the rock and its placement. The maximum test wave height was about 5 ft. Extrapolation above this value should be done with caution. The thickness of the riprap should be about

$$T = 20\, \frac{W_A}{62.4\, S}$$

where

T is the riprap thickness in inches. The gradation of the riprap should range from not larger than 4 W_{50} to not less than $W_{50/8}$. The test slopes used in the wave tank ranged from 2.5 to 5, and the expression for W_A is subject to the wave period. The expression given is for the most critical case. The riprap thickness should not be less than 12 in.

The configuration of the reservoir is significant as regards the maximum wave height. Shoal areas, islands, ridges and bends in the reservoir tend to reduce the wave height. The location of the dam just downstream of a sharp

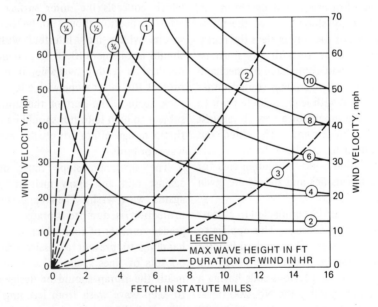

NOTE: FROM PLATES IV & V "REVISED WAVE FORECASTING GRAPHS AND PROCEDURE." SCRIPPS INSTITUTE OF OCEANOGRAPHY WAVE REPORT NO. 73, MARCH 1948 (PREPARED FOR U. S. NAVY HYDROGRAPHIC OFFICE).

Figure 6-4. Wave height as a function of wind velocity, fetch, and wind duration.

bend of the reservoir may result in a significant reduction in riprap costs.

The upstream riprap should extend from the crest of the dam to at least 5 ft or more below the low pool. Below this elevation the slope should be protected by less expensive riprap or by a sacrificial berm to protect against waves of the first filling or abnormally low pools.

Soil cement slope protection has been used on a few dams as a result of a test section at Bonny Reservoir by the U.S. Bureau of Reclamation. Design of such protection is based upon judgment and experience to date. Initially mixed in place soil cement was used but plant mix is now used because closer control can be obtained. The soil cement layer impedes drainage during reservoir drawdown. Drains under the layer with frequent outlets are used to prevent a build-up of excessive water pressure against the layer. Another possibility is to place the impervious zone as an inclined blanket adjacent to the soil cement layer and omit the drainage. The water in the inclined impervious zone will tend to drain downwards and to the downstream. The soil cement layer is brittle and subject to cracking from shrinkage and settlement. Use of soil cement must be based upon economics—not only construction cost but maintenance.

APPURTENANT STRUCTURES

Outlet Works

The location of the outlet works is important to the earth dam for reasons of cost and for convenience of construction. Because the stream must be diverted prior to closure of the embankment, the outlet works is one of the first items to be constructed. If possible, a tunnel should be used since its operation is not dependent upon the safety of the embankment against sliding. In some cases sluices in an ogee spillway are utilized; however, they have a disadvantage in that the pool cannot be completely emptied if necessary. A third and common solution is to use a conduit. This should be located on the strongest foundation at the site, often at or near the toe of an abutment. In no case should the conduit be located in the closure section of the dam. If the conduit is located on plastic soils, then the design should include structural collars and extendible water stops at each joint to prevent damage to the conduit and dam when the conduit stretches as the foundation deforms under the load of the dam. The need for seepage collars around the conduit and the impervious zone is subject to a difference of opinion among designers. A maximum of not over three is suggested. More than this is not necessary and they render compaction of the backfill difficult. The outer sides of the conduit should be such that the fill will always settle onto and not away from the conduit. The excavation for the conduit should be such as to permit a maximum use of tractor pulled rollers for fill compaction. Hand tamping should be kept to a minimum. Because of the protrusion of the axial and pull bars of the rollers, the fill adjacent to the conduit should be sloped away from it so that the roller can work right up to the conduit. In no case should a zone of noncompacted soil be permitted between the conduit and the compacted fill. If the downstream part of the dam is of random or impervious fill, then the horizontal drainage blanket should be wrapped around the downstream conduit to collect any seepage or leakage along the conduit. A pipe drain outlet should be provided on the abutment side of the downstream toe area adjacent to the conduit. The slopes of the conduit excavation should not be too steep to permit good compaction of the adjacent backfill.

If the conduit is located in a rock cut along the impervious zone, then a concrete plug should be provided between the rock and the conduit to prevent seepage through the adjacent rock. This rock is often badly fractured by blasting and excavation leaves a very rough surface which renders compaction of the soil fill difficult. The plug should be at least 50 ft long and should extend to the original rock surface. Concrete seepage collars should be used above the original rock surface if it is below the top of the conduit.

The control tower and outlet headwalls are often located within the embankment toes to reduce the length of the conduit. Plastic soils may be subject to significant horizontal deformation. This will create a condition of high earth pressures against the sides of the ridged concrete structures resting upon firm foundations. A minimum horizontal pressure coefficient is the at rest value and if the compaction water contents are excessive, the coefficient may approach the passive failure condition. These pressures can be sufficiently high to cause horizontal deformation and damage to the concrete structure.

The service bridge piers should be constructed only after the earthfill has been completed. The horizontal straining of the earthfills have in some cases been sufficient to produce excessive movements of the piers resulting in pre-ordered bridge steel being of unsuitable length.

Under no condition should the outlet works be located over a fault in an active seismic area because of the impossibility of determining that the fault is completely dead.

The stability of any slope, man-made or natural, should be carefully studied to be sure that it will be stable under conditions of pool raising, drawdown and seismic action.

Spillway

Excavation for the spillway is often complicated by emergence of the ground water and high cuts in weak materials. Studies should be made as to the suitability of such excavation for use in the embankment. If such is possible the slopes may be economically flattened. In no case should the instability of the cuts be permitted to reduce the spillway capacity.

In some cases the spillway must be located in erodible materials. For these conditions the spillway control works should be located as far upstream as feasible so that should downstream erosion occur, sufficient time will be available for corrective action. Should the material between the dam and the spillway be erodible, then adequate protection must be provided in the form of heavy revetment and concrete retaining walls. The discharge of the spillway should never impinge upon the downstream embankment toe, where necessary the flow must be diverted by substantial protective works.

Miscellaneous Requirements

The excavation slopes for all items should be given careful consideration. They should be designed by the engineer where significant to the safety of the project construction or operation. Excavation pay lines, in lieu of excavation lines, should be designated on the plans only for minor items.

Care must be exercised when the dam embankment abuts a concrete structure, especially an ogee spillway. The end of the spillway should have a batter not steeper than 10 vertical on 1 horizontal so that the earthfill will settle onto the concrete. In no case should horizontal chamfered joints be carried into the contact area of the earth fill since they may form the start of a seepage pipe. A suitable juncture is to wrap the earth fill around the end of the spillway. The upstream part of such a wraparound must be protected not only against wave erosion but also against currents. The downstream part of the wraparound should contain a drainage zone to collect and lead away any through or contact seepage. The compaction of the wraparound fill must be equal to or better than that of the embankment. The fill should be sloped adjacent to the concrete to allow clearance of the roller drawbar and axial. Power hand tamping should be used with thin lifts where clearance is prevented by the conformation of the concrete structures.

CONSTRUCTION AND MAINTENANCE

General

The design of an earth dam is not completed until completion of construction. In some cases it extends into the operational phase where significant maintenance changes are required. During construction, the character of the foundation and borrow areas are exposed by the stripping of the foundation and abutments, the excavation of the cutoff trench or drainage trench and the opening up of the borrow areas. The weather will also influence the water content of the foundation, the borrow areas and the ground water table. These factors may result in changed conditions which require modifications to the design plans and specifications.

During construction it is important that representatives of the design staff visit the site to discuss both the design concepts and the significance of any changed foundation and borrow area conditions with the resident engineer's staff. Such visits are a standard procedure in the Corps of Engineers, U.S. Army on civil projects. It is highly important that the resident staff be the best available. The competency of the resident engineers are as important as that of the designers and, in fact, even more so. It is poor practice to economize on the resident staff. A poor staff may damage a good design by not enforcing the requirements of the plans and specifications or noting changed conditions. A good staff can improve on a poor design by noting the errors and the changed conditions so that proper modifications may be made.

Care must be exercised to see that damage to the environment is minimized; that cut soil slope and borrow areas are reshaped and seeded; that pollution and sedimentation of the stream is minimized; and that excessive damages to the vegetation is prevented.

Visual Observations

One of the most important control means during construction and operation is intelligent visual inspection by members of the resident staff. The staff should be alert to changed conditions such as excessively soft areas of the soil foundation, jointing, faulting and excessive seepage of the rock foundation, movement of the fills such as slumps or bulging, or cracking of fills or cut slopes. Such observations often give indications that distress or failure is impending. The appearance of muddy water in the downstream drains, wet areas, or cave-ins are also indications of trouble during operations.

Construction Control

The control of compaction is obtained by enforcing the plans and specification requirements regarding water content, lift thickness, type of rollers and number of passes. An intelligent inspector can quickly learn to judge the compliance of these items by visual observations. To aid in developing such abilities, it is important that the taking and testing of representative samples be intensified early in the use of each different type of soil. There, specimens should be tested for density, water content and soil gradation. As the staff gains experience, the number of samples may be reduced. The chief value of such sampling and testing is to confirm or correct the inspector's judgment.

It is also important to take record samples to determine if the compacted fills possess the required shear strength, permeability and compressibility. These undisturbed samples must be carefully taken, packaged and transported to the laboratory for testing. While it is highly desirable that such tests be made at the project, economics often dictate that it be done in some distant place.

The inspection of riprap placement is very important not only because of the high unit cost but also of the serious damage to the dam which can result from riprap failure. When the placement of riprap starts, a representative sample of the acceptable gradation should be obtained and placed in a convenient location for ready reference by the resident staff and the contractor's foreman. Care must be taken to prevent excessive segregation of the rocks.

Construction Records

Complete construction records must be maintained. Such records will include the number and type of equipment in operation, the location of work, weather conditions, location of test samples. Records must be kept of the location of piezometers and movement markers; significant conditions of the foundation soil and bedrock and seepage. Finally, it is important that a record of the as-built drawings, construction photographs and a report on the foundation and construction be compiled. Copies of these items should be filed at the designers office and also at the project site. The site storage must be suitable to protect the records against fire, weather and vandals. After completion of the project, detailed inspections should be made periodically. The inspections should cover the degree of operational maintenance, of the slope protection, drain, relief wells, seepage and erosion, sinkholes and the condition of the field instruments.

INSTRUMENTATION

General

The embankment and foundation of an earth dam are instrumented to obtain data so that the safety and the action of the structure may be assessed during construction, during the operation of the project, and to obtain research data for use in future design or operational studies. Detailed information on the field instruments, installation and operations is given in Ref. 20. The frequency of reading during construction is dependent upon the conditions existing in the structure. In normal cases monthly readings during construction are satisfactory. For critical cases readings may be weekly or more often. For operational cases readings should be taken initially at significant changes of the pool. Suggested schedules are given in Ref. 21. The data should be reduced and plotted, summarized and evaluated promptly.

Types of Instruments

The type of instrumentation depends upon the size and complexity of the project. The devices in common use are:

1. Piezometers.
2. Surface movement monuments.
3. Settlement gages.
4. Inclinometers.
5. Internal movement and strain indicators.
6. Pressure cells.
7. Seismic accelerometers.
8. Movement indicators at conduit joints and other concrete structures.

Piezometers

The safety of a dam is affected by the hydraulic pressures developed in the dam and foundation by seepage and by compression of the soil structure. Piezometers are used to check such conditions and are useful for the determination of the effectiveness of drainage provision during pool operation and the development of excess pore pressures during construction. The development of excess pore pressures are symptomatic of the failure to develop increasing shear strength with increasing soil stresses. Knowledge of this condition is of great importance during construction. Piezometer should be located at critical cross sections in a layout both vertically and horizontally to develop the excess pore pressure and seepage pressure pattern. Various types of piezometers are available; electrical and gas operated devices permit the installation to be completed without further interruption to construction. They are subject to mechanical breakdown and to breakage of leads by soil straining. In no case should a lead be carried completely through the impervious zones. Open-end tubes, such as the Casagrande type, are very reliable and if properly designed and installed have a small time lag. One problem difficult to solve is the damage to the tubes by consolidation of the dam and foundation. Another difficulty is the interference of installation with construction.

Surface Movement Monuments

These are used to determine settlement and horizontal movement of the dam. They are installed at intervals along the outer slopes and along the crest of the dam. The vertical and horizontal spacing depends upon the size of the project and the magnitude of the straining problem. These monuments should be installed as soon as possible during construction.

Settlement Gages

These are plates with telescoping casing to permit sounding down to the plate at intervals to determine settlement of the soil below and also to permit an estimate of the amount and rate of consolidation. Data of this nature is needed to estimate the camber of the embankment crown and for correlation with excess pore pressures developed by slow soil consolidation.

Inclinometers

These are devices which permit the measurement of internal horizontal movements. A sounding device is used which measures inclination from the vertical by electrical means. These data permit the determination of horizontal movement knowing the vertical distance between inclination readings. These devices are useful for cases where slide may be impending and to obtain research data for future soil straining problems.

Internal Movement Devices

These consist of plastic tubes with telescopic joints. A metal plate surrounds each joint. An electrical sounder is moved along the tube and locates the plates by electrical induction. The distance to the plates is measured by a tape towed by the sounder; distances are referred to an external point whose vertical and horizontal movement are also determined by surveying.

Pressure Cells

A need exists to determine the internal stresses in a soil mass and especially at the interface with concrete structures. Data to aid improvements to present analytical methods are obtained from soil pressure cells. These generally use an oil-filled cell with a flexible face plate. Straining of the cell is indicated electrically using resistance wires or else induction coils which have been previously calibrated. These devices indicate only normal pressures which are influenced to some extent by the presence of the cell itself. There exists a need for devices to measure shear in the soil or at structural interfaces. Such devices exist but are yet in the experimental stage. Since pressure cells are operated by electrical means, trouble occurs frequently from moisture intruding into the systems and also from the creep of the devices.

Seismic Accelerometers

Strong motion, self-triggering recording accelerometers are frequently installed in important dams in active seismic areas. The data obtained is useful in improving analytical methods to determine the reaction of dams and foundations to seismic actions. A minimum installation should consist of one device on bedrock, one at the crown of the dam, and one at the toe. Additional devices are desirable. The instruments should have an auxiliary source of battery power in case the commercial electric system is disabled. For further details see Refs. 20 and 22.

Structural Movement Monuments

These devices are installed in conduits on both sides of joints and are generally bronze plugs. Similar devices are also set into retaining walls and control towers for similar purposes.

Seepage Measurements

Seepage may be measured by V weirs at points where seepage emerges or placed at intervals along a seepage collection ditch. Discharges from relief wells can be measured by similar weirs fitted to the well outlet, or by a flow meter designed and calibrated for the well size.

Plans and Records

The instrumentation for a dam and its foundation should be designed prior to construction. Changed conditions may require modification to the system. As-built plans of the devices and location should be kept at the design office and at the project site if possible. The data should be promptly reduced and the results plotted for review. The data should be taken and used for specific purposes to either check on the safety of the structure or to increase knowledge as to how the structures react to seepage and stresses induced by the pool and the dam itself.

REFERENCES

1. "Earth and Rock-Fill Dams, General Design and Construction Considerations," EM1110-2-2300,1 March 1971, Department of the Army, Corps of Engineers.
2. "Stability of Earth and Rock-Fill Dams," EM1110-2-1902, 1 April 1970. Department of the Army, Corps of Engineers.
3. "Seepage Control," EM1110-2-1901, Department of the Army, Corps of Engineers.
4. "Freeboard Allowances for Waves in Inland Reservoirs," Saville, T., McClendon, E. W., and Cochran, A. L. A.S.C.E., Waterways and Harbors Division, Journal, Vol. 88, Paper 3138, No. WW2, May 1962, pp. 93–124.
5. "Computations of Freeboard Allowances for Waves in Reservoirs," ETL1110-2-8. Department of the Army, Corps of Engineers.
6. "Embankment Pore Pressures During Construction," Clough, G. W. and Snyder, J. W., Technical Report No. 3-722, May 1966, U.S. Army Engineers Waterways Experiment Station, CE, Vicksburg, Miss.
7. "Review of Soils Design, Construction and Prototype Analysis, Blakely Mt. Dam, Arkansas," Technical Report No. 3-439, October 1956, U.S. Army Engineers Waterways Experiment Station, Vicksburg, Miss.
8. "Subsurface Investigation, Geophysical Explorations," EM1110-2-1802, September 1948, Department of the Army, Corps of Engineers.
9. "Subsurface Investigations–Soils," EM110-2-1803, March 1954, Department of the Army, Corps of Engineers.
10. "Soil Sampling," EM1110-2-1907, 31 March 1972. Department of the Army, Corps of Engineers.
11. "Subsurface Exploration and Sampling of Soils for Civil Engineering Purposes," Hvorslev, M. J., November 1948, U.S. Army Engineering Waterways Experiment Station, Vicksburg, Miss.
12. "Laboratory Soils Testing," EM1110-2-1906, 30 November 1970, Department of the Army, Corps of Engineers.
13. "West Point Project, Chattahoochee River, Alabama and Georgia; Construction of Slurry Trench Cutoff," May 1968, U.S. Army District, Savannah, CE, Savannah, Ga.
14. "Foundation Cutoff Wall for Allegheny Reservoir Dam," Fuquay, G. A., A.S.C.E. Soil Mechanics and Foundation Division, Journal, Vol. 93, No. SM3, May 1967, pp. 37–60.
15. "Relief Well Design," Civil Works Engineering Bulletin 55-11, 28 June, 1955, Department of the Army, Corps of Engineers, Office, Chief of Engineers.
16. "Pavement Design for Frost Conditions," TM5-818-2, Department of the Army, Corps of Engineers.
17. "Settlement Analysis," EM1110-2-1904, January 1953, Department of the Army, Corps of Engineers.
18. "Embankments," (For Earth Dams) CE-1306, April 1967 with amendments. Department of the Army, Corps of Engineers.
19. "Wave Tank Studies for the Development of Criteria for Riprap," Beene, R. W. and Ahrens, J. P. Proceedings, Q42, R15, Commission Internationale des Grands Barrages, Onzième Congrès des Grands Barrages, Madrid, 1973.
20. "Instrumentation of Earth and Rockfill Dams," EM1110-2-1908, 31 August 1971, Department of the Army, Corps of Engineers.
21. "Periodic Inspection and Continuing Evaluation of Completed Civil Works Structures," ER1110-2-100, 22 April 1968, Department of the Army, Corps of Engineers.
22. "Proceeding, Seismic Instrumentation Conference on Earth and Concrete Dam," Nov. 1969, Department of the Army, Corps of Engineers, published by the Waterways Experiment Station, June 1970.

7
Design of Rockfill Dams

KARL V. TAYLOR
Executive Engineer
Bechtel Incorporated
San Francisco, California

I. INTRODUCTION

Rockfill dams originated in California about the middle of the nineteenth century as a result of the need to impound water for mining operations in remote areas. Suitable rock was abundant and rock-handling operations were familar to the miners. Most of these early dams were small and consisted of a loose, dumped rockfill forming the mass of the dam, and an impervious upstream facing, often constructed of timber, with a layer of placed rock between the facing and the rockfill. The term "rockfill" dam strictly applied to this type of structure until about the mid-1900's. However, the increased use of rockfill in a variety of configurations of dams led to confusion in nomenclature, and in 1958, a definition was proposed[1] for a rockfill dam as "one of which at least half of the material in the maximum cross section is comprised of loose rock placed by dumping," regardless of the thickness of lifts or the manner of compaction of the materials. In the "Symposium on Rockfill Dams" by The American Society of Civil Engineers in 1960, a rockfill dam was defined as "one that relies on rock, either dumped in the lifts or compacted in layers as a major structural element." Included are rockfill dams with (1) impervious face membranes, (2) sloping earth cores, (3) thin central cores, and (4) thick central cores." The ASCE definition is used in this chapter, with the further qualification that "rock" shall include angular fragments such as produced by quarrying or occuring naturally as talus, and subangular or rounded fragments such as coarse gravel, cobbles, and boulders.

Although the history of rockfill dams is short compared to that of other types of dams, the development of this type of dam during the last several decades has been rapid in the United States and abroad. Prior to 1925 only eight rockfill dams in the United States and four in other countries were higher than 100 ft (30.5 m)[1,2]. Dix Dam, built in 1923–25, raised the maximum height limit to 275 ft (83.9 m). In 1931 Salt Springs Dam was built to a height of 328 ft (100 m), and in 1967 New Exchequer

Dam was raised to a height of 479 ft (146 m). About 1940, the design and construction of rockfill dams began to undergo changes and significant departures from past practices. The 250-ft (76.25 m) high sloping earth core Nantahala Dam was completed in 1942, and the 400-ft (122 m) high central core Mud Mountain Dam in 1941. Since that time the heights of sloping and central core dams in the United States have continued to increase and in 1976 were within a 550 to 750-ft (168 to 229 m) range.

II. EARLY ROCKFILL DAMS WITH IMPERVIOUS FACINGS

The original or "true" rockfill dam with an upstream impervious face was defined by J. D. Galloway in his paper entitled "The Design of Rock-Fill Dams, "published in 1939[3] as follows: "Rock-fill dams, as now built, are composed of three elements: a loose rockfill forming the mass of the dam; an impervious face next to the water; and a rubble cushion between the two."

"The characteristic that differentiates rockfill dams from other types is that the element resisting the thrust of the water pressure is of loose rock of varying sizes, placed as a fill at an appropriate site. In almost every case the rock is dumped loosely in position, and there is no attempt at orderly arrangement of the individual rocks; nor is there any other material introduced to bind the rocks together. The mass of rock is somewhat consolidated when placed in position and further consolidation takes place by settlement under load and the action of the weather. Resistance to the forces imposed by water is obtained from the weight of the mass of rock in the dam. There can be no arch action; nor can there be any action such as the cantilever offset of a gravity masonry dam. As the mass of loose rock permits the free passage of water, it is necessary to provide the dam with an impervious element to make it watertight. Several different arrangements have been tried. A combination of an earthfill backed by a rockfill has been used, but this arrangement is not properly a rockfill dam. In a few cases a diaphragm has been placed in the center of the dam, examples being the old Lower Otay Dam in California with its steel diaphragm encased in concrete (1894), or the Crane Valley Dam (California) with its masonry core wall. The most usual arrangement is to place the impervious element as a facing on the water side of the dam. This facing in the earlier dams was made of timber, but in the larger and more recent structures, it has been made of concrete, usually reinforced. Sheet steel has also been used. On one structure, the Movena Dam in California, the facing was made of uncoursed rubble masonry set in concrete. Asphalt concrete has also been proposed for this purpose."

Reference 3 as of 1939 lists 29 important rockfill dams in the United States, eleven of which were in service or under construction before 1900. Seven of the eleven dams had timber facings; one had a concrete facing; one a rubble facing, and two had steel cores. Three of the eleven were higher than 100 ft (30.5 m). Of the eighteen built between 1900 and 1932, twelve had concrete facings; two, timber facings; one had a concrete face, and one was rubble faced. One of these dams was 328 ft (100 m) high, two more were higher than 270 ft (82.4 m), and eight ranged from 125 to 167 ft (38.1 to 50.9 m) high.

In the intervening 35 years to 1974 many impervious face rockfill dams have been built in the United States and abroad. In the development of this type of structure, changes in details have evolved but the basic concepts remain essentially unaltered. In the following paragraphs the three basic elements of the original type of rockfill dam will be discussed, together with some of the major changes in design which have taken place in recent years.

Materials for Dumped Rockfill

The dumped rock mass was almost invariably composed of angular rock fragments obtained from local quarries, spillway cuts, foundation excavation, or talus deposits. There has in the past been a considerable diversity of opinion amongst rockfill dam designers as to the optimum size and gradation of rock for dumped rockfill. For example, Galloway[3] states: "The nature of rockfill is one upon which difference of opinion will develop. It is believed that the fill should be composed of individual rocks of fairly uniform size, one rock bearing directly on another, usually expressed as "rock to rock." Any wide divergence in size will cause excessive and unequal settlements, something to be avoided wherever possible. To illustrate the idea, if one selects a sample from a mass of broken stone of fairly uniform size, it will be found extremely difficult to force the sample into the mass by added pressure. On the contrary, if a large rock, which may rest upon small gravel or sand, is given a load, it will settle into the mass by displacing the smaller grains. In the usual method of dumping rock into the fill, the larger ones roll to the bottom and the smaller ones remain at the top. A graded fill results." However, in a discussion of Galloway's paper, Peterson[3] states, "In the writer's opinion an unsegregated quarry-run mass, with excess fines wasted, will give the highest degree of contact, rock-to-rock, and the greatest resulting fill density. Such a fill will have ample rock-to-rock contact and will still have sufficient voids to provide adequate drainage." This opinion coincides with that of most authorities on rockfill dams at the present time.

The specifications for Cogswell Dam[4], a 200-ft high structure completed in 1935, provided for "the material for the rockfill to consist of three Classes A, B, and C of large rock with maximum size on the downstream face and toe, and with derrick-placed rock, commonly known as packed rock, immediately below the facing. The character of all rock was to be sound, hard, durable, angular quarried rock, weighing not less than 160 lb /cu ft (2562.9 kg/m^3); to be unaffected by air and moisture and of such toughness as to withstand dumping without undue shattering or breakdown; and to have a minimum compressive strength of 5000 lb/sq in. (351.5 kg/cm^2). Class A rock, for general use throughout the main fill of the dam, was to be a well graded mixture, 40 percent of which was to vary in weight from quarry chips to 1,000 lb (373. 2 kg), 30 percent from 1,000 to 3,000 lb (373.2 to 1119.6 kg), and the remaining 30 percent from 3,000 to 14,000 lb (1119.6 to 5224.8 kg). In addition, this mixture was not to contain more than 3 percent of its total weight in quarry dust and the maximum dimension of any piece was not to be more than three times its minimum dimension."

"Class B rock was to be selected extra large rock, one-half of which to weigh not less than 14,000 lb (5224.8 kg) and the other half not less than 6,000 lb (2239.2 kg) each. The greatest dimension of each piece was not to be more than 4 times its least dimension. This rock was to be placed at the downstream toe and on the downstream face of the dam."

"Class C rock was to vary in weight from quarry chips to 14,000 lb (5224.8 kg), the relative proportion of the various sizes to be regulated according to the requirements of placing to result in a packed rockfill of maximum density." Figure 7-1 shows the maximum section of Cogswell Dam.[4] The grading of rockfill for Salt Springs Dam, completed in 1931,[5] was "from fines to 25-ton (22,679 kg) rocks, with the average large rock exceeding three tons (2721.5 kg)." "Field notes and photographs indicate that the rock coming from the quarry varied from large clean rock just after a quarry blast to smaller rocks with many fines when cleaning up the quarry." The maximum section of Salt Springs Dam[5] is shown in Fig. 7-2.

For Lower Bear River Dam No. 1 built in 1952,[5] "the rock is gray, fine-grained granodiorite that has a compressive strength of 15,000 to 20,000 psi (1054.5 to 1406.0 kg/cm^2). Quarry run rock of predominately large size, 1-10 tons, (0.91 to 9.07 mt.) was specified. Fines of 4 in. and less in size were specified to be less than 5 percent by weight. As constructed, more than 50 percent of the rock is estimated to be 1 to 20-ton size and many loads

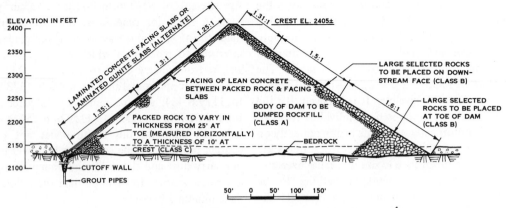

Figure 7-1. Cogswell Dam—maximum section. (After Baumann[4])

with more than five percent fines were used in the fill."

At Wishon Dam, a concrete face dam, 261 ft (79.6 m) high, completed in 1958,[6] "rock was specified to be sound, hard and durable with definitions for each term. At least, 50 percent (by weight) of the rock in the dam was specified to range from 0.5 to 10 tons (0.45 to 9.07 mt) and over." The specifications for rock (basalt) to be placed in Lemolo No. 1 Dam, 120 ft (36.6 m) high, completed in 1954,[7] shown in Figure 7-3, "required that it should be hard and durable and that is should be free from cracks or incipient jointing which would cause it to shatter when dumped in the fill." "The specifications further required that the grading be such as to produce a dense fill with rock sizes ranging from one to 10 tons predominating in the fill. It was further required that not more than 5 percent of the fill should consist of fines less than 4 in. (10.2 cm) in size."

From these and other samplings of the literature, it is apparent that the majority of designers of impervious faced dumped rockfill dams, after about 1925, preferred a reasonably well graded mix of rock ranging in size from about 4 in. (10.2 cm) to a weight of 10 tons (9.07 mt) or more, for the main rock mass of the dam.

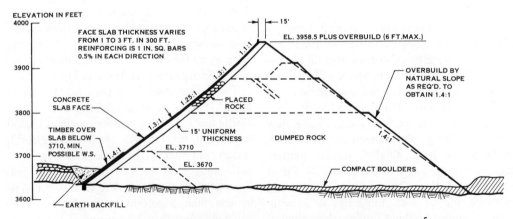

Figure 7-2. Salt Springs Dam—maximum section. (After Steele and Cooke[5])

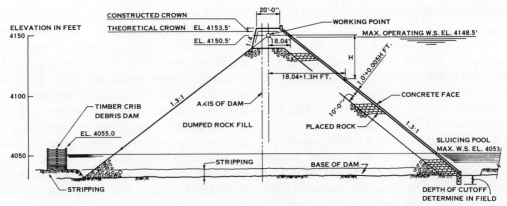

Figure 7-3. Lemolo No. 1 Dam. (After Boyle and Barrows[7])

Placement of Dumped Rockfill

Opinions as to the proper method of placing and compacting dumped rock have varied widely amongst designers, and over the years. Referring again to the references cited above, at Cogswell Dam "Loose rockfill was to be placed in lifts not to exceed 25 ft (7.63 m). It was to be spread approximately level and parallel to the axis of the dam. Placing the fill from the abutments toward the center and thus leaving a gap at or near the center was not permitted. Due to scarcity of stream flow (1932) sluicing for consolidation of the loose rockfill was not provided." However, when the dam was about 80 percent complete, it was subjected to a heavy rainstorm which yielded 15.07 in. (38.3 cm) of rain at the dam. "The application of this natural lubrication to the dry rockfill caused the latter to settle, especially that part dumped in the gap at the right abutment. Immediate settlement there amounted to some 4 percent of the height. Vertical settlement was accompanied by slight bulging in the upstream lower half of the dam, resulting in damage to the laminated facing and sub-slab, particularly near the abutments." "The storm reversed the water supply from deficient to abundant. Thus, water in ample quantity was now available for artificial sluicing of the rockfill. This was accomplished by drilling holes through the facing and sub-slab . . . , and pumping clear water into the rockfill to hasten final settlement. After several months of this sluicing operation the afore-mentioned maximum settlement of 4 percent had increased to 6 percent. The average residual settlement amounted to 4.5 percent."

At San Gabriel Dam,[4] "all loose rock was placed by the use of 2 cu yd (1.53 m^3) of water to 1 cu yd (0.76 m^3) of rock. Initial settlement thereby produced averaged 6 percent. This checks the afore-mentioned settlement at Cogswell Dam due to rain and sluicing after dry placing of rock."

Salt Springs Dam[5] was constructed by dumping principally in three main lifts, progressing outward from each abutment. In each lift the rate of progress from each abutment varied so that the "valleys" formed by the approaching faces of the dumped fills would not all be in the same general zone of the dam. The three main lifts were each across the full thickness of the dam and were 170 to 185 ft (51.9 to 56.4 m), 75 ft (22.9 m) and 65 ft (19.8 m) high. As the lower lifts were placed the materials were sluiced by low pressure hose streams, with about 1 cu yd of water to 4 cu yd of fill. More water was applied to the upper lifts but at best the authors of Ref. 5 considered the sluicing nominal, and in retrospect considerably deficient in comparison with that at Wishon Dam in 1957-58 which was also constructed in high lifts but sluiced with high pressure monitors applying a volume of water 3 times as great as the volume of rock being dumped.

The dumped rockfill at Lemolo No. 1 Dam[7] was placed in three lifts, and all the fill was sluiced as it was dumped using 3 volumes of water to each volume of rock.

Paradela Dam is a 361-ft (110.1 m) high concrete faced rockfill dam built in Portugal in the late 1950's.[8] The dumped rockfill was placed in four lifts of approximately 190, 98, 28, and 12 ft (58, 30, 8.5 and 3.7 m) height, working outward principally from one abutment. The sluicing water used during the dumping was equivalent in volume to 4 times the volume of the dumped rock, and was applied by remotely controlled monitors which could direct the water jets directly upon the rock dumped from the trucks.

These few examples selected from many illustrate the fact that nearly all authorities agree that rock fill dumped in high lifts must be sluiced with copious quantities of water by high pressure jets as it is placed to minimize the post-construction settlements which otherwise may be intolerably high. Not all authorities agree, however, on what is actually accomplished by the sluicing. Some contend that the fines are washed from the points of contact between the larger rocks, thus allowing rock-to-rock contact, and consequently a more firmly keyed together and compacted rock mass. Others seem to believe that by sluicing the fines into the voids between the larger rocks during placement, the bulk density of the rock mass is improved and the compressibility is reduced.

Lubrication of the rock mass by the sluicing water has also been credited with the acceleration of settlement by allowing the rock to adjust and consolidate more rapidly as the load is increased. Another concept[9] is that the softening effect of partially saturating the points and edges of the rock allows them to break and crush more readily than if the rock had been dry. It has long been known that the strength of saturated rock is significantly less than the strength of rock in a dry condition.

Settlement of a dumped rockfill occurs principally because of the crushing and breaking of the individual rocks at points of contact with adjacent rocks. The corners and edges are particularly susceptible to crushing under load. It therefore seems logical that if the fines are washed from the contact points between rocks, and the edges and corners of the rocks at the contact points are weakened by partial saturation, and the rock mass is made more mobile by lubrication, then the maximum amount of settlement can occur as the loading of the fill is increased during construction and less will appear as post-construction settlement. It is unlikely that the partial filling of the voids between large rocks by sluicing fines into them would have a significant effect in reducing the total settlement except perhaps at the bottom of the fill.

The question naturally arises as to how much water should be applied during the sluicing operation. The amount of water required to saturate the edges and corners of the rocks is much less than the amount of water which experience indicates is required to accelerate the settlement of the fill. When the rock is sluiced by high pressure jets as it is dumped, the mechanical action of the impinging jet does, in fact, wash fines into the larger interstices of the rock mass. This mechanical action is, however, effective only to a limited depth below the surface of the rock mass because the energy of the jet is quickly dissipated by impact on the surface rocks. The sluice water percolates downward through the rock mass, carrying some fines with it and maintaining the fill in a moistened state. This is significant because as load is added by dumping on the advancing lift face, rocks already in place tend to break and crush at the corners and transfer load to adjacent units, which in turn undergo the same action. This chain reaction is manifest in the settlement of the rockfill, but it requires time for it to progress. Large quantities of water are necessary to maintain the lubrication and rock-softening effect on the rocks already in place back of the advancing fill face.

The dumping of rockfill in high lifts causes

segregation of the materials. The fines are deposited near the top of the slope of the dumping face and the large rocks slide and roll to the bottom. Thus there is a tendency toward a natural sorting action as each lift is placed, causing the bottom zone to be composed of the large, coarse rock, the top zone to contain much of the fine material, and the intermediate zone to contain rock graded between these limits. However, depositing the rock at the crest of the slope causes surficial slides to occur which have the beneficial effect of somewhat remixing the materials and creating a more homogeneous fill. As large rocks move down the slope of a high lift some of the sharp corners and edges are removed and, moreover, the drag and the vibration created by the movement of the large rocks tend to move much of the fine material down slope. Despite these effects, however, each high lift tends to remain a more or less sorted or zoned mass. The tendency toward segregation of fines at the top of a dumped rockfill lift has the effect of interposing strata or bodies of fines within the rock mass of a dam if construction is accomplished with multiple lifts. Every effort is made during placement to avoid such stratification of fines by special sluicing, light blasting, etc., at the top of each lift but some segregation of fines is unavoidable. At such an interface between lifts, coarse rock forming the bottom of the upper lift may be in contact with fine material collected at the top of the lower lift. Some settlement could be caused by the penetration of the large rocks into the finer materials upon which they rest. The application of large quantities of sluicing water is beneficial in this case also to accelerate settlement during construction.

In short, dumped rockfill should be sluiced with high pressure jets directed so as to maximize the benefits of the localized mechanical action of the jets in washing out fines, but also copious quantities of water must be delivered to lubricate the rock and enhance the corner breakage to accelerate the settlement of the fill. Experience indicates that jets operating at 100 to 200 psi (7.0 to 14.0 kg/cm^2) and delivering a volume of water equivalent to 2 to 4 times the volume of rock will achieve good results.

Settlement of Dumped Rockfill

The settlement of a dumped rockfill under its own weight occurs mostly during construction, but continues on at a decreasing rate, sometimes for many years, after completion of the fill. The settlement is mostly in the vertical direction but when fills are placed in "V"-shaped valleys, there is a very significant component of movement of the rock mass from the abutments toward the center of the valley. This results in a tendency for the fill to elongate at the abutments and compress at the center, and the design of the facing must take this action into account. A still further lateral movement of the fill has been noted in the direction opposite to the direction of advancement of the high lifts used in placing the material. The reason for this appears to be that the impact of large rocks rolling and sliding down the steep face of the lift causes a relatively high degree of compaction, generally in the plane of the slope of the lift but less so in the direction normal to that surface, and therefore, the lateral component of settlement will be greater normal to the lift face than parallel to it.

The settlement of dumped rockfill during construction depends on several factors, the most important of which are the height of the lifts, thoroughness of the sluicing, hardness of the rock, and the pattern of lift placement. Most of the settlement takes place during the dumping and sluicing, and not more than about one percent of the height of the lift should take place after the completion of the lift.

If concrete facing slabs are placed concurrently with the raising of the rockfill, as is customary for economic reasons, it must be expected that construction settlements will affect the slabs already in place, particularly near abutments and may cause some damage to the slabs even before the dam is completed. However, the cost of repairs is usually minor compared to the cost of delaying construction of the slabs until completion of the rockfill when a large proportion of the total settlement will have occurred.

At the first filling of the reservoir, additional settlement of the rockfill will occur due to the weight of water on the facing, and in addition there usually will be an appreciable horizontal movement in the downstream direction of the upper portion of the fill. These horizontal components of movement can amount, on the average, to about 50 percent or more of the vertical movement induced by the water loading.

The post construction settlement of rockfill continues for a number of years at an ever-decreasing rate. Most of the settlement occurs after the first complete filling of the reservoir, but continues at a slowing rate as the dam ages and the reservoir level fluctuates. The "ultimate" post-construction settlement of the crest, defined as the settlement when the annual rate is equal to or less than 0.20 percent of the height, of a well-constructed rockfill built with high lifts and adequate sluicing may be expected to be in the order of 0.5 percent to 1.0 percent of the height of the section, for fills under about 300 ft (91.5 m) in height.

In Ref. 10 Lawton and Lester reported investigations of the settlement of a number of dumped rockfill dams and found that the crest settlement, "S," is related to height approximately as follows:

$$S = 0.001 H^{3/2}$$

(see also, Refs. 5, 6, 8, 11 and 16)

Because of the crest settlement, the dams are overbuilt in height or cambered in proportion to the height of the section and the anticipated percent of settlement. For dams in the 300-ft (91.5 m) and above height range, 3.5 to 7.0 ft (1.06 to 2.14 m) of camber is customary, or approximately 1 percent to 2 percent of the height.

Some dams are curved upstream in plan so that the horizontal movements induced by the water loading will tend to cause axial compression in the fill and reduce the total movement.

Cooke and Steel[5,6] made detailed measurements of the effects of settlement on the concrete facing at Salt Springs and Wishon Dams. It was their observation that under the influence of water pressure and settlement of the rockfill the resultant movement of the facing slabs was essentially normal to the slope. At Salt Springs Dam after 27 years of service the maximum movement of about 5.5 ft (1.7 m) at the center of the dam occurred at about 0.4 of the height of the dam. The settlement contours indicate a shallow bowl-shaped depression of the slab extending from the abutments to the afore-mentioned maximum depression at the center. As a resultant of the vertical, horizontal, and lateral movements of the fill the facing slabs in the lower three-quarters, approximately, of the dam were in compression, but in the upper part of the face the horizontal joints opened about $\frac{1}{2}$-in. to 1-in. (1.27 to 2.54 cm). The lateral movement closed all vertical joints except those well toward the abutment ends of the facing, which opened up as much as several inches. Also, some shifts of the slabs in the lateral direction were noted which could be associated with the direction of dumping the rockfill.

Placed Rock

The slopes of impervious face dumped rockfill dams are unusually steep as compared to other types of fill dams. The downstream slopes are customarily constructed at approximately the angle of repose of the dumped rock; i.e., 1.3 H to 1.0 V or 1.4 H to 1.0 V, but some are as steep as 1.25 H to 1.0 V and a few are flatter than the angle of repose where the foundation conditions or the conservatism of the designer seemed to so warrant. Upstream slopes are constructed no flatter than the angle of response of the rock and in many cases they are appreciably steeper than the repose angle. Some of the larger dams have been constructed with compound slopes which become steeper than the angle of response near the top of the dam. Many of the old impervious face or "deck" type dams still in service in the Sierra Nevada

Mountains of California have upstream slopes as steep as 0.75 H to 1.0 V. Such steep slopes are made possible by the use of a layer or zone of "placed rock" on the face of the slope. This placed rock layer, which is in reality a type of rough dry rubble masonry, forms a semirigid mass and retains the toe of the loose dumped rock as the fill is brought up to grade (see Fig. 7-3).

The most important function, however, of the placed rock zone is to transmit uniformly the thrust of the water on the facing to the supporting mass of the main rockfill. It is imperative therefore that intimate contact be established between the downstream face of the placed rock and the main fill.

The thickness of the placed rock layer was established on the basis of experience and judgment. Some have been built with a constant thickness for the full height of the dam; e.g., 15 ft (4.6 m) normal to the face at Salt Springs Dam and 10 ft (3.05 m) normal to the face at Lemolo No. 1 Dam. Most placed rock sections, however, are tapered as a function of height of dam, for instance, thickness, "t" = 10 ft + 0.05 H, where H = distance below the crest, as at Paradela Dam and Lower Bear River Dam. Large rock was used for the placed section, ranging in weight from about 0.5 ton (453.6 kg) to upward of 10 tons (9072 kg.). The rocks were individually and carefully placed to make good contact with the course below and with the minimum workable space laterally between units. All spaces or joints between rocks were chinked with small rock. Special attention was given to the contact between the placed rock and the dumped rockfill so that the water load on the facing would be transferred uniformly to the dumped rockfill without undue distortion of the placed rock layer and consequent damage to the concrete facing slab.

During construction, a substantial part of the dumped rockfill could be placed with slopes at the angle of repose to elevations well above the working level on the placed rock section. As the placed rock section was raised, rock was dumped on the upstream face of the fill and retained at the toe of the slope by the placed rock. By this method (or others) work on the rockfill, placed rock section, and the facing could be performed concurrently on portions of the dam.

Cutoff Wall

Foundations for most (but not all) dumped rockfill dams are rock, and it is customary to install a grout curtain and cutoff wall along the alignment of the upstream toe of the placed rock. The depth and thickness of the grout curtain and the configuration of the cutoff wall will depend on specific conditions at each site. The cutoff wall may be trenched deeply into the foundation, or it may be little more than an anchored slab acting as a grout cap, but in either case it forms the base or footwall of the concrete facing. Since this footwall is rigidly fixed to the foundation, some sort of hinge or type of articulation at the bottom joint between the footwall and facing slab must be provided to permit the facing to move as the rockfill settles. On high sections of dams with concrete slab facings one or more hinge slabs are provided adjacent to the footwall. Careful attention must be given to the design of the articulation system around the periphery of the facing to avoid damage to the facing as the rockfill settles under its own weight and the thrust of the reservoir water. A variety of treatments have been developed to accomplish the necessary peripheral articulation. Some which have been successfully used in the past are discussed in Refs. 4 through 7.

Concrete Facings

In order to attain the desired flexibility to allow the facing to accommodate to the movement of the rockfill, it has been the practice in the United States to construct concrete facings of individual panels or slabs about 50 or 60 ft square. Near the bottom of the facing and around the periphery the slabs are narrower so as to place more horizontal or sloping joints in areas where the bending due to settlement will be most severe.

The thickness of the facing slab usually increases with depth below the crest. For example, at Lower Bear River Dams, thickness of facing slabs in feet, "t" = 1 + 0.0063 H and at Paradela, "t" = 1 + 0.00735 H, where "H" is the depth in relation to the crest of the dam. The area of steel reinforcement in the slabs, in single or multiple layers in accordance with slab thickness, is about 0.5 percent of the area of the concrete section, in both horizontal and vertical directions. Horizontal joints about $\frac{1}{2}$ to $\frac{3}{4}$ in. (1.27 to 1.9 cm) wide, formed with wood filler strips and sealed by copper water stops, are often used in the upper approximately two-thirds of the facing. The lower horizontal joints, cutoff wall joints, and sloping joints are cold joints, coated with asphalt and also sealed with water stops. Vertical joints about $\frac{3}{4}$ to 1 in. wide near the abutments and 1 to 2 in. wide in the center areas of the facing are formed using asphalt board or similar plastic material and are sealed with water stops and asphaltic or rubberized sealing compound. Joint ribs are provided under all joints to permit movement of the panels without damage to the edges. These ribs are concrete members, cast in the placed rock section, with smooth top surfaces coated with asphalt to provide slip surfaces for the edges of the panels.

Depending on the anticipated amount of settlement which must be accommodated, other designs of concrete facings, for example, continuously reinforced facings (Bucks Creek Dam, 1928) or laminated concrete facings (Cogswell Dam, 1932), have been utilized. However, the articulated panel type of facing such as that installed on dams in 1940-50, has stood the test of time and has generally proved to be satisfactory.

The settlement of the rockfill will create high compressive stresses in the concrete facing panels. Sometimes the facing is designed to be slightly concave so as to avoid the possibility of the panels being forced outward or away from the dam face when the compressive loads are applied. Also, the concavity will compensate for the increasing gross settlement from the bottom to the top of the fill.

The volume of leakage past a concrete-faced dumped rockfill dam would, of course, be related to the amount of slab cracking and the efficiency of the cutoff, but aside from the value of the water lost it would rarely if ever, be significant, particularly with regard to the stability of the structure. Reported leakages have been noted in the literature ranging from practically none to amounts well in excess of 100 cfs (2.83 m^3/sec).

III. MODERN CONCRETE-FACED ROCKFILL DAMS

As a result of experiences with the earlier concrete-faced rockfill dams discussed above, a number of changes in design treatment and construction practice have evolved in recent years.

1. Compacted rockfill, rolled in thin layers, has practically superseded dumped rock placed in high lifts. The use of compacted rock is not an innovation. It was incorporated to some extent even in quite early rockfill dams, such as San Ildefonso Dam, Mexico, a 189-ft (57.6 m) high rockfill structure designed in 1937, San Gabriel Dam No. 1 in 1937, the transitions of numerous earth core dams including Mud Mountain Dam in 1939, and later in Quoich Dam in Scotland.[144] Compacted rockfill is now almost universally used in the main rockfill sections of modern dams. The reasons for this are several fold:

   The post construction settlement and distortion of rolled rockfill is much less than that of dumped fill. The uniformity of the fill can be controlled and a high degree of compaction can be achieved during construction rather than relying on gravity, water forces, and time, as in the case of dumped fill.

   A wider range of rock materials, with respect to both physical properties and gradation, may be used for rolled fill than for dumped rock. Soft rock which would not survive dumping in high lifts can be

DESIGN OF ROCKFILL DAMS   329

used in compacted fills, and since there is little or no tendency to segregate when placed in thin lifts, the gradation is less critical and the elimination of a high percentage of fines would not be required insofar as placement problems are concerned, but of course, it may be necessary in zones where high permeability is a requisite.

The development of modern construction equipment, particularly the vibratory roller, has made handling and compacting rockfill in thin lifts the preferred procedure.

Interference with concurrent work in adjacent areas of the dam is less when rolled fill methods are used.

2. The upstream dry rubble masonry or placed-rock zone is no longer used. The upstream slopes are constructed at or about the normal angle of repose of the rock, and a compacted bedding layer of small rock is placed on the upstream face to support the facing. This avoids the problems of handling selected large rock and affords better and more uniform support to the facing than can be accomplished by placed rock. The cost of additional materials to build the upstream face at a flatter angle than that which could be constructed using placed rock is more than offset in labor cost.

3. Cutoff walls deeply entrenched into the rock have largely been superseded by anchored footwalls which support and seal the periphery of the facing and serve as a grout cap for curtain grouting.

4. The present trend is toward the elimination of deformable joint fillers in vertical joints of the facing, and very limited use of horizontal joints. Both contraction and construction joints are provided with single or multiple water-stops of pre-molded rubber or plastic and expandable metal shapes. Facings are placed by slip-forming[12] and horizontal panel widths are usually determined accordingly.

Some or all of these design and construction modernizations have been utilized in the two dams selected as examples in the following brief discussions.

Cethana Dam in Australia is a 361-ft (110 m) high concrete faced rockfill dam, completed in 1971.[13] The upstream and downstream slopes are 1.3 H to 1.0 V. Fig. 7-4 shows the cross

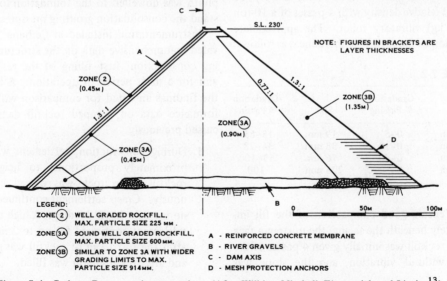

Figure 7-4. Cethana Dam—maximum section. (After Wilkins, Mitchell, Fitzpatrick and Liggins[13])

section of the dam. The main rockfill was sound quartzite graded approximately within the following limits:

TABLE 7-1

Size	Gradation % Passing	Size	Gradation % Passing
0.60 mm	0–10	76 mm	22–100
4.8 mm	0–40	305 mm	60–100
19 mm	0–65	610 mm	100

The rock was placed in layers 36 in. (0.9 m) thick and rolled with 4 passes of a 10-ton (9072 kg) vibratory roller. Immediately preceding and during compaction all rockfill was sprayed with water, the volume applied being not less than 15 percent of the rockfill volume. This procedure produced a relative density of the rockfill close to 100 percent. Within the downstream one-third of the section, lifts 53 in. (1.35 m) thick were used and the gradation limits for the rockfill were wider than for the main rockfill. A zone about 10 ft (3.0 m) wide in the upstream part of the main fill was composed of specially selected rock and compacted in 18 in. (0.45 m) lifts.

At the upstream face of the fill a zone 9.8 ft (3.0 m) wide of well graded rockfill was placed and compacted in 18 in. (0.45 m) lifts to 100 percent relative density with 4 passes of a 10-ton (9072 kg) vibratory roller. The approximate gradation of the rockfill in this zone is as follows:

TABLE 7-2

Size	Gradation % Passing	Size	Gradation % Passing
0.07 mm	0–5	19 mm	15–65
0.60 mm	0–18	38 mm	34–78
4.8 mm	0–40	76 mm	53–100
9.5 mm	8–50	225 mm	100

To ensure good compaction of the fill immediately beneath the facing, the upstream face of the rockfill was initially given 4 passes of the roller without vibration, and the slope was trimmed to remove high spots. It was then given 4 passes at half vibration and low areas were filled. To avoid displacement of stones by roller vibration, traffic, and rain, a bitumen emulsion treatment was applied. The final compaction was attained by 8 passes of the roller with full vibration.

The concrete facing was slip formed[12] in panels 40 ft (12.2 m) wide with horizontal contraction joints except near the periphery where intermediate vertical joints, terminating in horizontal contraction joints, were introduced to control cracking of the slab in zones where horizontal tensile strains were expected. These horizontal joints contained wood filler strips, but no filler of any kind was used in the vertical joints. The concrete facing was tapered according to the formula "t" $= 0.3 + 0.002\ h$, where "t" = thickness and "h" = head, in meters. The reinforcement was 5 percent in each direction except in zones close to the abutments where expected tension called for an increase in the steel area. The plinth or footwall at the lower periphery of the facing was not entrenched but cast on the rock surface after removal of weathered and open jointed rock. The minimum contact surface on the foundation rock was $\frac{1}{20}$ of the head on sound rock and $\frac{1}{10}$ of the head on poor rock, but not less than 3.0 m. The plinth was dowelled to the foundation to withstand the consolidation grouting pressures.

Instrumentation installed at Cethana[14] provided comprehensive data on the structure during construction, first filling of the reservoir, and for a short period of operation. A few of the findings are cited for comparison with performance data on dumped rockfill dams discussed previously.

1. During construction, settlement was approximately proportional to height of rockfill when rock was placed continuously. Creep settlement continued with no increase in fill loading. A high rate of settlement continued for up to 2 months after the top 12 m of rockfill was placed and after the reservoir was filled.

2. The vertical and horizontal deflections of the crest during a period of about 11 months following the commencement of filling the reservoir were:

   a) Maximum vertical = 2.7 in. (69 mm)
   b) Horizontal downstream maximum = 1.6 in. (41 mm)
   c) Horizontal transverse = 0.71 in. (18 mm) toward center from left and 0.31 in. (8 mm) toward center from right.

3. The deflection of the membrane was essentially normal to the face and was a maximum of about 5.0 in. (13 mm) at about the lower 0.4 point of the slope.
4. The maximum perimetric joint opening was about $\frac{1}{2}$ in. or 11.5 mm.
5. After filling the reservoir the strains in the facing were compressive, the maximum being 207×10^{-6} in the slope direction and 290×10^{-6} in the transverse direction.
6. The leakage past the dam at full reservoir was 1.24 cfs (0.035 m^3/s).

New Exchequer Dam is a concrete-faced rockfill dam, 490 ft high, (149.5 m) completed in 1966. The dam is unique in that it incorporates the old 310 ft (94.55 m) high concrete gravity arch dam in the upstream toe as a retaining wall for the rockfill of the new dam, and as the lower part of the upstream impervious membrane on the face of the new dam. This required that a carefully designed joint be constructed between the old dam and the new facing on the rockfill.

The 490 ft high fill is considerably higher than any previously constructed rockfill dam with a concrete facing. The settlement of the fill was a matter of prime concern. Since the records of dumped rockfills indicated settlements in the order of 1 percent, it was decided to compact most of the rockfill in the dam to reduce the settlement to a tolerable amount.

The slopes of the dam, both upstream and downstream, are 1.4 H to 1.0 V. The rockfill was placed in 4 zones, as follows:

1. The zone immediately under the concrete facing consists of 15 in. (38.1 cm) maximum size rock compacted in 2-foot (0.61 m) lifts by a 10-ton (9072 kg) vibratory roller.
2. An upstream zone adjacent to the old dam and extending to the top of the new dam, varying in width from about 200 ft (61 m) at the bottom to about 40 ft (12.2 m) at the top, was constructed of 4 ft. maximum size rock placed in 4 ft. (1.22 m) lifts and compacted by 10-ton (9.07 mt) rollers.
3. The main body of the dam consists of 4 ft maximum size rock placed in 10-ft lifts and compacted by the hauling and grading equipment.
4. The downstream slope section of the dam consists of the largest rock which could be placed with a minimum of 50 percent larger than 12 in. (30.48 cm). The fill material was dumped in lifts up to 60 ft (18.3 m) high but no compaction was specified. All the fill except the zone immediately under the facing was sluiced with high pressure jets.

The New Exchequer Dam rockfill represents a compromise or transition between the traditional practice of dumping rock in high lifts without compaction and the more recent trend toward heavy mechanical compaction of the fill in relatively thin lifts.

The concrete facing was constructed of blocks 60 ft (18.3 m) wide and varying in length in the direction of the slope from 21 ft (6.41 m) at the bottom of the new facing to 50 ft (15.25 m) at the top. The thickness of the slabs varied from 34 in. to 18 in. (86.36 to 45.72 cm). All joints were provided with copper water-stops.

Measurements made during construction and during the first filling of the reservoir showed normal movements of the facing slabs, with the vertical joints near the center tending to close and those near the abutments tending to open. After the reservoir was filled the maximum crest settlement was 1.5 ft (0.46 m) or about 0.3 percent of the height. The maximum horizontal downstream movement of the crest was 0.4 ft (12.2 cm) or about 30 percent of the associated vertical movement. The crest settlement was only about one-third of that which normally would have been expected for a dumped rockfill, but about 5 times that of the fully compacted rockfill of Cethana Dam. The settlement of the facing itself formed the characteristic bowl-shaped depression with a maximum depth of about 2.0 ft (61 cm) normal to the slope at a point 0.3 to 0.4 of the height above the toe.

During the first two fillings of the reservoir the leakage through the dam increased from 12.5 cfs (0.35 m^3/sec) to a maximum of 490 cfs (13.72 m^3/sec). This was caused by the spalling and cracking of the face slabs at and near the junction of the new facing with the old dam. A supplementary zone composed of sand, gravel, clay, and bentonite was placed underwater in the "V" notch formed at the contact of the new facing with the downstream facing of the old dam. This material was placed to a depth of 20 to 25 ft. (6.1 m to 7.6 m) using a specially designed skip. The sealing blanket reduced the leakage to about 8 cfs (0.224 m^3/sec.)[27]

IV. ASPHALTIC CONCRETE DECK DAMS

Only a relatively few major dams in the United States have been constructed with facings of asphaltic concrete. The first use of asphaltic concrete facing was principally for slope protection. The acceptance of asphaltic concrete for the impervious membrane on rockfill dams in the United States lagged considerably behind the adoption of this material in other countries. Ghrib Dam, for example,[15] is a 135 ft (41.2 m) high rockfill dam in Algeria, built in 1936, which has an asphaltic concrete facing on a slope somewhat steeper than 45 degrees. The asphaltic concrete was confined between slabs of porous concrete, the bottom slab being 4.75 in. (12.1 cm) thick and the top, 4.0 in. (10.2 cm) thick. The success of this and other asphaltic concrete facings on dams in Algeria, Germany, Portugal, and elsewhere demonstrated that a properly designed facing of this material is durable and stable even on steep slopes and under severe climatic conditions.

The first major dam in the United States to be built with an asphaltic concrete facing as the impervious member was Montgomery Dam completed in 1957 for the City of Colorado Springs, Colorado.[17,18] This is a dumped rockfill dam, 113 ft (34.47 m) high with an upstream slope of 1.7 H to 1.0 V and a downstream slope of 1.5 H to 1.0 V. Since this dam was the first of its kind and no precedent existed for placing asphaltic concrete on the steep face of a dam where it would be completely exposed, consideration had to be given to (1) watertightness of asphaltic concrete, (2) stability and behavior on steep slopes, (3) effects of high temperature, freezing and thawing, and variations in types of aggregate and asphalt binder, (4) appropriate thickness of asphaltic concrete deck and manner of placing the materials, (5) nature and preparation of the base and (6) manner of making watertight connections of the asphaltic concrete to structures constructed of portland cement concrete. Exhaustive field and laboratory tests were undertaken to solve these problems, determine the proper asphaltic concrete mix and develop satisfactory construction procedures. This research led to the conclusion that asphaltic concrete could be placed on slopes as steep as 1.5 H to 1.0 V and remain stable at surface temperatures somewhat in excess of 140°F. However, the upstream slope of 1.7 H to 1.0 V was chosen in the interest of personnel safety and facility of machine operations in placing the asphaltic concrete.

It was concluded from the tests that the best asphaltic concrete mix would be one containing 8.5 percent (in terms of weight of dried aggregates) of steam refined, paving grade asphalt and on-site aggregates graded as follows:

TABLE 7-3

Size	Percent Passing Square Mesh Screen (Washed Sample)	Size	Percent Passing Square Mesh Screen (Washed Samples)
1½ in.	100	No. 10	47.2
¾ in.	87.5	No. 40	30.8
½ in.	75.9	No. 80	20.6
No. 4	59.2	No. 200	11.8

The maximum size of rock in the fill was limited to 5 tons (4536 kg) and the amount of fines (4.0 in. (10.2 cm) and smaller) could not exceed 10 percent. The rock was dumped in lifts not to exceed 40 ft. (12.2 m) in height and sluiced by streams operating at nozzle pressure of 60 p.s.i. (414 kN/m^2). The volume of water was twice the volume of sluiced rock. A layer of clean rock, completely free of all rock of 3 in. (7.62 cm) and smaller size, 10 ft (3.05 m) in horizontal width was placed on the upstream face of the rockfill and provided at the base with drain pipes leading to the downstream toe. The purpose of this drain zone was to insure that no water would be trapped in the rockfill which on reservoir drawdown could exert back pressure on the asphaltic concrete face. This zone was covered by a compacted layer, averaging 6 in. (15.24 cm) thick, of small stones (¾ - 3 in.) to form a smooth consolidated surface so as to avoid local settlements of the facing and provide a working surface for the paving equipment. The hot asphaltic concrete was placed by paving machine in three layers to a total thickness of 10.5 in. (26.7 cm). The cutoff wall and grout curtain were located slightly upstream of the theoretical heel of the dam and the asphaltic concrete facing was extended over and sealed to the top of the cutoff wall.

Completed in 1966 was a major rockfill dam with an asphaltic concrete facing or deck known as Homestake Dam.[19] This structure is a compacted (rather than dumped) rockfill dam, located in the high Colorado Mountains above Elevation 10,000 ft. The dam is curved in plan and consists of a zoned, rolled rockfill embankment having a maximum height of 250 ft (76.25 m), a crest length of 1996 ft (608.8 m) and a crest width of 20 ft (6.1 m). The dam is shown in section in Fig. 7-5. The upstream slope is 1.6 H to 1.0 V, and the downstream slope is 1.4 H to 1.0 V except for the top 30 ft (9.15 m) where the slope is 1.3 H to 1.0 V.

The mass of the rockfill (Zone 1) was specified to be clean, durable, selected quarry run granitic rock. The material is well graded from 36 in. (91.4 cm) maximum size of rock to 10

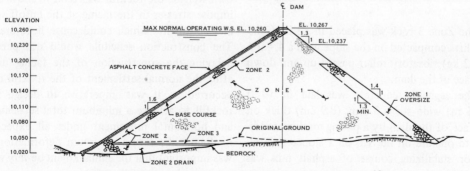

Figure 7-5. Homestake Dam. (After Burke, Forrest and Perley[19])

percent passing a No. 4 sieve. The rock was placed in 36-in. lifts and compacted with 4 passes of a 10 ton (9072 kg) vibratory roller. Oversize rock was pushed to the downstream slope into Zone 1 (Oversize), and no compaction was required. A narrow zone of smaller rock (Zone 2) was provided along the upstream face. Zone 2 rock gradation is:

TABLE 7-4

Size	Grading Percent Passing
18 in. (45.72 cm)	100
12 in. (30.48 cm)	70–100
6 in. (15.24 cm)	30–70
2 in. (5.08 cm)	0–5

This material was placed in 18 in. lifts and compacted with four passes of a 10 ton vibratory roller.

Zone 3 material was used as a base course under the asphaltic concrete facing and for a filter course above a rock drain which followed the low contours to the downstream toe. The gradation of the Zone 3 material is:

TABLE 7-5

Size	Grading Percent Passing
3 in. (7.62 cm)	100
2 in. (5.08 cm)	50–58
1 in. (2.54 cm)	20–50
$\frac{3}{8}$ in. (0.95 cm)	5–15
No. 4 (4.76 mm)	0–5

The Zone 3 rock was placed in 12 in. (30.5 cm) lifts compacted on the slope with a $4\frac{1}{2}$ ton (4082 kg) vibratory roller passed up and down the face of the dam.

The asphaltic concrete was laid in 10 ft (3.05 m) wide strips $3\frac{1}{2}$ in. (8.9 cm) thick on the face of the dam by a paving machine modified to operate on the 1.6 on 1 slope. A leveling or stabilizing course of asphalt mix was spread by a separate laydown machine prior to placing the asphaltic concrete. Compaction of each layer was obtained by four passes of a $4\frac{1}{2}$-ton (4082 kg) vibratory roller. The total thickness of the facing varied from 14 in. (35.6 cm) at the bottom to 7 in. (17.8 cm) at the top of the dam, as shown on Fig. 7-6.

In order to design an asphaltic concrete facing which would be dense, well graded and stable under the conditions of placement and use and yet be flexible and impervious with adequate strength to withstand water loads and wave action, extensive testing and mix design was necessary. These characteristics were achieved by using 8 percent (by weight of dry aggregate) of 50-60 penetration asphalt together with an aggregate gradation having a higher than normal percentage of fines. The gradation is:

TABLE 7-6

Size	Grading Percent Passing
$1\frac{1}{2}$ in. (3.81 cm)	100
No. 4 (4.76 mm)	48–75
No. 200 (0.074 mm)	7–15

The Marshall Stability ratings for this mix design are in the 1100 to 1200 range, lower than those required for highway pavements.

Although the asphaltic concrete facing was designed to be as flexible as possible and still fulfill the other requirements, the possibility of cracking as the result of differential movements of the rockfill was the subject of much investigation. Due to the water load and its own weight, the rockfill will tend to move and impose stresses in the facing at the cutoff wall and abutments which could cause it to crack. The construction schedule would not permit delaying the construction of the facing until after the normal settlement of the rockfill had occurred so it was imperative to design the rockfill to achieve a minimum total settlement and horizontal movement under all expected loadings. To insure minimum deformation it was important that the embankment be very well compacted, especially the upstream portion underlying the facing. A study was necessary

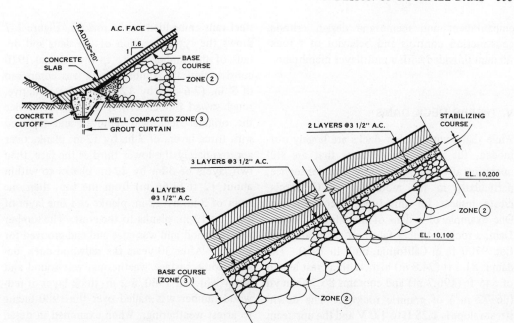

Figure 7-6. Homestake Dam—details of asphaltic face and upstream cutoff wall.[19])

to determine the modulus of deformation of the rockfill as related to density, confining pressure and loading, to establish the compaction requirements of the rockfill which must be met to avoid overstrain of the asphaltic concrete. The afore-mentioned compaction requirements resulted from this study but other measures also were taken to minimize differential settlements as follows:

1. The stepped contour of the foundation rock was smoothed to avoid abrupt changes in elevation of the cutoff wall.
2. A fillet of concrete was provided downstream of the cutoff wall to effect a gradual transition to the compacted rockfill. In addition, as shown in Fig. 7-6, a hinged concrete slab, tied to the cutoff wall, was provided to bridge a short distance from the cutoff.
3. Extreme care was taken to insure adequate compaction of the rockfill immediately adjacent to the junction of the facing and cutoff wall. The cutoff wall (Fig. 7-6) also served as a grout cap for a single row of grout holes on 15-ft (4.6 m) centers.

The holes were 90 ft (27.5 m) to 120 ft (36.6 m) deep and were slanted at various angles to intercept as many seams as possible.

Despite the precautions taken to avoid cracks, some minor cracking in the asphaltic concrete occurred during construction and was repaired. In 1969, about 13 cfs (0.37 m³/sec) of seepage past the dam was observed but was reported to be through fissures in the foundation rock, possibly opened by quarry blasting, but the asphaltic concrete facing was not involved.

Asphaltic concrete membranes for rockfill dams have been widely used elsewhere in recent years. In a number of cases these membranes are multilayered, and include a pervious drainage layer within the membrane. Reference 20 describes the design and performance of such membranes.

Reference 21 gives a theoretical and empirical approach to the design of multilayer membranes and presents research and test data on the materials and procedures for practical applications of the data. Reference 23 describes the

V. TIMBER DECK DAMS

While dams with timber decks are largely outmoded, the many timber dams that are still operating after 60-70 or more years of service, particularly in the mountains of California, create respect for this type of construction. One example of a dam of this kind is Hillside Dam, a rockfill-timber faced structure at Elevation 9700 ft in California, built in 1910. The dam is 81.5 ft (24.8 m) high, with a crest length of 645 ft (196.7 m) and contains 86,500 cu yd (66172 m^3) of granitic rockfill. The downstream slope is 1.25 H to 1.0 V and the upstream slope is 0.75 H to 1.0 V. The timber deck is spiked to stringers anchored by eye-bolts to steel rails embedded in the rockfill. Figure 7-7 shows the typical section of the dam and details of the timber deck. In the original 1910 construction the timber facing was composed of 3 in. (7.6 cm) by 12 in. (30.5 cm) native, rough-sawed timber. After 20 years of service the original deck was removed and replaced with three layers of 3-in. by 12-in. planks over approximately the lower third of the face, then two layers of 3-in. by 12-in. planks to within about 12 ft (3.66 m) from the top, then one layer of 3-in. by 12-in. planks and one layer of 2-in. by 12 in. planks to the crest. The lumber was redwood and was edge and end-grooved for splines. After 30 years the redwood deck, despite some surface weathering, was sound and watertight. In 1960, a 2 in. thick layer of redwood lumber was nailed over the 1930 facing to arrest weathering. When examined in detail by the author in 1971, the facing was found to be in excellent condition.[3,24]

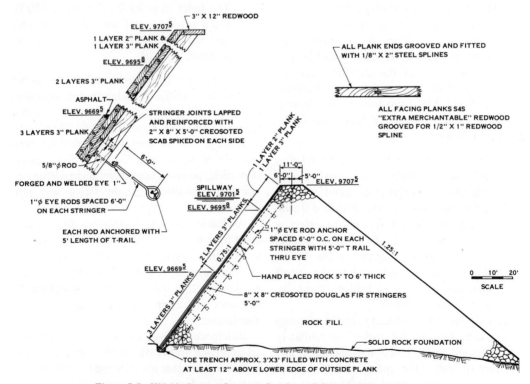

Figure 7-7. Hillside Dam. (*Courtesy Southern California Edison Company*)

VI. STEEL DECK DAMS

A few steel dams and steel-faced dams have been built in the United States and attest to the long life of the material in this service. For example, Ash Fork Dam in Arizona, a steel structure 184 ft (56.12 m) long and 46 ft (14.03 m) high, was built in 1898 and is still in service (1973). The owner reports that painting at intervals of about 8 years has been the only maintenance required.[24]

Skaguay Dam built in 1901 on Beaver Creek near Victor, Colorado, is a 70 ft (21.4 m) high rockfill dam with a steel facing on about a 60° slope.[3] The steel plate is $\frac{1}{2}$ in. (1.27 cm) thick at the bottom of the dam, $\frac{3}{8}$ in. (0.95 cm) at the middle of the face, and $\frac{1}{4}$ in. at the top, and was riveted and caulked.

In Ref. 25, it was reported in 1959:

"In the past 30 years the steel face has been given two complete treatments of cleaning, chipping, buffing, and painting. The last of these was in 1940."

In 1959 the owner further reported that through the years the steel face had become pitted but not to an alarming extent.

El Vado Dam near Santa Fe, New Mexico, constructed by the Bureau of Reclamation in 1935, is a compacted gravel fill embankment with a $\frac{1}{4}$ in. (0.64 cm) thick steel membrane on the upstream face. This dam and two other steel-faced earthfill dams in the United States are discussed in Ref. 15.

A number of notable steel-faced rockfill dams have been constructed in other countries; i.e., France, Venezuela, New Guinea, and Portugal.[15] The design and construction of Salazar Dam in Portugal is described in Ref. 26. The 184-ft (56.1 m) high Salazar Dam has a crest length of 630 ft (192.1 m), a crest width of 16.4 ft (5.0 m), and slopes of 1.25 H to 1.0 V on the upstream face and 1.4 H to 1.0 V between each of three berms, 6.6 ft (2.01 m), 8.2 ft (2.50 m) and 9.8 ft (3.0 m) wide on the downstream face.

The rockfill was dumped and partly hand-placed in the central embankment in layers approximately 9.2 ft (2.8 m) in depth. The upstream face zone of the embankment is composed of cement mortar masonry 6.6 ft (2.0 m) thick at the bottom of the dam and 2.6 ft (0.8 m) thick at the top. A zone of large rock underlies the masonry and a zone of similar material also forms the downstream toe fill. The downstream face is covered with dry masonry. The surfacing of the upstream face was a concrete layer with an average thickness of 8 in. (20.3 cm) on which the steel facing sheets were laid. The steel sheets are $\frac{5}{16}$ in. (0.8 cm) thick up to one-third of the dam height and $\frac{1}{4}$ in. (0.64 cm) thick from there to the top. Structural steel members, anchor-bolted to the cutoff wall, are used to secure and seal the lower edges of the steel facing. Expansion joints, reinforcing bars, and intermediate anchor bolts keep the plates from buckling and rupturing. The cutoff wall under the upstream part of the embankment, between the steel facing and foundation, contains an inspection gallery. A drainage galley, embedded in a drain formed by small stones, was constructed in the deepest zone of the foundation and ended in a downstream pumping well.

The foundation, composed of silicious and argillaceous shale has a low strength and, moreover, has a large fault 12 ft (3.66 m) thick in the river bed, passing transversely to the axis of the dam. These conditions called for a dam with a wide base and one which could undergo substantial settlement and distortion without serious damage. The steel-faced rockfill structure met these qualifications in a completely satisfactory manner.

The Aguada Blanca rockfill dam completed in 1970 in Peru is 148 ft (45 m) high, and has a metal facing. The design and construction are described in Ref. 145.

In view of the excellent performance of the relatively few steel-faced dams which have been built, it would appear that this type of structure has not received the consideration it merits. The steel membrane is quite flexible, distributes the water pressure to the rockfill uniformly, and is, of course, completely impermeable and wave resistant. If excessive distortions of the rockfill do occur and cause rupture of the facing, it is quite easily repaired, and if such repairs are necessary, they probably can be made at the convenience of the owner or operator of the dam because the escape of water through fractured welds or ripped plates is not likely to greatly increase with time due to erosion of rockfill embankment materials or progressive enlargement of the openings in the plates.

The principal causes of the lack of acceptance of steel facings on rockfill dams are probably:

1. Concern as to possibility of excessive damage due to corrosion,
2. Initial cost, and
3. The necessity that provision be made for occasional emptying of the reservoir so that maintenance inspection, repairs, and repainting of the steel facing can be performed.

As to Item (1), experience has shown that steel facings are remarkably resistant to damage from corrosion, particularly if cathodically protected. On none of the steel facings in service has this apparently been a serious problem. The initial cost (Item 2) is relatively high. When several of the above-mentioned dams were built (1940's-1950's) the costs were competitive with other facings. At 1974 prices, however, this would probably not be generally true. Item (3) above would relegate the steel-facing to special case situations where the reservoir would normally be drawn down annually, or where it could be periodically dewatered without substantial economic loss.

VII. ADVANTAGES AND DISADVANTAGES OF ROCKFILL DAMS WITH IMPERVIOUS FACINGS

Advantages

A dam with an impervious upstream facing makes more efficient use of material in resisting the thrust of the reservoir water that any other type of dam; i.e., it has the greatest possible mass of the embankment resisting the reservoir pressure.

An upstream faced dam develops a higher safety factor against shear failure per unit of material in the dam than any other type. Consequently, for a given safety factor the slopes can be steeper and the volume smaller than for other dams.

The thrust of the water acting on the inclined face is directed downward toward the foundation, creating an optimum condition with respect to foundation stability.

Inclined faced rockfill dams usually leak to some extent, but aside from the value of the water that may be lost, such leakage is not a matter of serious concern. Leaks through cracks, joints, or fissures in the facing would not reach sufficient volume of flow to cause significant seepage forces within the mass of the rockfill. Leps, in a comprehensive study of flow through rockfill[28] developed procedures for analyzing the stability of such rock masses against deep shear failure, and ravelling and progressive slide failures of the downstream slope during throughflow. The analyses applied to rockfill embankments without upstream facings. It is obvious from these studies that the amount of leakage which could conceivably escape through cracks and fissures in a reasonably intact facing could not jeopardize the stability of a rockfill dam.

Impervious faced rockfill dams are ideally suited to locations where the foundations and abutments consist of hard competent rock, and there is an adequate supply of coarse granular material which may be utilized as borrow for the rockfill. In this case, the base shear at the foundation contact would probably not govern the maximum allowable angle of the embankment slopes. Steep slopes and a minimum required volume of fill may permit economies in construction cost not attainable with other types of dams. Also, the facing protects the embankment from wave action and makes it unnecessary to provide other and often expensive slope protection. Because of the short base width of faced rockfill dams, the length and cost of conduits through the dam will be less than for flatter-sloped embankments.

The construction schedule may be shortened because the rockfill may be placed in advance of other work, such as foundation cutoff grouting along the upstream toe, and rockfill placement can continue during wet weather which would preclude placement of earthfill.

The impervious facing is accessable for repair, if needed. By contrast, if a leak developed in a core-type dam, the repair could be difficult and expensive.

Disadvantages

The position of the upstream impervious facing makes it vulnerable to damage caused by the vertical, horizontal, and lateral movements of the rockfill, particularly following the first filling of the reservoir. These movements cause spalling and cracking of the slabs and ripping of water stops of concrete facings, tears in sheets and fractures of welds in steel facings, and formation of fissures in asphaltic concrete facings. Spreading at the base of some dams is reported to have caused damage to their impervious facings. Settlement of rockfills often continues at a decreasing rate for many years and, consequently, damage to the facing resulting from that settlement may also continue progressively for a long time. Permanent repair of the facing might require the reservoir to be drawn down and taken out of service while the repairs are being effected.

The rate of deterioration of facing membranes depends on many factors, such as the kind of aggregates and mix design for portland cement and asphaltic concrete facings; the corrosion resistance of the steel in the particular environment of a steel facing; weather conditions; and the strains to which the facing may be subjected after installation. The service life of the membrane is, therefore, indeterminant. The facing could require extensive repairs or replacement in a relatively short time; i.e., 20 or 30 years, or it could last throughout a long lifetime of service of the dam as a whole.

With the possible exception of timber facings, the impervious membranes are constructed of manufactured materials which must be imported to the site. The base cost of the materials plus the cost of transportation and on-site handling may make the overall cost of the impervious-faced dam so high that it would not be competitive with a structure of greater volume constructed entirely of locally available materials.

VIII. EARTH CORE ROCKFILL DAMS

The use of compacted earth for the impervious element in high rockfill dams, as commonly practiced at present, received great impetus in the 1940s with the construction of the sloping core Nantahala Dam by the Aluminum Company of America, the 400 ft (122 m) high central core Mud Mountain Dam by the Corps of Engineers, U.S. Army, and the thick central core dams constructed by the Tennessee Valley Authority during this period. However, the combination of earth and rock in the building of low dams is as old as the art itself. Many earth and rock dams built in the United States from the middle to late 1800s are still in service (1973). Some of these are composite dams consisting of an upstream section comprising nearly one half of the dam, and a downstream section of rockfill with a zone of gravel and sand between the two parts. In a few instances, the importance of the protective filter zone was not recognized by early builders and the dams failed due to the piping of fine earth materials into the voids of the rockfill. Although rockfill forms a substantial part of composite dams, they are not generally classed as rockfill dams.

In earth core rockfill dams the impervious member is a relatively thin or narrow zone of compacted earth supported by rockfill, with filter zones of fine rock, gravel and sand on both sides of the core. Earth core dams are broadly classified as central core or sloping core dams, but the location of the core can vary from the central, vertical position to one paralleling the upstream slope of the dam section.

The central core rockfill dam was probably a logical development from earth dam practice, but the extremely sloping type of impervious earth core was a successful improvement designed to overcome some of the drawbacks to the uses of manufactured facings for rockfill dams.

Rockfill Dams with Sloping Earth Cores

The first extremely sloping core rockfill dam was Nantahala Dam, designed and constructed in 1941, under the direction of Mr. J. P. Growden for the Aluminum Company of America. The section of the dam[29] is shown in Fig. 7-8. The main rockfill is composed of quarried rock varying in size from 10 tons (9.07 mt) to dust. Selective dumping permitted placing most of the large rocks in the downstream section of the fill and most of the small rock in the upstream section so that the voids in the rockfill would become progressively larger from the upstream face to the downstream face. The rock was end-dumped from a series of high lifts averaging from 50 to 130 ft (15.3 m to 39.6 m) in height. The surface of each lift was thoroughly scarified and sluiced to remove fines. All the quarry rock dumped on the fill was sluiced with water, using monitors operating under 275 ft (83.9 m) of head and delivering a volume of water in excess of 4 times the volume of dumped rock.[29] Both the upstream and downstream slopes are at the natural angle of repose of the dumped rock.

The first filter layer adjacent to the rockfill is 14.75 ft (4.5 m) thick normal to the slope and is composed of selected rock ranging in size from 10 in. to 3 in. (25.4 cm to 7.6 cm). The second filter layer is 6.5 ft (2.0 m) thick and is composed of crushed rock not larger than 3 in. (7.6 cm) and not smaller than $\frac{1}{2}$ in. (1.3 cm). The third filter layer, adjacent to the core, is composed of crushed rock fines smaller than $\frac{1}{2}$ in. (1.3 cm). The sizes specified and the grading of the filter materials prevent the migration of finer materials into the voids of the larger rockfill. The thicknesses of the layers were designed so the materials could be placed without difficulty and the layers could adjust to the settlement of the rockfill without danger of rupture.

The availability of impervious core material was limited so the thickness of the core was established as the minimum to prevent excessive seepage. Laboratory tests determined that an impervious rolled fill, 29 ft (8.84 m) thick at the bottom where the head was 230 ft (70.2 m) and 13 ft (4.0 m) thick at the top would have a seepage rate of 2.75 cu ft (0.078 m^3)/min, which was considered satisfactory.

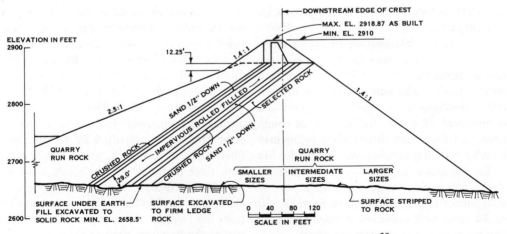

Figure 7-8. Nantahala Sloping Core Dam. (After Growdon[29])

The average gradation of the soil was:

TABLE 7-7

Size mm	Percent Passing
0.001	10
0.006	20
0.02	29
0.06	40
0.20	80
0.60	95
2.0	99
6.0	100

The average maximum dry density was 101 lb/cu ft (1617.8 kg/m^3). The angle of internal friction was 32° and cohesion was 0.1 ton/sq ft (0.098 kg/cm^2).

The material was compacted somewhat wetter than the optimum moisture content, to a dry density of 90 lb/cu ft (1441.6 kg/m^3). This resulted in a flexible core which could readily adjust to the settlement of the rockfill. At the base of the core the soil was compacted against clean, sound foundation rock to form a tight seal.

Above the core a 5 ft (1.53 m) layer of crushed rock fines and a second 5-ft thick layer of $\frac{1}{2}$ in. to 3-in. (1.27 to 7.62 cm) size crushed rock provides a reverse filter, over which was placed a layer of quarry run rock to armor the fill against wave action and stabilize it during sudden drawdown.

During the first 9 years after closure, the maximum vertical settlement of the crest was 2.16 ft (0.66 m) and the horizontal downstream movement was 1.01 ft (0.31 m). The flexible core apparently adjusted well to these movements, since the leakage was quite small and in the order of 2.75 cu ft/min (1298 cm^3/sec).

During the 15 years after the completion of Nantahala Dam, the Aluminum Company of America built 6 other sloping core dams similar in design to Nantahala Dam, but with some modifications of construction techniques. Discussions of these dams are given in Refs. (30) and (31), together with descriptions of several contemporary sloping core dams by discussers of the papers cited in the references. The performance of Kenney Dam in British Columbia, Canada, the largest of the sloping core rockfill dams at the time of its construction in 1952, is described in Ref. 10. Several other notable sloping core rockfill dams are the 200 ft (61 m) high Bersimis Dam and the 225 ft (68.6 m) high Desroches Dam in Canada, completed in 1955[39], and the 400 ft (122 m) high Brownlee Dam near Weiser, Idaho, completed in 1958.[40] Brownlee Dam is a particularly interesting structure because of the inclusion of large quantities of compacted small rock material in the dam, construction of the embankment on 110 ft (33.5 m) of river-deposited materials, and the diversion of flood waters over the partially completed embankment. The 394 ft (120 m) high Mont Cenis sloping core dam in France, constructed mostly with compacted rockfill, during 1963–68 is described in Ref. (32).

Rockfill Dams with Moderately Sloping Earth Cores

Moderately sloping earth core rockfill dams have gained considerable favor in recent years. There are a number of reasons for this, but in general, as will be discussed later, the moderately sloping core tends to eliminate some of the objectionable features of extremely sloping cores, yet retains some of their advantages.

Furnas Rockfill Dam, built in Brazil during 1960–63, is a 417 ft (127 m) high compacted rockfill structure with a narrow earth core, the upstream face of which slopes 0.70 H to 1.0 V.[33] There are transition zones upstream and downstream of the core, and random fill zones between the transition and the rockfill shells. The upstream slope is 2.0 H to 1.0 V and the downstream slope averages 1.8 H to 1.0 V.

The rockfill for the shells is principally quartzite, and due to the brittleness of the material, the fill contained a considerable amount of fines (smaller than 1.0 in.) averaging 35 percent of the total weight, with only 15 percent coarser than 6 in. (15.2 cm) in diameter. The rock was spread in 24 in. (60 cm) to 32 in. (80 cm) thick layers and compacted by two coverages of a D-8 tractor, without sluicing. Field densities average 1.56 t/yd^3 (1.85 t/m^3). The random zones, consisting of materials ranging from coarse rock to fines, were compacted in layers 19.7 in. (50 cm) to 11.8 in. (30 cm) thick, to a density of 1.67 t/yd^3 (1.98 t/m^3). The transition materials were spread in 11.8-in. (30 cm) thick layers and compacted by 2 tractor coverages. The core is sandy to silty clay of medium to high plasticity, compacted by rubber-tired rollers to an average density of 99.4 to 101.6 percent of the laboratory maximum density. Although the rock was not ideal for rockfill, the material was successfully utilized in the dam through carefully controlled compaction.

Two other examples of moderately sloping core dams are: Holjes Dam in Sweden (1961), a 265.6 ft (81 m) high structure with dumped and sluiced gravel shells[11], and Miboro Dam in Japan (1960), a 429.5 ft (131 m) high dam with dumped and sluiced granite rockfill shells.[34] The impervious core of Miboro Dam is composed of a mixture of clayey impervious material and disintegrated granite proportioned by mixing at the site.

Cougar Dam, a 445 ft (135.7 m) high moderately sloping core rockfill dam, located near Eugene, Oregon, was built by the Corps of Engineers, U.S. Army, during 1959-63.

The design of a rockfill embankment for Cougar Dam involved a choice between a central-core and a sloping-core section. The initial decision was for a sloping core section but this was later revised and a central core section was adopted. The principal factor in making the final decision was the uncertainty as to the amount of water-load settlement that could be effected in a rockfill embankment as high as the proposed dam and the susceptibility of a sloping core of compacted earth to rupture from the type of settlement that would occur.[44]

The original design of the Cougar Dam was essentially a sluiced rockfill with a zone of compacted rock on the upstream face to support a sloping compacted earth core which was protected and restrained by an upstream dumped rock shell. Since the dam would be considerably higher than any other sloping-core rockfill dam previously constructed, concern over the amount and effect of settlement—both laterally and normal to the slope—led to additional study and finally to a revised design.[47]

The final design was a compacted rockfill embankment with a narrow, centrally located moderately sloping compacted earth core. The base width of the core is equal to about 0.3 of the water head. The design called for 6 types or classes of rock, different in gradation or compaction requirements. In the early stages of construction the rockfill and spall zones immediately downstream of the core were placed in 18 in. and 12 in. (46 cm and 31 cm) lifts, respectively, and compacted with 4 coverages of a 50 ton (45,360 kg) 4-wheel pneumatic tire roller. Later, the 18 in. (46 cm) layers were compacted with a 5 ton (4536 kg) smooth-wheel vibrating roller. Still later, as a result of experimentation on test fills within the embankment, the lifts heights of the rockfill and spall zones were increased to 24-in. (61 cm) and compaction was accomplished by a 10 ton (9072 kg) vibrating roller. Coarser rockfill, with a maximum size of 24 in. (61 cm) was placed in 36 in. (91.4 cm) lifts and compacted with 2 coverages of a 60,000 lb (27.21 mt) tractor. Reference 47 gives an interesting account of a number of problems in the construction of Cougar Dam, and of construction methods and field tests.

An evaluation of post-construction performance is given in Ref. 64. As reported in 1967 the construction pore water pressures in the core had been generally 20 percent to 50 percent of the embankment pressures, and following the filling of the reservoir the fluctuations of the piezometric heads with reservoir operation were normal. The seepage through the dam, foundation and abutments ranged from about 0.5 cfs (0.014 m^3/sec) at low pool to about 3.0 cfs (0.085 m^3/sec) at full pool. Three years after the diversion tunnel was closed and impoundment began the maximum settlement of the crest was about 2.6 ft (0.8 m) at a point on the upstream shoulder and the maximum deflection was about 2.0 ft (0.6 m) on the downstream shoulder. During this period the crest spread laterally about 0.85 ft (0.26 m). When the reservoir first reached full pool elevation, about 10 months after closure, a transverse crack diagonally across the crest formed about 40 ft (12.2 m) from the left end of the dam, but no increase in seepage was detected. Several longitudinal cracks which were almost continuous for the full length of the high section of the dam, also formed at the interfaces between the core and gravel transition sections and between the transition and spalls section. These were caused by the differential settlement between the adjacent zones of materials in the dam. This type of cracking had been observed at other dams built prior to Cougar Dam (e.g. Mud Mountain Dam) and has occurred on still other later dams (see discussion of Round Butte Dam, following). Susceptibility to such cracking is one characteristic of high rockfill dams, and will be discussed in more detail in Section X.

The highest moderately sloping earth core rockfill dam in the United States is the 735-ft (225 m) high Oroville Dam in California which was completed in 1967. Since the shells are composed of coarse gravel and cobbles, some may contend that the dam does not properly belong in the rockfill category, but because such material must be handled and treated much as though it were quarried rock, it will be considered as rockfill for purposes of this discussion.

The source of the sand, gravel, and cobbles required for the more than 77,000,000 cu yd (59,000,000 m^3) of the dam embankment was the plentiful supply of dredge tailings resulting from the excavation, washing, separation, and redeposition of the original recent alluvium sediments by gold dredges which worked in the area from the 1880s to about the late 1930s. As the material was passed through the dredges and the gold extracted, the fines were separated from the gravel and were discharged at the rear of the dredges, with the gravels being deposited farther back and on top of the sand. A distinct boundary was thus formed between the gravel and cobbles used for the shells and the sands used for the transition zones.[35, 36]

Figure 7-9[35] shows the maximum section of Oroville Dam. Zone 1 is a relatively thin inclined core of clayey sand and gravel. The borrow areas contained about 33 percent to 56 percent of rock larger than No. 4 U.S. Standard Sieve size (4.76 mm), but by blending in the

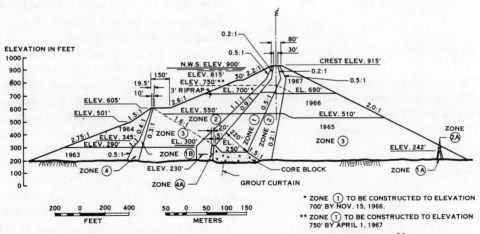

Figure 7-9. Oroville Dam—maximum section. (After Gordon and Wulff[35])

pit and removal of rock larger than 3.0 in. (76.2 mm) the following gradation for Zone 1 was produced:

TABLE 7-8

Size	Percent Passing
5 micron	5-20
No. 200 U.S. Std. Sieve (0.074 mm)	10-40
No. 4 U.S. Std. Sieve (4.76 mm)	40-70
3 in. (76.2 mm)	100

The impervious materials were placed in layers of 10-in. (0.25 m) compacted thickness, and compacted by four coverages (8 passes) of a towed 100-ton (91 mt) pneumatic tired roller. Field tests indicated that it was necessary to scarify between lifts to avoid horizontal laminations.

By mixing the coarse dredge tailings and underlying sands in proper proportions ideal transition materials for Zone 2 resulted, with high compacted density and shear strength and permeability between that of Zone 1 and Zone 3 materials.

The gradation of Zone 2 is:

TABLE 7-9

Size	Percent Passing
No. 200 U.S. Std. Sieve (0.074 mm)	0-5
No. 4 U.S. Std. Sieve (4.76 mm)	20-50
15 in. (380 mm)	100

The material was compacted in 15-in. (0.38 m) lifts by two coverages of an 8000-lb (3628 kg) towed smooth steel drum vibratory roller.

The gradation of the gravel and cobble shell (Zone 3) material is:

TABLE 7-10

Size	Percent Passing
No. 4 U.S. Std. Sieve (4.76 mm)	0-25
24-in. (610 mm)	100

The coarse material was readily compacted to the desired densities in 24-in. (0.61 m) layers with two coverages of the same vibratory roller used for the transition material. It was found by field trial that the application of water during compaction did not produce any significant improvement in final density, nor did sluicing alone produce densities equal to other methods of compaction. A 100-ton (91 mt) pneumatic tired roller produced densities nearly equal to those produced by the vibratory rollers, but the latter were preferred.

The dam was constructed in stages during 1962 through 1967 as indicated in Fig. 7-9. The mass concrete core block extending across the foundation provided[36]:

1. A retention block for the embankment placed upstream from the core block during the 1963 construction season so that the 1963-64 flood runoff would be passed without damage to the embankment placed,
2. A downstream toe for the embankment placed during the 1964 construction season, and
3. A solid foundation across the inner gorge for the impervious core.

The zoned section in the upstream shell was incorporated as a construction expediency in bringing the main embankment to Elevation 550 in 1965. The small impervious section near the downstream toe was installed to facilitate making measurements of seepage.

Extensive investigations were made to select the type of dam to be built at the Oroville site. "Early investigations[35] considered first a massive concrete dam section, which appeared warranted on the basis of preliminary foundation and materials exploration. Later foundation explorations, however, showed that for the height of dam considered, certain geologic complexities made a concrete dam less economical and structurally less desirable than an embankment type dam. Of the many potential designs considered (concrete gravity, multiple arch, and arch buttress; quarried rockfill with impervious core; and gravel fill with either vertical or inclined impervious core) the fill of dredger tailings with an inclined impervious core was the section finally selected. The section was chosen in favor of a slightly less expensive vertical core section because of the more favorable geometry provided for the settlement of the Zone 1 material with relation to Zones 2 and 3, which are both considerably less compressible than Zone 1 material under the high loads to be experienced within the embankment."

Moderately Sloping Vs. Extremely Sloping Earth Cores

Moderately sloping impervious cores are in general subjected to less distortion in a direction normal to the slope than extremely sloping cores as a result of settlement of the rockfill. Either type of sloping core minimizes the possibility of arching with the consequent hazard of horizontal cracking if the core material is more compressible than the transition and shell materials (see Section X). The optimum position of a sloping core with respect to the stability of the dam and hydraulic facturing of the core has been investigated and reported in Ref. (37 and 38), which also cite previous investigations in this regard. The results of elastic analyses based upon the finite element method indicate that the moderately inclined core has clear advantages with respect to arching (i.e., reduction of the maximum minor principal stress resulting from settlement of the core with respect to the shell) over a vertical symmetrical core. The optimum position of the core, in this respect, as defined by the inclination of the core-shell interface lies between 0.5 H to 1.0 V and 0.6 H to 1.0 V.[37]

If the shearing strength of the compacted core material is significantly less than that of the rockfill, as is often the case, then the upstream slope of the dam can be considerably affected by the position of the sloping core. If the slope is such that much of the potential slide surface would lie within the weaker zone of material then the upstream slope of the dam must be considerably flattened to have the same safety factor against shear failure than would be the case if the potential critical slip surface were substantially within the rockfill shell. The studies reported in Ref. (38) indicate that if the upstream inclination of the core is steeper than about 1.0 to 1.0 the core normally has no decreasing effect on the stability of the upstream part of the dam. In some cases the distribution of the vertical pressure on the foundation is a matter of some importance. For transmission of water and earth pressures from the core to the foundation, an inclination of the core of about 2.0 V to 1.0 H gives an even bottom pressure distribution below the fill downstream of the core.[38] Great differences in bottom pressure upstream and downstream of the core will arise for very flat cores and for vertical cores.

The use of very flat sloping cores will permit the placement of much or all the main rockfill prior to the placement of the core. If construction of the core can be delayed until the initial settlement of the rockfill has taken place, the post-construction distortion of the core will be greatly reduced. The ability to schedule construction of the main rockfill independently of the core construction may have some advantages in areas of heavy rainfall or where flood flows may have to be passed over the incomplete dam between construction seasons. In the latter case anchorage of the downstream facing rocks may be necessary (see Refs. 28 and 94 through 102).

Since the contact between the base of a flat sloping core and the foundation is located near the upstream toe of the dam, the contact pressure is less than that under a central or slightly sloping core. This could result in the flat sloping core contact being more susceptible to seepage and piping along the contact face than would be the case for cores near the center of the dam. The foundation grout curtain beneath the base of a flat sloping core must be installed with particular care. Should additional grouting be necessary to stop excessive leakage after the reservoir is filled, the location of such a grout curtain would make it difficult or impossible to perform the additional grouting without draining the reservoir, which in some cases would be infeasible. The contact under a slightly to moderately sloping core can usually, if necessary, be re-grouted from the top of the dam and without dewatering the reservoir (see also Refs. 73, 74 and 75).

Rockfill Dams with Central Earth Cores

The central earth core rockfill dam was probably the out-growth of practices developed for the design and construction of zoned earthfill dams. One of the most notable early high central core rockfill dams is Mud Mountain Dam, constructed by the Corps of Engineers, U.S. Army, on the White River about 50 mi southeast of Seattle. The dam was built during 1939 to 1941[41], to a height of 400 ft (122 m) with symmetrical upstream and downstream slopes of 1.75 H to 1.00 V for the top 150 ft (45.8 m) and 2.25 H to 1.00 V for the lower 250 ft (76.3 m). The central core is 30 ft (9.2 m) thick at the top and 150 ft (45.8 m) thick at the base. Transition zones on either side of the core are each 10 ft (3.05 m) thick at the top and 50 ft (15.25 m) thick at the base.

The shells are composed of andesite, dumped in 40-ft (12.2 m) lifts and sluiced with $1\frac{1}{2}$ cu yds (1.15m^3) of water per cu yd (0.765 m^3) of rock. The transitions are crushed fine-grained granite, compacted in 12-in. (30.5 cm) layers. The core is pit run sand and gravel under 6-in. (15.2 cm) size, mixed with 20 percent silty, glacial till-like material to reduce permeability. The core zone was constructed in 6-in. (15.2 cm) compacted layers.

Settlement of the crest of Mud Mountain Dam in the nine years following its completion measured 2.7 ft (0.82 m) at the maximum section, which was very close to the predicted settlement based on large scale consolidation tests. Horizontal movement during this period varied from zero to 0.90 ft (27.5 cm). In 1942, cracks appeared on the crest of the dam along the junction between the core and the upstream and downstream transition zones parallel to the axis. The cracks, which test pits revealed were about 6 ft (1.83 m) deep, were backfilled, but continued movement reopened the cracks to a width of about 4 in. (10.2 cm) some six years later. The shell zones had settled a total of about 18 in. (45.7 cm) below the core, but the differential movement was not considered to be serious.

During the years 1941 through 1950 the Tennessee Valley Authority constructed three central earth core rockfill dams, namely, Notteley Dam, 184 ft (56.1 m) high; Watauga Dam, 318 ft (97.0 m) high; and South Holston Dam, 285 ft (86.9 m) high.[42] These dams are distinguished by having very wide cores in proportion to their heights. This design resulted largely from there being an abundance of both earth and rockfill materials available at the sites, and to some degree as an expedient to facilitate construction. The slopes of the cores of the three dams varied between 1.2 H to 1.0 V and 0.75 H to 1.0 V. The average upstream and downstream slopes of all dams are 2.0 H to 1.0 V, except the upstream slope of South Holston Dam which varies from 2.2-2.1 H to 1.0 V because of the use of somewhat weaker sandstone in the rockfill. At all dams the face slopes were broken by berms spaced 30 ft (9.2 m) vertically, the actual slope between berms being 1.4 H to 1.0 V, the natural slope of the dumped rock. The rockfill was dumped from trucks on the 30-ft (9.2 m) deep lifts along the full length of the dams. All the rockfill was sluiced with high pressure streams at each point of dumping.

Filter zones were used between the core and rockfill on both slopes. Two zones were placed on all dams on the upstream side and two similar zones were placed on the downstream side at Watauga and South Holston but at Notteley there is only a single 3 ft (0.92 m) thick zone.

After 10 years the crest of Notteley Dam had settled 1.1 ft (0.34 m) or 0.6 percent of the height, in 8 years Watauga Dam settled 2.6 ft (0.79 m) or 0.8 percent of the height, and in $6\frac{1}{2}$ years South Holston Dam settled 2.85 ft (0.87 m) or 1.0 percent of the height. Differential settlement of the core and rockfill shells resulted in considerable cracking along the shoulders of the core at all three dams. The cracks were similar to those reported at Mud Mountain, but of less width, and were adjudged not to be detrimental to the stability of the structures.

Cherry Valley Dam, built in California during the years 1953 through 1955, is a central core rockfill structure, 315 ft (96.1 m) high and very similar in cross section to the wide core dams constructed by TVA.[43] The core slopes are 0.70 H to 1.0 V upstream and 0.75 H to 1.0 V downstream. The face slopes of the dam average 2.0 H to 1.0 V upstream and 1.33 H to 1.0 V downstream, but each slope is stepped by a series of berms 20 ft (6.1 m) wide and 30 ft (9.2 m) in vertical height. The transition or filter zones between the core and rockfill consist of a fine and a coarse layer totaling 20 ft (6.1 m) in thickness.

The core is composed of disintegrated granite, classified as sandy silt to silty sand, and was compacted in 9 in. (22.8 cm) layers with 12 passes of a 35 ton (31.8 mt) sheepsfoot roller. Quarried granitic rock was used for the shells, and it was placed by end dumping, generally in 30 ft (9.2 m) high lifts, and sluiced with high pressure monitors.

During the first two years of reservoir operation the maximum settlement of the crest, i.e., the core settlement, was 0.45 ft (14.9 cm). Settlement points were not established on the rockfill slope but the settlement of the central portion of the berm at about mid-height of the upstream slope was approximately 2.0 ft (0.61 m). Differential settlement between the core and shell caused the formation of cracks at the core-transition interface, similar to, but less extensive than those observed at Mud Mountain Dam and the TVA dams (see also Section X). On the first filling of the reservoir the maximum downstream deflection of the crest was 0.14 ft (4.27 cm) but on drawdown the rebound was 72 percent of the original movement. On the second filling the movement was 0.465 ft (14.18 cm), but the rebound on drawdown was 43 percent. At other locations on the dam the initial movement was downstream, but the points rebounded to a position upstream from their initial positions with an empty reservoir.

During the period 1950-1960 a number of notable central core rockfill dams were constructed or construction begun in countries other than the United States. Goschenen Dam in Switzerland[48] is a thin central core rockfill dam 510 ft (155 m) high, completed in 1960. The rockfill is unprocessed talus material containing a high percentage of fines. It was dumped wet in layers 6.56 ft (2.0 m) thick and not exceeding 9.84 ft (3.0 m) thick, with no sluicing and no compaction other than by hauling and spreading equipment. The core was designed as a thin member because it was necessary to manufacture the impervious material from local sand and gravel with the addition of imported clay amounting to 875 lb/cu yd (519 kg/m^3).

Ambuklao Dam in the Philippines was completed in 1955. It is a vertical thin core rockfill dam, 430 ft (131 m) high[49]. The material available from the quarry for the rockfill shells ranged from sand size to large boulders with approximately 60 percent smaller than coarse gravel, and was, therefore, not suitable for dumped rockfill construction. The finer rockfill was placed in a zone in the upstream shell adjacent to the core. The rockfill was placed in 24 in. (61 cm) layers and compacted by 50 ton (45.36 mt) rubber-tired rollers and sluicing. The sluicing operations were effective in creating a compact mass. A two-stage filter was provided at the upstream face of the core and a three-stage filter downstream. The core material was placed 5 to 10 percent above optimum moisture because of the long rainy season and high humidity. The impervious core and filter materials were placed in 18 in. (45.7 cm) layers and compacted with 4 passes of a 50 ton (45.36 mt) roller.

Derbendi Khan Dam in Iraq is a 410 ft (125 m) high central core rockfill dam, completed in 1961.[50] The core has side slopes of 0.3 H to 1.0 V, and the face slopes of the dam vary upstream, 2.25-1.75 H to 1.0 V and 2.0-1.75 H to 1.0 V downstream, but are formed in 10-m benches to facilitate construction of the dumped rockfill shells. The lifts for end-dumping the shell material were specified as 30 m (100 ft) maximum and 10 m (33 ft) minimum. The rock was sluiced, with the ratio of the volume of sluicing water to the volume of rockfill specified as 3 to 1. The core was specified to be placed in 10-cm lifts and compacted with 4 passes of a rubber-tired roller, or in 15-cm lifts and compacted with four passes of a sheepsfoot roller. A two-layer transition or filter was placed upstream of the core and a three-layer filter downstream.

Makio Dam in Japan, completed in 1961, is 348 ft (106 m) high with a thin central core, composed of volcanic ash and breccia, having slopes of 0.1 H to 1.0 V.[51] Thick transitions of alluvium have slopes of 1.5 H to 1.0 V. The shells are sandstone placed with no specific requirements for lift thickness or compaction in the upstream shell, but the downstream shell was to be placed in lifts approximately 2 m thick, dumped and bulldozed into place and compacted by hauling and spreading equipment without sluicing.

Among other notable central core rockfill dams built during the 1950-1960 period were Messaure Dam in Sweden, 330 ft (101 m) high[48] and Kajakai Dam in Afghanistan, 328 ft (110 m) high.[53,54] More recently (since about 1960) construction has been initiated or completed on a number of other high central core dams outside the United States, a few of which are noted in the following paragraphs.

Nurek Dam in Russia, when completed, will be the highest rockfill dam in the world.[55,56] This central core dam will be 985 ft (300 m) high, with external slopes of 2.25 H to 1.0 V upstream and 2.22 H to 1.0 V at the downstream face. The core, constructed of rocky clay with side slopes of 0.25 H to 1.0 V is founded on a concrete block under the highest part of the dam. The foundation block houses the galleries used for surface and deep grouting. The shells are mainly alluvial gravel and coarse gravel with thick surface blankets of oversize rock. A single stage filter is provided at the upstream face of the core, and a two-stage filter is provided between the core and the downstream shell. The dam is constructed in a region where seismic shocks might be as high as 8-9 magnitude. Extensive investigations were conducted with respect to the effects on settlement and loadings.

Gepatsch Dam in Austria, completed in 1964, is 503 ft (153 m) high and has a fill volume of 7.1 million m^3.[57,58] The rock fill shells were compacted in 2-m lifts with 4 passes of an 8.5-ton vibratory roller, and have face slopes of 1.5 H to 1.0 V. The very thin impervious core, with slopes of 0.125 H to 1.0 V, is divided into two zones, one of boulder clay and talus, and the other of similar material with the addition of one percent bentonite. Sand and gravel transition zones are provided between each face of the core and the shell. Reference 57 contains interesting descriptions of the placement of the dam fill, the control testing during construction, and the instrumentation installed in the dam.

El Infiernillo Dam in Mexico is 485 ft (148 m) high with a thin, vertical, central core and two-stage filter and transitions between each face of the core and the rockfill shells.[59] The relatively thin core, which has a ratio of water head to thickness of about 5, was adopted because of the shortage of suitable well graded plastic soils in the vicinity of the dam and the necessity for hauling such materials a distance of about 18 kilometers. The core was compacted in 6 in. (15 cm) lifts by a specially designed sheepsfoot roller at a moisture content between 2 and 6 percent above optimum. The sand filter zones were placed in 12 in. (30 cm) lifts, sprayed with water and compacted with 4 passes of a 3-ton vibratory roller. The transition zones were similarly compacted in 8 in. (20 cm) lifts. The rockfill shells were divided in two sections, both upstream and downstream from the core. The inner sections were of 18 in. (45 cm) maximum size rock compacted in 24 in. (60 cm) thick lifts with 4 passes of a heavy crawler tractor. The outer rockfill sections were composed of rock particles larger than 18 in. (45 cm) placed in layers 6.5 ft (2 m) thick.

Special emphasis was placed on materials and control tests. Two large triaxial test units were built for testing samples 90 cu ft (2.5 m^3) in volume at confining pressures of 14.22 psi (1.0 kg/cm^2) and 355.6 psi (25 kg/cm^2) and axial loads to 1763 t (1600 mt). The instrumentation installed in the dam consisted of surface reference monuments, crossarm settlement devices, inclinometers, piezometers, radial strain meters, and strong motion accelerometers. Performance records of El Infiernillo Dam are cited in Refs. 55 and 65.

In the United States a number of major central core rockfill dams have been constructed since 1960. Several of these more recent dams are described briefly in the following paragraphs.

Summersville Dam on the Gauley River in West Virginia is 390 ft (119 m) high and is one of the first high rockfill dams to be constructed of sandstone rock placed in thin lifts and compacted by rolling.[60] The dam is a multi-zoned structure, and in order to make the best use of available materials, a total of 9 zones was specified in the main section as designed. The central core is asymmetric and consists of an upstream zone of clay and a downstream zone of clay and sand. A transition zone of 3 in. (7.6 cm) maximum size sandstone spalls was specified adjacent to the upstream face of the core, and two similar zones of weathered and unweathered sandstone spalls at the downstream core face. Both the upstream and downstream shells were to consist of three zones of rolled sandstone rock with size limitations of 9 in. (22.9 cm), 18 in. (45.7 cm), 24 in. (61 cm), and an outer zone of oversize rock.

Although the dam had not been completed at the time the report on the design was prepared[60], the experience that already had been gained indicated that:

(a) "Sandstones and shales can be compacted to form a dense fill, meeting all design requirements for stability and imperviousness;

(b) "Substantial savings can be effected by making provisions for utilization of quantities of lower grades of rock, particularly when such rocks are readily available or form part of the necessary excavation for appurtenant structures;

(c) "Placement of rock embankment material in thin lifts is an economical and effective process for construction of high rockfill dams."

Carters Dam is located on the Coosawattee River in northwest Georgia. Construction was started on the 400 ft (122 m) high zoned rockfill dam in 1964.[61] The site materials were generally massive quartzite, argillite, and phyllite, with an overburden mantle of lean clay or silt and clay, overlying a zone of highly weathered and disintegrated rock. The preliminary design section of the dam consisted of an upstream good-quality rock shell with a face slope of 2.0 H to 1.0 V, separated from a transition zone by a coarse filter sloping 0.8 H to 1.0 V. In the center of the dam, downstream from the single transition zone, there was a thin impervious core sloping 0.3 H to 1.0 V upstream and 0.1 H to 1.0 V downstream. A filter separated the core from a large interior zone of only fair quality rock. Beneath and over the downstream slope of this interior zone there was to be a zone of hard, good quality rock with an outer, downstream slope of 2.0 H to 1.0 V. However, as a result of (1) findings from extensive test fill operations and experiences in early rock placement in the fill: (2) the development and availability of vibratory rollers of 5 tons weight at the outset of construction to 15 tons during the course of the work; and (3) exposure of the foundation and abutment foundation rock in the early construction stages, the final section was gradually changed over a period of about $2\frac{1}{2}$ years to one differing substantially from that envisioned in the preliminary design. The final section consists of a somewhat thicker, near-vertical central core with an upstream face sloping 0.3 H to 1.0 V, constructed of silts and clays with rock fragments limited to 4 in. (10.1 cm) in size, and compacted in 6 in. (15.2 cm) lifts. Transition zones, composed of phyllite spalls and weathered rock fines, well graded from 8 in. (20.3 cm) size to no more than 30 percent passing No. 4 sieve, flank the core on either side.

The upstream shell is composed of two zones, (1) a downstream zone of reasonably sound to slightly weathered quartzite, containing up to 40 percent argillite and phyllite, and (2) a large outer zone of reasonably well graded sound, fresh quartzite with not more than 30 percent of sound argillite and phyllite included. The maximum size rock was limited to that which could be placed in the specified 2 ft and 3 ft lifts without interfering with proper compaction. The downstream shell is similar but contains an additional zone adjacent to the downstream transition, composed of random moderately weathered rock which would not break down under compaction to the extent required to qualify for transition material. The outer slopes of the shells were steepened to 1.9 H to 1.0 V upstream and 1.8 H to 1.0 V downstream.

Carters Dam is an excellent example of the adaptability of compacted rock techniques to site conditions with respect to the economical utilization of available construction materials, and to foundation characteristics.

Round Butte Dam, a 440 ft (134.2 m) high rockfill structure on the Deschutes River in Oregon, was completed in 1964.[62] The design and the construction of the dam were strongly influenced by the geologic conditions at the site. The canyon of the Deschutes River is 1000 ft (305 m) deep. In past geologic time basalt flows nearly filled the ancestral canyon and erosion of this infilling left steep-cliffed intercanyon basalt benches clinging to the old canyon walls, and buried river channels in locations where the river eroded a new channel. The axis of the dam was located within a short reach of the river where the abutments would be the original canyon walls and thus avoid possible high seepage zones. The Deschutes formation at the site is approximately 1000 ft thick and consists of lava flows interbedded with horizontally stratified sediments of siltstone, sandstone, and lightly cemented sand and gravel. The lowest member of the formation is the 260 ft (79.3 m) thick highly jointed and multi-layered Pelton Basalt. Above the Pelton Basalt are about 700 ft (213.5 m) of sediments containing several basalt flows. It was concluded that the topography, abutment and foundation conditions and availability of materials at the site were favorable for the construction of a rockfill dam provided seepage through the Pelton Basalt could be controlled to an acceptable amount. This was successfully accomplished by an extensive array of drainage and grouting tunnels, and foundation grouting.

The materials available for construction of the dam within a distance of $7\frac{1}{2}$ mi from the site were rock from talus slopes and quarries, pumiceous sand and limited quantities of nonplastic silty sand. The dam was founded on 180 ft (54.3 m) of jointed, pervious rock and 260 ft (79.3 m) of stratified sediments generally impervious but containing strata of lightly cemented sand, and to a lesser extent, strata of gravel of varying permeability. A 50 to 100 ft (15.25 to 30.5 m) thick stratum of basalt also existed within the layer of sediments.

In order to minimize the cost of diversion and power tunnels the outer slopes were required to be as steep as possible, thus dictating the use of a narrow central core. As shown in Fig. 7-10 the resulting embankment section consisted of a narrow central core sloping slightly upstream protected by transition zones of natural and processed materials and supported by compacted rockfill shells. Adjacent to the downstream face of the core there is a substantial transition zone of natural sand, and between this zone and the rockfill are two transition zones of processed material. The upstream face of the core is also protected by three transition zones between it and the rockfill shell.

In order to obtain greater contact area between the core and the rock foundation, fillets were provided at the base of the core. Because

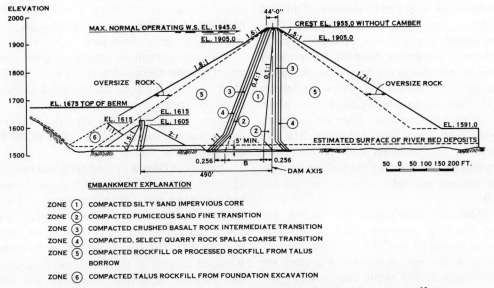

Figure 7-10. Round Butte Dam—maximum section. (After Patrick and Ferris[62])

of the presence of some weaker strata in the abutments, the embankment slopes were flattened at the abutments, thus providing a flairing of the shells.

The impervious core was placed in 12 in. (30.5 cm) layers, moisture conditioned, and compacted with a 50 ton (45.36 mt) rubber-tired roller. The fine, medium, and coarse transition zones were placed in 12 in. (30.5 cm) layers and compacted with a 10-ton vibratory roller. The rockfill was placed in 24 in. (61 cm) layers, and the oversize rock was pushed to the outer edges. The rockfill was compacted by 4 passes of a 10-ton vibratory roller, with no sluicing or moisture conditioning required.

As mentioned previously, the possibility of excessive seepage through the pervious basalt in the foundation and abutments was a major concern of the designers. To reduce the seepage through the Pelton Basalt, a deep grout curtain penetrating to a more impervious basalt formation was provided as well as shallow consolidation grouting beneath the contact areas of the core and transition zones. The exposed pervious basalt from the downstream face of the core to the upstream embankment slope was sealed by gunite, as were the most pervious sedimentary strata. The pervious strata underlying the downstream transition zones and rockfill shell were protected by processed transition material between the strata and the rockfill shell.

The post-construction behavior of Round Butte Dam is described in Ref. 63. The maximum settlement and horizontal deflection noted at a line of settlement markers set during construction at an elevation about 100 ft (30.5 m) below the final crest, were about 1.5 ft (0.46 m) some 16 months after the dam was completed and water impounded to Elevation 1945, the normal operating level. These settlement and deflection observations on the downstream slope, therefore, represent movement caused, in part, by additional loading resulting from embankment construction as well as movement caused by water loading.

On October 16, 1964, about $3\frac{1}{2}$ months after the dam was completed and the water level was still 45 ft (13.7 m) below maximum operating level, a longitudinal crack was observed in the approximate center of the 44 ft (13.4 m) wide crest of the dam. The crack was initially about 200 ft (61 m) long but lengthened to about 500 ft (152.5 m) in 14 days. Later another crack appeared in the area of the first transition zone upstream of the core. This longitudinal cracking was produced by the differential settlement between the principal zones of materials in the dam. It would appear that the first crack which developed in the core of the dam may have been caused by the downstream shell settling away from the upstream-sloping downstream face of the core. The later cracks in the transition zone resulted from downstream deflection of the core under reservoir pressure and settlement of the upstream shell.

About one month after the reservoir had been filled, a transverse crack in the core formed about 68 ft (20.7 m) from the rock contact at the right abutment. The crack was nearly vertical, and about one in. (2.54 cm) wide at the top, tapering to about $\frac{1}{16}$ in. (0.16 cm) at a depth of 16 ft (4.83 m). There was no free water in the crack even though it extended some 5 ft (1.53 m) below reservoir level, attesting to the self-healing action of core and transitions. The transverse crack resulted from the tensile stress in the brittle core material caused by the differential settlement between the maximum section of the dam and the abutment section and concentrated by the presence of a relatively abrupt change in foundation elevation directly beneath the location of the crack (see also Section X). The crack was repaired by filling it with a mixture of dry sand and bentonite. None of the afore-mentioned cracks, although a matter of concern at the time, have had any discernible adverse effects on the structural integrity of the dam.

The total amount of seepage past the dam which can be measured is in the order of 53 cfs, (1.5 m^3/sec) but an undetermined amount probably emerges into the river downstream of the dam. The total seepage appears to be well within the originally estimated range of 80 to 200 cfs (2.26 to 5.66 m^3/sec).

Don Pedro Dam is a zoned earth and rockfill structure, 585 ft (178.4 m) high above streambed, curved in plan, with a crest length of 1300 ft (396.5 m), a crest width of 40 ft (12.2 m), and a maximum base thickness of about 2800 ft (854 m). The dam is located in a narrow V-shaped canyon of the Tuolumne River near the town of La Grange, California, and was completed in late 1970.[66] The canyon is some 500 to 600 ft (152 to 183 m) deep and exposes a variety of metamorphic rocks of Jurassic Age. The rocks are massive and blocky, and the foundation is quite sound and competent. The dam is founded entirely on such bedrock.

Figure 7-11 shows the maximum developed cross section of Don Pedro Dam. The centrally located impervious earth core is composed of silty to clayey sand with a permissible gradation within the limits of 15 percent to 60 percent passing No. 200 sieve, 25 percent to 80 percent passing No. 50 sieve, 70 percent to 100 percent passing No. 16 sieve, with a maximum size of $1\frac{1}{2}$ in. (3.8 cm). The plasticity index varies

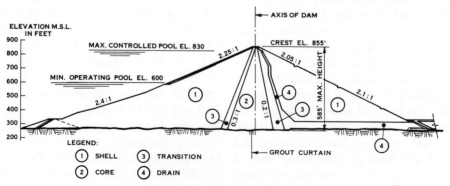

Figure 7-11. Don Pedro Dam—developed maximum section. (After Taylor[66])

from N.P. to 13 percent, and the liquid limit from 18.6 percent to 28 percent. The dry density of the compacted core material varies only slightly from the average of 125.5 pcf (2010.3 kg/m^3) and the permeability is generally less than 1×10^{-6} cm/sec. Although the core material is excellent in most respects, the low plasticity of most of the material causes the compacted material to tend to be brittle in nature. This property was one of the factors which called for extraordinary precautions to be taken during design to minimize the possibility of transverse cracks forming in the core, and to alleviate their effects if such cracks should develop. Some of the precautionary measures adopted wholly or in part because of the characteristics of the core and possible cracking potentialities were (1) the axis of the dam was curved upstream on a 3000 ft (915 m) radius, (2) special attention was given to the configuration of the rock-core contact faces at the abutments and to the convergence of the rock contours immediately downstream of the contact areas, (3) the core was widened or flaired at both abutments from the normal crest width of 20 ft (6.1 m) to 70 ft (21.4 m) at the rock contact to increase the length of the seepage path and reduce the gradient should some cracks form in the settlement induced tension zones near the top of the dam, (4) special instrumentation was installed to detect the onset of the development of tension zones near the top of the dam adjacent to the abutments and to locate precisely any cracks which might form, and (5) most importantly, rather thick transition zones of selected sandy gravel were placed adjacent to both the upstream and downstream slopes of the core. For a distance of 400 ft (122 m) out from each abutment the upper 130 ft (40 m) portions of these transition zones were widened to several times normal width. A principal function of the upstream transition zone is to supply materials of the proper gradation to induce self-healing if a transverse crack should form in the core. The downstream transition is designed to inhibit propagation of any crack initiated within the core and to act as an efficient filter with respect to the core materials and prevent them from being washed into the drain should there be a flow of water through a core crack. The flow capacity of the drain is far in excess of that required to provide for normal seepage, so as to be capable of handling any abnormal flow which could conceivably result from a crack through the core.

The shells of the dam are composed of coarse gravel and cobbles obtained from the abundant quantities of gold dredge tailings in reaches along the Tuolumne River. This material is similar in nature to the shell materials used at Oroville Dam, discussed previously. It is ideal for the purpose, being a relatively clean, washed gravel, well graded in size generally from about 20 percent passing $\frac{3}{8}$ in. (0.96 cm) sieve to 12 in. (30.5 cm) in diameter. The compacted dry density of the shell material averaged 130.2 pcf (2085.6 kg/m^3). In general, the shell material is highly pervious but due to the variation in fines in the borrow material and possible stratification in the fill, and further, because of the increase in fine material due to degradation of the coarser materials under high pressure near the base of the fill, it was deemed advisable to provide a definite internal drain zone of washed and screened shell material. This drain zone contains the only processed material in the fill.

The transition zone material is a sandy gravel obtained from a deposit of dredge tailings containing a much higher percentage of fines than the gravel used for the shells, with permissible limits of 20 percent to 55 percent of the material passing No. 4 sieve.

The compaction of the core material was accomplished by sheepsfoot rollers, but all other materials were compacted by vibratory rollers.

The instrumentation installed in Don Pedro Dam for monitoring its performance during construction, the first filling of the reservoir, and subsequent operation, consists of 43 pneumatic piezometers; seven special hydraulic settlement devices for measuring the relative movement between the various zones in the dam; eight inclinometers, including one in the upstream shell sloped at an angle of 40° with the vertical; seven crossarm settlement devices; 68 surface alignment and settlement markers, and four arrays of strainmeters at two elevations at each abutment to furnish a continuous record of the development of any tensile strains in these susceptible zones. There are 90 meters in the 4 lines installed at levels of 10 ft (3.05 m) and 45 ft (13.7 m) below the crest and extending some 250 ft (76.3 m) out from each abutment. This was the first installation of such meters in the United States.

Reservoir storage began in November, 1970. Because of certain early release requirements and heavy demands for irrigation water, the reservoir as of November, 1974 had not reached full pool elevation. The instrumentation data indicate that the dam is performing in a normal and completely satisfactory manner. The post-construction settlement of the core at the maximum section has been about 1.0 ft (0.31 m). The crossarms on the settlement devices were installed at a vertical spacing of 10 ft (3.05 m) as construction progressed. During construction, the incremental compression in the lowest 10 ft (3.05 m) lift near the maximum section of the core was 3.2 percent; at mid-height it was 2.6 percent, and 1 percent at 100 ft (30.5 m) below the crest. The maximum post-construction settlement measured in the shell has been less than 0.5 ft (0.15 m) (May, 1974). The strain meters near the top of the dam indicate the development of small amounts of tensile strain but no evidence of incipient crack formation.[66]

IX. ROCKFILL DAMS WITH CORES OF MATERIALS OTHER THAN EARTH

Ever since the earliest period of rockfill dam design and construction, consideration was given to the utilization of materials other than earth for cores of rockfill dams. These attempts met with only limited success. Concrete cores were installed in a number of these dams, for example, Crane Valley Dam in California. This dam was completed in 1910 to a height of 130 ft (39.6 m), and consists of a vertical central core of concrete about 6 ft (1.8 m) thick at the bottom and 1 ft (0.31 m) thick at the top, with an hydraulic-fill earth embankment forming the upstream shell and quarried rockfill, dumped in high lifts, forming the downstream shell. On the first filling of the reservoir the crest began to deflect downstream and during the next 4 years the estimated movement was about 9 ft (2.74 m). Measuring points were then installed, and during the next 38 years about 5 ft (1.53 m) of additional movement occurred at a steadily decreasing rate. Large quantities of additional rockfill were placed on the downstream slope to stop the movement. By means of observation in exploratory shafts, it was determined that a large horizontal crack had occurred in the core wall about 6 ft (1.83 m) above the base, and although the lower part of the wall had remained in place the portion above the crack had deflected downstream. The failure of the core wall was probably due to pressure from the hydraulic-fill upstream shell, but the fine-grained soil sealed the cracks in the wall and inhibited leakage through the dam.

Concrete core walls at other dams have cracked due to movement of the shells, but the damage was not generally as extensive as in the foregoing case. It is now recognized that the introduction of a thin, rigid concrete member into a mass of rockfill which inevitably undergoes settlement and deflection will generally result in damage to the concrete wall. For this reason the use of thin concrete core walls extending to the full height of the dam has largely been discontinued. However, various forms of concrete cutoff walls in pervious foundations and extending partially into impervious earth cores are extensively used in modern dams. Reference 67 describes such a cutoff wall which has a maximum depth of 210 ft (64 m) in pervious river alluvium and extends upward some 30 ft (9.2 m) into the impervious core of the dam to give the wall a total depth of 240 ft (73 m).

Steel diaphragms, centrally located in rockfill dams, have been used as impervious water barriers. However, concern as to the possible rate of deterioration of the steel and the virtual impossibility of repairing such diaphragms have limited their use to only a very few installations.

More recently, rockfill dam cores of bituminous materials have been used with considerable success. Reference 68 lists 15 dams in Germany, Ethiopia, Austria, and Ecuador, varying in height from 13 m to 54 m, built between the years 1954 and 1972, which have impervious cores of asphaltic concrete. The cores are relatively quite thin, generally within the range of 40 cm to 100 cm in thickness. Transition zones are required on both sides of asphaltic concrete cores to provide uniform support, to act as a normal filter in case of leakage through the core, and to provide a groutable zone if serious leakage does occur.

Considerations and investigations for the design of the 302 ft (92 m) high Finstertal Dam in Austria, are discussed in Ref. 69. To accommodate the terrain, the dam is uniquely S-shaped in plan. The shells are quarry-run and morainal material rockfill. Transition zones, approximately 3 m thick, of graded moraine or quarry materials with a maximum grain size of 0.2 m are placed on each side of the core. The slightly inclined asphaltic concrete core varies in thickness from 70 cm at the base to 40 cm at the crest.

Asphaltic concrete cores have been used in earth dams as well as in rockfill dams. The design of such cores in earth embankment dams is discussed in Refs. 70 and 71.

X. DESIGN AND CONSTRUCTION PRACTICES FOR MODERN ROCKFILL DAMS

Factors Influencing the Alignment and the Design of the Cross Section of the Dam

The site conditions which importantly influence the final configuration and composition of the component parts of a rockfill dam are often those which are dominant in the selection of the type of dam itself, i.e. whether it should be earthfill, rockfill, or concrete. Some of these factors are:

1. Topography.
2. Meteorological conditions.
3. Geologic conditions, as they relate both to the foundations and abutments and the kind and availability of construction materials. Regional seismicity would be included in the category of general geologic conditions.
4. Schedules.
5. Costs.

The matters of site selection, selection of type of dam, geologic and materials investigations, and runoff studies are treated in depth in other chapters, so the above listed factors are discussed briefly herein only as they relate to the design of rockfill dams.

Topography, of course, controls the maximum possible dam height and the length, but the topographic details are important in the location of the dam axis, the spillway, the diversion works, outlet control, and power facilities. The configuration of the valley walls should be considered in positioning the axis to avoid large vertical or overhang areas at the core contact with the abutments because of the difficulties of preparing the foundation and compacting the core material to effect a satisfactory seal at the interface. Also, it is now considered good practice by some designers to position the impervious core, if possible, so that the abutment contours in the contact areas will converge in the downstream direction, which would tend to cause the core to be forced more tightly against the rock when the water loading is applied. Topographic features (usually related to geologic conditions) such as deep ravines adjacent to abutment contacts, or narrow promontories utilized as abutments, which could create short seepage paths from the reservoir should be evaluated with respect to stability or cost of remedial treatment.

Because an abundance of rock at or near the site is, of course, a necessary prerequisite for construction of a rockfill dam, the majority of such dams are founded on rock of some type, but this is not a requirement for utilization of rockfill construction as, for example, at Brownlee Dam, which is founded on a deep bed of alluvium, or Gepatsch Dam on moraine and alluvial materials.

Most rock types are sufficiently strong to support the characteristically steep embankment slopes designed to develop the intrinsic shear strength potential of the rockfill. However, for foundations of high rockfill dams some varieties of rock, as for example, a very low strength shale, may be encountered which would be unacceptable because of the relatively flat embankment slopes which would be required to avoid overstressing the material in shear. Such materials occurring within the lithologic sequence for some distance below the foundation contact could form significant zones of weakness which must be considered in the stability analyses of the dam.

Dense, high-strength alluvial, morainal or similar granular materials are usually acceptable as foundation materials under the rockfill shells of dams. The high density is required to preclude undue settlement of the shell with respect to the core and to eliminate any possibility of liquefaction of the cohesionless foundation material under seismic vibration. The shear strength of the foundation materials should be comparable to that of the rockfill—otherwise, the strength of the foundation may govern the slope angles of the embankment and result in an uneconomical utilization of the rockfill. For most rock foundations, the geologic details affect the quality of the foundation more importantly than the rock type. Geologic features such as joints, fissures, bedding planes, faults, shear zones or pronounced schistosity may constitute serious defects in the foundation and require remedial treatment, special features incorporated within the embankment, or may dictate the selection of the type of dam section or the final alignment. Even so, the impact of geologic defects upon the competency of the foundation is generally much less severe for a rockfill dam than for a concrete structure because of the ability of the rockfill dam to adjust to differential settlements and to impose less concentrated loads on the foundation.

The availability and in-place cost of construction materials are obviously governing factors in the design of the dam. A paucity of earth core material, among other factors, may dictate the selection of an impervious-faced rockfill section. If an adequate amount of suitable earth core material is obtainable at reasonable cost a sloping or central core dam might be the preferred choice to minimize settlement or seepage problems. The physical characteristics of the available rock and the costs for quarrying, hauling and placing are, of course, outstanding considerations in determining whether the fill will be dumped in high lifts or compacted in thin lifts, and to a large measure will control the slopes and internal zoning of the section.

Rockfills can be designed to be stable when subjected to throughflow or overflow—clearly an important advantage of rockfill dams over earthfill structures subject to overtopping during construction. The mechanics of flow through rockfill have recently been developed by Leps[28] and Johnson.[94] Laboratory tests and observations of prototypes have shown that failure of an unreinforced rockfill subjected to throughflow would be initiated by ravelling and minor sliding at the downstream toe, followed by the headward progression of a succession of larger slides as support was removed and the hydraulic gradient steepened. The slides could (and have on occasion) cut through the crest of the dam and cause complete breaching of the structure. However, if the initial erosion is prevented by securely anchoring the larger rocks comprising the lower portion of the downstream slope the rockfill mass will remain stable under flow gradients far greater than those which would otherwise cause failure.

The susceptibility to failure by throughflow of an unreinforced rockfill slope is dependent upon the inclination of the slope, the maximum gradient, rate of discharge, and characteristics of the rockfill with respect to the density of the rock mass and the size, shape, gradation and specific gravity of the rock particles. The evaluation of these factors can be approached logically[28] but the design of an anchorage system to prevent the dislodgment of larger key rocks at the embankment toe, is largely empirical. Weiss[95] pioneered the concept of stabilizing the downstream rockfill slopes of dams to permit them to be safely overtopped by tieing down the rocks with a grid of steel bars anchored in place by tiebacks extending horizontally into the rock mass. Other designers have utilized and further developed this method. Slope stabilization by means of concrete slabs, asphaltic concrete membranes or long flat berms of heavy rock, has also been accomplished. (see Refs. 96, 97, 98, 99, 100. 101, 102).

It can be shown, at least theoretically, that under certain conditions pertaining to the embankment configuration, developed pore pressures, and characteristics of the rockfill, a deep-seated slide failure could occur in a rockfill embankment subjected to overtopping or throughflow.[28] While it does not appear from the published literature that a major failure of this type has actually occurred, designers of rockfill dams subject to throughflow or overflow should be cognizant of this possibility.

Climatic conditions could be a dominant factor in the choice of the type of dam section. Rockfill can be placed in areas where heavy rainfall would make the construction of an earthfill difficult or virtually impossible. Some types of rockfill construction can be accomplished at below-freezing temperatures which would preclude earthfill placement. Thus, the construction season is far less restricted with respect to rockfill rather than earth, and possibly concrete, construction. Moreover, in view of the above discussion on the effects of overtopping and throughflow, diversion problems during construction, particularly between construction seasons, may be minimized by the use of rockfill for the main body of a dam.

The final design of the dam section is never a straight-forward procedure, and in many cases is not really complete until the construction of the dam is finished. The influence of each of the afore-mentioned five major factors is varied, complex and interrelated to that of the others, but the relative magnitudes of the influences are unique to each situation.

Foundation Treatment

In 1972, the American Society of Civil Engineers Committee on Embankment Dams and Slopes of the Soil Mechanics and Foundations Division conducted a symposium on Foundation and Abutment Treatment For High Embankment Dams on Rock. The Committee Report and the papers contributed to the Symposium are listed herein as Refs. 86 through 93 (see also Ref. 103). Current and some past practices regarding foundation treatment for all types of embankment dams are discussed in depth in these referenced papers.

Excavation and Surface Preparation. As mentioned previously, it would appear that the majority of rockfill dams are founded on rock of some type. Stripping the entire foundation area to rock would generally be practiced where the overburden is shallow, i.e. where the depth is in the order of 10 to 15 ft (3.1 to 4.6 m). Where the foundation is stripped to rock under the core contact only the materials left in place should be equivalent in strength to the overlying embankment fill. This same criterion should apply generally in those cases where the depth of overburden or the character of the overburden materials influenced the choice of extending a cutoff only beneath the core to the bedrock or suitable impervious stratum.

All overhangs within the core contact area at the foundation and abutments must be eliminated. Vertical contact faces of limited height and width may on occasion be tolerated when it is difficult or infeasible to do otherwise, providing extraordinary care in selecting, placing and compacting the core materials in these areas is exercised. However, at most sites a sloping core contact would be provided even if considerable rock excavation was necessary. Such a sloping face facilitates machine compaction of the core material in contact with and immediately adjacent to the rock, thus forming a tight interface seal. More importantly, as the core settles the wedging action along such a sloping face tends to maintain or even increase the contact pressure between the core and rock face. Practical considerations relating to the topography of the abutments, volume of excavation required, the bedding, jointing and strength of the bedrock usually influence the limiting slopes selected for the core contact area. Commonly, adopted slopes are within the range of about 45° to 70° with the horizontal, but slopes as steep as about 85° are sometimes acceptable, and valley topography may fix abutment slopes at angles much flatter than 45°.

Rock excavation and trimming in core contact areas must be carefully controlled to prevent damage to the underlying rock. Controlled blasting with relatively light charges, under the direct supervision of knowledgeable, experienced personnel is the usual procedure but excavation by mechanical equipment has also been performed. Presplitting or line drilling is sometimes specified.

Often, trimming the rock surface to an acceptable contour by excavation alone is not feasible, in which case, the projections are trimmed and depressions below the desired surface contours are filled with "dental" concrete. On steep abutment slopes it is sometimes necessary to use forms to retain the dental concrete.

The requirements for preparation of the foundation beneath the rockfill shells are less severe than in the core areas. The rock or overburden left in place to form the foundation should be equivalent in strength and density to that of the overlying fills, but sometimes this will be infeasible due to geologic or economic considerations. Special trimming of rock is usually not required, but prominent rock projections or pronounced overhangs are customarily removed to facilitate placement and compaction of the rockfill adjacent to the abutments.

After the necessary excavation in the foundation and abutment areas has been completed, surface treatment of these areas, particularly the contact areas of the impervious core and filter zones, is generally required. The main purpose of the surface treatment in the core area is to prevent or reduce seepage along and adjacent to the contact between the core and foundation. All soil, loose rocks and debris remaining in the core area after the excavations are complete are removed by hand or machine cleanup. The area is then thoroughly cleaned by brooming and washing with air or water jets to completely remove all loose material. Surface and near-surface defects in the rock along which seepage could occur, such as open joints, seams, faults or shear zones are thoroughly cleaned and backfilled with concrete, slush grout, or mortar such as gunite or shotcrete. Such defects should be treated to the maximum practicable depth as influenced by the size and continuity of the defect, its attitude with respect to the axis of the dam, and the hydraulic head to which it may be subjected. Large surface defects such as solution channels, potholes and cavities may be cleaned and backfilled with concrete or mortar but impervious core material may be used if it can be properly compacted. Rock surfaces which are badly fractured, friable, or tend to spall or slake when exposed to air are commonly covered with a thin layer of slush grout. Asphalt or asphalt emulsion is also sometimes used to inhibit air or water slaking after cleanup and prior to embankment placement. Some engineers advocate covering the entire core contact area with slush grout and placing the first lift of the core while the grout is still plastic in order to effect a good bond between the fill and foundation.

The foundation beneath filter or transition zones is usually treated similarly to that of the core with respect to filling discrete voids and crevices in the rock to inhibit piping of the finer granular materials.

Special surface treatment of foundation areas beneath rockfill shells, other than machine cleanup is not normally required for most types of rock foundations. Shell foundations of other materials, such as residual alluvium, may require some treatment, e.g., vibratory compaction or placement of a protective transition zone, before the fill is placed. Such situations are quite varied and call for unique solutions.

Grouting. Foundation grouting for embankment dams is performed to (1) minimize seepage through the dam foundation, (2) reduce hydrostatic pressures under the downstream portion of the dam and (3) preclude the possibility of piping or internal erosion of materials comprising the body of the dam. Since rockfill shells are usually quite pervious and free draining, hydrostatic pressures under the downstream shell normally present no problem, but if the shell material contains a high percentage of fines which markedly reduce the permeability, special provision for drainage of the downstream shell may be required, perhaps by the use of a drainage blanket in contact with the foundation. Coarse-grained shell material would not pipe or erode due to any underseepage which could reasonably be expected to occur, but fine-grained core material must be protected against displacement by underseepage (and through-seepage) through the use of correctly designed and constructed filter zones and the grouting of the foundation rock.

Both blanket and curtain grouting procedures are commonly used in and adjacent to the core foundation area. Where the foundation rock is closely jointed or fractured, blanket grouting is used to consolidate the upper 10 to 30 ft (3.1 to 9.2 m) of rock so as to prevent piping of fines from the core into the rock crevices and to seal the near surface rock against undue loss of grout during the high pressure curtain grouting operation to follow. This grouting is usually accomplished through a pattern of holes drilled on 10 to 15 ft (3.1 to 4.6 m) centers to depths of about 15 to 30 ft (4.6 to 9.2 m).

Curtain grouting to reduce the deep underseepage through the foundation is normally performed by using single or multiple lines of grout holes approximately parallel to the axis of the dam and located within the upstream half of the core foundation contact area. The spacing of holes varies greatly depending upon the characteristics of the rock being grouted, but the split-spacing technique is nearly always used with the initial holes drilled at intervals of 10 to 100 ft (3.1 to 30.5 m) and subsequent intermediate holes at locations splitting the initial spacing into as many intervals as required for the grouting to effect closure of the curtain. Spacings such as: 40/20/10/5 ft (12.2/6.1/3.05/1.5m); 20/10/5/2$\frac{1}{2}$/1$\frac{1}{2}$ ft (6.1/3.05/1.5/0.76/0.46 m); or 100/50/25/12 ft (30.5/15.25/7.63/3.66 m) are not uncommon. The maximum depth as well as the spacing of curtain grout holes will depend upon the rock characteristics, but will usually be at least half of the height of the dam.

Multiple lines of curtain grout holes are often used where the nature of the rock jointing or fracturing would cause the length of travel of the grout to be limited or erratic. The lines are spaced about 10 to 15 ft (3.05 to 4.57 m) apart and the hole spacings staggered with respect to their locations in the lines. Angled holes are commonly used to intersect discrete faults, fissures, and cracks, or to close out or check the continuity of the grout curtain.

Cement is the most common grouting material with the ratio of water to cement being about 5 to 1 for the initial injection, and thickened gradually to perhaps 1 to 1 to close the hole. For special applications various chemical grouts, sand-cement mixtures, bentonite, etc., are sometimes used.

Grouting pressures must be kept low enough to avoid fracturing or moving the rock but high enough, in keeping with this important criterion, to achieve reasonable grout penetration in the rock crevices and fractures. It is better to err on the low side rather than the high, even at the expense of having to drill more intermediate holes to close the curtain. Practice varies somewhat as to the maximum allowable grout pressure but most specifications establish it between 0.5 psi (0.035 kg/cm^2) and 1.0 psi (0.07 kg/cm^2)/ft (0.305 m) of rock cover. Collar pressures are adjusted so the hydrostatic pressure of the grout column in the hole plus the collar pressure meet the adopted limit. Stage grouting with packers, starting at the lower depths and progressing upward is preferred since high pressures can be applied without subjecting the upper portions of the rock to damaging pressures. Grout caps, consisting of a trench in the rock and backfilled with concrete or a concrete slab anchored to the rock surface, are often used to permit higher grout pressures to be applied to the upper part of the rock. For rockfill dams with impervious facings this grout cap and the toe wall are usually combined into one structure. On some high dams the grout cap is enlarged sufficiently to contain a grouting gallery from which the foundation grouting operations are conducted while the construction of the dam is in progress. Such a gallery also facilitates additional post-construction grouting, if needed, which otherwise might be quite difficult.

Characteristics of Materials for Rockfill Dams

Early rockfill dams, such as those described briefly in Section II of this chapter, were designed on the basis of intuition, engineering judgment, and observation of the behavior of prototype structures. Many of these dams, standing today, are monuments attesting to the innate skills of their designers. Time has proved them to be safe structures but certainly the designers had no measure of how safe their creations really were in terms of safety factors or other quantified indices of stability as understood at present. The large particle size and harsh nature of the rock used in the dumped and sluiced fills for virtually all rockfill dams prior to about 1940 made the meaningful determination of the characteristics of the material difficult or impossible with the techniques available. However, the intuitive and judgmental approach to design was adequate since the consequences of partial or even complete failure of the structure were, in general, far less serious than they would be at present. Population growth, the development of communities below dams, and the increasing height, size and popularity of rockfill dams now demand that safety definitely be determined at least to a degree commensurate with that of earth dams designed in accord with the best practices recognized at the present state-of-the-art.

Most modern rockfill dams, as discussed in preceding sections of this chapter, consist of (1) an impervious core of compacted soil or sometimes other material such as asphaltic concrete, (2) transition zones, usually on both sides of the core, composed of relatively small sized rock particles or gravel and sand, with a well graded composition, (3) filter zones of clean but well graded sand, (4) internal drain zones, if required, of clean, pervious (often processed) rock particles, and (5) rockfill shells composed of coarse rock fragments or cobbles and gravel, ranging widely in particle size gradation but commonly with a maximum size of about 18 to 48 in. (45.7 to 122 cm), grading down to fines with perhaps 10 to 20 percent passing No. 4 sieve (4.76 mm). The rockfill materials are placed in lifts with a thickness commensurate with the maximum particle size, and compacted by rolling and vibrating with mechanical equipment and usually wetted but not often sluiced. The materials investigations, testing, design practices, and construction procedures for impervious earth cores are the same as those required for earth dams, and are well covered in the voluminous literature on the subject. Transition and filter materials usually may also be treated as soil. Rockfill materials (and sometimes drain rock), because of the large particle size, have certain characteristics which pose problems in testing, design, and handling which are not encountered in working with fine grained materials. The remainder of this section will therefore be devoted to discussions of rockfill materials.

As so very well expressed in Ref. 104, "a thorough understanding and appraisal of the physical properties of soils (and rockfill materials) is essential both to the use of the current methods of design and also as a key to further progress in the field. No stability analysis, regardless of how intricate and theoretically exact it may be, can be useful for design if an incorrect estimation of the shearing strength of the construction materials has been made. In many cases the errors arising from an improper appraisal of the soil properties can far exceed those resulting from the use of the more approximate methods of analysis." The physical properties of soils can be determined in the laboratory under conditions similar to those to which the materials will be subjected in the dam. Laboratory testing of rockfill materials, however, under simulated field or prototype conditions is virtually impossible. At the present time there is no laboratory equipment in the world which is capable of accepting test samples comprised of particles as large as a few feet. As early as 1956 experiments by Holtz and Gibbs[106] indicated that the diameter of a triaxial test specimen should be at least 6 times the diameter of the maximum size particle (see also, Ref. 154), and other investigations (Bishop and Green)[107] show that the height of the specimen should be about $2\frac{1}{2}$ times the diameter. Obviously, these constraints would require the fabrication and testing of enormous specimens to accommodate materials with particle sizes commonly used in rockfill. Leps[108] listed 10 large scale triaxial testing devices known to be in operation in the Western Hemisphere in 1970. The largest, at El Infiernillo Dam in Mexico,[109,111] could test specimens 44.5 in. (113 cm.) in diameter at a lateral pressure of 350 psi (24.6 kg/cm^2), and at the laboratories of Geo Testing, Inc, California, 2000 psi (140.6 kg/cm^2) lateral pressure could be applied to specimens 6 in. (15.24 cm) in diameter. The University of California, Berkeley, operates a facility capable of testing specimens 36 in. (91.4 cm) in diameter at a lateral pressure of 750 psi (52.7 kg/cm^2).[110] Fumagalli[112,113] reports on triaxial tests performed in Italy using specimens 3.94 in. (100 mm), 19.7 in. (500 mm) and 51.2 in. (1300 mm) in diameter. At present, therefore, triaxial test at confining pressures in the ranges comparable to those developed within prototype structures, considering even the few largest machines in the world, are limited to specimens containing particles with a maximum diameter of about 6 in. (15.24 cm) or less. This means that the determination of the characteristics of rockfill for most dams would require either (1) the laboratory testing of specimens with particle sizes very much smaller than in the prototype or (2) some alternative approach to the test method.

Laboratory Testing. The fabrication of specimens of testable size can be accomplished either by scalping out all particles of the rockfill material larger than the maximum permissible size for the test specimen or by modeling the gradation of the entire specimen so as to bear some adopted relationship to the prototype gradation. Because of the absence of the large particles the characteristics of such fabricated sample materials can differ greatly from those of the prototype materials. In order for the test data to be useful in predicting the behavior of the prototype these differences must be known, and much investigation by a number of very able researchers has been, and is at present directed toward establishing these relationships.

In 1957, Zeller and Wullimann[114] used scalped samples for testing the rockfill to be used in the shell zones of Goschenenalp Dam[79] in Switzerland. Four sieved fractions were separated and, "for each of these grain size distributions thus obtained the dependence of the shear strength "S" on the porosity "N" was investigated by a series of triaxial shear tests. Consequently, the relationships between the three elements, shear strength "S," porosity "N," and grain size distribution, could be established, enabling an extrapolation of the results to the conditions prevailing in the shell proper to be made." Although the uniformity coefficients of the specimens were different, trends could be established as to the effects of particle size and gradation on the shear strength of the rockfill material. This method (or variations thereof) was widely adopted and is commonly used by a number of laboratories.

The second method mentioned above for estimating the shear strength of rockfill by laboratory triaxial testing, i.e. modeling the fabricated sample, was developed by Lowe[77] in 1964 for the design and construction of Shihmen Dam in Taiwan. The test specimens were fabricated so that each rock particle would be one-eighth the size and similar in shape to the corresponding particle in the prototype material. The grain size distribution curve for the model would have the same shape as that of the prototype material and the same uniformity coefficient but shifted into the range of finer materials. It can be shown that for a regular packing of elastic spheres the strain under load and the maximum contact stress are independent of the particle size, and this was the basis for the assumption that if the materials for the model and prototype were alike in all respects except the size of particles, their shear strengths would be closely duplicated. However, for materials other than the ideal spheres there are other factors to be considered, i.e. the modulus of elasticity and surface texture, as well as shape factor, of the particles in model and prototype must be the same, and the very important effects of particle crushing at the points of contact must be evaluated (see Refs. 76, 111). The method of testing modeled specimens under either triaxial or plane strain conditions can have a marked effect on the correlation between laboratory and field behavior of rockfill materials.

A number of researchers (e.g. Seed, Chan, Marachi, Duncan, Becker, Marsal, Ref. 76, 104, 105, 110, 111) have investigated in detail these and other factors with respect to the correlation between sample and prototype behavior.

From an engineering point of view the objectives of testing programs on rockfill samples are to provide bases for estimating important characteristics of prototype materials so that these properties may be quantified for use in design analyses. The more salient properties are discussed briefly as follows:

1. Void ratio and relative density of the granular mass. If the effects of other variables are held constant, a decrease in void ratio and an increase in relative density increases the angle of internal friction of granular material. Void ratio (or density) has a more profound effect on shear strength of any given granular material than any other variable.

2. Physical characteristics of the rock particles. The mineral constituents affect the sliding friction between individual rock particles and hence to some degree the friction angle of the granular mass. The lithology and structural characteristics of the rock particles affect the strength of the individual fragments and influence the extent of breakage and crushing under load.

3. Gradation. When compacted to high density a well graded granular material will have a higher angle of friction than that of a uniformly graded material. Some studies have indicated, however, that at lower densities and at a given void ratio the reverse may occur. In order to maximize shear strength a granular material must be well graded and compacted to its maximum density.

4. Particle Size. It appears that particle breakage under load tends to increase as the particle size increases, probably due to a greater number of inherent flaws and cracks which may be present in the larger size particles. Shear strength of granular material significantly decreases as particle breakage and crushing at points of contact increases.

5. Partical shape. With all other factors remaining constant, the angle of internal friction of a granular material composed of angular fragments is higher than that of one comprised of rounded particles. Tests by several investigators[104] however, have shown that for samples composed of the two types of fragments, at equal void ratio the strength difference would be significant but if subjected to equal compactive effort the difference becomes very much smaller.

6. Moisture Content. Saturated rock parti-

cles are generally significantly weaker than dry ones. At points of contact the weakening due to moisture increases the breakdown and crushing of the rock which adversely affects the shear strength of the granular mass.
7. Confining Pressure. Shear strength, expressed as the angle of internal friction or principal stress ratio, decreases at a decreasing rate as the confining pressure applied to the granular mass increases. However, tests on widely different materials[110] indicate that further decrease in the angle of internal friction should not be appreciable for confining pressures above about 650 psi (45.7 kg/cm^2). Consequently, the Mohr envelope for shear stress vs. normal stress on the mean failure plane would usually be distinctively curved for the lower values of normal stress, with a high tangent angle near zero normal stress and decreasing gradually as the normal stress increased.

In Ref. 108, Leps presents a plot of the shearing strength of rockfill from about 100 tests on 15 varieties of rock placed at low, medium, and high density. While the effects of saturation are not indicated in the plot the overall influence of the other variables mentioned above is quite apparent. The friction angles of the materials plotted against the log of the normal pressure on the failure plane indicates a straight-line variation for high density, well graded, strong particles from 60° at normal pressure of 1.0 psi (0.07 kg/cm^2) to about 42° at 500 psi (35.2 kg/cm^2). Similarly, for low density, poorly graded, weak particles the variation is from 50° at 1.0 psi to about 32° at 500 psi. An average line between these two bounds indicates a friction angle varying from 55° at 1.0 psi to about 37° at 500 psi normal pressure.

The compressibility characteristics of granular materials are, in general, influenced by the same factors as those discussed in connection with the shear strength of the materials. High relative density connotes relatively low compressibility. The compressibility of an initially loose material is much higher than that of an initially dense material but under very high pressures where the effects of initial void ratio are no longer dominant, both initially loose and initially dense materials have about equal compressibilities. Generally, a uniformly graded granular material is more compressible than a well graded material but under high pressure the compressibilities tend to become equal because of the greater particle breakage or degradation of the uniform material. Angular materials are more compressible than well rounded ones, also because of greater particle breakage and crushing.

Compressibility is an important property of rockfill because the settlements experienced by rockfill dams are related directly to it. Differential settlements between the various zones of materials in dams are functions of the relative compressibilities of the materials. The mobilization of shear strength of granular material requires considerable deformation of the material. The analyses of stresses, stress patterns, interaction between zones, and the determination of the strength-stress ratios within an embankment dam necessitate the full evaluation of the stress-strain relationships for each of the materials in the structure.

Modulus of deformation and Poisson's ratio of the materials are two parameters needed in analyses of dams which involve stress-strain relationships as in the finite element method. The values of the modulus and Poisson's ratio for a cohesionless granular material are not constants, but are stress-dependent with a nonlinear variation. However, empirical relationships from triaxial tests provide data so that the values of the modulus and Poisson's ratio for any state of stress may be known.[76,104,110,111,136]

It is apparent from the above that the testing of scalped or modeled specimens of rockfill material using triaxial or plane strain equipment must be very carefully done if the data derived can be extrapolated to forecast, with confidence, the behavior of prototype materials. Most investigators agree that if the ratio of the nominal diameter of the maximum size particle to the diameter of the test specimen is less than about $\frac{1}{6}$ for well graded materials to about $\frac{1}{20}$ for uniform materials for triaxial testing, or about $\frac{1}{40}$ for maximum particle size to width of shear box for direct shear, the values of shear strength or deformation so determined would not be affected by specimen size. While there appears to be a large body of theoretical and experimental evidence indicating that the maximum contact stresses and strains between particles are not affected by properly modeling the particle size distribution and shapes; some researchers do not agree. However, at present, triaxial or plane strain testing of scalped or modeled specimens is the most widely adopted and generally acceptable approach toward obtaining essential data needed for the design of rockfills.

Large Scale Field Tests. As mentioned previously, methods other than those discussed above have been developed and applied for determining the shear strength of granular material. In 1956, Lewis[115] reported the results of a series of direct shear tests, using a 12 in. shear box, on many sizes of uniformly graded rock particles to about $\frac{1}{4}$ in. diameter. Because of noted deficiencies inherent in direct shear testing, some check tests were run triaxially and were found to yield values of friction angles somewhat higher than the direct shear tests. Also, "it appears that the stone size must not exceed about one fortieth of the width of the box" (see also Ref. 117).

Boughton reported in 1970[137] preliminary results of direct shear tests with laboratory scale apparatus which appear to overcome many of the deficiencies usually noted in direct shear testing. "The greatest obstacle to the analyses of the direct shear test is that there is insufficient data from the test results to determine the magnitude and directions of the principal stresses, even if the stresses within the specimen were uniform. However, with the conventional shear box, it is highly probable that the various boundary conditions introduce nonuniformity of stress from point to point in the sample. This nonuniformity of stress has been largely overcome by forming the specimen into an annular shape, and by shearing it in the plane of the annulus. Some preliminary tests have been carried out in the HEC laboratories using specimens formed in a horizontal annulus of center diameter 22 in. (55.9 cm) and of cross section 2 in. $\times$ 2 in. (5.08 $\times$ 5.08 cm). The vertical (normal) load was applied by hydraulic jack, and the tangential (shear) load by dead weights which tensioned wires attached to the top plate of the annulus, while the bottom plate was fixed." With this apparatus values of shear strength, Young's Modulus and Poisson's Ratio for use in stress-deformation (finite element) analyses could be obtained.

Sowers, (1971)[108] discussed the application of large scale direct shear testing of rockfill materials in contrast to laboratory triaxial tests. "... it is imperative that realistic shear test data are available to designers. Such a realistic test of rockfill must meet two requirements: (1) the material tested should be the same as that at the prototype embankment; and (2) the stresses imposed by testing should simulate those in the dam. The large scale triaxial tests ... do not meet the first requirement and the second test only partially." Large direct shear tests have been developed that fulfill most of the objectives of evaluating rockfill under conditions similar to those that will be experienced in the prototype. They permit full scale rock placement and compaction, and induce plane strain on the weakest surface, the interface between compacted layers. Although they introduce some difficulties not present in triaxial compression it is the writer's opinion that they offer a more valid representation of embankment behavior than the triaxial tests.

The first large direct shear test was developed by the TVA in its evaluation of sandstone for the South Holston Dam in 1948. The device consisted of a steel box 9.3 ft by 9.3 ft, 3 ft deep (2.8 $\times$ 2.8 $\times$ 0.92 m) supported above a saw-tooth rock surface. The rockfill, including particles 6 in. to 12 in. (15.2 to 30.5 cm) maximum size, was placed in the box and vibrated to densities comparable to those anticipated in the embankment. The box was loaded vertically with large concrete blocks that produced normal stresses as high as 10 psi (0.703 kg/cm^2) on the shear surface. The rockfill was sheared by jacking the box horizontally, thereby forcing a shear surface a foot or so above the saw-tooth rock foundation. Other large shear boxes were described; one designed for Alabama Power Co. in 1957, capable of exerting 27.5 psi (1.93 kg/cm^2) normal stress on the rockfill; and another designed for Duke Power Company in 1966 could apply a normal stress of 78 psi (5.48 kg/cm^2) on the failure surface. The unique feature of these devices and the testing built around them (compared to triaxial testing) was that they permitted placement and compaction of the fill in the same manner as in the embankment[108].

There are three shortcomings of the direct shear test. First, the failure surface is forced between the two halves of the box. However, as long as this is the layer interface, this is an advantage because that surface should exhibit the least particle interlocking and lowest strength. Second, the rigid box introduces stress concentrations as well as nonuniform strains. In loose materials the stress concentrations may induce progressive failure; in dense rockfill the effect is probably less, but its magnitude is not known. Finally, the maximum stresses are lower than available in some triaxial devices. The test results cited on weathered gneiss did not exhibit curvature of the Mohr envelope. Instead, the envelopes were straight and the angles of internal friction well below 45°. The strengths from the direct shear tests are somewhat higher than those from large scale triaxial testing and may yield either more realistic evaluations of safety or increased design economy,[108] (see also Ref. 148).

Jain and Gupta (1974)[116] report on the performance of large scale direct shear tests of compacted clay shale on rockfills for the Ramgana Project in India. "These tests afforded testing on in-situ samples 4 ft (1.2 m) square by 2.62 ft (0.8 m) deep, by sinking a hollow mild steel plate box in compacted rockfill containing sizes up to 8 in. (200 mm). Another similar setup was also devised that permitted tests to be performed on samples 2 ft (0.6 m) square by 1 ft (0.3 m) deep for testing materials containing sizes up to 4 in. (100 mm)." The hollow boxes, provided with knife edges, were pressed into the compacted rockfill by hydraulic jacks as the surrounding fill was simultaneously excavated. When the box was lowered completely the undisturbed rockfill in the box served as the specimen to be tested. Normal loads were applied by jacks reacting against movable loaded platforms, and shearing loads by other hydraulic jacks. Plots of peak horizontal stress vs. normal stress show straight-line relationships for all tests performed with the in-situ shear boxes, and with laboratory direct shear equipment on $\frac{3}{16}$ in. (4.8 mm) size material.

Large scale direct shear tests similar to these examples yield shear strength values which can be used in conventional rigid body analyses of the embankment but normally do not fully define the stress-strain relationships for the materials which are needed in the stress-deformation type of analysis discussed briefly in the following subsection.

Stability Analysis

Early designers of rockfill dams were apparently concerned principally with the ability of the structure to resist sliding along the base, but relied mostly on engineering judgment and experience in proportioning the section so as to be stable against sliding within the rock mass. Literature of the time contains statements to the effect that almost any rockfill dam with an impervious upstream face which would be practicable to build on a suitable foundation would necessarily have a relatively high safety factor against sliding because of the large mass involved. Galloway (1939)[3] states "On the assumption of a dam 200 ft high and a crest 15 ft wide, a downstream slope of 1 on 1.4 and a unit weight of 100 lb/cu ft for the loose rock and with the water load, ratios of height to base of 1:2.25, 1:2.5, and 1:3 would have sliding factors (ratios of weight of rock to water pressure) of 4.50, 5.14 and 6.45, respectively. These ratios are practically constant for all heights of dams." The 1920s and 1930s were periods of great activity in the development of soil mechanics and the application of the new science to the problems of slope stability of earth dams. Techniques were developed for expressing quantitatively the degree of stability or safety factor of the structures. The pertinent strength properties of earth materials could be determined by laboratory tests and applied as input parameters for any of a number of newly devised procedures for calculating the stability of earth dams. However, the large size, coarse materials then used almost exclusively in rockfill dam construction were not amenable to testing by soil laboratory methods to determine their strength properties—nor, indeed was it even generally considered necessary to attempt to do so. As the configurations of rockfill dams began to resemble earth dams with internal impervious cores, techniques developed for earth dams were utilized to investigate the stability of these dams also. However, the shearing strength of the principal member, the rockfill itself, had to be estimated from the angle of repose of the material or, in the case of important structures, from large scale field tests. It is only in comparatively recent years that insight has been gained as to the behavior of rockfill masses, through the research efforts of many investigators in the United States and abroad.

The various methods which have been developed for performing conventional static state stability analyses are fully treated in textbooks on soil mechanics, and the details of the procedures will not be repeated here. Static limit-equilibrium methods, such as the ϕ circle, sliding block or wedge, circular arc, slices, infinite slope, and logarithmic spiral methods are well known and have been applied in dam analyses for many years. Reference 118 presents an appraisal of the state of the art of limit-equilibrium analysis as of 1967. References 119 and 120 discuss the application of the sliding block analysis, for example, to the case of the sloping core dam with rockfill shells (see also Refs. 122 and 123). Though a number of useful modifications of the various static methods have been offered by researchers, the basic principles have not changed materially in the last two decades. The static limit-equilibrium concept has been in the past, and apparently continues at present (1976) to be the most widely used approach to the stability analysis of rockfill dams.

The basic premise of all static equilibrium analyses is that the dam will act as a rigid body and the shear strength of the material will be fully developed at all points simultaneously along a surface of incipient failure. Deformations of the dam and the relationships of strain to developed strength of the materials are not accounted for in the analyses as customarily performed. For embankments in general, and rockfill dams in particular, such assumptions are significant departures from reality, but the magnitude of the errors introduced in the analyses would depend upon the stress-strain relationships of the materials involved. This situation has long been recognized, but in view of the difficulties of determining such stress-strain relationships and including their effects in analyses, the designer perforce adopted the more tractable approach offered by the rigid body, limit-equilibrium, concept. For some years past, however, mathematical processes with the aid of the computer have made it feasible to compute displacements, stresses, and strength-stress ratios in embankments and foundations, providing the stress-strain relationships of the materials involved are known.

As mentioned above, the behavior of soils in the prototype can be forecast with acceptable accuracy by laboratory tests, but for rockfill materials the basic laws of similitude between test specimen and prototype characteristics are presently only in the process of being established. Sufficient progress has been made, however, to amply justify stress-deformation (e.g. finite element) analyses on all major rockfill dams. This by no means infers that rigid body analyses should be abandoned. They are very useful design tools, particularly for preliminary design, comparisons between alternative sections, etc., and in fact, into the final design stage of rockfill embankments, but before designs are finally adopted the stability of major, important rockfill dams should be thoroughly examined by stress-deformation (static and dynamic) analyses. It must be recognized that although the rigid body method of analysis in either the static or dynamic case leaves very much to be desired, embankments so designed have apparently been remarkably successful. A few earth dams have failed during earthquakes and a number have been damaged, but there appear to have been no instances of failure of a major rockfill dam due solely to shear slides which could be attributed to faulty slope design. In those cases where failure of or severe damage to earth dams has occurred during earthquakes, other factors in addition to shear displacement, such as liquefaction of the embankment or foundation materials seem to have been involved.[82,125,126,127] Whether the excellent safety record of rockfill dams is due to the inherent stability of the structures or to the fact that very few major dams have been subjected to violent earthquake shocks, is unknown. Nevertheless, rigid body analyses would offer no clue as to whether the embankments were actually only marginally safe or uneconomically conservative in design.

The deficiencies of the rigid body analysis become most apparent when the method is applied to the problem of assessing the stability of an embankment subjected to seismic vibrations (e.g. Refs. 82, 121, 126, 127). In the analysis of embankment stability so as to account for the dynamic forces induced by an earthquake it is commonly the practice to compute the minimum factor of safety against sliding when a static, horizontal force or system of such forces, attributable to seismic acceleration, is included in the computations. The horizontal force is expressed as the product of a seismic coefficient and the weight of the potentially sliding mass of material. The dynamic forces are replaced by a presumably equivalent static force and the analysis can then be treated as a static problem. The method therefore may more properly be termed a "pseudo-static" analysis.

The values of the seismic coefficients used for design in the United States typically lie in the range of 0.05 to 0.15 and appear to be based largely on judgment and custom[82,121] although many engineers relate adopted values to the anticipated maximum ground acceleration at the dam site.

There is ample evidence to indicate that most dams do not respond to seismic forces as rigid bodies. Inertia forces produce deformations, the magnitudes and patterns of which are uniquely characteristic of the structure. Maximum accelerations may momentarily reduce the safety factor of some portion of the dam below 1.0, particularly near the crest, causing a small amount of slumping or sliding but the movement ceases when the direction of the inertia force reverses or the force decreases. The cumulative effect of an application of a series of such forces during an earthquake may result in a very substantial sloughing, sliding or settlement of the dam. There is obviously no similarity or equivalence between a pseudo-static force applied for an unlimited time and a series of seismic impulses which vary in duration and direction.

In an effort to overcome some of the deficiencies inherent in the assumption of rigid body response, a number of investigators have envisioned the embankment as being composed of thin horizontal slices interconnected by elastic springs and viscous-damping devices.[121] Such a structure would respond as a deformable body, and comparisons between the computed distributions of accelerations induced by ground motion and the measured response of a number of dams indicate good agreement. The patterns of accelerations could be utilized to establish a corresponding pattern of maximum inertia forces acting within the embankment at any instant during an earthquake which could then be applied in a modified pseudo-static analysis of stability. While this viscoelastic approach would lead to a much better evaluation of dynamic stability than the conventional rigid body analysis, it also has some shortcomings. Since the analysis is based on the assumption that the response of the embankment to the ground motion is controlled only by the shearing action between horizontal slices, the vertical deformations within the dam resulting from ground motion that contribute significantly to the overall dynamic stress patterns are not included. Moreover, the deformation pattern in the transverse section near the upstream and downstream faces appears to be in error. In view of these deficiencies the shear slice method does not realistically measure dam behavior, but nevertheless it does have very considerable value in determining the general characteristics of the dynamic response.

In order to develop improved methods of analysis Clough and Chopra[128] in 1965 introduced the finite element approach for evaluating the response of an embankment constructed of linearly elastic, homogeneous, isotropic materials. Subsequently, the method was extended to include nonlinear, elastic-plastic materials. The principles of the finite element method are fully treated in a number of textbooks (e.g. Refs. 129 and 130) and numerous publications, and need not be discussed in detail here. Very briefly, the embankment section is idealized as an assemblage of discrete elements interconnected at their nodal points. If deformations within each element are assumed so as to satisfy certain conditions the stiffness properties can be calculated, and the stiffness of the complete assemblage can be obtained by adding the appropriate stiffness components of the individual elements connecting to each nodal point. The stiffness relationships of the assemblage are used in the derivation of equations for evaluating the response of the system.

Recently, (Jan. 1974) an analytical procedure that permits the use of a different damping ratio for each individual element was formulated. This procedure allows the incorporation of both stiffness and damping values that are strain-dependent, for each element.[146] The analysis by the finite element method involves such voluminous calculations that it is practicable only if performed with the aid of a digital computer. The method is a powerful tool by means of which the complete time-history of displacements, velocities, accelerations, strains and stresses at all nodal points in the system may be obtained. In applying the method to practical problems, such as assessing the stability of dams during earthquakes, the values obtained from the analyses will obviously be only as valid as are the input data upon which the computations are based. Both static and dynamic characteristics of the materials must be known, and accelerograms developed for credible earthquakes to be considered in dynamic finite element dam analyses. The synthesis of applicable design earthquake characteristics involves extensive seismological studies, discussion of which is not within the scope of this chapter. Reference 131 and 133 for example, should be consulted in this regard. Stress-strain relationships for soil, including shear strength, in triaxial, plane strain or simple shear states, for both static and dynamic conditions can readily be determined in the laboratory. The techniques for performing cyclic dynamic tests on soil are discussed in Ref. 138 and 139.

As discussed previously, due to equipment limitations, static and dynamic characteristics of rockfill materials can presently be estimated with far less certainty than for soil. Consequently, for rockfill dams the alternative is sometimes adopted of substituting in dynamic analyses well established input parameters from tests on a granular material known to be somewhat less competent than the prototype rockfill and thus determine a lower bound for the dynamic stability factors of the dam. This procedure was followed, for example, in the extensive dynamic investigations of New Don Pedro Dam[66] which were performed in 1969.[132]

Static as well as dynamic properties of the construction materials are required because, "experimental work has shown that the dynamic strength (of granular material) is a function of the static stresses existing in the material before the application of the dynamic loading. Consequently, in order to assess the dynamic strength of the dam, it is necessary to know the distribution of static stresses throughout the dam prior to the occurence of an earthquake."[132]

The static stresses can be determined either by (1) applying gravity loads simultaneously to the entire embankment or (2) simulating the placement of fill in successive layers. For zoned embankments, method (1) would yield incorrect displacements and stresses, therefore, "the final static stress distribution in the dam is arrived at by considering the actual changes of stress that occur in the dam throughout construction. This is done by analyzing the dam in increments corresponding to the addition of layers of material in the actual construction sequence, with the values of modulus and Poisson's ratio throughout the dam adjusted for the average stress conditions existing during the addition of that increment."[132] A static analysis of this type for Oroville Dam is also discussed in Ref. 134. A static finite element analysis of a concrete-faced rockfill dam is described in Ref. 137. An outline of the steps typically required in the dynamic analysis of an earth or rockfill dam is given in Ref. 133.

Means other than those outlined above have been used for investigating the stability of rockfill dams. In 1956, Clough and Pirtz[140] investigated the earthquake resistance of sloping and central core rockfill dams by means of models. The entire dam and component materials were modeled then subjected to simulated earthquake vibrations on a specially designed shaking table. The seismic stability of Oroville Dam embankment was investigated through the use of model techniques of this type.[35] The modeling approach has been used by a number of investigators to examine the behavior of structures under both dynamic and static conditions. A recent report (1970) on the use of static models to investigate the deformations in rockfill dams under hydrostatic loadings is given by Fumagalli, et al. in Ref. 113 (see also Ref. 156).

The stability studies for the design of high dams which will impound large reservoirs should include dynamic analyses of the structures despite the fact that they may be constructed in areas that have no prior history of significant earthquake activity. Recent evidence indicates a possibility that earthquakes of considerable magnitude can be created by the impoundment of large bodies of water. Studies of the mechanics of this phenomenon are being pursued by a number of investigators at the present time (1977).

Instrumentation

The use of instrumentation to monitor the performance of rockfill dams during the construction and post-construction periods is a fairly recent development. Even had adequate equipment been available it would have been generally impracticable to install measuring devices in the dumped rockfill embankments typical of the early construction. Measurements of settlement, deflection and lateral movement were, therefore, limited to observations of movements of surface monuments and markers installed after the embankment was completed, and with few exceptions[144,147] no data were available regarding internal stresses and deformations. With the advent of the compacted rockfill dam, instrument installation similar to that which had slowly been developing for earth dams became feasible and was widely adopted. By the 1960's, instruments for measuring internal vertical and horizontal displacements, stresses and strains, and total and fluid pressures, as well as surface monuments and markers, were usually installed in both rockfill and earth dams.

The following is a very brief (and by no means complete) description of instruments commonly installed in rockfill dams. Literature on the design and application of the various instruments is extensive and the reader should review, for example, Refs. (14; 15, Ch. 3; 35; 55; 58; 59; 63; 64; 65; 66; 81; 141; 142; and 143).

1. Piezometers measure the static pressure of the fluid (water or in some cases gas)

in the pore spaces of the soil or rockfill. There are a number of types of piezometers some of which are suitable for measuring pressures in materials with low permeability such as in the core of a rockfill dam; others are applicable to measurements in permeable shell, transition or drain materials; and some to either. The simplest is the open standpipe which is merely an observation well in which the water level is measured directly by a probe. It is not normally suitable for use in impervious soil because of the time lag required for the relatively large quantities of water to move into and out of the well. The Casagrande piezometer consists of a porous cell encased in sand and sealed in a bore hole at the desired location, and connected to a small diameter tube leading to the surface. Because of the small amount of water transfer required, the device is suitable for use in relatively impervious soils. Hydraulic piezometers, such as the type developed by the U. S. Bureau of Reclamation,[143] have been used many years. Small cells equipped with porous inserts and buried in the fill, are connected by long tubing lines to pressure gauges located in a housing, usually on the downstream slope of the dam. Two lines to each cell permit purging the system of accumulated gas. Pneumatic piezometers consist of a sealed cell equipped with a pressure sensitive diaphragm which closes a port connected to one of two tubing lines leading to the surface. Gas or fluid pressure applied to the line, equal to the pore pressure, opens the port and bypasses the gas to a return line. The applied gas pressure is therefore a measure of the pore pressure. The devices are sensitive, quick acting and durable, and offer considerable freedom in the location of readout equipments and minimum interference with construction. Electrical piezometers consist of cells with pressure sensitive diaphragms which actuate various types of electrical transducers and generate signals indicative of the pore pressures acting on the diaphragms. They also are quite sensitive and accurate but, as presently designed, possibly may not have sufficient durability to give reliable readings over a very long period of time when buried in embankments.

2. Devices for measuring internal vertical movements are quite varied but some of the more commonly used equipment may be mentioned as briefly as follows:

(a) Cross-arm settlement devices as developed by the USBR[143] consist of telescoping $1\frac{1}{2}$ in. and 2 in. pipe sections with horizontal crossarms fixed to the $1\frac{1}{2}$ in. sections at intervals of 5 to 10 ft. Sections are added as the embankment is raised, and the cross-arm sections furnish reference points within the embankment. The positions of the crossarm sections are periodically monitored by lowering a torpedo equipped with latching pawls down the pipe, engaging each of the crossarm sections and measuring the depths to the latching points. This simple and reliable device has been used for many years. Some types of inclinometer casings (discussed below) are equipped with telescoping joints which allows them to function in a manner similar to a crossarm settlement device. Friction between the casing and the soil rather than anchorage by crossarms is relied upon to telescope joints.

(b) Various versions of taut-wire devices have been developed to measure internal settlements. In essence these devices consist of an anchorage in the fill to which is attached a wire, protected by casing, leading to tensioning and read-out equipment at the surface for determining the vertical position of the anchorage. Some taut-wire extensometers activate potentiometers or other electrical devices connected to read-out circuitry which furnish continuous readings of the positions of the anchorages. Solid rods with anchors which can be expanded and set at bottoms of bore holes to provide reference points, have also been described in the literature.

(c) Hydraulic settlement devices of various kinds have been installed in dams. A device used at Oroville Dam consists of two small interconnected chambers arranged vertically with a float valve between them. Three tubing lines lead from the buried unit to the sensing equipment located at a lower elevation. The unit is flushed by pumping water into the lower chamber and out through a return line attached to the same chamber. Air is then admitted to the upper chamber to depress the water level in the lower chamber to the level of the return line outlet port. With the water level in the lower chamber thus established, the hydrostatic head in the water line is read by a sensitive gauge and the elevation of the chamber determined. A device developed for Don Pedro Dam consists of a small closed vessel containing a very sensitive pneumatic piezometer tip and connected by water lines to a circulating water tank and a fixed elevation weir tank. Water is circulated in the system then the flow is shut off, the top elevation of the water column communicating with the buried vessel drops to the weir level, and the pressure in the vessel is read by the piezometer, from which the elevation of the vessel can be determined. This device was especially designed to measure the differential settlement between zones within the dam. Various other settlement devices based on measuring the head of water between a point in the dam and a fixed outside reference point have been assembled and used successfully. A different version of a device operating on this same principle[55] consists of a torpedo-shaped vessel filled with water and connected by long flexible tubing to a manometer. As the torpedo is pulled through a tube buried in the fill and inclined slightly above the horizontal, the manometer is read, and the readings can be converted to elevations of the torpedo at all points along the horizontal tubes. In the Idel system metal plates also are anchored at intervals along the traverse tube, and a sonde which triggers an audible signal at the outside controls when centered at each metal plate, is pulled through the tube. This device, therefore, can measure both vertical and horizontal displacements in the dam.

3. Internal horizontal movement devices which measure either directly or indirectly the horizontal displacements within dams are in common use.
    (a) Taut-wire arrangements, similar in principle to those used to measure vertical movements, have been installed in a number of dams.[143]
    (b) Crossarm devices, using telescoping pipe sections, installed in a horizontal position, have also been utilized. Readings of displacements are taken by means of a special cable-operated torpedo.
    (c) Inclinometers are displacement sensing devices which have gained considerable favor during the past decade. A casing, usually of about 4 in. diameter aluminum tubing with 4 alignment grooves formed in the interior wall, is installed in a near ver-

tical position in the embankment. A torpedo containing electrical circuitry which can sense in one plane the deviation of the device from the vertical or a predetermined reference angle is lowered down the casing with guide wheels engaged in 1 of the 4 internal grooves. Deformations in the embankment acting normal to the axis of the casing deflect the casing from its originally installed position. The casing deflections are detected and registered as point to point changes in slope as the torpedo traverses the length of the casing. The torpedo guide wheels are engaged in each of the 4 casing grooves in turn, consequently, cross-checked deflection measurements in 2 orthogonal planes are obtained. The bottom of the casing is usually fixed in bedrock to serve as the reference point for the measured cumulative slope changes indicated by the torpedo. Normally, the grooved casing is placed in a near-vertical position but at Don Pedro Dam one unit was installed at a 40° angle from the vertical to detect deformations near the face of the upstream slope and also to permit readings to be taken when the slope is submerged. This innovation required a specially designed torpedo and very careful alignment of the casing grooves as sections of the casing were added during construction of the fill.

(d) Strain meters designed for embedment in embankments are essentially electrically operated gauging units containing linear potentiometers arranged to detect compression or extension within a gauge length of 10 to 15 ft (3.05 to 4.56 m). These instruments are a fairly recent development and have not as yet been installed in very many dams. The arrays of strain meters installed in 1969 at Don Pedro Dam are believed to be the first in the United States. Lines of strain meters were placed at 2 levels in the upper part of the dam, extending for several hundred feet out from each abutment for the purpose of detecting and measuring the extension which would be expected to occur in these zones of the embankment.

4. Other measuring devices installed in or on embankment dams are:
   (a) Stress meters, to measure pressures within an earth or rockfill mass or the pressure exerted by such materials against structures, have been installed in some dams. However, there are inherent problems associated with such gauges. The size and stiffness of the buried unit and the manner in which it is installed can very significantly affect the accuracy of its response. Several types of gauges are available, however, which apparently minimize the disturbance of the stress pattern caused by the presence of the gauge itself.
   (b) Surface monuments and alignment markers are installed on nearly all dams. Their vertical and horizontal positions are periodically determined by accurate surveys with reference to fixed monuments and bench marks located outside of the area of influence of the dam and reservoir. Various other methods for making surface deformation measurements, such as those based on the use of laser beams as reference lines, are used to a limited degree at present.
   (c) Seismographic recorders and seismoscopes have been installed on many dams in the last few years and the trend is toward the placement of such instruments generally on dams, particularly those located in areas with any known history of significant seismic activity.

Examples of the arrays of instruments installed in modern rockfill dams to monitor their performance are those at Oroville Dam (Section VIII, discussed in detail in Ref. 142, or at Don Pedro Dam, described briefly in Section VIII and Ref. 66 (see also Ref. 59 and 141).

Deformation and Cracking of Dams

During construction of dams deformations of the completed portions occur, the magnitudes and directions of which depend on a number of factors. Some of the most influential of these factors are (1) the profile of the valley in which the dam is built, (2) the configuration of the dam with respect to both plan and section, (3) internal zoning of the dam, (4) physical characteristics, including the stress-strain relationships and time-dependent effects, of the fill materials, (5) construction methods, practices, schedules, and construction control, (6) climatic condition in the construction areas. These deformations are important not only as early indices of safety, adequate design and proper construction, but their effects can be strongly manifest in the post-construction behavior of the dam. They are complex but are usually defined by three components, i.e. (1) settlement (vertical movement), (2) deflection (movement normal to the dam axis), and (3) lateral displacement (cross-valley movement). In view of the many influencing variables there can be no generally applicable "normal" pattern of deformation of fill dams, except perhaps in the broadest sense, but comparisons can be made between structures which have certain dominant features in common. Some values of observed deformations of specific dams have been cited in other sections of this chapter. The results of detailed observations, measurements, and analyses of deformations of a number of major earth and rockfill dams are given in Refs. 55, 81, 83, 85. Recent measurements of movements occurring during construction of a high rockfill dam in Britain are presented in Ref. 141.

The following are general comments regarding deformations in rockfill dams.

1. The post construction settlement profile along the crest of the dam will reflect the height of fill at each specific point. A dam in a V-shaped valley will show gradually increasing settlement toward the valley center. The differential settlement will create axial compression toward the center of the dam and axial extensions in zones in the upper part of the dam adjacent to each abutment. If the valley is U-shaped, differential settlement and shear strain near the abutments can be relatively higher than would be the case if the abutment slopes were more gradual.
2. Settlement is often accompanied by appreciable lateral movement of materials toward the center of the dam.
3. The maximum shear strains at the abutments, accompanying the vertical and lateral movements, do not usually occur at the foundation contact but in a zone some feet above it.
4. During the first filling of the reservoir it is not unusual for the crest of a central core dam to deflect upstream due to an accelerated settlement of the upstream shell but, as the reservoir rises toward full pool, the direction of the deflection reverses because of the water pressure, and the net movement will usually be downstream.
5. The first filling of the reservoir is a critical period for most dams because of changes in the deformation pattern, water loadings and effects of saturation on the properties of the in-place fill materials. The rate of filling has an important bearing on the severity of the effects on the dam—it may adjust to a slowly rising pool whereas a rapid filling might possibly cause some structural distress, such as the formation of cracks in the core or shell.
6. Differential movements between zones in dams usually occur, particularly those with central or moderately sloping cores, primarily because of the differences in stress-strain properties of the materials as placed in the structure. The relative

rates of consolidation as well as the compressibilities of the materials are significant. For example, the initial settlement of the shell and transition zones with respect to the core may impose a downward shearing stress on the core, but with passing time the more slowly consolidating core may tend to move downward with respect to the shell and the direction of the shearing stresses would reverse.

7. The action of load transfer between zones and arching in rockfill dams was first reported by Lofquist in 1951.[151] The matter of differential settlement between zones and the resultant redistribution of stresses within the dam has since been studied by a number of investigators (see Refs. 37, 55, 63, 64, 65, 81, 83, 84, 85, 124, 141, 150, 157, 158). It must be given careful consideration in the design of the structure because differential movements of this kind are one of the principal causes of cracking in dams, as will be discussed later.

8. As the shells settle the crest tends to spread laterally because of the formation of an extension zone at the top of the dam.

9. Impervious-membrane-faced or extremely sloping core dams of dumped rockfill settle and deflect downstream significantly more than compacted rockfill dams. Lateral or cross-valley movements are also usually quite significant and the direction and magnitude of movement may be influenced by the direction of dumping and height of lifts of the rockfill. The facing tends to deflect under water loadings so as to acquire a shallow bowl-shaped configuration. The peripheral areas of such facings or sloping cores are the most vulnerable to settlement damage.

The distortions which dams undergo may lead to cracking of several different kinds. Longitudinal cracks frequently form in the crests of high multi-zoned dams as a result of the differential settlement between the zones. Such cracks sometimes have the appearance of having been caused by shear sliding of the embankment but are not in fact related to shearing instability of the structure. Longitudinal cracks may extend over several hundreds of feet, paralleling the axis of the dam, and although objectionable, they do not seriously menace the safety of the dam unless, influenced by strains in other directions, they develop diagonal branches extending into the core. Near-vertical transverse or diagonal cracks into or through the core are matters of particular concern since they can form paths of concentrated seepage and, in some cases, lead to erosion of the core and jeopardize the safety of the dam. Transverse tensile cracks can be caused by differential settlement within the core creating shear strains in planes oriented transversely to the axis. Such strains and the resulting transverse cracks are sometimes caused by abrupt irregularities in the foundations of the embankment, particularly at the abutments where the top of the embankment tends to be in a state of extension rather than compression. If the core is nonplastic and brittle in nature the tensile forces in the zones near the abutments can crack the core even though there may be no abrupt offsets of any kind in the foundation (see Refs. 152, 158, 159). The depths to which the vertical transverse cracks can exist as open fractures depend upon the relationship of the unconfined compressive strength of the core material to the vertical loading imposed by the weight of the embankment and the hydraulic pressure which may propagate the crack by hydraulic fracturing.

As the core is deformed by differential settlement the minor principal stress in the zone of an incipient crack decreases and could even become zero if the major principal stress is less than the unconfined compressive strength of the soil. An open crack would form if the soil in this state were further deformed but even at lesser deformation fine closed hairline cracks could be created. Reservoir water can penetrate such fine cracks or thin zones where the minor principal stress is low and exerts a significant differential pressure on the crack or zone boundaries because the head losses in the crack or loosened zone may be substantially less than in the adjacent intact soil. The result is a widening and probably further propagation of the crack. The process of hydraulic fracturing has been known for some time and is, in fact, used to advantage in some drilling operations. It may be of interest to note that on occasion when drill holes were put down through an impervious core for supplemental foundation grouting or other purposes, at some depth the drilling fluid would suddenly be lost as though the drill had entered a void in the core. Subsequent investigation indicated the fluid loss to be the result of localized hydraulic fracturing due to the pressure of the drilling fluid and not due to a pre-existing void. It follows therefore that at considerable depth in the core where the vertical loading is high an incipient crack would be forced to remain closed but in an upper zone in the embankment where the compressive loading is less or the soil is in a state of lateral extension the cracks would be open.

Of most serious concern is the possibility of horizontal cracks forming in the core. Since they could not be observed at the surface they might remain undetected until excessive seepage through the core or other signs of distress made their presence known. The most likely mechanism which could contribute to horizontal cracking is that of arching. Under certain conditions this can happen when the core settles more than the shells. Through load transfer, very narrow, near-vertical cores with vertical or steeply sloped faces could be partially supported by the shells. Such arching could cause the minor principal stress in a part of the core to be greatly reduced and thus create conditions conducive to hydraulic fracturing and the formation of horizontal cracks. The cracks could conceivably extend, at least locally, completely through the core.

The matter of cracking of dams warrants the most careful consideration by the designer. A comprehensive analysis of the mechanism of cracking in dams is presented in Ref. 84.

A number of measures may be adopted in the design and construction of the dam to minimize cracking and to alleviate its effects should it occur. Some of these measures are listed briefly as follows:

1. Position the axis whenever practicable so that the foundation contours at the abutments downstream of the core will tend to converge, and curve the axis upstream so any downstream deflection will tend to increase rather than decrease the axial compression. This may be of some benefit particularly if a canyon site is narrow and the core of the dam is relatively rigid and brittle.
2. Provide ample transition zones of cohesionless material both upstream and downstream of earth cores. This is one of the most important safety measures that can be incorporated in the design. The upstream transition, in addition to forming a filter between the core and shell, will provide material to be washed into any substantial, open transverse crack in the core and suppress any leakage flow through the crack. The downstream transition is designed as a filter to prevent the sluicing of core material into the internal drain zone or downstream shell in the event of a through crack in the core. It also serves as a "crack-stopper" since a crack cannot propagate and remain open in the cohesionless filter material. At Don Pedro Dam (Figure 7-11) the upper part of the downstream transition was widened substantially adjacent to the core zone where cracking would be most likely to occur.
3. Provide drains, or assured drainage through the downstream shell, with ample capacity to accommodate any reasonably

conceivable flow of water that could issue through a core crack.
4. The ratio of the volume of impervious core material to the volume of rockfill established by the design of the dam will usually be governed by factors such as the availability of materials, cost of hauling and placing, climatic conditions, schedules, etc., other than considerations of cracking potential. If a low ratio is desired or mandatory a sloping core rather than a central core section might be preferred, but if a very narrow, steep-sided centrally located core is adopted special care in the design and the construction of the transitions, filters and drains is required. Locally increasing the core thickness at the abutments by gradually flairing the core slopes will provide some further safeguard against the effects of cracking and lengthen the seepage path at the foundation contact.
5. The rockfill should be placed at the maximum density that can practicably be attained to reduce post-construction distortions of the embankment, increase the strength of the fill material and therefore enhance the overall safety factor of the dam. Obviously, the selection, loading, placing, moisture control, and compaction of the core materials are equally important. The placement of more plastic or moist soil in the abutment areas than elsewhere in the core has often been done to aid in suppressing cracks.
6. Adequate instrumentation should be installed to monitor the overall performance of the dam, as suggested in this Section under "Instrumentation." For high dams, the instrument array should include strain meters in the normal tension zones in the core near each abutment to detect the onset of cracking and greatly facilitate the locating of any cracks which may form so prompt repairs (such as grouting with sand-bentonite mixtures) may be accomplished.

Construction Quality Control

The performance of any dam will depend not only upon the competence and thoroughness with which the design had been accomplished but equally, at least, upon the quality of the construction operations. If the dam is to perform as anticipated, the properties of the materials in place must conform to those upon which the design was based. Therefore, the objective of construction quality control is to obtain values of the significant properties of the fill materials comparable to those used in design.

Soils, for engineering purposes, are identified and classified as to particle size, gradation, Atterberg limits or limits of consistency with respect to moisture content, including dilatancy, dry strength, and toughness; and the physical properties of permeability, compressibility and shear strength of the soil mass as determined by laboratory tests for use in design. The quantified values of the latter 3 properties depend primarily on the particle size, gradation, moisture content and density of the compacted material. Field quality control, therefore, is required for assurance that at every location in the dam, materials of the proper classification, particle size, and gradation are placed and densified as specified. For soils the placement lift thickness, the distributed moisture content in relation to the optimum moisture content prior to compaction, and the compactive effort applied as measured by the number of passes of the specified type, size and weight of the compacting equipment, must be closely controlled. The degree of compaction achieved is measured by the relationship of the dry density of the compacted material to a maximum dry density as determined in the laboratory by one of several standard methods. The procedures for field quality control of earthwork operations have been developed through many years of earth dam construction, and although the details of methods and equipment may differ amongst individuals and engineering organizations, the basic requirements for such control are universally recognized. Literature dealing with both the technical soil mechanics and construction practice aspects of field control is voluminous and no attempt at summarization will be made here. As general references, however, the publications of the U.S. Bureau of Reclamation (e.g. Refs. 52 and 143) are suggested.

Field control procedures for rockfill operations are not as well standardized as those for earthwork. Rockfill materials are usually produced by quarrying, and the physical characteristics of the native rock, the shot hole size and pattern, the type and quantity of explosives used in blasting operations, and the handling of the shot-rock product—all influence the size and gradation of the rockfill materials available for placement in the dam. The interrelationship of these factors can only be determined on the basis of experience or by experimentation.

Further experimentation is necessary to establish the lift thickness; type, kind and weight of compaction equipment; compactive effort or number of passes, and amount of water to be applied to achieve optimum density of compacted rockfill. Test fills are customarily constructed using various lift thicknesses and numbers of passes of the compaction equipment which may be used on the fill to determine the best combination of these variables for each class of rockfill material. The revolutionary changes in the type, particle size, and gradation of acceptable rockfill materials, and in the placement techniques for densifying such materials, brought about by the introduction and continuing development of the vibratory roller, have made close field control for rockfill as necessary as it is for earthfill.

High relative density is a principal qualification for acceptable compacted rockfill, but the field measurement of in-place density is considerably more difficult and time-consuming than it is for earthfill because of the size of both the sample and the necessary equipment. The methods developed, for example, at Cougar Dam, Oroville Dam, and Don Pedro Dam for determining the relative density of the fill were generally similar, but differed somewhat as to the details of equipment and procedures. In brief, the method was as follows:

The surface of the compacted rockfill where the test was to be performed was leveled as much as possible by the removal of projecting sharp rock particles. A flanged steel ring 4 ft to 6 ft in diameter and about 6 in. deep was secured in place on the leveled area by means of sandbags or other weights or anchors. The inside face of the ring and the rock surface were lined with a flexible membrane of plastic sheets, and the container thus formed was filled to a marked level below the top of the ring with a measured volume of water. The water and plastic sheets were then removed and the rockfill material inside the ring was carefully removed to a depth of several feet and the surface of the excavation smoothed as much as possible. The plastic sheets were replaced, and the hole refilled with water to the previous level and the required volume measured; the difference between the 2 volumes of water being the volume of the excavation. The excavated material was placed in a container in the loosest possible state and the bulk volume at minimum density determined. The container and contents were then taken to the on-site laboratory, a known surcharge loading applied to the rock, and the container and rock vibrated for a predetermined length of time at an established frequency, using a vibrating table or an attached vibrator. The volume of the compacted rock was measured and the material weighed. The minimum density, maximum laboratory density, and the relative density of the material as placed in the fill could then be calculated. The sample was then screened to determine the particle size gradation, and, on occasion, standard soil tests were performed on the fine fraction.

A set of control tests for soil is usually made on samples, representing about 2000 to 5000 cu yd of placed fill, but tests on rockfill are normally made much more infrequently, with one density test representing perhaps 50,000 to 70,000 cu yd of rock. The important matter of quality control during construction is treated much more extensively in Ref. 78 (see also Refs. 35, 47, 64, 65, 141, 149, and 155).

REFERENCES

1. "Rockfill Dams: Review and Statistics" by John Snethlage, F. W. Scheidenhelm and Arthur N. Vanderlip, ASCE Journal of the Power Division Proc. Paper 1739, August 1958.
2. "Register of Dams in the United States," compiled and edited by T. Mermel, Chairman, Committee on Register of Dams, United States Committee of the International Commission on Large Dams.
3. "Design of Rockfill Dams" by J. D. Galloway, Transactions, American Society of Civil Engineers Vol. 104, Paper No. 2015, 1939.
4. "Rockfill Dams: Cogswell and San Gabriel Dams" by Paul Baumann, Transactions, American Society of Civil Engineers, Vol. 125, Part II, Paper No. 3064, 1960.
5. "Rockfill Dams: Salt Springs and Lower Bear River Concrete Face Dams" by I. C. Steele and J. B. Cooke. Transactions, American Society of Civil Engineers Vol. 125, Part II, Paper No. 3065, 1960.
6. "Rockfill Dams: Wishon and Courtright Concrete Face Dams" by J. Barry Cooke. Transactions, American Society of Civil Engineers, Vol. 125, Part II, Paper No. 3083, 1960.
7. "Rockfill Dams: Lemolo No. 1 Dam" by J. C. Boyle and W. R. Barrows. Transactions, American Society of Civil Engineers Vol. 125, Part II, Paper No. 3074, 1960.
8. "Rockfill Dams: Paradela Concrete Face Dam" by Luis Henrique Gomes Fernandes, Edgard de Oliveria, and Nuno de Vasconcelos Porto. Transactions, American Society of Civil Engineers, Vol. 125, Part II, Paper No. 3076, 1960.
9. "Discussion of Paper No. 3065, "Salt Springs and Lower Bear River Concrete Face Dams," (Ref. 5) by Dr. Karl Terzaghi. Transactions, American Society of Civil Engineers, Vol. 125, Part II.
10. "Settlement of Rockfill Dams," by F. L. Lawton and M. D. Lester. Transactions of the Eighth International Congress on Large Dams, Vol. III, Edinburgh 1964.
11. "Design and Construction of Holjes Dam," by E. Reinius. Transactions of the Eighth International Congress on Large Dams, Vol. III, Edinburgh 1964.
12. "Methods and Equipment for Slipforming of Concrete Faces on Rockfill Dams," by T. J. Szczepanowski. Transactions of the Eleventh International Congress on Large Dams, Vol. III, Madrid 1973.
13. "The Design of Cethana Concrete Face Rockfill Dam," by J. K. Wilkins, W. R. Mitchell, M. D. Fitzpatrick, and T. Liggins. Transactions of the Eleventh International Congress on Large Dams, Vol. III, Madrid 1973.
14. "Instrumentation and Performance of Cethana Dam," by M. D. Fitzpatrick, T. B. Liggins, L. J. Lack, and B. P. Knoop. Transactions of the Eleventh International Congress on Large Dams, Vol. III, Madrid 1973.
15. Earth and Earth-Rock Dams, by James L. Sherard, Richard J. Woodward, Stanley F. Gizienski, and W. A. Clevenger. John Wiley and Sons, Inc. 1963.
16. "Limit Height Criteria For Loose-Dumped Rockfill Dams," by P. F. Baumann. Transactions of the Eighth International Congress on Large Dams, Vol. III, Edinburgh 1964.
17. "Rockfill Dams: Montgomery Dam With Asphaltic Concrete Deck," by W. F. Scheidenhelm, John B. Snethlage, and Arthur Vanderlip. Transactions, American Society of Civil Engineers, Vol. 125, Part II, Paper No. 3078, 1960.
18. "Rock Dam In The Rockies," by Ray Day. An account of the construction of Montgomery Dam. Excavating Engineer, January, 1957, p. 34.
19. "Design and Construction of Homestake Dam," by Harris H. Burke, M. M. Forrest, and R. L. Perley. ASCE Water Resources Conference, Preprint 346, Denver 1966.
20. "Report On The Behavior Of Impervious Surface Of Asphalt," by H. W. Koenig and K. H. Idel. Transactions of the Eleventh International Congress on Large Dams, Vol. III, Madrid 1973.
21. "Empirical Research And Practical Design Of Rockfill Dams With Asphalt Facings," by T. Sawada, Y. Nakazima, and T. Tanaka. Transactions of the Eleventh International Congress on Large Dams, Vol. III, Madrid 1973.
22. "The Membrane of the Pozo De Los Ramos Dam," by J. A. Herreras. Transactions of the Eleventh International Congress on Large Dams, Vol. III, Madrid 1973.
23. "Ogliastro Reservoir Peripherical Rockfill Dam, with 90000 m.^{2} Upstream Bituminous Membrane," by G. Baldovin and A. Ghirardini. Transactions of the Eleventh International Congress on Large Dams, Madrid 1973.

24. "Slope Protection on Earth and Rockfill Dams," by Karl V. Taylor. Transactions of the Eleventh International Congress on Large Dams, Vol. III, Madrid 1973.
25. "Rockfill Dams: Steel-Faced Dam," by James L. Sherard. Transactions, American Society of Civil Engineers, Vol. 125, Part II, Paper No. 3079, 1960.
26. Discussion of Paper No. 3076, "Paradela Concrete Face Dam," by Armondo da Palma Carlos. Transactions, American Society of Civil Engineers, Vol. 125, Part II, 1960.
27. "General Paper No. 7," by the General Paper Committee of the United States National Committee on Large Dams. Transactions, Tenth International Congress on Large Dams, Vol. IV, Montreal 1970.
28. Flow Through Rockfill, by Thomas M. Leps. Embankment-Dam Engineering-Casagrande Volume. John Wiley and Sons 1973.
29. "Rockfill Dams: Nantahala Sloping Core Dam," by James P. Growdon. Transactions, American Society of Civil Engineers, Vol. 125, Part II, Paper No. 3066, 1960.
30. "Rockfill Dams: Dams with Sloping Earth Cores," by James P. Gorwdon. Transactions, American Society of Civil Engineers, Vol. 125, Part II, Paper No. 3069, 1960.
31. "Rockfill Dams: Performance of Sloping Core Dams," by James P. Growdon. Transactions, American Society of Civil Engineers, Vol. 125, Part II, Paper No. 3070, 1960.
32. "Essais De Mise En Oeuvre D'Enrochements Au MontCenis," by L. Pousse and J. Molbert. Transactions, Eighth International Congress on Large Dams, Vol. III, Edinburgh 1964.
33. "The Furnas Rockfill Dam," by F. H. Lyra and L. Queiroz. Transactions, Eighth International Congress on Large Dams, Vol. III, Edinburgh 1964.
34. "The Miboro Dam," by I. Asao. Transactions, Eighth International Congress on Large Dams, Vol. III, Edinburgh 1964.
35. "Design and Methods of Construction—Oroville Dam," by Bernard B. Gordon and Jack G. Wulff. Transactions, Eighth International Congress on Large Dams, Vol. IV, Communication C. 12, Edinburgh 1964.
36. "Oroville Dam," by Alfred R. Golze, Western Construction, April 1962, p. 25.
37. "Optimum Position Of The Central Clay Core Of A Rockfill Dam In Respect To Arching And Hydraulic Fracture," by M. Maksimovic. Transactions, Eleventh International Congress on Large Dams, Vol. III, Madrid 1973.
38. "Some Stability Properties Of A Dam Having An Inclined Core," by E. Reinius. Transactions, Eleventh International Congress On Large Dams, Vol. III, Madrid 1973.
39. "Rockfill Dams: Bersimis Sloping Core Dams," by F. W. Patterson and D. H. Macdonald. Transactions, American Society of Civil Engineers, Vol. 125, Part II, Paper No. 3081, 1960.
40. "Rockfill Dams: Brownlee Sloping Core Dam," by Torald Mundal. Transactions, American Society of Civil Engineers, Vol. 125, Part II, Paper No. 3082, 1960.
41. "Rockfill Dams: Mud Mountain Dam," by Allen S. Cary. Transactions, American Society of Civil Engineers, Vol. 125, Part II, Paper No. 3067, 1960.
42. "Rockfill Dams: TVA Central Core Dams," by George K. Leonard and Oliver H. Raine. Transactions, American Society of Civil Engineers, Vol. 125, Part II, Paper No. 3068, 1960.
43. "Rockfill Dams: Cherry Valley Central Core Dam," by H. E. Lloyd, O. L. Moore, and W. F. Getts. Transactions, American Society of Civil Engineers, Vol. 125, Part II, Paper No. 3075, 1960.
44. "Rockfill Dams: Design of Cougar Central Core Dam," by Paul Thurber. Transactions, American Society of Civil Engineers, Vol. 125, Part II, Paper No. 3085, 1960.
45. "General Paper No. 9, U. S. Committee on Large Dams," by the General Paper Committee of the United States Committee on Large Dams. Transactions, Eighth International Congress on Large Dams, Vol. IV, Edinburgh 1964.
46. "Design, Methods of Construction, and Performance of High Rockfill Dams," by J. Barry Cooke, General Reporter, Question 31, Transactions, Eighth International Congress on Large Dams, Vol. IV, Edinburgh 1964.
47. "Construction Experience—Cougar Dam," by Donald H. Basgen. Transactions, Eighth International Congress on Large Dams, Vol. III, Edinburgh 1964.
48. Discussion by J. Barry Cooke of Paper No. 3082, "Rockfill Dams: Brownlee Sloping Core Dam" (Ref. 40). Transactions, American Society of Civil Engineers, Vol. 125, Part II, 1960.
49. "Ambuklao Rock-Fill Dam, Design and Construction," by E. Montford Fucik and Robert F. Edbrooke. Transactions, American Society of Civil Engineers, Vol. 125, Part I, Paper No. 3057, 1960.
50. "Rockfill Dams: Derbendi Khan Dam," by Calvin V. Davis. Transactions, American Society of Civil Engineers, Vol. 125, Part II, Paper No. 3084, 1960.
51. Discussion by S. Sakurai by Paper No. 3084,

"Rockfill Dams: Derbendi Khan Dam" (Ref. 50). Transactions, American Society of Civil Engineers, Vol. 125, Part II, 1960.
52. "Design of Small Dams," United States Department of the Interior, Bureau of Reclamation-Denver, Colorado. Second Edition, 1973.
53. "Rockfill Dams: Design and Construction Problems," by D. J. Bleifuss and James P. Hawke. Transactions, American Society of Civil Engineers, Vol. 125, Part II, Paper No. 3072, 1960.
54. "Rockfill Dams: Kajakai Central Core Dam-Afghanistan," by Glenn F. Sudman. Transactions, American Society of Civil Engineers, Vol. 125, Part II, Paper No. 3073, 1960.
55. "Earth and Rockfill Dams," by Stanley D. Wilson and R. Squier. Seventh International Conference on soil Mechanics and Foundation Engineering, State of the Art Volume, Mexico 1969.
56. "Measures Providing Impermeability of the Nurek Dam," by V. I. Vutsel, P. P. Listrovoy, M. P. Malyshev, and V. I. Shcherbina. Transactions, Eleventh International Congress on Large Dams, Vol. III, Madrid 1973.
57. "The Gepatsch Rockfill Dam in the Kauner Valley," by H. Lauffer and W. Schober. Transactions, Eighth International Congress on Large Dams, Vol. III, Edinburgh 1964.
58. "The Interior Stress Distribution of the Gepatsch Dam," by W. Schober. Transactions, Tenth International Congress on Large Dams, Vol. I, Montreal 1970.
59. "El Infiernillo Rockfill Dam," by R. J. Marsal. Transactions, Eighth International Congress on Large Dams, Vol. III, Edinburgh 1964.
60. "Design and Construction of Summersville Dam," by J. N. Barnes. Transactions, Eighth International Congress on Large Dams, Vol. III, Edinburgh 1964.
61. "Rockfill Design-Carters Dam," Journal of the Construction Division, American Society of Civil Engineers, 92, (No. CD3), Proc. Paper No. 4906, September 1966, pp. 51-74.
62. "Design of Round Butte Dam," J. G. Patrick and W. R. Ferris. Pre-print—Water Resources Engineering Conference, American Society of Civil Engineers, Denver, Colorado, May 17, 1966.
63. "Post-Construction Behavior of Round Butte Dam," by James G. Patrick. Journal of the Soil Mechanics and Foundations Division, American Society of Civil Engineers, 93, No. (SM 4), July 1967.
64. "Evaluation of Cougar Dam Embankment Performance," by Robert J. Pope. Journal of the Soil Mechanics and Foundations Division, American Society of Civil Engineers, 93, (No. SM 4), July 1967.
65. "Performance of El Infiernillo Dam, 1963-1966," by Raul J. Marsal and Luis Ramirez de Arellano, Journal of the Soil Mechanics and Foundations Division, American Society of Civil Engineers, 93, (No. SM 4), July 1967.
66. "New Don Pedro Project," by Karl V. Taylor, Paper No. 3 CANCOLD/USCOLD Joint Technical Meeting, Vancouver, British Columbia, October 1971.
67. "Concrete Diaphragm Wall, Bighorn Dam," by D. J. Forbes, J. L. Gordon, and S. E. Rutlege. Transactions, Eighth International Congress on Large Dams, Vol. III, Madrid 1973.
68. "Asphaltic Concrete Cores, Experiences and Developments," by A Lohr and A Feiner. Transactions, Eleventh International Congress on Large Dams, Vol. III, Madrid 1973.
69. "Considerations and Investigations for the Design of a Rockfill Dam with a 92 m. High Bituminous Mix Core," by W. Schober. Transactions, Eleventh International Congress on Large Dams, Vol. III, Madrid 1973.
70. "The Vertical Asphaltic Concrete Core of the Earth-Fill Dam Eberblaste of the Zemm Hydroelectric Scheme, by Hans Kropatschek and Kurt Rienossl. Transactions, Tenth International Congress on Large Dams, Vol. I, Montreal 1970.
71. "Embankment Dams with Asphaltic Concrete Cores," by K. Rienossl. Transactions, Eleventh International Congress on Large Dams, Vol. III, Madrid 1973.
72. Discussion of Paper No. 2015, "Design of Rockfill Dams," (Ref. 3) by F. Gomez-Perez and M. Jinich. Transactions, American Society of Civil Engineers, Vol. 104, 1939.
73. "The Effect of Regional Conditions in Japan on Design and Construction of Impervious Elements of Rockfill Dams," by M. Takahashi and K. Nakayama. Transactions, Eleventh International Congress on Large Dams, Vol. III, Madrid 1973.
74. "Design of Loose Fill Dam Slopes by the Method of Characteristics," by S. V. Romero. Proceedings, Seventh International Congress on Soil Mechanics and Foundation Engineering, Vol. 2, Mexico 1969.
75. "Influence of Core Position on Stability of Rockfill Dam Founded on Sand, Gravel, and Boulder Deposits in Seismic Zone," by K. Y. Murthy, J. N. Srivastava, and S. K. Bhatia. Transactions, Eleventh International Congress on Large Dams, Vol. III, Madrid 1973.
76. Mechanical Properties of Rockfill, by Raul J. Marsal. Embankment-Dam Engineering-Casagrande Volume. John Wiley and Sons 1973.
77. "Shear Strengh of Coarse Embankment Dam Materials," by John Lowe III. Transactions, Eighth International Congress on Large Dams, Vol. III, Edinburgh 1964, pp. 745-761.
78. Field Tests for Compacted Rockfill, by George

E. Bertram. Embankment-Dam Engineering-Casagrande Volume. John Wiley and Sons 1973.

79. "Essais Sur Place Et Mesures De Controle Des Digues De Goschenenalp Et De Mattmark," by B. Gilg. Transactions, Eighth International Congress on Large Dams, Vol. III, Edinburgh 1964.

80. "Studies and Methods in Designing and Building High Fill Dams," by H. Press. Transactions, Eighth International Congress on Large Dams, Vol. III, Edinburgh 1964.

81. Deformation of Earth and Rockfill Dams, by Stanley D. Wilson. Embankment-Dam Engineering–Casagrande Volume. John Wiley and Sons 1973.

82. Stability of Earth and Rockfill Dams during Earthquakes, by H. Bolton Seed. Embankment-Dam Engineering–Casagrande Volume. John Wiley and Sons 1973.

83. "Deformations and Stability of Rockfill Dams," by A. A. Nitchiporovitch. Transactions, Eighth International Congress on Large Dams, Vol. III, Edinburgh 1964.

84. Embankment Dam Cracking, by James L. Sherard. Embankment-Dam Engineering–Casagrande Volume. John Wiley and Sons 1973.

85. "Design, Construction Methods and Performance of Rockfill Dams over 80 Meters High," by A. Lohr. Transactions, Eighth International Congress on Large Dams, Vol. III, Edinburgh 1964.

86. "Foundation and Abutment Treatment for High Embankment Dams on Rock," by The Committee on Embankment Dams and Slopes, American Society of Civil Engineers–Soil Mechanics and Foundations Division, Proceedings, ASCE, Vol. 98, No. SM 10, Paper No. 9269, October 1972.

87. "Foundation and Abutment Treatment for Rockfill Dams," by Richard C. Acker and Jack C. Jones. Proceedings, American Society of Civil Engineers–Soil Mechanics and Foundations Division, Vol. 98, SM 10, Paper 9303, October 1972.

88. "Abutment and Foundation Treatment for High Embankment Dams on Rock," by Reginald A. Barron. Proceedings, American Society of Civil Engineers–Soil Mechanics and Foundations Division, Vol. 98, SM 10, Paper No. 9270, October 1972.

89. "Current Practice in Abutment and Foundation Treatment," by Harris H. Burke, Charles S. Content, and Richard L. Kulesza. Proceedings, American Society of Civil Engineers–Soil Mechanics and Foundations Division, Vol. 98, SM 10, Paper No. 9268, October 1972.

90. "Foundations and Abutments-Bennet and Mica Dams," by Harold K. Pratt, Robert C. McMordie and Robert M. Dundas. Proceedings, American Society of Civil Engineers–Soil Mechanics and Foundations Division, Vol. 98, SM 10, Paper No. 9290, October 1972.

91. "Foundation Treatment for Embankment Dams on Rock," by Elmer W. Stroppini, Donald H. Babbitt, and Henry E. Struckmeyer. Proceedings, American Society of Civil Engineers–Soil Mechanics and Foundations Division, Vol. 98, SM 10, Paper No. 9274, October 1972.

92. "Foundation Practices for Talbingo Dam, Australia," by B. J. Wallace and J. I. Hilton. Proceedings, American Society of Civil Engineers–Soil Mechanics and Foundations Division, Vol. 98, SM 10, Paper No. 9273, October 1972.

93. "Treatment of High Embankment Foundations," by Fred C. Walker and R. W. Bock. Proceedings, American Society of Civil Engineers–Soil Mechanics and Foundations Division, Vol. 98, SM 10, Paper No. 9272, October 1972.

94. "Flow Through Rockfill Dam," by Horace A. Johnson. Proceedings, American Society of Civil Engineers–Journal, Soil Mechanics and Foundations Division, 97, No. (SM 2), February 1971, Paper No. 7914. Discussion by Thomas Kluber and Herbert Breth, S, M, and F Div., 97, No. (SM 11), November 1971; Closure, S, M, and F Div., 98, No. (SM 5), May 1972.

95. "Construction Techniques of Passing Floods over Earth Dams," by Andrew Weiss. Transactions, American Society of Civil Engineers, Vol. 116, 1951, Paper No. 2461.

96. "The Stability of Overtopped Rockfill Dams," by J. K. Wilkins. Proceedings, Fourth Australian Conference on Soil Mechanics and Foundation Engineering 1963, p. 1.

97. "Rockfill Structures Subject to Water Flow," by Alan K. Parkin, David H. Trollope, and John D. Lawson, Journal, American Society of Civil Engineers–Soil Mechanics and Foundations Division, 92, (SM 6), November 1966, Paper No. 4973. Discussion: J. K. Wilkins, (SM 3), May 1967; Sergio Giudici, (SM 5), September 1967; Closure: (SM 4), July 1968.

98. "The Sloughing, Overtopping, and Reinforcement of Rockfill and Earth Dams," by A. D. W. Sparks. Transactions, Ninth International Congress on Large Dams, Vol. IV, p. 327, Istanbul, 1967.

99. "Reinforcement of Rockfill Dams in South Africa," by H. N. F. Pells. Proceedings, Seventh International Conference on Soil Mechanics and Foundation Engineering, Vol. 2, p. 345, Mexico City 1969.

100. "Rockfill Designed to Withstand Overflow," by Hans Werner Koenig and Karl Heinz Idel. Transactions, Tenth International Congress on Large Dams, Vol. I, p. 259, Montreal 1970.

101. "Experience in the Design and Construction of Reinforced Rockfill Dams," by Ninham Shand and P. J. N. Pells. Transactions, Tenth International Congress on Large Dams, Vol. 1, p. 291, Montreal 1970.
102. "Design of Overflow Rockfill Dams," by F. Hartung and H. Scheuerlein. Transactions, Tenth International Congress on Large Dams, Vol. 1, p. 587, Montreal 1970.
103. Preparation of Rock Foundations for Embankment Dams, by William F. Swiger. Embankment-Dam Engineering—Casagrande Volume. John Wiley and Sons, New York 1973.
104. "Strength and Deformation Characteristics of Rockfill Materials," by N. Dean Marachi, C. K. Chan, H. Bolton Seed, and J. M. Duncan. Department of Civil Engineering—Institute of Transportation and Traffic Engineering—University of California, Berkeley. Report No. TE 72-3 to State of California Department of Water Resources, September 1969.
105. "Strength and Deformation Characteristics of Rockfill Materials in Plane Strain and Triaxial Compression Tests," by E. Becker, C. K. Chan, and H. Bolton Seed. Department of Civil Engineering—Institute of Transportation and Traffic Engineering—University of California, Berkeley. Report No. TE 72-3 to State of California Department of Water Resources, October 1972.
106. "Triaxial Shear Tests on Pervious Gravelly Soils," by W. G. Holtz and H. J. Gibbs. Proceedings, American Society of Civil Engineers-Journal, Soil Mechanics and Foundations Division, 82, No. (SM 1), January 1956, Paper No. 867.
107. "The Influence of End Restraint on the Compression Strength of a Cohesionless Soil," by A. W. Bishop and G. E. Green. Geotechnique, 15, pp. 244-266, 1965.
108. "Review of Shearing Strength of Rockfill," by Thomas M. Leps. Proceedings, American Society of Civil Engineers-Journal, Soil Mechanics and Foundations Division, 96, No. (SM 4), July 1970, Paper No. 7394. Discussion: Thomas F. Thompson, 97, (SM 1), January 1971; Raul J. Marsal, 97, (SM 3), March 1971; Don C. Banks and Bruce N. MacIver; 97, (SM 5), May 1971; George F. Sowers, 97, (SM 5), May 1971; Closure: 97, (SM 12), December 1971.
109. "Large Scale Triaxial Testing of Granular Materials," by Raul J. Marsal. American Society of Civil Engineers—Structural Engineering Conference, Miami, January 31, 1966, Preprint No. 303.
110. "Evaluation of Properties of Rockfill Materials," by N. Dean Marachi, Clarence K. Chan, and H. Bolton Seed. Proceedings, American Society of Civil Engineers—Journal, Soil Mechanics and Foundations Division, 98, No. (SM 1), January 1972, Paper No. 8672. Discussion: Mosaid Al-Jussaini, 98, (SM 12), December 1972; Raymond J. Frost, 98, (SM 12), December 1972.
111. "Large Scale Testing of Rockfill Materials," by Raul J. Marsal. Proceedings, American Society of Civil Engineers—Journal, Soil Mechanics and Foundations Division, 93, No. (SM 2), March 1967, Paper No. 5128. Discussion: Josef Brauns and Hans Leussink, (SM 6), Nov. 1967; Sidney F. Hillis and Nigel A. Skermer, (SM 1), Jan. 1968; Closure: (SM 4), July 1968.
112. "Tests on Cohesionless Materials for Rockfill Dams," by Emanuele Fumagalli. Proceedings, American Society of Civil Engineers—Journal, Soil Mechanics and Foundations Division, 95, (SM 1), January 1969, Paper No. 6353. Discussion: J. K. Wilkins, SM 2, March 1970; Closure: (SM 1), January 1971.
113. "Laboratory Tests on Materials and Static Models for Rockfill Dams," by E. Fumagalli, B. Mosconi, and P. P. Rossi. Transactions, Tenth International Congress on Large Dams. Vol. 1, p. 531, Montreal 1970.
114. "The Shear Strength of the Shell Materials for the Goschenenalp Dam, Switzerland," by J. Zeller and R. Wullimann. Proceedings, Fourth International Conference on Soil Mechanics and Foundation Engineering, Vol. II, p. 399, London 1957.
115. "Shear Strength of Rockfill" by J. G. Lewis. Proceedings, Second Australia-New Zealand Conference on Soil Mechanics and Foundation Engineering, January 1956, pp. 50-51.
116. "In-situ Shear Test for Rock Fills," by Surendra P. Jain and Ramesh C. Gupta. Proceedings, American Society of Civil Engineers, Journal of the Geotechnical Engineering Division, 100, No. (GT 9), September 1974, Paper No. 10823.
117. "Large Scale Shear Tests," by E. Schultze. Transactions, Fourth International Conference on Soil Mechanics and Foundation Engineering, Vol. I, p. 193, London 1957.
118. "Stability Analysis of Embankments," by John Lowe III. Proceedings, American Society of Civil Engineers—Journal, Soil Mechanics and Foundations Division, 93, (SM 4), July 1967, Paper No. 5305.
119. "Stability of Sloping Core Earth Dams," by Hassan A. Sultan and H. Bolton Seed. Proceedings, American Society of Civil Engineers—Journal, Soil Mechanics and Foundations Division, 93, (No. SM 4), July 1967, Paper No. 5307.
120. "Stability Analyses For a Sloping Core Embankment," by H. Bolton Seed and Hassan A. Sultan. Proceedings, American Society of Civil Engineers—Journal, Soil Mechanics and Founda-

tions Division, **93**, (No. SM 4), July 1967, Paper No. 5308.
121. "Slope Stability During Earthquakes," by H. Bolton Seed. Proceedings, American Society of Civil Engineers–Journal, Soil Mechanics and Foundations Division, **93**, (No. SM 4), July 1967, Paper No. 5319.
122. "Use of Computers For Slope Stability Analysis," by Robert V. Whitman and William A. Bailey. Proceedings, American Society of Civil Engineers–Journal, Soil Mechanics and Foundations Division, **93**, (SM 4), July 1967, Paper No. 5327.
123. "Special Problems In Slope Stability," by Willard J. Turnbull and Mikael J. Hvorslev. Proceedings, American Society of Civil Engineers–Journal, Soil Mechanics and Foundation Division, **93**, (SM 4), July 1967, Paper No. 5328.
124. "Analysis of Embankment Stresses and Deformations," by Ray W. Clough and Richard J. Woodward III. Proceedings, American Society of Civil Engineers–Journal, Soil Mechanics and Foundations Division, **93**, (SM 4), July 1967, Paper No. 5319.
125. "Potentially Active Faults in Dam Foundations," by James L. Sherard and Lloyd S. Cluff. A.S.C.E. National Structural Engineering Meeting, April 9-13, 1973, San Francisco. Preprint No. 1948.
126. "Earthquake Considerations in Earth Dam Design," by James L. Sherard. Proceedings, American Society of Civil Engineers–Journal, Soil Mechanics and Foundations Division, **93**, No. (SM 4), July 1967, Paper No. 5322.
127. "Earthquake-Resistant Design of Earth Dams," by H. Bolton Seed. Department of Civil Engineering–Institute of Transportation and Traffic Engineering–University of California, Berkeley.
128. "Earthquake Stress Analysis in Earth Dams," by Ray W. Clough and Anil K. Chopra. Proceedings, American Society of Civil Engineers–Journal, Engineering Mechanics Division, **92**, (No. EM2), April 1966.
129. The Finite Element Method in Structural and Continuum Mechanics, O. C. Zienkiewicz in collaboration with Y. K. Cheung. McGraw-Hill Publishing Company, Ltd., London 1967.
130. Introduction to the Finite Element Method, by Chandrakant S. Desai and John F. Abel. Van Nostrand Reinhold Company, New York 1972.
131. "Characteristics of Rock Motions During Earthquakes," by H. Bolton Seed, I. M. Idriss, and F. W. Kiefer. College of Engineering, University of California, Berkeley. Report No. EERC 68-S, September 1968.
132. "New Don Pedro Dam–Dynamic Investigations," by Bechtel Corporation, San Francisco. For Turlock Irrigation District and Modesto Irrigation District. August 1969. Unpublished.
133. "Seismic Stability of Upper San Leandro Dam," by Bernard B. Gordon, David J. Dayton, and Khosrow Sadigh. A.S.C.E. National Structural Engineering Meeting, April 9-13, 1973, San Francisco. Preprint No. 2025.
134. "Nonlinear Finite Element Analysis of Stresses and Movements in Oroville Dam," by F. H. Kulhawy and J. M. Duncan. Department of Civil Engineering–Institute of Transportation and Traffic Engineering–University of California, Berkeley, January 1970.
135. "Response of Earth Banks During Earthquakes," by I. M. Idriss and H. Bolton Seed. Proceedings, American Society of Civil Engineers–Journal, Soil Mechanics and Foundations Division, **93**, (No. SM 3), May 1967, Paper No. 5232.
136. "Nonlinear Analysis of Stress and Strain in Soils," by James M. Duncan and Chin-Yung Chang. Proceedings, American Society of Civil Engineers–Journal, Soil Mechanics and Foundations Division, **96**, (No. SM 5), September 1970, Paper No. 7513.
137. "Elastic Analysis for Behavior of Rockfill", by Neville O. Boughton. Proceedings, American Society of Civil Engineers–Journal, Soil Mechanics and Foundations Division, **96**, (No. SM 5), September 1970, Paper No. 7532.
138. "Liquefaction of Saturated Sands During Cyclic Loading," by H. Bolton Seed and Kenneth L. Lee. Proceedings, American Society of Civil Engineers–Journal, Soil Mechanics and Foundations Division, **92**, (No. SM 6), November 1966. Paper No. 4972. Discussion: Reginald A. Barron, (SM 2), March 1967; Bing Cheng Yen, (SM 3), May 1967; Yoshiaki Yoshimi, (SM 5), September 1967, W. L. Schroeder and Robert L. Schuster, (SM 5), September 1967; Closure: (SM 3), May 1968.
139. "Cyclic Stresses Causing Liquefaction of Sand," by Kenneth L. Lee and H. Bolton Seed. Proceedings, American Society of Civil Engineers–Journal, Soil Mechanics and Foundations Division, **93**, (No. SM 1), January 1967, Paper No. 5058. Discussion: Bing C. Yen, (SM 5), September 1967; O. Clarke Mann, (SM 5), September 1967; Closure: (SM3), May 1968.
140. "Earthquake Resistance of Rock-Fill Dams," by Ray W. Clough and David Pirtz. Proceedings, American Society of Civil Engineers–Journal, Soil Mechanics and Foundations Division, **82**, (SM 2), April 1956, Paper No. 941.
141. "Constructional Deformations in Rockfill Dam," by Arthur Penman and Andrew Charles. Proceedings, American Society of Civil Engineers–

Journal, Soil Mechanics and Foundations Division, **99**, (No. SM 2), February 1973, Paper No. 9560 (also **100**, (GT 8), Aug. 74). Discussion: Nigel A. Skermer, (SM 8), August 1973; David J. Naylor and D. Barrie Jones, (SM 12), December 1973; Daniel Resendiz, **100**, (GT 3), March 1974; Closure: **100**, (GT 10), October 1974.

142. "Performance Instrumentation Installed In Oroville Dam," by John E. O'Rourke. Proceedings, American Society of Civil Engineers—Journal, Geotechnical Engineering Division, **100**, (No. GT 2), February 1974.

143. "Earth Manual." United States Bureau of Reclamation, Denver, Colorado.

144. "Quoich Rockfill Dam," by C. M. Roberts. Transactions, Sixth International Congress on Large Dams, Vol. III, pp. 101-121, New York 1958.

145. "Aguoda Blanca Rockfill Dam with Metal Facing," by Piero Sembenelli and Marco Fagiolo. Proceedings, American Society of Civil Engineers—Journal, Geotechnical Engineering Division, **100**, (No. GT 1), January 1974, Paper No. 10270.

146. "Seismic Response By Variable Damping Finite Elements," by I. M. Idriss, H. Bolton Seed, and Norman Serff. Proceedings, American Society of Civil Engineers—Journal, Geotechnical Engineering Division, **100**, No. (GT 1), January 1974, Paper No. 10284.

147. "Descriptions of Some Swedish Earth and Rock Fill Dams with Concrete Core Walls and Measurements of the Movements and Pressure in the Filling Material and Core Walls," by G. Westerberg, G. Pira, and J. Hagrup. Transactions, Fourth International Congress on Large Dams, Vol. I, pp. 67-97, New Delhi 1951.

148. "Large Scale Pre-construction Tests on Embankment Materials for an Earth-Rockfill Dam," by G. F. Sowers and C. E. Gore. Proceedings, Fifth International Conference on Soil Mechanics and Foundation Engineering, Vol. 2, pp. 717-720. Paris 1961.

149. "Materials Exploration and Field Testing at Oroville Dam," by A. L. O'Neill and R. G. Nutting. Symposium on Soil Exploration—American Society for Testing and Materials, Philadelphia, STP-351 1964, pp. 96-107.

150. "Stresses and Deformations in Cores of Rockfill Dams," by E. Nonveiller and P. Anagnosti. Proceedings, Fifth International Conference on Soil Mechanics and Foundation Engineering, Vol. 2, pp. 673-680, Paris 1961.

151. "Earth Pressures in a Thin Impervious Core," by B. Lofquist. Transactions, Fourth International Congress on Large Dams, Vol. 1, pp. 99-109, New Delhi 1951.

152. "Behavior of Compacted Soil in Tension," by Addanki V. Gopala Krishnayya, Zdenek Eisenstein, and Norbert R. Morgenstern. Proceedings, American Society of Civil Engineers—Journal, Geotechnical Engineering Division, **100**, No. (GT 9), September 1974, Paper No. 10828.

153. "An Analysis of Stresses and Deformations in the Wide Clay Core of a Rockfill Dam," by P. Anagnosti. Proceedings, Sixth International Conference on Soil Mechanics and Foundation Engineering, Vol. 2, pp. 447-450, Montreal 1965.

154. "End Restraint Effects in the Triaxial Test," by W. H. Peroloff and L. E. Pombo. Proceedings, Seventh International Conference on Soil Mechanics and Foundation Engineering, Vol. 1, pp. 327-333, Mexico 1969.

155. "Nuclear Radiation in Construction Control of Earth and Rockfill Dams," by L. Bernell and K. A. Sherman. Proceedings, Seventh International Conference on Soil Mechanics and Foundation Engineering, Vol. 2, pp. 285-289, Mexico 1969.

156. "Centrifugal Model Test of a Rockfill Dam," by M. Mikasa, N. Takada and Y. Yamada. Proceedings, Seventh International Conference on Soil Mechanics and Foundation Engineering, Vol. 2, pp. 325-333, Mexico 1969.

157. "Strains and Stresses in Earth and Rockfill Dams," by A. A. Nitchiporovitch and A. P. Sinitsyn. Proceedings, Seventh International Conference on Soil Mechanics and Foundation Engineering, Vol. 2, pp. 335-343, Mexico 1969.

158. "Stresses in Narrow Cores and Core Trenches of Dams," by G. E. Blight. Transactions, Eleventh International Congress on Large Dams, Vol. III, Madrid 1973.

159. "Tensile and Flexural Strength of Earth Dam Materials," by A. Kezdi. Transactions, Eleventh International Congress on Large Dams, Vol. III, Madrid 1973.

160. "The Impervious System of the Mattmark Dam," by B. Gilg. Transactions, Eleventh International Congress on Large Dams, Vol. III, Madrid 1973.

Design of Concrete Dams

MERLIN D. COPEN
Head, Concrete Dams Section (retired)
ERNEST A. LINDHOLM
Supervisory Civil Engineer
Concrete Dams Section
GLENN S. TARBOX
Supervisory Civil Engineer
Concrete Dams Section
U.S. BUREAU OF RECLAMATION
Department of the Interior
Denver, Colorado

I. GENERAL

A. Introduction

Concrete dams may be categorized into three principal types according to their particular physical form and the features of their design. The three types are: arch, gravity, and buttress.

An arch dam is curved upstream in plan and transmits a major part of the imposed load to the canyon walls by horizontal thrust. A single curvature arch dam is curved in plan only. A double curvature arch dam is curved in plan and elevation with undercutting at the heel and, in most instances, downstream overhang near the top of the dam.

A gravity dam resists the applied load primarily by its weight. Gravity dams are usually straight in plan but are sometimes curved to take advantage of the topography of a site. In cross section gravity dams are roughly triangular.

Buttress dams depend principally upon the weight of the water in addition to the concrete weight for stability. They are composed of two structural elements: a sloping water-supporting deck, and buttresses which support the deck. Buttress dams are usually classified according to the type of water-supporting deck. A slab-and-buttress dam is one whose deck is comprised of flat slabs supported on transition sections at the upstream edge of the buttresses. Multiple-arch dams consist of a series of arch segments supported by buttresses. A massive-head buttress dam is formed by flaring the upstream edges of the buttresses to span the distance between the buttress walls. The terms diamond-head and round-head, which refer to the shape of the enlargement at the upstream face, more fully describe this type.

B. Definitions

Most of the descriptive terms which define the various types of concrete dams and their pertinent features are well recognized and applied in a similar manner throughout the engineering field. In some localities, however, and in many technical books and articles dealing with these subjects, there are variations in the interpretation of some of these terms. In order that the terms encountered throughout this chapter may be fully understood, definitions are given below.

A *plan* is an orthographic projection on a horizontal plane, showing the principal features of a dam and its appurtenant works with respect to topography and available geologic data.

A *profile* is a developed elevation of the intersection of a dam with the original ground surface, rock surface, or excavated surface along the axis of the dam, the upstream face, the downstream face, or other designated location.

A *section* is a representation of a dam as it would appear if cut by a plane.

Dams less than 100 ft in height are considered to be low dams.

Dams from 100 to 300 ft high are designated as medium-high dams, and those *higher than 300 ft* as high dams.

The *structural height* of a concrete dam is the difference in elevation between the top of the dam and the lowest point in the excavated foundation area, exclusive of such features as narrow fault zones. The *top of the dam* is the crown of the roadway or the level of the walkway if there is no roadway.

The *axis of a dam* is a vertical reference plane coincident with the upstream edge of the top of the dam. In an arch dam, the axis is curved horizontally.

The *length of a concrete dam* is the distance measured along the axis at the top of the main body of the dam from abutment to abutment. If a spillway is located adjacent to the dam on its abutment, the length of the spillway is not included in the length of the dam.

The *volume* of a concrete dam includes all of the body of the dam and appurtenances not separated from the dam by specific joints. Where a power plant is constructed on the downstream toe of the dam, the limit of the dam is the downstream face projected to the excavated foundation surface.

Definitions which apply specifically to arch dams include the following:

An *arch element* is that portion of a dam bounded by two horizontal planes spaced 1 ft apart.

The *extrados* is the curved upstream surface of arch elements.

The *intrados* is the curved downstream surface of arch elements.

A *section of an arch* is a portion of an arch element which is selected for ease of computation. The section must have a constant extrados radius but may be variable in thickness.

A *voussoir* is a smaller segment of an arch section which, for ease of computation, is assumed to have constant thickness.

The *abutment* of an arch element is the surface, at either end of the arch, which contacts the rock of the canyon wall.

An *arch centerline* is the locus of all midpoints of the thickness of an arch section.

The *central angle* of an arch is the angle bounded by lines radiating from the arch extrados center to points of intersection of the arch centerline with the arch abutments.

The *thickness of an arch* at any point is the length of a line normal to the extrados from the upstream to the downstream face which passes through the point.

The *length of an arch* is the distance along a curve which is concentric with the extrados and passes through the midpoint of the arch thickness at the crown.

A *cantilever element* is that portion of an arch dam which is contained within two vertical planes normal to the extrados and spaced 1 ft apart at the axis. Cantilever elements of arch dams, other than the constant-center type, are bounded by warped surfaces since extrados centers vary with the elevations of the arches.

The *crown cantilever* is a cantilever element located at the point of maximum depth in the canyon.

The *height of a cantilever element* is the vertical distance between its base elevation and the top of the dam.

Upstream projection is the horizontal distance from the extrados to the axis measured along a line normal to the extrados.

Downstream projection is the horizontal distance from the intrados to the axis measured along a line normal to the extrados.

Reference plane is a vertical plane which passes through the crown cantilever and the center for the axis radius.

The *line of centers* is the loci of centers for circular arcs used to describe a face of a dam or a portion thereof. The number of lines of centers necessary to describe an arch dam may vary from 1 for a circular dam with uniform thickness arches to 6 for a three-centered dam with variable thickness arches. Polycentered dams may require more lines of centers if more than three arch segments are used to describe each face of the arch elements. Lines of centers are located with respect to the reference plane.

A *fillet* is an increase in thickness of a dam beginning near and extending to the abutments of the arches. Fillets are usually placed at the downstream face but may also be used at the upstream face.

Abutment pads are mass concrete structures placed between arch dams and their abutments. They are used to reduce load intensity and/or distribute the load more uniformly on the foundation rock, reduce effects of foundation irregularities, and assist in obtaining more symmetry for the dam. The pads, usually trapezoidal in cross section, may be monolithic with the dam or separated from it by peripheral joints.

A *constant-center arch dam* is one in which the upstream and downstream faces are described by radii whose centers are coincident with the axis center at all elevations. The line of centers is a vertical straight line. The arches are of uniform thickness and all cantilevers have identical shape, varying only in base elevation. The thickness of a constant-center dam may be modified near the abutments of the arches and the bases of the cantilevers by the use of fillets.

Variable-center dams include all classes of arch dams whose arch centers for either or both extrados and intrados vary in location with respect to the axis center at different elevations. Arch elements may be of uniform or variable thickness, with or without fillets. Cantilever elements vary in shape and thickness according to the variation in curvature between arches and according to the types of arches used. Single-centered, two-centered, three-centered, polycentered, and constant-angle dams are all variable centered dams.

Single-centered dams are formed by one set of lines of centers on the reference plane defining both sides of the dam. The faces of each arch element are described by concentric circular arcs. This type of arch dam is usually designed for U- or V-shaped canyons having near symmetrical profiles and crest length to height ratios of about 2:1 or less.

Two-centered dams are formed by two sets of lines of centers; one set for each side of the dam. Both sets are co-planar on the reference plane. The faces of the arch elements are described by two circular arcs compounded at the reference plane. Two-centered dams are usually designed for sites with pronounced non-symmetry.

Three-centered or elliptically shaped arch dams are made up of arch elements whose extrados and intrados are both formed by three radii. Usually a shorter radius curve is used in the central part of the arch and longer radius curves near the abutments. Arch elements may be either uniform or variable in thickness.

Polycentered dams are composed of multi-centered arch elements which may be defined to produce essentially any arch shape desired. The horizontal elements may be either uniform or variable in thickness.

Definitions which apply specifically to gravity dams include the following:

A *beam element* is a portion of a gravity dam bounded by two horizontal planes 1 ft apart. For analytical purposes the edges of the elements are assumed to be vertical.

A *cantilever element* is a portion of a gravity dam bounded by two vertical planes, normal to the axis and 1 ft apart.

Twisted structures refer to vertical elements

with the same structural properties as the cantilever elements and horizontal elements with the same properties as the beams. The twisted structure resists only torsion in the vertical and horizontal planes.

The *abutment of a beam element* is the surface, at either end of the beam, which contacts the rock of the canyon wall.

C. Selection of Site and Type of Dam

Although the selection of the type of dam is discussed in Chapter 5, a few essential observations are appropriate at this point.

The site finally selected should be that site where the dam and appurtenances can be most economically constructed consistent with environmental and technical considerations. The foundations of the dam should be sound and relatively free of faults, shears, etc. A narrow site will minimize the quantity of concrete required for the dam, with a resultant low cost for the dam alone. However, a wider site that can more economically and efficiently accommodate the appurtenant structures, such as spillways, outlets, power plants, etc., may be the least costly overall.

Narrow canyons with sound rock for foundations are generally perferred for concrete dams. The narrower canyons are especially adaptable to arch dams. A crest length to height ratio of 5 to 1 or less should always be considered for an arch dam. A number of techniques are available, which, if properly applied, may make wider sites feasible for an arch dam. These techniques include polycentered arches, use of formed joints at the abutments to reduce or eliminate tensile stresses, prestressing, control of temperature change and others. At locations where earth materials are difficult to obtain or where environmental factors render earth or rockfill construction unsatisfactory, a well designed concrete dam may prove to be entirely satisfactory. A concrete dam should always be considered if a rock foundation is available.

The foundation rock need not be perfect to provide adequate support for concrete dams.

For example, an ideal foundation for an arch dam would be approximately three times as deformable as concrete. This deformability may result from a relatively soft rock, such as sandstone, or from rock containing numerous joints. Highly variable rock foundations may accommodate massive-head buttress dams or slab-and-buttress dams. All rock foundations for concrete dams must be stable when subjected to dam and reservoir loadings, otherwise the site must be abandoned.

Low gravity and buttress dams may be considered for other than rock foundations provided adequate precautions are taken. Loads from the dam must be kept within the allowable bearing capacity of the foundation material. Aprons and/or cutoffs must be provided to reduce uplift pressures and prevent undermining of the dam by piping. Because of the limited sliding resistance, measures such as reduced uplift and cutoff walls, may be necessary to ensure sliding stability. Care must be taken to prevent undermining of the dams by erosive forces.

Additional site selection considerations are discussed with the layout and design of the particular type of dam.

II. DESIGN CRITERIA

A basic philosophy of concrete dam design is that dams should be designed according to rational and consistent criteria to ensure safe, economical, functional, durable, and easily maintained structures. Established design criteria should be updated through a continuing program of review and evaluation to assure that design practices remain consistent with developing technologies and sound engineering practice.

Dams designed for loading combinations whose simultaneous occurrence is highly improbable result in overly conservative, uneconomical structures. Care should also be taken to prevent unintentional allowances for safety beyond specified safety factors. Dams so designed may have a margin of safety greater than intended which can also lead to uneconomical designs.

Design data for analyses must be as accurate and complete as possible. The data should be based on field tests, laboratory tests, and behavioral measurements taken from prototype structures.

The relationship between construction practices and design criteria is important. The responsibility for insuring that a dam is built with uniformly good quality concrete lies with those in charge of construction. The best of materials and design practice become less effective without properly performed construction procedures.

A. Material Properties

The two basic materials encountered in concrete dam design are the concrete and the foundation material.

1. Concrete Properties. The concrete properties which must be accurately determined include compressive, tensile, and shear strength; the instantaneous and sustained modulus of elasticity; Poisson's ratio; and the coefficient of thermal expansion, thermal conductivity, specific heat, and diffusivity. These properties are determined by laboratory tests of the design concrete mix.

Concrete strengths should be determined by tests of the full mass mix in cylinders of sufficient size to accommodate the size aggregate to be used. The compressive, tensile, and shear strength tests should be made as companion test series on specimens of equal age since concrete strengths vary with age. The shear strengths should be determined for several different normal stresses which correspond to the expected operating normal loads.

The elastic modulus of concrete varies with age. The instantaneous modulus of elasticity is the ratio of stress to strain which occurs immediately with the applied load. The sustained modulus of elasticity reflects the added long-term deformation which is the result of creep in the concrete. The test cylinders used to determine the elastic properties of instantaneous and sustained modulus of elasticity and Poisson's ratio should be the same size and cured accordingly as the cylinders used for strength tests.

The effects of temperature change on concrete dams can be significant. Consequently, the accurate determination of thermal properties is important.

The effects on strength and elastic properties of cyclic-type dynamic loads which alternate between tension and compression have not been fully investigated. The instantaneous modulus of elasticity, as previously discussed, can be used for dynamic analyses, however.

2. Foundation Properties. The foundation properties which must be accurately determined include deformation modulus, shear strength, permeability, compressive strength, and Poisson's ratio. The deformation modulus is defined as the ratio of applied stress to elastic plus inelastic strain. The effective deformation modulus is a composite of deformation moduli for all materials within a particular part of a foundation. The analysis of a concrete dam should include the effective deformation modulus and variation over the entire contact area of the dam with the foundation. The contribution to inelastic strain from joints, shears, and faults should be evaluated and included in the deformation modulus determination.

Shear resistance within a foundation and between a dam and its foundation depends upon the cohesion and internal friction in the foundation and on the bond between the concrete and foundation at the dam contact. Although shear resistance has traditionally been expressed as varying linearly with respect to normal load according to Coulomb's equation, the relationship is often nonlinear, particularly for materials other than intact rock. The variation of shear resistance versus normal load should be determined using laboratory and/or in-situ tests. The values of shear resistance selected for design computations should be commensurate with the expected level of normal load and should also be carefully considered with respect to scale effects. The scale effects result when tests are performed on small samples and the results are extrapolated to in-situ conditions. If treatment, such as grouting, drainage or backfill concrete is used to beneficiate the foundation, its effects on the physical properties of the foundation should be considered in the analyses.

The permeability of the foundation including joints, shears, fault zones, and solution cavities is necessary to determine pore pressures for analyses of stresses and stability. The exit gradient for materials near the surface or adjacent to openings such as drain holes should be determined to check the possibility of piping. If foundation grouting, drainage, or other treatment is involved, their effects on the pore pressure distributions should be included.

Compressive strength of the foundation rock or the bearing pressure of other foundation materials can be an important factor in determining dam thicknesses at the dam foundation contact. Allowable compressive strengths and bearing pressures should be determined for each foundation material to be loaded when a foundation is nonhomogeneous.

B. Loads

The loads commonly considered in the design of a concrete dam are: (1) dead weight, (2) temperature, (3) reservoir and tailwater, (4) internal hydrostatic pressure, (5) silt, (6) ice, and (7) earthquake. Other loads, where appropriate, should be considered at the discretion of the designer.

1. Dead Load. The dead load is the weight of concrete plus appurtenances such as gates and bridges. For gravity and buttress dams, all the dead load is assumed to be transmitted vertically to the foundation without transfer by shear between adjacent blocks. For arch dams, all the dead load imposed on the structure before closure of the contraction joints or cooling slots is assumed to be transmitted vertically to the foundation as for gravity dams. Dead load imposed after grouting is assumed to be distributed vertically and horizontally similar to other loads.

2. Temperature. Volumetric increases in gravity dams due to temperature rise are considered unrestrained if the contraction joints are ungrouted and therefore there is no lateral transfer of load. If the joints are grouted, horizontal thrusts, caused by volumetric increases due to rising temperature, will produce load transfer across the joints. The load transfer will increase twist effects and the loads at the abutments.

An arch dam, exposed to temperature variations, will experience temperature loads whenever there is a volumetric change which is restrained by adjacent blocks and the abutments. Concrete temperatures to be expected during operation of a dam are determined from studies of the site which include the effects of ambient air temperatures, reservoir water temperatures, and solar radiation.

Studies can be made for different weather conditions to obtain the range of temperatures and gradients from average to extreme values. The combination of the mean daily variation, the mean weekly variation and the mean monthly variation in air temperature will produce temperatures considered indicative of usual weather conditions.

3. Reservoir and Tailwater. Reservoir and tailwater loads are obtained from reservoir operation studies and tailwater curves. These studies are based on operating and hydrologic data such as reservoir capacity, storage allocations, stream flow records, flood hydrographs, and reservoir releases for all purposes. Design reservoir elevations can be derived from these operation studies which will reflect the full range of water surfaces from the minimum to the maximum.

4. Internal Hydrostatic Pressure. The internal hydrostatic pressure has only a small effect within an arch dam and is normally not included in the design analyses. Formed drains placed in the dam should be utilized, however, to minimize any internal hydrostatic pressures as a safeguard.

The distribution of internal hydrostatic pressure along a horizontal section through a gravity dam is assumed to vary linearly from full reservoir pressure at the upstream face to zero or tailwater pressure at the downstream face and to act over the entire area of the section. The pressure distribution should be adjusted to reflect the size, location, and spacing of formed drains if they are used. Experimental and analytical studies indicate that drains set in from the upstream face at 5 percent of the maximum reservoir depth and spaced laterally twice that distance will reduce the average pressure at the drains to approximately tailwater pressure plus one-third the differential between tailwater and reservoir water pressures.

5. Silt. Horizontal silt pressure, including the effect of water, is assumed equivalent to a fluid weighing 85 lb/cu. ft (1361.6 kg/cu m). Vertical silt pressure including the effect of water, is assumed equivalent to a soil with a wet density of 120 lb/cu. ft (1922.2 kg/cu m).

6. Ice. Ice pressures should be computed according to an accepted method of analysis[1] if the designer considers that ice will form to appreciable thickness and remain for a long duration. An estimate of ice pressure may be taken as 10,000 lb/linear ft (14,882 kg/linear m) of contact with the dam for ice depths of 2 ft (0.61 m) and greater.

7. Earthquake. Concrete dams are elastic structures and may be excited to resonance by seismic ground accelerations. The determination of seismic loads on a dam require knowing first, the input ground motion at the site and, secondly, the dynamic response of the dam to vibratory motions. There are methods available for estimating the magnitudes and locations of earthquakes to which a dam may be subjected.[2] Concrete dams should be designed to remain elastic when subjected to the operating basis earthquakes which are events that could happen occasionally during the useful life of the dam. A concrete dam should also be designed to withstand the maximum credible earthquake. In this instance, some structural damage can be permitted but any sudden release of the reservoir must be prevented. A maximum credible earthquake is one having a magnitude usually larger than any historically recorded event and although its probability of occurrence is very small, it is considered to be the most severe expected earthquake which would cause maximum ground motion at the dam.

Whenever the ground acceleration at the dam exceeds 0.05 g, a dynamic analysis of the dam should be made using, at a minimum, a response spectrum analysis and if the stress variations with time are critical, a time history analysis.

C. Load Combinations

Concrete dams should be designed for all reasonable combinations of the loads discussed in the previous section. Combinations of loads whose simultaneous occurrence is highly improbable should not be considered as reasonable. As the probability that a particular combination of loads will occur decreases, the required safety factor can also be reduced. Most load combinations can be categorized as usual, unusual, or extreme.

For example, normal design reservoir elevation plus the usual concrete temperatures occurring at that time and the appropriate dead loads, tailwater, ice, and silt is typical of a *usual* type load combination. An *unusual* type combination might be the maximum reservoir elevation plus minimum usual temperatures and the appropriate dead loads, tailwater, and silt. The example of usual load combination above coupled with a maximum credible earthquake would be considered an *extreme* load combination.

D. Cracking

Tensile stresses large enough to cause cracking are not allowed in designs for new dams under static load combinations. There are, however, some instances when cracking in concrete dams occurs and must be dealt with, e.g., older dams subjected to changed loading conditions and new dams exposed to maximum credible earthquake loads.

Wherever computed tensions exceed the tensile strength of the concrete, cracking should be assumed to exist in an arch dam and the depth of crack assumed to extend to the point of zero stress within the section. The resulting changes in the static flexibility of the dam due to cracking must be computed and the analyses repeated to reflect the changed flexibilities.

Horizontal cracking is assumed to occur in a gravity dam wherever the vertical normal stress exceeds the specified minimum stress as discussed under Factors of Safety. The depth of crack is assumed to extend horizontally to the point where the compressive stress computed without uplift equals the internal hydrostatic pressure.

E. Factors of Safety

The following discussions of safety factors are based on the presumption that all uncertainties concerning applied loads and the load-carrying capacity of the dam be resolved as far as practicable by field or laboratory tests, through exploration and inspection of the foundation, good concrete control, and good construction procedures. It is also presumed that an adequate instrumentation system be installed to assess the behavior of the dam and its foundation and to monitor the continued safe operation of the structure.

Foundation sliding stability safety factors are based on the assumption that the shear stress is distributed uniformly over any given plane being analyzed. The safety factors required for foundation studies are slightly larger than those for the dam because defining the foundation properties is more difficult.

The values of safety factors given should be considered as minimum values.

1. Arch Dams. The safety factors presented herein for the dam were developed to be compatible with the theory and assumptions contained in the Trial Load Method of Analysis and the computer program Arch Dam Stress Analysis System (ADSAS).

Compressive Stress. The maximum allowable compressive stress for concrete in the dam should be less than the specified compressive strength divided by 3.0, 2.0, and 1.0 for Usual, Unusual, and Extreme Loading Combinations, respectively. In no case should the compressive stress exceed 1500 psi (105.46 kg/cm^2) for Usual Loading Combinations or 2250 psi (158.19 kg/cm^2) for Unusual Loading Combinations.

The maximum allowable compressive stress in the foundation should be less than the foundation compressive strength divided by 4.0, 2.7, and 1.3 for Usual, Unusual, and Extreme Loading Combinations, respectively.

Tensile Stress. Tensile stresses should be avoided whenever practicable. Limited tensile stresses may, however, be allowed in restricted locations at the designer's discretion. Any tensile stresses allowed should not exceed 150 psi (10.55 kg/cm^2) for Usual Load Combinations or 225 psi (15.82 kg/cm^2) for Unusual Load Combinations.

Concrete should be assumed to crack if its tensile strength is exceeded due to an Extreme Load Combination including a maximum credible earthquake. If, after the effects of any cracking are included, the stresses do not exceed the concrete compressive strength and structural stability is assured, the dam can be considered safe against sudden release of the reservoir despite the damage sustained.

Shear Stress. The maximum allowable average shear stress on any plane within the dam should be less than the concrete shear strength divided by 3.0, 2.0, and 1.0 for Usual, Unusual, and Extreme Loading Combinations, respectively.

Sliding Stability. The shear-friction factor of safety is a measure of stability against sliding or shearing. The equation for the shear-friction factor of safety (Q) is as follows:

$$Q = \frac{CA + (\Sigma N + \Sigma U) \tan \phi}{\Sigma V} \quad (1)$$

where

C = unit cohesion
A = area of planes considered
ΣN = summation of normal forces acting on planes

ΣU = summation of uplift forces acting on planes
$\tan \phi$ = coefficient of internal friction
ΣV = summation of driving shear forces
Use consistent units on all input variables.

The shear-friction factor of safety can be used to determine sliding stability at the concrete-rock contact or any plane within an arch dam in lieu of checking the allowable shear stress. The minimum allowable values of Q are 3.0, 2.0, and 1.0 for the Usual, Unusual, and Extreme Loading Combinations, respectively.

The factor of safety against sliding on any plane of weakness within the foundation is also assessed using the shear-friction and its value should be not less than 4.0, 2.7, and 1.3 for the Usual, Unusual, and Extreme Loading Combinations, respectively.

2. Gravity Dams. Safety factors for gravity dams are based on the Gravity Method of Analysis. The buttress portions of buttress dams are considered to be gravity-type structures and their safety should be evaluated according to these criteria.

Structural elements of buttress dams such as the water bearing deck, footings, stiffeners, and corbels should be designed according to reinforced concrete design codes.

Compressive Stress. The criteria as discussed for arch dams applies equally to gravity dams.

Tensile Stress. The critical stress associated with gravity dams is generally the vertical normal stress acting at the upstream face. The following equation for the vertical normal stress at the upstream face (σz_u) provides a criterion which ensures that the allowable tensile stress will not be exceeded.

$$\sigma z_u = p \cdot w \cdot h - \frac{f_t}{s} \quad (2)$$

where

σz_u = minimum allowable vertical normal stress at the upstream face

p = a reduction factor to account for drains ($p = 1.0$ without drains and 0.4 with drains)
w = unit weight of water
h = depth below reservoir surface
f_t = tensile strength of lift surfaces
s = safety factor
Use consistent units on all input variables.

The compressive stress calculated for comparison with σz_u should be computed without the effect of internal hydrostatic pressure.

The value of the safety factor, s, should be 3.0, 2.0, and 1.0 for the Usual, Unusual, and Extreme Loading Combinations, respectively. The value of σz_u must be greater than zero for all Usual Loading Combinations and if the upstream stress is less than σz_u due to an Extreme Load Combination including a maximum credible earthquake, cracking should be assumed to occur. If, after the effects of any cracking are included, the stresses do not exceed the concrete compressive strengths and structural stability is assured, the dam can be considered safe against sudden release of the reservoir despite the damage sustained.

Sliding Stability. The criteria for the arch dam and its foundation holds for the gravity dam.

III. DESIGN OF ARCH DAMS

A. Description

The design of a complex, highly indeterminate structure like an arch dam must include analysis as an integral part of the design procedure. The designer conceives a design and uses analytical methods to determine the stress distributions throughout the structure. The design is continually improved by alternately modifying the structure and checking the results of the analysis until the design objectives are achieved within the allowable design criteria.

The order in which the topics are presented in this section does not necessarily represent the sequence that would be followed for design and analysis but does represent a logical chronology of steps beginning with an initial design.

Section III-D is a discussion of a layout procedure and some general design philosophy. Methods used for comprehensive stress and stability analysis of concrete arch dams are discussed in Section III-E.

B. Required Data

The principal data which should be on hand before preparing a design layout are: A topographic map of the proposed location; geological data on rock types; depth of overburden; the location of faults and jointing patterns; operating reservoir water surface and tailwater elevations; probable sediment accrual in the reservoir; the size and location of required openings in the dam for spillways, outlets, etc; climatological data for studies on temperature variations within the dam; laboratory and possibly in situ test data for determining the strength and elastic properties of the rock and concrete; and seismological data for the region.

C. Level of Design

There are three levels of design, reconnaissance, feasibility, and final. The level of a design determines the degree of refinement to which the design is taken. A reconnaissance design is made in conjunction with field investigations during project planning to estimate the concrete volume required for use in determining the feasibility of the project. Reconnaissance designs may be based on previous designs which are similar in height and shape of profile and for which the stresses are satisfactory. Another means of determining the information is by using nomographs constructed from empirical formulas derived by statistical analyses of several analytical studies completed for previously designed dams. A more detailed discussion describing the use of nomographs is given in Section III-D-1.

Feasibility designs are used in the selection of the final location and as a basis for requesting construction funds. Feasibility designs are made in greater detail than reconnaissance designs because a closer approximation to final design is required. Analyses should be made of the adopted structure for the usual and the most severe loading conditions expected in actual service. The best of the alternative designs should have stresses distributed as uniformly as possible within allowable limits and have a minimum cost.

Final designs are used to develop specifications drawings and construction drawings. Stress analyses should be made which include all normal operating loading conditions and dynamic loading conditions due to earthquake ground accelerations. The temperature distribution applied during final design is determined using detailed analysis and is based on finalized temperature and operating data. The final design will be one which best satisfies the requirements for acceptable stresses most economically.

D. Layouts

A layout drawing for an arch dam includes a plan, profile, and section along the reference plane. The drawing is made to describe the dam geometrically and to locate the dam in the site. The necessary data for analyzing the structure are also obtained from the layout drawing.

The primary objective in making a layout for an arch dam at a particular site is to determine the arches which will fit the topographic and geologic conditions most advantageously, provide for the installation of adequate facilities for reservoir operation, and distribute the load with the most economical use of materials within stress limitations. The load distribution, and the stresses resulting from such distribution, depend largely on the shape of the canyon, length and height of dam, shape of dam, thicknesses and loading conditions.

To produce a satisfactory design, the engineer conceives and constructs a design layout, makes a stress analysis for the design, reviews the results to determine appropriate changes in the design shape which will improve the stress distributions, and draws a new design layout incorporating the changes. The process is repeated until a design is achieved which has (1) a reasonably uniform varying distribution of stress, (2) a compressive stress level throughout as nearly as equal as practicable to the defined allowable limits, and (3) a minimum practicable volume of concrete.

It is very difficult to design an arch dam which has compressive stresses throughout that are near the maximum allowable. A good design is usually a compromise solution which yields a dam that has some very low compressive stresses and may even have limited zones of tensile stresses, provided they are within the allowable defined limits. Nevertheless, the three objectives of a good design, as stated above, still represent the goal toward which to strive.

1. Procedure. A single centered, variable thickness arch dam will be used for the purpose of this discussion. The procedure for laying out other types of arch dams differs only in the way the arches are defined. The following dimensions are parameters used in empirical formulas to determine initial values of the axis radius and thicknesses of the crown cantilever; H, structural height in feet (vertical distance from the crest of the dam to the lowest assumed point of the foundation); L_1, straight line distance in feet at crest elevation between abutments excavated to assumed sound rock; L_2, straight line distance in feet between abutments excavated to assumed sound rock and measured at an elevation $0.15H$ ft above the base.

The empirical formulas referred to can be used to determine initial values of axis radius, crown cantilever dimensions, and estimated concrete volumes. In order to facilitate the use of such formulas, nomographs have been constructed and are shown in Figs. 8-3-1 through 8-3-4.[3]

The first step in making a layout is to draw a tentative axis for the dam in plan on transparent paper. The paper is overlaid on the site topography and shifted about until an optimum orientation for the axis is achieved, i.e., the angle of incidence to the surface contours at the crest elevation is approximately equal on each side. The following formula can be used as a guide to selecting a tentative radius for the axis:

$$R_{axis} = 0.6 L_1$$

Because empirical formulas are only a guide to choosing R_{axis}, the designer should make appropriate adjustments in R_{axis} so that the central angle of the top arch and intersection of the axis with the topography are satisfactory. The top arch and the axis are discussed interchangeably because the extrados and the axis have the same radius by definition.

The magnitude of the central angle of the top arch is a controlling value which influences the curvature of the entire dam. Objectionable tensile stresses will develop in arches of insufficient curvature, such a condition being apt to occur in the lower elevations of a dam having a V-shape profile. The largest central angle practicable should be used, and consideration given to the fact that the bedrock topography may be inaccurately mapped and the arch abutments may need to be extended to points of somewhat deeper excavation than originally planned. Owing to limitations imposed by topographic conditions and foundation requirements, it will be found that, for most layouts, the largest practicable central angle for the top arch varies between 90° and 110°.

The next step after the axis has been located is to define the reference plane and crown cantilever section. The crown cantilever is usually located at the point of maximum depth. The reference plane for a single centered dam is a vertical plane which passes through the crown cantilever and the R_{axis} center. Ideally, the reference plane should be at the midpoint along the axis. This seldom occurs, however, because most canyons are not symmetrical about their low point.

After the crown cantilever and reference plane have been located, the thicknesses and

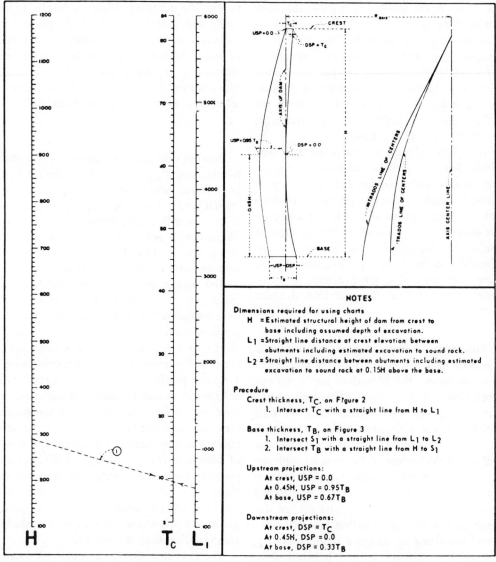

Figure 8-3-1. Nomograph for obtaining crest thickness and projections on crown cantilever. (*Courtesy Bureau of Reclamation*)

shape of the crown cantilever should be determined. Proportioning the crown cantilever is facilitated by considering separately the top thickness, intermediate thickness and base thickness. Estimates for crown cantilever thicknesses may be computed using empirical equations. The equations are to be used as guides and only for initial layouts.

1. Crest thickness, in feet,

$$T_c = 0.01[H + 1.2L_1]$$

2. Base thickness, in feet,

$$T_B = \sqrt[3]{0.0012\, HL_1 L_2 \left(\frac{H}{400}\right)^{H/400}}$$

DESIGN OF CONCRETE DAMS 397

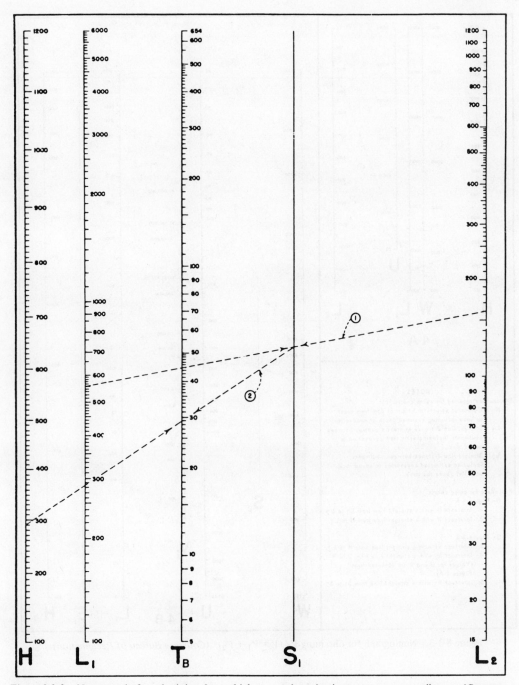

Figure 8-3-2. Nomograph for obtaining base thickness and projections on crown cantilever. (*Courtesy Bureau of Reclamation*)

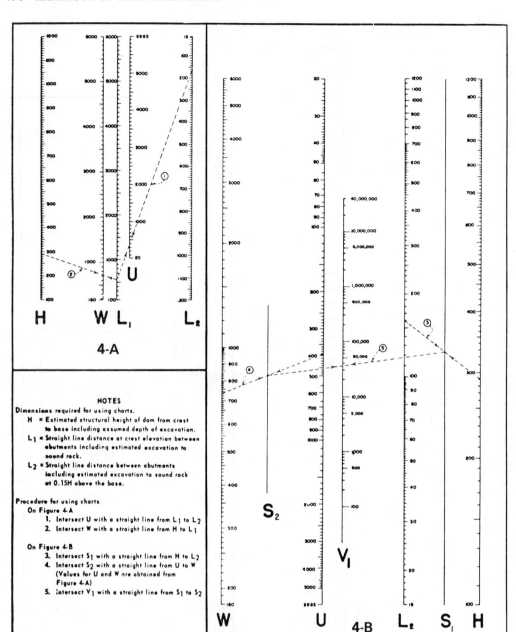

Figure 8-3-3. Nomograph for obtaining V_1 ($V = V_1 + V_2$). (*Courtesy Bureau of Reclamation*)

3. Thickness at $0.45\,H$, in feet,
$$T_{0.45\,H} = 0.95\,T_B$$
Values for T_c and T_B can be read directly from the nomograph in Figs. 8-3-1 and 8-3-2.

After values for thickness have been determined, they can be divided into upstream and downstream projections according to the following:

DESIGN OF CONCRETE DAMS 399

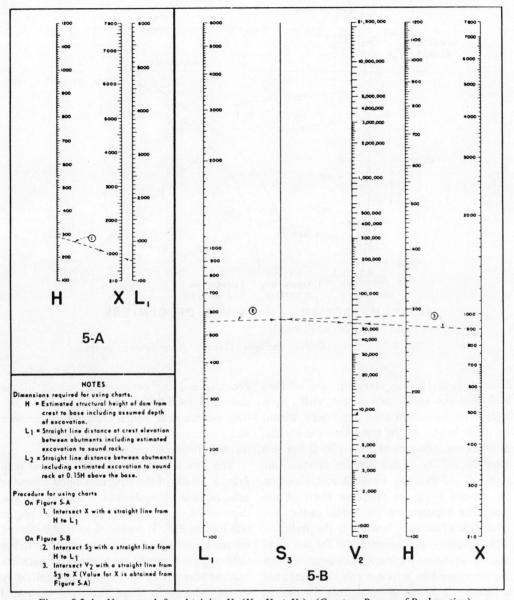

Figure 8-3-4. Nomograph for obtaining V_2 ($V = V_1 + V_2$). (*Courtesy Bureau of Reclamation*)

	Upstream projections	Downstream projections
At crest	0.0	T_c
At 0.45 H	0.95 T_B	0.0
At base	0.65 T_B	0.33 T_B

The crown cantilever can be constructed after the controlling thicknesses have been determined as is shown in Fig. 8-3-5.

The next step is to draw the arches in plan at convenient elevations for the stress analysis.

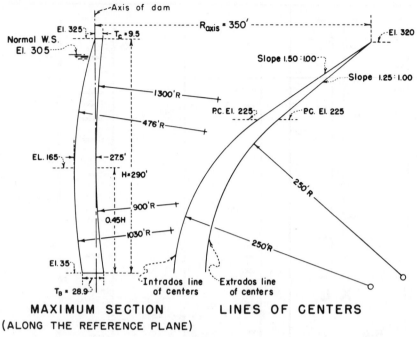

Figure 8-3-5. Crown cantilever and lines of centers for a preliminary design.

Usually 5 to 10 arch elevations are selected such that the entire dam is represented by a system of horizontal elements evenly spaced over the height of the structure. Usually the intervals are not greater than 100 ft nor less than 20 ft. The radius centers defining the intrados and extrados for each arch elevation are plotted along the reference plane of the dam. The locations of the radius centers are a function of the canyon width at the particular arch elevation, the required rise for the arch, and the abutment thickness. The radius centers are determined by selecting trial positions until a location is found which defines the desired arch shape. To ensure that the dam is smooth in both the horizontal and vertical directions, the arch radii centers must lie along the reference plane in plan and be connected by smooth continuous curves in elevation called lines of centers. Slight adjustments in the trial radii center locations are usually necessary as a result. Fig. 8-3-5 and 8-3-6 show examples of how the arch radii centers are defined. The plan is completed by drawing in the arch abutment contacts and the perimetrical contact of the dam and foundation as shown in Fig. 8-3-6. The perimetrical contact should be smooth and continuous. This may also require adjustments in the radius centers.

The final step in constructing a layout is to draw a profile of the dam developed along the axis. Surface irregularities such as ridges, depressions, or undulations in the profile will be revealed. Pronounced anomalies should be removed by reshaping the affected arches until a smooth profile is obtained. Smoothness may be thought of as a continuous uniformly varying change in slope along the excavated surface.

2. Factors Affecting Layouts. Consideration of all factors affecting the layout should be a part of the four steps of designing a dam:

    layout
    analysis
    evaluation
    modification

DESIGN OF CONCRETE DAMS

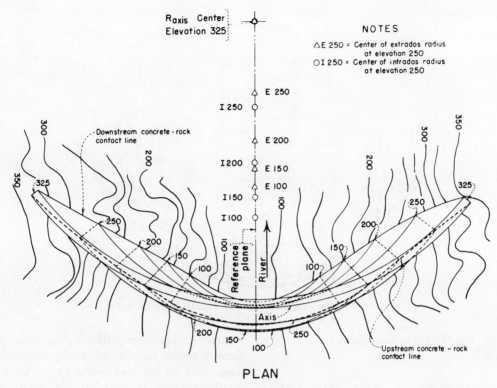

Figure 8-3-6. Plan for a preliminary design.

Length to Height Ratio. The length-height ratios of dams may be used as a basis for comparison of proposed designs with existing designs. Such comparisons should be made in conjunction with the relative effects of other controlling factors such as central angle, shape of profile, and type of layout. The length-height ratio also gives a rough indication of the economic limit of an arch dam as compared with a dam of gravity design. Generally, the economic limit of an arch dam occurs for a maximum ratio between 4 to 1 and 6 to 1, depending somewhat on the height of dam and local conditions. Even if the length-height ratio for an arch dam falls within the economic range, the combined cost of dam and spillway may be such that another type of dam would be more economical.

Symmetry. Although not a necessity, a symmetrical or nearly symmetrical profile is desirable from the standpoint of stress distribution. A region of stress concentration is likely to exist in an arch dam having a nonsymmetrical profile, a condition tending toward an uneconomical section compared with that of a symmetrical dam. In some cases improvements of a nonsymmetrical layout by one or a combination of the following methods may be warranted: Deeper excavation in appropriate places, constructing artificial abutments, reorienting and relocation.

When a nonsymmetrical canyon is encountered such that a single-centered arch cannot be satisfactorily fitted to the site, a two-centered scheme can be used to define the dam. This type of layout is constructed by using two separate sets of lines of centers, one for each side of the dam. To maintain continuity, however, each set of lines of centers must lie along the reference plane. In some cases the axis radius (R_{axis}) may be different on each

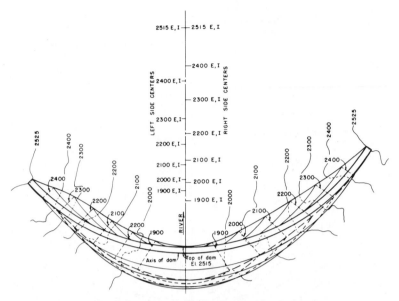

Figure 8-3-7. Two-centered dam with uniform thickness arches.

side. Fig. 8-3-7 is an example of a two-centered arch.

Canyon Shape. In dams constructed in U-shaped canyons, the lower arches have approximately the same chord length as those near the top. In such cases, use of a variable-thickness arch layout will normally give a relatively uniform stress distribution. Undercutting on the upstream face may be desirable to eliminate areas of tensile stress at the bases of cantilevers.

In dams having narrow V-shape profiles, the lower arches are relatively short, and the greater portion of the load is carried by arch action. From the standpoint of avoiding excessive tensile stresses in the arch, a type of layout should be used which will provide as much curvature as possible in the arches. This may be accomplished by using variable-thickness arches with a variation in location of centers to produce greater curvature in the lower arches. Fig. 8-3-8 shows an example of a two-centered variable thickness arch dam.

Assuming for comparison that factors such as central angle, height of dam, and shape of profile are equal, the arches of dams designed for wider canyons would be more flexible in relation to cantilever stiffness than those of dams in narrow canyons, and a proportionately larger part of the load would be carried by cantilever action. In dams for wide canyons in which there is a tendency for cantilever stresses to be greater than arch stresses, it is desirable to obtain the maximum possible advantage from dead weight by using a crown section having both faces curved, with undercutting at the base of the upstream face and overhang at the top of the downstream face. The layout would normally be a variable-thickness or polycentered arch.

Arch Shapes. In most cases, uniform-thickness arches may be used in the upper part of the dam since the thinner, longer arches are more flexible and do not carry as much of the load as those in the lower portion of the dam. The need for additional thickness at the abutments will vary with each layout but is normally not required in the upper few arches.

In the most efficient and economical design, the stresses approach uniform values close to the established allowable limits. Variable thickness arches will have a more uniform stress distribution than arches with fillets, since the thickness

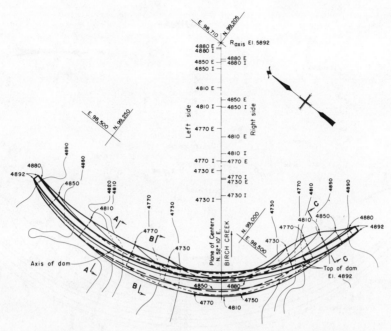

Figure 8-3-8. Two-centered dam with variable thickness arches.

varies gradually without any change in curvature. They also furnish adequate thickness for cantilevers with bases near the midheight of the dam. The angle of intersection of the intrados and a line parallel with the corresponding surface contour should not be less than 30° to insure stability of the arch abutment. If the angle of intersection is less than 30°, a special study should be made to evaluate the abutment stability. For sites where abutment thickening is desirable, short radius fillets may be added to uniform thickness arches on the downstream face. The length of radii used at one side need not be equal to that used at the other side. The fillet centers at each side of the dam should fall on smooth curves in plan to avoid irregularly warped surfaces. The locus of points of tangency between the intrados of the arches and the fillets at each of the dam, called the trace of beginning of fillets, also requires a smooth curve.

Fillet radii should have enough length to ensure that the resultants of arch forces are directed safely into the abutment rock, and that curvatures at the downstream face of both the arch and cantilever elements are not so great as to produce excessive stresses parallel with the face of the dam. For this reason, the angle of intersection between fillet and arch abutment should be not greater than 45°. Fillets should be laid out, as a general rule, so that the traces of beginning of fillets will intersect the top arch at its abutment and intersect approximately the three-fourths points of the arches in the region of greatest arch abutment stresses, that is at about one-half to three-fourths of the height of the dam above the base. Fig. 8-3-9 shows an example of short radius fillets used with arches of uniform thickness.

Three-centered or elliptical arches can be used advantageously in wide U- or V-shaped canyons. Elliptical arches have the inherent characteristic of conforming more nearly to the line of thrust for wide sites than do circular arches. Consequently the concrete is stressed more uniformly throughout its thickness. Because of smaller influences from moments, elliptical arches require little if any variable thickness. The direct benefit is reduced required volume and subsequently increased economy. Fig. 8-3-10 shows a typical three-centered arch.

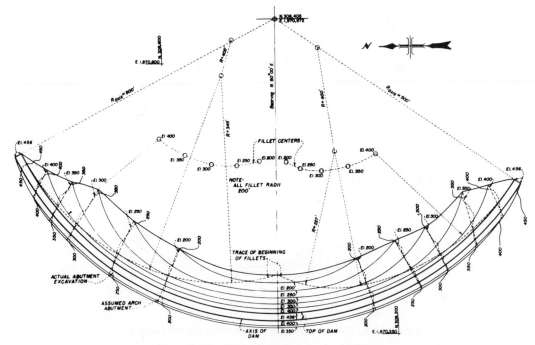

Figure 8-3-9. Uniform thickness arch with fillets.

Arch Abutments. Care must be taken to assure that the arch abutments are adequately keyed into sound rock, and that there is sufficient rock mass to withstand the applied loads. The directions of joint systems in the rock should be given careful consideration in making the layout to insure stable abutments under all loading conditions.

Full-radial arch abutments (normal to the axis) are advantageous for good bearing against the rock, but where excessive excavation at the extrados would result from the use of full-radial abutments, and the rock has the required strength and stability, the abutments may be reduced to half-radial as shown in Fig. 8-3-11(a). Where excessive excavation at the intrados would result from the use of full-radial abutments, greater than radial abutments may be used as shown in Fig. 8-3-11(b).

In such cases, shearing resistance should be carefully investigated. Where excessive excavation would result from the use of either of these shapes, special studies may be made for determining the possible use of other shapes having a minimum of excavation. These special studies would determine to what extent the arch abutment could vary from the full-radial and still fulfill all requirements for stability and stress distribution.

Analytical Results. The dam as defined by the layout is analyzed for stresses and deflection due to the applied loads. Stress analyses are made using methods discussed in Sections III-E-1 through III-E-4. The results of an analysis serves two purposes. The designer can evaluate the adequacy of the design and if improvement is required, he can utilize the analysis to ascertain the appropriate modifications to be made.

Evaluation requires a thorough examination of all the analytical output. The following represents the type of information to be reviewed; crown cantilever description; intrados and extrados lines of centers, geometrical statistics, dead load stresses and stability of blocks during construction, radial, tangential, and angular deflections, loading distributions, arch and cantilever stresses, and principal stresses. If any aspect of the design is either

DESIGN OF CONCRETE DAMS 405

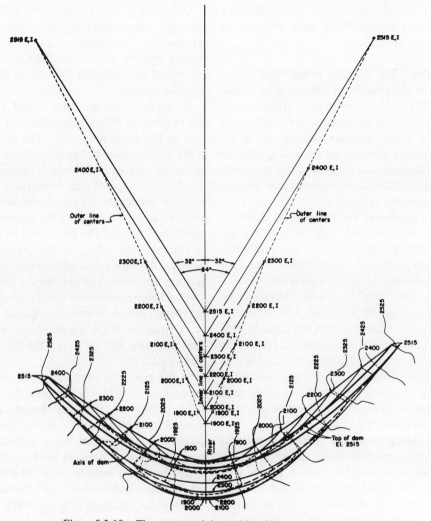

Figure 8-3-10. Three-centered dam with uniform thickness arches.

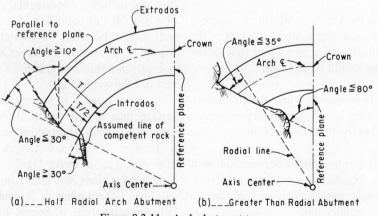

Figure 8-3-11. Arch abutment types.

incorrect or does not comply with established criteria then modifications must be made to improve the design.

3. Shaping. The primary means of effecting changes in the behavior of the dam is by adjusting the shape of the structure. Whenever the overall stress level in the structure is below the allowable limits, concrete volume can be reduced thereby utilizing the remaining concrete more efficiently and improving the economy. Following are several examples of how a design can be improved by shaping:

When cantilevers are too severely undercut they are unstable and tend to overturn upstream during construction. The cantilevers must then be shaped to redistribute the dead weight such that the sections are stable.

If an arch exhibits tensile stress on the downstream face at the crown, one alternative would be to reduce the arch thickness by cutting concrete from the downstream face at the crown while maintaining the same intrados contact at the abutment. Another possibility would be to stiffen the crown area of the arch by increasing the horizontal curvature which increases the rise of the arch.

Load distribution and deflection patterns should vary smoothly from point to point. Often when an irregular pattern occurs it is necessary to cause load to be shifted from the vertical cantilever elements to the horizontal arches. Such a transfer can be effected by changing the stiffness of the cantilever relative to the arch.

Shaping is the key to producing a complete and balanced arch dam design. The task of the designer is to determine where and to what degree the shape should be adjusted.

4. Design Techniques. Four design techniques, pads, thrust blocks, joints and prestressing are discussed in the following sections:

Pads. A pad can be utilized between the arch shell and the foundation as a transition section. Three particular uses for pads are discussed to illustrate their application.

Occasionally the abutments at a dam site are especially rugged and irregular or the site is unsymmetrical. To produce a smooth uniformly varying keyway would necessitate an excessive amount of excavation in the case of the highly irregular canyon.

An intermediate structure or pad, however, can be used to effect transition from the arch dam to the foundation rock. The pad surface at the dam interface is constructed to a smooth, uniformly varying profile which coincides with the desirable type of dam-foundation contact. The pad surface at the foundation interface is constructed to conform to the contour of the excavated keyway. The pad should be designed to eliminate the effects of the irregular abutment on the dam yet simultaneously produce a uniform distribution of load on the foundation rock. These requirements usually precipitate a pad design which resembles a spread footing, i.e., a structure which is generally more massive than the dam and consequently has a larger relative stiffness.

In the case of a nonsymmetrical site, a pad can be used to reduce or eliminate the nonsymmetry. This use of a pad requires that the physical mass of the pad be enough larger than the arch at their contact that a definite change in stiffness occurs between the two structures. If this condition does not exist, the arch dam behaves as a continuous arch structure from abutment to abutment and the effects of nonsymmetry will persist.

An abrupt change in stiffness between the dam and its pad is fundamental to the concept of a pad. If this condition is not satisfied, the arch would be more appropriately described simply as a uniform thickness arch or in the case of some thickening, either a variable thickness or fillet-type arch.

In some cases it is advantageous to form a surface near the perimetrical contact of the dam and its foundation which has no resistance to tension. This is a technique developed to deal with the excessive tensile stresses which can develop along the foundation at the dam extrados. The surface or formed joint usually requires some type of special treatment, e.g., painting to prevent bond, installation of special shear resistors or leakage preventive devices. Construction of a pad, the surface of which is the perimetrical joint, is useful in this situation.

Some dam foundations are completely adequate to support a dam safely but have a low bearing capacity. A pad can be used in situations such as this in a manner directly analogous to a spread footing. The loads imposed on the pad by the dam are redistributed sufficiently to reduce the stress on the rock within allowable limits.

Other words used for pad in the nomenclature of arch dam design are plinth, sockle, and pulvino.

Thrust Blocks. A thrust block constitutes an artificial abutment. A thrust block by virtue of the purpose it serves behaves as a gravity structure and therefore must be designed to resist all applied loads simply by the stability of its mass. Resultant forces should be considered in the analysis because their magnitude and direction represent the combined effects of all the applied forces.

A thrust block can be used near the top of a dam to reduce the length of the arches in canyons that widen rapidly. They can also be used at the top of the dam to reduce nonsymmetry and to provide an artificial abutment in cases where no natural rock abutment exists. Thrust blocks have been used to form artificial foundations in wide valleys to support the abutments of multiple arches.

Thrust blocks can be incorporated into the spillway design of a dam by forming either a gated or free overflow section over the top of the thrust block. Waterways often times take the form of conduits and these too can be conveniently located through a thrust block.

Joints. Joints as used here pertain to non-bonded surfaces that cannot transmit tensile forces. Normally an arch dam can be shaped to eliminate or substantially reduce the tensile stresses to within acceptable limits. Occasionally, however, other techniques are necessary to provide relief for excessive tensions. Both vertical and horizontal joints can be designed and located so that areas within the dam that may be subjected to tension and possible subsequent cracking are relieved of such forces.

For example, the top arches of a very thin dam exposed to severe temperature drops usually experience horizontal tensile stresses. This occurs when the arch contracts, thus trying to become shorter and the fixed abutments provide a restraint. Vertical contraction joints can be used to advantage in this situation to eliminate tensile stresses while permitting the arch to function normally as it expands with temperature rise and goes into compression.

Another application of formed joints is near the base of a dam along the extrados. Some situations exist when excessive vertical tensile stresses occur in the cantilever. In this case a horizontal joint can be formed which extends partially into the dam from the upstream face. Once again, as for the arch, the joint relieves tensile stresses yet functions in compression for a corresponding type loading condition. This type of joint is usually ended at the upstream side of a gallery. The gallery allows for periodic inspections and acts as an intercept for any seepage which might occur through the seals installed across the joint.

Prestressing. The technique of prestressing has been used for several years in concrete arch dam design. The particular application referred to is prestressing the arch such that it expands upstream against the water when the reservoir is filled. This is accomplished by lowering the temperature of the dam after it is constructed but prior to filling the reservoir. The dam is cooled by use of embedded cooling systems or simply by exposure to the ambient air. The temperature to which the dam is cooled is lower than the predetermined usual operating temperature. The vertical contraction joints between the construction blocks open as the dam is cooled. After reaching the prescribed temperature, the open contraction joints are filled with a cement and water grout. When the grout is set, the dam can act monolithically in compression. As the temperature of the concrete rises to the operating temperature, it is accompanied by an upstream expansion of the dam which in effect partially offsets the load from the reservoir.

More recent adaptations of prestressing have been performed using jacks and cables. Vertical cables are installed extending from inside a dam into the foundation rock. In this manner, the cantilevers can be prestressed into compression, thus compensating for tensile stresses which would exist under normal loads.

Flat jacks, as pictured schematically in Fig. 8-3-12, have also been used to prestress arches horizontally by placing the jacks in vertical joints and pressurizing them to induce the prestressing loads. Varying patterns of prestress can be produced in an arch depending on where and how many jacks are used. For example, jacks placed at an arch crown will produce uniform compression at the crown and a pattern at the abutment varying from tension at the intrados to compression at the extrados.

Flat jacks used in conjunction with a cable produce a uniform compressive prestress around the entire length of an arch. The jacks are placed at the arch abutment and a cable or cables embedded in the arch and anchored at the abutment. As the jacks are pressurized, the arch expanding upstream is resisted by the cables. The opposing action of the jacks and cables produces a net compression around the complete length of the arch.

Summary. There are two fundamental principles in the design of an arch dam. The first is to keep the design simple and to insure that all surfaces vary smoothly without abrupt changes in direction. The second is to remember that the structure is a continuum and therefore the behavior of the entire dam must be considered whenever any change in shape is contemplated.

E. Analysis

The Trial-Load Method is a three-dimensional method developed by the Bureau of Reclamation to perform stress analyses on concrete arch dams. The method was originally developed prior to 1940.[4,5] and has been expanded and improved during the ensuing years.[6] The reliability of the method has been confirmed by extensive research measurements.[7-11] The chief limitation of the trial load method is its complexity and the large amount of required time and labor. The development and application of the Trial Load Method is too extensive for inclusion in this volume but a brief discussion appears in Section III-E-1.

The advent of computers made possible the development of a computerized version of the Trial-Load Method named ADSAS, which is the acronym for Arch Dam Stress Analysis System. The excellent accuracy of ADSAS has been demonstrated by comparisons with results of trial-load method obtained without using computers. The average required time for a static analysis using ADSAS after programming the computer has been reduced to less than a minute. ADSAS is discussed more fully in Section III-E-2.

The most recent structural analysis method to be introduced in arch dam design is the finite

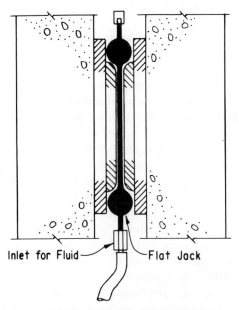

Figure 8-3-12. Schematic sketch of flat jack.

element method. A discussion of a new three-dimensional finite element computer program developed for the Bureau of Reclamation by the University of California is given in Section III-E-3.

In addition to the actual stress analysis of an arch dam, there are some additional analyses which must be performed to determine certain loadings on the dam as well as foundation deformation effects.

Sections III-E-4, III-E-5, and III-E-6 are devoted to discussions of temperature analysis, dynamic analysis, and foundation analysis, respectively.

1. The Trial Load Method. The trial-load method is based on the assumption that the water load is divided between arch and cantilever elements; that the division may or may not be constant from abutment to abutment for each horizontal element; and that the true division of load is the one which causes equal arch and cantilever deflections at all points in all arches and cantilevers instead of at the crown cantilever only. Furthermore, the method assumes that the distribution of load must be such as to cause equal arch and cantilever deflections in all directions; that is, in tangential and rotational directions as well as in radial directions. To accomplish the preceding agreement, it is necessary to introduce internal, self-balancing trial-load patterns on the arches and cantilevers.

A complete trial load analysis is obtained by properly dividing the radial, tangential, and twist loads between the arch and cantilever elements until agreement is reached for each of the three axial and three rotational movements for each arch-cantilever node point. The accuracy of this analysis is limited only by the exactness of the basic assumptions, the number of horizontal and vertical elements chosen, and the magnitude of error permitted in the slope and deflection adjustments.

Theory of the Trial Load Method. A comparatively elaborate analysis is required if a dependable estimate of stress distribution in an arch dam is to be obtained, because of the redundant nature of this type of structure. The requirements for a correct solution of the stress problem may be inferred from the Kirchhoff uniqueness theorem[12] in the theory of elasticity. The implied requirements are the following:

1. The elastic properties of the body must be completely expressible in terms of two constants: Young's modulus and Poisson's ratio.
2. If the volume of the body in the unstressed state is divided into small elements by passing through it a series of intersecting planes or surfaces, each of the elements so formed must be in equilibrium under the forces and stresses which act upon it.
3. Each of the elements described above must deform in such a way as the body passes into the stressed state that it will continue to fit with its neighbors on all sides.
4. The stresses or displacements at the boundaries of the body must conform to the stresses or displacements imposed.

Under these conditions Kirchhoff proved that it is impossible for more than one stress system to exist. If follows, therefore, that if the actual structure conforms to requirement (1), and a stress system is obtained which meets requirements (2), (3), and (4), this stress system is the one which must exist in the structure under the assumed conditions. A stress system meeting the above requirements may be obtained by the trial-load method.

A complete analysis by the trial load method is voluminous and time-consuming. An illustration of this analysis in full detail is beyond the scope of this book. The design engineer is referred to "Design of Arch Dams."[6]

2. ADSAS (Arch Dam Stress Analysis System). The need for an acceptably accurate, comprehensive method of analyzing arch dams resulted in the development of the trial-load method discussed in Section III-E-1. The chief limitations of the trial-load method are its complexity and the protracted computations required to make an analysis. Programming the computations for the electronic computer and linking them together to form the Arch Dam Stress Analysis System (ADSAS) permits the stress analysis of an arch dam to be made in a very short time and at low cost. In general, ADSAS follows the same procedures used in the trial-load analysis with some exceptions which help to make the method more adaptable to the computer. The significant changes from the trial-load method to ADSAS are discussed in the following paragraphs.

In a trial-load analysis the water load is not applied directly to either arch or cantilever elements, but is divided between them as a part of the radial adjustment loading. In the computer solution, however, all the external initial loads, including the water loading, are applied to the cantilever elements. Initial deformations of the cantilever elements are computed for these loads. Geometrical continuity is then attained by applying equal but opposite loading to arch and cantilever elements for the radial adjustment as well as for the tangential and twist adjustments.

Arch quarter points are used in the trial-load method with unit loads (1 kip) peaking at the abutments. The computer program locates the load points on the arches by using the intersections of cantilever elements with the arches. Unit loads are peaked at these points and varied linearly to zero at adjacent points on each side.

Unit arch loads for each section of the arch between node points are calculated for radial, tangential, and twist uniform and triangular loads.

Figure 8-3-13 is a sketch of the radial loads on Section B-C.

Tangential and twist loads are obtained in a similar manner. Computations for uniform thickness and variable thickness arches are described in Section III-E-1. Arch constants, load constants, and forces are transferred to other arch points on the same side of the arch. Total forces and deflections due to each unit load on each arch section are then computed for the entire arch.

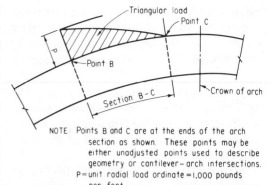

NOTE: Points B and C are at the ends of the arch section as shown. These points may be either unadjusted points used to describe geometry or cantilever-arch intersections.
P = unit radial load ordinate = 1,000 pounds per foot.

Figure 8-3-13. Unit radial load patterns on an arch section.

The triangular load for an arch section containing a point of compound curvature, as encountered with a polycentered arch, is obtained by combining a triangular load on Section O-C with a uniform load and a triangular load on Section B-O as shown in Fig. 8-3-14. The magnitude of the triangular load ordinate is also demonstrated in Fig. 8-3-14. The uniform load for Section B-C is obtained by combining the uniform loads on Sections O-C and B-O.

Unit arch loads peaking at node points and extending linearly to zero at the adjacent node points on either side as shown in Fig. 8-3-15 are obtained by combining the loads on two adjacent sections. The triangular load on the section nearest the crown (Section B-C) is added to the uniform load minus the triangular load on Section A-B.

The differences outlined above facilitate the generation of a set of simultaneous equations for each adjustment, radial, tangential, and twist. Matrix methods are used to solve each set of equations for the loading distributions required to maintain geometric continuity throughout the structure. The direct solution for loads replaces the trial and error procedure used in the trial-load method to determine load distributions. The present capabilities of ADSAS offer several options for geometrical configurations and loading combinations.

Geometry. The approach used for the geometrical description of the arch dam is the "solid

DESIGN OF CONCRETE DAMS 411

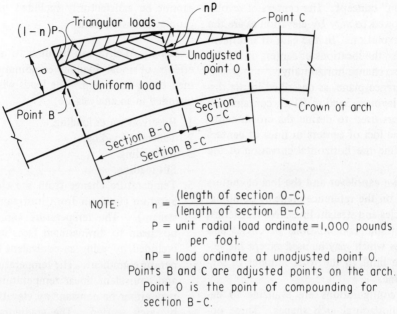

NOTE: $n = \dfrac{\text{(length of section O-C)}}{\text{(length of section B-C)}}$

P = unit radial load ordinate = 1,000 pounds per foot.

nP = load ordinate at unadjusted point O.
Points B and C are adjusted points on the arch.
Point O is the point of compounding for section B-C.

Figure 8-3-14. Unit triangular load pattern on an arch section with a point of compound curvature.

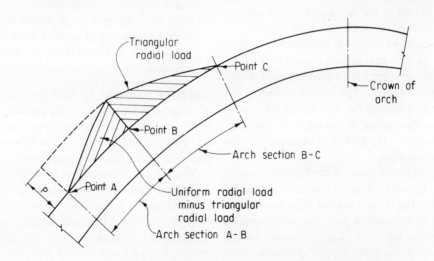

NOTE: The unit radial load peaking at point B is composed of a unit triangular radial load on section B-C plus a unit uniform radial load on section A-B minus a unit triangular radial load on section A-B.
The triangular load may be either as shown on figure 8-3-13 or figure 8-3-14.

Figure 8-3-15. Unit triangular load peaking at a node point.

of revolution" concept. The centers of revolution are allowed to vary by elevation as are the angles of revolution. In the case of multicentered designs, the locations of centers of revolution may also change horizontally.

The reference plane, as described in the discussion of layouts, Section III-D, contains the radius centers used to define the crown cantilever and the loci of centers or lines of centers used to define the horizontal curvation of the arches.

The crown cantilever and the loci of centers are defined on the reference plane by combinations of circles and straight lines. No restriction is placed on the combinations of circles and straight lines which may be used except the relatively large limit to the total number of segments per line.

Several configurations are available to describe the horizontal arch shapes. These options are listed below:

1. Single-centered dam. A configuration usually used in narrow, symmetrical sites. The loci of centers are the same for both sides of the dam and lie on the reference plane.
2. Two-centered dam. A configuration which may be used in relatively narrow sites which are too assymetrical to permit a good single-centered design. The loci of centers differ from one side of the dam to the other with the compounding of curvature at the reference plane. The loci of all centers lie on the reference plane.
3. Three-centered dam. A configuration which may be used to advantage in some wider sites. The three-centered dam has shorter radii in the central part and longer radii in the outer segments. Loci of centers for the central portion lie on the reference plane although those for the outer portion do not.

Abutment pads can be modeled on either or both faces of the dam. In addition, geometric data can be punched on cards to be used as input for the arch, cantilever, and abutment segments of ADSAS. These may be changed to modify the geometry when such modifications cannot be satisfactorily included in the basic geometric descriptions.

Loading. ADSAS can be used to analyze the effects of most loading combinations which may occur on a dam. The loads which can be included in an analysis are:

1. Reservoir water loading.
2. Tailwater loading.
3. Ice loading.
4. Silt loading.
5. Temperature change from the closure temperature (uniform from upstream to downstream). The temperature variation from upstream to downstream faces may also be included by using an equivalent linear temperature gradient. The temperature changes and equivalent linear temperature gradients may either be constant by elevation or vary by arch section. The gradients may also vary by cantilever and elevation.
6. Radial and tangential loads may be applied to node points to include the effects of dynamic response to earthquake.
7. Dead load stresses are included with total stress and are also computed for a check of stresses due to concrete weight during construction.

Additional Options. In addition, the effects of stage construction can be simulated by a series of studies. The results from these studies are then combined to give total stresses in the dam when a construction program is included.

The effects of variations in foundation deformation moduli both vertically and horizontally can also be included in the analysis.

Analytical models which simulate openings of the upper vertical contraction joints which interrupts arch action can also be studied. Loads are carried by cantilever action only in the portion of the dam where the arch action is considered not to occur.

Horizontal cracking can be included in the analyses which simulate cantilever cracking due to excessive tensile stress or the modeling of an intentionally formed horizontal joint used in a design to relieve tensile stresses.

Output. The normal output from ADSAS is a print of the following items:

1. Input data.
2. Summary sheet of physical properties and loading data.
3. Geometric properties at the reference plane as well as for arch and cantilever elements.
4. Volume of the dam.
5. Foundation constants.
6. Dead load stresses during construction.
7. Loads and movements for final tangential, twist, and radial adjustments.
8. Arch and cantilever forces.
9. Lists of stresses.
10. Stress maps and summary of principal stresses along the foundation.

The normal output can be supplemented by optional prints of data to be used for checking.

For special studies, stresses from several studies can be stored on tape for later use when certain combinations of studies are superimposed, e.g., stage construction and earthquake modal analyses.

Limitations. Although the analysis system has a large capacity and many capabilities, there are some limitations. The possible geometrical configurations are limited to those described earlier unless input to the arch, cantilever, and abutment segments are to be by punched cards. If cards are used, almost any shape can be approximated.

At present, the analysis system does not have the capability of analyzing the effects of nonradial abutments or thrust blocks.

The maximum number of arch and cantilever node points is limited. The system will analyze a dam with up to 12 arch elements and 25 cantilever elements (one at each arch abutment plus the crown cantilever) or 10 arch elements and 25 cantilever elements (one at each arch abutment, one at the crown, and two on each side between the crown and the lowest arch abutment).

Only one combination of loading conditions can be handled at one time. Any change in shape requires a separate analysis.

ADSAS requires a machine capacity of 64,000 word storage for the maximum size analysis. A machine capacity of 32,000 words can be used with some reduction in maximum problem size.

3. Finite Element Method. The finite element method is based on the principle that a continuous body can be replaced and modeled as an assemblage of discrete elements connected at their corner nodes. Although it may be extremely difficult to describe the behavior of the continuum mathematically, the behavior of each individual structural element is definable mathematically. The solution of the finite element model becomes a piecewise solution for the continuum. The method has become an accepted and widely used means of stress analysis.

Recent literature contains numerous examples of specialized uses of the finite element method. A principle reason for the ready acceptance and widespread use of the method is that it provides a better solution to many engineering problems which have previously been neglected or solved by overdesigning or making crude approximations. The ability to model complex geometry and to include variations in physical properties was very difficult prior to the finite element method and modern high-speed digital computers. The finite element method permits a very close approximation of the actual geometry and extensive material property variations simply and inexpensively. The formulation and theory of the finite element method is given in several publications.[13,14]

Two-dimensional finite element models can be used to analyze many problems associated with concrete dam design. Analysis of the dam, the dam's foundation, or the interaction of the two can be made economically even when a fine mesh is used to attain greater accuracy. Many times three-dimensional effects can be satisfactorily approximated by making two-dimensional analyses in one or more planes.

A three-dimensional finite element model can be used when the structure or loading is not a case of plane stress, plane strain, or is axially symmetric. The complexity and cost of performing three-dimensional analysis are much greater than ADSAS. Application of the three-dimensional analysis will, however, become easier and more economical as newer computers are built with larger storage capacity and faster execution times.

There is also a two-dimensional finite element analysis used to determine stresses and deflections for two-dimensional structures of arbitrary shape.[15] The system can optionally generate element meshes, loads, and material properties. The analyzed structure can be loaded with concentrated forces, gravity, temperatures, displacements, and accelerations as a percentage of the acceleration due to gravity. Variable material properties in compression and tension can be included. For example, the redistribution of load if tensile stresses are limited or not allowed can be determined by an iterative procedure. Openings in the structure can be modeled and the effect of shear stiffness in the third dimension can be included. The normal stress and shear stress acting on an arbitrary plane can be determined and the factor of safety computed for a given angle of internal friction and unit cohesion. In addition to printed output, the program produces optional microfilm prints of the entire grid or portions of it with material numbers shown. The plots can also have printed on them the horizontal, vertical, shear, or principal stresses.

There is a computer program[16] for a three-dimensional finite element analysis developed especially for the analysis of arch dams. The program was developed by the University of California at Berkeley through a research contract between the Bureau of Reclamation and the University.

The dam can be modeled using eight noded isoparametric hexahedron elements or either of two shell elements as shown in Fig. 8-3-16. Multiples of the 3D element can be used through the dam thickness to determine nonlinear stress distributions or to apply nonlinear temperature variations. The 3D element is used to define the foundation in all cases. Elastic orthotropic material properties can be specified

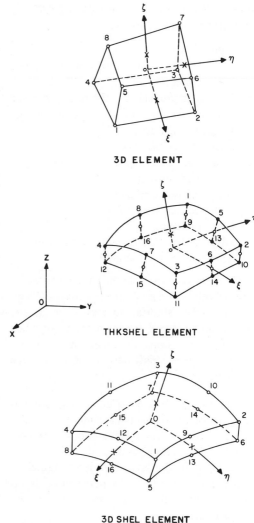

Figure 8-3-16. Element types available in ADAP.

for the 3D element and the material properties for both the THKSHEL and 3D-SHEL elements are restricted to elastic isotropy. The temperature in a 3D element is a function of the nodal point temperatures and the natural coordinates η, ζ, ξ. The temperature distribution within either shell element varies linearly through its thickness and quadratically in the surface directions.

The program contains mesh generation options for single-centered, two-centered, and three-centered dams. Any structure of arbitrary geometry can be analyzed, however, by describing the mesh to the program externally. The program can handle gravity, hydrostatic pressure, uniform pressure, temperature, and earthquake loads. An earthquake analysis is performed by first computing the natural frequencies and mode shapes of the dam. The stresses due to 3 translational components of earthquake ground motion can be computed using either a time history response or the response spectrum method and mode superposition.

The output from a static analysis includes nodal deflections and element stresses. Output from a dynamic analysis includes mode shapes and frequencies plus deflections and stresses.

The program offers a very accurate method of analysis provided a fine enough mesh is used. It can also be used to simulate stage construction and openings in the dam. Further, the dam and the foundation can by analyzed as a unit.

4. Temperature Analysis. Temperature analyses are necessary to determine the temperature distributions which will exist within dams and appurtenant structures during the construction operations and later throughout the operating life of a project. The information required to perform a temperature analysis includes weather data describing the climatic conditions, projected reservoir water temperatures and operation schedules, thermal properties of the materials involved, the effects of solar radiation, and the proposed construction schedule.

Air Temperature Amplitudes. Air temperatures expected to occur during the life of the structure are estimated based on recorded temperatures which have occurred at or in the vicinity of the dam site. The U.S. Department of Commerce, National Weather Service, has compiled and published recorded temperature data in their "Climatography of the United States" by states. The data can be adjusted for latitudinal distance to the site and differences in elevation. An increase of 1.4° latitude is assumed to decrease the temperature 1°F, and an increase of 250 ft in elevation is assumed to decrease the temperature 1°F. Figure 8-3-17 is a typical curve of annual climatic temperature data.

The methods developed are based on theories which assume the air temperature variations to be sinusoidal although the actual temperature cycles are not truly sine waves. The two principal temperature cycles used are the mean daily and the mean annual variations. The amplitudes for the two assumed sine curves are computed from the mean annual, the mean monthly, the mean monthly maximum, and the mean monthly minimum air temperatures. The period for the two sine waves is 24 hr for the daily variation and 8760 hr for the annual.

There are three categories of air temperature amplitudes that generally suffice for most design situations. The three categories are Mean Temperature Conditions, Usual Weather Conditions, and Extreme Weather Conditions. The Usual and Extreme categories include the effects of the maximum and minimum ever recorded temperatures.

A third sinusoidal temperature cycle is assumed to account for the maximum and minimum ever recorded temperatures. This cycle is fictitious but does have some empirical basis. This fictitious cycle is associated with

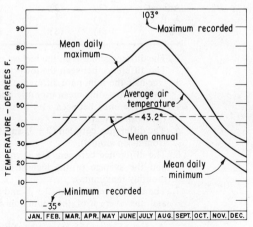

Figure 8-3-17. Typical curve of annual climatic temperature data.

barometric pressure variations which indicate a 1- to 3-week period. The third temperature cycle, therefore, has been arbitrarily set as 15 days or 365 hr for the Extreme Weather Conditions and 7 days or 168 hr for the Usual Weather Conditions. For the extreme Weather Conditions, the amplitudes are computed such that the actual maximum and minimum recorded air temperatures are accounted for when the daily, bi-weekly, and annual cycles are added together. The amplitude of the sine wave for the Usual Weather Conditions is computed to represent the temperatures halfway between the mean monthly maximum or minimum and the maximum or minimum ever recorded, respectively. Table 8-3-1 describes the computation of all the air temperature amplitudes required.

Figure 8-3-17 shows a typical plot of the climatic data for a dam site.

Reservoir Water Temperatures. Reservoir water temperatures vary with depth and are assumed to have an annual cycle. The factors which affect reservoir temperatures include the temperature of the water entering the reservoir, air temperatures, and reservoir operations. The best means of estimating what reservoir temperatures to expect during the normal operation of a dam is to examine temperature records for nearby lakes and reservoirs that have similar depths and inflow-outflow conditions. When

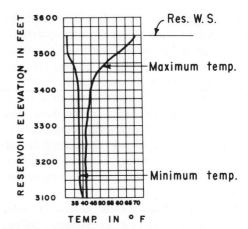

Figure 8-3-18. Typical curve of annual reservoir operation temperatures.

TABLE 8-3-1. Amplitudes of Air Temperatures

Period	Extreme weather conditions		Usual weather conditions		Mean temperature conditions	
	Above mean	Below mean	Above mean	Below mean	Above mean	Below mean
Annual	(1)	(2)	(1)	(2)	(1)	(2)
A, B, C	(4)	(5)	(6)	(7)	–	–
Daily	(3)	(3)	(3)	(3)	(3)	(3)

1. The difference between the highest mean monthly and the mean annual.
2. The difference between the lowest mean monthly and the mean annual.
3. One-half the minimum difference between any mean monthly maximum and the corresponding mean monthly minimum.
4. The difference between (1 + 3) and (the highest maximum recorded minus the mean annual).
5. The difference between (2 + 3) and (the lowest minimum recorded difference from mean annual).
6. The difference between (1 + 3) and (the difference between the mean annual and the average of the highest maximum recorded and the highest mean monthly maximum).
7. The difference between (2 + 3) and (the difference between the mean annual and the average of the minimum recorded and the lowest mean monthly minimum).
   A. Use 15-day (365-hr) cycle for Extreme Weather Conditions.
   B. Use 7-day (168-hr) cycle for Usual Weather Conditions.
   C. Not applicable for Mean Temperature Conditions.

DESIGN OF CONCRETE DAMS 417

there are no data available, an estimate can be made based on the exposure air temperatures, the operation and capacity of the reservoir, and the characteristics of the river inflow. The water surface temperature essentially follows the mean monthly air temperatures down to 32°F but not below. A system to measure air and reservoir water temperatures should be installed on all dams to collect data that can be compared against assumed values used in the design studies. Figure 8-3-18 shows a plot of typical reservoir temperatures variations.

Solar Radiation Effects. The mean concrete temperatures obtained from air and water temperatures require adjustments due to the effect of solar radiation on the surface of the dam. The downstream face, and the upstream face when not covered by reservoir water, receive an appreciable amount of radiant heat from the

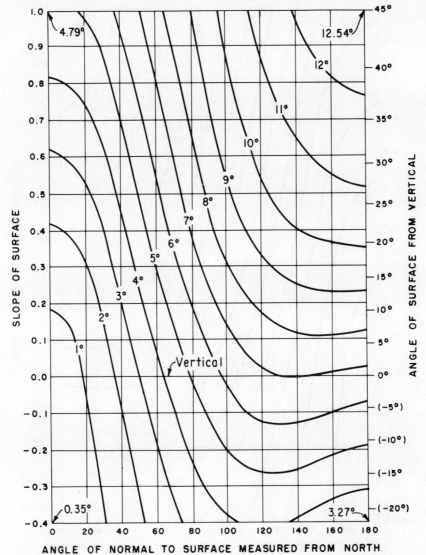

Figure 8-3-19. Solar radiation curves for 30°–35° latitudes. (*Courtesy Bureau of Reclamation*)

418 HANDBOOK OF DAM ENGINEERING

sun and this has the effect of warming the concrete surface above the surrounding air temperature. The amount of this temperature rise above the surrounding air temperatures has been recorded at the faces of several dams in the western portion of the United States. These data were then correlated with theoretical studies which take into consideration varying slopes, orientation of the exposed faces, and latitudes. Figures 8-3-19, 20, 21, and 22 give values of temperature increase due to solar radiation for various latitudes, slopes, and orientations.

It should be noted that the curves give a value for the mean annual increase in temperature and not for any particular hour, day, or month.

An example computation for solar radiation

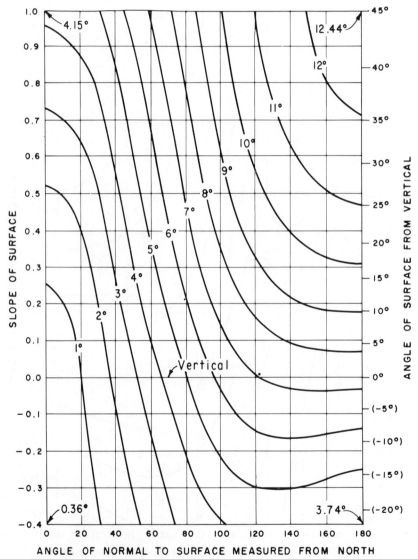

Figure 8-3-20. Solar radiation curves for 35°–40° latitudes. (*Courtesy Bureau of Reclamation*)

DESIGN OF CONCRETE DAMS 419

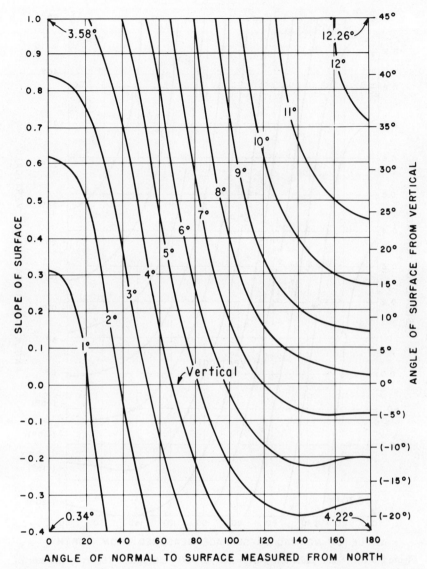

Figure 8-3-21. Solar radiation curves for 40°–45° latitudes. (*Courtesy Bureau of Reclamation*)

effects on an arch dam is given in Fig. 8-3-23. The temperature rises shown on the graph should be corrected by a terrain factor which is expressed as the ratio of actual exposure to the sun's rays to the theoretical exposure. This is required because the theoretical computations assumed a horizontal plane at the base of the structure and the effect of the surrounding terrain is to block out certain hours of sunshine. Although this terrain factor will actually vary for different points on the dam, an east-west profile of the area terrain, which passes through the crown cantilever of the dam, will give a single factor which can be used for all points and remain within the limits of accuracy of the method itself.

Heat Flow Analysis. The application of variational calculus combined with the finite element technique provides a powerful method of

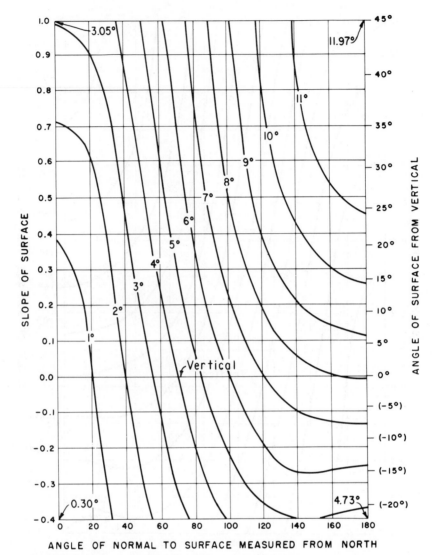

Figure 8-3-22. Solar radiation curves for 45°-50° latitudes. (*Courtesy Bureau of Reclamation*)

analysis for the determination of temperature distributions in complex structures of arbitrary geometry.[17] A computer program [18] has been developed, which utilizes the two-dimensional finite element technique for transient or steady-state temperature analyses of plane and axisymmetric solids.

This program has several optional capabilities which include: simulation of concrete heat of hydration; convection heat transfer across boundaries; sinusoidal external temperature variation for daily, weekly, biweekly, and annual cycles; simulation of stage construction according to concrete lifts; simulation of cooling by embedded cooling coils; and plotting the finite element grid with material numbers and element temperatures on computer microfilm output.

The program HEATFL can be used to determine placing temperatures, required cooling

DESIGN OF CONCRETE DAMS 421

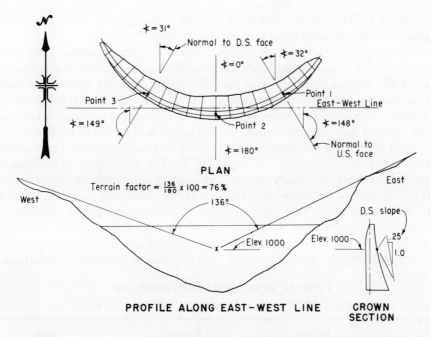

Figure 8-3-23. Example solar radiation computation.

water temperatures, cooling coil spacing, cooling periods, height of construction lifts, width of construction blocks, insulation requirements, operating temperature distributions in the dam, temperature rise during construction, and temperature gradients near surfaces and openings.

Cooling. Cooling is sometimes necessary to partially offset the heat of hydration occurring during construction and also to bring the concrete mass to a specified grouting temperature. Cooling can be accomplished naturally by exposure to climatic conditions. However, this method is limited to whatever the minimum air temperature at a site may be and is also difficult to control.

Refrigeration has become an important means of accomplishing cooling in mass concrete dams. Refrigeration is used to reduce the placing temperature of a mix by lowering the temperatures of various mix components. Refrigeration is also used to chill the coolant circulated through temperature control coils embedded at the lift surfaces in a dam.

The computations necessary to simulate the effect of cooling coils is integrated into the computer program HEATFL as discussed in the preceding section. The required input data to include cooling are obtained from curves which define the temperature change in the coolant as a function of the thermal properties, coil dimensions and spacing, and the volume rate of flow.[19] Typical computations necessary to determine the required tons of refrigeration to chill the coolant or to lower mix temperatures are discussed below.

Concrete mix cooling is usually accomplished by one or more of the following: water immersion, water spray or air blast cooling of large aggregate, use of refrigerated water, substitution of ice for part of the mix water, and cooling of sand and cement. The method used should be designed for each particular construction project.[20]

There are some empirical constants which generally apply in computing refrigeration requirements. The mix temperature leaving the batch plant should be 3°F below the required placing temperature to allow for heat gains during transit to the point of placement. The temperature of portland cement is assumed to be 150°F. The temperature of stockpiled coarse aggregate will generally lag behind the exposure air temperatures and, as a result, will be 5°F less than the mean monthly maximum and 3°F more than the mean monthly minimum. The following is an example problem to demonstrate typical refrigeration computations:

TYPICAL REFRIGERATION PROBLEM
GIVEN DESIGN DATA FOR 1 CUBIC YARD OF CONCRETE

Material	Dry weight (lb/cu yd)	Surface moisture (%)	Specific heat	Surface moisture pounds	Btu's per 1°F	Initial temperature °F	Total Btu's
3"-6"	804	0.5	0.23	4	186	72	13,392
1½"-3"	761	1.2	0.23	9	177	72	12,744
¾"-1½"	550	1.5	0.23	8	128	72	9,216
No. 4-¾"	605	1.0	0.23	6	141	72	10,152
Sand	812	6.0	0.23	49	198	70	13,860
Cement	231	–	0.23	–	53	150	7,950
Pozzolan	92	–	0.23	–	21	80	1,680
Water	162	–	1.0	–	162	60	9,720
TOTAL					1,066		78,714

Number of mixers = 4
Mixer capacity (cu yd) = 4
Hp per mixer = 50
Mix time (min) = 3
Placing temp. (°F) = 50
Placing rate (cu yd/day) = 4,000

Heat generated by mixers = $\dfrac{50 \times 2{,}545 \text{ (Btu/hr/hp)} \times 3}{4 \times 60}$ = 1,591 Btu's

Initial temperature of 1 cu yd = $\dfrac{78{,}714 + 1{,}591}{1{,}066}$ = 75°F

Required refrigeration for 47°F = $\dfrac{(1{,}066)(4{,}000)(75 - 47)}{288{,}000 \text{ (Btu/ton/day)}}$ = 415 tons

Assumed losses:

Heat leakage = 5% of required refrigeration

Water leakage = $\dfrac{2\% \text{ of water wt circulated/day} \times \text{temperature change}}{288{,}000 \text{ (Btu/ton/day)}}$

$$\text{Aggregate absorption} = \frac{2\% \text{ of Agg Wt/day} \times \text{temperature change}}{288{,}000 \text{ (Btu/ton/day)}}$$

$$\text{External heat from pumps} = \frac{\text{Bhp (Brake horsepower)} \times 61{,}068 \text{ (Btu/day/hp)}}{288{,}000 \text{ (Btu/day/hp)}}$$

Total (tons) = 415 + losses

If ice is to be used as part of the mix water, a typical computation might be as follows:

Assume coarse aggregate (¾″–6″) is cooled to 47°F.

$$\begin{aligned}
\text{Aggregate:} \quad & 2{,}115 \times 0.23 \times (72 - 47) = 12{,}161 \\
\text{Surface moisture:} \quad & 21 \times 1.0 \times (72 - 47) = \phantom{12{,}}525 \\
& \text{Btu/cu yd} = 12{,}686
\end{aligned}$$

Assume remaining heat is removed by mix water in the form of ice and water at 35°F.

Remaining heat = 1,066 (75 − 47) − 12,686 = 17,162 Btu's

Let I equal pounds of ice/cu yd. Then pounds of water at 35°F equals (162 − I)/cu yd.

$$\begin{aligned}
\text{I}[144^* + (47 - 32)] + (162 - \text{I})(47 - 35) &= 17{,}162 \\
147\text{I} + 1{,}944 &= 17{,}162 \\
\text{I} &= 104 \text{ lb}
\end{aligned}$$

Tons of ice per day = 208

*Ice requires 144 Btu's/lb to melt.

5. Dynamic Analysis.

The Time-History Method of earthquake analysis is presented in this section. The method may be applied in general to any structure which acts elastically under the influence of an earthquake. The method of analysis utilizes lumped masses, generalized coordinates, and mode superposition. Brief discussions of the theory and development of equations are given here. More detailed discussions of dynamic analysis are available in textbooks[21] and other publications.[22,23]

The first step is to discuss the basic concept of motion and to develop some useful expressions used in dynamic analysis.

The equations of motion are most easily developed for a one-degree-of-freedom system. Such a system can be demonstrated by having a mass, m, affixed to a rigid support by a spring, with stiffness, k, as shown in Fig. 8-3-24.

The vibrating system is of the simplest type because only one coordinate is necessary to describe the motion of the mass. Thus the term one degree of freedom.

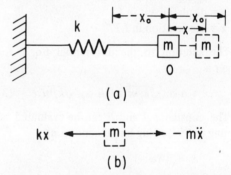

Figure 8-3-24. Simple one-degree-of-freedom system.

If the mass m in Fig. 8-3-24a is pulled to the right a distance x_0 and released, it will be acted upon by a restoring force in the spring toward its original position. Intuitively the motion will be confined to a range $\pm x_0$ on either side of the equilibrium position and can be demonstrated experimentally as well. Each back and forth movement requires the same increment of time and the motion would continue indefinitely if there was no loss of energy by friction. This type of motion, under the influence of an elastic restoring force and no friction is called simple harmonic motion. This type of motion is also called periodic because it repeats itself in equal time intervals. The period of the motion represented by T, is the time required for one complete vibration. The frequency, equal to the number of complete vibrations/sec, is the reciprocal of the period or

$$f = \frac{1}{T} \text{ cycles/unit time} \qquad (1)$$

Consider now the equation of equilibrium for the mass, m, in Fig. 8-3-24(a). Dimensionally, F, L, and T represent force, length, and time. According to D'Alembert's Principle

$$m\ddot{x} + kx = 0 \qquad (2)$$

where

$$m = \frac{w}{g} \text{ in } FL^{-1}T^2$$

w = weight in F
g = acceleration due to gravity in LT^{-2}
k = spring stiffness in FL^{-1}
x = displacement in L
$\ddot{x}$ = acceleration in LT^{-2}

The solution to the differential Eq. 2 for time t is

$$x = A \sin \sqrt{k/m}\, t + B \cos \sqrt{k/m}\, t \qquad (3)$$

The constants A and B can be evaluated assuming the following initial condition:

$$x = x_0$$

and

$$\dot{x} = \dot{x}_0 \text{ at the time } t = 0.$$

Substitution of the two initial conditions into Eq. 3 produces values of A and B as given below:

$$A = \dot{x}_0 \sqrt{m/k}$$

$$B = x_0$$

Rewriting Eq. 3 gives

$$x = \dot{x}_0 \sqrt{m/k} \sin \sqrt{k/m}\, t + x_0 \cos \sqrt{k/m}\, t \qquad (4)$$

This equation shows that harmonic or free vibration as it is also called may be started by either initial velocity or by an initial displacement.

Now using Eq. 4 and assuming there is no initial velocity, it is possible to calculate the time for the mass to go from its initial position $x = x_0$ to the extreme left position $x = -x_0$. The distance traveled is one-half a complete oscillation and therefore requires a time equal to one-half a period, $T/2$. From Eq. 4

$$t = \sqrt{\frac{m}{k}} \cos^{-1} \frac{x}{x_0}$$

or

$$\frac{T}{2} = \sqrt{\frac{m}{k}} \cos^{-1} \frac{-x_0}{x_0} = \sqrt{\frac{m}{k}} \cos^{-1}(-1)$$

The angle whose cosine is -1 is $180°$ or π. Hence

$$\frac{T}{2} = \sqrt{\frac{m}{k}} \pi$$

or

$$T = 2\pi \sqrt{\frac{m}{k}} \text{ time/cycle} \qquad (5)$$

Since frequency $f = 1/T$

$$f = \frac{1}{2\pi} \sqrt{\frac{k}{m}} \text{ cycles/unit time} \qquad (6)$$

Equations 5 and 6 can be used to find the period or frequency of a body given its mass when vibrating under the influence of an elastic restoring force with known stiffness. It is interesting to note that period does not depend on the amplitude of vibration but on the mass and stiffness of the structure.

Another useful expression demonstrates that ω, the angular velocity in radians/second of a point traveling around the circumference of a circle and the frequency f, the number of complete revolutions of the point per second are related by

$$\omega = 2\pi f \text{ radians/unit time} \qquad (7)$$

From Eq. 6 it is also evident that

$$\omega = \sqrt{\frac{k}{m}} \qquad (8)$$

The next step is to describe the displacement response of the single degree of freedom system subjected to an arbitrary forcing function defined as

$$P(t) = P'f(t) \qquad (9)$$

From Newton's Second Law, $F = ma$, the following expression can be written for time $t = \tau$

$$P(\tau) = ma \qquad (10)$$

Equation 10 can be written in the impulse momentum form as

$$P(\tau)d\tau = m(d\dot{x})$$

or

$$P'f(\tau)d\tau = m(d\dot{x})$$

and

$$d\dot{x} = \frac{P'f(\tau)d\tau}{m} \qquad (11)$$

The displacement of mass, m, due to the load P' applied statically would be $x_{ST} = P'/k$ or $P' = kx_{ST}$. If this expression for P' is substituted into Eq. 11 it can be written as

$$d\dot{x} = \frac{kx_{ST} f(\tau)d\tau}{m}$$

and remembering that $\omega^2 = k/m$ then

$$d\dot{x} = \omega^2 x_{ST} f(\tau)d\tau \qquad (12)$$

The displacement of a single-degree-of-freedom system due to an initial velocity is

$$x = \frac{\dot{x}_0}{\omega} \sin \omega t$$

as given in Eq. 4.

Because each incremental velocity change can be considered independently, then summed, Eq. 4 can be used to find the displacement at some later time $t = \tau$ by

$$dx = \frac{d\dot{x}}{\omega} \sin \omega(t - \tau)$$

Substitution for $d\dot{x}$ from Eq. 12 gives

$$dx = \omega x_{ST} f(\tau) \sin (t - \tau)d\tau \qquad (13)$$

The displacement at time t is obtained by integration of the incremental displacement between $\tau = 0$ and $\tau = t$.
Thus

$$x = x_{ST} \omega \int_0^t f(\tau) \sin \omega(t - \tau)d\tau \qquad (14)$$

is the displacement of a system initially at rest.

As is true for any type of motion, the displacement can be expressed as the static displacement multiplied by a dimensionless dynamic or magnification factor $D(t)$.

In Eq. 14 the magnification factor is

$$D(t) = \frac{2\pi}{T} \int_0^t f(\tau) \sin \frac{2\pi}{T} (t - \tau)d\tau \qquad (15)$$

The integral in Eq. 15 is called the Duhamel Integral and is fundamental to dynamic analysis.

When dealing with earthquake accelerations, the response of a structure is excited by a particular ground motion. Therefore, in addition to initial velocities, displacements and forcing functions, movements of a system's support can cause vibration of the structure. If the support of the system in Fig. 8-3-25 is subjected to a ground displacement $x_g(t)$, then the equation of motion is

$$m\ddot{x} + k(x - x_g) = 0 \qquad (16)$$

In Eq. 16, the mass term involves the total displacement x and the stiffness term the relative displacement of the mass to the ground $x - x_g$. Letting u equal the relative displacement $x - x_g$ and substituting into Eq. 16 gives

$$m(\ddot{u} + \ddot{x}_g) + ku = 0$$

or

$$m\ddot{u} + ku = -m\ddot{x}_g(t) \qquad (17)$$

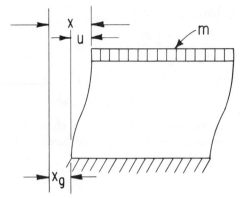

Figure 8-3-25. Relative displacement of mass due to a ground displacement.

This very important relationship shows that relative motion can be treated like an ordinary forced vibration problem where the forcing function is expressed in terms of the ground acceleration.

The solution to Eq. 17 for relative displacement is

$$u(t) = \frac{T}{2\pi} \int_0^t \ddot{x}_g(t) \sin \frac{2\pi}{T} (t - \tau) d\tau \quad (18)$$

A useful relationship can be derived from Eq. 16 by noting that $\ddot{x} = -\omega^2 u$. This states that absolute acceleration and relative displacement are related as in simple harmonic motion.

To this point, the effects of damping, c, have not been included in the equations of motion. Damping is present in all moving systems, however, and it affects the response of the system. There are different types of damping, but in structural design, viscous damping is usually assumed and is represented as a resisting force proportional to the velocity. This type of damping is known as the dashpot effect.

The equation of motion for free vibration including damping is

$$m\ddot{x} + c\dot{x} + kx = 0 \quad (19)$$

where c is the damping factor with dimensions of FTL^{-1} and $\dot{x}$ is the velocity in LT^{-1}. The solution to this differential equation is of the form

$$x = A_1 e^{\alpha_1 t} + A_2 e^{\alpha_2 t} \quad (20)$$

Substitution of Eq. 20 into Eq. 19 yields expressions for the two roots.

$$\alpha_1 = \frac{-c}{2m} + \sqrt{\frac{c^2}{4m^2} - \frac{k}{m}} \quad (21)$$

$$\alpha_2 = \frac{-c}{2m} - \sqrt{\frac{c^2}{4m^2} - \frac{k}{m}} \quad (22)$$

For brevity let $\beta = c/2m$ and of course $\omega^2 = k/m$. There are three distinct cases which must be investigated

$\beta^2 - \omega^2 = 0$ Case 1
$\beta^2 - \omega^2 > 0$ Case 2
$\beta^2 - \omega^2 < 0$ Case 3

For Cases 1 and 2, the two roots, α_1 and α_2 are negative. Substitution of the values into Eq. 20 shows that for both cases the motion is aperiodic, meaning not periodic. Physically this means that after an initial displacement a mass would gradually return to its starting position but never go beyond and have to return. Actually Case 1 is the limiting case for Case 2. The motion described in Case 1 is called aperiodic motion and the damping necessary to cause such motion is called critical damping c_{cr}. The motion described by Case 2 is said to be over damped because it is less than critical. Figure 8-3-26 demonstrates critical and over critical motion.

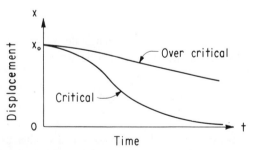

Figure 8-3-26. Curves showing critical and over critical damping.

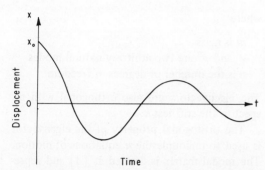

Figure 8-3-27. Curve showing decay of damped periodic motion.

For Case 3 $c < c_{cr}$ a system is said to be underdamped. This represents the most common case and applies to the type motion associated with an arch dam. By letting $\omega_d^2 = \beta^2 - \omega^2$ and substituting values of α_1 and α_2 into Eq. 20 the solution for x is

$$x = e^{-\beta t}(C_1 \sin \omega_d t + C_2 \cos \omega_d t) \quad (23)$$

A plot of Eq. 23 is shown in Fig. 8-3-27 and clearly demonstrates how the amplitude of the periodic motion decays exponentially.

The damped period, T_d, is somewhat greater than the undamped or natural period, T_n, as shown by the following:

$$T_d = \frac{2\pi}{\omega_d} = \frac{2\pi}{\omega_n} \frac{1}{\sqrt{1 - (\beta/\omega_n)^2}} \quad (24)$$

However, since values of β are usually small, the natural period can be used with little inaccuracy.

The corresponding form of Eq. 14 including damping can be written as

$$x = x_{ST}\omega \int_0^t f(\tau) e^{-\beta(t-\tau)} \sin \omega(t-\tau) d\tau \quad (25)$$

and for relative displacement

$$u(t) = \frac{1}{\omega} \int_0^t \ddot{x}_g(\tau) e^{-\beta(t-\tau)} \sin \omega(t-\tau) d\tau \quad (26)$$

Having developed the basic equations for a one-degree-of-freedom system, it is possible, using matrix algebra, to expand the equations to define the response of a multidegree-of-freedom structure. There are two methods to solve the governing equation:

1. Modal Analysis
2. Direct Integration of the differential equation of motion

The method presented in this section, as stated at the beginning, is modal analysis where the contributions of each mode of vibration are superimposed to find the total response of a system.

A mode of vibration is the time variation of the configuration a system assumes during a particular motion. For a multidegree-of-freedom system, there exist as many distinct modes of vibration as there are degrees of freedom. Each mode of vibration is harmonic, i.e., it is periodic and definable as a sine or cosine function.

There is also a unique frequency associated with each mode of vibration. When a multidegree-of-freedom system is allowed to vibrate freely, it will vibrate according to one of its natural modes and at an associated frequency called a natural frequency. Any irregular mode of vibration can be described by the sum of all the normal modes each of some particular frequency.

To perform a modal analysis, the mode shapes and frequencies of the system must first be determined.

The equation of motion can be generalized to obtain n equations of motion for a system with n degrees of freedom as follows:

$$[M]\{\ddot{x}\} + [K]\{x\} = \{F(t)\} \quad (27)$$

where

$[M]$ is a diagonal matrix of masses
$[K]$ is a square matrix of stiffnesses
$\{\ddot{x}\}$, $\{x\}$, and $\{F(t)\}$ are vectors of acceleration, displacement, and force, respectively.

For the case of free vibration, $\{F(t)\} = 0$.

Equation 27 represents a characteristic value problem and solutions for this homogeneous equation exist only for certain characteristic frequencies. The frequencies turn out to be the natural frequencies for each degree of freedom and are called eigenvalues. The solutions to the equation are simple harmonic motions given by

$$\{x\} = \{\varphi\} \sin \omega t \qquad (28)$$

where $\{\varphi\}$ is a vector of the amplitudes of vibration for each respective mass. Substituting Eq. 8 into Eq. 27 gives

$$[K]\{\varphi\} = \omega^2 [M]\{\varphi\} \qquad (29)$$

The existence of nontrivial solutions of this homogeneous equation requires that the characteristic determinant vanish:

$$|[K] - \omega^2 [M]| = 0 \qquad (30)$$

The expansion of the determinant 30 yields a characteristic polynomial having n roots for ω^2. These values for ω^2 are the eigenvalues or natural frequencies.

The amplitudes $\{\varphi\}$ cannot be evaluated numerically although the ratios of the amplitudes can be determined for each natural frequency. These ratios represent mode shapes and are called eigenvectors.

There are several methods for evaluating eigenvalues and eigenvectors from the standard form of the equation[24]

$$\lambda\{\varphi\} = [R]\{\varphi\} \qquad (31)$$

Equation 29 can be rewritten as

$$\omega^2 \{\varphi\} = [M]^{-1}[K]\{\varphi\} \qquad (32)$$

By comparing Eqs. 31 and 32 the following is evident:

$$\lambda = \omega^2$$

and

$$[R] = [M]^{-1}[K]$$

A very useful relationship exists between eigenvectors and is defined as orthogonality. Orthogonality, simply stated, means the following:

$$\sum_{i=1}^{n} m_i \varphi_i' \varphi_i'' = 0 \qquad (33)$$

where

m is mass
φ' and φ'' are two arbitrary natural modes
n is the number of degrees of freedom

The eigenvectors are also orthogonal with respect to the stiffness k.

The orthogonal property of the eigenvector is used to uncouple the n equations of motion. The modal matrix is defined as $[A]$ and represents the assembly of the n eigenvectors. The square matrix $[A]$ is used to introduce new generalized coordinates, $\{q\}$, via the transformation

$$\{x(t)\} = [A]\{q(t)\} \qquad (34)$$

Substituting Eq. 34 into the Equation of motion with damping gives

$$[M][A]\{\ddot{q}\} + [C][A]\{\dot{q}\} + [K][A]\{q\} = \{F\} \qquad (35)$$

Premultiplying by $[A]^T$ gives

$$[\overline{M}]\{\ddot{q}\} + [\overline{C}]\{\dot{q}\} + [\overline{K}]\{q\} = \{\overline{Q}\} \qquad (36)$$

where

$$[\overline{M}] = [A]^T [M][A] \qquad (37a)$$

$$[\overline{K}] = [A]^T [K][A] \qquad (37b)$$

$$[\overline{C}] = [A]^T [C][A] \qquad (37c)$$

$$\{\overline{Q}\} = [A]^T \{F\} \qquad (37d)$$

The orthogonal property of the eigenvectors causes the generalized quantities in Eq. 37 to be diagonal matrices with the restriction that the damping matrix be proportional to the stiffness and/or mass matrices. Thus the n equations have been uncoupled. Equation 36 can be simplified by dividing through by $\overline{M}$. The ith equation becomes

$$\ddot{q}_i + \frac{\overline{C}_i}{\overline{M}_i} \dot{q}_i + \frac{\overline{K}_i}{\overline{M}_i} q_i = \frac{\overline{Q}_i}{\overline{M}_i} \qquad (38)$$

If η is set equal to C/C_{cr}, the damping factor, and remembering that $C_{cr} = 2m\omega$, then

$$C = 2\eta m\omega \qquad (39)$$

Equation 38 can be rewritten as

$$\ddot{q}_i + 2\eta_i \omega_i \dot{q}_i + \omega_i^2 q_i = \frac{\overline{Q}_i}{\overline{M}_i} \qquad (40)$$

which is the equation of motion with generalized coordinates, damping, and some arbitrary forcing function.

The equation of motion is completed by adding the effects of support motion. In this case the forcing function is replaced by the product of the mass and the ground acceleration $\{F\} = [M]\ddot{x}_g(t)$. The results then become relative displacements $\{u\}$ rather than absolute displacements $\{x\}$. Returning to the generalized force vector $\{\overline{Q}\} = [A]^T \{F\}$ and substituting for F the expression becomes

$$\{\overline{Q}\} = [A]^T [M] \ddot{x}_g \qquad (41)$$

For the ith mode

$$\overline{Q}_i = \{\varphi\}_i^T \{M\} \ddot{x}_g \qquad (42)$$

and the uncoupled equation becomes

$$\ddot{q}_i + 2\eta_i \omega_i \dot{q}_i + \omega_i^2 q_i = \Gamma_i \ddot{x}_g \qquad (43)$$

where

$$\Gamma_i = \frac{\{\varphi\}_i^T \{M\}}{\overline{M}_i} \qquad (44)$$

The term Γ is called a participation factor.

The solution to this single-degree system is

$$q_i = \frac{\Gamma_i}{\omega_i} \int_0^t \ddot{x}_g e^{-\eta_i \omega_i (t-\tau)} \sin \omega_i (t-\tau) d\tau \qquad (45)$$

The n single-degree equations are usually solved using numerical methods and the results from each mode are superimposed using Eq. 40.

After the displacements are known the inertia forces can be determined. Remember, it was pointed out earlier that absolute acceleration and relative displacement are related by

$$\ddot{q}_i(\text{absolute}) = \omega_i^2 q_i(\text{relative}) \qquad (46)$$

Also, by removing the ith eigenvector from the transformation matrix $[A]$ and using Eq. 40.

$$\{\ddot{x}\} = \{\varphi\}_i \ddot{q}_i \qquad (47)$$

Thus

$$\{\ddot{x}\} = \{\varphi\}_i \omega_i^2 q_i$$

Therefore the inertia force vector for the ith mode is

$$\{F\}_i = [M]\{\ddot{x}\} = [M]\{\varphi\}_i \omega_i^2 q_i \qquad (48)$$

The total inertia force vector for all modes is

$$\{F\} = [M][A][E]\{q\} \qquad (49)$$

where

$[E]$ = matrix of ω^2 values.

Having briefly shown the method of modal analysis and some of the basic development, its specific application to an arch dam can be discussed.

To begin, the dam is divided into a system of horizontal arches and vertical cantilevers. A mass point is coincident with a nodal point or the intersection of an arch and a cantilever. The mass associated with each point is equivalent to the weight divided by the acceleration due to gravity of that volume of concrete contained within the space halfway to an adjacent node in every direction.

The total mass of each associated volume is assumed to be concentrated or lumped at the node point. Hence, the term "lumped mass." The movement of each mass can be described by a set of coordinate displacements and rotations. Each displacement or rotation is considered a degree of freedom. The number of mass points is directly proportional to the degrees of freedom for a particular structure. Therefore, the more mass points used increases the degrees of freedom. The accuracy of a dynamic analysis improves as the degrees of freedom are increased. However, it is advisable to determine the number of degrees of freedom beyond which no significant improvement in accuracy can be obtained and to assume that number is sufficient.

The equation of motion for an n degree of freedom arch dam is

$$[M]\{\ddot{u}\} + [C]\{\dot{u}\} + [K]\{u\} = -[M]\ddot{x}_g(t) \qquad (50)$$

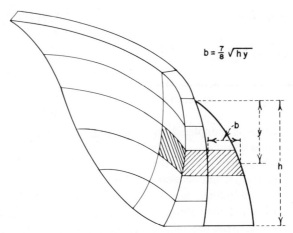

Figure 8-3-28. Dimensional sketch of water mass assumed to be accelerated with dam.

The mass matrix, $[M]$, is composed of all the lumped masses. Each lumped mass includes the concrete enclosed by the intersecting surfaces of the arch and the cantilever and the corresponding mass of the water assumed to be accelerated with the dam. The water mass is estimated using a formula developed by Westergaard.[25] The radially measured dimensions, b, of the water assumed to be moving with the dam is shown in Fig. 8-3-28.

The damping matrix $[C]$ contains the damping factors $c/c_{cr} = \eta$ for each mode.

Little data are available on the damping factor for concrete dams. Measurements during the forced vibration tests at Monticello Dam[26] indicated a damping factor of from 2 to 3 percent. Other measurements of damping on monolithic concrete structures indicate a similar range of 2 to 3 percent.[27,28]

The stiffness matrix, $[K]$, is equal to the inverse of the flexibility matrix, $[D]$. The flexibility matrix is more easily constructed than the stiffness matrix so it is used and the inversion is performed by computer as is the entire dynamic analysis. An element, d_{ij}, of the flexibility matrix is the deflection at i due to a unit load at j. Both radial and tangential deflections are used for the flexibility matrix. A number of methods may be used to obtain the matrix.[29] At a minimum, bending and shear deflections should be taken into account for the cantilevers and bending and rib shortening deflections for the arches.

The term $\ddot{x}_g(t)$ is the ground acceleration and is taken from an accelerogram which is the acceleration record for a particular earthquake. The vectors $\{u\}$, $\{\dot{u}\}$, and $\{\ddot{u}\}$ are the relative displacements, velocity, and accelerations respectively for each node or mass point.

In modal analysis the displacement vector is expressed in terms of the eigenvectors or mode shapes as

$$\{u(t)\} = q_1(t)\{\varphi\}_1 + q_2(t)\{\varphi\}_2 + \ldots + q_n(t)\{\varphi\}_m \quad (51)$$

where $q_k(t)$ are the generalized coordinates which come from the transformation

$$\{u\} = [A]\{q\} \quad (52)$$

The matrix $[A]$ is the collection of all the eigenvectors

$$[A] = [\{\varphi\}_1 \{\varphi\}_2 \ldots \{\varphi\}_m] \quad (53)$$

The mode shapes, $\{\varphi\}_k$, are dimensionless and q has the dimensions of length.

Remembering Eq. 38 through 44, the uncoupled equation for the kth mode becomes

$$\ddot{q}_k + 2\eta_k \omega_k \dot{q}_k + \omega_k^2 q_k = \Gamma_k \ddot{x}_g \quad (54)$$

where the dimensionless participation factor

$$\Gamma_k = \frac{\{\varphi\}_k^T \{M\}}{\overline{M}_k},$$

ω_k is the angular frequency and $\overline{M}_k$ is the generalized mass. All other terms are as previously defined. The solution to Eq. 54 is as given in Eq. 45.

The inertia force for a particular node, i, and mode, k, is

$$F_{i_k} = m_i \varphi_{i_k} \Gamma_k \omega_k$$

$$\cdot \int_0^t \ddot{x}_g(\tau) e^{-\eta_k \omega_k (t-\tau)} \sin \omega_k (t-\tau) d\tau \quad (55)$$

The total force at mass point i then is

$$F_i = a_1 L_{i_1} + a_2 L_{i_2} + \ldots + a_m L_{i_m} \quad (56)$$

where

$$L_{i_k} = m_i \varphi_{i_k} \Gamma_k \quad \begin{array}{l} i = 1, n \\ k = 1, m \end{array}$$

$$a_k = \omega_k \int_0^t \ddot{x}_g(\tau) e^{-\eta_k \omega_k (t-\tau)} \sin \omega_k (t-\tau) d\tau$$

n = number of nodes
m = number of modes

Using summation notation

$$F_i = \sum_{k=1}^m a_k L_{i_k} \quad i = 1, n \quad (57)$$

It is impractical to make a complete stress analysis for the loading pattern occurring at each interval of time. However, the forces, F, can be determined for a unit acceleration by substituting a value of 1 for $\ddot{x}_g$ in the coefficient terms, a_k.

Any reliable method of stress analysis, such as ADSAS, may then be used to determine the stresses, S_i, from the forces, F_i. These unit stress patterns are stored on magnetic tape and can be used to determine total stresses due to any earthquake.

New coefficient values designated a_k^* are computed at selected intervals of time (usually 0.01 second) using actual ground accelerations from a digitized record of an earthquake. To determine the stress history for the earthquake, the a_k^* values at each interval of time are multiplied by the corresponding S_i values and summed.

The equation for total stress at node i is

$$S_i = a_1^* S_{i_1} + a_2^* S_{i_2} + \ldots + a_k^* S_{i_k} \quad (58)$$

or in summation notation

$$S_i = \sum_{k=1}^m a_k^* S_{i_k} \quad i = 1, n \quad (59)$$

6. Foundation Analysis. The foundation or portions of it must be analyzed for stability whenever the rock against which the dam thrusts has a configuration such that direct shear failure is possible or whenever sliding failure is possible along faults, shears, and joints. Associated with stability are problems of local overstressing in the dam due to foundation deficiencies. The presence of weak zones causes problems for one of two reasons: (1) differential displacement of rock blocks on either side of weak zones, and (2) bridging by the dam when the width of a weak zone represents an excessive span for the dam to bridge over. To reduce local overstressing the zones of weakness in the foundation must be strengthened to distribute the applied forces without causing excessive differential displacements and to insure that the dam is not overstressed due to bridging. Analyses can be performed to determine the geometric boundaries and extent of the necessary replacement concrete to be placed in weak zones to limit overstressing.

Stability Analyses. Methods available for stability analysis are generally either two dimensional or three dimensional. The two-dimensional methods are:

1. Rigid section method
2. Finite element method

The three-dimensional methods are:

1. Rigid block method

2. Partition method
3. Finite element method

Each method produces a shearing force and a normal force. The normal force is used to determine the shearing resistance as described in Section II-D. The factor of safety against sliding is then computed by dividing the shearing resistance by the shearing force.

Two-dimensional Methods. A problem may be considered two dimensional if there is little variation in physical and geometrical properties of a cross section through the rock mass for a considerable length and if the end conditions are either negligible or are free faces offering no resistance.

Rigid Section Method. The rigid section method offers a simple method of analysis. The assumption that no deformation of the section occurs allows a static solution. The resultant of all loads acting on the section is resolved into a shearing force, V, parallel to the sliding plane and a force, N, normal to the sliding plane as shown in Fig. 8-3-29. The normal force is used to determine the shearing resistance, and the shear friction factor of safety is determined by dividing the resisting force by the sliding force. This method may also be used when two or more features form the sliding surface. In such cases the cross section can be divided into discrete areas by drawing vertical lines through the section at every point of juncture between the features forming the sliding plane. Each subdivision is analyzed for stability beginning at the highest end. Any excess shearing force not resisted by a subdivision is passed on to the adjacent subdivision. If a net excess of shearing force occurs after considering all the subdivisions, the section is unstable, of course. This procedure is similar to the method of slices in soil mechanics except the sliding surface may have abrupt changes in direction.

The Finite Element Method. The finite element method differs from the rigid section method in essentially two ways. The study section is permitted to deform and loads are applied more accurately. Results from the analysis are stress distributions in the section. The distribution is used to determine the variation in normal load and subsequently the resisting force and shearing force along the sliding plane. The shear friction factor of safety can then be computed along the plane to determine a measure of the stability.

Three-dimensional Methods. A typical three-dimensional stability problem is a four-sided wedge with two exposed faces and two sliding surfaces. The wedge shown in Fig. 8-3-30 is used to illustrate the various methods.

Rigid Block Method.[30] The following assumptions are made for this method:

1. All forces may be combined into one resultant force.
2. No deformation within the rock mass can take place.
3. Sliding on a single plane can occur only if the shear force on the plane is directed toward an exposed (open or free) face.

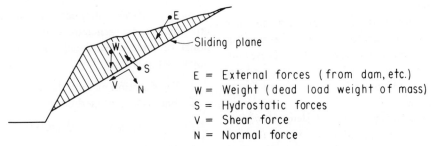

E = External forces (from dam, etc.)
W = Weight (dead load weight of mass)
S = Hydrostatic forces
V = Shear force
N = Normal force

Figure 8-3-29. Vector forces acting on a rigid section.

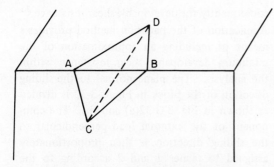

Figure 8-3-30. Typical four sided wedge.

4. Sliding on two planes can occur only in the direction parallel with the intersection of the two planes and toward an exposed face.

5. No transverse shear forces are developed, i.e., shear on the planes normal to the sliding direction.

The rigid block analysis proceeds in the following manner:

1. The planes forming the block are defined and the intersection of the planes form the edges of the block.

2. The areas of the faces of the block and the volume of the block are computed.

3. The hydrostatic forces normal to the faces, if any, are computed.

4. The resultant of all forces is computed.

5. The possibility of sliding on one or two planes is checked.

6. The factor of safety against sliding is computed for all cases where sliding is possible.

To determine whether the rock mass will slide on one or two planes, as stated in Step 5, a test is applied to each possible resisting plane. If the resultant vector of all forces associated with the rock mass has a component normal to a plane and causes compression, then it offers resistance to sliding. If only one plane satisfies the criteria, the potential sliding surface will be one plane and if two planes satisfy the criteria, then the potential sliding surface is comprised of two planes.

Sliding on three planes is impossible according to the assumptions of the rigid block method. If an analysis of a block having more than two resistant surfaces is desired using the rigid block method, the block must be subdivided into smaller blocks and each smaller block analyzed. Any excess shear load not resisted by a particular smaller block is applied to an adjacent block. The technique is similar to that suggested for the two-dimensional rigid section method when the sliding surface consists of more than one geologic feature.

The resultant force for the case of a single sliding plane is resolved into a normal and a shear force. For the case of sliding on two planes the resultant is divided into a shear force along the intersection line of the two planes and a resultant force normal to the intersection line. Forces normal to the two planes are then computed such that they are in equilibrium with the resultant normal force. Figure 8-3-31 shows a section through the sliding mass normal to the intersection line of the two planes with the resultant normal load balanced by normals to the two sliding planes. As a result of assumption (5), these normal loads are the maxi-

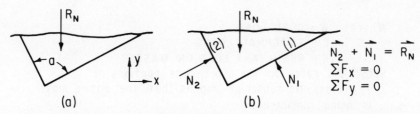

R_N = The portion of the resultant normal to the direction of movement.

N = The normal load on the face indicated by the subscript.

Figure 8-3-31. Section through mass sliding on two planes drawn normal to sliding direction.

mum that can occur and the resulting shear resistance developed is a maximum.

The shearing resistance developed for either a single plane or two planes is computed using the normal forces acting on the planes and the methods discussed in Section II-D.

Partition Method. The rigid block method assumes no deformation of the mass. Because of this restriction no shear load is developed on the sliding planes transverse to the direction of sliding. The development of shear in the transverse direction decreases the normal load and consequently the developable shear resistance.[31] Application of the partition method provides a means of including an approximation of the minimum developable shear resistance within the analysis. The plane normal to the sliding direction of the block in Fig. 8-3-32 is divided as shown in Fig. 8-3-32(a) and (b). The component of the external load perpendicular to the sliding direction is then proportionately assigned to planes 1 and 2 according to the ratio of projected areas of the planes with respect to the direction of loading as shown in Fig. 8-3-32(d). All the forces on each plane are

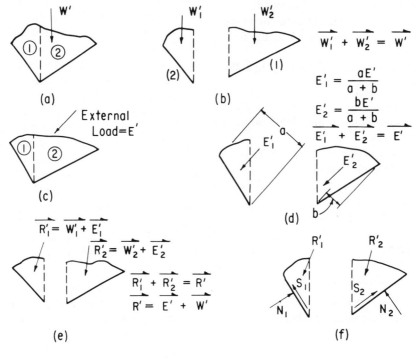

Where: W = DEAD LOAD
E = EXTERNAL LOADS
R = RESULTANT LOAD ON MASS
Subscripts refer to the appropriate portions of the mass. No subscript implies that the entire mass is being considered.
Planes are normal to the direction of sliding.
Loads resolved into the plane normal to direction of sliding are indicated with a prime.

Figure 8-3-32. Schematic showing vector forces acting on a mass using the partition method.

then combined to form a resultant on that plane (Fig. 8-3-32(e)). Each resultant is assumed to be balanced by a normal force and a shear force acting on the sliding plane (Fig. 8-3-32(f)). The normal forces are then used to determine the combined resistance of the block to sliding.

The shearing force tending to drive the block in the direction of sliding is determined as described for the rigid block method. The computation of the resistance according to the partition method utilizes the information obtained from the rigid block analysis and, therefore, requires very little additional computation. The shear resistance determined by the rigid block method is an upper bound and that determined by the partition method is considered a lower bound. As the angle between the planes in Fig. 8-3-32(a) increases, the results from the two methods converge. The correct answer lies between the upper and lower bounds and is a function of the deformation properties of the sliding mass and host mass of rock. More importantly, the correct answer is a function of the sliding and deformation characteristics of the geologic features which form the sliding surface. The effect of these properties on the resistance developed can be approximated and included in the analysis by using a three-dimensional finite element program with planar failure zone elements.

The partition method can readily be extended to multifaced blocks. Just as the section normal to the direction of sliding was partitioned so can a section parallel with the direction of sliding as shown in Fig. 8-3-33. The two resulting blocks A and B are further subdivided and treated as discussed previously for the four-sided wedge. Block A is considered first because it tends to move toward Block B. Any excess shear load from Block A must be applied to B as an external loading.

Finite Element Method. A computer program developed by Mahtab[32] allows representation of rock masses by three-dimensional solid elements and representation of the sliding surface by two-dimensional planar elements. The planar elements are given properties in two directions of deformation, compression (normal stiffness) and shear (shear stiffness).

The ratio of the normal stiffness to the shear stiffness significantly influences the amount of load which will be taken in the normal and transverse shear directions. If the normal stiffness is much larger than the shear stiffness as is the case for a joint with a slick coating, the solution approaches that given by rigid block method. However, as the shear stiffness increases with respect to the normal stiffness, more load is taken by transverse shear and the solution approaches the results computed by the partition method.

The three-dimensional finite element method allows another important refinement in the solution of stability problems. Since deformations are allowed, the state of stress on all planes of a multifaced block can be computed and thereby stress concentrations can be evaluated and located.

The three-dimensional finite element method should be used when the upper and lower bounds determined by the other two methods are significantly different. The method should also be used if there is considerable variation in material properties either in the sliding planes or in the rock mass.

Differential Displacement Analysis. The problem of relative deflection or differential displacement of masses or blocks within the foundation arises due to variation in the foundation materials. Typical problems that may occur are displacement of the mass necessary to mobilize frictional shearing resistance, displacement of a mass due to deformation of a low modulus zone against which the mass bears, and displacements larger than that of adjacent rock of a zone isolated by geologic features incapable of transmitting shear load. The displacements may be estimated from (1) shear strength versus displacement curves developed from specimen testing in situ or in the laboratory; (2) by model testing; (3) by hand computed analytical models; and (4) by two or three-dimensional finite element methods.

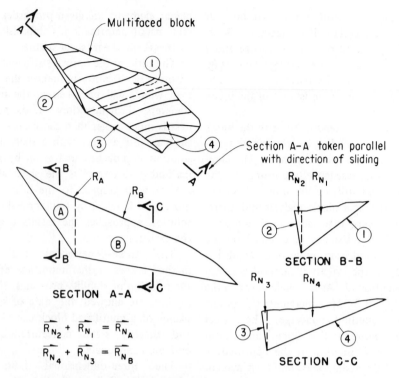

NOTE: Circled numbers refer to faces.
Circled letters refer to blocks.
R_{N_1} = The portion of the resultant assigned to a face. The subsubscript indicates the face number.
R_{N_A} = The portion of the resultant normal to the direction of movement of a block. The subsubscript refers to the block.
R_A = Resultant external load acting on block A.

Figure 8-3-33. Partition method applied to multifaced blocks.

Although the method used depends on the particular problem, it should be noted that the finite element method offers considerable advantages over the other procedures. The finite element method allows accurate material property representation, gives stress distributions, and provides a means of including the effects of remedial treatment used to beneficiate the foundation.

Analysis of Stress Concentrations Due to Bridging. A stress concentration may occur in the dam when a zone of low modulus material is

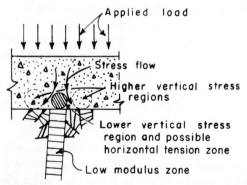

Figure 8-3-34. Stress concentration due to a low modulus zone.

encountered in the foundation as shown in Fig. 8-3-34.

To reduce such stress concentration, a portion of the weak material in the low modulus zone may be replaced with concrete. The depth of replacement required is that which causes the stresses in the dam and foundation to fall within allowable limits. The plane stress, plane strain finite element method as discussed in Section III-E-3 is a good method for solving problems of this type.

IV. DESIGN OF GRAVITY DAMS

The term, gravity dam refers to solid concrete or masonry dams of roughly triangular cross section which depend primarily on their own weight and cohesion with the foundation for stability. The dams are usually straight, but may be slightly curved in plan. In spite of their impressive bulk gravity dams usually require the same factor of safety as thin arch dams. This fact should be kept in mind at all stages of design and construction.

A. Layout

The initial step in designing a new dam is to lay out the structure based on previous experience with similar dams. After the layout is completed, a stress and stability analysis of the structure is made to determine the stress distributions and magnitudes and the stability. If the results of the analyses indicate the stresses are not within allowable limits, stress distributions are not satisfactory because of stress concentrations, or the dam is not stable, modifications must be made by reshaping the structure to improve the design. The design of a gravity dam is accomplished by making successive layouts, each one being progressively improved based on the results of the stress and stability analysis of the preceding layout.

The upstream face of gravity dams is usually made vertical to concentrate the concrete weight at the upstream face where it acts to resist reservoir water loading. Except where additional thickness is required at the crest, the downstream face usually has a constant slope from the top of the dam to the base. The base width as determined by both stress and stability requirements determines the slope. The slope will provide an adequate section to meet the stress and stability requirements at the higher elevations unless a large opening is formed in the dam.

The thickness of the dam at the crest is usually determined by roadway or other access requirements for the nonoverflow portion. However, it should be adequate to withstand all possible loadings including ice pressures and impact of floating objects. When additional crest thickness is used, the downstream face is usually vertical from the downstream edge of the crest to an intersection with the sloping downstream face.

A batter may be used on the lower part of the upstream face to increase the base thickness and thereby improve the sliding safety at the base. However, unacceptable stresses may develop at the heel of the dam because of the change in moment arm for the concrete weight about the center of gravity of the base. If a batter is used, the stresses and stability should be checked where the batter intersects the vertical upstream face. The dam should also be analyzed at any other changes in slope on either face.

If an overflow spillway is incorporated in the dam, the layout of the spillway section should be similar to the nonoverflow section. The curves describing the spillway crest and the junction with the energy dissipator are designed to meet hydraulic requirements discussed elsewhere. The slope joining these curves should be tangent to each and, if practicable, parallel to the downstream slope on the nonoverflow section. An upstream batter may be used on the spillway section under the same conditions as for the nonoverflow section.

Figure 8-4-1 is a typical layout drawing showing a plan, sections, and profile for a gravity dam.

The maximum water surface elevation should not exceed the top of the nonoverflow section of the dam. A solid concrete parapet wall can

438 HANDBOOK OF DAM ENGINEERING

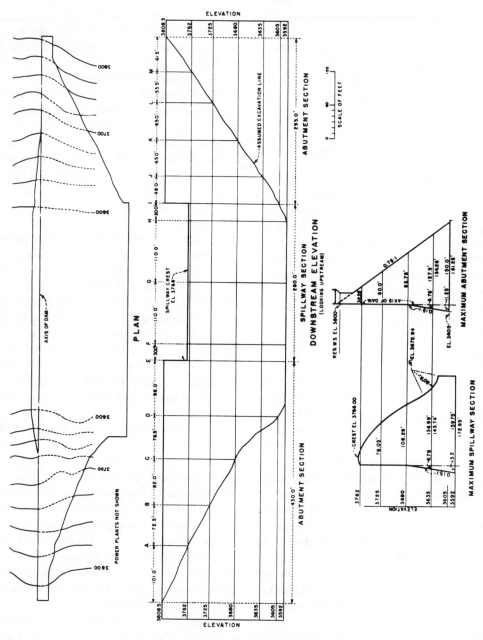

Figure 8-4-1. Typical layout for a gravity dam.

be constructed on top of the dam to provide freeboard. Height of the parapet will depend upon the anticipated height of the waves.

B. Methods of Analysis

Gravity dams can be analyzed by the Gravity Method, Trial-load Twist Analysis, or the Beam and Cantilever Method, depending upon the configuration of the dam, continuity between the blocks, and the degree of refinement required. The Gravity method is used when blocks are not made monolithic by keying and grouting the joints between them. Thus, each block acts independently and the load is transmitted to the foundation by cantilever action and is resisted by the weight of the cantilever. Trial-load twist analysis and the beam and cantilever analysis are used when the blocks are keyed and grouted together to form a monolith because part of the load is transmitted to the abutments by beam action. The gravity analysis may be used, however, as a preliminary analysis for keyed and grouted dams. The Gravity Method is presented in this section and discussion of the Trial-Load Twist and Beam and Cantilever methods are given in other publications.[33]

C. Gravity Method of Analysis

The Gravity Method is applicable to the general case of a gravity section with vertical upstream face and a constant downstream slope and to the cases where there is variable batter on either or both faces.

The formulas shown for calculating stresses are based on the assumption of a straight-line distribution of vertical stress and a parabolic distribution of horizontal shear stress on horizontal planes. These formulas provide a direct method of calculating stresses at any point within the boundaries of a transverse section of a gravity dam. The assumptions are substantially correct, except for horizontal planes near the base of the dam where foundation yielding affects the stress distributions. The stress changes which occur due to foundation yielding are usually small in dams of low or medium height but they may be important in high dams. Stresses near the base of a high masonry dam should therefore be checked by the Finite Element Method or other comparable methods of analyses.

Uplift pressures on a horizontal section are usually not included with the contact pressures in the computation of stresses, but are considered in the computation of stability factors.

The analysis of overflow sections presents no added difficulties. Usually, the dynamic effect of overflowing water is negligible and can be omitted. Any additional head above the top of the section can be included as a surcharge load on the dam.

1. **Assumptions.** Design criteria were given in Section II. The following are assumptions peculiar to the gravity analysis:
   (a) The concrete in the dam is a homogenous, isotropic, and uniform elastic material.
   (b) There are no differential movements which occur at the dam site due to water loads on the reservoir walls and floors.
   (c) All loads are carried by the gravity action of vertical, parallel-side cantilevers which receive no support from the adjacent elements on either side.
   (d) Unit vertical pressures, or normal stresses on horizontal planes, vary uniformly as a straight line from the upstream face to the downstream face.
   (e) Horizontal shear stresses have a parabolic variation across horizontal planes from the upstream face to the downstream face of the dam.

2. **Notations.** Notations used in the analyses are listed below.
   (a) Properties and Dimensions
   O = origin of coordinates, at downstream edge of section.
   ϕ = angle between face of element and the vertical.

T = horizontal distance from upstream edge to downstream edge of section.

c = horizontal distance from center of gravity of section to either upstream or downstream edge, equal to $T/2$.

A = area of section.

I = moment of inertia of section about center of gravity, equal to $T^3/12$.

w_c = unit weight of concrete or masonry.

w = unit weight of water.

h or h' = vertical distance from reservoir or tailwater surface, respectively, to section.

p or p' = reservoir water or tailwater pressure, respectively, at section. It is equal to wh or wh'.

(b) Forces and Moments

W_c = dead-load weight above base of section under consideration.

M_c = moment of W_c about center of gravity of section.

W_w or W'_w = vertical component of reservoir or tailwater load, respectively, on face above section.

M_w or M'_w = moment of W_w or W'_w about center of gravity of section.

V or V' = horizontal component of reservoir or tailwater load, respectively, on face above section. This is equal to $wh^2/2$ for normal conditions.

M_p or M'_p = moment of V or V' about center of gravity of section, equal to $wh^3/6$ or $wh'^3/6$.

ΣW = resultant vertical force above section, equal to $W_c + W_w + W'_w$.

ΣV = resultant horizontal force above section, equal to $V + V'$.

ΣM = resultant moment of forces above section about center of gravity. It is equal to $M_c + M_w + M'_w + M_p + M'_p$.

U = total uplift force on horizontal section.

(c) Stresses

σ_z = normal stress on horizontal plane.

σ_y = normal stress on vertical plane.

$\tau_{zy} = \tau_{yz}$ = shear stress on vertical or horizontal plane.

$a, a_1, a_2, b, b_1, b_2, c_1, c_2, d_2$ = constants.

σ_{p1} = first principal stress.

σ_{p2} = second principal stress.

ϕ_{p1} = angle between σ_{p1} and the vertical. It is positive in a clockwise direction.

(d) Subscripts

u = upstream face.

d = downstream face.

w = vertical water component.

p = horizontal water component.

(e) Notations for Horizontal Earthquake

p_E = pressure normal to face.

$\alpha = \dfrac{\text{horizontal earthquake acceleration}}{\text{acceleration of gravity}}$

z = depth of reservoir at section being studied.

h = vertical distance from the reservoir surface to the elevation in question.

C_m = a dimensionless pressure coefficient.

W_{wE} or W'_{wE} = change in vertical component of reservoir water load or tailwater load on face above section due to horizontal earthquake loads.

M_{wE} or M'_{wE} = moment of W_{wE} or W'_{wE} about center of gravity of section.

V_E = summation of horizontal inertia forces of incremental concrete weights above section.

M_E = moment of V_E about center of gravity of section.

V_{pE} or V'_{pE} = summation of changes in horizontal components of incremental reservoir or tailwater loads on face above section due to horizontal earthquake loads.

M_{pE} or M'_{pE} = moment of V_{pE} or V'_{pE} about center of gravity of section.

ΣW = resultant vertical force above section, equal to $W_c + W_w + W'_w \pm W'_{wE} \pm W'_{wE}$.

ΣV = resultant horizontal force above section, equal to $V + V' \pm V_E \pm V_{pE} \pm V'_{pE}$.

ΣM = resultant moment of forces above horizontal section about center of gravity. It is equal to $M_c + M_w + M'_w + M_p + M'_p \pm M_E \pm M_{wE} \pm M_{wE} \pm M_{pE} \pm M'_{pE}$.

The algebraic signs of the terms with subscript E in the earthquake equations for ΣW, ΣV, and ΣM depend upon the direction assumed for the horizontal earthquake acceleration of the foundation. No notation is given for vertical earthquake shock.

3. **Forces and Moments Acting on Cantilever Elements.** Forces acting on the cantilever element, including uplift, are shown for normal loading conditions in Fig. 8-4-2. Reservoir and tailwater pressure diagrams are shown for the portion of the element above the horizontal cross section OY. Positive forces, moments, and shears are indicated by the directional arrows.

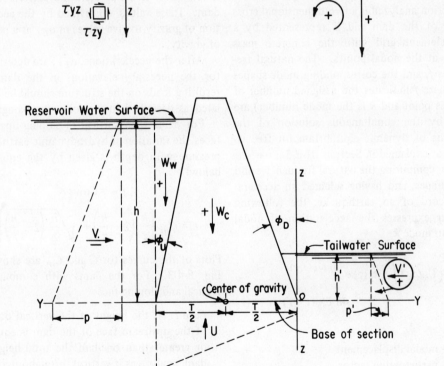

(a) – VERTICAL CROSS-SECTION

Figure 8-4-2. Section through a dam showing loads acting on dam.

Hydrodynamic and concrete inertia forces acting on the cantilever element for a horizontal earthquake shock with the foundation acceleration acting in an upstream direction are in addition to those shown in Fig. 8-4-2. For a foundation acceleration acting in a downstream direction, the direction of these forces must be reversed.

Forces and moments for static loads are computed for each section by determining areas and moment arms of the triangular pressure diagrams and the area and eccentricity of vertical sections. However, to evaluate the quantities V_{pE}, V'_{pE}, W_{wE}, and W'_{wE} for hydrodynamic effects of earthquake shock, it is necessary to first determine the dynamic response of the structure. A discussion of the fundamentals of dynamic analysis is given in Section III-E-5.

The method used to analyze the dynamic response of a gravity dam is a two dimensional solution similar to that proposed by Chopra.[34] The section analyzed is a two-dimensional cross section of the dam and is represented by a finite element grid with the concrete mass lumped at the nodal points. The natural frequencies f_j and the corresponding mode shapes φjk (where j indicates the assigned number of the mass point and k is the mode number) are found by the simultaneous solution of the equations of dynamic equilibrium for free vibration as explained in Section III-E-5.

After computing the natural frequencies and mode shapes, and having selected an acceleration record of an earthquake, the following equation expresses the acceleration of nodal point j in mode k,

$$\ddot{x}_{j_k} = \varphi_{j_k} \Gamma_k \omega_k \int_0^t \ddot{x}_g(\tau) e^{-\eta_k \omega_k (t-\tau)} \sin \omega_k (t - \tau) \, d\tau \quad (1)$$

where

φ_{j_k} = modal displacement
Γ_k = participation factor
ω_k = angular frequency
$\ddot{x}_g$ = acceleration due to gravity
τ = time

t = a particular time $\tau = t$
η = damping force

Little data are available on the damping in concrete gravity dams. Chopra[34] indicates that a reasonable assumption for η in a concrete gravity structure is 0.05.

The total acceleration at a particular node j due to all modes is

$$\ddot{x}_{j_\text{total}} = \sum_{k=1}^{m} \ddot{x}_{j_k} \quad (2)$$

Equation (1) is evaluated at chosen increments of time, usually 0.01 seconds. At the end of each time increment, the acceleration at all node points for all modes considered are summed using equation (2). The response history is scanned for the time of maximum acceleration at the top of the dam. The vertical distribution of the average acceleration values at that time is the acceleration pattern for the dam. These values are divided by the acceleration of gravity to give values of α_{E1} as a percent of gravity.

After the accelerations, α_{E1}, are determined for the necessary elevations in the dam, the resulting loads on the structure should be calculated as described in the following paragraphs.

For dams with vertical or sloping upstream faces, the variation of hydrodynamic earthquake pressure with depth is given by the equations below:

$$p_E = C \alpha w z \quad (3)$$

$$C = C_m \left[\frac{h}{z} \left(2 - \frac{h}{z}\right) + \sqrt{\frac{h}{z} \left(2 - \frac{h}{z}\right)} \right] \quad (4)$$

Plots of the curves for C and C_m are shown in Fig. 8-4-3. For the dams with combination vertical and sloping face:

Case 1: If the height of the vertical portion of the upstream face of the dam is equal to or greater than one-half the total height of dam, analyze as if vertical throughout.

Case 2: If the height of the vertical portion of the upstream face of the dam is less than one-half the total height of the dam, use the

DESIGN OF CONCRETE DAMS 443

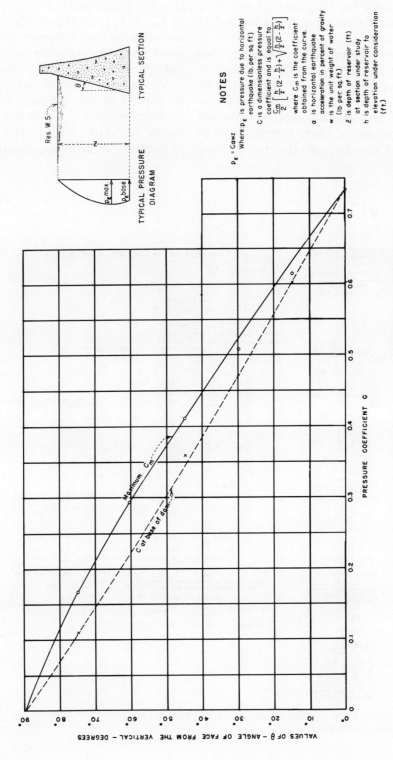

Figure 8-4-3. Pressure coefficients for constant sloping faces for horizontal earthquakes.

pressure which would occur assuming that the upstream face has a constant slope from the water surface elevation to the heel of the dam.

The inertia forces for concrete in the dam should be computed for each increment of height using the acceleration factor through the centroid for that increment. The inertia force to be used for a particular elevation in the dam is the summation of all the incremental forces above that elevation and the total of their moments about the center of gravity at the elevation being considered.

V_{pE} or V'_{pE} and M_{pE} or M'_{pE} should also be computed for each increment of elevation selected for the study and the totals obtained by summation because of the nonlinear response.

Vertical earthquake loads are produced when the mass of the dam and water are accelerated in direct proportion to the values of α for vertical earthquake motion. This is equivalent to increasing or decreasing the density of con-

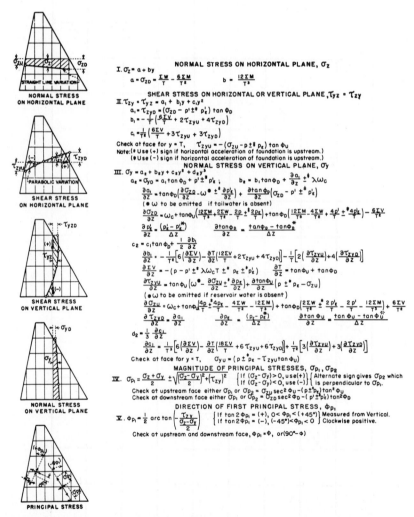

Figure 8-4-4. Stresses in straight gravity dams, including effects of tailwater and horizontal earthquake.

crete and water, depending on the direction of the shock.

4. **Stress and Stability Equations.** A summary of the equations for stresses computed by the Gravity Method is given in Fig. 8-4-4. These equations include those for normal stresses on horizontal planes, shear stresses on horizontal and vertical planes, normal stresses on vertical planes, and direction and magnitude of principal stresses for any point within the boundaries of the cantilever element.

The shear-friction factor is expressed as follows:

Shear-friction factor for horizontal planes

$$= \frac{\begin{pmatrix}\text{horizontal}\\\text{area}\end{pmatrix}\begin{pmatrix}\text{unit}\\\text{shear}\\\text{resistance}\end{pmatrix} + \begin{pmatrix}\text{weight}-\\\text{uplift}\end{pmatrix}\begin{pmatrix}\text{coefficient}\\\text{of internal}\\\text{friction}\end{pmatrix}}{\text{horizontal force}}$$

Shear-friction factors are computed at each elevation for which stresses are calculated in the cantilever element. All possible conditions of loading should be investigated. It should be noted that a large margin of safety against sliding is indicated by high shear-friction factors. The allowable minimum values of this factor for use in design are given in Section II.

A factor of safety for overturning is not usually tabulated with other stability factors. Before bodily overturning of a gravity dam can take place, other failures may occur such as crushing of the toe material and cracking of the upstream material with accompanying increases in uplift pressure and reduction of the shear resistance. However, it is desirable to provide an adequate factor of safety against the overturning tendency. This may be accomplished by specifying the maximum allowable stress at the downstream face of the dam. Because of their oscillatory nature, earthquake forces are not considered as contributing to the overturning tendency. A factor of safety for overturning may be calculated if desired by dividing the total resisting moments by the total moments tending to cause overturning about the downstream toe.

Overturning safety factor

$$= \frac{\text{moments resisting}}{\text{moments overturning}} \quad (6)$$

V. DESIGN OF BUTTRESS DAMS

Traditionally buttress dams have been classified according to three categories; flat-slab type, massive-head type, and multiple-arch type. There is, however, another form of buttress dam which is a hybrid developed by combining arch dam design with gravity dam design. The resulting structure has been named a double curvature multiple arch dam. An example of this type is shown in Fig. 8-5-1. The gravity buttresses should be analyzed as discussed in Section IV and the double curvature arch sections according to the methods discussed in Section III-E. The layout of such a scheme is the prerogative of the designer, of course. The primary objective, presuming the foundation is adequate throughout, is to determine the optimum number and arrangement of arch and buttress sections. Inventive and resourceful schemes can produce designs that are economically competitive with other types of concrete dams.

There are special design considerations associated with the three traditional types of buttress dams which are not discussed elsewhere. This section, therefore, is devoted to the discussion of the design and analysis for three types.

A buttress dam consists essentially of two principal structural elements: (1) a sloping upstream deck that supports the water, and (2) the buttresses or vertical wall that support the deck and transmit the load to the foundations.[35,36,37] Secondary structural elements include footings, stiffeners, lateral braces or struts between buttresses, corbels or other transition members for transferring the water load to the buttresses, cutoff walls, and grout curtains. Figure 8-5-2 shows the different types

446　HANDBOOK OF DAM ENGINEERING

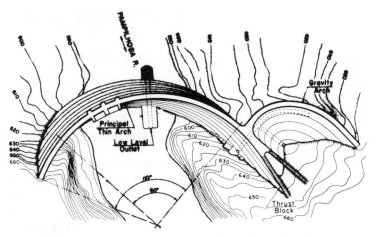

Figure 8-5-1. Santa Luzia Dam, Portugal. ("Dams of the World," Portland Cement Assn.)

of buttress dams and the principal structural elements.

The deck of the buttress dam is usually built on a slope of approximately 45°, thus utilizing the vertical component of the water load as a stabilizing force against sliding or overturning of the dam. As a result of this feature, a buttress dam requires much less material and is therefore much lighter in weight than other types of concrete dams of equal height. Accordingly, buttress dams may sometimes be built on foundations that do not have sufficient strength to support other more massive types of dams.

A. Layout

The preparation of a layout for a buttress dam involves two separate considerations:

- Selection of the most suitable type; namely, flat-slab, multiple-arch, or massive-head, depending largely upon such local conditions as topography, geology, and accessibility
- A study of the possible economics obtainable within the type by varying buttress spacing, upstream slope, varying the strength of concrete, etc.

The effects of variations within the type may best be studied by determining actual quantities and comparative cost for a number of designs involving changes in the variable factors, then selecting the most economical set.

1. Selection of Type. The choice of the type of buttress dam is usually determined on the basis of tentative designs and cost estimates and consideration of the site conditions. In making the choice, consideration is given to the following general advantages and disadvantages of each type.

Flat-Slab Type. Since the units of this type are structurally independent, ordinary foundation deformations or settlements have little or no effect on the distribution of foundation stresses. It is therefore often preferred, if otherwise satisfactory, for construction on jointed or faulted foundations where unequal subsidences or deformations may be expected. Although extensively used for low dams because of its simplicity and economy of construction, the flat-slab design is by no means restricted to low dams. The principal disadvantages of the flat-slab type are the complete dependence upon the tension steel of the slab

DESIGN OF CONCRETE DAMS 447

for support of the imposed loads and the difficulty of transmitting the slab loads to the buttresses without causing objectionable stress concentrations in the corbels.

Multiple-Arch Type. The multiple-arch type is, in general, most suitable for the higher buttress dams with buttress spacings of 50 to 60 ft. The dam is a continuous structure in which

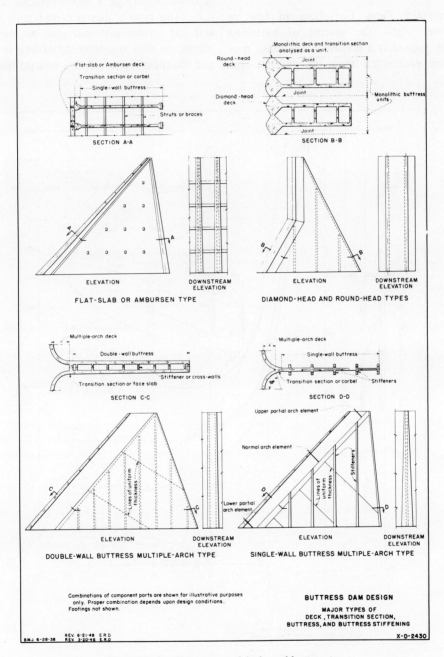

Figure 8-5-2. Major types of decks and buttresses.

the stability of each unit may be dependent upon the stability of adjacent units. With 180° arches the lateral thrust is small, and the stability of one unit may not be dependent upon the stability of adjacent units.

Massive-Head Type. For dams of low or moderate height, the spacing of buttresses may be such that a massive-head design may be more economical than the multiple-arch and may compare favorably with the flat-slab type. Since the head is designed so that dead weight and water induce little or no tension stress in the structure, very little or no reinforcement is required with the massive-head type which may be a factor favoring its selection. This type of dam is considerably heavier and has greater sectional area on horizontal planes, so that resistance to sliding is higher and the shear-friction factor is larger than with

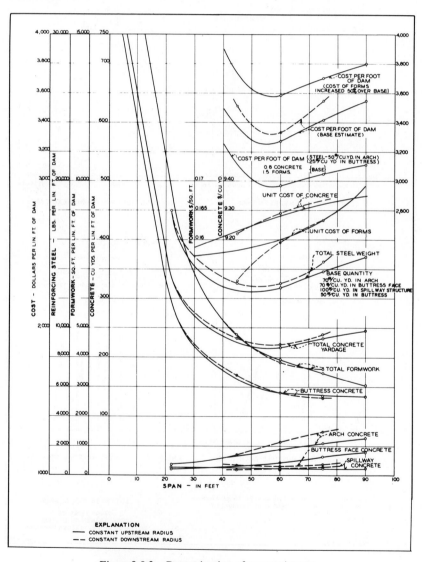

Figure 8-5-3. Determination of economic span.

other types. Since the units are structurally independent, small unequal foundation settlements may be permissible.

2. Buttress Spacing. The economic spacing of buttresses is a complex problem in that it involves, in addition to economy of material, the relative cost of formwork, excavation, and labor. The variation in local conditions determines the spacing for each individual case.

The most economical buttress spacing from the standpoint of the buttresses alone is one in which the minimum thickness of concrete is fully utilized to carry the load, since any concrete is uneconomical when stressed below its normal working stress. The cost of formwork in the buttresses decreases with wider spacing. The minimum concrete and reinforcing-steel quantities in the deck will be obtained by close buttress spacing.

The solution of the problem can best be approached by trial. Curves representing quantities of concrete, formwork, and reinforcement for a typical study of buttress spacing in a multiple-arch dam 250 ft high are shown in Fig. 8-5-3. Knowing their relative cost in any particular geographical location, the cost per foot of dam is readily computed. This study indicates an economical buttress spacing of from 50 to 60 ft for this particular structure.

Foundation conditions, especially the extent and character of the required excavations and the presence of faults or fractures in the rock, are sometimes an important factor in the determination of buttress spacing and must always be taken into consideration. In some cases, the foundation conditions determine the spacing.

3. Upstream Slope of Dam. The upstream slope of a buttressed dam is controlled by the relation between the requirements of stability and intensities of foundation pressure. By varying the slope, the stability factors and foundation pressure may be altered within reasonable limits. Slopes of approximately 45° have high stability and a low sliding factor and are, therefore, often used.

When buttresses of uniform thickness are used with a slope of 45°, tension develops parallel to the upstream face. However, buttresses tapered in plan and elevation will eliminate this tension. In the tapered section, the center of gravity is moved upstream, giving a greater upstream dead load moment which results in higher compression at the upstream edge of the buttress.

The best slope to use is found by trial. For economy the slope should be as steep as will provide satisfactory sliding factors and stresses.

B. Design of Deck and Transition Section

The three types of deck considered here are the flat-slab, multiple-arch, and massive-head (diamond-head or round-head). These are illustrated in Fig. 8-5-2.

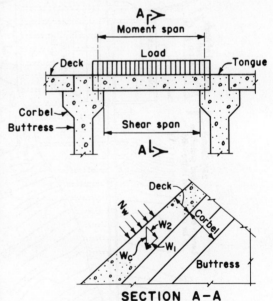

N_w = Water load
W_C = Weight of concrete
W_1 = Weight of concrete to slab
W_2 = Weight of concrete to buttress

Figure 8-5-4. Deck loading assumptions.

450 HANDBOOK OF DAM ENGINEERING

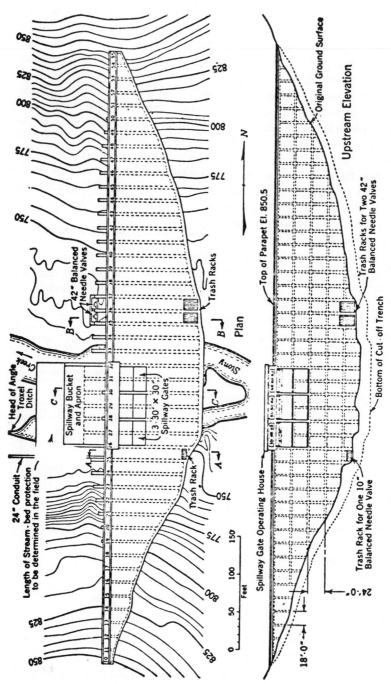

Figure 8-5-5. Example of slab and buttress dam, Stoney Gorge Dam. *(Courtesy Bureau of Reclamation)*

DESIGN OF CONCRETE DAMS 451

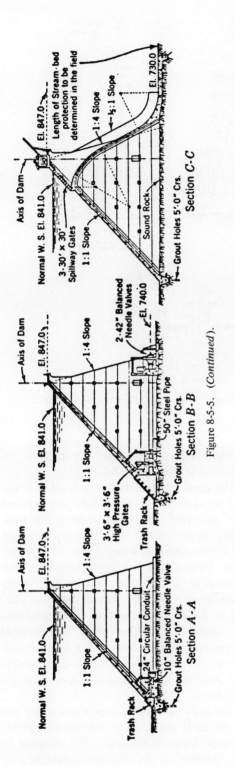

Figure 8-5-5. (Continued).

1. Flat-Slab Deck. The flat-slab deck is designed as a simply supported slab or beam using reinforced concrete design procedures. The deck loading assumptions are shown in Fig. 8-5-4. The slab should be designed for the water load, N_w, plus the portion of the concrete weight, W_c, that is carried by the buttress. Special precautions should be taken to limit the deflection of the slab to prevent cracking of the concrete. This could expose the reinforcement and lead to its disintegration which would, in turn, cause a failure of the dam.

Stony Gorge Dam, Fig. 8-5-5, is an example of the slab and buttress dam. Details of the slab are shown in Fig. 8-5-6.

To minimize the effect of temperature change, provision should be made to allow for the expansion and contraction of the deck both parallel with the axis of the dam and parallel with the upstream face. This can be done by coating the contact between the buttress and the slab with a bituminous mastic or other suitable material. The mastic provides a small clearance for expansion and allows the slab to contract without breaking the water seal.

The effects of deck slab movement along the corbel must also be included in the design of the buttresses. It may be desirable to obtain the benefits of having the full weight of the concrete on the buttress. This can be accomplished by placing the slab in sections and providing keys in the corbel. The keys transfer the portion of the weight that would normally be carried by the slab to the buttress. A compressible material and a waterstop is placed between the sections to eliminate the effect of expansion on the slab on the buttresses while providing a water seal.

The details for attaching the deck to the buttress for Stony Gorge Dam as shown on Fig. 8-5-6 are typical for dams with slab decks. A tongue or extension of the buttress separates the slabs. Keys may be provided in the tongue to secure the slab to the buttress but permit expansion and contraction of the deck. Corbels are used to support the edge of the slab. The bearing surface area of the corbel must be

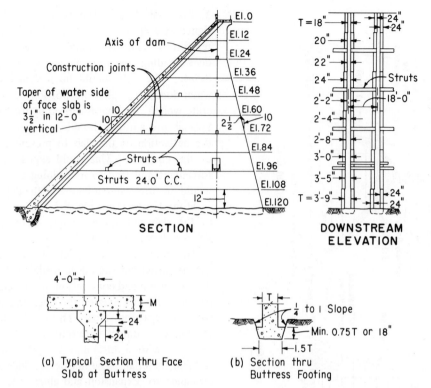

Figure 8-5-6. Details of slab and buttresses for Stoney Gorge Dam. (*Courtesy Bureau of Reclamation*)

sufficiently large so that the maximum bearing pressure will not exceed the allowable. The loading diagram on the corbel is usually assumed to be triangular with the maximum ordinate at the outer edge.

A temperature drop in the slab will tend to pull the corbel from the buttress. This may require additional reinforcement in the corbel. The corbel is designed using reinforced concrete design procedures.

2. Multiple-Arch Deck. Three principal factors enter into the design of a multiple-arch deck. These are: (1) the slope of the upstream face, which gives the variation of load between the crown and the arch abutment; (2) the temperature range expected in the concrete due to the seasonal variation in water temperature; and (3) the economical central angle. Bartlett Dam, Fig. 8-5-7, is an example of a multiple-arch buttress dam. Details of the arches are shown on Fig. 8-5-8.

The economical central angle must be evaluated for each dam. It will usually vary between 150° and 180°, the variation being due principally to the effects of temperature changes and of the upstream slope. Values close to the lower limit of central angle will be found economical in localities where the yearly temperature variation is small and the 180° will be found to be the most economical central angle in localities where the seasonal temperature range is larger. The longer length of the 180° arch gives a more flexible member with resulting lower stress of a given quantity of concrete.

The most economical central angle may not be the most desirable to use due to possible eccentric buttress loading. The variation in the direction of the resultant should be considered, since the failure of one arch should not tend to produce failure of the entire structure. The thickness of the arch may be controlled by dead load and temperature and not by the water load since the eccentricity of loading for dead load is usually greater than that for water load.

The use of arches other than circular can be used in multiple-arch dams and may prove advantageous under some conditions. It may be found that these special shapes can be made applicable to only one set of temperature and load conditions. The circular arch can be economically designed to meet many variations in load and temperature. An advantage of the circular-arch type is that layout work in the field and office is less complex than for other shapes.

The deck is assumed to be made up of a series of unit arches in planes normal to the axis of the arch barrel. The unit arches are analyzed using the methods outlined in Section III-E to determine the moments, shears, and thrust. After these have been determined, the stresses and reinforcement requirements are determined using reinforced concrete design procedures for beams subject to combined bending and axial loads.

The effect of abutment or transition rotation should be included, when appropriate, in determining the moments, shears, and thrust. Generally, on a single-wall buttress, the moments from the arches are approximately equal in magnitude but opposite in direction. Therefore, rotations are usually small and may be neglected. However, on a double-wall buttress unless the transition is designed to prevent rotations, the rotation is usually of such magnitude that it must be considered.

The analyses indicate the minimum amount

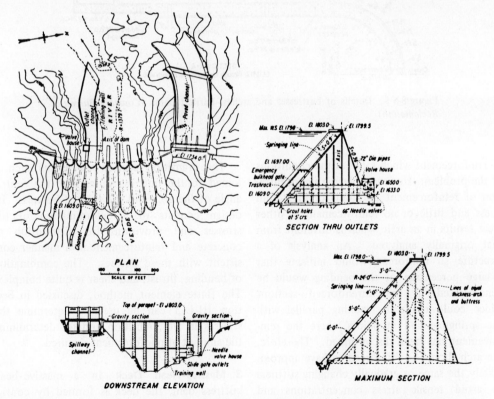

Figure 8-5-7. Example of multiple-arch buttress dam—Bartlett Dam. (*Courtesy Bureau of Reclamation*)

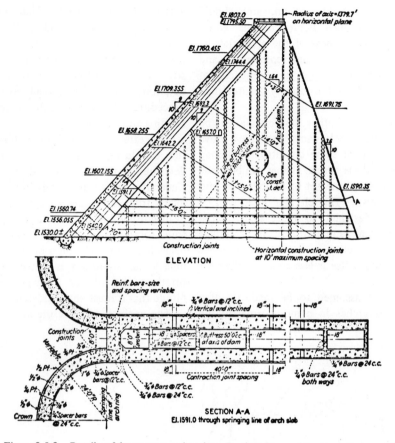

Figure 8-5-8. Details of buttresses and arches—Bartlett Dam. (*Courtesy Bureau of Reclamation*)

of reinforcement which satisfies the conditions of the problem. However, placing a concentration of reinforcement at points of maximum stress and little or no reinforcement in other areas results in an arch which is different from that originally analyzed. An analysis of a structure so reinforced would indicate that a large percentage of the bending would be concentrated in the unreinforced sections which could result in cracking parallel with the springline and at points where the reinforcement has been discontinued. Therefore, the arch should be reinforced to have approximately the same or gradually changing stiffness to avoid tensile stress concentrations and possible cracking.

The transition section where the arch is joined to the buttress varies with the central angle of the arch and the type of buttress. Its design consists of a determination of the stresses and provisions for distribution of concrete and reinforcement in a manner consistent with these stresses. The combination of bending, thrust, and shear is quite complex. The finite element method, discussed in Section III-E-3 can be used to determine the stresses in the transition and for determining the amount of reinforcement required.

3. Massive-Head Deck. In a massive-head buttress dam, the deck is formed by contiguous massive heads which replace the slab and corbels used in the other types. The massive-head deck is designed in two forms, the round-head, and the diamond-head (see Fig. 8-5-2). In each form the deck is so designed that the resultant water pressure on the overhanging sides is directed toward the center of the buttress. Pueblo Dam, Fig. 8-5-9 and -10, is an example of a massive-head buttress dam. Details of the diamond head are shown on Fig. 8-5-11.

Massive-heads are shaped to minimize tensile stresses in the head which in turn minimizes or eliminates the need for reinforcement. For a properly shaped head, tension should occur only as a minor principal stress near the center of the head. Stress distribution and magnitudes can be determined using the finite element method as discussed in Section III-E-3. Tensile stresses at the upstream face due to bending should be avoided since this is

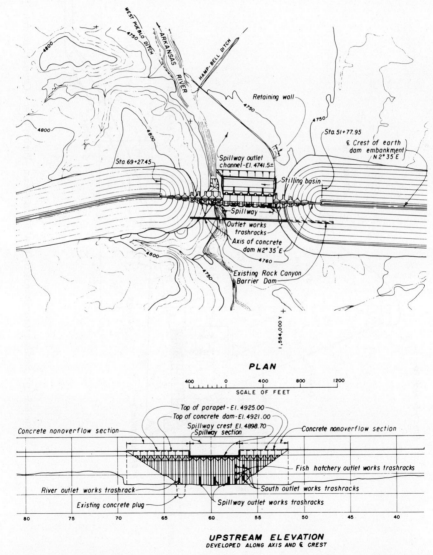

Figure 8-5-9. Massive-head buttress dam—Pueblo Dam. (*Courtesy Bureau of Reclamation*)

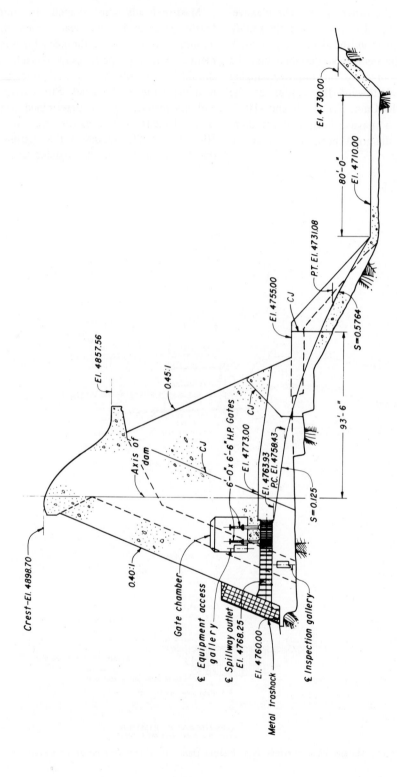

Figure 8-5-10. Section through massive-head buttress dam—Pueblo Dam. (*Courtesy Bureau of Reclamation*)

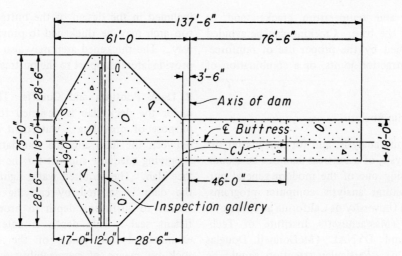

Figure 8-5-11. Detail of massive-head for buttress dam—Pueblo Dam. (*Courtesy Bureau of Reclamation*)

an area of existing tensile stress due to hydrostatic pore pressures. If excessive tensile stresses occur, which could cause cracking, the use of reinforcement should be considered.

C. Design of Buttresses

The basic principles of buttress design and analysis are applicable to all types of buttress dams. When properly applied, the principles can be extended to include double-wall or hollow buttresses as well as the usual single-wall or solid buttress.

The design of a buttress involves two steps: (1) selection of a trial buttress, and (2) analysis of this buttress to determine if it satisfies stress and stability requirements. The proportions of the trial buttress may be determined from previous experience or from comparisons with other dams that have been constructed.

The stress and stability analysis are made in a manner similar to that used for gravity dams as discussed in Section IV. The design element is taken as the width of the buttress and all applied loads including those from the deck are considered. In addition to meeting the stress and stability requirements for gravity dams, the buttress must conform to reinforced concrete design.

The discussion of foundation analysis given in Section III-E-6 is also pertinent to buttress dams, particularly the references to gravity dams.

To prevent buckling, the buttresses are considered to be bearing walls and the allowable loads are in accordance with reinforced concrete design procedures. Bracing or struts can be constructed between adjacent buttresses to prevent buckling by reducing the unsupported length. The effect of temperature change must be considered in the design particularly if the dam is of any length. Stiffeners or counterforts may also be placed on the sides of the buttresses monolithic with it to increase its resistance to buckling.

For double-wall buttresses, diaphragm or curtain walls are placed between the walls. The effective width of the buttress for buckling is then the outside to outside dimension of the walls.

Concrete in the buttresses as in other structures is subject to volume change due to temperature variation and shrinkage. Because the base is rigidly attached to the foundation, shrinkage and temperature cracks tend to form near the base. Cracking can be avoided or controlled by the proper use of reinforcement, contraction joints, or a combination of the two.

D. Dynamic Analysis

The dynamic analysis, and for that matter the stress analysis, of a buttress dam can be accomplished using one of the modern general purpose structural analysis computer programs, e.g., SAP (University of California at Berkeley), STRUDL (Massachusetts Institute of Technology) and DYNAL (McDonnell Douglas Corporation). Particular attention should be directed to properly modeling the connection between the buttresses and the deck. The interaction between buttresses should also be considered carefully. The method for determining the ground acceleration at the site is the same as that referred to in Section III-E-5.

E. Miscellaneous Design Features

1. Top of Dam. The design features encountered at the top of the dam involve no special problems in the case of flat-slab or massive-head dams. Slabs or decks for walkways and roadways can be supported on the buttress similar to the deck slab.

Conditions encountered at the top of multiple-arch dams, however, may be more complex. Where no provisions are required for walkway or roadway, the individual arches may be carried upward as continuous arch-barrels to the required elevation. In this case, no special treatment is necessary other than the considerations of temperature, required minimum thickness, and possible overtopping. Roadways and walkways can be constructed on top of multiple-arch dams as separate elements capable of free movement by adding piers on top of the buttresses for their support. The roadways, walkways, and piers would be designed using design procedures for reinforced concrete design and the loads from the piers included in the design of the buttresses. The top arch may be thickened to provide a walkway. The thickened arch may also be used to provide lateral support to the buttresses.

2. Deck Footings and Cutoffs. The footing under the deck slab is designed for the dual purpose of providing a watertight connection between the dam and the foundation and of supplying adequate support for the base of the face slab. The required water tightness under the dam is secured by carrying the cutoff trench to sufficient depth to provide an efficient seal. This depth is determined by examination and study of the foundation rock by means of permeability and leakage tests of test borings. Special attention must also be given to planes of weakness and pervious strata which may cross the section. For dams on soft or pervious foundation the cutoff must be designed to prevent piping. The connection between the deck and the footing presents no special problems in the flat-slab or massive-head dams. Structural adequacy and safety against shear and sliding are the design requirements for these structures.

There are two principal types of footing for the multiple-arch dam. The first is of monolithic construction. This type introduces secondary stresses which can be evaluated by making a trial-load analysis at the base of the dam and making provision for bending along lines parallel to the arch springline. When there is considerable difference in the elevations of the footings of adjacent buttresses, the closure is effected by means of partial arches and will necessitate the use of more reinforcement than is required in the symmetrical arches higher in the structure. The partial arches will be subject to high temperature stress in locations where the annual variation is great. This will require that a minimum length of arch be established either by excavation or by building up the footing adjacent to the lower buttress.

The second type of footing is constructed as a separate gravity structure, allowing the arch to deflect independently of the footing.

This lack of restraint will eliminate secondary stresses at the base of the arch barrel. Watertightness is secured by waterstops, and the joint is filled with mastic or other suitable material. It may be desirable to use this type of footing in locations where temperatures vary through a large range. The footing is carried to an elevation defined by a plane normal to the arch springline. The footing itself should have sufficient mass to resist the applied forces as an independent unit.

For dams on soft foundations, spread footings can be used beneath the buttresses to spread the load into the foundation. In extreme cases, continuous slabs can be placed under the buttresses for footings. Such slabs must be adequately drained similar to gravity dams to prevent uplift pressures on the slab.

3. Spillways. The design of a spillway on a slab and buttress dam involves no special problems. Figure 8-5-5 is an example of a spillway on a slab and buttress dam. The crest and downstream aprons are designed as a slab between the buttresses in the same manner as the deck slabs. The slabs will, of course, be shaped to accommodate the crest and apron shape as determined by the hydraulic compu-

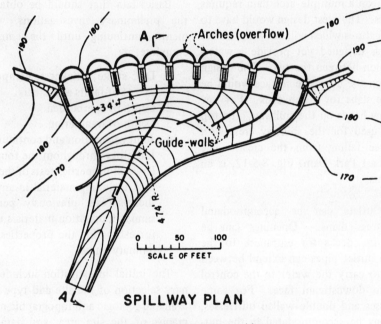

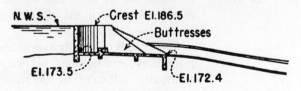

Figure 8-5-12. Multiple-arch spillway at East Park Dam. (*Courtesy Bureau of Reclamation*)

tations. The slabs are supported by corbels attached to the buttresses.

For the massive-head type of dams, an overflow crest can be constructed on top of the heads and buttresses similar to that shown in Fig. 8-5-9. Piers could be constructed on the weir and the crest gated if desired. In Fig. 8-5-9, a flip bucket directs the water downstream and away from the dam, letting it free fall into the stilling pool. As an alternative, slabs could have been constructed on the downstream face to provide an apron to carry the water to the stilling pool. These slabs would be designed similar to those for a slab deck.

A spillway on a multiple-arch dam requires special designs. The crest design would have to incorporate features which will allow the arches to function as designed yet provide a watertight connection between the two. The design of an apron on the downstream face would be similar to the slabs for a slab deck. For low dams with low heads on the spillway, the top arch can be used for the spillway crest with the water free falling from the crest. The spillway at East Park Dam, Fig. 8-5-12, is an example of this.

4. Outlets. Outlets can be accommodated through buttress dams. Openings can be provided in the decks for entrances to the outlets. The outlet pipes can extend between the buttress to carry the water to the control works at the downstream face. For some thick buttresses and double-walled buttresses, the outlet may be accommodated in the buttress. Openings in the buttresses and decks will require reinforcement to carry the stresses around the openings.

VI. DAM FOUNDATIONS

A. Geologic Investigation

The purpose of a geologic foundation investigation is to provide the data necessary to properly evaluate a foundation. A properly sequenced and organized foundation investigation for a dam will provide all the data necessary to evaluate and analyze the foundation at any stage of the site development.

The collection, study, and evaluation of foundation data is a continuing program from the time of the preliminary investigation to the completion of construction. Preliminary investigation provides the data required for the selection of the site and the type of dam. The data collection continues on a more detailed basis through the design phase. Data are also collected continuously during construction as the foundation is exposed, to correlate with previously obtained information and to evaluate the possible need for design changes.

Basic data that should be obtained during the preliminary investigations, with refinements continuing until the construction is complete, are:

1. Dip, strike, thickness, composition, and extent of faults and shears.
2. Depth of overburden.
3. Depth of weathering.
4. Joint orientation and continuity.
5. Lithology throughout the foundation.
6. Physical properties tests of the foundation rock including material in any faults and shears. Tests previously performed on similar foundation materials may be used for estimating the properties in the preliminary phase.

The initial investigation includes a preliminary selection of the site and type of dam. All available geologic and topographic maps, photographs of the site area, and data from field examinations of natural outcrops, road cuts, and other surface conditions should be utilized in the selection of the site and preliminary evaluation of the foundation.

The amount of investigation necessary during preliminary investigations will vary with the anticipated difficulty of the foundation. In general, the investigation should be sufficient to define the major geologic conditions with emphasis on those which will affect design.

The geologic history of a site should be thoroughly studied, particularly where the geology is complex. A study of the history may assist in recognizing and adequately investigating hidden but potentially dangerous foundation conditions.

Diamond core drilling during preliminary investigations may be necessary in more complex foundations and for the foundations for larger dams. The number of drill holes required will depend upon the areal extent and complexity of the foundation. Some foundations may require as few as three or four drill holes to define an uncertain feature. Others may require substantially more drilling to determine foundation treatment for a potentially dangerous foundation condition.

As the investigations proceed, the location of the dam is finalized and the basic design data are confirmed. The geologic mapping and sections are reviewed and supplemented by additional data such as new surveys and additional drill holes. The best possible topography should be used. In most cases, the topography is easily obtained by aerial photogrammetry to almost any scale desired.

The additional drill holes are incorporated into a drilling program which takes advantage of knowledge of special conditions revealed during the preliminary investigations. The drill holes become more specifically oriented and increased in number to better define the foundation conditions and to determine the foundation treatment required.

Laboratory testing of specimens is usually minimized until the decision on the site for construction of the dam has been made. Test specimens are used to determine physical properties of the foundation rock. Physical properties of joint or fault samples may be estimated for preliminary and intermediate stage investigations by using conservative values from past testing of similar materials. The similarity of materials can be judged from the cores retrieved from the drilling.

Final design data are required prior to the preparation of the construction specifications. A detailed foundation investigation is conducted to support the final design data. This investigation involves as much exploration work as necessary to ascertain fully foundation conditions. It may also involve detailed mapping of surface geology, exploratory dozer trenches and exploratory openings such as tunnels, drifts, and shafts in addition to core drilling. Exploratory openings are usually oriented to cross particular geologic features or discontinuities. The number and depth of the openings depend on the type of rock and the extent of geologic discontinuities such as faults, shears, and prominent joint sets. These openings provide the best possible means of examining the foundation.

In addition to test specimens for determining the physical properties, specimens may be required for final design to use in determining the shear strength of the rock types, healed joints and open joints. This information may be necessary to determine the stability of the foundation and is discussed under the shear friction factor in Section II-E.

Permeability tests should be performed as a routine matter during the drilling program.[38,39] The information from these tests can be utilized in establishing flow nets which will aid in studying uplift conditions and establishing drainage systems.

The geology as encountered in the excavation during construction should be defined and compared with the preexcavation geology. Geologists and engineers should consider carefully any geologic change and check its relationship to the design of the structure.

As-built geology drawings should be developed even though revisions in design may not be required by changed geologic conditions, since operation and maintenance problems requiring detailed foundation information may develop.

B. Testing

In-Situ Testing. In-situ shear tests[40] are more expensive than laboratory tests and usually only comparatively few are made. The advantage of a larger test surface is that fewer in-situ tests are required. The in-situ tests can be supplemented by laboratory tests. In either case, the shearing strength relative to both horizontal and vertical movement should be obtained.

Foundation deformation tests can be performed within the exploratory shafts and tunnels to determine values of deformation modulus. The deformation tests applicable to rock masses are the radial jacking test[41] and the uniaxial jacking test.[42] For softer and less massive materials, the plate gouge test[43] may be used.

Laboratory Testing. Laboratory tests have been standardized and the methods and test interpretations should not vary substantially from one laboratory to another. A major problem in laboratory testing is the difficulty of obtaining representative samples. Further, sample size is often dictated by the laboratory equipment with a corresponding effect on test results. Laboratory tests usually performed are:

Physical Properties Tests

(1) Compressive strength
(2) Elastic modulus
(3) Poisson's ratio
(4) Bulk specific gravity
(5) Porosity
(6) Absorption

Shear Tests

(1) Direct shear } Performed on intact specimens and those with healed joints
(2) Triaxial shear
(3) Sliding friction — Performed on open joints

Other Tests

(1) Solubility
(2) Petrographic analysis

C. Design

1. Effective Deformation Modulus. A concrete dam is considered to be homogeneous, elastic, and isotropic but its foundation is, in general, heterogeneous, inelastic, and anisotropic. This variation in the foundation can affect the distribution of loads within the dam. Thus, the foundation deformation characteristics should be evaluated over the entire extent of the dam contact.

The reaction of the foundation to the loads from the dam control, to some extent, the stresses within the dam. Conversely, the response of the dam to the external loads and the foundation determines the stresses on the foundation. The proper determination of the interaction between the dam and the foundation requires an accurate knowledge of the deformation characteristics of the foundation.

Deformation modulus is defined as the ratio of stress to elastic plus inelastic strain and is used to define the deformation characteristics of the foundation. Deformation modulus differs from elastic modulus in that an elastic modulus is the ratio of stress to elastic strain only. An effective deformation modulus (EDM) is one which, when substituted for the various deformation moduli in a composite foundation, produces an equivalent behavior.

To determine an EDM, the deformation moduli of all foundation materials must first be determined and then, using appropriate methods of analysis, such as the finite element method as described in Section III-E, the EDM can be evaluated for the composite of materials.

Basic to the idea of an EDM is the fact that there is an EDM for any discrete block of foundation regardless of how arbitrarily or systematically a block is determined. If several sections were taken through the foundation normal to the dam contact and the EDM determined for each section, for example, it is apparent that there could be as many values of EDM as sections. Therefore, to properly represent the variation in the EDM along the concrete to rock contact an adequate number of sections should be analyzed to describe the EDM distribution. If the EDM is highly variable, the load distribution in the dam may be adversely affected and high load and stress concentrations could be induced. In such cases it is necessary to design treatment to reduce the variability of the foundation EDM. A discussion of treatment design follows in Section VI-C-3.

2. Stability. The stability of a dam foundation as referred to here is its ability to develop resisting forces such that a new state of equilibrium is established after the system's initial equilibrium is disturbed by external forces.

Regardless of how well a dam is designed, if the foundation on which it stands is not stable, the design is worthless. Therefore, the stability of the foundation must of necessity be included in the design.

The primary condition which constitutes instability is the sliding failure of a rock mass within the foundation or between the dam and the foundation. There are two other conditions usually discussed in reference to foundation stability although neither one is actually a sliding mechanism. The first is differential movements which may occur across a geologic discontinuity like a shear or due to a sudden change in deformatiom modulus in adjacent materials for example. The second is stress concentrations which can occur where zones of weak material in the foundation are bridged over by the dam.

Methods of stability analysis are given in Section III-E.

3. Treatment

Foundation Shaping. The entire area to be occupied by the base of the concrete dam should be excavated to firm material capable of withstanding the loads imposed by the dam, reservoir, and appurtenant structures. Considerable attention must be given to blasting operations to assure that excessive blasting does not shatter, loosen, or otherwise adversely affect the suitability of the foundation rock. All excavations should conform to the lines and dimensions shown on the construction drawings where practicable; however, it may be necessary or even desirable to vary dimensions or excavation slopes due to local conditions.

Foundations such as shales, chalks, mudstones, and siltstones may require protection against air and water slaking, or in some environments, against freezing. Such excavations can be protected by leaving a temporary cover of several feet of unexcavated material, by immediately applying a minimum of 12 in. of pneumatically applied mortar to the exposed surfaces, or by any other method that will prevent damage to the foundation.

Shaping for Gravity and Buttress Dams. If the canyon profile for a gravity dam site is relatively narrow with steep sloping walls, each vertical section of the dam from the center towards the abutments is shorter in height than the preceding one. Consequently, sections closer to the abutments will be deflected less by the reservoir load and sections closer toward the center of the canyon will be deflected more. Since most gravity dams are keyed at the contraction joints, the result is a torsional effect in the dam that is transmitted to the foundation rock.

A sharp break in the excavated profile of the canyon will result in an abrupt change in the height of the dam. The effect of the irregularity of the foundation rock causes a marked change in stresses in both the dam and foundation, and in stability factors. For this reason, the foundation should be shaped so that a uniformly varying profile is obtained free of sharp offsets or breaks.

Generally, a foundation surface will be horizontal in the transverse (upstream-downstream) direction. However, where an increased resistance to sliding is desired, particularly for structures founded on sedimentary rock foundations, the surface can be sloped upward from heel to toe of the dam. The foundation excavation for Pueblo Dam (Fig. 8-6-1), a massive head buttress-type gravity dam, is an example of an excavation sloped in the transverse direction. Figure 8-6-1 also represents a special type of situation wherein the foundation excavation is shaped to the configuration of the massive head butress.

Shaping for Arch Dams. Although not necessary, a symmetrical or nearly symmetrical profile is desirable for an arch dam from the standpoint of stress distribution. However, asymmetrical canyons frequently have to be chosen

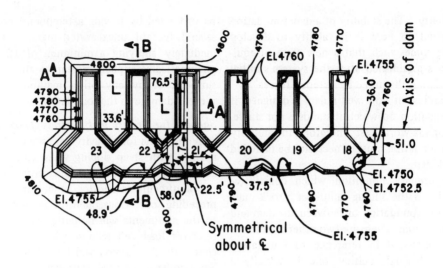
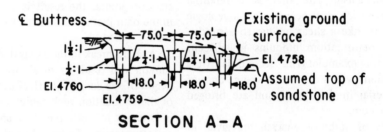
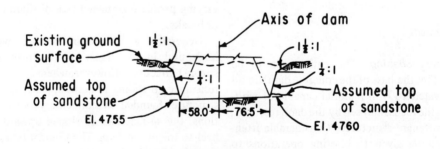

Figure 8-6-1. Foundation shaping excavation for a massive head buttress dam. (*Courtesy Bureau of Reclamation*)

as arch dam sites. The nonsymmetry may introduce stress problems, but these can be overcome by proper design. Abutment pads between the dam and its foundation may be used to overcome some of the detrimental effects of nonsymmetry or foundation irregularities. Thrust blocks are sometimes used at nonsymmetrical sites. However, the primary use of a thrust block is not to provide symmetry, but simply to establish an artificial abutment where a natural one did not exist. Overexcavation of a site to achieve symmetry on a nonsymmetrical profile is not recommended. In all cases, the foundation should be excavated in such a way as to eliminate sharp breaks in the excavated profile, since these may cause stress concentrations in both the foundation rock and the dam. The foundation should also be excavated to radial or part radial lines as explained in Section III-D.

Dental Treatment. Very often the exploratory drilling or final excavation uncovers faults, seams, and shattered or inferior rock extending to such depths that it is impracticable to attempt to clear such areas out entirely. These conditions require special treatment in the form of removing a portion of the weak material and backfilling the resulting excavations with concrete. This procedure of reinforcing and stabilizing such weak zones is frequently called "dental treatment."

Theoretical studies have been made to develop general rules for guidance as to how deep transverse seams should be excavated. These studies, based upon foundation conditions and stresses at Shasta and Friant Dams, have resulted in the development of the following approximate formulas:

$$d = 0.002bH + 5 \quad \text{for} \quad H \geq 150 \text{ ft}$$
$$d = 0.3d + 5 \quad \text{for} \quad H < 150 \text{ ft}$$

where

H = height of dam above general foundation level in ft,
b = width of weak zone in ft, and
d = depth of excavation of weak zone below surface of adjoining sound rock in ft.

(In clay gouge seams, d should not be less than 0.1 H.)

These guides provide a means of approach to the question of how much should be excavated, but final judgment must be exercised in the field during actual excavation.

Although the preceding rules are suitable for application to foundations with a relatively homogeneous rock foundation with nominal faulting, some dam sites may have several distinct rock types interspersed with numerous faults and shears. The effect of rock-type anomalies complicated by large zones of faulting on the overall strength and stability of the foundation requires a definitive analysis. The finite element method of analysis, as discussed in Section III-E can be used in evaluating the foundation. This method provides a way to combine the physical properties of various rock types, and geologic discontinuities such as faults, shears, and joint sets into a value representative of the stress and deformation in a given segment of the foundation. The method also permits substitution of backfill concrete in faults, shears, and zones of weak rock, and thus evaluates the degree of beneficiation contributed by the "dental concrete."

Data required for the finite element method of analysis are: dimensions and composition of the lithologic bodies and geologic discontinuities, deformation moduli for each of the elements incorporated into the study, and the loading pattern imposed on the foundation by the dam and reservoir.

"Dental treatment" may also be required to improve the stability of rock masses. A safety factor for a particular rock mass can be calculated by inputting the following data; the shearing strength of faults, shears, joints and intact rock, pore water pressures induced by the reservoir and/or ground water, the weight of the rock mass, and the driving forces induced by the dam and reservoir.

For an arch dam foundation the geologic discontinuities can also affect both the stability and the deformation modulus. From a practical viewpoint, however, the foundation modulus need not be known accurately if, in a quantitative sense, the ratio of the foundation modulus E_f to the concrete modulus E_c of the dam is known to be greater than one-fourth. Moreover, canyons with extensive zones of highly deformable materials and consequently E_f/E_c ratios less than one-fourth should not be arbitrarily dismissed as potential arch dam sites. The deformation modulus of such zones can be improved by removing sheared material, gouge, and/or inferior rock and replacing the same with backfill concrete. A case in point is the foundation analysis for a dam located in northeast California. Here the basic volcanic rock was interspersed with faults and shears, most of which had continuous seams of gouge that varied from paper thin to several feet in thickness, and with other rock anomalies each with individual deformation moduli. The resulting anisotrophy required a definitive analysis to obtain the existing foundation modulus, and then a determination of which and how much of the geologic discontinuities needed to be treated to obtain acceptable deformation moduli throughout the foundation. The finite element method of analysis was used to evaluate the foundation.

The approximate and analytical methods discussed above will satisfy the stress, deformation, and stability requirements for a foundation, but they may not provide suitable protection against piping. Since faults and seams may contain material conducive to piping, upstream and downstream cutoffs may be required to eliminate this condition. For this reason upstream and downstream cutoff shafts must be excavated in each seam and backfilled with concrete. The dimension of the shaft perpendicular to the seam should be equal to the width of the weak zone plus a minimum of 1 ft on each end to key the concrete backfill into sound rock. The shaft dimension parallel with the seam should be at least one-half of the other dimension. In any instance a minimum shaft dimension of 5 ft each way should be used to provide working space.

The depth of cutoff shafts required to control piping may be computed by constructing flow nets or by the methods outlined by Khosla.[44] These two methods are particularly applicable for medium to high dams. For low head dams, the weighted creep method for determining cutoff depths may be used.[45]

Other adverse foundation conditions may be due to horizontally bedded clay and shale seams, caverns, or springs. Procedures for treating these conditions will vary and will depend upon field studies of the characteristics of the particular condition to be remedied.

Grouting. The principal objectives of grouting in a rock foundation are to establish an effective barrier to seepage under the dam and to consolidate the foundation. Spacing, length, and orientation of grout holes and the procedure to be followed in grouting a foundation are dependent on the height of the structure and the geologic characteristics of the foundation. Since the characteristics of a foundation will vary for each site, the grouting plan must be adapted to suit existing conditions.

Grouting operations may be performed from the surface of the excavated foundation, from the upstream fillet of the dam, from the top of concrete placements for the dam, from galleries within the dam, and from tunnels driven into the abutments, or any combination of these locations. The general plan for grouting the foundation rock of a dam provides for preliminary low-pressure, shallow consolidation grouting to be followed by high-pressure, deep curtain grouting. As used here, "high pressure" and "low pressure" are relative terms. The actual pressures used are usually the maximum that will result in filling the cracks and voids as completely as practicable without causing any uplift or lateral displacement of foundation rock.

Consolidation Grouting. Low-pressure grouting to fill voids, fracture zones, and cracks at and below the surface of the excavated foundation is accomplished by drilling and grouting relatively shallow holes. The extent of the area grouted and the depth of the holes will depend on local conditions.

Usually for structures 100 ft and more in height, a preliminary program will call for lines of holes parallel to the axis of the dam extending from the heel to the toe of the dam and spaced approximately 10 to 20 ft apart. Holes are staggered on alternate lines to provide better coverage of the area. The depths of the holes vary from 20 to 50 ft depending on local conditions and to some extent on the height of the structure. For structures less than 100 ft in height, and depending on local conditions, grouting will be applied only in the area of the heel of the dam.

The upstream line of holes should lie at or

near the heel of the dam to furnish a cutoff for leakage of grout from the high-pressure holes drilled later in the same area. Consolidation grout holes are usually drilled normal to the excavated surface unless it is desired to intersect known faults, shears, fractures, joints, and cracks. Drilling is usually accomplished from the excavated surface, although in some cases drilling and grouting to consolidate steep abutments has been accomplished from the tops of concrete placements in the dam to prevent "slabbing" of the rock. In rarer cases, consolidation grouting has been performed from foundation galleries within the dam after the concrete placement has reached a certain elevation. This method of consolidation grouting requires careful control of grouting pressures and close inspection of the foundation to assure that the structure is not being disbonded from the foundation.

In the execution of the consolidation grouting program, holes usually 1½ in. in diameter (EX drill bit size) are drilled and grouted 40 to 80 ft apart before split-spaced intermediate holes are drilled. The amount of grout which the intermediate holes accept determines whether additional intermediate holes should be drilled. This split-spacing process is continued until grout "take" for the final closure holes is negligible, and there is reasonable assurance that all groutable seams, fractures, cracks, and voids have been filled.

Water-cement ratios for grout mixes may vary widely depending on the permeability of the foundation rock. Starting water-cement ratios usually range from 8:1 to 5:1 by volume. Most foundations have an optimum mix which should be determined by trial in the field by gradually thickening the starting mix. An admixture such as sand or clay may be added if large voids are encountered.

Consolidation grouting pressures vary widely and are dependent in part on the characteristics of the rock, i.e., its strength, tightness, joint continuity, stratification, etc.; and on the height of rock above the stage being grouted. Grout pressures as high as practicable but which, as determined by trial, are safe against rock displacement, are used in grouting. These pressures may vary from a low of 10 lb/sq. in to a high range of 80 to 100 lb/sq. in. A common rule of thumb is to increase the collar pressure by 1 lb/sq. in./ft of hole above the packer. If the take is small the pressure may be increased.

Curtain Grouting. Construction of a deep grout curtain near the heel of the dam to control seepage is accomplished by drilling deep holes and using higher grouting pressure. Tentative designs will usually specify a single line of holes drilled on 10-ft centers, although wider or closer spacing may be required depending on the rock condition. To permit application of high pressures without causing displacement in the rock or loss of grout through surface cracks, this grouting is usually carried out subsequent to consolidation grouting and after some of the dam concrete has been placed. Grouting will usually be accomplished from galleries within the dam and from tunnels driven into the abutments especially for this purpose. However, when no galleries are provided, as is the case for most low gravity dams, and some thin arch dams, high-pressure grouting is done from curtain holes located in the upstream fillet of the dam before reservoir storage is started.

The alignment of holes should be such that the base of the grout curtain will be located on the vertical projection of the heel of the dam. If drilled from a gallery that is some distance from the upstream face, the holes may be inclined as much as 15° upstream from the plane of the axis. If the gallery is near the upstream face, the holes will be nearly vertical. Holes drilled from foundation tunnels may be inclined upstream or they may be vertical depending on the orientation of the tunnel with the axis of the dam. When the holes are drilled from the upstream fillet, they are usually vertical or inclined downstream. Characteristics of the foundation seams may also influence the amount of inclination.

To facilitate drilling, pipes of 2-ft minimum

length are embedded in the floor of the gallery or foundation tunnel, or in the upstream fillet. When the structure has reached an elevation that is sufficient to prevent movement of concrete, the grout holes are drilled through these pipes and into the foundation. Although the tentative grouting plan may indicate holes to be drilled on 10-ft centers, the usual procedure will be first to drill and grout holes approximately 40 ft apart, or as far apart as necessary to prevent grout from one hole leaking into another drilled but ungrouted hole. Also, leakage into adjacent contraction joints must be prevented by prior grouting of the joints. Intermediate holes, located midway between the first holes, will then be drilled and grouted. Drilling and grouting of additional intermediate holes, splitting the spaces between completed holes, will continue until the amount of grout accepted by the last group of intermediate holes indicates no further grouting is necessary.

The depth to which the holes are drilled will vary greatly with the characteristics of the foundation and the hydrostatic head. In a hard, dense foundation, the depth may vary from 30 to 40 percent of the head. In a poor foundation the holes will be deeper and may reach as deep as 70 percent of the head. During the progress of the grouting, local conditions may determine the actual or final depth of grouting. Supplementary grouting may also be required after the waterload has come on the dam and observations have been made of the rate of seepage and the accompanying uplift.

Usually the foundation will increase in density and tightness of seams as greater depths are reached, and the pressure necessary to force grout into the tight joints of the deep planes may be sufficient to cause displacements of the upper zones. Two general methods of grouting are used, each permitting the use of higher pressures in the lower zones.

(1) Descending stage grouting consists of drilling a hole to a limited depth or to its intersection with an open seam, grouting to that depth, cleaning out the hole after the grout has taken its initial set, and then drilling and grouting the next stage. To prevent backflow of grout during this latter operation a packer is seated at the bottom of the previously grouted state. This process is repeated, using higher pressures for each succeeding stage until the final depth is reached.

(2) Ascending stage grouting consists of drilling a hole to its final depth and grouting the deepest high-pressure stage first by use of a packer which is seated at the top of this stage. The packer limits grout injection to the desired stage and prevents the grout from rising into the hole above the packer. After grouting this stage, the grout pipe is raised so that the packer is at the top of the next stage which is subsequently grouted using somewhat lower pressure. This stage process is repeated, working upward until the hole is completely grouted. Ascending stage grouting is becoming more generally used, as it reduces the chances for displacement of the foundation rock, gives better control as to the zones of injection, and expedites the drilling.

The discussion for consolidation grouting concerning grout pressures applies in general to curtain grouting. An exception is that higher initial collar pressures are permitted depending on the height of concrete above the hole.

Foundation Drainage. Although a well-executed grouting program may materially reduce the amount of seepage, some means must be provided to intercept the water percolating through and around the grout curtain which, if not removed, may build prohibitive hydrostatic pressures on the base of the structure. Drainage is usually accomplished by drilling one or more lines of holes downstream from the high-pressure grout curtain. The size, spacing, and depth of these holes are assumed on the basis of judgment of the physical characteristics of the rock. Holes are usually 3 in. in diameter (NX drill bit size). Spacing, depth, and orientation are all influenced by the foundation conditions. Usually the holes are spaced on 10-ft centers with depths dependent on the grout

DESIGN OF CONCRETE DAMS 469

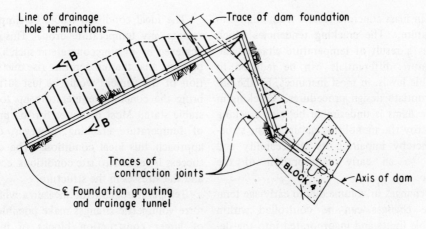

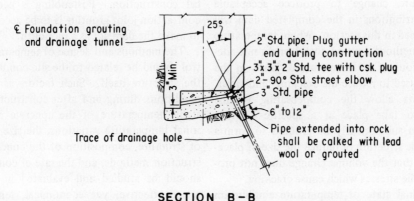

Figure 8-6-2. Foundation grouting and drainage tunnel.

curtain and reservoir depths. As a general rule, hole depths vary from 20 to 40 percent of the reservoir depth and 35 to 75 percent of the deep curtain grouting depth.

Drain holes should be drilled after all foundation grouting has been completed in the area. They can be drilled from foundation and drainage galleries within the dam, or from the downstream face of the dam if no gallery is provided. Frequently drainage holes are drilled from foundation grouting and drainage tunnels excavated into the abutments as shown in Fig. 8-6-2.

In some instances where the stability of a rock foundation may be benefited by reducing the hydrostatic pressure along planes of potentially unstable rock masses, drainage holes have been introduced to alleviate this condition. A collection system for such drainages should be designed so that flows can be gathered and removed from the area.

VII. TEMPERATURE CONTROL AND JOINTS

Mass concrete structures undergo volumetric changes which, because of the large dimensions involved, concern the designer. These volumetric changes can cause cracking in the dam because of the restraint against volumetric change.[6,33,46] They also affect the stresses in the completed dam. The largest volumetric change in mass concrete results from change in temperature. The cracking tendencies which occur as a result of temperature changes and temperature differentials can be reduced to acceptable levels, in most instances, by the use of appropriate design procedures. Cracking in concrete dams is undesirable because cracking can destroy the monolithic nature of the structure, thereby impairing its serviceability and leading to an early deterioration of the concrete.

The changes in volume due to early-age temperature changes can be controlled within reasonable limits and incorporated into the design of the structure. Methods of controlling temperature change to produce acceptable stress distributions in the completed dam will be discussed in this section. Where temperature control methods are too costly or do not adequately control the temperature, joints in the dam are used to reduce the cracking tendencies. The joints allow the compensating volume change to take place at a specific location which can subsequently be grouted to form a monolithic structure or reduce the size of placement so that the volume change does not produce tensile stresses which cause cracking.

The final state of temperature equilibrium depends upon site conditions, and little if any degree of control over the subsequent periodic volumetric changes can be affected. The effects of these periodic temperature changes must, therefore, be included in the analysis of the dam as discussed in the design criteria.

A. Temperature Control

Measures commonly used to control temperature changes in early-age concrete include precooling, postcooling, or a combination of the two and by the use of a concrete mix designed to limit the heat of hydration.[47] These measures will reduce the peak temperature which otherwise would have been attained. Proportionately, this reduction in peak temperature will reduce the subsequent volumetric change and the accompanying crack-producing tendencies.

The ideal condition would be simply to eliminate any temperature drop. This could be achieved by placing concrete at such a low temperature that the temperature rise due to hydration of the cement would be just sufficient to bring the concrete temperature up to its final stable state. Most measures for the prevention of temperature cracking, however, can only approach this ideal condition. The degree of success is related to site conditions, economics, and the stresses in the structure.

Temperature control measures which minimize volumetric changes make possible the use of larger construction blocks or monoliths, thereby resulting in a more rapid and economical construction. Postcooling is necessary if contraction joint grouting is to be accomplished to make the dam act as a monolith.

The methods and degree of temperature control should be related to the site conditions and the structure itself. Such factors as exposure conditions during and after construction, final stable temperature of the concrete mass, seasonal temperature variations, the size and type of structure, composition of the concrete, construction methods, and the rate of construction should be studied and evaluated in order to select effective, yet economical, temperature control measures. The construction schedule and design requirements must also be studied to determine those procedures necessary to produce favorable temperature conditions during construction. Such factors as thickness of lifts, time interval between lifts, height differentials between blocks, and seasonal limitations on placing of concrete should be evaluated. Study of the effect of these variables will permit the determination of the most favorable construction schedules consistent with the prevention of cracking from temperature stresses.

One of the most effective and positive temperature control measures is precooling which reduces the placing temperature. The method or combination of methods used to reduce concrete placing temperatures will vary with the degree of cooling required and the contractor's equipment. For some structures, sprinkling and shading of the coarse aggregate piles may be the only precooling measures required. Evaporative cooling, associated with sprinkling, can also reduce the placing temperature but is restricted to areas with a low relative humidity. Insulating and/or painting the surfaces of the batching plant, waterlines, etc., with reflective paint can also be beneficial.

Mixing water can be cooled to varying degrees, the more common temperatures being from 32° to 40° F. Adding slush or crushed ice to the mix is an effective method of cooling because it takes advantage of the latent heat of fusion of ice. Cooling of the coarse aggregates to about 35° F can be accomplished in several ways. One method is to chill the aggregate in large tanks of refrigerated water for a given period of time or spraying with cold water. Relatively effective cooling of coarse aggregate can also be attained by forcing refrigerated air through the aggregate while the aggregate is draining in stockpiles, while it is on a conveyor belt, and while it is in the bins of the batching plant. Sand may be cooled by passing it through vertical tubular heat exchangers. Cold air jets directed on the sand as it is transported on conveyor belts can also be used. Immersion of sand in cold water is not practical because of the difficulty in removing the free water from the sand after cooling.

Use of the above treatments has resulted in concrete placing temperatures of 50° F in a number of instances. Concrete placing temperatures as low as 45° F have been attained, but these can usually be achieved only at a considerable increase in cost.

Postcooling reduces the peak temperatures of the concrete due to the heat of hydration. It also is used to reduce the concrete temperature to the desired contraction joint grouting temperature.

A typical layout of the embedded cooling system used for postcooling is shown in Fig. 8-7-1. The system consists of pipe or tubing placed in grid-like coils over the top surface of each lift of concrete after the concrete has hardened. Coils are formed by joining together lengths of thin-wall metal pipe or tubing. The number of coils in a block depends upon the size of the block and the horizontal spacing. Supply and return headers, with manifolds to permit individual connections to each coil, are normally placed on the downstream face of the dam. In some instances, cooling shafts, galleries, and embedded header systems can be used to advantage.

The velocity of flow of the cooling water through the embedded coils is normally required to be not less than 2 ft/sec, or about 4 gal/min for the commonly used 1-in. pipe or tubing. Cooling water is usually pumped through the coils, although a gravity system has at times been used. When river water is used, the warmed water is usually wasted after passing through the coils. River water having a high percentage of solids should be avoided as it can clog the cooling systems. When refrigerated water is used, the warmed water is returned to the water coolers in the refrigerating plant, recooled, and recirculated.

The design of a cooling system requires a study of each structure, its environment, and the maximum temperatures which are acceptable from the standpoint of crack control. The temperature effects of various heights of placement lifts and such layout variables as size, spacing and length of embedded coils should be investigated. Variables associated with the operation of the cooling systems, such as rate of water circulation and the temperature differential between the cooling water and the concrete being cooled, are studied concurrently. All of these factors should be considered in arriving at an economical cooling system which can achieve the desired temperature control.

Figure 8-7-2 shows the effect of cooling water temperature, length of cooling tubing, and the spacing of coils on the concrete temperatures. These studies were made using four sacks of Type II cement/cu yd, a diffusivity of 0.050 ft^2/hr, a flow of 4 gal/min through 1-in. outside-diameter pipe, 5-ft placement lifts, and assumed a 3-day exposure of each lift. The design of a cooling system is given in detail in other sources.[6,33] See also p. 422.

Postcooling is also used to reduce the temperature of the concrete so that the contraction

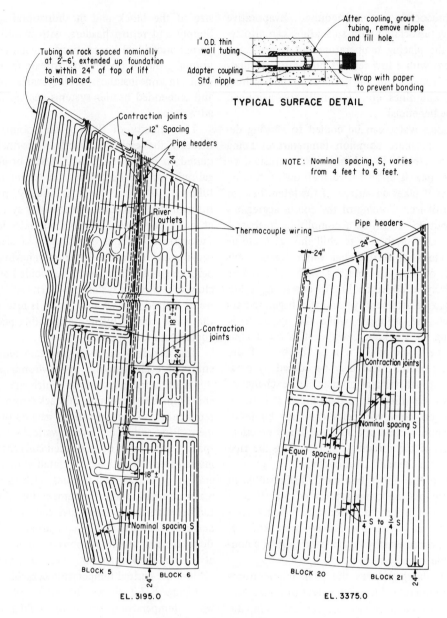

Figure 8-7-1. Typical cooling coil layouts from Glen Canyon Dam. (*Courtesy Bureau of Reclamation*)

joints between blocks will open sufficiently to be grouted. In some cases, such as arch dams, additional cooling may be required so that the joints may be grouted at a specified temperature as discussed in the design criteria. Figure 8-7-3 shows a typical temperature history of artificially cooled concrete.

Use of a concrete mix designed to limit the heat of hydration can significantly reduce the peak temperatures.[48,49] Since the heat gener-

DESIGN OF CONCRETE DAMS 473

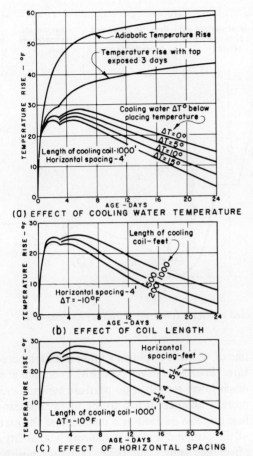

Figure 8-7-2. Artificial cooling of concrete. (*Courtesy Bureau of Reclamation*)

ated within the concrete is directly proportional to the amount of cement used/cu yd, the selected mix should provide the required strength and durability with the lowest cement content.

The heat-producing characteristics of cement play an important role in the amount of temperature rise. Although cements are classified by type, the heat generation within each type may vary widely because of the chemical compounds in the cement. Types II and IV were developed for use in mass concrete construction. Type II cement is commonly referred to as modified cement, and is used where a relatively low heat generation is desirable. Type IV cement is a low-heat cement characterized by its low rate of heat generation during early age. The various types of cement are shown in Table 8-7-1. The effect of the type of cement on the adiabatic temperature rise can be seen in Fig. 8-7-4. Also shown on the figure is the effect of heat loss to the surface on the temperature rise.

Normally, specifications for portland cement do not state within what limits the heat of hydration shall be for each type of cement. They do, however, place maximum percentages on certain chemical compounds in the cement. They further permit the purchaser to specifically request maximum heat of hydration requirements of 70 to 80 cal/g at ages 7 and 28 days, respectively, for Type II cement; and 60

TABLE 8-7-1. Types of Portland Cement

Type	TYPICAL COMPOUND COMPOSITION, PERCENTAGE*								Typical uses
	C_3S	C_2S	C_3A	C_4AF	$CaSO_4$	Free CaO	MgO	Ignition loss	
Type I	49	25	12	8	2.9	0.8	2.4	1.2	General concrete construction.
Type II	46	29	6	12	2.8	0.6	3.0	1.0	Concrete subject to moderate sulfate attack. Concrete requiring a moderate heat of hydration.**
Type III	56	15	12	8	3.9	1.3	2.6	1.9	Concrete requiring high early strength.
Type IV	30	46	5	13	2.9	0.3	2.7	1.0	Concrete requiring a low heat of hydration.
Type V	43	36	4	12	2.7	0.4	1.6	1.0	Concrete subject to severe sulfate attack.

*For complete physical and chemical requirements of each type, see Comparison of Requirements for Portland Cement, Portland Cement Association, MS204.04T, May, 1974.
**By invoking optional provisions for controlling heat of hydration.

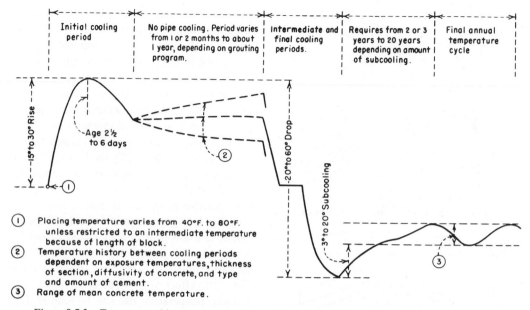

Figure 8-7-3. Temperature history of artificially cooled concrete. (*Courtesy Bureau of Reclamation*)

to 70 cal/g at ages 7 and 28 days, respectively, for Type IV cement.

In most instances, Type II cement will produce concrete temperatures which are acceptable. In the smaller structures, Type I cement will often be entirely satisfactory. Other factors being equal, Type II cement would normally be used because of its better resistance to sulfate attack, better workability, and lower permeability. Type IV cement is used only where an extreme degree of temperature control is required. For example, it would be beneficial near the base of long blocks where a high degree of restraint exists. Concrete made with Type IV cement requires more curing than concrete made with other types of cement and extra care is required at early ages to prevent damage to the concrete from freezing during cold weather. Often, the run-of-the-mill cement from a plant will meet the requirements of a Type II cement, and the benefits of using this type of cement can be obtained at little or no extra cost. Type IV cement, because of its special composition, is obtained at premium prices.

Pozzolans are used as a replacement for part of the cement in concrete for several reasons, one of which is to obtain a lower peak temperature. Pozzolans develop heat of hydration at a much lower rate than do portland cements and more of the heat is lost to the surface. Pozzolans can also be used as a replacement for part of the portland cement to improve workability, effect economy, and obtain a better quality concrete. The more common pozzolans used in mass concrete include calcined clays, diatomaceous earth, volcanic tuffs and pumicites, and fly ash.

Other temperature control methods include:

1. Use of shallow placement lifts so that a greater percentage of the heat generated is lost to the surface.

2. Use of water curing which reduces the surface temperature to about that of the water instead of the prevailing air temperature. In areas of low humidity, the effect of evaporative cooling may result in lowering the surface temperature even lower.

3. Use of retarding agents which retard the rate of heat generation allowing more heat to be lost to the surface.

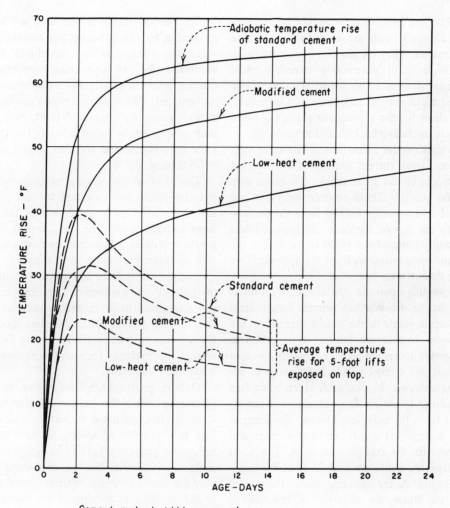

Figure 8-7-4. Effect of type of cement on temperature rise in mass concrete. (*Courtesy Bureau of Reclamation*)

B. Joints

Joints are used to limit the size of concrete placements so that tensile stresses which occur as a result of temperature changes will not cause cracking. They are also used to limit the size of placement to facilitate construction.

Contraction and expansion joints are provided in concrete structures to accommodate volumetric changes which occur in the structure after placement. Contraction joints are provided in a structure to prevent the formation of tensile cracks in the structure as the structure undergoes a volumetric shrinkage due to a temperature drop. Expansion joints are provided in a structure to allow for expansion (a volumetric increase due to temperature rise).

A construction joint in concrete is defined as the surface of previously placed concrete upon or against which new concrete is to adhere when the previously placed concrete has attained its initial set and hardened to such an extent that the new concrete cannot be incorporated integrally with the earlier place concrete by vibration. They are placed in concrete structures to facilitate construction, to reduce initial shrinkage stresses, to permit installation of embedded metalwork in blockouts at a later date, or to allow for the subsequent placing of other concrete, including backfill and second-stage.

To control the formation of cracks in mass concrete dams, current practice is to construct the dam in blocks separated by transverse contraction joints. These contraction joints are vertical and normally extend from the foundation to the top of the dam. Transverse joints are usually normal to or radial to the axis of the dam and are continuous from the upstream face to the downstream face.

Depending upon the size of the structure, it may also be necessary to provide longitudinal contraction joints in the blocks formed by the transverse contraction joints. If longitudinal contraction joints are provided, construction of the dam will consist of placing a series of adjoining columnar blocks, each block being free to undergo its own volume change without restraint from the adjoining blocks. The longitudinal contraction joints are also vertical and parallel with the axis for arch dams. The joints are staggered a minimum of 25 ft at the transverse joints. As the joint approaches the face of the dam either the direction of the joint is changed from the vertical to effect a perpendicular intersection with the face, with an offset of 3 to 5 ft, or the joint is terminated at the top of a lift when it is within 15 to 20 ft of the face. In the latter case, strict temperature control measures will be required to prevent cracking of the concrete directly above the terminated joint.

Typical transverse contraction joints can be seen on Figs. 8-7-5, 8-7-6 and 8-7-7, and a typical longitudinal contraction joint can be seen on Fig. 8-7-8. Contraction joints should be constructed so that no bond exists between the concrete blocks separated by the joint. Reinforcement should not extend across a contraction joint.

Contraction joints in the dam are usually grouted after the concrete has undergone its volumetric change so that the blocks will act monolithically. In some cases, transverse contraction joints in gravity and buttress dams are not grouted. Since stresses would not be transmitted across an ungrouted joint, each block must, therefore, be stable in itself. Contraction joint grouting details are also shown in Figs. 8-7-5 through 8-7-8.

The location and spacing of transverse contraction joints are governed by the physical features of the dam site, details of the structures associated with the dam, results of temperature studies, placement methods, and the probable concrete mixing plant capacity.

Foundation defects and major irregularities in the rock are conducive to cracking and this can sometimes be prevented by judicious location of the joints. Consideration should be given to the canyon profile in spacing the joints so that the tendency for such cracks to develop is kept to a minimum.

Outlets, penstocks, spillway gates, or bridge piers may affect the location and/or spacing. Consideration of other factors, however, may lead to a possible relocation of these appurtenances to provide a spacing of joints which is more satisfactory to the dam as a whole. Probably the most important of these considerations is the permissible spacing of the joints determined from the results of concrete temperature control studies. If the joints are too far apart, excessive shrinkage stresses will produce cracks in the blocks. On the other hand, if the joints are too close together, shrinkage may be so slight that the joints will not open enough to permit effective grouting.

Contraction joints should be spaced close enough so that, with the probable placement methods, plant capacity, and the type of concrete being used, batches of concrete placed in a lift can always be covered while the concrete is still plastic. For average conditions, a spacing of 50 ft has proved to be satisfactory. In dams where pozzolan and retarders are used, spacings up to 80 ft have been acceptable. An effort

DESIGN OF CONCRETE DAMS 477

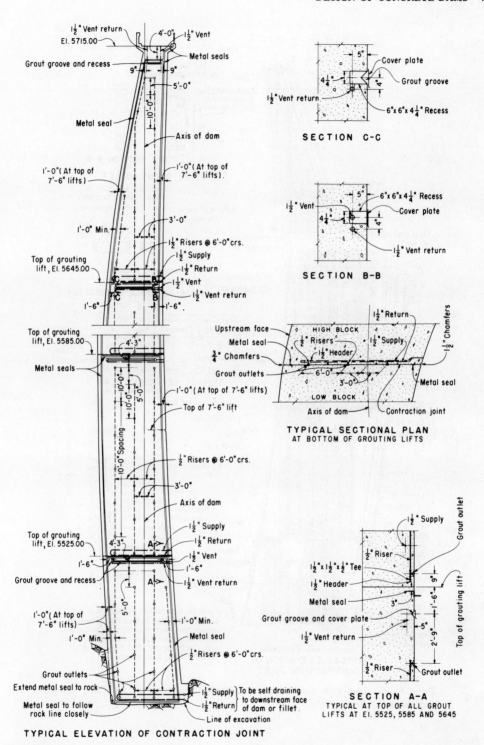

Figure 8-7-5. Transverse contraction joint details from East Canyon Dam. (*Courtesy Bureau of Reclamation*)

478 HANDBOOK OF DAM ENGINEERING

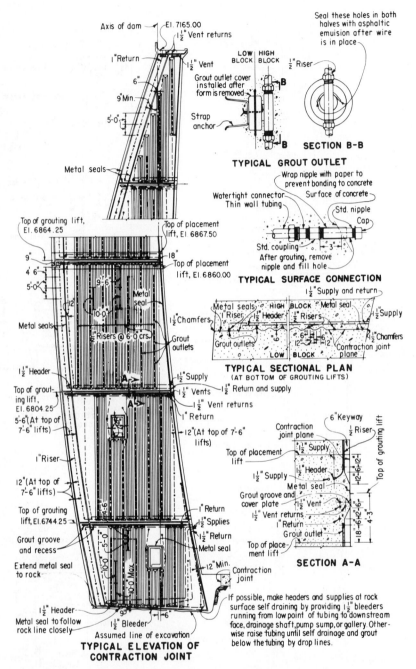

Figure 8-7-6. Transverse contraction joint details from Morrow Point Dam. (*Courtesy Bureau of Reclamation*)

DESIGN OF CONCRETE DAMS 479

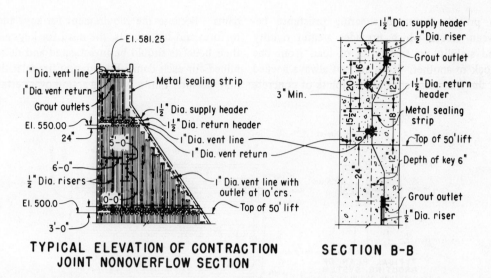

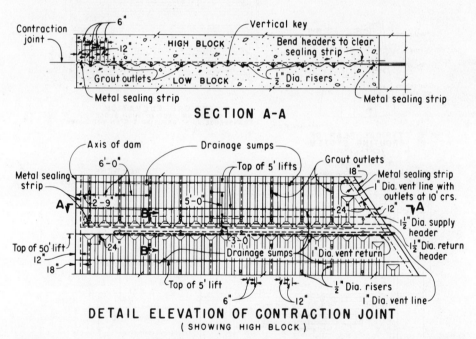

Figure 8-7-7. Transverse contraction joint details from Friant Dam. (*Courtesy Bureau of Reclamation*)

should be made to keep the spacing uniform throughout the dam.

The practice of spacing longitudinal joints follows, in general, that for the transverse joints, except that the lengths of the blocks are not limited by plant capacity. Depending on the degree to which artificial temperature control is exercised, spacings of 50 to 200 ft may be employed.

Keys in contraction joints are used primarily

480 HANDBOOK OF DAM ENGINEERING

to provide increased shearing resistance between blocks which provides greater rigidity and stability in transferral of load from one block to another. Keys are not always needed in the transverse contraction joints of concrete dams. Because the requirement for keys adds to form and labor costs, the need for keys and their benefits should be investigated and determined for each dam. Foundation irregularities may be such that a bridging action over certain

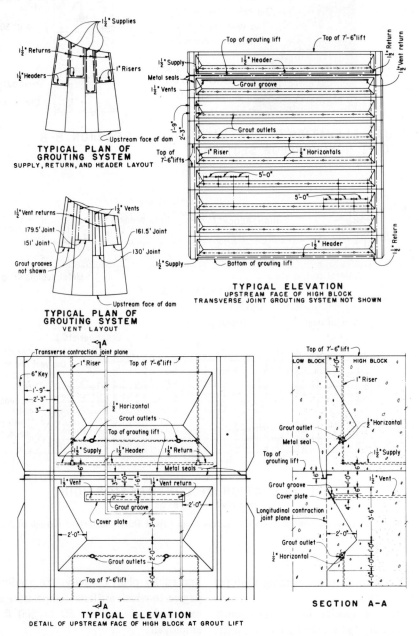

Figure 8-7-8. Longitudinal contraction joint details from Glen Canyon Dam. (*Courtesy Bureau of Reclamation*)

portions of the foundation would be desirable. Keys can be used to lock together adjacent blocks to help accomplish this bridging action.

Shear keys are important accessories in longitudinal contraction joints and are provided to maintain stability of the dam by increasing the resistance to vertical shear. The key faces are inclined to make them conform approximately with the lines of principal stress for full waterload. Details of the shape and dimensions of keys used on Glen Canyon Dam are shown on Fig. 8-7-8.

In some cases, a construction joint can be utilized to reduce the tension stresses caused by the volumetric change due to the heat of hydration.

By placing a portion of the structure and permitting the heat of hydration to dissipate before placing the adjacent portion, the tensile stresses due to the heat of hydration are those associated with the portion of the structure rather than the entire structure. An example of the use of the construction joint is placing the stem of a massive head buttress dam, and allowing it to cool before placing the massive head.

Lift lines in the dam, which are a special construction joint, permit loss of heat to the surface thereby reducing the peak temperatures. Shallower lifts permit a more rapid dissipation of heat to the surface and consequently experience a smaller temperature rise.

Expansion joints are provided in concrete structures primarily to accommodate volumetric change due to temperature rise. In addition, these joints frequently are installed to prevent transfer of stress from one structure to another. Notable examples are (1) powerplants constructed adjacent to the toe of the dam, wherein the powerplant and the mass of the dam are separated by a vertical expansion joint; and (2) outlet conduits encased in concrete and extending downstream from the dam, in which case an expansion joint is constructed near the toe of the dam separating the encasement concrete from the dam.

Like contraction joints, expansion joints are constructed so that no bond exists between the adjacent concrete structures. A corkboard, mastic, sponge rubber, or other compressible-type filler usually separates the joint surfaces to prevent stress or load transferral. The thickness of the compressible material depends on the magnitude of the anticipated deformation induced by the load.

VIII. OPENINGS IN DAMS

Openings in dams can be divided into two categories: the gallery system and waterways. Gallery systems provide access to the interior of the dam for inspection of the dam, for operation of equipment controls located in the dam, and for maintenance of the dam and equipment.

Waterways, such as spillways, outlets, and penstocks can usually be incorporated within concrete dams. Their effect on the design of the dam will be discussed here.

A. Gallery Systems

The gallery systems are comprised of the various vaults, galleries, adit, shafts, and other access openings, provided in the dam. Galleries are formed in the dams for a variety of reasons.[6,33] Some of the more common are to provide:

(1) Drainage for water percolating through the upstream face or seeping through the foundation.
(2) Space for drilling and grouting the foundation.
(3) Space for headers and equipment used in artificially cooling the concrete blocks and grouting contraction joints.
(4) Access to the interior of the structure for observing its behavior after completion.
(5) Access to, and room for, mechanical and electrical equipment such as that used for the operation of gates in the spillways and outlet works.
(6) Access through the dam for control cables and/or power cables.
(7) Access routes for visitors.

Typical gallery layouts are shown in Figs. 8-8-1 and 8-8-2. Because of space limitations, galleries are not usually formed in very thin arch dams and low head dams such as diversion dams.

Galleries are usually rectangular, 5 ft wide by 7½ ft high with a 12-in.-wide gutter along the upstream face for drainage. The 4-ft width is a comfortable width for walking and the 7½-ft height corresponds with the 7½-ft placement lift in mass concrete. Experience has shown that this size of gallery will provide adequate work area and access for equipment for normal maintenance except where special equipment is required such as at gate chambers. Galleries as narrow as 2 ft have been used in thin arch dams; however, a minimum of 3 ft is recommended.

To protect against a leakage crack developing between the upstream face of the dam and

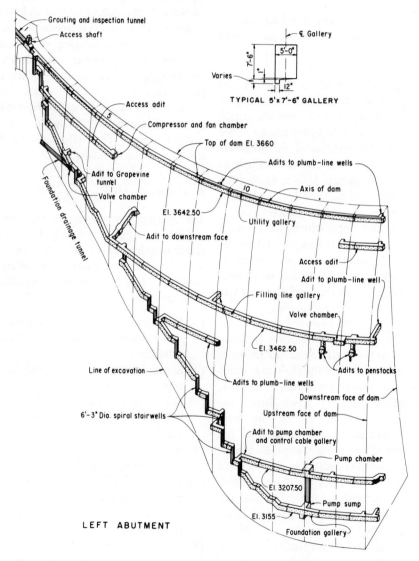

Figure 8-8-1. Gallery layout for Yellowtail Dam. (*Courtesy Bureau of Reclamation*)

DESIGN OF CONCRETE DAMS 483

a gallery, the minimum distance between the gallery and the upstream face of the dam is usually 5 percent of the reservoir head. A minimum of 5 ft clear distance should be used between galleries and the faces of the dam and contraction joints to allow room for placement of mass concrete and to minimize stress concentrations.

Chambers or vaults may be formed in the dam to provide rooms for mechanical and electrical equipment. For example, gate chambers may be used in connection with spillways and outlets to facilitate the installation of gates and valves associated with these features. Transformer chambers may be required to provide room for transformers supplying power for lighting and operation of equipment in the dam.

Elevators are installed in dams to provide access between the top of the dam and the

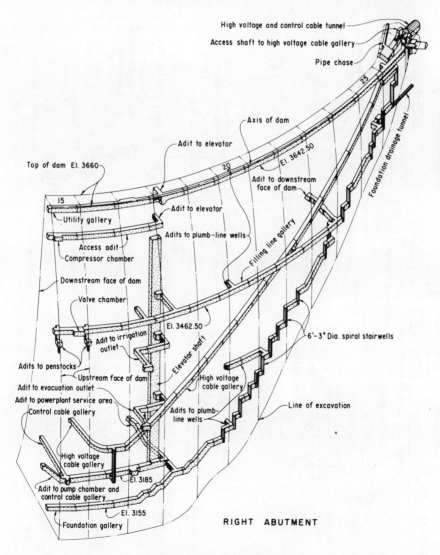

Figure 8-8-1. (*Continued*).

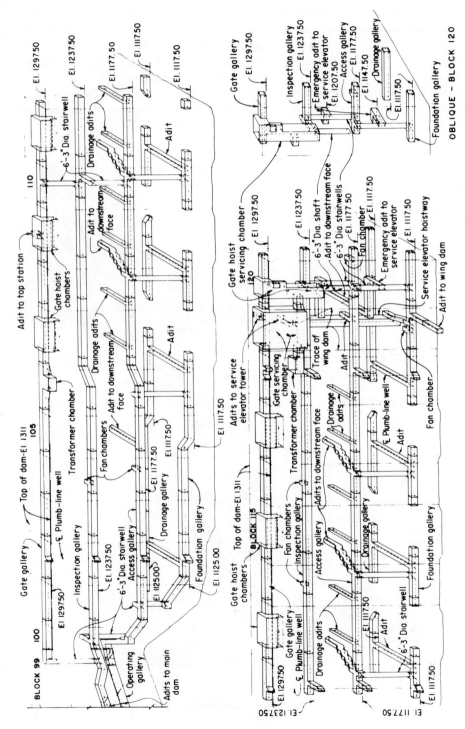

Figure 8-8-2. Gallery system for Grand Coulee Forebay Dam. *(Courtesy Bureau of Reclamation)*

gallery system, equipment and control chambers, and powerplant when adjacent to the dam. The elevator shaft is formed within the mass concrete, with a tower at the crest of the dam. The shaft has connecting adits which provide access to the various galleries and other openings. Stairways and/or emergency adits to the gallery system should be incorporated between elevator stops to provide an emergency exit.

The tower provides a sheltered entrance at the top of the dam and houses the elevator operating machinery and equipment. The tower may be designed to provide space for utilities, storage, and offices.

In general, reinforcement is not required around galleries in a dam. However, it may be required around large openings, openings whose configuration produces high tensile stress concentrations, and openings which are located in areas where the surrounding concrete is in tension due to loads on the dam or temperature or shrinkage. Reinforcement may also be required where conditions are such that a crack would begin at the gallery and propagate through the dam to the reservoir.

Stresses around openings can be determined using the finite element method for various loading assumptions such as dam stresses, temperature, and shrinkage loads. Reinforcement is usually not required if the magnitudes of tensile stresses in the concrete around the opening are less than 10 percent of the compressive strength of the concrete.

In areas of high stress or where the stresses are such that a crack once started could propagate, reinforcement should be used. If unreinforced, such a crack could propagate to the surface where it would be unsightly and/or admit water to the gallery. It could also threaten the structure's safety. The stresses determined by the finite element analysis can be used to determine the amount of reinforcement required around the opening to control the cracking.

In some cases, reshaping or relocating the gallery can reduce or eliminate the tensile stresses.

Five-inch-diameter drains are formed in the mass concrete to intercept water which may be seeping into the dam along joints or through the concrete. By intercepting the water, the drains minimize the hydrostatic pressure which could develop within the dam. They also minimize the amount of water that could leak through the dam to the downstream face where it would create an unsightly appearance.

The drains are usually about 10 ft from the upstream face and are parallel to it. They are spaced at approximately 10-ft centers along the axis of the dam. The lower ends of the drains are taken to the gallery or connected to the downstream face near the fillet through a horizontal drain pipe or header system if there are no galleries. The tops of the drains are usually located in the crest of the dam to facilitate cleaning when required. Where the top of the dam is thin, the drains may be terminated at about the level of the normal reservoir water surface. A 1½ in. pipe then connects the top of the drain with the crest of the dam and can be used to flush the drains.

B. Waterways

Waterways through the dam are often the most economical. They can be accommodated provided stress concentrations around the openings are not excessive and the integrity of the dam is maintained. Where feasible, the openings should be designed to minimize the stress concentrations around the openings.

Reinforcement and/or metal liners are usually placed around the openings. It is important to prevent cracks from forming which would allow water from the waterways to percolate into the interior of the dam increasing pore pressure. The reinforcement around the openings should be designed to withstand the stresses due to the stresses in the dam in addition to those resulting from the waterway itself.

IX. INSTRUMENTATION

Instruments should be installed at selected locations throughout the dam and its foundation so that measurements can be taken to determine the structural behavior of the dam and foundation. Of primary importance is the obtaining of information by which the structural safety of the dam can be determined. Of secondary importance is the use of the information obtained to provide better criteria for use in the design of future concrete dams. Instruments can be used to determine temperature, strain, stress, hydrostatic pore pressure, and deformation for loading conditions on the dam at the time the instrument measurements are taken. Results of these measurements taken at a particular time can be compared with corresponding analytical values.

Two principal methods of measurement are used to obtain the necessary information. The first method involves several types of instruments that are embedded in the mass concrete to measure the behavior of the dam and appurtenances. These instruments indicate strain, stress, contraction joint opening, temperature, and pore pressure. Instruments are also installed at the rock-concrete contact of the abutments and the base of the dam for determining deformation of the foundation.

The second method involves several types of precise surveying instruments. Measurements are made on targets on the downstream face and on the top of a dam, through galleries and vertical wells in the dam, in tunnels in the dam abutments, and in tape gage deformation wells in the foundation. The measurements are used to determine the deformation of the dam and its foundation. Figures 8-9-1, 8-9-2, and 8-9-3 show typical installations of instruments for concrete dams.

Instruments should be located so that the data obtained can be used to make comparisons with the structural analysis of the dam. For example, strain meters should be located to correspond to points where the stresses were

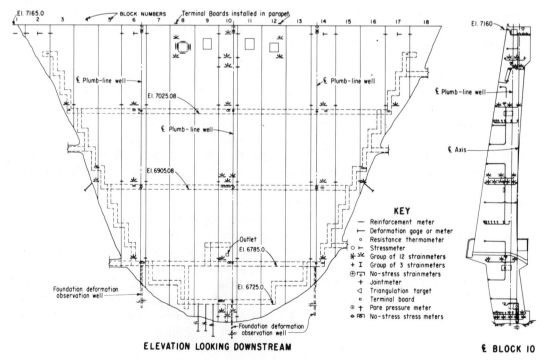

Figure 8-9-1. Instrument layout for Morrow Point Dam. (*Courtesy Bureau of Reclamation*)

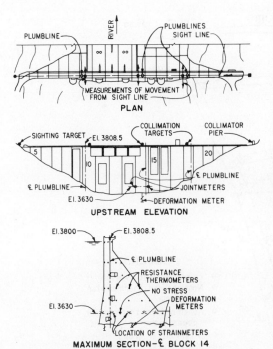

Figure 8-9-2. Instrument layout for Canyon Ferry Dam. (*Courtesy Bureau of Reclamation*)

computed in the design analysis. Locations of maximum stresses, as predicted by the structural analysis, should be instrumented. It should also be recognized that design assumptions may not accurately predict the behavior in the completed dam and enough instrumentation should be provided so that a complete answer may be obtained and not a partial answer limited by the number of points studied. Deformation meters and survey points should be located where the measured deformation can be compared with anticipated deformation.

The number of instruments will vary with size and complexity of the structure, the larger and more complex structures requiring more instruments. Arch dams, because of their more complex configuration, usually require more instruments than gravity and buttress dams.

A sufficient number of instruments should be installed at a location to obtain the stress distribution across the dam. Similarly, sufficient locations should be instrumented to obtain stress patterns for the dam. Of particular concern would be areas of anticipated stress concentration. These areas should be instrumented.

Temperature changes in massive concrete structures cause changes in stress which influence the behavior of the dam. Thus it is important that the temperature history of the concrete be obtained for use when comparing measured stress and deformations with those obtained by analytical methods. Thermometers at the upstream face can also be used to determine the temperature stratification of the reservoir.

Layouts of instrumentation are discussed in detail in other publications.[6-33]

A. Embedded Instruments

Embedded instruments used to obtain measurements in a concrete dam may be selected from several types presently available on the commercial market. One type, the Carlson elastic wire instrument shown in Fig. 8-9-4, is commonly used in the United States and is available in patterns suited to most purposes.[50] They are dual-purpose instruments and measure temperature as well as the function for which they were designed. The instrument takes advantage of the fact that the electrical resistance of steel wire varies directly with the temperature and with the tension in the wire. These instruments have been proved reliable and stable for measurements extending over long periods of time. Installations have been made in many Bureau of Reclamation dams and experience with the instruments covers a period of many years. Their operation, description, and the manner in which they are installed appear in other publications.[51,52,53]

The vibrating wire instrument, Fig. 8-9-5, is in common use in European countries. This instrument makes use of the fact that an increase in tension increases the frequency of vibration of the wire when plucked. The frequency is measured by use of a magnetic circuit.

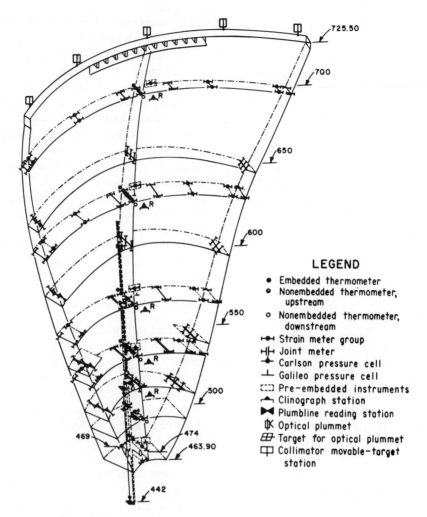

Figure 8-9-3. Instrument layout for Viant Dam. (*Courtesy SADE*)

Embedded instruments are usually constructed so that measurements can be made by electrical readouts. For convenience, cable leads are usually taken to terminal boards and instrument measurements from an area of the dam are made at one location.

Groups of 12 strain meters are usually installed for determining the stress at a particular location. Eleven meters are supported by a holding device or spider and installed in a cluster as shown on Fig. 8-9-6. The twelfth strain meter is placed vertically adjacent to the

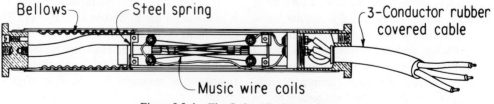

Figure 8-9-4. The Carlson Resistance Meter.

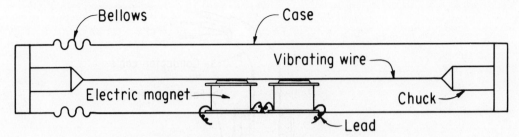

Figure 8-9-5. Schematic diagram of vibrating wire instrument.

cluster. Nine meters provide complete information on strain for any direction in the mass. Three meters are duplicates and provide backup and a check on the readings from other meters. In some areas, fewer strain meters may give the information desired at a particular location. For example, if the strain in one direction is all that is desired, one meter, oriented in that direction, will provide the data. A duplicate meter gives the assurance of mutual checking. Varying numbers of meters may be installed at a particular location to provide the data required.

Stress meters, Fig. 8-9-7, are used to determine compressive stresses. They are usually used for special applications such as determining vertical stress at the base for comparison with the results from strain meters. They are also used in arches for determining horizontal compressive stresses parallel to the direction of thrust in the thinner arch elements near the top of a dam.

Joint meters are used to measure contraction joint openings. The meters are similar to the strain meter but because of the greater anticipated movement, the range of the meter is increased and the sensitivity is reduced. The joint meter can be adapted to measure foundation deformation as shown in Fig. 8-9-8. Other applications of the meters may be made as conditions to be investigated require.

For locations where only temperature measurements are desired, resistance thermometers are used. Temperature measurements of a special nature and short duration such as for concrete cooling operations are made with thermocouples.

B. Precise Survey Methods

Measurements made by precise survey methods are used to determine the deformation of the dam and its foundation. These measurements involve plumblines, tangent line collimation,

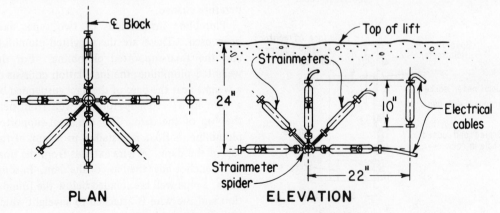

Figure 8-9-6. Layout for group of 12 strain meters.

490 HANDBOOK OF DAM ENGINEERING

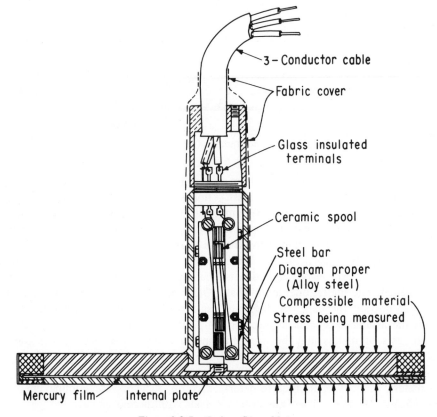

Figure 8-9-7. Carlson Stress Meter.

precise leveling, tape gages, and triangulation of deflection targets on the face of the dam. Over a period of several years, results from these measurements show the range of deformation of a structure during the cyclic loading conditions of temperature and water to which the dam is subjected.

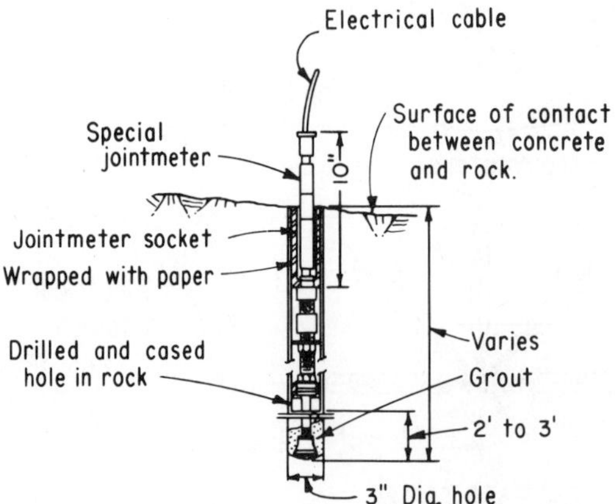

Plumblines are a convenient and relatively simple way to measure the manner in which a dam deforms due to the water load and temperature change.

Plumbline installations of two types have been used. These are the weighted plumbline and the float-supported plumbline. For the weighted plumbline, the installation consists of a weight near the base of the dam suspended by a wire in a vertically formed well from near the top of the dam. For the float-supported plumbline, a float is installed in a tank at the top of the dam. A wire extends from the float to an anchor near the base of the dam. In some cases, a pipe well is extended below the foundation and the wire is attached to a weight which is lowered into the well.

Figure 8-9-8. Foundation deformation meter.

DESIGN OF CONCRETE DAMS 491

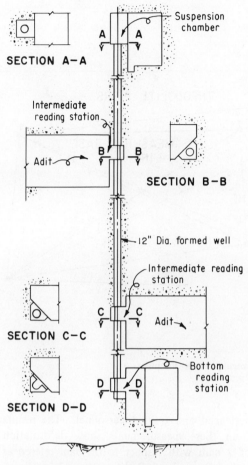

Figure 8-9-9. Typical plumb line well.

Wherever feasible, reading stations are located at various elevations to measure the deflected position of the dam over the full height of the structure. A typical well with reading stations is shown on Fig. 8-9-9. In some large arch dams with a curved upstream face, it is necessary to use more than one plumbline at the same section because of the curvature of the dam. By summing the deflections of the plumblines, total deformations may be determined.

Measurements are made with a micrometer slide device having either a peep sight or a microscope located in the reading stations. The measured movements indicate deformation of the structure with respect to the fixed end of the plumbline.

Triangulation, leveling, and collimation measurements are used to determine deformation of the dam with respect to off-dam reference points. Collimation measurements are made at the top of the dam with a theodolite or jig-transit. An instrument pier is constructed on one abutment and at a higher elevation than the dam. A reference target is located on the opposite abutment at about the same elevation. They are located so that the line of sight between them passes through locations on the top of the dam where measurements are to be made. On arch dams, because of their curvature, more than one target and/or pier might be required. By making the line of sight normal to the desired measurements, the deformation can be measured directly. A movable measuring target is located on the dam. Differences in the position of the movable target from the line of sight indicate the

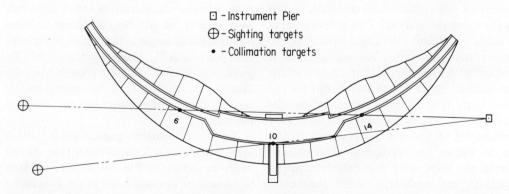

Figure 8-9-10. Layout for collimation measurements—Morrow Point Dam. (*Courtesy Bureau of Reclamation*)

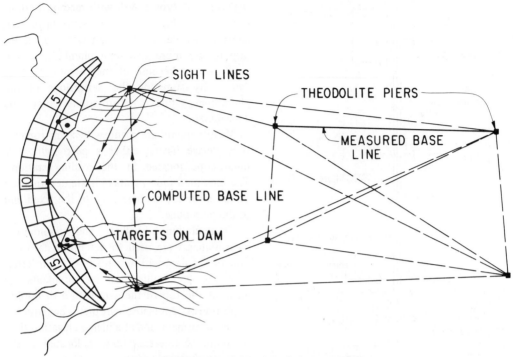

Figure 8-9-11. Layout for triangulation measurement.

deformation. Usually three or four locations on the dam are measured. The results are correlated with plumbline measurements to provide data for charting the behavior of the structure. A typical layout for a collimation system is shown on Figs. 8-9-2 and 8-9-10.

A more elaborate installation is a system of triangulation targets on the face of the dam. Targets are usually located so that data can be correlated with other instrumentation as well as the structural analysis. This system requires a net of instrument piers and a baseline downstream from the dam. The configuration is laid out to provide the greatest strength of the geometrical figures and to afford sight lines to each target from as many instrument piers as is feasible.[54] The nature of the terrain and the topography of the area are governing factors in the size of the net layout. The measurements are made using first-order equipment, methods, and procedures which require experienced and trained personnel. The results from these measurements show deformation of a dam with respect to off-dam references and deformation of the canyon downstream from a dam in the streamwise and cross-stream directions. The layout of a system and locations of items of equipment are shown on Fig. 8-9-11.

Leveling measurements are used to determine vertical displacements of a structure with respect to off-dam references. These measurements employ first-order equipment and procedures.[55] Base references for the measurements should be far enough from the dam to assure that they are unaffected by vertical displacement caused by the dam and reservoir.

Tape deformation gages, which utilize invar-type tapes and micro-meter-type reading heads as shown on Fig. 8-9-12, are installed in tunnels or in cased holes in the foundation

DESIGN OF CONCRETE DAMS 493

C. Uplift Pressure Measurements

A system of piping is usually installed in several blocks at the contact between the foundation rock and the concrete of the dam as shown on Fig. 8-9-13. The piping is installed to determine the hydraulic pressures present at the base of the dam due to percolation or seepage of water along underlying foundation seams or joint systems. Measured values of uplift pressure also may indicate the effectiveness of foundation grouting and drainage.

Uplift pressures are determined by pressure gages or by soundings. When a pipe is under pressure, the pressure is measured by a Bourdon-type pressure gage attached through a gage cock to the pipe. The gages can be calibrated to measure the pressure in any units desired. When there is no pressure in the pipe, the water level is determined by sounding.

Porous-tube piezometers, Fig. 8-9-14, can also be used to measure the uplift pressures. The piezometers can be located at the contact between the dam and foundation or at selected locations throughout the foundation. The plastic tubing from the porous-tube piezometer runs to a convenient location where they can be probed to determine the level of water in the tubing. They are particularly useful in small dams such as diversion dams. Figure 8-9-14 shows a typical layout for a diversion dam.

Another system for measuring uplift pressure is to install pore pressure cells at the locations to be investigated. Electrical cables may be routed from the cell locations to appropriate reading stations where measurements can be obtained. The installation of pore pressure cells is particularly applicable to installations beneath concrete apron slabs. A typical pressure cell installation is shown on Fig. 8-9-15.

Pressure cells can also be installed at selected locations within the dam to determine the concrete pore pressures. Porous-tube piezometers may also be utilized by installing them in holes drilled in the concrete.

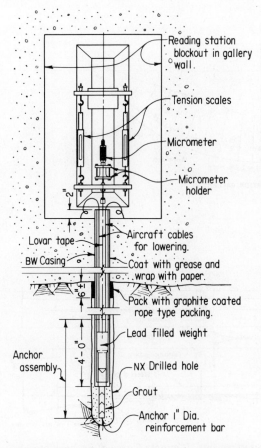

Figure 8-9-12. Typical tape deformation gage installation.

of the dam. These gages extend beyond the contact between the rock and the concrete. They show length change over their depths into the rock in the same manner as the deformation meter shows the amount of deformation caused by the weight of the dam and by the loading on the dam. Reading heads are located at selected intervals along the invar tape so that the deformation of various portions can be determined. If cased holes are used, several holes of different depths are used and by comparing the deformations for the various depth holes. Deformation for the various portions of the foundation can then be determined.

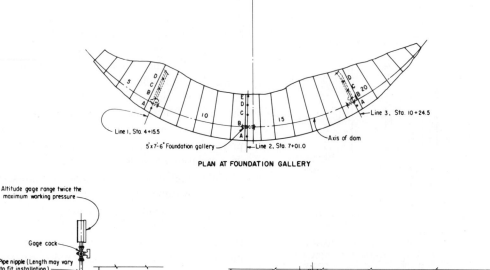

Figure 8-9-13. Layout of piping system for uplift measurements. (*Courtesy Bureau of Reclamation*)

D. Other Instruments

At some large dams in Europe, devices are used to measure the change in angle or rotation of reference axes with respect to horizontal or vertical planes. These measurements are made at selected locations in the galleries. Instruments employed for measuring rotations are called clinometers and inclinometers. Clinometers, which measure changes in angles relative to horizontal planes, are more widely used than inclinometers.

Clinometers have two basic components, a precision level and a rigid frame about 1 m in length. The level, mounted at midspan on the frame, is pivoted at one end and fitted with a micrometer screw at the other. Measurements of changes in rotation are made by supporting each end of the frame on reference points attached to supports which are embedded in the dam, and then centering the bubble by means of the micrometer screw.

Inclinometers make use of a pendulum and measure the angle of rotation from the vertical. The pendulum is attached to a wall by means of a bracket. Strain meters are installed to measure the strain developed as the pendulum tries to remain vertical and the bracket rotates with the dam. The strain is a measure of the rotation.

DESIGN OF CONCRETE DAMS 495

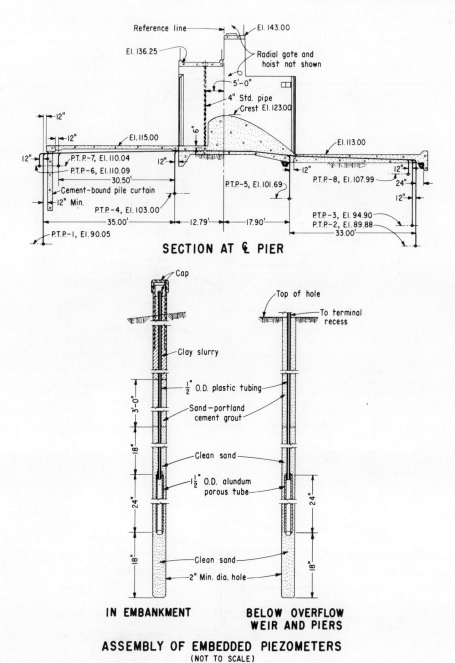

Figure 8-9-14. Layout of porous tube piezometers for uplift measurements. (*Courtesy Bureau of Reclamation*)

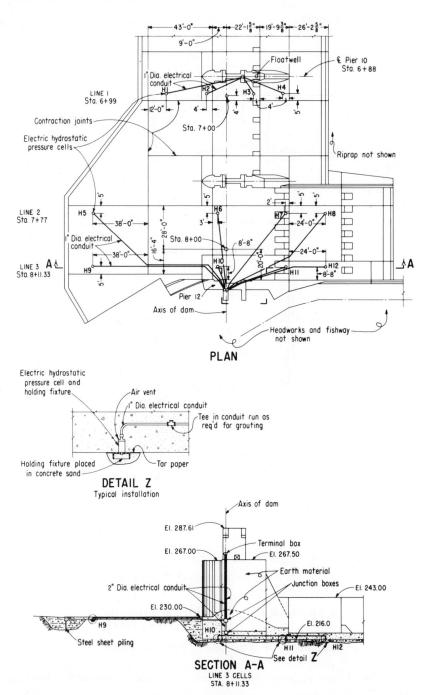

Figure 8-9-15. Layout of pressure cells for uplift measurements. (*Courtesy Bureau of Reclamation*)

REFERENCES

1. Monfore, G. E., and Taylor, F. W., "The Problem of an Expanding Ice Sheet" Bureau of Reclamation Memorandum, March 18, 1948.
2. Boggs, H. L. et al., "Method for Estimating Design Earthquake Rock Motions" Engineering and Research Center, Bureau of Reclamation, Denver, Colorado, November 1972.
3. "Guide for Preliminary Design of Arch Dams," Engineering Monograph No. 36, 1966, Bureau of Reclamation, Denver Federal Center, Denver, Colorado 80225.
4. Glover, R. E., "Fundamentals of the Trial-Load Method for the Design of Arch Dams," presented to faculty of the Graduate College in the University of Nebraska, April 30, 1936 in partial fulfillment of requirements for the professional degree of Civil Engineer.
5. Howell, C. H. and Jaquith, A. C., "Analysis of Arch Dams by Trial-Load Method," ASCE Proceedings, January 1928 (Trans. ASCE, Vol. 93, 1929, p. 1191).
6. Bureau of Reclamation, "Design of Arch Dams," 1976.
7. The Engineering Foundation, "Report on Arch Dam Investigations," Vols. I and III, 1927 and 1933.
8. Larned, A. I., and Merrill, V. S., "Actual Deflections and Temperatures in a Trial-Load Arch Dam," Trans. ASCE, Vol. 99, 1934, pp. 897-961.
9. Bureau of Reclamation, "Model Tests of Boulder Dam," Boulder Canyon Project Final Reports, Part V, Bulletin 3.
10. Bureau of Reclamation, "Comparison of Analytical and Structural Behavior Results for Flaming Gorge Dams," Research Report No. 14, 1968.
11. Kramer, M. A., and Jones, K., "Comparison of Analytical and Structural Behavior Results for Morrow Point Dam," REC-ERC-72-8, Bureau of Reclamation.
12. Love, A. E. H., "Mathematical Theory of Elasticity," fourth edition, 1927, par 118.
13. Clough, R. W., "The Finite Element Method in Plane Stress Analysis," ASCE Conference Papers (Second Conference on Electronic Computation, September 1960).
14. Zienkiewicz, O. C., "The Finite Element in Structural and Continuum Mechanics," McGraw-Hill, London, 1967.
15. The computer program 220-PLNSTR, contained in BRECS (Bureau of Reclamation Engineering Computer System).
16. The computer program 220-ADAP, contained in BRECS (Bureau of Reclamation Engineering Computer System).
17. Wilson, E. L. and Nickell, R. E., "Application of the Finite Element Method of Head Conduction Analysis," Nuclear Engineering and Design 4 (1966), 276-286, North Holland Publishing Company, Amsterdam.
18. 220-HEATFL, contained in BRECS (Bureau of Reclamation Engineering Computer System).
19. "Control of Cracking in Mass Concrete Structures," Engineering Monograph No. 34, Bureau of Reclamation, 1965, Denver, Colorado.
20. Kinley, Frederick B., "Refrigeration for Cooling Concrete Mix," Air Conditioning, Heating, and Ventilating, March 1955.
21. Myklestad, N. O., Vibration Analysis, McGraw-Hill Book Company, Inc., New York, New York, 1944.
22. Earthquake Response of Structure, by Ray W. Clough, Chapter 12, "Earthquake Engineering," Prentice-Hall, Englewood Cliffs, New Jersey, 1970.
23. "Earthquake Engineering for Concrete and Steel Structures," Proceedings of a conference with Ray W. Clough and Bureau of Reclamation staff members, Denver, Colorado, March 1963.
24. Fadeev, D. K. and Faddeeva, V. N., "Computational Methods of Linear Algebra," W. H. Freeman and Company, San Francisco and London, 1963.
25. Westergaard, H. M., "Water Pressures on Dams During Earthquakes," Transactions, American Society of Civil Engineers, Vol. 98, 1933.
26. Rouse, George C. and Boukamp, Jack G., "Vibration Studies of Monticello Dam," A Water Resources Technical Publication Research Report No. 9, Bureau of Reclamation, 1967.
27. Keightly, W. O., Housner, G. W., and Hudson, D. E., "Vibration Tests of Encino Dam Intake Tower," Engineering Research Laboratory, California Institute of Technology, Pasadena, California, July 1961.
28. Cozart, C. W., "The Response of an Intake Tower at Hoover Dam to Earthquakes," Bureau of Reclamation Report REC-ERC-71-50, Denver, Colorado, 1971.
29. Dungar, R. and Severn, R. T., "Dynamic Analysis of Arch Dams," Paper No. 7, Symposium on Arch Dams, Institution of Civil Engineers, March 1968.
30. Londe, P., "Une Methode d'Analyze o'trois dimensions de la stabilite d'une reve rocheme," Annls Ponts Chaus. No. 1 37-60, 1965.
31. Guzina, Bosko and Tucovic, Ignjat, "Determining the Maximum Three Dimensional Stability of a Rock Wedge," Water Power, October 1969.
32. Mahtab, M. A. and Goodman, R. E., "Three-Dimensional Finite Element Analysis of Jointed Rock Slopes," Final Report to U.S. Bureau of

Reclamation, Contract No. 14-06-D-6639, December 31, 1969.
33. Bureau of Reclamation, "Design of Gravity Dams," 1976.
34. Chopra, A. K., and Chakrabarti, P., "A Computer Solution for Earthquake Analysis of Dams," Report No. EERC70-5, Earthquake Engineering Research Center, University of California, Berkeley, California, 1970.
35. Design Standards No. 2, Treatise on Dams, Chapter 11, Buttress Dams, USBR.
36. Handbook of Applied Hydraulics, C. V. Davis, 1942, Section 6, p. 191.
37. Engineering for Dams, W. P. Creager, J. D. Justin, and Julian Hinds, 1944, Vol. II, Chapter 14, Buttressed Concrete Dams, p. 558.
38. "Field Permeability Tests in Boreholes," Earth Manual, Bureau of Reclamation.
39. "Drill Hole Water Tests Technical Instructions," report, Bureau of Reclamation, 1972.
40. "Morrow Point Dam Shear and Sliding Friction Tests," Concrete Laboratory Report No. C-1161, Bureau of Reclamation.
41. Wallace, G. B., Slebir, E. J., and Anderson, F. A., "Radial Jacking Tests for Arch Dam," Tenth Rock Mechanics Symposium, University of Texas, Austin, Texas, 1968.
42. Wallace, G. B., Slebir, E. J., and Anderson, F. A., "In Situ Methods for Determining Deformation Modulus Used by the Bureau of Reclamation," American Society for Testing and Materials, Denver, Colorado, 1969.
43. Wallace, G. B., Slebir, E. J., and Anderson, F. A., "Foundation Testing for Auburn Dam," Eleventh Symposium on Rock Mechanics, University of California, Berkeley, California, 1969.
44. Khosla, "Design of Dams on Permeable Foundations," Central Board of Irrigation, India, September 1936.
45. "Design of Small Dams," United States Department of the Interior, Bureau of Reclamation, U.S. Government Printing Office, Washington, D.C. 20402, 1973.
46. "Control of cracking in concrete structures" ACI Committee 224, ACI Journal Procedures Vol. 69, Dec. 1972, pp. 717–753.
47. "Control of cracking in mass concrete structures," Engineering Monograph No. 34, Water Resources Technical Publication, Bureau of Reclamation, 1965.
48. Concrete Manual, Seventh Edition, 1963, U.S. Bureau of Reclamation.
49. ACI Manual of Concrete Practice, 1973, Part 1.
50. Technical Bulletin Series, Bulletins 16 through 23, Terrametrics Division of Earth Sciences, Teledyne Co., Golden, Colo., 1972.
51. Raphael, J. M., and Carlson, R. H., "Measurement of Structural Action in Dams," James J. Gillick and Co., Berkeley, Calif., 1956.
52. Technical Record of Design and Construction, "Flaming Gorge Dam and Powerplant," Bureau of Reclamation, 1968.
53. Technical Record of Design and Construction, "Glen Canyon Dam and Powerplant," Bureau of Reclamation, December 1970.
54. "Manual of Geodetic Triangulation," Special Publication No. 247, Coast and Geodetic Survey, Department of Commerce, Washington, D.C., 1950.
55. "Manual of Geodetic Leveling," Special Publication No. 239, Coast and Geodetic Survey, Department of Commerce, Washington, D.C., 1948.

9

Design of Spillways and Outlet Works

Abstracted (with permission of U.S. Bureau of Reclamation) from "Design of Small Dams," Second Edition 1973.

CARL J. HOFFMAN
Engineer (retired)
Principal author

I. Spillways

A. GENERAL

1. Function of Spillways

Spillways are provided for storage and detention dams to release surplus or flood water which cannot be contained in the allotted storage space, and at diversion dams to bypass flow exceeding those which are turned into the diversion system. Ordinarily, the excess is drawn from the top of the pool created by the dam and conveyed through an artificial waterway back to the river or to some natural drainage channel.

The importance of a safe spillway cannot be overemphasized; many failures of dams have been caused by improperly designed spillways or by spillways of insufficient capacity. Ample capacity is of paramount importance for earthfill and rockfill dams, which are likely to be destroyed if overtopped, whereas concrete dams may be able to withstand moderate overtopping. Usually, increase in cost is not directly proportional to increase in capacity. Very often the cost of a spillway of ample capacity will be only moderately higher than that of one which is obviously too small.

In addition to providing sufficient capacity, the spillway must be hydraulically and structurally adequate and must be located so that spillway discharges will not erode or undermine the downstream toe of the dam. The spillway's bounding surfaces must be erosion resistant to withstand the high scouring velocities created by the drop from the reservoir surface to tailwater, and usually some device will be required for dissipation of energy at the bottom of the drop.

The frequency of spillway use will be determined by the runoff characteristics of the drainage area and by the nature of the development. Ordinary riverflows are usually stored in the reservoir, diverted through headworks, or released through outlets, and the spillway is not required to function. Spillway flows will result during floods or periods of sustained high runoff when the capacities of other facilities are exceeded. Where large reservoir storage is provided, or where large outlet or diversion capacity is available, the spillway will be utilized infrequently. At diversion dams where storage space is limited and diversions are relatively small compared to normal river flows, the spillway will be used almost constantly.

2. Selection of Inflow Design Flood

(a) General Considerations. When floods occur in an unobstructed stream channel, it is considered a natural event for which no individual or group assumes responsibility. However, when obstructions are placed across the channel, it becomes the responsibility of the sponsors either to make certain that hazards to downstream interests are not appreciably increased or to obligate themselves for damages resulting from operation or failure of such structures. Also, the loss of the facility and the loss of project revenue occasioned by a failure should be considered.

If danger to the structures alone were involved, the sponsors of many projects would prefer to rely on the improbability of an extreme flood occurrence rather than to incur the expense necessary to assure complete safety. However, when the risks involve downstream interests, including widespread damage and loss of life, a conservative attitude is required in the development of the inflow design flood. Consideration of potential damage should not be confined to conditions existing at the time of construction. Probable future development in the downstream flood plain, encroachment by farms and resorts, construction of roads and bridges, etc., should be evaluated in estimating damages and hazards to human life that would result from failure of a dam.

Dams impounding large reservoirs and built on principal rivers with high runoff potential unquestionably can be considered to be in the high-hazard category. For such developments, conservative design criteria are selected on the basis that failure cannot be tolerated because of the potential damages which could approach disaster proportions. However, small dams built on isolated streams in rural areas where failure would neither jeopardize human life nor create damages beyond the sponsor's financial capabilities can be considered to be in a low-hazard category. For such developments design criteria may be established on a much less conservative basis. There are numerous instances, however, where failure of dams of low heights and small storage capacities has resulted in loss of life and heavy property damage. Most small dams will require a reasonable conservatism in design, because a dam failure must not present a serious hazard to human life.

(b) Inflow Design Flood Hydrographs. Chapter 2 discusses the determination of floodflows which may be used as inflow design floods. The procedures presented permit the derivation of inflow flood hydrographs. (See 2-11 of Chapter 2.)

3. Relation of Surcharge Storage to Spillway Capacity

Streamflow is normally represented in the form of a hydrograph, which charts the rate of flow in relation to time. A typical hydrograph representing a storm runoff is illustrated in Fig. 2-16 of Chapter 2. The flow into a reservoir at any time and the momentary peak can be read from the curve. The area under the curve is the volume of the inflow, because it represents the product of rate of flow and time.

Where no storage is impounded by a dam, the spillway must be sufficiently large to pass the peak of the flood. The peak rate of inflow is then of primary interest and the total volume in the flood is of lesser importance. However, where a relatively large storage capacity above normal reservoir level can be made available economically by a higher dam, a portion of the flood volume can be retained temporarily in reservoir surcharge space and the spillway capacity can be reduced considerably. If a dam could be made sufficiently high to provide storage space to impound the entire volume of the flood above normal storage level, theoretically no spillway other than an emergency type would be required, provided the outlet capacity could evacuate the surcharge storage in a reasonable period of time in anticipation of a recurring flood. In such a case the maximum reservoir level would depend entirely on the volume of the flood and the rate of inflow would be of no concern. From a practical standpoint, however, there will be relatively few sites that will permit complete storage of an inflow design flood by surcharge storage. Such sites usually will be off-channel reservoirs; that is, reservoirs which are supplied by canal and which have small tributory drainage areas.

In many reservoir projects, economic considerations will necessitate a design utilizing surcharge. The most economical combination of surcharge storage and spillway capacity requires flood routing studies and economic studies of the costs of spillway-dam combinations, subsequently described. However, in making these studies, consideration must be given to the minimum size spillway which must be provided for safety. The inflow design flood hydrographs determined by the methods given in chapter 2 are for floods resulting from runoff from rain. Normally, such floods will have the highest peak flows but not always the largest volumes. When spillways of small capacities in relation to these inflow design flood peaks are considered, precautions must be taken to insure that the spillway capacity will be sufficient to (1) evacuate surcharge so that the dam will not be overtopped by a recurrent storm, and (2) prevent the surcharge from being kept partially full by a prolonged runoff whose peak, although less than the inflow design flood, exceeds the spillway capacity. To meet these requirements, the minimum spillway capacity should be in accord with the following general criteria:

(1) In the case of snow-fed perennial streams, the spillway capacity should never be less than the peak discharge of record that has resulted from snowmelt runoff. (This value may have to be estimated from a study of records on the stream itself or nearby streams.)

(2) The spillway capacity should provide for evacuation of sufficient surcharge storage space so that in routing a succeeding flood the maximum water surface does not exceed that obtained by routing the inflow design flood. In general, the recurrent storm is assumed to begin 4 days after the time of peak outflow obtained in routing the inflow design flood.

(3) In regions having an annual rainfall of 40 inches or more, the time interval to the beginning of the recurrent storm in criterion (2) should be reduced to 2 days.

(4) In regions having an annual rainfall of 20 inches or less, the time interval to the beginning of the recurrent storm in criterion (2) may be increased to 7 days.

4. Flood Routing

The accumulation of storage in a reservoir depends on the difference between the rates of inflow and outflow. For an interval of time Δt, this relationship can be expressed by the equation:

$$\Delta S = Q_i \Delta t - Q_o \Delta t$$

where:

ΔS = storage accumulated during Δt,
Q_i = average rate of inflow during Δt, and
Q_o = average rate of outflow during Δt.

The rate of inflow versus time curve is represented by the inflow design flood hydrograph; the rate of outflow is represented by the spillway discharge versus reservoir-elevation curve; and storage is shown by the reservoir storage versus reservoir-elevation curve. For routing studies, the inflow design flood hydrograph is not variable once selection of the inflow design flood has been made. The reservoir storage capacity also is not variable for a given reservoir site, so far as routing studies are concerned. The spillway discharge curve is variable: It depends not only on the size and type of spillway but also on the manner of operating the spillway (and outlets in some instances) to regulate the outflow.

The quantity of water a spillway can discharge depends on the type of the control device. For a simple overflow crest the flow will vary with the head on the crest, and surcharge will increase with an increase in spillway discharge. For a gated spillway, however, outflow can be varied with respect to reservoir head by operation of the gates. For example, one assumption for an operation of a gate-controlled spillway might be that the gates will be regulated so that inflow and outflow are equal until the gates are wide open; or an assumption can be made to open the gates at a slower rate so that surcharge storage will accumulate before the gates open wide.

Outflows need not necessarily be limited to discharges through the spillway but might be supplemented by releases through the outlets. In all such cases the size, type, and method of operation of the spillway and outlets with reference to the storages or to the inflow must be predetermined in order to establish an outflow-elevation relationship.

If equations could be established for the inflow design flood hydrograph curve, the spillway discharge curve (as may be modified by operational procedures), and the reservoir storage curve, a solution of flood routing could be made by mathematical integration. However, simple equations cannot be written for the flood hydrograph curve and the reservoir storage curve, and such a solution is not practical. Many techniques of flood routing have been devised, each with its advantages and disadvantages. These techniques vary from a strictly arithmetical integration method to an entirely graphical solution. Mechanical and electronic routing machines have been developed, and digital computers have been employed.

A rough approximation of the relationship of spillway size to surcharge volume can be obtained without making an actual flood routing, by drawing an arbitrary outflow-time curve below the inflow hydrograph, such that the outflow peak crosses the declining leg of the hydrograph at a selected spillway discharge; and then measuring the area between the curves. This area, translated to the product of the time and discharge scales, then will represent the surcharge volume.

5. Selection of Spillway Size and Type

(a) General Considerations. In determining the best combination of storage and spillway capacity to accommodate the selected inflow design flood, all pertinent factors of hydrology, hydraulics, design, cost, and damage should be considered. In this connection and when applicable, consideration should be given to such factors as (1) the characteristics of the flood hydrograph; (2) the damages which would result if such a flood occurred without the dam; (3) the damages which would result if such a flood occurred with the dam in place; (4) the damages which would occur if the dam or spillway were breached; (5) effects of various dam and spillway combinations on the probable increase or decrease of damages above or below the dam (as indicated by reservoir backwater curves and tailwater curves); (6) relative costs of increasing the capacity of spillways; and (7) use of combined outlet facilities to serve more than one function, such as control of releases and control or passage of floods. Service outlet releases may be permitted in passing part of the inflow design flood unless such outlets are considered to be unavailable in time of flood.

The outflow characteristics of a spillway depend on the particular device selected to control the discharge. These control facilities may take the form of an overflow weir, an orifice, a tube, or a pipe. Such devices can be unregulated or they can be equipped with gates or valves to regulate the outflow.

After a spillway control of certain dimensions has been selected, the maximum spillway discharge and the maximum reservoir water level can be determined by flood routing. Other components of the spillway can then be proportioned to conform to the required capacity and to the specific site conditions, and a complete layout of the spillway can be established. Cost estimates of the spillway and dam can then be made. Estimates of various combinations of spillway capacity and dam height for an assumed spillway type, and of alternative types of spillways, will provide a basis for selection of the economical spillway type and the optimum relation of spillway capacity to height of dam.

To make such a study requires many flood routings, spillway layouts, and spillway and dam estimates. Even then, the study is not necessarily complete since many other spillway arrangements could be considered. A comprehensive study to determine alternative optimum combinations and minimum costs may not be warranted for the design of some dams. Judgment on the part of the designer would be required to select for study only the combinations which show definite advantages, either in cost or adaptability. For example, although a gated spillway might be slightly cheaper than an ungated spillway, it may be desirable to adopt the latter because of its less complicated construction, its automatic and trouble-free operation, its ability to function without an attendant, and its less costly maintenance. See Fig. 9-1.

(b) Combined Service and Auxiliary Spillways.
Where site conditions are favorable, the possibility of gaining overall economy by utilizing an auxiliary spillway in conjunction with a smaller service-type structure should be considered. In such cases the service spillway should be designed to pass floods likely to occur frequently and the auxiliary spillway control set to operate only after such small floods are ex-

Figure 9-1. Left abutment spillway, Navajo Dam, Colorado. (USBR)

ceeded. In certain instances the outlet works may be made large enough to serve also as a service spillway. Conditions favorable for the adoption of an auxiliary spillway are the existence of a saddle or depression along the rim of the reservoir which leads into a natural waterway, or a gently sloping abutment where an excavated channel can be carried sufficiently beyond the dam or other structures to avoid the possibility of damage to the dam.

Because of the infrequency of use, it is not necessary to design the entire auxiliary spillway for the same degree of safety as required for other structures; however, at least the control portion must be designed to forestall failure, since its breaching would release large flows from the reservoir. For example, concrete lining may be omitted from an auxiliary spillway channel excavated in competent rock. Where the channel is excavated through less competent material, it might be lined but terminated above the river channel with a cantilevered lip rather than extending to a stilling basin at river level. The design of auxiliary spillways is often based on the premise that some damage to portions of the structure from passage of infrequent flows is likely to occur. Minor damage by scour to an unlined channel, by erosion and undermining at the downstream end of the channel, and by creation of an erosion pool downstream from the spillway may be expected. Damage to the main dam would not be involved.

An auxiliary spillway can be designed with a fixed crest control, or it can be stoplogged or gated to increase the capacity without additional surcharge head. "Fuse plug" dikes which are designed to breach and wash out when overtopped often are substituted for some or all of the gates. Their advantage over gates is that, if properly designed, breaching become automatic whenever overtopping occurs; furthermore, they are cheaper to install and to maintain. Since the chance of their failure from overtopping is contingent on the occurrence of infrequent floods, their cost for replacement is too problematical for evaluation. By dividing the dike into short sections of varying height so that they are not all simultaneously overtopped, smaller floods might be passed with the failure of one or several of the sections, with total failure occurring only as the probable maximum flood is approached. The breaching of one section at a time will minimize the flood wave brought about by sudden failure of the dike.

(3) Emergency Spillways. As the name implies, emergency spillways are provided for additional safety should emergencies not contemplated by normal design assumptions arise. Such situations could be the result of an enforced shutdown of the outlet works, a malfunctioning of spillway gates, or the necessity for bypassing the regular spillway because of damage or failure of some part of that structure. An emergency might arise where flood inflows are handled principally by surcharge storage and a recurring flood develops before a previous flood is evacuated by the small service spillway or the outlet works. Emergency spillways would act as auxiliary spillways if a flood greater than the selected inflow design flood occurred.

Under normal reservoir operation, emergency spillways are never required to function. The control crest is, therefore, placed at or above the designed maximum reservoir water surface. The freeboard requirement for the dam is based on a water surface determined by assuming an arbitrary discharge which might result from a possible emergency. Usually, an encroachment on the freeboard provided for the designed maximum water surface is allowed in considering the design of an emergency spillway.

Emergency spillways are provided primarily to avoid an overtopping of the main dam embankment because of an emergency condition. Therefore, to be effective the emergency spillway must offer resistance to erosion greater than does the dam itself. Emergency spillways are often formed by lowering the crest of a dike section below that of the main embankment, by utilizing saddles or depressions along the reservoir rim, or by excavating channels through ridges or abutments. The exit channel of an emergency spillway should be a sufficient dis-

DESIGN OF SPILLWAYS AND OUTLET WORKS 505

Figure 9-2. General view Oroville Dam on Feather River, California. Highest in U.S.A. (DWR)

tance from the dam to preclude damage to the embankment or appurtenances should the spillway operate. See Figs. 9-2 and 9-3.

B. DESCRIPTION OF SERVICE SPILLWAYS

1. Selection of Spillway Layout

A composite design of a spillway can be prepared by properly considering the various factors influencing the spillway size and type, and correlating alternatively selected components. Many combinations of components can be used in forming a complete spillway layout. After the hydraulic size and outflow characteristics of a spillway are determined by routing of the design flood, the general dimensions of the control can be selected. Then, a specific spillway layout can be developed by considering the topography and foundation conditions, and by fitting the control structure and the various components to the prevailing conditions.

Site conditions greatly influence the selection of location, type, and components of a spillway. The steepness of the terrain traversed by the spillway control and discharge channel, the class and amount of excavation and the possibility for its use as embankment material, the chances of scour of the bounding surfaces and the need for lining, the permeability and bearing capacity of the foundation, and the stability of the excavated slopes must all be considered in the selection.

The adoption of a particular size or arrangement for one of the spillway components may influence the selection of other components. For example, a wide control structure with the crest placed normal to the centerline of the spillway would require a long, converging transition to join it to a narrow discharge channel or to a tunnel; a better alternative might be the selection of a narrower gated control arrangement. Similarly, a wide stilling basin may not be feasible for use with a cut-and-cover conduit

Figure 9-3. Oroville Dam spillway in action. (DWR)

or tunnel because of the long, diverging transition needed.

A spillway may be an integral part of a dam such as an overflow section of a concrete dam, or it may be a separate structure. In some instances, it may be combined as a common discharge structure with the outlet works or integrated into the river diversion plan for economy. Thus, the location, type, and size of other appurtenances are factors which may influence the selection of a spillway location or its arrangement. The final plan will be governed by overall economy, hydraulic sufficiency, and structural adequacy.

The components of a spillway and common types of spillways are described and discussed herein. Hydraulic design criteria and procedures are discussed in Chapter 2.

2. Spillway Components

(a) Control Structure. A major component of a spillway is the control device, for those reservoirs where it regulates and controls the outflows. This control limits or prevents outflows below fixed reservoir levels, and it also regulates releases when the reservoir rises above that level. The control structure may consist of a sill, weir, orifice, tube, or pipe. The discharge-head relationship may be fixed as in the case of a simple overflow crest or unregulated port, or it may be variable as with a gated crest or a valve-controlled pipe. The control characteristics of a closed conduit might change with the stage relationship. In a culvert spillway, for example, the entrance will act as a weir for low heads when it is not submerged and as an orifice when submerged. As the amount of submergence increases, the flow will be controlled by the conduit acting as a tube, and finally, for greater submergence, the conduit will flow full and the flow will be governed by pressure pipe characteristics. See Figs. 9-4 and 9-5.

Control structures may take various forms in both positioning and shape. In plan, overflow crests can be straight, curved, semicircular,

DESIGN OF SPILLWAYS AND OUTLET WORKS 507

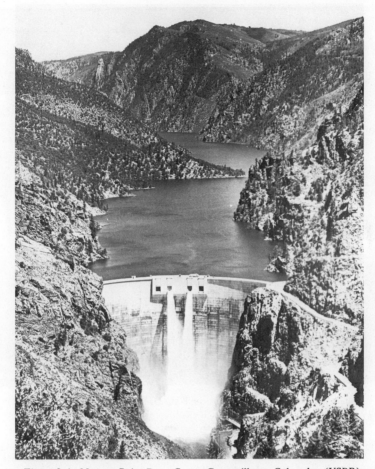

Figure 9-4. Morrow Point Dam, Center Gate spillway, Colorado. (USBR)

U-shaped, or round. Orifice controls can be placed in a horizontal, inclined, or vertical position. Tubes or pipes may be placed vertically, horizontally, or inclined and can be straight, curved, or follow any profile. They can be circular, square, rectangular, horseshoe, or other shape in cross section.

An overflow can be sharp crested, ogee shaped, broad crested, or of varied cross section. Orifices can be sharp edged, round edged, or bellmouth shaped, and can be placed so as to discharge with a fully contracted jet or with a suppressed jet. They may discharge freely or discharge partly or fully submerged. Tubes may have entrance corners which are sharp edged, rounded, or bellmouthed; and they can be of uniform size or be divergent or convergent. Tubes or pipes can operate freely discharging or they can flow full for part or all of their length and can be of uniform or changing size, with the control placed at some point between the inlet and the downstream end. Pipes can be of uniform or changing size, with the control placed either at the downstream end or at some intermediate point along the length. Pipes can flow full under pressure for their entire length or they can flow full and partly full, respectively, above and below their control point.

(b) Discharge Channel. Flow released through the control structure usually is conveyed to the streambed below the dam in a discharge channel or waterway. Exceptions are where the dis-

Figure 9-5. Upstream inlets, Morrow Point Dam spillway, Colorado. (USBR)

charge falls free from an arch dam crest or where the flow is released directly along the abutment hillside to cascade down the abutment face. The conveyance structure may be the downstream face of a concrete dam, an open channel excavated along the ground surface, a closed cut-and-cover conduit placed through or under a dam, or a tunnel excavated through an abutment. The profile may be variably flat or steep; the cross section may be variably rectangular, trapezoidal, circular, or of other shape; and the discharge channel may be wide or narrow, long or short.

Discharge channel dimensions are governed primarily by hydraulic requirements, but the selection of profile, cross-sectional shapes, widths, length, etc., is influenced by the geologic and topographic characteristics of the site. Open channels excavated in the abutment usually follow the ground surface profile; steep canyon walls may make a tunnel desirable. In plan, open channels may be straight or curved, with sides parallel, convergent, divergent, or a combination of these. A closed conduit may consist of a vertical or an inclined shaft leading to a nearly horizontal tunnel through the abutment or to a cut-and-cover conduit under or through the dam. Occasionally a combination of a closed conduit and an open channel might be adopted, such as a culvert under an embankment emptying into an open channel leading down the abutment slope. Discharge channels must be cut through or lined with material which is resistant to the scouring action of the accelerating velocities, and which is structurally adequate to withstand the forces from backfill, uplift, waterloads, etc.

(c) Terminal Structure. When spillway flows fall from reservoir pool level to downstream river level, the static head is converted to kinetic energy. This energy manifests itself in the form of high velocities which if impeded result in large pressures. Means of returning the flow to the river without serious scour or erosion of the toe of the dam or damage to adjacent structures must usually be provided.

In some cases the discharge may be delivered at high velocities directly to the stream where the energy is absorbed along the streambed by impact, turbulence, and friction. Such an arrangement is satisfactory where erosion-resistant bedrock exists at shallow depths in the channel and along the abutments or where the spillway outlet is sufficiently removed from the dam or other appurtenances to avoid damage by scour, undermining, or abutment sloughing. The discharge channel may be terminated well above the streambed level or it may be continued to or below streambed.

Upturned deflectors, cantilevered extensions, or flip buckets can be provided to project the jet some distance downstream from the end of the structure. Often, erosion in the streambed at the point of contact of the jet can be minimized by fanning the jet into a thin sheet by the use of a flaring deflector.

Where severe scour at the point of jet impingement is anticipated, a plunge basin can be excavated in the river channel and the sides and bottom lined with riprap or concrete. For small installations, it may be expedient to perform a minimum of excavation and to permit the flow to erode a natural pool; protective riprapping or concrete lining may be later provided to halt the scour. In such arrangements an adequate cutoff or other protection must be provided at the end of the spillway structure to prevent it from being undermined.

Where serious erosion to the streambed is to be avoided, the high energy of the flow must be dissipated before the discharge is returned to the stream channel. This can be accomplished by the use of an energy dissipating device, such as a hydraulic jump basin, a roller bucket, a sill block apron, a basin incorporating impact baffles and walls, or some similar energy absorber or dissipator.

(d) Entrance and Outlet Channels. Entrance channels serve to draw water from the reservoir and convey it to the control structure. Where a spillway draws water immediately from the reservoir and delivers it directly back into the river, as in the case with an overflow spillway over a concrete dam, entrance and outlet channels are not required. However, in the case of spillways placed through abutments or through saddles or ridges, channels leading to the spillway control and away from the spillway terminal structure may be required.

Entrance velocities should be limited and channel curvatures and transitions should be made gradual, in order to minimize head loss through the channel (which has the effect of reducing the spillway discharge) and to obtain uniformity of flow over the spillway crest. Effects of an uneven distribution of flow in the entrance channel might persist through the spillway structure to the extent that undesirable erosion could result in the downstream river channel. Nonuniformity of head on the crest may also result in a reduction in the discharge.

The approach velocity and depth below crest level have important influence on the discharge over an overflow crest. A greater approach depth with the accompanying reduction in approach velocity will result in a larger discharge coefficient. Thus, for a given head over the crest, a deeper approach will permit a shorter crest length for a given discharge. Within the limits required to secure satisfactory flow conditions and nonscouring velocities, the determination of the relationship of entrance channel depth to channel width is a matter of economics.

Outlet channels convey the spillway flow from the terminal structure to the river channel below the dam. In some instances only a pilot channel is provided, on the assumption that scouring action will enlarge the channel during major spills. Where the channel is in a relatively nonerodible material, it should be excavated to an adequate size to pass the anticipated flow without forming a control which will affect the tailwater stage in the stilling device.

The outlet channel dimensions and its need for protection by lining or riprap will depend on the influences of scour on the tailwater. Although stilling devices are provided, it may be impossible to reduce resultant velocities below the natural velocity in the original stream, and some scouring of the riverbed, therefore, cannot be avoided. Further, under natural conditions the beds of many streams are scoured during the rising stage of a flood and filled during the falling stage by deposition of material carried by the flow. After creation of a reservoir the spillway will normally discharge clear water and the material scoured by the high velocities will not be replaced by deposition. Consequently, there will be a gradual retrogression of the downstream riverbed, which will lower the tailwater stage-discharge relationship. Conversely, scouring where only a pilot channel is provided may build up bars and islands downstream, thereby effecting an aggradation of the downstream river channel which will raise the tailwater elevation with respect to discharges. The dimensions and erosion-protective measures at the outlet channel may be influenced by these considerations.

3. Spillway Types

(a) General. Spillways are ordinarily classified according to their most prominent feature, either as it pertains to the control,to the discharge channel, or to some other component. Spillways often are referred to as controlled or uncontrolled, depending on whether they are gated or ungated. Commonly referred to types are the free overall (straight drop), ogee (overflow), side channel, open channel (trough or chute), conduit, tunnel, drop inlet (shaft or morning glory) baffled apron drop, culvert, and siphon.

(b) Free Overall (Straight Drop) Spillways. A free overfall or straight drop spillway is one in which the flow drops freely from the crest. This type is suited to a thin arch or deck overflow dam or to a crest which has a nearly vertical downstream face. Flows may be free discharging as will be the case with a sharp-crested weir control, or they may be supported along a narrow section of the crest. Occasionally the crest is extended in the form of an overhanging lip to direct small discharges away from the face of the overfall section. In free overfall spillways the underside of the nappe is ventilated sufficiently to prevent a pulsating, fluctuating jet.

Where no artificial protection is provided at the base of the overfall, scour will occur in most streambeds and will form a deep plunge pool. The volume and depth of the hole are related to the range of discharges, the height of the drop, and the depth of tailwater. The erosion-resistant properties of the streambed material including bedrock have little influence on the size of the hole, the only effect being the time necessary to scour the hole to its full depth. Where erosion cannot be tolerated, an artificial pool can be created by constructing an auxiliary dam downstream from the main structure, or by excavating a basin which is then provided with a concrete apron or bucket.

If tailwater depths are sufficient, a hydraulic jump will form when a free overfall jet falls upon a flat apron. It has been demonstrated that the momentum equation for the hydraulic jump may be applied to the flow conditions at the base of the fall to determine the elements of the jump.

A free overfall spillway which will be effective over a wide range of tailwater depths can be designed for use with earthfill dams [1, 2, 3, 4].[2] This type of structure is not adaptable for high drops on yielding foundations, because of the large impact forces which must be absorbed by the apron at the point of impingement of the jet. Vibrations incident to the impact might crack or displace the structure, with danger from failure by piping or undermining. Ordinarily, the use of this structure for hydraulic drops from head pool to tailwater in excess of 20 feet should not be considered.

---

[2] Numbers in brackets refer to the Bibliography D—Spillways, pages 525-526 of this chapter.

(c) Ogee (Overflow) Spillways. The ogee spillway has a control weir which is ogee or S-shaped in profile. The upper curve of the ogee ordinarily is made to conform closely to the profile of the lower nappe of a ventilated sheet falling from a sharp-crested weir. Flow over the crest is made to adhere to the face of the profile by preventing access of air to the under side of the sheet. For discharges at designed head, the flow glides over the crest with no interference from the boundary surface and attains near-maximum discharge efficiency. The profile below the upper curve of the ogee is continued tangent along a slope to support the sheet on the face of the overflow. A reverse curve at the bottom of the slope turns the flow onto the apron of a stilling basin or into the spillway discharge channel.

The upper curve at the crest may be made either broader or sharper than the nappe profile. A broader shape will support the sheet and positive hydrostatic pressure will occur along the contact surface. The supported sheet thus creates a backwater effect and reduces the efficiency of discharge. For a sharper shape, the sheet tends to pull away from the crest and to produce subatmospheric pressure along the contact surface. This negative pressure effect increases the effective head, and thereby increases the discharge.

An ogee crest and apron may comprise an entire spillway, such as the overflow portion of a concrete gravity dam, or the ogee crest may only be the control structure for some other type of spillway. Because of its high discharge efficiency, the nappe-shaped profile is used for most spillway control crests. See Fig. 9-6.

(d) Side Channel Spillways. The side channel spillway is one in which the control weir is placed along the side of and approximately parallel to the upper portion of the spillway discharge channel. Flow over the crest falls into a narrow trough opposite the weir, turns an approximate right angle, and then continues into the main discharge channel. The side channel

Figure 9-6. Shasta Dam, California. Spillway and outlet works in center of concrete dam. (USBR)

design is concerned only with the hydraulic action in the upstream reach of the discharge channel and is more or less independent of the details selected for the other spillway components. Flows from the side channel can be directed into an open discharge channel or into a closed conduit or inclined tunnel. Flow into the side channel might enter on only one side of the trough in the case of a steep hillside location, or on both sides and over the end of the trough if it is located on a knoll or gently sloping abutment. The "bathtub" type of side channel spillway, illustrates the latter type.

Discharge characteristics of a side channel spillway are similar to those of an ordinary overflow and are dependent on the selected profile of the weir crest. However, for maximum discharges the side channel flow may differ from that of the overflow spillway in that the flow in the trough may be restricted and may partly submerge the flow over the crest. In this case the flow characteristics will be controlled by a constriction in the channel downstream from the trough. The constriction may be a point of critical flow in the channel, an orifice control, or a conduit or tunnel flowing full.

Although the side channel is not hydraulically efficient nor inexpensive, it has advantages which make it adaptable to certain spillway layouts. Where a long overflow crest is desired in order to limit the surcharge head and the abutments are steep and precipitous, or where the control must be connected to a narrow discharge channel or tunnel, the side channel is often the best choice.

(e) Chute (Open Channel or Trough) Spillways.
A spillway whose discharge is conveyed from the reservoir to the downstream river level through an open channel, placed either along a dam abutment or through a saddle, might be called a chute, open channel, or trough type spillway. These designations can apply regardless of the control device used to regulate the flow. Thus, a spillway having a chute-type discharge channel, though controlled by an overflow crest, a gated orifice, a side channel crest, or some other control device, might still be called a chute spillway. However, the name is most often applied when the spillway control is placed normal or nearly normal to the axis of an open channel, and where the streamlines of flow both above and below the control crest follow in the direction of the axis.

The chute spillway has been used with earthfill dams more often than has any other type. Factors influencing the selection of chute spillways are the simplicity of their design and construction, their adaptability to almost any foundation condition, and the overall economy often obtained by the use of large amounts of spillway excavation in the dam embankment. Chute spillways have been constructed successfully on all types of foundation materials, ranging from solid rock to soft clay.

Chute spillways ordinarily consist of an entrance channel, a control structure, a discharge channel, a terminal structure, and an outlet channel. The simplest form of chute spillway has a straight centerline and is of uniform width. Often, either the axis of the entrance channel or that of the discharge channel must be curved to fit the alinement to the topography. In such cases, the curvature is confined to the entrance channel if possible, because of the low approach velocities. Where the discharge channel must be curved, its floor is sometimes superelevated to guide the high-velocity flow around the bend, thus avoiding a piling up of flow toward the outside of the chute.

Chute spillway profiles are usually influenced by the site topography and by subsurface foundation conditions. The control structure is generally placed in line with or upstream from the centerline of the dam. Usually the upper portion of the discharge channel is carried at minimum grade until it "daylights" along the downstream hillside to minimize excavation. The steep portion of the discharge channel then follows the slope of the abutment.

Flows upstream from the crest are generally at subcritical velocity, with critical velocity occurring when the water passes over the control. Flows in the chute are ordinarily maintained at supercritical stage, either at constant or accelerating rates, until the terminal structure is reached. For good hydraulic performance, abrupt vertical changes or sharp convex or concave vertical curves in the chute profile should be avoided. Similarily, the convergence or divergence in plan should be gradual in order to avoid cross waves, "ride-up" on the walls, excessive turbulence, or uneven distribution of flow at the terminal structure.

(f) Conduit and Tunnel Spillways. Where a closed channel is used to convey the discharge around or under a dam, the spillway is often called a tunnel or conduit spillway, as appropriate. The closed channel may take the form of a vertical or inclined shaft, a horizontal tunnel through earth or rock, or a conduit constructed in open cut and backfilled with earth materials. Most forms of control structures, including overflow crests, vertical or inclined orifice entrances, drop inlet entrances, and side channel crests, can be used with conduit and tunnel spillways. See Fig. 9-7.

With the exception of those with orifice or drop inlet entrances, tunnel and conduit spillways are designed to flow partly full throughout their length. With the drop inlet or orifice control, the tunnel or conduit size is selected so that it flows full for only a short section at the control and thence partly full for its remaining length. Ample aeration must be provided in a tunnel or conduit spillway in order to prevent a make-and-break siphonic action which would result if some part of the tunnel or conduit tends to seal temporarily because of an exhaustion of air caused by surging of the water jet, or by wave action or backwater. To guarantee free flow in the tunnel, the ratio of the flow area to the total tunnel area is often limited to about 75 percent. Air vents may be provided at critical points along the tunnel or conduit to insure an adequate air supply which will avoid unsteady flow through the spillway.

Tunnel spillways may present advantages for damsites in narrow canyons with steep abut-

Figure 9-7. Spillway operation, Hoover Dam, Colorado River, 1941. (USBR)

514 HANDBOOK OF DAM ENGINEERING

ments or at sites where there is danger to open channels from snow or rock slides. Conduit spillways may be appropriate at damsites in wide valleys, where the abutments rise gradually and are at a considerable distance from the stream channel. Use of a conduit will permit the spillway to be located under the dam near the streambed.

(g) Drop Inlet (Shaft or Morning Glory) Spillways. A drop inlet or shaft spillway, as the name implies, is one in which the water enters over a horizontally positioned lip, drops through a vertical or sloping shaft, and then flows to the downstream river channel through a horizontal or near horizontal conduit or tunnel. The structure may be considered as being made up of three elements; namely, an overflow control weir, a vertical transition, and a closed discharge channel. Where the inlet is funnel-shaped, this type of structure is often called a "morning glory" or "glory hole" spillway. See Fig. 9-8.

Discharge characteristics of the drop inlet spillway may vary with a range of head. The control will shift according to the relative discharge capacities of the weir, the transition, and the conduit or tunnel. For example, as the heads increase on a glory hole spillway, the control will shift from weir flow over the crest to tube flow in the transition and then to full pipe flow in the downstream portion. Full pipe flow design for spillways except those with extremely low drops is not recommended.

A drop inlet spillway can be used advantageously at dam sites in narrow canyons where the abutments rise steeply or where a diversion tunnel or conduit is available for use as the downstream leg. Another advantage of this type of spillway is that near maximum capacity is attained at relatively low heads; this characteristic makes the spillway ideal for use where the maximum spillway outflow is to be limited. This characteristic also may be considered disadvantageous, in that there is little increase in

Figure 9-8. Morning glory type spillway, Monticello Dam, California. (USBR)

capacity beyond the designed heads, should a flood larger than the selected inflow design flood occur. This would not be a disadvantage if this type of spillway were used as a service spillway in conjunction with an auxiliary or emergency spillway.

Additional information on the design and performance of drop inlet spillways is given in the references listed in the bibliography [5, 6, 23].

(h) Baffle Apron Drop Spillways. Baffled aprons or chutes are used in flow ways where water is to be lowered from one level to another and where it is desirable to avoid a stilling basin. The baffle piers partially obstruct the flow, dissipating energy as the water flows down the chute so that the flow velocities entering the downstream channel are relatively low. Advantages of baffled aprons include: economy, a low terminal velocity of the flows regardless of the height of the drop, downstream degradation which does not affect the spillway operation, and there are no requirements for initial tailwater depth in order for the stilling action to be effective.

The chute is normally constructed on a slope of 2:1 or flatter, extending below the outlet channel floor. Chutes having slopes steeper than 2:1 should be model tested [11] and their structural stability checked. The lower end of the chute should be constructed far enough below the channel floor to prevent damage from degradation or scour.

Design capacities of baffled aprons have varied from less than 10 second-feet to over 80 second-feet per foot of width. At Conconully Dam in north central Washington, the spillway baffled apron was designed to discharge up to 78 second-feet per foot of width and to operate effectively at 150 second-feet per foot of width.

(i) Culvert Spillways. A culvert spillway is a special adaptation of the conduit or tunnel spillway. It is distinguished from the drop inlet and other conduit types in that its inlet opening is placed either vertically or inclined upstream or downstream, and its profile grade is made uniform or near uniform and of any slope. The spillway inlet opening might be sharp edged or rounded, and the approach to the conduit might have flared or tapered sidewalls with a level or sloping floor. If it is desired that the conduit flow partly full for all conditions of discharge, special precautions are taken to prevent the conduit from flowing full; if full flow is desired, bellmouth or streamlined inlet shapes are provided. Special hooded inlets are sometimes added to facilitate the flow passing from part full to full flow conditions as well as to prevent the formation of vortices which would interfere with the full flow action [24].

Culvert spillways operating with the inlet unsubmerged will act similarly to an open channel spillway. Those operating with the inlet submerged, but with the inlet orifice arranged so that full conduit flow is prevented, will act similarly to an orifice-controlled drop inlet spillway, or to an orifice-controlled chute spillway. Where priming action is induced and the conduit flows full, the operation will be similar to that of a siphon spillway. When the culvert spillway is arranged to operate as a siphon, recognition must be taken of the disadvantages of siphon flow, especially those listed as items (4), (5), and (6) in section (j), following.

When culvert spillways placed on steep slopes flow full, reduced or negative pressures prevail along the boundaries of the conduit. Where negative pressures are large, there is danger of cavitation to the surfaces of the conduit or of its collapsing. Where cracks or joints occur along the low-pressure regions, there is the possibility of drawing in soil surrounding the conduit. Culvert spillways, therefore, should not be used for high-head installations where large negative pressures can develop. Further, the transition flow phenomenon, when the flow changes from part-full to full stage, is attended by rather severe pulsations and vibrations which increase in magnitude with increased fall of the culvert. For these reasons, culvert spillways should not be used for hydraulic drops exceeding 25 feet.

For drops not exceeding 25 feet, culvert spillways offer advantages over similar types because of their adaptability for either part-full or full flow operation and because of their simplicity and economy of construction. They might be placed on a bench excavated along the abutment on a relatively steep sidehill location, or they can be placed through the main section of the dam to discharge directly into the downstream river channel. As is the case with a drop inlet or siphon spillway, a principal disadvantage of the culvert spillway is that because its capacity does not substantially increase with increase in head, it does not provide a factor of safety against underestimation of the design flood. This disadvantate would not apply if the culvert type were used as a service spillway in conjunction with an auxiliary or emergency spillway.

(j) Siphon Spillways. A siphon spillway is a closed conduit system formed in the shape of an inverted U, positioned so that the inside of the bend of the upper passageway is at normal reservoir storage level. The initial discharges of the spillway, as the reservoir level rises above normal, are similar to flow over a weir. Siphonic action takes place after the air in the bend over the crest has been exhausted. Continuous flow is maintained by the suction effect due to the gravity pull of the water in the lower leg of the siphon.

Most siphon spillways are composed of five component parts. These include an inlet, an upper leg, a throat or control section, a lower leg, and an outlet. A siphon-breaker air vent is also provided to control the siphonic action of the spillway so that it will cease operation when the reservoir water surface is drawn down to normal level. Otherwise the siphon would continue to operate until air entered the inlet. The inlet is generally placed well below the normal reservoir water surface to prevent entrance of ice and drift and to avoid the formation of vortices and drawdowns which might break the siphon action. The upper leg is formed as a bending convergent transition to join the inlet to a vertical throat section. The throat or control section is generally rectangular in cross section and is located at the crest of the upper bend of the siphon. The upper bend then continues to join a vertical or inclined tube which forms the lower leg of the siphon. Often the lower leg is placed on an adverse slope, to provide a more positive priming action by forming a flow curtain which seals across the leg. The lower leg can be terminated so as to discharge vertically or along the face of a concrete dam, or it may be provided with a lower bend and diverging outlet tube to release the flow in a horizontal direction. The outlet flow can be free discharging or submerged, depending on the arrangement of the lower leg and on tailwater conditions.

A siphon spillway can be formed of concrete or from steel pipe. Because of the negative pressures prevalent in the siphon, the structure wall thicknesses should be sufficiently rigid to withstand the collapsing forces. Joints must be made water-tight, and measures must be taken to avoid cracking of the structure from movement or settlement. In order to prevent absolute pressures within the conduit from approaching cavitation or collapsing pressures, the total drop of the siphon should be limited to a maximum of 20 feet.

The principal advantage of a siphon spillway is its ability to pass full-capacity discharges with narrow limits of headwater rise. A further advantage is its positive and automatic operation without mechanical devices or moving parts.

In addition to its higher cost, as compared with other types, the siphon spillway has a number of disadvantages, including the following:

(1) The inability of the siphon spillway to pass ice and debris.
(2) The possibility of clogging the siphon passageways and siphon breaker vents with debris or leaves.
(3) The possibility of water freezing in the inlet legs and air vents before the reservoir rises to the crest level of the spillway, thus preventing flow through the siphon.

(4) The occurrence of sudden surges and stoppages of outflow as a result of the erratic make-and-break action of the siphon, thus causing radical fluctuations in the downstream river stage.

(5) The release of outflows in excess of reservoir inflows whenever the siphon operates, if a single siphon is used. Closer regulation which will more nearly balance outflow and inflow can be obtained by providing a series of smaller siphons, with their siphon breaker vents set to prime at gradually increasing reservoir heads.

(6) The more substantial foundation required to resist vibration disturbances, which are more pronounced than in other types of control structures.

As is the case with other types of closed conduit structures, a principal disadvantage of the siphon spillway is its inability to handle flows materially greater than designed capacity although the reservoir head exceeds the design level. Consequently, the siphon spillway is best suited as a service spillway to be used in conjunction with an auxiliary or emergency structure.

4. Controlled Crests

(a) General. The simplest form of control for a spillway is the free or uncontrolled overflow crest which automatically releases water whenever the reservoir water surface rises above the crest level. The advantages of the uncontrolled crest are the elimination of the need for constant attendance and regulation of the control devices by an operator, and the freedom from maintenance and repairs of the devices.

A regulating gate or other form of movable crest must be employed if a sufficiently long uncontrolled crest or a large enough surcharge head cannot be obtained for the required spillway capacity. Such devices will also be required if the spillway is to release storages below the normal reservoir water surface. The type and size of the selected control device may be influenced by such conditions as discharge characteristics of a particular device, climate, frequency and nature of floods, winter storage requirements, flood control storage and outflow provisions, the need for handling ice and debris, and special operating requirements. Whether an operator will be in attendance during periods of flood and the availability of electricity, operating mechanisms, operating bridges, etc., are factors which will influence the type of control device employed. See Fig. 9-9.

Many types of crest control have been devised. The type selected for a specific installation should be based on a consideration of the factors noted above as well as economy, adaptability, reliability, and efficiency. In the classification of movable crests are such devices as flashboards, stoplogs, bear-trap gates, tilting hinged-leaf gates, and drum gates. Regulating devices include stoplogs, needle beams, bulkheads, vertical or inclined rectangular lift gates, roller gates, and radial gates.

For simplicity of design and operation, only the less complicated control devices are considered appropriate for spillways for small dams. Such devices as flashboards, stoplogs, rectangular gates, and radial gates should be utilized wherever possible, since they can be easily fabricated or obtained commercially.

(b) Flashboards and Stoplogs. Flashboards and stoplogs provide a means of raising the reservoir storage level above a fixed spillway crest level, when the spillway is not needed for releasing floods. Flashboards usually consist of individual boards or panels supported by vertical pins or stanchions anchored to the crest; stoplogs are boards or panels spanning horizontally between grooves recessed into supporting piers. In order to provide adequate spillway capacity, the flashboards or stoplogs must be removed before the floods occur, or they must be designed or arranged so that they can be removed while being overtopped.

Various arrangements of flashboards have been devised. Some must be placed and removed manually, some are designed to fail after being overtopped, and others are arranged to

Figure 9-9. General view, Grand Coulee Dam on Columbia River, Washington, showing spillway in operation, power plants on both sides. (USBR)

drop out of position either automatically or by being manually triggered after the reservoir exceeds a certain stage. Flashboards provide a simple economical type of movable crest device, and they have the advantage that an unobstructed crest is provided when the flashboards and their supports are removed. They have numerous disadvantages, however, which greatly limit their adaptability. Among these disadvantages are the following: (1) they present a hazard if not removed in time to pass floods, especially where the reservoir area is small and the stream is subject to flash floods; (2) they require the attendance of an operator or crew to remove them, unless they are designed to fail automatically; (3) if they are designed to fail when the water reaches certain stages their operation is uncertain, and when they fail they release sudden and undesirably large outflows; (4) ordinarily they cannot be restored to position while flow is passing over the crest; and (5) if the spillway functions frequently the repeated replacement of flashboards may be costly.

Stoplogs are individual beams or girders set one upon the other to form a bulkhead supported in grooves at each end of the span. The spacing of the supporting piers will depend on the material from which the stoplogs are constructed, the head of water acting against the stoplogs, and the handling facilities provided for installing and removing them. Stoplogs which are removed one by one as the need for increased discharge occurs are the simplest form of a crest gate.

Stoplogs may be an economical substitute for more elaborate gates where relatively close spacing of piers is not objectionable and where removal is required only infrequently. Stoplogs which must be removed or installed in flowing water may require such elaborate hoisting mechanisms that this type of installation may prove to be as costly as gates. A stoplogged spillway requires the attendance of an operating crew for removing and installing the stoplogs. Further, the arrangement may present a hazard to the safety of the dam if the reservoir is small and the stream is subject to flash floods, since the stoplogs must be removed in time to pass the flood.

(c) Rectangular Lift Gates. Rectangular lift gates span horizontally between guide grooves in supporting piers. Although these gates may be made of wood or concrete, they are often made of metal (cast iron or steel). The support guides may be placed either vertically or inclined slightly downstream. The gates are raised or lowered by an overhead hoist. Water is released by undershot orifice flow for all gate openings.

For sliding gates the vertical side members of the gate frame bear directly on the guide members; sealing is effected by the contact pressure. The size of this type of installation is limited by the relatively large hoisting capacity required to operate the gate because of the sliding friction that must be overcome.

Where larger gates are needed, wheels can be mounted along each side of the rectangular lift gates to carry the load to a vertical track on the downstream side of the pier groove. The use of wheels greatly reduces the amount of friction and thereby permits the use of a smaller hoist. Rubber or belting is used along the sides to seal the openings between the upstream leaf plate and the sides of the pier.

(d) Radial Gates. Radial gates are usually constructed of steel or a combination of steel and wood. They consist of a cylindrical segment which is attached to supporting bearings by radial arms. The face segment is made concentric to the supporting pins so that the entire thrust of the waterload passes through the pins; thus, only a small moment need be overcome in raising and lowering the gate. Hoisting loads then consist of the weight of the gate, the friction between the side seals and the piers, and the frictional resistance at the pins. The gate is often counterweighted to partially counterbalance the effect of its weight, which further reduces the required capacity of the hoist.

The small hoisting effort needed to operate radial gates makes hand operation practical on small installations which otherwise might require power. The small hoisting forces involved also make the radial gate more adaptable to operation by relatively simple automatic control apparatus. Where a number of gates are used on a spillway, they might be arranged to open automatically at successively increasing reservoir levels, or only one or two might be equipped with automatic controls, while the remaining gates would be operated by hand or power hoists.

Small radial gates which may be operated either automatically or by hoist operation are available commercially. These gates are fabricated from structural steel members and have either a corrugated-metal or plate-steel faceplate.

C. STRUCTURAL DESIGN DETAILS

1. General

The structural design of a spillway and the selection of specific structural details follow the determination of the spillway type and arrangement of components and the completion of the hydraulic design.

Usually the foundation material of a spillway is not competent to resist the destructive action of high-velocity flows; therefore, a nonerodible lining ordinarily must be provided along the spillway waterway. This lining may be of wood, steel, handlaid grouted riprap, rubble masonry, or concrete. Such lining serves to prevent erosion, reduces friction losses by providing smooth bounding surfaces for the channel (which also permits smaller hydraulic sections), and provides a relatively watertight conveyance channel for directing flow past the dam. Economy and durability most often favor concrete as the appropriate lining material for water conveyance structures.

A spillway can be constructed on almost any foundation capable of sustaining applied loads without undue deformation. Although it is not usually advisable, a spillway can be placed on the face of an earthfill dam or through the dam, provided precautions are taken in the selection of design details to accommodate settlement and to prevent leakage from the structure. The type of walls, linings, and associated structures of a spillway and the details of the design will depend on the nature of the foundation. For instance, the details of the design for a spillway founded entirely in rock will differ from one constructed on clay. Structural details will differ according to foundation bearing capabilities, settlement or heave characteristics, and permeability and seepage features. Although concrete walls, linings, and associated structures may be adequate to withstand normal hydrostatic and earth loadings, they must also be arranged to allow for movements due to temperature change, and for unequal or large foundation settlements and heavings because of frost action. Provisions must also be made for handling leakage from the channel or underseepage from the foundation which might cause saturation of the underlying materials and large uplifts against the structure undersurfaces.

Subsequent sections discuss the structural designs of open channel spillways, including crest structures, walls, channel linings, and miscellaneous details. The structural designs of spillway conduits and tunnels are similar to those for outlet conduits and tunnels which are discussed later in this chapter.

2. Crest Structures and Walls

Spillway control structures and overflow crests against which reservoir heads act are essentially overflow dams, and spillway abutment structures or flanking dikes are similar to concrete nonoverflow dams or earthfill embankments. The design of earthfill dams is described in chapter 6; the design of overflow and nonoverflow concrete dams is discussed in Chapter 8.

The nature or type of confining side walls for open channel spillways will depend on the material upon which they are founded and on the loading to which they will be subjected.

For spillway channels excavated in rock or firm material, and where sloping of the wall faces is permissible, lining placed directly against the excavated slopes may offer sufficient stability for forming the channel sidewalls. Otherwise, self-supporting retaining walls of the gravity, cantilever, or counterforted type will be required. A monolithic flume-type section with the walls continuous with the floor and heels may also be considered.

The design of a gravity or reinforced concrete retaining wall is similar to that for a gravity dam, in that the stability against sliding and overturning and the magnitude and distribution of the foundation reaction resulting from the weight and applied loads must be determined. Methods of analyzing gravity structures for stability are discussed in Chapters 6, 7 and 8.

Earth loadings are assumed on the basis of equivalent fluid pressures, based on cohesionless soil, as given by Rankine. Wall footings must be safeguarded against frost heaving, and the wall panels must be articulated to provide for adjustments in the event of foundation yielding or unequal settlement. To avoid differential settlement in soft or yielding foundations, wall footing dimensions should be selected to minimize foundation load concentrations and to provide nearly uniform bearing reactions across the base areas.

Inlet channel and chute walls may be subject to various combinations of loading. When flow is occurring through the spillway, hydrostatic loads on the channel side of the wall tend to offset the loads caused by backfill. If, however, the fill has shrunk away from the walls, the wall members may be subject to full channel-side waterload before the deflection is sufficient to gain support from the backfill. This condition is more likely to exist where the wall leans into the backfill. On the other hand, when the reservoir is drawn down below the spillway level, there is no spill through the structure and the walls will be subject to full backfill loads without any support from waterloads. The structural design of wall members must recognize all these possibilities of loading. In the case of the assumption that the backfill may not be tight against the wall to help support the wall against water pressures, an increase in allowable stresses may be considered.

When permeable backfill is placed behind stilling basin walls or when the back of the wall is partly exposed to tailwater, the water pressure resulting from tailwater will need to be added to the backfill loading. For higher spillway discharges, the water level inside the basin will be depressed by the profile of the jump and an unbalanced hydrostatic load acting to overturn the walls will occur. The design loading assumptions must recognize this condition of unbalanced pressures as well as the increased uplift forces when sliding and overturning analyses are considered.

3. Open Channel Linings

Floor pavings are provided primarily to form a reasonably watertight protective surfacing over the channel to prevent erosion or damage to the foundation. During spillway flows, the floor is subjected to hydrostatic forces due to the weight of the water in the channel, to boundary drag forces due to frictional resistance along the surface, to dynamic forces due to flow impingement, to uplift forces due to reduction of pressure along the boundary surface, or to uplift pressure caused by leakage through joints or cracks. When there are no spills, the floor is subject to the action of the elements including expansion and contraction due to temperature variations, alternate freezing and thawing, and weathering and chemical deterioration; to the effects of settlement and buckling; and to uplift pressures brought about by underseepage or high ground-water conditions. Since it is not always possible to evaluate the various forces which might occur nor to make the lining heavy enough to resist them, the thickness of the lining is most often established on a more or less arbitrary basis; and underdrains, anchors, cutoffs, etc., are utilized to stabilize the floor.

To provide a relatively watertight lining which will withstand reasonable weathering and abrasion, and which will hold up against ordinarily experienced forces caused by expansion, contraction, frost heave, and settlement of the foundation, a nominal minimum thickness of 8 inches is recommended for small spillways when the lining is placed directly on rock. When the lining is placed directly on earth or on an intervening gravel layer, a somewhat thicker slab should be provided to forestall cracking or buckling if expansion and contraction can move or displace the slab.

When a spillway channel is excavated in rock, the paving slab is cast directly on the excavated surface. Anchor bars grouted into holes drilled into the rock can be provided to tie the paving to the foundation. A slab which is bonded to the foundation may not move as the result of expansion and contraction. Instead, numerous cracks which in effect divide the slab into a series of small individual blocks will occur. Reinforcement therefore must be provided to tie these individual blocks together and to distribute the cracking and minimize the crack openings. The anchorage provided increases the effective weight of the slab against displacement by the amount of foundation rock to which the anchors can be tied. Depth and spacing of anchors will depend on the nature of the bedrock, its stratification, jointing, weathering, etc. The anchor should be of sufficient size to hold the weight of the foundation to which it is anchored without exceeding the yield stress of the steel. A gridwork of underdrains laid with open joints in gravel-filled trenches is provided to prevent a buildup of uplift under the paving. When leakage through the joints is to be minimized, metal or rubber waterstops are provided.

When a spillway channel is excavated through earth, the paving slab may be cast directly on the excavated surface, or an intervening pervious blanket may be required, depending on the nature of the foundation as related to its permeability, susceptibility to heaving from frost action, and heterogeneity as it may affect differential settlement. Because the slab is not bonded to the foundation, it is subject to movement from expansion or contraction, and it must be restrained from creeping when it is constructed on a slope. This restraint is best achieved by cutoffs which can be held in a more or less fixed position with respect to the slab and to the foundation, or by tying the slab to walls, piles, or similar rigid members of the spillway structure. Since the slab is relatively free to move upon the foundation, the movement will take place from the fixed edges and the paving should be reinforced sufficiently to permit its sliding without cracking of the concrete or yielding of the reinforcement. To assist further in holding the slab to the foundation, bulb anchors are sometimes employed. The anchor in this instance in effect ties the slab to a cone of earth, the volume of which will depend on the anchor depth and spacing and on the angle of internal friction of the soil.

A pervious gravel blanket is often provided between the slab and the foundation when the foundation is sufficiently impervious to prevent leakage from draining away, or where it is subject to capillarity which will draw moisture to the underside of the lining. The blanket serves not only as a free-draining medium but also aids in insulating the foundation from frost penetration. The thickness of the blanket thus depends on the climate at the site and on the susceptibility of the foundation to frost heaving. A gridwork of underdrains laid with open joints in gravel and bedded on a mortar pad to prevent the foundation material from being leached into the pipe is provided as a collecting system for the seepage. The network of drainage pipe empties into one or more trunk drains which carry the seepage flows to outlets through the channel floor or walls.

In stratified foundations, ground water or seepage can cause uplift on layers below the floor lining, and drainage holes are sometimes augered into the underlying material and backfilled with gravels to relieve the underpressure.

When watertightness of the paving against exterior water heads is required, metal or rubber waterstops are installed to seal the joints. Such seals are provided in floor slabs upstream from the control structure if watertightness is desired to increase the percolation path under the structure. They are commonly provided at transverse joints along concave curved portions of the downstream channel where the dynamic pressures on the floor cause a high head for introducing water into the joint. Seals may be desirable along longitudinal joints in a stilling basin on a permeable base. Differential heads resulting from the sloping water surface of the jump can cause a circulating flow under the slab if leakage is allowed to enter the joint at the downstream end of the basin and to flow out of the joint at the upstream end.

Lateral joints over which flow velocities are high are arranged so that the upstream edge of the lower slab cannot heave without moving the lower edge of the upper slab a like amount; further, the lower slab edge is constructed about one-half inch lower than the upper slab edge. These provisions are made to forestall a high buildup of dynamic head at the joint which would result if the surface of the downstream slab were to project above the surface of the upstream slab. The dynamic head could introduce water at high pressure under the slab, which would result in uplift or dislodgement.

Contraction joints are generally placed from 25 to 50 feet apart in both the floor and walls. Joints are also provided where angular changes of the floor surface occur and where they are required to avoid reentrant angles in the slab which often cause cracking of the paving. The use of joint fillers in contraction joints should be minimized because deterioration of the filler will result in an open joint which is difficult to maintain. If joints are provided at the indicated spacings, the contraction or expansion movements may not be severe and filler material in the joint may not be necessary. Floor slabs can be constructed in alternate panels; the initial placement shrinkage of the concrete may then afford sufficient joint opening for subsequent expansion. Keyed joints in thin floors and walls which may be subjected to differential movement are unsatisfactory, since inequalities in deflection across the joint will place high stress on the keys or keyways and cause them

DESIGN OF SPILLWAYS AND OUTLET WORKS 523

to spall. An unkeyed joint with slip dowels is a better detail.

Water normally will stand in a stilling basin whose floor is at a lower level than the river channel. With this condition the foundation under the basin will be permanently saturated. When the water is lowered in the basin, the floor can be subjected to an uplift equal to the tailwater head or higher if the pressure is augmented by head from a higher source. During times of spillway discharge, the water weight in the basin will be reduced because of the slope of the jump profile, and at the upstream end of the basin the uplift will far exceed the downward weight. The basin floor must be heavy enough to withstand this unbalanced waterload, unless an adequate drainage system is installed to relieve the uplift pressure when jump sweep-out occurs. Since a drainage system cannot be considered entirely effective because of the possibility of clogging or silting of the drain outlets, the floor slab is usually made sufficiently heavy to resist the flotation effect on the floor.

For design, the stilling basin floor is considered to be a free body in static equilibrium with foundation reactions balancing active loads. Uplift forces caused by hydrostatic head on the bottom of the slab are counterbalanced by the weight of the concrete and the effective weight

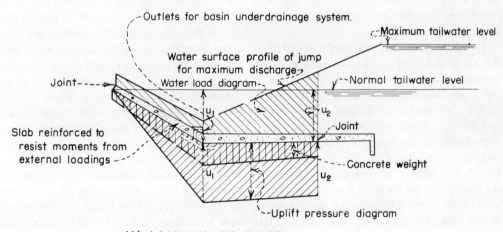

(A) LOADINGS ON FLOOR

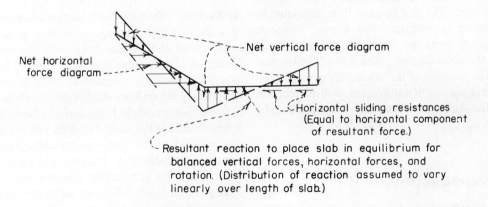

(B) FORCE DIAGRAMS

Figure 9-10. Illustration of uplift forces acting on a stilling basin floor.

of the water in the basin. Differential horizontal hydrostatic forces are opposed by the sliding resistance of the horizontal leg of the slab on the foundation. Equilibrium against rotation is achieved by equating any unbalanced forces with a foundation reaction force positioned so that the moments will be zero about any point.

An illustrative force diagram for a typical basin analysis is shown in Fig. 9-10. Note that the dynamic force which will occur because of impingement of the incoming high-velocity flow on the horizontal apron is not included. Where chute blocks or dentates are used at the upstream end of the apron, reduced pressure zones opposite the blocks will occur. Because the reduced pressures and the impingement forces tend to offset each other, they are both neglected in the analysis.

It will be seen that the weight required in the floor will vary in direct relation to the amount of assumed uplift. If uplift equal to the head from maximum tailwater is considered, the required floor thickness will be almost prohibitive. Therefore, an uplift based on a lesser tailwater condition, is generally assumed and reliance is placed on the drainage system for relief of greater uplift pressures. Ordinarily the outlets for the basin under-drainage system are located in the sills near the upstream end of the basin floor. The head on these outlets will be reduced when the jump occurs, which facilitates relief of hydrostatic pressures under the slab. The stilling basin floor slab must be designed to withstand the internal moments resulting from the external loadings indicated by the force diagram. The slab will be thickest at the junction of the sloping leg and the horizontal apron. Reinforcement will be required in the slab to withstand the computed internal stresses.

4. Miscellaneous Details

(a) Cutoffs. One or more cutoffs are generally provided at the upstream end of a spillway for various purposes. They may form a watertight curtain against seepage under the structure, or they may be used to increase the path of percolation under the structure and thus reduce uplift forces. Cutoffs also can be used to intercept permeable strata in the foundation so as to minimize seepage and prevent a buildup of uplift pressure under the spillway or adjacent areas. When the cutoff trench for the dam extends to the spillway, it is generally joined to the upstream spillway cutoff to provide a continuous barrier across the abutment area. In jointed rock the cutoff acts as a grout cap for a grout curtain which is often extended across the spillway foundation.

A cutoff is usually provided at the downstream end of a spillway structure as a safeguard against erosion and undermining of the end of the structure. Cutoffs at intermediate points along the length of a spillway are sometimes provided as barriers against water flowing along the contact between the structure and the foundation and to lengthen the path of percolation under the structure. When the spillway is a conduit under the dam, the cutoff takes the shape of collars placed at intervals around the conduit barrel. Wherever possible, cutoffs in rock foundations are placed in vertical trenches. In earth foundations where the cutoffs must be formed in a trench with sloping sides, care must be taken to compact the trench backfill properly with impervious material to obtain a reasonably watertight barrier.

(b) Backfill. When a spillway is placed adjacent to a dam so that the impervious zone of the embankment abuts the spillway walls, the wall backfill is actually the impervious zone of the dam and is similarly compacted. Backfill elsewhere along the spillway walls ordinarily should be free-draining material to minimize hydrostatic pressures against the walls. Backfill other than that adjacent to the dam may be either compacted or uncompacted. The choice of backfill material and the compaction methods used in placing such material will affect the design loadings on the walls.

(c) Riprap. When the spillway approach channel is excavated in material that will be eroded by the approach velocities, a zone of riprap is often provided immediately upstream from the inlet lining to prevent scour of the channel floor and side slopes adjacent to the spillway concrete. The riprap is generally a continuation of that along the upstream face of the dam, is of similar size and gradation, and has similar bedding. Riprap is normally used in the outlet channel adjacent to the downstream cutoff to prevent excessive erosion and undermining of the downstream end of the structure. To resist scour from high exit velocities, the riprap should be the largest size possible and should be bedded on a graded material. The voids should be filled with spalls to prevent the underlying material from washing out, which would cause the riprap to settle or to be displaced.

D. BIBLIOGRAPHY—SPILLWAYS*

[1] Donnelly, C. A., and Blaisdell, F. W., "Straight Drop Spillway Stilling Basin," University of Minnesota, Saint Anthony Falls Hydraulic Laboratory, Technical Paper No. 15, series B, November 1954.

[2] Morris, B. T., and Johnson, D. C., "Hydraulic Design of Drop Structures for Gully Control" Trans. ASCE, vol. 108, 1943, p. 887.

[3] Moore, W. L., "Energy Loss at the Base of a Free Overfall," Trans. ASCE, vol. 108, 1943. p. 1343.

[4] Rand, Walter, "Flow Geometry at Straight Drop Spillways," ASCE Proceedings, Paper No. 791, September 1955.

[5] Peterka, A. J., "Morning-Glory Shaft Spillways," Trans. ASCE, vol. 121, 1956, p. 385.

[6] Bradley, J. N., "Morning-Glory Shaft Spillways: Prototype Behavior," Trans. ASCE, vol. 121, 1956, p. 312.

[7] Bureau of Reclamation, "Studies of Crests of Overfall Dams," Bulletin 3, pt. VI, Hydraulic Investigations, Boulder Canyon Project Final Reports, 1948.

[8] Hinds, Julian, "Side Channel Spillways," Trans. ASCE, vol. 89, 1926, p. 881.

[9] Bradley, J. N., and Peterka, A. J., "The Hydraulic Design of Stilling Basins," ASCE Proceedings, vol. 83, October 1957, Journal of the Hydraulics Division, No. HY5, Papers No. 1401 to 1406, inclusive.

[10] Warnock, J. E., "Experiments Aid in Design at Grand Coulee," Civil Engineering, vol. 6, 1936, p. 737.

---
*Direct reference is not made in the text to all the publications listed here.

[11] Bureau of Reclamation, "Hydraulic Design of Stilling Basin and Bucket Energy Dissipators," Engineering Monograph No. 25, 1964.

[12] Bradley, J. N., and Peterka, A. J., "Hydraulic Design of Stilling Basins: Small Basins for Pipe or Open Channel Outlets—No Tailwater Required (Basin VI)," ASCE Proceedings, vol. 83, October 1957, Journal of the Hydraulics Division, No. HY5, Paper No. 1406.

[13] Doddiah, D., Albertson, M. L., and Thomas, R. A., "Scour From Jets," Proceedings, Minnesota International Hydraulics Convention (Joint Meeting of International Association for Hydraulics Research and Hydraulics Division, ASCE), Minneapolis, Minn., August 1953, p. 161.

[14] Veronese, Alessandro, "Erosioni Di Fondo A Valle Di Uno Scarico," (Downstream Bed Erosion Due to a Discharge), Annali dei Lavori Pubblici, vol. 75, No. 9, September 1937, p. 717.

[15] Wagner, W. E., "Morning Glory Shaft Spillways: Determination of Pressure-Controlled Profiles," Trans. ASCE, vol. 121, 1956, p. 345.

[16] Mavis, F. T., "The Hydraulics of Culverts," Pennsylvania State College Engineering Experiment Station, Bulletin 56, vol. XXXVII, No. 7, February 12, 1943.

[17] Straub, L. G., Anderson, A. G., and Bowers, C. E., "Importance of Inlet Design on Culvert Capacity, Culvert Hydraulics," Highway Research Board of the National Academy of Sciences, National Research Council, Publication 287, Research Report No. 15-B, 1953, p. 53.

[18] Karr, M. H., and Clayton, L. A., "Model Studies of Inlet Designs for Pipe Culverts on Steep Grades," Engineering Experiment Station, Oregon State College, Bulletin No. 35, June 1954.

[19] Schiller, R. E. Jr., "Tests on Circular Pipe Culvert Inlets," Culvert Flow Characteristics, Highway Research Board of the National Academy of Sciences, National Research Council, Publication 413, Bulletin No. 126, 1956, p. 11.

[20] Shoemaker, R. H. Jr., and Clayton, L. A., "Model Studies of Tapered Inlets for Box Culverts," Culvert Hydraulics, Highway Research Board of the National Academy of Sciences, National Research Council, Publication 287, Research Report No. 15-B, p. 1.

[21] Blaisdell, F. W., "Hydraulics of Closed Conduit Spillways—Part 1—Theory and Its Application," University of Minnesota, Saint Anthony Falls Hydraulic Laboratory, Technical Paper No. 12, series B, January 1952, revised February 1958.

[22] Blaisdell, F. W., "Hydraulics of Closed Conduit Spillways—Parts II through VII—Results of Tests on Several Forms of the Spillway," University of Minnesota, Saint Anthony Falls Hydraulic Laboratory, Technical Paper No. 18, series B, March 1958.

[23] Blaisdell, F. W., "Hydraulics of Closed Conduit Spillways—Part VIII—Miscellaneous Laboratory Tests and Part IX—Field Tests," University of Minnesota, Saint Anthony Falls Hydraulic Laboratory, Technical Paper No. 19, series B, March 1958.

[24] Blaisdell, F. W., and Donnelly, C. A., "Hydraulics of Closed Conduit Spillways—Part X—The Hood Inlet," University of Minnesota, Saint Anthony Falls Hydraulic Laboratory, Technical Paper No. 20, series B, April 1958.

[25] Beichley, G. L., "Hydraulics Design of Stilling Basin for Pipe or Channel Outlets," Bureau of Reclamation, Research Report No. 24, 1971.

[26] U. S. Bureau of Reclamation, "Design of Small Dams," 2d Edition, 1973.

II. Outlet Works

A. GENERAL

1. Functions of Outlet Works

An outlet works serves to regulate or release water impounded by a dam. It may release incoming flows at a retarded rate, as in the case of a detention dam; divert incoming flows into canals or pipelines, as in the case of a diversion dam; or release stored waters at such rates as may be dictated by downstream needs, evacuation considerations, or a combination of multiple-purpose requirements.

Outlet works structures can be classified according to their purpose, their physical and structural arrangement, or their hydraulic operation. An outlet works which empties directly into the river could be designated a river outlet; one which discharges into a canal could be classed as a canal outlet; and one which delivers water into a closed pipe system could be termed a pressure pipe outlet. An outlet works may be described according to whether it consists of an open channel or closed conduit waterway, or whether the closed waterway is a conduit in cut-and-cover or in tunnel. The outlet works may also be classified according to its hydraulic operation, whether it is gated or ungated or, for a closed conduit, whether it flows under pressure for part or all of its length or only as a free-flow waterway.

Occasionally the outlet may be placed at a higher level to deliver water to a canal, and a bypass extended to the river to furnish necessary flows below the dam. Such flows may be required to satisfy prior right uses downstream from the site; or they may be required for the maintenance of a live stream for abatement of stream pollution, preservation of aquatic life, or stock watering purposes. For dams constructed to provide reservoirs principally for recreation or fish and wildlife, a fairly constant lake level is desired and an outlet works may be needed only to release the minimum flows which will provide a live stream below the dam.

In certain instances the outlet works of a dam may be used in lieu of a service spillway in conjunction with an auxiliary or secondary spillway. In this event the usual outlet works installation might be modified to include a bypass overflow, so that the structure can serve as both an outlet works and a spillway.

An outlet works may also act as a flood control regulator, to release waters temporarily stored in flood control storage space or to evacuate storage in anticipation of flood inflows. Further, the outlets may serve to empty the reservoir to permit inspection, to make needed repairs, or to maintain the upstream face of the dam or other structures normally inundated. The outlets may also aid in lowering the reservoir storage when it is desired to control or to poison scrap fish or other objectionable aquatic life in the reservoir.

2. Determination of Required Capacities

Outlet works controls are designed to release water at specific rates, as dictated by downstream needs, flood control regulation, storage considerations, or legal requirements. Delivery of irrigation water is usually determined from project or farm needs and is related to the consumptive use and to any special water requirements of the irrigation system. Delivery for domestic use can be similarly established.

Releases of flows to satisfy prior rights must generally be included with other needed releases. Minimum downstream flows for pollution abatement, fish preservation, and other companion needs may often be accommodated through other required releases.

Irrigation outlet capacities are determined from reservoir operation studies and must be based on a consideration of a critical period of low runoff when reservoir storages are low and daily irrigation demands are at their peak. The most critical draft from the reservoir, considering such demands (commensurate with remaining reservoir storage) together with prior rights or other needed releases, generally determines the minimum irrigation outlet capacity. These requirements are stated in terms of discharge at either a given reservoir content or water surface elevation. Occasionally outlet capacity requirements are established for several reservoir contents or alternate water surfaces. For example, outlet requirements may be set forth as: 20 second-feet capacity at reservoir content 500 acre-feet, and 100 second-feet capacity at reservoir content 3,000 acre-feet.

Evacuation of waters stored in an allocated flood control storage space of a reservoir can be accomplished through a gated spillway at the higher reservoir levels or through an outlet at the lower levels. Flood control releases generally can be combined with the irrigation outlet releases if the outlet empties into a river instead of into a canal. The capacity of the flood control outlet is determined by the required time of evacuation of a given storage space, considering the inflow into the reservoir during this emptying period. The combined flood control and irrigation releases ordinarily must not exceed the safe channel capacity of the river downstream from the dam and must allow for any anticipated inflows immediately below the dam. These inflows may be the natural runoffs or may result from releases from other storage developments along the river or from adjacent developments on tributaries emptying into the river.

If an outlet is to serve as a service spillway in releasing surplus inflows from the reservoir, the required discharge for this purpose may fix the outlet capacity. Similarly, for emptying the reservoir for inspection or repair, the volume of water to be evacuated and the allotted emptying period may be the determining conditions for establishing the minimum outlet capacity. Here again, the inflow into the reservoir during the emptying period must be considered. The capacity at low reservoir level should be at least equal to the average inflow expected during the maintenance or repair period. It can, of course, be assumed that any required repair work might be delayed until service demands are light and that it will be done at times of low inflow and at seasons favorable to such construction.

An outlet works cut-and-cover conduit or tunnel often may be utilized for diverting the riverflow during the construction period, thus avoiding the necessity for supplementary installations for that purpose. The outlet structure size dictated by this use rather than the size indicated for ordinary outlet works requirements may determine the final outlet works capacity.

3. Outlet Works Position in Relation to Reservoir Storage Levels

The establishment of the intake level and the elevations of the outlet controls and the conveyance passageway, as they relate to the reservoir storage levels, are influenced by many considerations. Primarily, in order to attain the required discharge capacity, the outlet must be placed sufficiently below minimum reservoir operating level to provide head for effecting outlet works flows.

Outlet works for small detention dams are generally constructed near riverbed level since permanent storage space, except for silt retention, is ordinarily not provided. (These outlet works may be ungated in order to retard the outflow while the reservoir temporarily stores the bulk of the flood runoff, or they may be gated in order to regulate the releases of the temporarily stored waters.) If the purpose of the dam is only to raise and divert incoming flows, the main outlet works generally is a headworks or regulating structure at a high level, and a sluiceway or small bypass outlet is provided to furnish water to the river downstream or to drain the water from behind the dam during off-season periods. For dams which impound waters for irrigation, domestic use, or other conservation purposes, the outlet works must be placed low enough to draw the reservoir down to the bottom of the allocated storage space; however, it might be placed at some level above the riverbed, depending on the elevation of the established minimum reservoir storage level.

It is usual practice to make an allowance in a storage reservoir for inactive storage for sediment deposition, fish and wildlife conservation, and recreation. The positioning of the intake still then becomes an important consideration since it must be high enough to prevent interference from the sediment deposits, but at the same time low enough to permit either a partial or a complete drawdown below the top of the inactive storage.

The size of an outlet conduit for a required discharge varies according to an inverse relationship with the available head for producing the discharge. This relationship may be expressed by the following equation:

$$H_T = K_1 h_v, \text{ or } H_T = K_2 \frac{Q^2}{a^2}$$

where

H_T = the total available head for producing flow,
Q = the required outlet works discharge, and
a = the required area of the conduit.

The above relationship for a particular design is illustrated in Fig. 9-11. In this example, if the head available for the required outlet works discharge is increased from 1.6 to 4.6 feet, the corresponding conduit diameter can be decreased from 6 to 4.75 feet. This shows that the conduit size can be significantly reduced if the inactive storage level can be increased. The reduction in active storage capacity resulting from increasing the inactive storage level 3 feet would have to be compensated by adding an equivalent amount of capacity to the top of the pool. By referring to the reservoir capacity curve, Fig. 9-12, it will be apparent that for equivalent storages (represented by *de* and *gh*) the 3 feet of head represented by ordinate *cd* added to obtain a reduced outlet works size would require a much

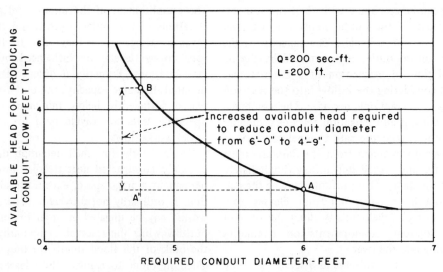

Figure 9-11. Relation of conduit size to available head.

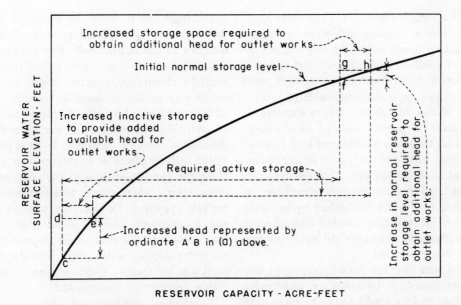

Figure 9-12. Relation of conduit size to normal storage level.

smaller increase (represented by the ordinate *fg*) in the height of the dam. Thus, economic studies can be utilized to determine the proper outlet size in relation to the minimum reservoir storage level.

Where an outlet is placed at riverbed level to accommodate the construction diversion plan or to drain the reservoir, the operating sill can be placed at a higher level to provide a sediment and debris basin and other desired inactive storage space, or the intake can be designed to permit raising the sill as sediment accumulates. During the construction period, a temporary diversion opening can be formed in the base of the intake for handling diversion flows and later closed with a plug. For emptying the reservoir, a bypass can be installed around the intake at riverbed level, either emptying into the lower portion of the conduit or passing under it. Delivery of water to a canal at a higher level can be made by a pressure riser pipe connecting the conduit to the canal.

4. Conditions Which Determine Outlet Works Layout

The layout of a particular outlet works will be influenced by many conditions relating to the hydraulic requirements, to the site adaptability and the interrelation of the outlet works to the construction procedures, and to other appurtenances of the development. Thus, an outlet works leading to a high-level canal or into a closed pipeline might differ from one emptying into the river. Similarly, a scheme in which the outlet works is used for diversion might vary from one where diversion is effected by other means. In certain instances, the proximity of the spillway may permit combining some of the outlet works and spillway components into a single structure. As an example, the spillway and outlet works layout might be arranged so that discharges from both structures will empty into a common stilling basin. In the same way components of an outlet works system may be combined with penstocks serving a hydropower system.

The topography and geology of a site may have a great influence on the layout selection. Some sites may be suited only for a cut-and-cover conduit type of outlet works, while at other sites either a cut-and-cover conduit or a tunnel can be selected. Unfavorable foundation geology, such as deep overburdens or inferior foundation rock, will obviate the selection of a tunnel scheme. On the other hand, sites in narrow canyons with steep abutments may make a tunnel outlet the only choice. Because of confined working space and excessive costs where hand construction methods must be employed, it is not practicable to make a tunnel smaller than about 6 feet in diameter. If constructed of precast material or if cast-in-place with the inside bore formed by a prefabricated liner, a cut-and-cover conduit can be constructed to almost any size. Thus the minimum size dictated by construction conditions, as compared to the size established by hydraulic requirements, will have considerable influence on the choice of alternative cut-and-cover conduit or tunnel schemes.

Some sites favorable for a tunnel outlet may have unfavorable portal conditions which make it difficult to fit the inlet and exit structures to the remainder of the outlet works. In this situation, a central tunnel with cut-and-cover conduits leading to and away from the tunneled portion of the outlet may prove to be feasible.

If water is to be taken from a reservoir for domestic use, special consideration must be given to the positioning of the intake. To assure the proper quality of the water, it may be necessary to draw from different levels of the reservoir at different seasons of the year or to restrict the draft to specific levels, depending on the reservoir stage. To prevent silt from being carried into the outlet system, intake locations at low points or pockets in the reservoir must be avoided. Similarly, intakes must not be placed at points in the reservoir where stagnant water or algae can accumulate or where prevailing winds will drift debris or undesirable trash to the intake entrance.

5. Arrangement of Outlet Works

An outlet works for a low dam, whether it is to divert water into a canal or release it to the river, often may consist of an open canal or a cut-and-cover structure placed at the dam abutment. The structure might consist of a conventional open flume or rectangular channel with a gate similar to that used for ordinary spillway installations, or it might be regulated by a submerged-type gate placed to close off openings in a curtain or head wall. Where the outlet is to be placed through a low earthfill embankment, a closed-type structure might be used which may consist of single or multiple units of buried pipe or box culverts placed through or under the embankment. Flow for such an installation could be controlled by gates placed at the inlet or placed at an intermediate point along the conduit, such as at the crest of the embankment, where a shaft would be provided for gate operation. Downstream from the control structure, the channel would continue to the canal or to the river where, depending on the exit velocities which might prevail for the particular installation, a stilling basin or similar stilling device might be employed.

For higher earthfill dams where an open channel outlet structure would not prove feasible, the outlet might be carried through, under, or around the dam as a cut-and-cover conduit or through the abutment as a tunnel. Depending on the position of the control device, the conduit or tunnel could be free flowing, flowing under pressure for a portion of its length, or flowing under pressure for its entire length. Intakes might be arranged to draw water from the bottom of the reservoir, or the inlets sills might be placed at some higher reservoir level. Dissipating devices similar to those described in this chapter could be utilized at the downstream end of the conduit. The outlet works also may discharge into the spillway stilling basin. Depending on the method of control and the flow conditions in the structure, access to the operating gates might be by bridge to an upstream intake tower, by shaft from the crest level of the dam, by walkway within the conduit or tunnel with entrance from the downstream end, or by a separate conduit or tunnel access adit. See Fig. 9-13.

For a concrete dam the outlet works installation is usually carried through the dam as a formed conduit or a sluice, or as a pipe embedded in the concrete mass. Intakes and terminal devices can be attached to the upstream and downstream faces of the dam. Often the outlet

DESIGN OF SPILLWAYS AND OUTLET WORKS 531

Figure 9-13. Outlet works and stilling basin, Blue Mesa Dam, Colorado. (USBR)

is formed through the spillway overflow section, using a common stilling basin to dissipate both spillway and outlet works flows. Where an outlet works conduit is installed in the nonoverflow section of the dam or where an outlet must empty into a canal, a separate dissipating device will, of course, be necessary. Instead of a large single conduit, multiple smaller conduits might be utilized in a concrete dam to provide a less expensive as well as a more feasible arrangement for handling the outlet works releases. The conduits might be placed at a single level, or for added flexibility they may be positioned at several levels. Such an arrangement would reduce the cost of the control gates, because of the lower heads in the upper level gates.

Where a diversion tunnel is utilized during the construction of a concrete dam, it is often feasible to convert the tunnel into a permanent outlet works by providing outlet sluices or conduits through the tunnel plug. Ordinarily, the diversion tunnel for a concrete dam will be in good quality rock and will therefore require a minimum of lining protection. Further, the outlet portal of the tunnel will generally be located far enough downstream from the dam so that no dissipating structure will be needed, or at most only a deflector will be required to direct the flow to the downstream river channel. See Fig. 9-14.

6. Location of Outlet Works Controls

(a) General. Where an outlet works is ungated, as will be the case with many detention dams, flow in the conduit will be similar to that in a culvert spillway. Where water must be stored and the release regulated at specific rates, control gates or valves will need to be installed at some point along the conduit.

Gates and valves for outlet works are categorized according to their functional use in the structure. Operating gates and regulating valves are used to control and regulate the outlet works flow and are designed to operate in any position from closed to fully open. Guard or emergency gates are designed to be utilized only to effect closure in the event of failure of the operating gates or where unwatering is

Figure 9-14. Testing outlet works, Hoover Dam, Colorado River, Arizona-Nevada, power plant in foreground. (USBR)

desired either to inspect the conduit below the guard gates or to inspect and repair the operating gates. Occasionally slots are provided at the conduit entrance to accommodate stoplogs or bulkheads so that the conduit can be closed off during an emergency period. For such installations, guard gates may or may not be provided, depending on whether or not the stoplogs can be placed readily if an emergency arises during normal reservoir operating periods.

The control gate for an outlet works can be placed at the upstream end of the conduit, at an intermediate point along its length, or in some instances at the lower end of the structure. Where flow from a control gate is released directly into the open as free discharge, only that portion of the conduit upstream from the gate will be under pressure. Where a control gate or valve discharges into a closed pressure pipe, the control will serve only to regulate the releases; full pipe flow will occur in the conduit both upstream and downstream from the control gate. For the pressure-pipe type, the location of the gate or valve will have little influence on the design insofar as internal pressures are concerned. However, where a control discharges into a free-flowing conduit, the location of the control gate becomes an important consideration in the design of the outlet. The effects of locating the control at various positions in a conduit are discussed in the following subsections. See Fig. 9-15.

(b) Control at Upstream End of Conduit. For an outlet works with an upstream control discharging into a free-flow conduit, part full flow will occur throughout the length of the structure. Ordinarily, with normal operating heads,

DESIGN OF SPILLWAYS AND OUTLET WORKS   533

Figure 9-15. External outlet works base, Monticello Dam, California. (USBR)

flow at supercritical stage will result. The structural design of the conduit and the safety and practical aspects of the layout will then be concerned only with the effects of external loadings and outside water pressures on the structure. Along the upstream portion of the conduit and extending until sufficient rock cover is available over a tunnel or until an adequate thickness of impervious embankment is obtained over a cut-and-cover conduit, practically full reservoir head will be exerted against the outside of the conduit barrel. The conduit walls must therefore be designed to withstand such pressures, and details of design must be selected to preserve the watertightness of the conduit. For a cut-and-cover conduit where settlement of the structure (due to foundation consolidation with increasing embankment load) must be anticipated, special care must be taken in design details to prevent the cracking of the conduit barrel and to seal any formed joints, since cracks and open joints will invite excessive leakage or piping of surrounding embankment material into the conduit.

With the controls placed at the upstream end of a conduit, fishscreens, stoplog slots, trashracks, guard gates, and regulating gates or valves can all be combined in a single intake structure. This arrangement will simplify outlet works operation by centralizing all control features at one point. Further, the entire conduit may be readily unwatered for inspection or repair. The intake will consist of a tower rising from the base of the outlet conduit to an operating deck placed above maximum reservoir water level, with the tower located in the reservoir area near the upstream toe of the dam. Access to the structure operating deck will then be possible only by boat, unless an access bridge is provided from the reservoir shore or from the crest of the dam.

(c) Control at Intermediate Point Along Conduit. Where a control gate is placed at an intermediate point along a conduit and discharges freely into the downstream section or where the flow is conveyed in a separate downstream pipe, the internal pressure upstream from the control will be approximately equal to full reservoir head. The structural design and safety aspects of the upstream portion will then be concerned with the effects of both the external loadings and the internal hydrostatic pressure acting on the conduit shell. The watertightness of the conduit in the extreme upstream section will be of less importance because the external and internal hydrostatic pressures will closely balance, and leakage into or out of the conduit will be minimized. However, the external pressure around the conduit will normally diminish with increasing distances from the reservoir. At downstream portions of the pressure conduit, there may be an excess of internal pressure which could cause leakage through joints or cracks into the material surrounding the conduit barrel. The flow from such leaks might follow along the outside of the conduit to the section not under pressure where piping through joints could occur. Where a pressure conduit is carried through an embankment, the development of piping with eventual failure of the dam is a possibility. Where such a conduit comprises a tunnel, leakage through seams in the rock might saturate the hillside overburden above the tunnel and cause a sloughing or landslide on the abutment.

To minimize the possibilities of failures such as those described above, it is normal practice to limit the length of the pressure portion of a cut-and-cover conduit to that part of the outlet upstream from the crest of the dam, or, in some instances, to approximately the upstream one-third of the dam only. Where there is concern regarding the watertightness of a pressure conduit in the upstream portion of a dam, but there are compelling reasons why the control cannot be located near the upstream end of the conduit, that portion upstream from the control may be provided with a steel liner.

For a tunnel installation, except for the possibilities of leakage discussed previously, the location of the control gate is not as critical as it is for cut-and-cover outlets. However, the pressure portion of the tunnel ordinarily should not extend downstream beyond a point where the weight of the column of rock above the tunnel or the side resistance to a blowout is less than the internal pressure forces, unless the tunnel lining is properly reinforced to withstand the internal pressure and a waterproof liner is provided to prevent a buildup of hydrostatic pressures outside the lining.

There may be cases where neither pressure nor free flow is desirable, either for a portion of a conduit or for its entire length. Such instances may occur where it is expected that excessive settlement or movement of the conduit will occur and that cracking and opening of joints cannot be avoided. In this situation, to forestall serious leakage that would occur if a free-flow or pressure conduit were used, a separate smaller pipe can be installed inside the larger conduit to convey the flow. The control gate or valve could be installed at the upper end of the pipe, at some intermediate location, or at the downstream end. If the control gates are not placed at the upstream end, guard gates might be provided at the upstream end of the pipe to effect closure in the event of a leak or failure along any part of the pipe.

Where a control gate discharges into a free-flow conduit, an access and operating shaft extending from the conduit to a level above high water surface in the reservoir will be required. For a cut-and-cover conduit under an earthfill dam, the location of the control gates is usually selected so that the operating shaft is positioned immediately upstream from the crest of the dam. Where flows in the downstream portion of the conduit are carried in a separate pipe, a control chamber is usually provided at the upstream end of the pipe.

The control gates or valves for a conduit or sluice through a concrete dam can be positioned at any point, either upstream to afford free flow in the sluice or at the downstream end to provide pressure pipe flow. Where the sluices are placed in the overflow section of the dam, upstream gates controlling the entrance or valves operated from an interior gallery in the dam are ordinarily employed. Where the outlets are placed in the nonoverflow section, either upstream gates or downstream valves are utilized.

B. OUTLET WORKS COMPONENTS

1. General

For an open channel outlet works or for a conduit-type outlet where part full flow prevails, the control gates or valves are the determining factors which establish the outlet works capacity. Where an outlet works operates as a pressure pipe, the size of the waterway as well as that of the control device determines the capacity.

The overall size of an outlet works is determined by its hydraulic head and the required discharge capacity. The selection of the size of some of the component parts of the structure, such as the tunnel, is dictated by practical considerations or by collateral requirements such as diversion. Since the capacity of a closed system outlet is influenced by the hydraulic losses through the components, the sizes of various features can be changed in relation to one another for a given capacity. For example, a streamlined inlet may permit the installation of a smaller gate for a given size conduit, or a larger gate may allow the use of a smaller conduit. Or, for a given discharge, enlargement of the upstream pressure conduit of a closed pipe system may permit reduction in the size of the downstream pressure pipe and consequently in the size of the downstream conduit. The determination of the best overall layout to achieve economy in the design may, therefore, require alternative studies involving various trial sizes of the different components of the outlet works.

When the type of waterway is chosen and the method of control is established, the associated structures to complete the layout can be selected. The type of intake structure will depend on its location and function and on the various appurtenances such as fishscreens, trashracks, stoplog arrangements, or operating platforms which must be furnished. A means for dissipating the energy of flow before returning the discharge to the river may have to be provided. This might be accomplished by a deflector lip, a stilling basin, or a similar dissipator device. Gate chambers, control platforms, or enclosures may be required to provide operating space and protective housing for the control devices. An outlet works also may require an entrance channel to lead diversion flows or flows when the reservoir is low to the intake structure, and an outlet channel to return releases to the river.

2. Waterways

(a) Open Channels. Open channel waterways for outlet works are similar to those for a canal headworks structure, a sluiceway through a dam, or an ordinary spillway. The waterway will usually consist of a channel or flume placed through the embankment to carry the flow from the reservoir to a canal or to the downstream river level. Details of the design are comparable to those for a gated orifice-controlled spillway.

(b) Tunnels. Because of its inherent advantages, a tunnel outlet works is preferred where abutment and foundation conditions will permit its utilization and if it is economical compared with other types. A tunnel is not in direct contact with the dam embankment, and therefore it provides a much safer and more durable layout than can be achieved with either a cut-and-cover conduit or an open channel structure. A minimum of foundation settlement, differential movement, and structural displacement will be experienced with a tunnel which has been bored through competent abutment material, and seepage along the outer surfaces of the tunnel lining or leakage into the material surrounding the tunnel will be less serious. Furthermore, there is less likelihood that failure of some portion of the tunnel would cause failure of the dam than if a cut-and-cover conduit passing under or through the dam were to fail.

Ordinarily, pressure tunnels in competent rock do not require lining reinforced to withstand full internal hydrostatic pressures, since the surrounding rock normally can assume such stresses. If the rock cover has sufficient weight and enough side resistance to prevent blowouts, only an unreinforced lining is necessary to provide watertightness in seamy rock and smoother surfaces for better hydraulic flow.

Where pressure tunnels are placed through less competent foundations, such as jointed or yielding rock, the tunnel lining must be designed to withstand external hydrostatic and rock loadings in addition to internal hydrostatic pressures. At the extreme upstream end of an outlet works tunnel, where external hydrostatic pressures may nearly balance the internal pressures, the lining will need to be reinforced to withstand rock loads only. However, if provision is made for unwatering the tunnel by use of intake gates, bulkheads or stoplogs, an unbalanced hydrostatic condition will exist. At the downstream portions of the tunnel, where outside water pressures diminish, the design of the tunnel lining will need to consider both external loads from the rock and internal water pressures.

For free-flow tunnels in competent rock, lining might be provided only along the sides and bottom to form a smooth waterway. In less competent material, lining of the complete cross section may be necessary to prevent caving. For that portion of a free-flow tunnel immediately adjacent to the reservoir or just downstream from a pressure tunnel, cognizance must be taken of the possibility of hydrostatic pressure buildup behind the lining due to leakage through the walls of the pressure tunnel or to seepage from the reservoir. Ordinarily, such external water pressure can be reduced by grouting and by providing drain holes through the lining of the free-flow tunnel.

The need for lining a tunnel in which an independent pipe is installed depends entirely on the competency of the rock to stand unsupported. Since such a tunnel is used to house the pressure pipe and provide access to an upstream gate, lining sufficient to avoid rock falls might be provided for protection of the pipe and operating personnel.

For a pressure tunnel a circular cross-sectional shape is the most efficient, both hydraulically and structurally. For a free-flow tunnel a horseshoe shape or a flat bottom tunnel will provide better hydraulic flow, but it is not as efficient as the circular shape for carrying external loads. For small tunnels under only moderate heads the horseshoe-shaped pressure tunnel and either the horseshoe or the flat-bottomed free-flow tunnel may be permissible, depending on foundation conditions. As discussed in Section IIA4 of this chapter, it is not practicable to provide tunnels much smaller than about 6 feet in diameter. The structural design of tunnels, including reinforcement of linings, is discussed in Section IIC2.

(c) Cut-and-Cover Conduits. If a closed conduit is to be provided and if foundation conditions are not suitable for a tunnel, or if the required size of the waterway is too small to justify the minimum-size tunnel, a cut-and-cover conduit must be used. Since such a conduit passes through or under the dam, conservative and safe designs must be used. Numerous failures of earthfill dams caused by improperly designed or constructed cut-and-cover outlet conduits have demonstrated the need for conservative procedures.

A conduit should be placed on the most competent portion of the dam foundation. Details of the design must allow for expected settlement, shrinkage, and lateral or longitudinal displacement without interfering with the continuity of the structure which must provide a safe and leakproof waterway.

Where bedrock occurs at the site, every attempt should be made to place the entire conduit on such a foundation. If this is not physically or economically feasible, the structure should be located where overburden is shallow so there will be a minimum of foundation settlement. If a uniform foundation exists and it is determined that foundation settlement will not be excessive, the excavation for the conduit should be to exact grade and the conduit supported on undisturbed material. Where the conduit foundation in its natural state is not suitable, the unsuitable material should be excavated to a depth where a material competent to support the load is reached, and the excavation should be refilled with compacted material of desired stability and impermeability. Unsuitable foundation materials include those which are so permeable as to permit excessive seepage, those subject to excessive settlement on loading, and those subject to settlement on saturation of the foundation by the reservoir. These materials are described in chapter 3. In all cases, regardless of the nature of the foundation, the contact of the conduit with the foundation must provide a watertight bond, free of void space or unconsolidated areas.

Cut-and-cover conduits must be designed of sufficient strength to withstand the load of the fill overlying the structure. Pressure conduits must also be designed to resist an internal hydrostatic pressure loading equal to full reservoir head. Design loadings for conduits are further discussed in section IIC3.

The adaptability of a cut-and-cover conduit and the desirability of utilizing such a conduit as a pressure pipe or as a free-flow waterway are discussed in section IIA6. Since a cut-and-cover conduit in most instances must be constructed before the embankment, conduit settlement will follow the foundation settlement resulting from the embankment loading. The conduit settlement therefore will be maximum at the point of highest fill and will diminish toward each end. Structure details must be selected to allow for such settlement, and conduit profiles must be adjusted to take account of the drop in grade near the center of the dam. Joint treatment and reinforcement requirements are discussed in section IIC3. See Fig. 9-16.

3. Controls.

(a) Control Devices. Selection of the outlet works arrangement for small dams should be

Figure 9-16. Outlet works conduit under construction, Silver Jack Dam, Colorado. (USBR)

based on the use of commercially available gates and valves or relatively simple gate designs, rather than on the use of special devices which will involve expensive design and fabrication costs. Cast-iron slide gates, which may be used for control and guard gates, are available for both rectangular and circular openings and for design heads in excess of 50 feet. Simple radial gates are available for ordinary surface installations, and top-seal radial gates can be secured from manufacturers on the basis of simple designs and specifications. For low heads, commercial gate valves and butterfly valves are suitable for control at the downstream end of pressure pipes if they are designed to operate under free discharging conditions. They are also suitable as inline guard valves for wide-open operation, and they can be adopted for inline control valves if air venting of the pipe is provided immediately downstream from the valve.

For higher head installations, specially designed gates and valves have been developed and utilized. [13], [14] Usual upstream controls are the rectangular leaf gates equipped either with sliding bronze bearing seats or with fixed-wheel or roller train side mountings. Inline installations include: a simple rectangular high-pressure leaf gate or a leaf gate with a ring follower, operating within a bonneted metal housing; rectangular leaf fixed-wheel gates operating in slots in either wet or dry wells; gate valves, needle valves or tube valves; and butterfly valves. At the downstream end of an outlet pipe, regulating gates or valves usually employed are: the high-pressure slide gate or wheeled jet-flow gate, contained within metal housings; the needle, tube or hollow-jet valves; and the Howell-Bunger fixed-cone valve.

(b) Arrangement of Controls. Flows through low-head outlet works can be controlled by various devices. A surface radial gate may be installed in an open channel.

Upstream gate controls for conduits are generally placed in a tower structure, with the gate hoists mounted on the operating deck. With this arrangement the tower must extend above the maximum water surface.

If controls are located at some intermediate point along the conduit, slide gates or top-seal radial gates can be used, operating in a wet wall shaft which extends vertically from the conduit level to the level of the crest of the dam.

A variation of the slide gate control which will eliminate the need for a wet well shaft is possible. In this instance watertight bonnets are provided over the gate slots and the gates are operated either from a dry shaft or from an operating chamber located above the conduit level. Watertight bushings are provided where the gate stems extend through the bonnets.

Valves also can be used as controls at intermediate points along conduits. A dry well is provided and the valve is placed in a length of pipe whose upstream end is encased in a concrete plug. If the flow is carried by separate pipe in a conduit sufficiently large to afford access along the pipe from the downstream end, a domed chamber can be used rather than a dry well shaft.

If a concrete dam utilizes a slide gate control on its upstream face, the gate frame and stem guides can be mounted directly on the concrete face and the hoist can be placed on a platform cantilevered from the crest of the dam. If the gate is placed at an intermediate point along a conduit formed through the concrete dam, the gate can be operated either in a wet well with the hoist placed at the crest of the dam, or from a gallery if the watertight bonnet cover arrangement is provided over the gate well. Inline valves can also be operated from the gallery or from a chamber formed inside the dam. A control valve placed on the end of the conduit at the downstream face of the dam can be operated from a platform extending from the face of the dam.

(c) Control and Access Shafts. Where a free-flow conduit is provided downstream from the control devices, access for operating is usually from a shaft located directly over the controls. If the wet well arrangement is utilized, a shaft of sufficient width and breadth to accommodate the several wells must be provided. When the type of controls permits dry well installations, only sufficient space to provide operating room at the bottom of the shaft is needed. A smaller access shaft, either directly above or offset from the chamber, and just large enough to permit passage of removable and replaceable gate parts, will then be needed.

The operating or access shaft for a tunnel outlet works can be sunk into the undisturbed hillside and lined with concrete as necessary to maintain the shaft walls intact. Where such a shaft is used for access and ventilation only, a minimum of wall lining will be needed. Where an access shaft is to be used for a wet well arrangement, adequate lining to make the shaft reasonably watertight will be required. If a cut-and-cover conduit scheme is used, the shaft must be constructed through the dam embankment. The structural design must consider the possibility of settlement and lateral displacement as a result of movement of the embankment. Where a wet well shaft is employed, care must be taken in the design to prevent cracks and the opening of joints which would permit leakage from the interior of the shaft into the surrounding embankment. The walls of the wet well shaft must be designed to resist internal hydrostatic pressure from full reservoir head in addition to the external embankment loading. If a shaft extends through the embankment and projects into the reservoir, external hydrostatic loads must also be considered. The protruding portion of the shaft constitutes a tower which is subject to the ice loads discussed in section IIB4.

(d) Control Houses. A housing around the outlet controls is sometimes provided where operating equipment would otherwise be exposed or where adverse weather conditions will prevail during operating periods. A house is sometimes provided to enclose the top of an access shaft, although the controls may be located elsewhere. Such houses are usually made sufficiently large to accommodate auxiliary equipment such as ventilating fans, heaters, flow measuring and recording meters, air pumps, small power-generator sets, and equipment needed for maintenance.

4. Intake Structures

In addition to forming the entrance into the outlet works, an intake structure may accommodate control devices. It also supports necessary auxiliary appurtenances (such as trashracks, fish screens, and bypass devices), and it may include temporary diversion openings and provisions for installation of bulkhead or stoplog closure devices.

An intake structure may take on many forms, depending on the functions it must serve as noted above, on the range in reservoir head under which it must operate, on the discharge it must handle, on the frequency of reservoir drawdown, on the trash conditions in the reservoir which will determine the need for or the frequency of cleaning of the trashracks, on reservoir ice conditions or wave action which could affect the stability, and on other such considerations. An intake structure may either be submerged or extended as a tower to some height above the maximum reservoir water surface, depending on its function. A tower must be provided if the controls are placed at the intake, or if an operating platform is needed for trash raking, maintaining and cleaning of fish screens, or installing stoplogs. Where the structure serves only as an entrance to the outlet conduit and where trash cleaning ordinarily will not be required, a submerged structure can be adopted.

The conduit entrance can be placed vertically, inclined, or horizontally, depending on intake requirements. Where a sill level higher than the conduit level is desired, the intake can be a drop inlet similar to the entrance of a drop inlet spillway. A vertical entrance is usually provided for inlets at the conduit level. In certain instances at small installations where the gate is placed and operated on the upstream slope of a low dam, an inclined entrance can be adopted. In most cases conduit entrances should be rounded or bellmouthed to reduce hydraulic entrance losses.

The necessity for trashracks on an outlet works depends on the size of the sluice or conduit, the type of control device used, the nature of the trash burden in the reservoir, the utilization of the water, the need for excluding small trash from the outflow, and other factors. These factors will determine the type of trashracks and the size of the openings. Where an outlet consists of a small conduit with valve controls, closely spaced trashbars will be needed to exclude small trash. Where an outlet involves a large conduit with large slide gate controls, the racks can be more widely spaced. If there is no danger of clogging or damage from small trash, a trashrack may consist simply of struts and beams placed to exclude only the larger trees and such floating debris. The rack arrangement will also depend on accessibility for removing accumulated trash. Thus, a submerged rack which seldom will be unwatered must be more substantial than one which is at or near the surface. Similarly, an outlet with controls at the entrance where the gates can be jammed by trash protruding through the rack bars must have a more substantial rack arrangement than if the controls are not at the entrance.

Trash bars usually consist of thin, flat steel bars which are placed on edge from 3 to 6 inches apart and assembled in rack sections. The required area of the trashrack is fixed by a limiting velocity through the rack, which in turn depends on the nature of the trash which must be excluded. Where the trashracks are inaccessible for cleaning, the velocity through the racks ordinarily should not exceed 2 feet per second. A velocity of up to approximately 5 feet per second may be tolerated for racks which are accessible for cleaning.

Trashrack structures also may take on varied shapes, depending on how they are mounted or arranged on the intake structure. Trashracks for a drop inlet intake are generally formed as a cage surmounting the entrance. They may be arranged as an open box placed in front of a vertical entrance or they may be positioned along the front side of a tower structure.

At some reservoir sites it may be desirable or required to screen the inlet entrance to prevent fish from being carried through the outlet works. Because small openings must be used to exclude fish, the screen can easily become clogged with debris. Provisions must therefore be made for periodically removing the screens and cleaning them by brooming or water jetting.

Where the control is placed at an intermediate point along a conduit, some means of unwatering the upstream pressure section of the conduit and the intake is desirable to make inspections and needed repairs. Stoplog or bulkhead slots are generally provided for this purpose in the intake or immediately downstream from the intake. In intake towers containing control devices, the stoplog slots are placed upstream from the controls. A circular, flat bulkhead which can drop down over the entrance is generally provided for a drop inlet structure. The bulkhead can be stored on supports near the top of the structure. Closure can then be effected under water by lowering the bulkhead with a cable winch operated from a barge, or from the top of the structure if the reservoir is low enough to expose the upper portion.

For an intake structure with inlet sill above the invert of the conduit, it may be desirable for various reasons to draw the reservoir down below that level. In such an instance a bypass can be provided near the base of the structure to connect the reservoir to the conduit downstream. In other instances, where flow must be maintained while installing or maintaining the control gates and outlet pipes or while repairing or maintaining the free-flow conduit concrete, it may be desirable to carry a separate pipe under or alongside the conduit to bypass it entirely. In either case, the bypass inlet can be placed in the intake structure and usually can be controlled by a slide gate mounted on one of the faces of the structure and operated from some higher level.

Where winter reservoir storage is maintained

and the surface ices over, the effect of such conditions on the intake structure must be considered. Where reservoir surface ice can freeze around an intake structure, there is danger to the structure not only from the ice pressures acting laterally but also from the uplift forces if a filling reservoir lifts the ice mass vertically. These effects must be considered when the advantages or disadvantages of a tower are compared with those of a submerged intake.

If a tower is constructed where icing conditions present a hazard, ice may be prevented from forming around the structure by the subsurface release of air. The air causes the slightly warmer water at lower depths to rise and mix with the colder surface water, thus preventing freezing. If an insufficient supply of warmer water is available, as when the approach channel to the tower is shallow or the reservoir storage is small, the compressed air may actually cause freezing around the structure.

5. Terminal Structures and Dissipating Devices

The discharge from an outlet, whether through gates, valves, or free-flow conduits, will emerge at a high velocity, usually in a nearly horizontal direction. For a free-flow conduit, deflector devices might be employed to direct the high-velocity flow away from the outlet structure and past the downstream toe of the dam if erosion-resistant bedrock exists at shallow depths in the downstream channel. Where softer foundations exist, a dissipating device might be provided to absorb the energy of flow before it is returned to the river or canal. The flow from valves at the end of an outlet will generally be in the form of a jet, which can be discharged into the river, into a plunge basin downstream from the outlets, or into a hydraulic jump-type basin. See Fig. 9-17.

Where an outlet is terminated as a submerged pipe, a stilling well dissipator is some-

Figure 9-17. Outlet works, Friant Dam, California. (USBR)

times employed to dissipate the flow energy. This device consists of a vertical water-filled well in which dissipation is achieved by turbulence and by diffusion of the incoming flow. The incoming flow can be directed horizontally into the well near the bottom, or it may be directed vertically downward into the well through a vertical pipe and released near the bottom. In both cases the flow rises upward and emerges out of the top of the well.

Terminal structures for free-flow conduit outlet works are essentially the same as those for spillways. The design of the basins to dissipate jet flow and the design of stilling wells to dissipate submerged pipe flow are discussed in Section IB2(c) of this chapter.

6. Entrance and Outlet Channels

An entrance channel leading to the outlet works intake and an outlet channel to deliver flow to the river downstream, are often required with a tunnel or cut-and-cover conduit layout. An entrance channel may be required to convey diversion flows to a conduit placed in an abutment, or to deliver water to the outlet works intake during low reservoir stage. Outlet channels may be required to convey discharges from the end of the outlet works to the river downstream or to a canal. All such channels should be excavated to stable slopes and to dimensions which will provide nonscouring velocities. Entrance channel velocities are usually made less than those through the trashracks, and the entrance channel is often widened near the intake structure to permit a smooth, uniform flow into all trashrack openings.

The outlet channel dimensions and the need for lining or riprap protection will depend on the nature of the material through which the channel is excavated. Occasionally a control or a measuring station is placed in the outlet channel, in which event the selection of the grade and cross section of the channel becomes an important consideration. The effects of aggradation or degradation of the main river channel must be considered in selecting the outlet works outlet channel dimensions.

C. STRUCTURAL DESIGN DETAILS

1. General

The same types of structures may be used for either spillways or outlet works. Spillways utilize open channels more often than do outlet works, and the structural design details for open channels are therefore discussed in part I-C of this chapter. The details of the design of the walls, open channel linings, and floors, discussed as spillway structures, are applicable to these structures when used for outlet works. Also, the headworks of open channel outlet works are similar to gated crest structures for spillways as regards structural design details.

On the other hand, closed conduit waterways are more commonly used for outlet works than they are for spillways, and the design details of closed conduits are therefore discussed in this section. The design details are the same in either case.

A closed conduit waterway might be a cast-in-place cut-and-cover culvert or conduit, a precast or prefabricated pipe, or a tunnel bored through the abutment. Waterways for a spillway will most often be free flowing, while those for outlet works may either flow full under pressure or partly full. The security of earthfill and rockfill dams depends to a large degree on the safety of the spillway and outlet structures, especially when conduits pass through the embankment. In those cases where all or part of a conduit is under internal pressure due to reservoir head, any leakage or failure of the conduit may cause openings through the dam which may gradually be enlarged until partial or complete failure results. There is also the danger of seepage along the contact surfaces between the conduit and the earthfill, which may result in serious damage. A third danger is the possibility of structural collapse of the conduit which would be almost certain to result in failure of an earthfill dam. These facts emphasize the importance of using durable materials, conservative design procedures, proper design details, and construction methods that will insure safe structures.

Replacement of a conduit through either an earthfill or rockfill dam is usually a difficult and expensive operation which can be avoided by the use of permanent material, such as cast iron for small size pipes and reinforced concrete cast-in-place conduit or precast concrete pipe for larger sizes. For small reservoirs with comparatively low heads, exceptions may be made only where the possible damage from failure is of little or no consequence. In such cases, where it is economically advantageous, the use of iron or steel pipe protected by galvanizing or bituminous coating or by some other rust-resisting treatment may be justifiable. It should be recognized that buried steel pipe is vulnerable to deterioration from electrolytic action, even where rust-resisting treatments or protective coatings are provided. Limited service with a possibility of eventual failure must therefore be expected. The use of iron or steel pipe as a watertight liner in a concrete conduit is permissible. However, unencased metal pipe should not be used if loss of life or serious damage to property will result from failure of the dam which might follow deterioration of the pipe.

Conduit joints must be made watertight to prevent leakage into the surrounding embankment. Joints of concrete cast-in-place conduits must be sealed with waterstops, and rubber-gasketed joints must be used for precast concrete pipe. For metal pipe, couplings are required which will remain watertight after movement or settlement of the pipe. Corrugated metal pipe with riveted seams should not be used for conduits unless the seams are welded or adequately calked and sealed with durable materials. If used for pressure conduits, corrugated metal pipe should be pretested for watertightness with pressures equal to twice the operating head.

When the outlet conduit consists of prefabricated pipe, whether of reinforced concrete, cast iron, or steel, the methods of bedding the pipe and backfilling around it should be such as to insure, insofar as possible, against unequal settlement and to secure the most uniform possible distribution of load on the foundation. When filling around these structures, extreme care should be taken to secure tight contact between the fill and the conduit surface and to obtain proper densities of the earthfill material. This is important not only for the prevention of seepage along the conduit, but also to insure that the fill develops a lateral restraint on the structure which will prevent excessive stresses in the conduit shell.

When the outlet consists of precast reinforced concrete or metal pipe, it should be set carefully on a good foundation and well bedded in concrete. The concrete bedding not only aids in distributing the conduit load on the foundation, but also guarantees against uncompacted zones and void spaces under the pipe which could induce leakage along the undersurface of the structure. Void spaces or inadequate compaction of impervious materials at the invert of pipes have been the cause of numerous failures of small earthfill dams. The practice of supporting pipes on piers or collars without a concrete bedding should be avoided, because the greater foundation reaction at the concentrated support points will cause unequal stress distribution in the pipe. Furthermore, if the foundation settles from under the conduit between piers, the unsupported conduit will sag and crack. If the conduit is sufficiently strong to sustain the fill load, the earth shrinking away from the underside will leave voids which will permit the free passage of water.

Details of designs for cut-and-cover conduits are discussed in Section IIC3 of this chapter.

2. Tunnel Details

Linings are provided in tunnel waterways for both hydraulic and structural reasons. The smooth boundary surfaces reduce frictional resistance and permit a smaller diameter tunnel for a required capacity. Lining also is used to prevent saturation of the surrounding ground by seepage. Structural lining is used to support the tunnel walls against raveling or yielding ground.

Where the purpose of the lining is to provide a smooth hydraulic flow surface or to reduce seepage, the thickness may be determined by requirements for shrinkage, temperature change, and concrete placement. For ordinary linings where reasonably stable ground is encountered and where a minimum of tunnel support is required, an average lining thickness between ¾ and 1 inch per foot of tunnel diameter is ordinarily used. The minimum thickness usually provided is 6 inches. Yielding ground or areas through water-bearing strata may require thicker linings to resist external rock loads and hydrostatic pressures. A full circular lining is the most efficient shape to withstand such external loads.

Where the tunnel lining is to be reinforced, it must be made sufficiently thick both to accommodate the reinforcement mat and to provide sufficient room for placing the concrete in the confined space behind the forms. A minimum thickness of 6 inches is suggested for tunnel linings with a single layer of reinforcement. Where two layers of reinforcement are required a minimum thickness of 9 inches may be desirable. In either case, the contractor may need to provide additional space outside the reinforcement to accommodate a concrete placement discharge pipe.

The portions of a tunnel which must be reinforced and the amount of reinforcement required depend on the tunnel shape, external and internal loadings, requirements for watertightness, and many geological factors. For a nonpressure tunnel, reinforcement may be required to resist external loads due to unstable ground or to grout or water pressures. Pressure tunnels with high internal hydrostatic loads must have lining reinforced sufficiently to withstand bursting where inadequate cover or unstable supporting rock prevails.

A suggested general guide for determining reinforcement requirements in tunnels is as follows:

(1) A pressure tunnel should ordinarily be reinforced whenever the depth of cover is adequate to withstand the unbalanced internal pressure head or whenever leakage control is important. The reinforcement should be sufficient to provide the required structural strength and leakage control for the maximum internal hydrostatic pressure reduced by a conservative estimate of the external hydrostatic pressure that is expected along the length of the tunnel. Restraint from the surrounding rock should only be considered when properties of the rock are known and a complete analysis is made. Where there are provisions for unwatering, the external pressure head should be the maximum that can be expected along the length of the tunnel.

(2) The transition from a pressure tunnel to a nonpressure tunnel should be specially reinforced to prevent excessive cracking which would permit leakage from the pressure portion of the tunnel to enter behind the lining of the nonpressure portion. Reinforcement for the pressure portion for a distance upstream from the junction equal to five times the diameter of the tunnel should be based on full internal hydrostatic head with no allowance for restraint from the surrounding rock. The nonpressure portion of the tunnel should be reinforced for a distance downstream from the control equal to from two to five times the tunnel diameter, for an external static head equal to the internal head just upstream from the control.

(3) A nominal amount of both longitudinal and circumferential reinforcement should be provided near the portals of both pressure and nonpressure tunnels to resist loads resulting from loosened rock above the tunnel or from sloughing of the portal cuts. This reinforcement should extend back from the portals for a distance equal to at least twice the tunnel diameter.

(4) If in competent rock, the tunnel, other than at the portals and at the transition from pressure to nonpressure, may be unreinforced where the rock cover is adequate to withstand the unbalanced internal pressure head. If in unstable ground, the lining should be reinforced to support probable rock loadings. Methods of estimating loadings for tunnel supports given in the publication, "Rock Tunneling With Steel Supports" [5], can be used to estimate requirements for the reinforced lining. Where the properties of the rock are known, an analytical determination can be made of the stresses that will theoretically exist in the rock surrounding the excavated tunnel and permanent supports and concrete lining, either acting separately or in combination, can be designed.

Permanent supports consist of steel ribs, steel lagging, steel-liner plates, shotcrete, reinforcement sheets, rock bolts with or without chain link fabric, or a combination of these. The choice of one or a combination of these types of supports should depend upon geologic conditions, groundwater levels, excavation methods to be used, length of time between excavating and placing permanent lining, and economic factors.

Supporting ribs, rock bolts, or other support types must be capable of holding up large blocks of material whose natural support was removed in excavating the tunnel. The lagging must be closely spaced where the rock is highly fractured or slacks off in small pieces; elsewhere it may be more widely spaced or even omitted. Methods of assuming and computing the size of supports are given by Proctor and White [5]. Loadings will be based on the nature of the ground encountered and, unless the exact underground conditions are known beforehand, the design of the ground-support system can only be approximated. Because of the uncertainties that may exist, the required size and spacing of supports are often determined by trial as the work progresses. In permanently supported sections of the tunnel, all spaces outside of the lagging or liner plates should be filled as completely and compactly as possible with clean gravel and rock spalls, and the spaces thoroughly filled with grout after the lining has been placed.

For tunnels through jointed rock or where seepage is to be minimized, the areas surrounding the tunnel are usually grouted both to consolidate the material and to fill open fissures in the rock and the voids between the lining and the rock. This grouting is accomplished by drilling holes through the lining, or through pipes placed in the lining for this purpose, into the surrounding rock and then injecting grout under pressure. Permissible grouting pressures will depend on the nature of the surrounding ground and on the lining thickness. For small tunnels, rings of grout holes are spaced at about 20-foot centers, depending on the nature of the rock. Each ring consists of grout holes distributed at about $90°$ around the periphery, with alternate rings placed on vertical and $45°$ axes.

Drainage holes are often provided in other than pressure tunnels to relieve external pressures caused by seepage along the outside of the tunnel lining. The drainage holes also are spaced at about 20-foot centers, at intermediate locations between the grout hole rings. At successive sections, one vertical hole is drilled near the crown alternating with two drilled horizontal holes, one in each side wall. In free-flow tunnels, drainage holes are provided only above the water surface; if flow through the tunnel is conveyed in a separate pipe, the horizontal holes are drilled near the invert.

3. Cut-and-Cover Conduit Details

(a) General. The design of a cut-and-cover conduit to be constructed through or under an earthfill embankment must include details which will provide for movement and settlement without excessive cracking or leakage. To obtain a safe structure, the following factors must be considered:

(1) Provide devices to minimize seepage along the contact of the conduit and the impervious embankment.

(2) Provide details to forestall cracking which might result in leakage of water into the fill surrounding a pressure conduit and to prevent piping of embankment material into a free-flow conduit.
(3) Select and treat foundation to minimize differential settlement which is a cause of cracking.
(4) Provide a structure to safely carry the loads to which the conduit will be subjected.

Selection of designs and details to accomplish the above purposes is discussed in this section.

(b) Cutoff Collars. Foundation preparation and compaction around conduits must be equivalent to foundation preparation for the dam and to compaction of the impervious earthfill. Projecting fins or cutoff collars are provided to minimize seepage along the contact between the outside surface of the conduit and the embankment. These collars should be made of reinforced concrete, generally from 2 to 3 feet high, 12 to 18 inches wide, and spaced from 7 to 10 times their height along that portion of the conduit which lies within the impervious zone of the dam. The length of the percolation path along the contact is thereby increased by 20 to 30 percent.

For a conduit on an earth foundation, the collar should completely encircle the conduit barrel. Where the foundation is sound rock, good contact along the base may be expected and the collars need extend only to be keyed into the rock foundation. The collars should be separated from the conduit to avoid introducing concentrated stresses into the conduit walls, which would alter the normal stress in the barrel. This is accomplished by constructing the collars with watertight fillers between the collars and the barrel. The structural separation permits lateral slipping of the collar on the barrel, eliminates secondary stresses in the conduit which would otherwise be caused by the stiffening effect of the collars, and avoids the introduction of torsional stresses in the conduit if horizontal movement or displacement of the embankment should occur. The joint filler material can be several layers of graphite-coated paper if only slight movement is expected, or premolded bituminous fillers where greater movement is expected.

Although cutoff collars usually are located between joints in the conduit, there are cases where collars have been constructed to span the joints. When so located they also serve as watertight covers for the joints. Where the collar is not placed at a conduit joint or where it is placed over a joint which is restrained from movement by keyways or by reinforcement extending across it, the collar ordinarily will not be subjected to large lateral loadings. In such cases it will need to be only strong enough to resist the superimposed fill load. When a collar covers a joint designed to permit differential movement, either the collar must be designed sufficiently strong to restrain such movement, or the collar must adjust to the movement without losing the watertight contact.

(c) Conduit Joints. Conduits constructed on rock or competent earth foundations may be subjected only to small settlement and longitudinal movements. Cast-in-place conduits on such foundations can be made more or less monolithic and, except for movement caused by initial setting shrinkage and by temperature expansion and contraction, should experience little cracking or joint opening. In such a design, major cracking is avoided by placing the conduit in short sections (usually 12 to 16 feet) and by liberal use of longitudinal reinforcement placed across construction joints to form a continuous structure. During construction, the adjoining sections of the barrel are not constructed until after the major volume change in the first-placed section due to initial setting shrinkage has taken place. Waterstops of metal or rubber are placed across the joints to provide a watertight seal.

Where considerable settlement and lateral or longitudinal adjustment of the foundation is expected, the conduit may be constructed as an articulated structure. The individual portions of the structure must be free to move without causing uncontrolled cracking which would permit leakage through the conduit walls. For such designs the reinforcement is not carried continuously across the joints, so that the individual sections are free to move longitudinally. Waterstops are provided to prevent leakage through the joints. Differential lateral displacement of the conduit sections at the joints is ordinarily restrained by a bell-and-spigot joint or by a reinforced collar encircling a plain joint. Rubber-gasketed joints similar to those utilized for precast concrete pipe can be adopted for joining individual lengths of conduit. These joints can be used in cast-in-place conduits by embedding short sections of precast pipe which contain the joint detail. Specifications for pipe and pipe joints can be found in ASTM specifications, designation C361-75, AWWA specifications C-302-74, or in the Bureau of Reclamation publication, "Standard Specifications for Reinforced Concrete Pressure Pipe," February 1, 1969.

Where precast pipe used for outlet works conduits is bedded on a concrete base, the base should be reinforced with the longitudinal reinforcement continuous through any transverse joints. Differential lateral displacement of precast pipe conduit sections at the joints is thus restrained.

(d) Design Loads. Embankment loads on conduits may vary over a wide range depending on many factors relating to the foundation, method of bedding, flexibility or rigidity of the conduit; and to soil characteristics of the embankment such as angle of internal friction, unit, weight, homogeneity, consolidation properties, cohesiveness, and moisture content. All possible combinations of these various factors must be considered to evaluate their overall effect. The loads must be considered not only as they may occur during construction but also as they may be altered after embankment completion, reservoir loading, and embankment saturation.

The "Marston Theory" of embankment pressures is usually adopted for precast conduits under relatively low fills. This theory is discussed in many bulletins published by the Iowa State College Experiment Station and is abstracted in various handbooks [6, 7] which contain bibliographies of publications dealing with this subject. On the basis of the Marston theory, the vertical load on a conduit is considered to be a combination of the weight of the fill directly above the conduit and the frictional forces acting either upward or downward due to the adjacent fill. A settlement of adjacent fill greater than the overlying fill induces frictional forces acting downward which increase the resultant load on the conduit; a greater settlement immediately above the conduit will result in an arching condition which reduces the load on the conduit. Thus a conduit laid in trench excavated in a compact natural soil will practically never receive the full weight of the backfill above it, because of the development of arching action when the backfill starts to settle. On the other hand, if the conduit is placed so that it projects in whole or in part above the natural ground surface, the embankment load which may come upon it can in some cases be as much as 50 percent greater than the weight of the fill directly above it.

For cast-in-place conduits under relatively high fills, where the conduit is placed in cut so that neither a full trench nor a complete projecting condition exists, a loading assumption which averages the extremes noted above is assumed. For this case the load on the conduit is assumed to be the weight of the column of fill directly above it. The load over that portion of a conduit under the upstream part of the dam includes both the weight of the saturated fill and the weight of the reservoir water above the fill. The conduit barrel is designed on the basis of a given factor of safety, considering that the unit horizontal lateral load on the conduit is one-third of the unit vertical load. The design is then checked on the basis of a reduced factor of safety considering no horizontal lateral load exists. The vertical reaction of the base of the conduit is taken equal to the vertical load plus the weight of the conduit. On an earth foundation, the base reaction is assumed to be distributed uniformly across the width of the conduit; on a rock foundation it is assumed to be distributed triangularly, varying from twice the average unit reaction at the outside edges to zero at the center of the base. External hydrostatic pressures are assumed to act equally in all directions, vertically downward as an increased load, upward as uplift, and laterally on the sides of the conduit.

Procedures for designing concrete box culverts and circular conduits are comprehensively discussed in "Concrete Culverts and Conduits" [8]. Cast-in-place conduits are designed on procedures using moment, thrust and shear coefficients from Beggs Deformeter stress analyses [9].

D. BIBLIOGRAPHY—OUTLET WORKS*

[1] King, W. H., "Handbook of Hydraulics," fourth edition, McGraw-Hill, New York, N.Y., 1954.
[2] Creager, W. P., and Justin, J. D., "Hydroelectric Handbook," second edition, John Wiley & Sons, Inc., New York, N.Y., 1950.

---

*Direct reference is not made in the text to all the publications listed here.

[3] "Friction Factors for Large Conduits Flowing Full," Engineering Monograph No. 7, Bureau of Reclamation, 1965.
[4] Rouse, Hunter, "Engineering Hydraulics," John Wiley & Sons, Inc., New York, N.Y., 1950.
[5] Procter, R. V., and White, T. L., "Rock Tunneling with Steel Supports," Commercial Shearing & Stamping Co., Youngstown, Ohio, 1946.
[6] "Concrete Pipe Handbook," American Concrete Pipe Association, 228 North La Salle Street, Chicago, Ill., 1951.
[7] "Handbook of Drainage and Construction Products," Armco Drainage & Metal Products, Inc., Middletown, Ohio, 1958.
[8] "Concrete Culverts and Conduits," Portland Cement Association, 33 West Grand Avenue, Chicago, Ill.
[9] Phillips, H. B., "Beggs Deformeter Stress Analysis of Single-Barrel Conduits," Engineering Monograph No. 14, Bureau of Reclamation, 1968.
[10] "Building Code Requirements for Reinforced Concrete," (ACI 318-63), American Concrete Institute, Detroit, Mich.
[11] "Supplement for ACI 318-63 Code," Bureau of Reclamation, Denver, Colo., 1968.
[12] "Ultimate-Strength Design Handbook," Special Publication No. 17, American Concrete Institute, Detroit, Mich.
[13] Kinzie, P. A. "High Pressure Reservoir Outlets," Dams and Control Works, U.S. Bureau of Reclamation, 1938, pp. 117–204.
[14] Halliday, B. A., and Hoffman, C. J., "Types of Gates, Valves, and Control Equipment Used for Bureau of Reclamation Spillways and Outlets," Transactions of the International Congress on Large Dams, 4th Congress, 1951, Vol. 2, pp. 169–194.

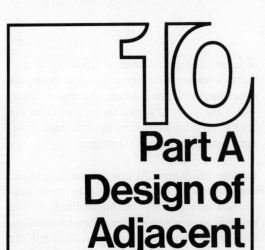

10

Part A
Design of Adjacent Power Plants

CHARLES H. FOGG
Chief, Hydroelectric Design Branch (retired)
Southwestern Division
U.S. Army Corps of Engineers
Dallas, Texas

Part B
Design of Adjacent Locks

JOHN MATHEWSON
Chief, Design Branch
Engineering Division (retired)
Nashville District
U.S. Army Corps of Engineers
Nashville, Tennessee

Part A: Design of Adjacent Power Plants—Charles H. Fogg

RELATION OF POWER PLANTS TO DAMS

The Case for Hydroelectric Power

The use of coal as a source of electrical energy can be planned far into the future. Natural gas, on the other hand, has been given a limited life. Oil will be available for many years, at a price. Nuclear-fueled thermal power is forecast for gradual increase. Why then, in a handbook planned for long time future use, is there a section on hydroelectric power? The answer is that such power is the natural complement of thermal power. The flexibility, low operating cost, and environmental advantages of hydroelectric power combine to make it the most desirable supplier of peak load power.

Forecast. Implicit in the foregoing is the assumption that power generation will continue to grow. Forecasts made prior to the development of earnest environmental opposition appear to need adjustment as to timing, at least. As to quantity, the consensus seems to be that maintenance of the status quo is extremely unlikely; that man can solve the environmental problems and, simultaneously, serve valid human needs by providing a continually increasing supply of convenient energy.

Purpose. The purpose of this portion of the handbook is, therefore, to provide useful guidance for the process of combining dams and hydroelectric power plants; having been assured that a need for such activity will exist in the future.

Basic Design Principles. The design of hydroelectric power plants is subject to three basic principles: safety, acceptable costs, and acceptable environmental disturbance. The designer undertakes to provide optimum safety combined with reasonable costs and minimal environmental upset. Like the solution of most engineering problems, the planning and arrang-

549

ing of a dam and a power plant is a matter of optimizing compromises—of doing the best you can with what you have available.

Toronto. The Toronto power plant at Niagara Falls is cited at this point as an example of the foregoing. Built in 1906, the portion above ground appeared as a monumental stone building adorned with Greek columns. Inside at ground level one found the generators at the top of iron shafts more than 100 ft (30.5 m) long. At the bottom of the shafts were the turbines and the water passages leading back to the river below the falls. The water reached the turbines in vertical penstocks from intakes on the river side of the superstructure. The arrangement was superbly compact. The designers had certainly done well with what they had available. Even from the environmental standpoint they could not be faulted. People loved monumental buildings in that era.

Elements of a Power Plant

In order to successfully combine a power plant with a dam, a competent knowledge of the elements of a power plant is required. The following paragraphs discuss these elements and their effect on the arrangement of a dam and power plant. The major physical parts of a hydroelectric power plant are, fairly obviously, the intake water passages, the powerhouse, and the discharge water passages.

Intake Water Passages

The intake water passages to a plant may be as simple as the short, culvert-like openings to a low head unit's wheel or as complex as the tunnel system feeding the very high head Kemano plant.[1] Simple or complex, the planning of the water passages is a major factor in the development of a safe and economic project. Design of the water passages (intakes, penstocks, tunnels, tailraces, surge tanks and chambers) of a power installation involves interdependent economic, hydraulic, and structural evaluations. For very low heads the economics of these passages has minimal effect on the project as a whole. However, as heads increase, the effect increases. At medium and higher heads, where the plant is separately located from the intake, maximum economy in water passage design is generally vital to the project.

Keys to Economy. At low and medium heads, compactness is the customary key to an economical arrangement. At higher heads no one key will always provide the optimum solution. Site geology and terrain will usually limit the choice of arrangements. At some locations, short pressure tunnels and long tailraces will be optimum. At others, the optimum layout will be obtained with a surface plant at the end of the intake tunnel. At higher heads, too, pressure rise, speed rise, and surge relief all become factors in the determination of the best arrangement.

Intake Water Passage Studies. At low heads the intake passage will usually consist of the transition section between the trash racks and the spiral case. Maximum compactness is available and is limited only by the desired velocity gradient between the racks and the case. At heads above 100 ft (30.5 m), the powerhouse will usually be a separate structure from the dam. Where a concrete gravity dam is used, suitable compactness can be obtained with penstocks fed through an intake on the upstream side of the dam. Examples are Krasnoyarsk[2] and Grand Coulee.[3] Reasonable compactness of intake water passages can also be obtained with arch dams where valley width is adequate for placing the plant in the valley adjacent to the dam. Glen Canyon[4] and Mossyrock[5] illustrate this solution. Tunneling must be resorted to for most other arch dams and for most embankment dams. At some sites, however, it may be feasible to excavate a forebay deep into one of the abutments and to use exposed penstocks on the downstream slope of the abutment. Another alternative, the cut-and-cover conduit, is more rarely used than formerly. One such conduit, however, is the very carefully designed TVA Tims Ford Dam[6] penstock placed in a trench and encased in concrete beneath a rockfill dam. A measure of compactness was achieved at two recently constructed U.S.S.R. plants having embankment dams. At Ust-Khantaisk[7] and at Vilyui[8] side openings in downstream-gated spillway channels serve as intakes to the units. The powerhouse at Ust-Khantaisk is underground, while at Vilyui it is in an open cut. When a tunnel through an abutment is used for the water passage, two choices are open for location of the powerhouse. Place it underground as at Oroville[9] or at the end of the tunnel on the surface as at Nurek.[10] Occasionally, the neck of an oxbow in a stream can be tunneled economically with substantial gain in net head. Such a layout was developed for the Broken Bow project of the Corps of Engineers on the Mountain Fork River in southeastern Oklahoma. At some locations the intake passages can be used for diversion purposes with an economic gain by the double service.

Surge Tanks. When the water passages become long enough to develop significant pressure rise and speed rise values, the need for surge tanks (chambers) must be investigated. The study should be very complete and careful since surge provision is extremely expensive. The post design studies which eliminated surge chambers from the first two lines for the Nurek[10] plant are witness of the never-say-die effort to avoid these costly appendages.

Powerhouses

As an element of the total power arrangement, powerhouse location is the factor of most concern to the engineer planning the project.

Low Heads. For power heads ranging up to about 100 ft (30.5 m), the intake water passages, the powerhouse, and the discharge water passages will be a structural unit replacing a like length of dam.

Higher Heads. Above 100 ft (30.5 m) of net head the powerhouse is seldom incorporated into the dam even in the case of gravity structures. Stability and framing problems become more complex with increasing height. With either gravity or arch structures, however, the powerhouse will be located immediately downstream if space is available. This meets the criteria for maximum practicable compactness of arrangement. When space is not available, and the powerhouse must be located either underground or at the portal of the intake water tunnel, an evaluation of the site conditions is necessary to reach a decision as to the best location for it.

Underground Plants. In Chapter XXII of Ref. 11 Dr. Charles Jaeger presents the case for the underground power plant. The reader is also referred to the technical literature describing such plants. The arguments for going underground are persuasive and many plants have been so planned in the last decade. Although military and climatic reasons supported early decisions to go underground, there are sound technical reasons such as space for larger plants; less costly pressure water passages; less environmental disturbance; turbine settings conducive to use of Kaplans at higher heads than customary; and possibly economic gain by avoidance of surge tanks.

Surface Plants Not Adjacent to Dams. Surface plants may be used by preference or because geologic conditions are unfavorable. In both cases, the record indicates that if the geology will permit tunneling, the water will be conveyed underground to a point near the plant. Surface penstocks (from intake to plant) have, however, been unavoidable at some locations and at some projects have been used by preference.

Discharge Water Passages

The discharge water passage provides for return of the flow to the stream after passing through the turbine. In most surface plant layouts it consists of the unit's draft tube and a short tailrace. In an underground layout the discharge passage requires technical and economic evaluation in comparison with the pressure passage on the upstream side of the turbine. Dr. Jaeger's study[11] discusses the hydraulic and constructional aspects of the problem.

River Basin Developments

Whole river basin complexes of dams and power plants have been arranged and constructed in the past, such systems are under study and design now, and there will be similar opportunities in the future for like developments. Accordingly, it will be desirable to note the nature of the major problems to be solved. Compromise is the key that unlocks the solutions. The power engineer sees economy in a few high dams. The navigation planner finds economy in standardization of structures. The geologist states that dams can be built here, here, and here, but not there; that this dam and this dam can be concrete, but that that dam must be embankment. The environmentalist points out that certain lands must be flooded each year but that certain other areas should be exempted from being submerged. Each field of interest presents its claim to the right to dominate the development. Resolution of these claims has, in the past through compromise, resulted in some truly outstanding achievements. Few would assert, however, that the results have been perfect, and most would agree that better projects should be designed in the future.

Future Planning

Having agreed that there is a future for hydroelectric power and that not all past planning was perfect, what improvements are needed in the future? Environmental concerns are being met and will continue to contribute to future planning. In addition, there are technical problems needing fundamental and innovative solutions.

Low Cost Concrete. Engineers have begun an earnest search for means of dramatically reducing the cost of concrete dams. If this search is successful to the point that concrete dams become generally competitive with embankment dams, the cost of associated power plant elements will also be dramatically reduced. Compact arrangements are customary with gravity structures and are frequently possible with arch dams.

Low Cost Tunneling. Comparisons of mining costs and power plant tunneling costs have been made that indicate an opportunity exists to drastically reduce the cost of this feature of plant layouts.

Siteless Pumped Storage. Reduction in tunneling costs will advance the day when pumped storage can be economically provided, regardless of the levelness of the plain at the site. When, also, the problems of the use of sea water as the pumped storage fluid have been solved,[12] seaside plants will become practicable.

Creative Arrangements. A few new arrangements have appeared in recent years that evinced innovative thinking. For example, the cluster type surface plant[13] wherein four units are encircled by a cylinder that economically resists the pressures of the deep settings of pumped storage units. The U.S.S.R. Toktogul plant[14] houses four units in a squarish structure to save space in a narrow valley. The spillway discharges over the roof of this plant.

Power Complexes

The most obviously desirable improvement in planning would be the combining of thermal and hydroelectric power sources at a single location with, of course, due consideration for the environmental effects of such an arrangement.

Tidal Developments

Although only minimal construction has resulted thus far from tidal power studies, the tides are a vast natural resource ready to be harnessed whenever the right combination of program, structures, and equipment can be assembled.

Inter-Basin Complexes

Large scale inter-basin transfers of water are inevitable in the future. Dam and power engineers must be ready with innovative planning to keep such works economic and environmentally satisfying.

REFERENCES

1. Huber, W. G., "Kemano Power Plant," Civil Engineering, February 1953, p. 50 and June 1953, p. 52.
2. Hydrotechnical Construction, eleven articles on Krasnoyarsk, September 1972, p. 819.
3. Arthur, H. G., "Proposed Third Power Plant at Grand Coulee Dam," ASCE Pow. Div. Jl., March 1967, p. 15.
4. Judd, Samuel, "Glen Canyon Power Plant," ASCE Pow. Div. Jl., July 1961, p. 41.
5. Moore, E. T., "Design and Construction of Mossyrock Penstocks," ASCE Pow. Div. Jl., January 1971, p. 49.
6. Day, G. E., and Bechner, H. S., "Steel Penstock Design for Tims Ford Dam," ASCE Pow. Div. Jl., January 1971, p. 203.
7. Plotnikov, V. M., and Golyshev, A. I., "The Underground Ust-Khantaisk HES," Hydrotechnical Construction, September 1971, p. 803.
8. Glushkin, Y. E., and Demidov, A. N., "Vilyui HES," Hydrotechnical Construction, February 1970, p. 118.
9. Golze, A. R., "Edward Hyatt (Oroville) Underground Power Plant," ASCE Pow. Div. Jl., March 1971, p. 419.
10. Pokrovskii, B. M., et al., "Regulation of the Nurek HES without Surge Shafts in the Headrace Tunnels," Hydrotechnical Construction, November 1971, p. 1012. Polonskii, G. A., and Polonskii, A. G., "Construction of the Nurek HES Penstocks," Hydrotechnical Construction, November 1972, p. 1028.
11. Hydro-Electric Engineering Practice, Editor, J. G. Brown, Three Volumes, 1958, Blackie & Son, Ltd., London.
12. Yoshimoto, Tajio, "Atashika," *Water Power*, February 1972, p. 57.
13. Patrick, J. G., "The Cluster Layout-A New Concept for Pumped Storage," *Water Power*, September 1972, p. 339.
14. Kuzmin, K. K., and Berezinskii, S. A., "Toktogul HES," Hydrotechnical Construction, May 1972, p. 428.

DAM AND POWER PLANT ARRANGEMENTS

If experience is the best teacher, the experience of others should be the next best. In this section the experience of others in the art of combining dams and power plants will be set forth. The projects described in the following are presented in the order of increasing head for convenient reference. However, the reader must not expect to find a "ready-to wear" arrangement for a given head. Experience has taught that attempts to bodily transfer complete designs have failed to achieve the anticipated gains. Differences in site conditions work their way up through the remainder of the plant.

Low Head Arrangements

Low head is defined as the range from zero to about 100 ft (30.5 m). The arrangements in this range are normally of the integral type. The water passages and power equipment are housed in structural blocks which usually comprise monoliths in a concrete dam. The power plant and a spillway will be flanked by concrete or embankment sections extending to the abutments. There are literally hundreds of these structures astride the flatland streams of the U.S.A., Britain, France, Germany, and European U.S.S.R. Many are on navigable rivers and to minimize interference of power discharge with tows, the lock will generally be found at the opposite end of the spillway from the power plant. Representative plants include most of the Tennessee River installations of the TVA, the Corps of Engineers plants on the lower Columbia River, and also the Corps of Engineers plants along the Arkansas. Among the latter are the Ozark and Webbers Falls projects which have the innovative inclined-axis units.[15]

Non-Standard Low Head Plants

Because most of the low head arrangements are somewhat standardized, it might be more stimulating to examine two plants that varied from the standard.

Sam Rayburn Plant. Although many of the low head plants of the U.S.S.R. are founded on soft materials, experience in the U.S.A. has been limited to stronger foundations with very few exceptions. Accordingly, when the location for the Sam Rayburn plant provided only dense sand as foundation material, previous experience was of little value. This two-unit plant to be located on the Angelina River in southeastern Texas was part of a Corps of Engineers project. The main dam was to be an embankment type with a high-level, weir-type spillway located in a broad saddle some distance away from the river valley. The power plant was assigned a location in a narrower saddle between the spillway and the dam. Initial attempts to develop a layout included an embankment section pierced by cut-and-cover conduits running from an upstream intake structure to the plant at the toe of the embankment. This arrangement would not have been unusual of itself but speed rise and pressure rise calculations indicated that a surge tank must be provided. This appeared to the designers to be an intolerable burden for a low head plant. Means to obviate the requirement were sought without success until an entirely different concept was developed. By utilizing the space below the erection bay for flood sluices, a monolithic rectangular structure was conceived that extended from headwater to tailwater. This structure was supported on a very heavily reinforced concrete base. The water passages of the units were now sufficiently short to avoid the need for surge relief. Concreting was performed on a schedule devised to secure approximately uniform settlement of the total structure. The arrangement has a pleasing appearance and no structural or settlement problems have arisen since the work was completed in the late 1950s.

Wells. The arrangement of the elements of this dam[16] and power plant is unique in the U.S.A. It is termed a "hydrocombine" by the designers. The spillway and power plant share the same monoliths. The use of this arrangement was considered to save on the order of $15,000,000. An extremely low level of foundation over much of the site spurred studies for maximizing use of what higher level foundation existed. The "combine" was the result of these studies. Another unusual feature of this Douglas County, Washington, Public Utility District dam on the Columbia River is the arrangement for handling draft tube gates. The gates are stored in a sealed tunnel below the spillway bucket. They are moved, installed, and serviced in the dry by depressing tailwater. Air locks are used for personnel and materials. The project has a capacity of 815,000 kva in 10 units. This "hydrocombine" is an example of the full cooperation of dam and power plant engineers.

Medium Head Arrangements

Included in this category will be plants ranging upward in head from 100 ft (30.5 m) to 500 ft (152 m). These are the plants found in foothill terrain on relatively large streams. The giants in capacity and in size of equipment are in this group. There are also many plants in this category that are interesting for features other than capacity.

Plyavinsk. This U.S.S.R. plant in Latvia on the Daugava River is a modern example of meeting the challenge of "soft ground" as a foundation. Ten units at a rated head of 131 ft (40 m) will have an output of 825,000 kw. Like many earlier Soviet plants, the generator room roof is sealed for use as a spillway. Discussion of some of the construction and operational problems is contained in Ref. 17 cited for this plant.

Boundary. This city of Seattle, Washington project[18] has one of the early U.S.A. underground powerhouses. The layout of the project might be termed classic. An almost natural forebay was available upstream of the left abutment and a broad area of the river lies immediately downstream of the dam site. With minimal forebay excavation an intake structure could be constructed to provide individual intakes for the six-unit plant. The pressure tunnels drop directly to the units in the power cavern located a very short distance upstream from the almost vertical downstream face of the abutment. The units discharge individually into the broad area of the river. The entire layout is remarkably compact for a tunnel type project.

Mossyrock. This plant[5] is owned by the City of Tacoma, Washington and is located on the Cowlitz River, south of Tacoma. Like Boundary this project has an arch dam, but unlike Boundary adequate space was available immediately downstream for a three-unit powerhouse to contain the 150,000-kw generators. The penstocks leave the dam on a downslope at a low level and continue some distance in the open before entering the powerhouse. Although elaborate provisions for movement of the dam were required in the penstock design, the total arrangement is obviously economical in respect to penstock and power plant costs. Maximum compactness of penstock installation was achieved and foundation interference of the dam and plant was avoided by skewing the plant to match the right wing of the dam.

Bull Shoals. The Bull Shoals Corps of Engineers plant on the White River in northern Arkansas is typical of the medium head plants associated with concrete gravity dams. This eight-unit plant required very little innovative effort of the planner and arranger. It sits close to the toe of the portion of the dam that adjoins the left end of the spillway. The left abutment rises steeply from the powerhouse portion of the dam while from the right end of the spillway the dam extends to the more gently sloping right abutment. At the time of design (the late 1940s), it had become good practice to separate the powerhouse foundation from the toe of the dam to allow for differential movement of the two structures. At some contemporary plants, flexible penstock joints were also provided. At Bull Shoals the 18 ft (5.5 m) diameter penstocks were given an unsupported span of about 30 ft (9.2 m) between structures for flexing needs. Being designed in the period before water temperature problems had become a major design factor, single level intakes, set below the bottom of the power pool, were provided. Although the powerhouse and dam were separated by several feet at foundation level, the sloping face of the dam was used to support upper floors of the service bay of the plant. The downstream ends of the beams for these floors were free to move on their beam seats and a 1-inch (2.54 cm) expansion joint space was provided for such movement for both beams and floors. The plant has four 40,000 kw units and four 45,000 kw units. The rated head is 190 ft (58 m).

Glen Canyon. This dam,[4] constructed by the Bureau of Reclamation, is upstream from the Grand Canyon of the Colorado River, near Page, Arizona. At first glance, it somewhat resembles the downstream Hoover Dam. The power plant, however, is a rectangular structure transverse to the stream and as near the arch dam as practicable. Many layout and evaluation studies were made in achieving this final compact arrangement. The eight penstocks are carried on concrete columns between the dam and the plant. Provisions for movement of the dam with respect to the penstocks are incorporated in the penstocks as they leave the dam. Joints in the dam were coordinated with the required spacing of the power units. Full capacity of the power plant is 900,000 kw.

Estreito. This project,[19] with its six units rated 160,000 kva each, is part of a very large power supply system being developed on the Rio Grande and Parana Rivers in south central Brazil. The layout includes a chute spillway through the right abutment, a pair of cut-and-cover diversion conduits in the river valley beneath the earth and rockfill dam, and a power plant just downstream of the left abutment. Relatively short penstocks extend up the abutment slope to the intake structure which is at the end of a rather long excavated forebay. Speed rise and pressure rise values were found to be within very reasonable limits with this arrangement. Each penstock is provided with an expansion joint as it leaves the intake structure and has an anchor block integral with its entrance to the powerhouse.

Ludington. This pumped storage installation[20] on the eastern shore of Lake Michigan won the 1973 A.S.C.E. Engineering Achievement Award. The plant belongs to the Consumers Power Company and to the Detroit Edison Company. The 1,872,000-kw plant will meet a need for peaking capacity of the Michigan Power Pool. There are many interesting and ususual features in the design and these have been well-covered in the technical literature. Of particular note in the context of penstocks and power plants are the following items. The powerhouse rests on clay and has been provided with a heavily reinforced substructure. Each of the penstocks, which lie along the slope between the reservoir and the powerhouse, has two sets of articulating joints which provide flexibility at the connection to the powerhouse and at the junction with the encased portion at the top of the slope.

Guri. This project[21] on the Caroni River in eastern Venezuela has been carefully planned for staged development to an ultimate capacity of 6,500,000 kw. The present dam height of 360 ft (110 m) supports over 2,000,000 kw and the future height of 518 ft (158 m), together with an additional powerhouse, will provide for the scheduled increase in capacity. To permit enlarging the gravity dam, about 165 ft (50 m) of each penstock is exposed in the initial development. In lieu of installing expansion or articulation joints, each penstock is designed as a "pipe bent" between the fixed end in the dam and the fixed end in the power plant. The penstocks leave the dam horizontally and turn downward through an angle of about 60 degrees on their way to the plant.

Aswan. Aswan[22] has received world wide attention for several reasons. The only concern here is what dam and power plant arrangers can learn from the development. The original planning called for an underground plant in the left abutment and water tunnels in the right.

The project as constructed was arranged by the Soviet engineers as follows. A long, open cut forebay in the right abutment leads to short tunnels which supply a surface powerhouse. The tunnels also connect to pressure spillway outlets through the powerhouse. These changes probably illustrate the fact that engineers feel most comfortable using elements that have been found satisfactory in their own experience. The power plant has a capacity of 2,100,000 kw in 12 units.

Krasnoyarsk. This giant of a plant,[2] 6,000,000 kw in 12 units, will reign as the world's largest until Grand Coulee's third power plant raises Coulee's output above that number. The arrangement of dam and powerhouse at Krasnoyarsk is standard for medium heads in a broad valley. The powerhouse is close to the toe of the dam which is gravity type. The turbines require an enormous flow and two 24.6 ft (7.5 m) penstocks are joined just upstream of each spiral case to provide the required quantity. This feature is the major unconventional element in a project that was otherwise sheer extrapolation from experience in respect to equipment size and capacity.

Grand Coulee. The original Grand Coulee with its two power plants adjoining the gravity dam was conventional in aspect. The modified Grand Coulee will have the third power plant close to its dam but the total arrangement is no longer strictly conventional. The planners were asked to find space for a plant where none existed. The technical literature reveals that the investigation was very thorough. Space was eventually found. While the layout is no longer conventional, it has a logical appeal that doubtless has had an economic bonus because of the simplicity of the modifications to the original arrangement. Desirable compactness for a very large volume of unit flow was achieved by bringing the reservoir close to the power plant. Many papers are available in the technical literature to furnish details of the Third Power Plant.

High Heads

At high heads, tunnels or surface penstocks become inevitable as the intake water passages. Oroville, Mica, and Nurek embankment dams have pushed the record for this type of dam up to about 1000 ft (305 m). Most of the combinations of dams and power plants in this range, however, utilize an "upper reservoir" feeding a tunnel or penstock system. Machine costs per kw decrease with increasing head but are offset by the rising costs of the water passages. Reversible-unit pumped storage plants are clustered at the 1000 ft (305 m) level. The range of Francis units extends upward several hundred feet above this level and overlaps the range of the Pelton units. The latter wheels are available up to 6000 ft (1830 m) heads.

Oroville. Oroville[9] was selected by the A.S.C.E. as the Outstanding Engineering Achievement of the year 1969. This 770 ft (235 m) embankment dam and the associated Edward Hyatt underground power plant were constructed as part of the California Water Plan. Only three features of the work will be mentioned here as the technical literature has many articles on the project. The choice of an underground plant was one of the earlier large U.S.A. ventures into this type of construction. The design of the water intake was even more of a pioneering effort. In order to withdraw water of desired temperature from a stratified reservoir, a movable intake opening was devised. This project was therefore in the van in meeting the need for selective withdrawals for ecological requirements. In conjunction with the downstream Thermolito regulating project, a total capacity of 725,000 kw is available. Three of the six units in the Edward Hyatt plant are reversible for pump back service to maximize the power revenue. The pumping feature, in conjunction with the use of the project's two diversion tunnels as tailrace water passages, resulted in unique hydraulic conditions downstream of the units. Analytical and model investigations were performed to insure satisfactory operation of the tailraces for both generating and pumping.

Mica. Mica,[23] located on the Columbia River in British Columbia some 280 mi (450 km) upstream from the U.S.A.-Canadian border, is an extremely large embankment dam. The planned hydroelectric installation, 2,610,000 kw, also makes it one of the larger power projects. The six-unit powerhouse is underground in the right abutment. The forebay has been notched deeply into the abutment, the six pressure tunnels are relatively short, and the two tailrace tunnels are quite long. This is a fairly typical arrangement with high embankment dams.

Northfield Mountain. This 1,000,000-kw pump back project[24] in western Massachusetts was selected as typical of pumped storage plants with power heads in the range of 800 ft (244 m) to 1200 ft (366 m). The arrangement has a short high pressure intake shaft and a single, long, plant discharge water passage. Unit surge chambers are, however, provided near the draft tube exits. The regulating orifices are rectangular rather than circular as they also serve as draft tube gate slots. The gates are handled by a crane traversing a gallery at the top of the surge shafts. An important factor in selecting an underground plant at the chosen site was minimal disturbance of the environment.

Nurek. Nurek[10] on the Vaksh River in Central Asia is well known for the record height, 984 ft (300 m) of its embankment type dam. The capacity of the power plant (which is a surface plant) is 2,700,000 kw. Three long tunnel systems feed the nine-unit plant. The initial design included surge chambers but later studies reached the conclusion that they could be omitted from the first two lines. The arrangement of the project is generally conventional, with the diversion tunnels through one abutment (the left) and the pressure power supply tunnels through the other.

Churchill Falls. This Labrador Peninsula project[25] has very large units (500,000 kw) and a very large total capacity, 5,225,000 kw. The catchment area for the project is extensive and the arrangements for control of the flows from the several lakes in the catchment area are of special interest to planners. The required relatively large flow per unit is supplied to the underground plant by 10 individual pressure tunnels. Two very long tailrace tunnels receive the discharge flow as the water leaves a unit-connecting surge chamber just downstream from the draft tube exits.

Castaic. This plant[26] uses the water flowing in the West Branch of the California Aqueduct but provides pumped storage service to the Los Angeles Department of Water and Power. The head of 1075 ft (328 m) and the plant's six 200,000 kw units make the installation typical of this kind of plant. What is not so standard is the intake water passage arrangement. The designers found that the combination of a tunnel with surface penstocks best served their needs. The upper reservoir (Pyramid) supplies water through the 7.5-mi (12.1 km) Angeles Tunnel which ends at a portal some 750 ft (229 m) upslope from the power plant. On the upstream side of the portal a 476-ft (145 m) high surge tank provides desired hydraulic control. On the downstream side of the portal there is a wye branch, each leg of which supplies three exposed penstocks. The turbine settings are so deep that, although the plant's superstructure is a conventional indoor arrangement, there is little of it to see. Normal high water in the lower reservoir is at the same level on the outside of the plant's wall that the crane runway is on the inside. This condition is typical for surface plants of high head reversible unit installations.

Kemano. This project[1] will serve to illustrate the possibilities of arrangements where the terrain will provide very high heads. Kemano takes water from the eastern side of the Coast Range in central British Columbia in Canada and discharges it into the Kemano River on the western side. The head is 2500 ft (762 m), the turbines are impulse type, and the total capacity, 1,700,000 kw. The sum of the tunnel, penstock, and tailrace lengths from the intake at Tahtsa Lake to the Kemano River is on the order of 11 mi (17.7 km).

River Basin Developments

Several river basin developments will be briefly discussed and referenced for convenient study.

Tennessee. Development of the Tennessee River basin by the Tennessee Valley Authority is the basin development best known to engineers the world over and needs no description here. Publications covering all aspects of the design and construction are available. Headquarters of the TVA is at Knoxville, Tenn.

Kama-Volga. A description of the projects of the Kama-Volga Cascade in the U.S.S.R. is available in the publication, "Hydroelectric Power Stations of the Volga and Kama Cascade Systems," G. A. Russo, Editor, published for the National Science Foundation and the Department of the Interior, Washington, D.C. by the Israel Program for Scientific Translations. The structures are founded on what the designers termed "soft ground," this being clay or sand. A considerable amount of general power plant design and test information is contained in this 350-p. publication.

Missouri. The power projects on the Missouri River, including Fort Peck, Garrison, Oahe, Big Bend, Fort Randall, and Gavins Point, while not constituting a complete development of the power in the basin, are the result of a coordinated plan. Information on these projects is available from the Corps of Engineers, U.S. Army, Department of Defense, Washington, D.C.

Arkansas. The McClellan-Kerr Navigation Project on the Arkansas River in Arkansas and Oklahoma included a number of power projects. The main river and major tributaries have been placed under control for the purposes of flood control, navigation, and power production. Information on individual projects and the work as a whole may be obtained from the Corps of Engineers, U.S. Army, Department of Defense, Washington, D.C.

Coosa. The basin developments cited above are under Government control. Development of the Coosa River in Alabama cited here was carried on by a private power company. The undertaking is discussed in a paper by Richard S. Woodruff in the A.S.C.E. Power Division Journal for July 1961. The paper, entitled, "Private Power Development of an Entire River," describes the planning, design, and construction of the work.

James Bay. The first four plants sited for this basin development[27] in northern Quebec will utilize 1200 ft (366 m) of head and develop 10,190,000 kw in 44 generating units. These plants are to be constructed in a 288-mile (464 km) stretch upstream from the mouth of La Grande River at Fort George. The plants will be guaranteed to produce 68 kWh $\times 10^9$ at 79 percent capacity factor. The first dam near Fort George will be a low head structure with an integral power plant. The upstream dams will be earth and rockfill. Two of the plants will be underground. The four plants are scheduled to be placed in service between 1980 and 1985.

Power Complexes

A pioneer development of this type is under design and construction in the northwestern part of South Carolina. The Duke Power Company was the winner of the Edison Award in 1972 for its Keowee-Toxaway power complex in the headwaters of the Savannah River. A total potential of 8,000,000 kw includes conventional hydroelectric power, pumped storage power, and nuclear power. The initial conventional and pump back hydroelectric capacity in two plants amounts to 750,000 kw and the nuclear in one plant to 2,658,000 kw. Two more nuclear plants are planned adjacent to the same lake as the initial plant. Future high head pumped storage will utilize the other (higher level) existing reservoir as a lower pool. Environmental interests are receiving attention and action. Particular emphasis is being paid to water temperature control and to the support of recreation usage.

Pumped Storage Arrangements

In the preceeding paragraphs attention has been given to illustrating typical pumped storage projects. Because of the rapid development of this field and the indicated future need, additional reference material is cited here for the convenience of the dam and power engineers engaged in this work.

a. Vol. No 4, April 1972, Hydrotechnical Construction, as translated from the Russian for publication by the A.S.C.E., pp. 311-359.

b. Vol. No. 22, March 1970, *Water Power*, published by IPC Electrical-Electronic Press, Ltd., Dorset House, Stamford Street, London. Four articles, with the first beginning on p. 89.

Unusual Arrangements and Solutions

In addition to having information on all the usual and conventional procedures for coordinating dams and power plants, it may be worthwhile to know about some of the more unusual arrangements.

Lake Taps. Construction of a headwater dam has been obviated entirely at some sites and the work greatly reduced at others by tapping an existing lake under water.

Yaupi. This power project in Peru northeast of Lima was extended for water storage by tapping the Huangush Alto Lake. A complete description of this very satisfactory operation is contained in *Water Power* magazine for August 1968, pp. 319-323.

Snettisham. Information on the lake tapping effort in connection with providing headwater for this Corps of Engineers designed plant in

Alaska may be obtained from their headquarters in Washington, D.C.

Askara. This project in a fjord on the west coast of Norway, about 100 mi (161 km) north of Bergen, is described in *Water Power* magazine, July 1971, beginning on p. 254. The tapping arrangements were rather complex as two taps were performed at different levels to feed one supply tunnel.

Shintoyone. Construction of the draft tube outlets for this Japanese plant required an unusual solution. Caissons were used to avoid drawing down the existing lower reservoir of this 1,125,000-kw pumped storage plant. The work is described in *Water Power* magazine, March 1972, p. 85.

Qattara Depression. Power engineers have long been intrigued by the power potential inherent in the difference in level between the Mediterranean Sea and this depression in northern Egypt. The concept depends on continuous natural evaporation of the power discharge; thus always maintaining a dependable head. A review of the studies and schemes involved in attempts to develop an economic arrangement is contained in *Water Power* magazine, January 1973, pp. 27-30. Additional studies are presented in the June-July and August 1975 issues of *Water Power*.

REFERENCES

15. Shannon, A. V., and Fogg, C. H., "Advantages of Inclined-Axis Hydroelectric Units," ASCE Pow. Div. J1., March 1971, p. 395.
16. Patrick, J. G., "Development of Wells Hydroelectric Project," ASCE Pow. Div. J1., March 1971, p. 267.
17. Taukach, A. D., "Plyavinsk HES," Hydrotechnical Construction, February 1970, p. 101. See also the February 1976 issue.
18. Pospisil, Jaroslav, and Hayes, M. D., "Design of Boundary Arch Dam," ASCE Pow. Div. J1., January 1970, p. 73.
19. Bendixen, B., "Brazil's 900 MW Project at Estreito," *Water Power*, March 1969, p. 87.
20. Comninellis, Eustratios, "Ludington Pumped Storage Project," ASCE Pow. Div. J1., May 1973, p. 69.
21. Hasen, Hans, and P. Palacios H., "Guri Hydroelectric Project Designed for Staged Development," Civil Engineering, December 1971, p. 55.
22. Aleksandrov, A. P. et al., "Major Victory of Arab and Soviet Hydrobuilders," Hydrotechnical Construction, April 1971, p. 305.
23. Meidal, P., and Webster, J. L., "Mica: One of the World's Largest Structures," *Water Power*, June 1973, p. 201.
24. Gunwaldsen, R. W., and Ferreira, A., "Northfield Mountain Pumped Storage Project," Civil Engineering, May 1971, p. 53.
25. Wermenlinger, D., et al., "Churchill Falls Power Facilities," ASCE Pow. Div. J1., March 1971, p. 515.
26. Haase, C. D., "Castaic Pumped Storage Project in California," *Water Power*, September 1971, p. 333.
27. Amyot, P., Laliberté, C., and Larocque, G. S., "Meeting Quebec's Power Needs for the Eighties," *Water Power*, July 1976, p. 27.

RELEVANT POWER PLANT DESIGN CRITERIA

Subject Matter

Water passage and power plant design criteria relevant to the associated dam are discussed in the following paragraphs. References are provided to competent sources of detail design information.

Water Passages

The design of the water passages for a low head plant usually involves only a determination of the proportions of the transition section to suit the planned velocity gradient from rack to spiral and reinforcement of the transition to carry the imposed loadings. Selection of a water passage system that includes penstocks (tunnels) and possibly surge relief, involves design as part of the selection process. Valid economic comparisons require cost data based on sound quantity and material estimates. Complete design of the alternative arrangements may be necessary to reach a decision.

Sizing the Water Passage. In the technical literature there are many charts and graphs depicting economic solutions for penstock size determinations. These are of value for preliminary studies. The engineer will, however, generally find it necessary to base the final evaluation on the criteria and materials available to the project being planned.

Pressure Rise and Speed Rise. An integral part of the complex problem of sizing and designing the water passages will be a determination of the pressure rise phenomena associated with the selected mode of operation of the units. There will also be a determination of the speed rise characteristics of the units resulting from reinforcement of the inherent machine characteristics by the pressure effect. Preliminary and refined methods of making these determinations are available in textbooks and in the technical literature. The technical literature in this specialized field is extensive in quantity and will provide guidance for solution of most of the problems to be encountered. The following note illustrates the magnitude of the continuing interest in the exploration and solution of such problems. For its symposium in Rome in the fall of 1972, the International Association for Hydraulic Research chose as one topic, "Current problems associated with hydraulic machinery for pumped storage power plants." Sub-topics included, "Transient Processes" and "Unsteady Behavior."

Surge Tanks and Chambers. Pressure rise and speed rise determinations will conclude with a decision that for the given operating conditions and machine characteristics, a surge tank or chamber either should be an adjunct of the system or (in the happier case) will not be required. If surge relief is indicated, and the need is marginal, an effort should usually be made to eliminate the need. The operating conditions can frequently be changed, the water passage sizes can often be changed, and the machine characteristics are subject to some adjustment from normal or book value. Preliminary surge tank proportions can be obtained by use of the procedures set forth in the Creager and Justin handbook referenced below.

Nonembedded Penstocks. Study of the "Nonembedded Penstocks" portion of the "New Design Criteria for USBR Penstocks" referenced below will provide the designer with guidance for such penstocks. The study should include the discussions of the criteria, and, for a broader view, should include the paper entitled, "Penstock Codes–U.S. and Foreign Practice," also referenced below. The publication, "Welded Steel Penstocks," referenced in the same paragraph will yield design guidance on associated matters.

Embedded Penstocks. Guidance for design of such penstocks will be found in the papers and publication noted above.

Power Plants

The relation of the design of a power plant to design of the associated dam is primarily one of location. When the two are wedded, as in a low head integral layout, there is only one set of criteria: that established for the dam. Different criteria may be acceptable for other locations where failure of the plant would not endanger the dam.

Low Head Dams. The power plant portion of low head dams will be designed to the same degree of safety as the dam. Criteria specifically applicable to such plants will be found in the publications and technical manuals of the Tennessee Valley Authority and the U.S. Army Corps of Engineers.

Surface Plants Located Close to Gravity Dams. Some of the criteria for such plants may be less conservative than for the dam itself. The first consideration is to coordinate the two designs to insure that no interference exists, either as to assumptions or physically. If the face of the dam is used to support part of the plant, adequate differential movement provisions need to be included in the design. Dam monolith joints will be arranged to suit penstock locations fixed by unit spacing. Frequently, the tailwater level selected for automatic flooding of the powerhouse (to insure stability) will be lower than the maximum tailwater level assumed for dam design. The costs of cleanup and dry-out are expected to be less than the alternative costs of full protection.

Surface Plants Close to Arch Dams. Locating plants close to arch dams requires coordination of dam and plant design in three major areas: location of the penstocks through the dam; allowance in the penstock arrangements for differential movement of dam and plant; and avoidance of the foundation of the dam by the foundation of the plant. Glen Canyon[4] and Mossyrock[5] illustrate solutions to these problems.

Surface Plants Close to Embankment Dams. Surface plants in close association with fill dams are frequently located at the downstream end of tunnels through one of the abutments. The plant's foundation will usually be quite independent of the dam. Orientation of the plant, however, should of course be such as to direct the power discharge away from the toe of the embankment.

Underground Plants. An underground plant might be considered an element of the water passage system. At locations where an underground plant is feasible, economic and other evaluations pertinent to the decision will be made in conjunction with the water passage evaluation studies. Criteria for safe design of the cavern for personnel will also provide adequately against possible hazard to the dam.

Major Power Plant Design References

Guidance for detail design of hydroelectric power plants will be found in the following works:

Hydro-Electric Handbook by W. P. Creager and J. D. Justin, 2nd Edition 1950, published by John Wiley & Sons, New York.

Hydro-Electric Engineering Practice, Editor, J. G. Brown, published in three volumes in 1958 by Blackie & Son, Ltd., London.

Handbook of Applied Hydraulics. C. V. Davis, Editor-in Chief, K. E. Sorenson, Co-Editor, Third Edition, published 1969 by McGraw-Hill Book Company, New York.

Supplementary Power Plant Design References

Design information for hydroelectric power plants is featured in the following publications.

Power Division Journal of the American Society of Civil Engineers.

Civil Engineering magazine, published by the American Society of Civil Engineers.

Water Power magazine, published by IPC Electrical-Electronic Press, Ltd., London.

Hydrotechnical Construction, a periodical translated from the Russian for publication by the American Society of Civil Engineers.

Penstock Design References

The following publications furnish detail design information on both embedded and nonembedded penstocks.

Penstock Codes–U.S. and Foreign Practice, a paper by A. E. Eberhardt in the ASCE Power Division Journal for April 1966, p. 137.

New Design Criteria for USBR Penstocks, a paper by H. G. Arthur and J. J. Walker in the ASCE Power Division Journal for January 1970, p. 129. Discussion in Journal for January 1971, p. 248 and closure in Journal for July 1971, p. 715.

Welded Steel Penstocks, Engineering Monograph No. 3, USBR, Revised 1966. (This reference is furnished because the parts not superseded by the reference next above provide guidance on many matters associated with the design of penstocks.)

An Assessment of Penstock Designs, an article by Andrew Eberhardt, *Water Power* for June, July and August 1975.

Part B: Design of Adjacent Locks— John Mathewson

A. GENERAL

1. Introduction

A navigation lock, or flight of locks, will be required in conjunction with a dam wherever

DESIGN OF ADJACENT LOCKS   563

there is existing navigation, or where provisions for navigation are a requirement in obtaining a permit for the dam construction. The essential parts of a lock consist of an open chamber with gates at both ends, and some means for admitting water to the chamber from the upper water level and discharging it from the chamber to the lower water level. The location of the lock with relation to the dam structures depends on site conditions. The lower approach to the lock should normally be along one side of the river below the project and not too near the power plant discharge. The relation between the upstream approach and the reservoir shoreline is not so critical, especially in a broad, deep pool. The lock itself of course acts as a portion of the dam, and must be designed for stability under the same loading conditions. Adjacent sections of the dam may be concrete gravity structures or may be of earth or rockfill construction, as is most practicable. The lock rarely forms one of the dam abutments, since its floor must be low enough to reach below minimum tailwater, and at least a short section of dam will be required to carry the crest level to high ground. Approach channels of adequate width and depth to and from the lock may require excavation by the project layout. A typical project, containing hydro-electric power plant, spillway, navigation lock, and adjacent earth embankment, is shown in Fig. 10B-1.

2. Scope

This discussion is primarily concerned with locks as appurtenant structures in connection

Figure 10B-1. Cordell Hull Lock and Dam, Cumberland River.

564 HANDBOOK OF DAM ENGINEERING

with multi-purpose projects, and not as the principal feature of a purely navigational development. Emphasis is therefore placed on medium to high lift locks, with lesser detail on other types. The author wishes to make full acknowledgement of material on locks and lock design contained in Engineering Manuals of the Corps of Engineers, U.S. Army, and in Technical Reports of the Waterways Experiment Station, Vicksburg, Miss. These, together with other publications, are listed in the bibliography at the end of this chapter, and are referred to in the text by raised numbers.

3. General Planning

The basic planning of a lock structure is interdependent with the planning of the overall dam project, discussed in other chapters of this book. It is assumed here that maps and topographic data, hydrologic and hydraulic information, and subsurface data have already been obtained in connection with planning for the adjacent dam. It is further assumed that the total lift, or height from maximum reservoir level to minimum tailwater, has been determined as a part of the dam planning. Other aspects of general planning for the lock include visibility and ease of approach, avoidance of a location which produces crosscurrents in the approaches, or shoaling; location with respect to river banks and stream bed; and accessibility with respect to initial construction, maintenance and operation. The relative importance of these factors depends greatly on the type and volume of existing or potential traffic. If either of these includes a significant number of large commercial tows or vessels, their needs should be carefully considered. A detailed discussion of the above factors is contained in EM-2-2601.[1]

4. Lock Characteristics

Lock Size. The lock should be large enough to accommodate the traffic which can reasonably be anticipated during the life of the structure, taking into account any limiting restrictions on size of tows due to sharp bends, narrow channels or shallow depths whose correction are not feasible. The dimensions of a lock chamber are determined by a balance between economy in construction and the average or probable size of tow likely to use the lock. The Corps of Engineers has certain standard lock sizes, as listed in Table 1. These sizes do not include locks to be used by small craft only.

Table 1. Standard Lock Sizes.

The following are usable dimensions; they are not to be encroached upon by lock gates, sills, or other lock features.

Lock Width × Length, Feet. (Meters).

56 × 400 (17.08 × 122m)
56 × 600 (17.08 × 183m)
84 × 600 (25.62 × 183m)
84 × 800 (25.62 × 244m)
84 × 1200 (25.62 × 366m)
86 × 675* (26.23 × 205.9m)
86 × 500* (26.23 × 152.5m)
86 × 360* (26.23 × 109.8m)
80 × 800** (24.40 × 244m)
110 × 600 (33.55 × 183m)
110 × 800 (33.55 × 244m)
110 × 1200 (33.55 × 366m)

*Locks in the Columbia River–Snake River system
**Locks in the St. Lawrence Waterway system

Depth. The effective depth of a lock is measured from the top of the sill to the water surface of the respective pool which it connects. The minimum depth over the sills should be at least 3 ft more than the existing or projected maximum draft of vessels using the lock. This will allow for the "squat" which always occurs when a tow which occupies a large proportion of the available waterway cross section passes over a sill, even at very low speeds. Additional depth should be provided over the lower sill if there is any probability of a future lowering of the tailwater from changed downstream conditions.

Height. The height of lock walls depends on the importance of navigation, and the extent to which it must be maintained during high reservoir stages. If considerable commercial use is anticipated, the walls should be at least 10 ft above normal pool so as to properly guide empty barges along the walls and not subject the top of walls to an overhanging barge rake. Top of walls should also be at least 2 ft above the maximum pool at which navigation is to be maintained. If maximum design reservoir level exceeds the height to which the lock walls can be justified, it is possible to raise the upstream gate bay and gates to a height which will contain this pool, and construct the remainder of the lock to a lower level. This construction has the disadvantage of a lesser convenience in operation, since lock operators will have to go up and down between these levels for every lockage, even though the necessity for the extra height may be very infrequent.

Lift. Locks have been constructed in this country with lifts up to 108 ft in one chamber. Where site conditions permit, it will probably be more economical to provide the total lift of the project in one chamber. It will certainly be less expensive to the users, and will entail less maintenance and operating costs, but the design will be more complicated. If the total lift exceeds the limits for a single lock, or foundation conditions are not suitable, multiple locks may be used. These may be in a continuous flight, as the Gatun Locks at Panama, or they may be separated by short approach channels.

Lock Types. Locks may be separated into those designed for inland waterway barge tows, and those designed for ships. They may be further divided into the following types:

1. Gravity or Mass Concrete. This type of lock is composed of wall sections which resist applied loadings by their massiveness. Base widths are made sufficient to prevent overturning, sliding, or overstressing the foundation materials. Intermediate and top widths are proportioned to provide a wall section which will withstand wall stresses and accommodate all necessary appurtenances, such as the filling and emptying system, gate recesses, anchorages for movable structures, operating equipment, etc. Walls may be supported on bedrock, soil, or piles. Advantages of this type of lock include:

a. Simplicity of design and detail.
b. Economy in construction.
c. Low maintenance cost.
d. Resists impact and abrasion of moving barges.

Some disadvantages are:

a. Relatively heavy lock walls, which must be properly designed to suit the supporting capacity of the foundation materials.
b. Possible unequal settlement of adjacent or opposite monoliths.
c. Consequent damage to or misalignment of the movable structures and their operating machinery.

Where site conditions require a considerable depth of sound rock to be excavated, a facing of reinforced concrete may be constructed adjacent to the vertical rock face to form the lower portion of the lock walls. This concrete is anchored to the rock by use of steel dowels grouted into drilled holes, and the gravity-type walls are continued from the top of this facing to the top of the wall level.

2. Dry-Dock Reinforced Concrete. This type consists of relatively thin walls constructed integrally with a thick floor slab. Walls and floor are designed to act together as a monolith, each being heavily reinforced to distribute the loads. Walls are designed as cantilevers above the floor, while the floor itself is designed as a beam on an elastic foundation. This requires a knowledge of the foundation modulus, the determination of which depends on the characteristics of the soil and the shape and size of the loaded area. The advantages of this type of lock include:

a. Lower, more uniform foundation loads, permitting construction on relatively weaker foundations, and where piling cannot penetrate to a stable subgrade.
b. Rigidity against differential settlement or wall rotation.
c. Chamber can be unwatered for inspection or maintenance without fear of a blowout.

Disadvantages of this type are:
a. Higher design and construction costs.
b. Types of filling and emptying systems are restricted.

3. Steel-Sheet Piling. This type is not suitable for an entire lock, but is used in combination with one or more of the other types of construction. Use of steel-sheet piling is restricted to those portions of the lock walls between the gate bays, and to the lock approach walls. A Z-type pile section is often used for this type of construction due to its high section modulus and strong interlocking properties. Walls are designed by the conventional methods of analysis for retaining walls. They may be provided with a horizontal waler, and tie rods with turnbuckles and dead-man anchorages. Concrete struts may be provided across the floor of the lock chamber to take the reaction of the walls and prevent their inward movement. The lock chamber face can be provided with horizontal fenders, of timber or steel, at levels where tows usually rub. The offsets formed by the Z-piles provide spaces for recessed mooring hooks and ladders. Special anchorages at the top of the lock walls can be provided for check posts to facilitate tying up barges during locking operations. This type of lock wall is useful for comparatively small lock chambers, where traffic is not heavy and less costly construction is indicated. The lock filling and emptying system will of course have to be confined to the concrete gate bays.

4. Reinforced Concrete. This type has separate reinforced concrete retaining walls with footings, as distinct from the dry-dock type. As in the case of the steel-sheet piling type, use of the retaining walls is limited to that portion of the lock chamber between gate bays, and to approach walls. The walls may be either cantilevered or counterforted. Spread footings, forming a portion of the lock floor, should be proportioned to satisfy the condition for stability, with keyways to prevent sliding.

Filling and Emptying Systems

1. General. As stated in the introduction to this section, means must be provided for leading water from the upper pool into the lock chamber, and from the chamber to lower pool. These means constitute the lock filling and emptying system. Various types of systems have been used, depending on the type of lock and the height of lift involved.

2. Low and Moderate Lifts. The most common system involves the use of longitudinal culverts. This system is particularly adapted to the gravity or mass concrete type of lock. It consists essentially of a longitudinal culvert of constant size in each wall, each with suitable intakes from upper pool, a filling valve, a series of chamber ports, an emptying valve, and a discharge manifold into the lower pool. Many types of valves have been used; vertical lift stoney gates, butterfly valves, cylinder valves, and tainter gates, both direct and reversed. Chamber ports are usually rectangular in shape, spaced at intervals which are staggered in the two walls; however, several locks have been constructed using a multiple system of small circular ports, arranged in two or three horizontal rows (see p. 574). For steel-sheet pile locks, or others having concrete gravity walls limited to the gate bays, a system of short loop culverts may be used. Two filling culverts extend from the upper pool around the upper gates into the lock chamber. A similar pair of culverts extend from the lock chamber around the lower gates and discharge into the lower pool. Other systems include filling around partially-opened upper lock gates (when sector type gates are used), or filling over or under the upper gates, when tainter type gates are used. Whatever type of system is used, the various parts must be proportioned so that during the filling or emptying operation, surges or surface turbulence in the lock chamber will not endanger the safety of the craft being locked through. The usual rule is that hawser stresses in the lines securing the craft to the lock walls should not exceed 5 tons for a tow of any size moored at random in the chamber, nor exceed 10 tons for deep-draft ships.

3. High Lifts. Longitudinal culverts are always used for high lifts. The simple wall ports used in low or medium lifts are replaced by

laterals extending across the lock chamber below floor level. The flow is discharged into the lock chamber through a number of ports in each lateral. In early lock design utilizing the bottom lateral system, the individual ports were in the roof of each lateral. This works satisfactorily with a deep water cushion, but is not suitable for the shallower cushion available in barge locks. More effective energy dissipation can be obtained by locating the ports in the sides of each lateral, so that adjacent laterals will discharge into the common trench or box between them. If ports in adjacent laterals are staggered, an even better stilling action is obtained. The width of each lateral should decrease from its culvert connection to the opposite wall so as to produce a uniform flow through all ports. Two types of lateral systems have been used; the intermeshed type, and the split type. In the intermeshed type, laterals from one culvert alternate with the laterals from the opposite culvert. The entire system is contained in about the middle third of the chamber, and produces excellent results if the tow is placed symmetrically over the laterals. However, unsymmetrically placed tows will experience much higher hawser stresses during filling operations. This can be overcome by the split lateral system, where one culvert feeds a set of laterals in the upstream half of the lock chamber, while the second culvert feeds a similar set of laterals in the downstream half of the chamber. Since each set of laterals receives the same amount of water, longitudinal currents in the lock chamber are held to a minimum, and hawser stresses will be about the same, regardless of the location of the tow. A third type of bottom filling system has recently been developed, using longitudinals in the lock floor, connected to the wall culverts. This system is described on p. 575.

B. HYDRAULIC DESIGN

1. General

The following discussion of hydraulic design assumes the use of the longitudinal culvert filling and emptying system, since this is by far the most common type. Nor is it in sufficient detail to contain all information needed for the complete hydraulic design of an actual lock. The intent is to set forth the general guidelines for design, with references to more detailed works for additional information. One of the initial steps in design is the selection of a tentative filling time. From the standpoint of expediting river traffic, the shortest possible operating time would be desirable. However, practical limitations, such as the maximum size of the culvert which can be accommodated in the lock walls, maximum allowable rate of rise of the water surface in the lock chamber, allowable surges and turbulence in the lock, cavitation pressures at the valves, and surging in the lock approaches will impose restrictions on the filling time. The selection of a tentative filling time is usually based on a study of operating experience and the results of hydraulic model tests on similar locks.

An equation for the total time required to either fill or empty a lock chamber, which may be used in selecting the culvert size is contained in EM1110-2-1604,[2] as follows:

$$T - kt_v = \frac{2A_L(\sqrt{H+d} - \sqrt{d})}{CA_c\sqrt{2g}}.$$

Where

T = total operating time in seconds
kt_v = a constant (assumed as 0.5 for design) times valve operating time in seconds
A_L = area of water surface in lock chamber, in square feet
A_c = total cross sectional area of culverts, in square feet
d = measured overtravel, in feet (overfilling or overemptying)
C = average coefficient of filling (or emptying) system
H = height from lower to upper pool, in feet.

Valve operating time usually varies from 2 min to as much as 8 min for a filling cycle. The time selected will depend on the rate of rise vs. cushion depth in the lock chamber required to maintain hawser stresses within desired limits.

568 HANDBOOK OF DAM ENGINEERING

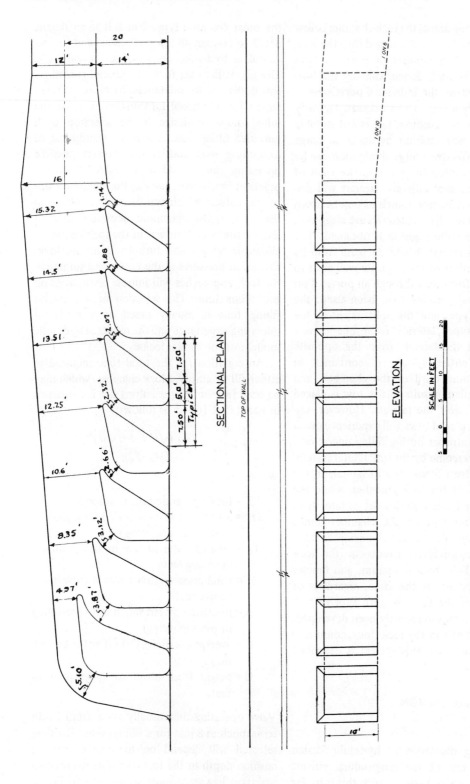

Figure 10B-2. Culvert intakes: (a) single (b) twin. (*Courtesy Corps of Engineers, U.S. Army*)

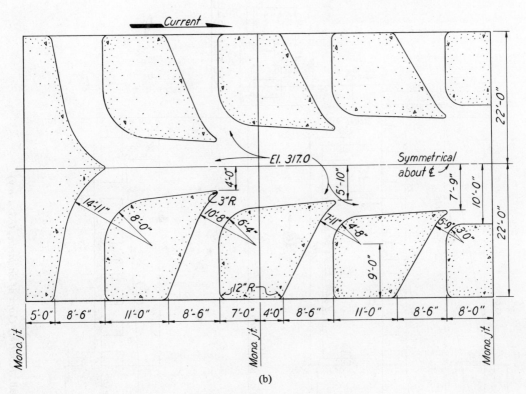

(b)

Figure 10B-2. (*Continued*)

The value of the average coefficient "C" can best be determined by a model test of the proposed lock. It can be approximated from other similar locks; however, for preliminary design, "C" for the filling cycle will be between 0.7 and 0.85, while that for the emptying cycle will be between 0.5 and 0.7. The amount of overtravel "d" varies inversely with the area of the lock chamber, and directly with the mass of water flowing through the culverts during a lock operation. If the culvert valves remain open, the overtravel will constantly reverse itself as a damped oscillation until the residual energy is expended through losses in the system. In practice, however, the lock gate opening will begin during the first cycle of overtravel, and residual excess head will then be dissipated immediately. Total overfilling varies from 0.6 to 1.3 ft, with an average of about 1 ft. Overemptying is usually somewhat less than overfilling.

In the hydraulic analysis of a lock filling and emptying system it is usually necessary to develop graphs showing the variation of water surface elevation in the lock chamber and the rate of flow in the culverts throughout the filling and emptying cycles. From these graphs, a rate of rise vs cushion depth can be developed for the entire cycle. Detailed instructions for development of these graphs are contained in EM-1110-2-1604.[2]

2. Intake Manifolds

Manifolds designed for flow in only one direction can be made more efficient than those subject to flow in both directions. The area of the intake at the face of the lock wall should be considerably larger then the culvert area to reduce velocities and entrance head losses. This also avoids damage to trashracks from impact of floating drift or ice, minimizes vortex formation, and lessens the tendency to draw air into the system. For structural reasons, several smaller intakes are preferable to one

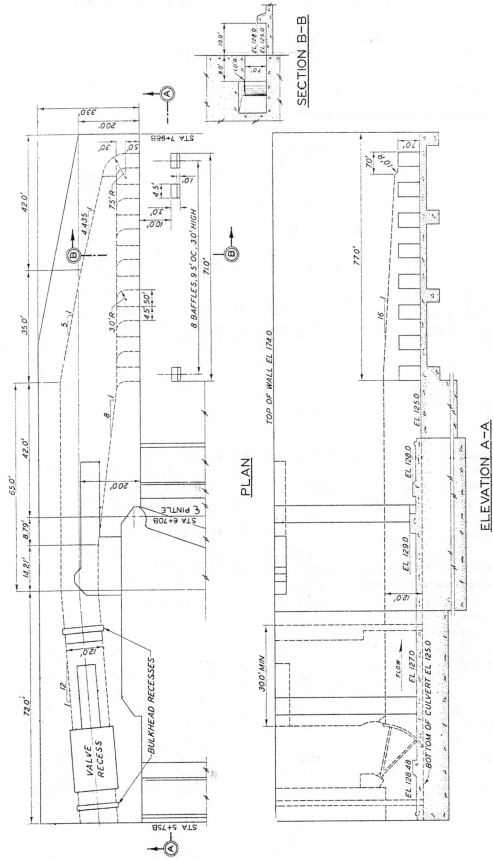

Figure 10B-3. Culvert discharge manifold: In Lockwall. (*Courtesy Corps of Engineers, U.S. Army*)

or two large ones. Intakes should be submerged below minimum pool a distance not less than the velocity head at the face of the wall to avoid vortices. Examples of single and twin manifolds located in an approach wall are shown in Figs. 10B-2a and 2b. Openings at the face of the wall are of uniform shape and size. Since each successive port in the downstream direction is subject to increased pressure differential, each throat is made smaller to obtain approximately equal flow. The culvert converges between successive ports, but expands abruptly immediately downstream from each port. This is done to increase the pressure in the culvert to that of the inflow from each succeeding port at its point of confluence, thus reducing impact losses. Curvatures at port entrances and throats, as well as general layout of the manifolds shown in Figs. 10B-2a and 2b are also applicable to designs for different numbers of ports.

3. Discharge Manifolds

These should be designed to obtain an efficient emptying operation and distribute the outflow from the lock at locations and velocities which will not imperil any craft in the immediate lock approach area. They may be located either in the lock walls, or as laterals in the lock approach floor. The latter type is particularly effective in preventing spiral motion of water from a one-culvert operation. Whenever practicable, the best type of discharge manifold is one which diverts the entire flow outside of the lock approach. Craft can then tie up to the lower approach walls in entire safety. For such a layout each culvert should be provided with a stilling basin having a sloping end sill in lieu of a manifold. This is to prevent high velocities from extending all the way across the lower pool and producing reflected eddy currents in the lower approach. If this type of emptying system results in a head differential between the lock chamber after emptying and the lower approach, the design of the lower gates and operating machinery may be affected. In extreme cases it may be necessary to provide an auxiliary emptying valve to equalize the residual head

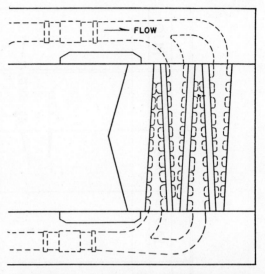

Figure 10B-4. Culvert discharge manifold: Laterals in lower approach. (*Courtesy Corps of Engineers, U.S. Army*)

differential. Figures 10B-3, 4 and 5 illustrate examples of each of the discharge manifolds discussed above.

4. Lock Chamber Wall Ports

Side wall ports in the lock chamber are rectangular, usually venturi-shaped with the larger area at the chamber face. They should have a rounded entrance at the culvert face for increased efficiency in filling, and a smaller radius at the wall face for improved entrance conditions during lock emptying. Dissipation of energy during filling is improved by staggering the ports in the two walls, so that the jets issuing from one wall pass between those from the opposite wall. Thus the jets travel twice as far horizontally and are more diffused when deflected upward by the opposite wall. Spacing of ports on 28-ft centers in each wall of a 110-ft wide lock will allow the jets to pass one another with only slight intermixing at their boundaries. Somewhat closer spacing may be used on narrower locks. Ports should be inclined downwards slightly for best results. With the 28-ft spacing, ports should be located in about the middle half of the lock chamber, symmetrically

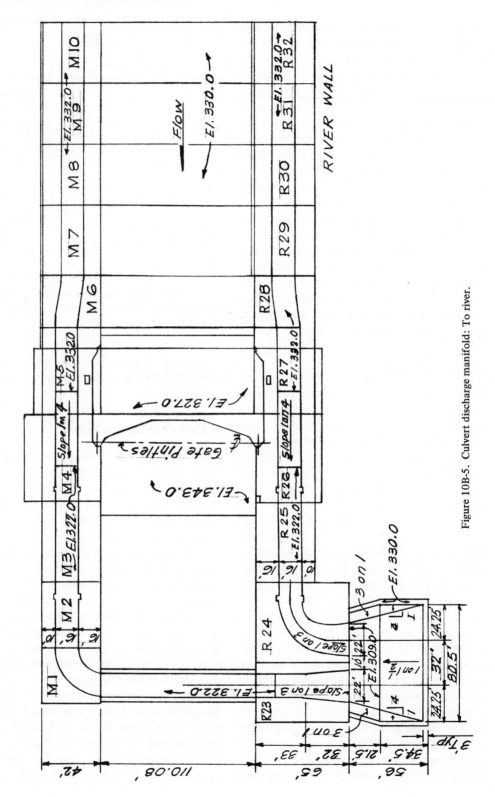

Figure 10B-5. Culvert discharge manifold: To river.

DESIGN OF ADJACENT LOCKS 573

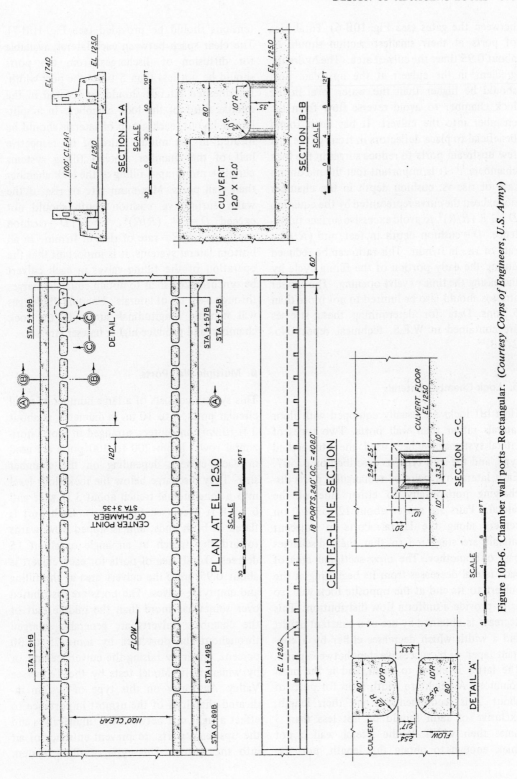

Figure 10B-6. Chamber wall ports—Rectangular. (*Courtesy Corps of Engineers, U.S. Army*)

between the gates (see Fig. 10B-6) Total area of ports at their smallest section should be about 0.95 times the culvert area. The hydraulic gradient in the culvert at the upstream port should be higher than the water level in the lock chamber to avoid reverse flow from the chamber into the culvert. It has been found beneficial to place deflectors in front of the first few upstream ports to reduce surges in the lock chambers.[11] It is important that the maximum rate of rise vs. cushion depth in the chamber not exceed the curve represented by the equation $D = 1.8\ (R/R)^2$ to avoid excessive surface turbulence. D = cushion depth in feet, and (R/R) = rate of rise in ft/min. This ratio can be reduced during the early portion of the filling cycle by increasing the time of valve opening. The hawser stresses should also be limited to not more than 5 tons. Data for determining these stresses are contained in W.E.S. technical report No. 2-713.[11]

5. Lock Chamber Laterals

High lift locks are usually equipped with floor laterals rather than wall ports. Two types of lateral systems have been used: the intermeshed type, and the split type, as described on p. 567. Each lateral has pairs of rectangular ports discharging horizontally to either side of the lateral. Pairs are spaced about 12 to 14 ft on centers along the lateral. Ports in adjacent laterals are staggered so that their discharges pass one another. The cross-sectional area of each lateral decreases from its beginning at the culvert to its end at the opposite lock wall, so as to provide a uniform flow distribution. This decrease is obtained by use of a constant height and a width which decreases either by a constant taper, or by successive steps between ports. The latter is easier to form, and is the most common. A height to width ratio for ports of about 2 is recommended, and their length, exclusive of radii should be not less than 3 times their width. If the lateral wall is not thick enough to obtain this length, port extensions should be provided (see Fig. 10B-7). The clear space between each lateral, available for diffusion of discharges from the ports should be not less than 5 times the port width. Intermeshed laterals should be located in the middle third of the lock chamber. In a split-lateral layout, each group of laterals should be located in the middle third of its respective half of the chamber. Lateral filling systems permit a more rapid filling of the lock chamber than wall ports. Maximum rate of rise of the water surface vs. cushion depth should not exceed $D = 0.4\ (R/R)^2$, where D = cushion depth, and R/R = rate of rise, in ft/min. In all bottom lateral systems, it is important that the operation of the filling valves in each culvert be synchronized so as to obtain equal discharges through each set of laterals. Unequal discharges will increase longitudinal surges in the lock chamber, and produce higher hawser stresses.

6. Multiple Wall Ports

This system consists of a large number of small circular ports, 8 to 10 in. in diameter, at about 3 ft (0.9m) on centers, arranged in 2 or 3 horizontal rows. From 200 to 300 ports are used in each culvert, depending on the chamber size. They discharge below the lock floor level into a longitudinal trench about 3 ft wide and 6 to 8 ft deep (see Fig. 10B-8). Each port is flared at both ends, and is sloped downwards toward the trench at an angle of about 15 degrees. Total area of ports for each culvert is about 0.95 times the culvert area at the filling and emptying valves. The ports are distributed over somewhat more than the middle half of the chamber. Culverts are generally enlarged throughout the port area by some 25 to 30 percent, either by raising the culvert ceiling, or by widening it. Model tests by the Tennessee Valley Authority on this type of system indicated that it is of the utmost importance to effect a tight seal between the filling valves and the upstream ports, to prevent entrance of air into the system. This type of filling system lends itself to a site where the lockwall foundations are considerably lower than the lock floor. The trench can be either excavated in the rock, or formed by concrete retaining walls. This system has been used by the TVA on 5 locks on the Tennessee River with lifts up to 60 ft, (18.3m), with excellent results.

7. Bottom Longitudinals

In an effort to overcome the weakness of bottom lateral systems to nonsynchronous operation of the culvert filling valves, the bottom longitudinal filling and emptying systems were developed. The aim is to admit water simultaneously through a number of ports, all equidistant from the center of the lock in travel time. To do this, the main culverts are led to the center of the lock, thence through large laterals to two or more longitudinals running upstream and downstream in the floor of the lock for most of its length. These longitudinals are provided with pairs of side ports similar to those in the bottom lateral system. The ports are arranged in a symmetrical pattern to obtain an even flow distribution throughout the chamber (see Fig. 10B-9). One of the first projects in this country using bottom longitudinals was the Millers Ferry lock. An improved system for a lock having a lift of 103 ft (31.4m) was developed for the Lower Granite lock. In this project, the main culverts were increased in area by about 50 percent to improve pressures below the valves. The distributing laterals and longitudinal sub-culverts are sized to be proportional to their discharge, so as to maintain a constant velocity. Area of ports in each longitudinal is about 0.85 times that of the sub-culvert, and total area of ports is about 1.2 times the culvert area at the filling valves. A more detailed description of the Lower Granite system is contained in a paper entitled "Filling System for Lower Granite Lock" by George C. Richardson.[3] Other papers dealing with lock filling systems are "U.S. Development of Hydraulic and Structural Designs for Locks,"[4] and "Hydraulic Design of Columbia River Navigation Locks."[5] The principal advantages of the bottom longitudinal system is that, since the main lateral at the center of the lock is connected to both main culverts, the lock can be filled or emptied through one main culvert (assuming the other is inoperative due to culvert repair or other cause) and still obtain a balanced filling or emptying operation. Also, with both main culverts in use, a faster filling of the chamber can be obtained without exceeding allowable hawser stresses.

8. Hydraulic Characteristics of Culvert Valves

The tainter type culvert valve has been adopted in the majority of locks recently built in the United States, although vertical lift (stoney) valves, butterfly valves, and cylinder valves have been used in the past. Until the advent of high-lift locks, the tainter valves had been installed with the skinplate upstream, and the arms in compression. As lifts increased, it was found that the pressure gradient just downstream of the valve dropped below the top of the culvert, and allowed large volumes of air to be drawn into the system. The explosive release of air into the lock chamber during lock filling caused disturbances hazardous to small craft and also decreased the efficiency of the system. Reversing the valve, so that the skinplate was downstream, and the arms were in tension, and with only a small clearance between skinplate and the downstream face of the valve pit, increased the pressure below the valve and eliminated the air problem at the valve pit. This arrangement also converts the valve pit into a surge chamber, which relieves water-hammer stresses on the valve in case of a sudden closing due to mechanical failure. A more detailed discussion of lock valves, including pressure gradients, possible cavitation tendencies, vibration problems, and head losses during valve operation, is contained in EM 1110-2-1604.[2] A short discussion on valves in other types of filling systems is also contained in the same referenced work.

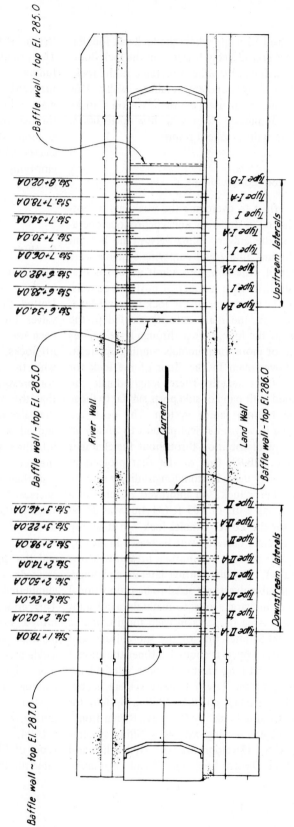

PLAN

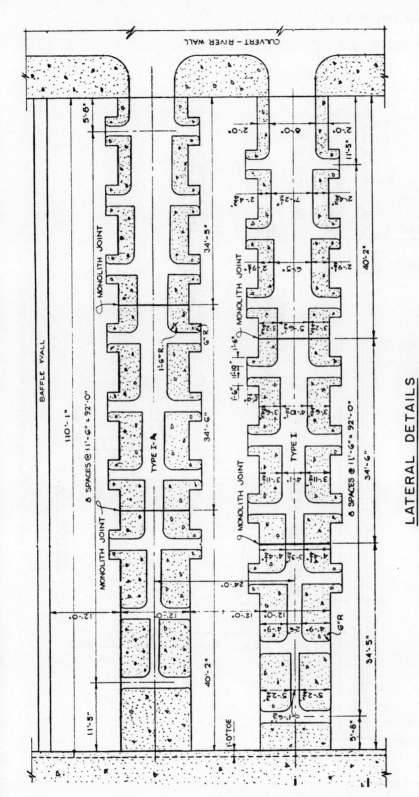

Figure 10B-7. Bottom lateral system—Split type.

578 HANDBOOK OF DAM ENGINEERING

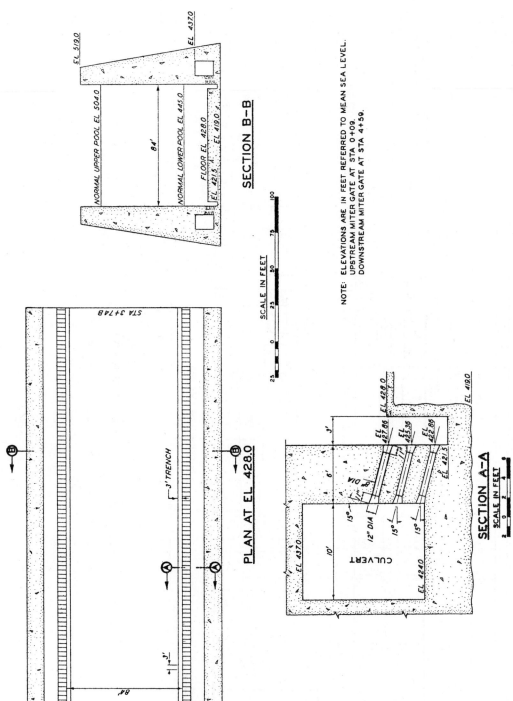

Figure 10B-8. Chamber wall ports—Multiple-port system.

DESIGN OF ADJACENT LOCKS 579

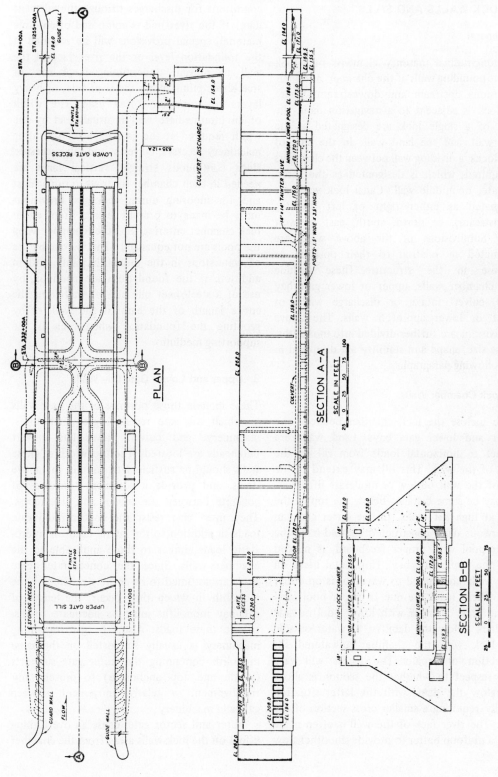

Figure 10B-9. Bottom longitudinal system. (*Courtesy Corps of Engineers, U.S. Army*)

C. LOCK WALLS AND SILLS

1. General

The longitudinal masonry elements of a lock are the bounding walls of the chamber, and their extensions upstream and downstream. Where the lock is adjacent to a navigation dam, the walls of a single lock are designated as the river wall, and the land wall. In the case of dual locks, a dividing wall between the chambers is required, which is designated as the intermediate, or middle wall. Canal lock walls are designated as either right or left (looking downstream), or north, south, east or west walls. Subdivisions of the above walls are designated in relation to their position or purpose in the structure. These include lock chamber walls, upper or lower gate bay walls, culvert intake or discharge walls, or upper or lower approach walls. The above subdivisions are further divided into monoliths, whose size, shape and stability are discussed in the following paragraphs.

2. Lock Chamber Walls

These inclose the lock chamber between the upper and lower gate bays. Land walls are subject to horizontal loads from fill on the back of the wall. This fill may extend to the top of the wall on low or moderate lift locks, or may be some distance below the top of the wall at high lift locks. In the latter case, an impervious dam, constructed to hold back the upper pool with proper freeboard, is needed near the upper gate bay. Horizontal loads on river walls are limited to water loads on either side. These loads are due to upper pool level in the lock chamber with lower pool outside the lock below the dam, or to water outside the lock with the chamber unwatered for inspection or repairs. The forces will vary with respect to whether the section is above or below the dam, with the latter situation usually requiring a smaller cross section of the wall. The river face of the wall is often made with a uniform batter to provide smoother flow conditions for discharges through the adjacent dam. If the river bed is composed of erodible material, special provisions will be required at the foundation level or the river face of the wall to prevent undermining it. A row of steel-sheet piling along the river side, or heavy layers of stone are often used. The width of an intermediate wall is usually set by the width required at the gate bays for operating machinery recesses, culverts, and valve recesses, since continuous straight vertical faces are needed in each chamber for tows to rub against and for mooring during lockages. Provisions must be made to prevent leakage from either lock chamber entering the other chamber when the pools are not equalized. This is done by use of waterstops in the monolith joints, and in addition, if the foundations are pervious, by use of a steel-sheet pile cut-off wall along the entire length of the base of the wall, or by grouting the foundation when rock is the supporting medium.

3. Upper and Lower Gate Bay Walls

These include those portions of the lock walls in which the gate recesses, gate anchorages, machinery, and culvert valves and culvert bulkheads are located. The top width of these walls should be sufficient to contain the above items, and provide a sufficient thickness of concrete between the culvert and gate recess. They must also resist the concentrated gate loads in addition to the lateral earth or hydrostatic loads similar to those applicable to the chamber walls. Since the concentrated gate loads are assumed to be entirely resisted by the monolith in which they occur, the length of gate-bay monoliths must be adequate to distribute these loads. Also, the gate operating machinery is usually supported on the same monolith containing the miter gate supports (pintle and top anchorage) to prevent any misalignment *or* relative movement between gate and machinery.

Miter and sector gates cause an overturning effect on the lock walls away from the chamber when they are loaded, and toward the chamber when they swing clear in an unwatered lock. These dead loads must be taken into account in computing wall stability. Gate bay monoliths are examined for stability as a complete unit, with all forces calculated in both the transverse and longitudinal axes, and the resultant located with respect to the kern area of the base.

4. Culvert Intake and Discharge Walls

Intake walls extend immediately upstream from the upper gate bays, and provide space for the culvert intake ports. They also provide space for the emergency bulkhead recesses, bulkhead handling equipment and storage when required, crossovers and risers, and other incidental equipment. Culvert intake walls are usually the same height as the chamber walls. The upstream emergency dam may be located either upstream or downstream from the culvert intakes. If downstream, then the filling valves or upstream valve bulkheads must be used to hold back the upper pool when unwatering the lock. Any part of the intake wall downstream from the emergency dam must be designed to resist lateral loads similar to those on the chamber walls during unwatered conditions. Only the land intake wall, however, will be required to withstand unbalanced lateral loads during normal operating conditions.

Discharge walls are immediately below the lower gate bays, and extend a sufficient distance to provide for the culvert discharge manifold or diffuser system, when these systems are used. The top of these walls may be lower than the chamber walls, whenever maximum navigable tailwater is appreciably lower than upper pool level. When the downstream maintenance bulkhead units are located downstream of the discharge manifold or diffuser system, the lateral loads to be resisted by the discharge walls are the same as for the chamber walls during unwatered conditions. Only the land discharge wall, however, will be subjected to unbalanced lateral loads during normal operating conditions. These unbalanced loads may be reduced by limiting the height of backfill.

5. Approach Walls

The primary purpose of approach walls is to provide protection for ships or tows, and the lock facilities, particularly as these craft approach the lock, by guiding them safely into the lock chamber. They also provide a means for checking the speed of an approaching tow, or correcting its alignment by putting out lines to check posts along the wall. Approach walls provide mooring space for the separated portion of a tow which is too long for a single lockage. In the past, the longer approach wall was located on the landward side of the lock. This was due to the general desire of pilots to operate close to the bank of the river when stages are such that adverse currents may exist. There are, however, other important considerations. If the spillway is located adjacent to the lock, there may be a current, or "draw" toward the dam. Prevailing winds may add to the tendency to move a tow toward the dam. In such cases, the long wall should be located on the riverward side of the lock. In canals, it may be desirable to locate the long approach wall on the side toward which the prevailing wind blows. Ship locks often have the long wall to port, since most ships berth more easily on that side. For locks serving barge tows, both the upstream and downstream long approach walls should be on the same side of the lock if possible; otherwise it would be necessary for a tow narrower than the lock chamber to "set over" from one wall to the other, a slow and tedious process. The length of the longer approach walls are generally made equal to the usable length of the lock chamber, unless conditions dictate a longer wall. Such a condition might be a rocky bank where it is impossible for tows or ships to nose safely into the natural bank during emergencies. In such a situation, it may be possible to install mooring piers or cells in lieu of a continuous wall. Length of the short walls opposite the longer walls varies with specific locations, and no general rules can be given. The long approach walls should generally be straight line extensions of the lock chamber wall. Approach walls should be able to absorb impact and withstand abrasion from moving vessels; however it is uneconomical to provide walls that cannot be damaged by the force of a heavily loaded tow, since local failure at the point of impact is not a serious matter compared to failure of a main lock wall, and will not put the lock out of operation. Numerous types of approach walls have been used, the more common being as follows:

Concrete gravity walls, continuous for the full length of wall, and founded on rock, soil, or piles. This type is the most expensive, since it usually requires cofferdam protection during construction. If the long upstream approach wall is located on the riverward side of the lock, and the dam spillway is adjacent to the lock, the approach wall should have openings below the navigation draft line to distribute flow toward the spillway and prevent the riverward flow or "draw" around the upstream end of the wall which would occur with a solid wall.

Timber cribs, filled with rock or sand, and capped with a concrete wall. The timber, of course, must not extend over minimum pool level, and if the wall is constructed on a soil foundation, piles will be required inside the crib to support the concrete wall. This type of wall is cheap, but excessive settlement at a relatively early age has often occurred, and its use is no longer recommended.

Cantilever or tied-back steel-sheet piling is applicable to land side approach walls where backfill is to be placed to the top of the wall, and an earth foundation exists. The wall is set back to allow for fenders, usually of wood. This type is relatively low in cost, but is subject to severe denting from heavy tows out of control.

Steel-sheet piling in a double row, connected by diaphragms or tie rods, filled with free-draining material, and capped with concrete. This type can often be employed on a riverward wall, where a single row of piling is not applicable.

Isolated steel-sheet pile cells, filled with sand, capped with concrete, and surmounted by a concrete wall which provides a continous rubbing surface for tows.

Where upper pools are very deep, and an approach wall founded on competent materials would be of excessive height, *floating approach walls* have been used. The Tennessee Valley Authority has floating upper guard walls at several Tennessee River locks. These floating walls are hinged to the upper end of the main lock walls through a wheeled guide operating in a slot, similar to a floating mooring bitt, except much heavier. A shock-absorbing device is incorporated into the connection. The upstream end of each wall is anchored by cables fastened to deadmen on the river bottom.

Isolated guides or mooring facilities, such as concrete piers, steel-sheet pile cells, or timber pile clusters, spaced at regular intervals, are applicable to both upstream and downstream approaches to locks in canals, where currents are not severe.

6. Lock Sills

These elements of a lock form the fixed portion of the damming surface under the service gates or temporary closures. The top of the sill in relation to the water surface controls the draft of vessels which can use the lock. A discussion of sill levels is contained on p. 564. All sills must resist the lateral forces, consisting of both earth and hydrostatic pressures, from the bottom of the gate or closure structure to the sill foundation. Where vertically framed miter gates are used, the sill must also resist the lateral load from the bottom girder of each gate leaf, and provide proper contact with the gate seal to close off any flow under the gate. Horizontally framed miter gates do not transmit any water loads to the sill; however, proper sealing must be provided. Sills for sector gates, rolling gates, or tainter gates must be designed to suit the particular requirements of these gates. In addition to the normal operating gates, locks are often equipped with temporary closures, used to permit unwatering the lock for periodic inspection and repairs. They are also used to stop the flow through the lock chamber should the regular gates become inoperative. In order that the temporary closures seal at the bottom, and have a base for support, sills are provided for them. The upper closure sill is located far enough upstream of the upper gate sill to permit a full swing of the miter gate leaves, and may be set at a slightly higher elevation than the gate sill to protect the latter from possible impact from tows or vessels. The closure sill may be a separate structure, or it may be combined with the gate sill, depending on the size of the gate sill required for stability. The lower closure sill is downstream from the lower gate sill, and can be much closer to it, since gate swing is not involved. The lower closure sill is usually at the same level as the adjacent gate sill, since the sealing face of the latter is not as subject to impact damage as the upper gate sill. The gate and closure sills are more often combined into a single structure at the lower end of the lock.

The principal types of closures include sectional bulkheads, poiree dams, and needle dams. Bulkhead sills are not required to resist any lateral forces from the bulkheads other than friction between the bottom section and the sill, and need only to be designed to support the weight of the bulkheads and the lateral hydrostatic and earth loads on the sill itself. Poiree dam sills must resist the full hydrostatic load on the dam as well as horizontal loads on the sill. The sill contains the structural anchorage to which the poiree A-frames are attached, and supports the needle damming surface. The water load is transferred from the A-frames to the sill through the anchorages in the form of a couple and horizontal shear. Needle dam sills are generally used on narrow locks, and these sills must resist the lower reaction of the vertical needles, while the top reaction is carried into the lock walls through the horizontal needle beam. Horizontal loads on the sill itself, from the top to the foundation must of course also be resisted.

7. Design Criteria

Gravity lock walls will be proportioned by methods similar to those used in the design of gravity dams.

Overturning Stability. Walls will be analyzed for dead loads, horizontal and vertical hydrostatic loads (including uplift) and horizontal and vertical loads from backfill where present. Horizontal earth pressures below assumed saturation levels will be computed at saturated values. The resultant must fall within the kern area of the base for all normal loading conditions when active earth pressures are assumed. Few rock foundations yield enough to develop active pressure values ("arching active") condition. Land walls on rock should therefore be designed for "at rest" pressures. Table 2 furnishes criteria and check lists for stability analyses of gravity lock chamber walls and sills for various load conditions.

Sliding Stability. Safety against sliding is achieved when the potential resistance of the foundation to sliding exceeds the total force tending to produce sliding. The degree of sliding safety is represented by the following shear-friction equation (see EM 1110-2-2602):[6]

$$S_{s-f} = \frac{fV + sA}{H}$$

where

V = summation of vertical forces
f = tangent of internal friction angle
s = unit shearing strength of foundation at zero normal load
A = area of potential failure plane considered. Any portion of the base not in compression should be excluded from A.
H = net horizontal driving force
S_{s-f} = shear-friction factor of safety

TABLE 2. Navigation Lock on Rock Foundation Stability Analysis Criteria

Loading Condition	Case	Earth Pressure Condition	Overturning Direction	Minimum Base Area in Compression	Minimum S_{s-f}
Land Wall:					
Normal Operating:					
Lower pool in chamber	Ia	Active	Towards lock	100 percent	Not controlling
Lower pool in chamber	Ib	At-rest	Towards lock	75 percent	4.0
Maintenance-Chamber unwatered	IIa	At-rest	Towards lock	75 percent	$2\,2/3$
Emergency-Case Ia or Ib with earthquake	IIb	At-rest	Towards lock	*	$2\,2/3$
Construction	III	At-rest	Towards lock	75 percent	$2\,2/3$
Middle Wall:					
Normal Operating:					
Upper pool in landward lock, lower pool in riverward	Ia	Not appl.	Towards river	100 percent	4.0
Upper pool in riverward lock, lower pool in landward	Ib	Not appl.	Towards land	100 percent	4.0
Maintenance:					
Upper pool in landward lock, lower pool in riverward	IIa	Not appl.	Towards river	75 percent	$2\,2/3$
Upper pool in riverward lock, lower pool in landward	IIb	Not appl.	Towards land	75 percent	$2\,2/3$
River Wall:					
Normal Operating:					
Upper pool in chamber, lower pool outside (1)	Ia	Not appl.	Towards river	100 percent	4.0
Lower pool in chamber, upper pool outside (2)	Ib	Not appl.	Towards land	100 percent	4.0
Maintenance:					
Lock chamber unwatered, lower pool at max. maint. stage	IIa	Not appl.	Towards lock	75 percent	$2\,2/3$
Emergency:					
Case Ia or Ib with earthquake	IIb	Not appl.	Towards lock	*	$2\,2/3$
Gate Sills:					
Normal Operating:					
Upper pool upstream from gate, lower pool downstream	I	Not appl.	Downstream	100 percent	4.0
Emergency:					
Same as I with earthquake	II	Not appl.	Downstream	75 percent	$2\,2/3$
Maintenance: (Upper gate sill only)					
Upper pool upstream of gate, lock chamber unwateredq	III	Not appl.	Downstream	75 percent	$2\,2/3$

*Resultant to fall within base; max. pressure allowable. (1) If dam is upstream from section. (2) If dam is downstream from section.

Note: See text for explanation of terms used.

Minimum factor of safety should not be less than 4 for normal operating conditions, including at-rest earth pressures. For maintenance or earthquake conditions, the factor of safety may be reduced to 2.67.

Internal Stability. Internal stresses in a gravity lock wall are seldom critical, except in regions of discontinuity as around openings and at points of high concentrated load transfers such as gate anchorages. Components of lock walls subject to high localized stresses should be designed as reinforced concrete.

8. Loadings

Principles of Load Determination. In the design of lock walls, sills, and other related elements, the basic design loadings may be difficult to determine. Because of the varying characteristics of soils, and different properties of foundation materials at each lock site, no definite rule can be applied. Water stages for high, normal and low discharges have a decided effect on the design of the lock structure, since their frequency and duration determine (1) extent of saturation of backfill on land walls, (2) hydrostatic pressures to be considered active on all walls during normal and unwatered conditions, (3) intensity of uplift pressures, and (4) loads from lock gates or temporary closures transferred to lock walls or sills.

Saturation Levels. No definite rule can be followed in determining the level of ground water in the backfill adjacent to a lock landwall. It is well established, however, that the saturation level varies between upper and lower pools, the degree of variation depending on the physical characteristics of the backfill material and the perviousness of the foundation. Since the thrust produced by the backfill is one of the most important to be considered in lock wall stability, and depends a great deal on the saturation line, the latter should not be determined entirely by arbitrary assumptions but should be given careful consideration including soil tests if necessary. Where practicable, backfill material should be pervious and free-draining. If sufficient pervious material is not available for the entire backfill, then a pervious zone immediately adjacent to the lock wall may be used. If such pervious backfill is used, it may be desirable to provide an impervious core between the lock wall and natural ground near the upper gate bay to prevent seepage of upper pool ground water to lower pool. Lowering of the saturation line behind the lock wall can be further assured by installation of a drain line in the pervious fill. This drain would extend from the downstream edge of the impervious core to lower pool, and should be at a level slightly above normal tailwater. If wall stability depends on the action of this drain, it should be provided with facilities for inspection and maintenance to assure its continued effectiveness.

Uplift. Fluctuating water levels in a lock chamber cause the problem of uplift to be complex. There is little knowledge to date as to how rapidly uplift under a wall builds up as the chamber is filled, or how rapidly it decreases when the chamber is emptied. The only conservative approach is to assume uplift on lock walls on the same basis as for dams. In all cases, uplift should be assumed to act on 100 percent of the base area. Where effective drainage is not provided, the following uplift assumptions are recommended:

CASE I—Upper pool in lock chamber:
   River Wall—uplift will vary from 100 percent of lower pool head at the riverward face to 100 percent of upper pool head at the chamber face.
   Land Wall—uplift will vary from 100 percent of the head represented by the saturation level at the landward face, to 100 percent of upper pool head at the chamber face.
CASE II—Lower pool in lock chamber:
   River wall—Uniform uplift of 100 percent of lower pool head over the entire base.
   Land wall—Uplift will vary from 100 percent of saturation level head at

landward face to 100 percent of lower pool head at chamber face.

CASE III—Maintenance:

River wall—Uplift will vary from 100 percent of lower pool head at maximum maintenance level to 100 percent of pumped-down level at chamber face.

Where adequate drainage, relieving to tailwater, is provided near the chamber face under the river wall, the total uplift, for the condition of upper pool in the chamber may be reduced as follows: Assume uplift varies from 100 percent of tailwater head at the riverward face, to tailwater plus 50 percent of the difference between headwater and tailwater at the chamber face. Drainage under the land wall is not advantageous in reducing uplift. The most effective drainage is the reduction of the saturation level. Uplift on lock walls supported on piles in an impervious foundation should be assumed as described above, unless the cutoff is at an appreciable distance from the high water face. In this situation, 100 percent of upper pool head should be used for the entire portion of the base on the high water side of the cutoff. Uplift under walls supported on piles driven into pervious materials should be assumed as 100 percent of upper pool head on the high water side of the cutoff, and 100 percent of lower pool on the low water side. Except for earthquake loadings, any portion of the base not in compression should be assumed to be subjected to a uniform uplift equal to 100 percent of the adjacent pool or saturation level. Uplift for earthquake loadings may be assumed to be equal to the same loading without earthquake. Hydrostatic pressures within the body of the lock wall may be assumed to vary from 100 percent of tailwater head at the low water face to 100 percent of the tailwater plus 50 percent of the difference between headwater and tailwater at the high water face. Because of the adverse effects of even minor movements of gate sills, all sill blocks should be analyzed for stability and internal pressures resulting from maximum differential heads. Uplift on the base, and internal hydrostatic pressures should be assumed to vary from 100 percent of high water head to 100 percent of low water head. A uniform uplift of 100 percent of high water head should be assumed active on any portion of the plane not in compression.

Earth pressures. Careful investigation of available backfill materials and methods of backfilling is of primary importance. The dry, naturally drained, and submerged weights and corresponding angle of internal friction of the proposed backfill material for the conditions anticipated to result from the proposed field placement method should be adopted for the design of the lock wall. At-rest pressures should be used for gravity sections on rock foundations as discussed on p. 583. Values for these pressures should be determined for the various conditions of the backfill (drained, saturated, or submerged) by accepted soil analysis methods.

Foundation pressures. Allowable foundation pressures for the lock walls should be determined in the same manner as for the adjacent dam.

Gate loads. The wall monoliths which support the lock gates are considered as individual units in their design, because of the concentrated gate thrusts that must be distributed over their entire areas. Gate dead loads are important in wall stability analysis for the lock unwatered condition, since they add to the overturning forces toward the lock chamber. Live load forces are used for normal operating conditions; lock chamber at lower pool for the upper gate, and upper pool in chamber for the lower gate. Computation of these forces, their location, and direction are discussed on p. 588.

Miscellaneous Loads

1. Earthquake. It is assumed that the consideration of earthquake forces has already been made for the design of the adjacent dam. Obviously, the same intensities and force distribution should be used for the lock walls as for

the concrete portions of the dam, which are supported on similar foundations.

2. Tow Impact. Lock walls, and particularly approach walls, need to be designed with allowances for impacts from tows or vessels, and line pulls which may be generated by the tow during a lockage. These forces vary considerably; the following summarizes forces allowed on approach walls of a 110 × 1200 ft lock on the Ohio River:

a. Impact of 100 kips per monolith, applied 4 ft above normal pool, acting *away* from the lock chamber. (Normal operating condition).
b. Pull of 20 kips per monolith, applied 10 ft above normal pool, acting *toward* the lock chamber. (Normal operating condition).
c. Same as *a* above, except applied 4 ft above *maximum* navigation pool. (Emergency operating condition).
d. Same as *b* above, except applied at top of wall. (Emergency operating condition).
e. Same as *a* above, except acting *toward* the lock chamber. (Emergency operating condition applicable to upstream riverward approach wall only). This force represents a downbound tow which has missed its approach and strikes the back of the wall.

3. Ice Loads. These are not normally included in wall stability analysis. However, approach walls, particularly in the upper pool, are sometimes subjected to moving ice and floating debris, and their effects should not be neglected. For projects where ice conditions are severe, and the ice sheet is short or can be restrained or wedged between structures, its magnitude should be estimated, with consideration given to the locality and available records of ice conditions. A unit pressure of not more than 5,000 lb/sq ft applied to the contact surface of the structure at normal pool level is recommended in EM 1110-2-2602.[6] Ice thickness assumed for design in the U.S. will normally not exceed 2 ft.

4. Wind Loads usually need not be included in the wall analysis, except where major portions of the walls are not backfilled, or where projections extend above their tops. Where included, wind forces should be placed in the most unfavorable direction, and should be assumed at 30 lb/sq ft (1.26 kg/m^2) unless past records indicate that other assumptions are warranted.

9. Reinforcing of Gravity Walls

General. Concrete gravity walls are reinforced primarily at discontinuities in the section, such as culverts, galleries, valve pits, and gate recesses, and at points of concentrated load applications such as gate pintles and anchorages, line hooks, and check posts. Any localized thin or cantilevered section should, of course be designed as reinforced concrete.

Culverts. Longitudinal wall culverts in modern locks may be quite large, and their proper reinforcement is important. Lock walls containing culverts are of two types. One type has relatively thin walls, 4 or 5 ft (1.2 to 1.5 m) thick enclosing the culvert, with a cantilevered projection above to effect a retaining wall for the lock chamber. This type will normally occur on locks with lifts varying from 5 to 15 ft (1.5 to 4.6 m), with only minor variations in water levels, and may be founded on piles. This wall should be analyzed as a purely reinforced concrete structure by rigid frame analysis incorporating moment distribution in accordance with the relative rigidity of the frame parts. The second type is used for high lifts, requiring considerable bottom width for stability, and resulting in much greater and probably unequal concrete thickness along the culvert sides. This type does not lend itself to a moment distribution analysis, since the accuracy of the results depends on the assumptions made for the frame dimensions, which are at best only approximate. The method recommended for a lock with thick walls consists of calculating stresses produced by the unbalanced horizontal loads on the culvert walls, considering the walls as beams fixed at their bottoms and guided at their tops. To these stresses are added those calculated from the wall stability analyses. A detailed description of this method is contained in EM 1110-2-2602.[6] The finite element method of analysis may be used in lieu of the above if desired. From the total stresses due to horizontal and vertical loads, for all loading cases, the necessity for and the amount of reinforcing steel can be determined. Whenever the analysis indicates tension from hydraulic and other applied loads, reinforcement should be provided for the total tensile force, without reliance on tension in the concrete. Culvert floors should be analyzed as restrained beams resisting 100 percent uplift from the foundation with the culvert dry. Roofs should be able to support one concrete lift 7 days old plus one fresh lift, at working stresses compatible with 7-day strength. Lifts are normally 5 ft in height. Minimum transverse reinforcement of #7 bars at 12 in. centers should be provided around all water passages where analysis does not require a greater amount. Such minimum steel is recommended because of shrinkage stresses and unpredictable stress concentrations at corners.

Valve Recesses. Walls of valve shafts should be designed as fixed beams resisting internal and external horizontal water loads. The walls are assumed to span horizontally, transferring their loads to the end blocks. Each valve shaft should be investigated for loadings to which it can be subjected. For example, upper valves may be holding back the upper pool while the lock chamber is unwatered. The shaft walls would then be subjected to an *internal* hydrostatic head equal to the height from the upper pool to the culvert floor. Again, a sudden accidental closure of the valve might cause water in the shaft to rise as high as the top of the lock wall, subjecting the side walls of the valve shaft to an *internal* pressure equal to the height from the lower pool to top of the lock walls. Maximum *external* pressure would occur when the valve is unwatered for inspection or repair, and would be a hydrostatic head equal to the difference between culvert floor and upper pool. These are all considered emergency conditions, and reinforcing steel provided on the basis of unit stresses one third above normal. Culvert bulkhead recesses should be reinforced with sufficient steel to control shrinkage cracks.

Load Concentrations. A grillage of reinforcing steel is usually placed under each gate pintle to distribute the vertical load from the lock gate. Likewise, reinforcing is used around the top anchorage of each lock gate to distribute horizontal gate reactions into the surrounding concrete. Line hooks and check posts are provided with reinforcing steel to distribute their pulls into the mass of the lock walls.

D. LOCK GATES AND CULVERT VALVES

1. General

The lock gates form the upstream and downstream ends of the lock chamber. They must be capable of being opened to allow a vessel or tow to enter or depart from the chamber, and of being closed to hold back the upper pool to permit the vessel or tow to be raised or lowered from one water level to the other. A number of types of gates have been used for this purpose, the most common by far being the miter gate. Other types include rolling gates, sector gates, vertical lift gates, and tainter gates.

The culvert valves permit water to be admitted to the lock chamber from the upper pool, or to be discharged from the chamber into the lower pool. Again, a number of types of valves have been used for this purpose, including vertical lift, or stoney valves, butterfly valves, and tainter type valves, both direct and reversed. Each of these types has its place, but modern medium and high lift locks exclusively use reverse tainter type valves. Detailed discussion of culvert valves in this section will therefore be limited to this type; however, the general principles involved are applicable to the other types.

2. Lock Gate Types

Miter Gates. A miter gate consists of two leaves, each supported by a lock wall, which close against each other at an angle pointed upstream against the water load. When open, they lie flush with the face of the lock walls, so as not to obstruct passage of vessels or tows. Leaves are supported on the bottom by a hemispherical ball (pintle) and socket, and held at the top by a pair of anchorages. Early barge locks in this country had gates made of wood. As the locks were narrow, such construction was feasible. As barges increased in size, and locks became wider, the wooden gates were replaced by steel gates. Gate parts were riveted or bolted together. With the improvement of the arc-welding process, riveted gates were replaced by welded ones; and today, practically all miter gates are of all-welded construction. This has virtually eliminated the gradual sagging of the leaf formerly experienced with riveted joints, and has greatly decreased periodic adjustments which were necessary to bring the leaves back into proper sealing postion. Modern miter gates have proved to be the most rugged, trouble free, simplest operating type of gate developed and their use is highly recommended unless some special condition indicates another type.

Types of Miter Gates

Vertically Framed. In a vertically framed leaf, the skinplate transfers its water load to a series of vertical beams extending from the bottom to the top of the leaf. These beams in turn transfer their reactions to the horizontal top and bottom girders. The bottom girder transfers its load directly into the lock sill, against which it bears when the gate is in the closed position. The top girders of each leaf bear against each other at their outer, or miter ends and against thrust blocks in each lock wall at their inner, or quoin ends, thus forming a three-hinged arch which transfers all the top reactions of the vertical beams into the lock walls. Approximately two thirds of the total water load is transferred to the lock sill, which must be designed for this added load. Vertically framed gates are seldom used where the ratio of leaf height to length exceeds 0.50. This limits their use to very low lift locks, or to upper gates, regardless of lift, where a high sill can be used.

Horizontally Framed. As miter gates increased in height, vertical framing became impractical. The horizontally framed leaf was developed, wherein the skinplate transfers its water loads to a series of horizontal beams, each of which, with its corresponding member of the other leaf, forms an arch and transfers the load from its portion of the skinplate into the adjacent lock wall. The total load on the gate is therefore transmitted into the lock walls, and no load is carried by the sill. Thrust blocks on the gate leaves and in the lock walls are continuous from top to bottom of the gate, and the bottom seal is usually a "J" type rubber seal suspended below the bottom girder.

Rolling Gates. This is a horizontally framed gate, constructed to span the lock chamber as a single structure. It requires a slot in each lock wall: a shallow slot in the river wall to develop a bearing surface, and a slot equal to the width of the gate in the land wall to receive the gate in its fully opened position. The bottom of the land wall recess and the lock sill are provided with rollers on which the gate is moved. Water loads are transmitted horizontally into each lock wall through vertical bearing beams in each slot. This type of gate was more commonly used with relatively narrow locks, and is not suitable for a modern wide lock due to the length of recess required in the land wall. It is also not suited to twin locks adjacent to the open river. A rolling gate is subject to deposition of silt in the area of the rollers, and requires considerable maintenance of the underwater rollers.

Tainter Gates. This type is applicable to an upper gate, when the upper sill is high enough above the chamber floor to permit the gate to be completely lowered behind the sill so as not to be an obstruction to the passage of vessels or tows across the sill at minimum upper pool level. The gate must extend from sill level to maximum upper pool with some freeboard allowance. Side frames and trunnions must be recessed into the lock walls unless they are lower than the sill when the gate is lowered, and must not extend into the usable lock length when raised. Tainter gates may require guard piers in the lock chamber for protection against impact of up-bound tows which may not be able to stop in time, and to prevent tows from mooring too close to the gate so it can not be fully lowered. Piers must be spaced so that a barge cannot pass between them, and their downstream face will form the upstream limit of the usable lock chamber. Lifting chains or cables and hoisting machinery must not encroach on the lock width. Hoists cannot be operated by a common drive shaft unless they are located above the navigation clearance line, but must normally be electrically synchronized to assure that the gate remains level throughout its entire operating range.

Vertical Lift Gates. A few locks have been constructed with a vertical lift lower gate in lieu of the miter type. This gate is framed similarly to a rolling gate, but is lifted vertically instead of being rolled horizontally. This eliminates the need for the deep land wall recess, but requires that lifting machinery be set on towers high enough so that the gate can be lifted above the downstream navigation clearance line to allow vessels and tows to pass under it. Operating time is much longer than for a miter gate.

3. Miter Gate Design (Horizontally Framed)

General. Horizontally framed miter gate leaves are generally of the shape shown in Fig. 10B-10. The "spring line" of each girder is a straight line through the center of the miter and quoin contact blocks. The "thrust line" is an approximately circular arc through the quoin contact of each pair of matching girders and their common miter contact which form the hinges of the arch. Formerly, gates were often constructed with curved girders whose center of gravity followed the the thrust line so as to avoid bending stresses due to eccentricity. With rising labor costs, the additional fabrication cost of the curved girders more than offset the additional material required for straight girders. Also, lock walls for curved gates must be wider to allow for the deeper gate recesses.

It should be noted that the center of leaf rotation, or "C/L Pintle" shown in Fig. 10B-10 is offset from the spring line. This is done so that the quoin contact block on the gate moves away from the matching wall contact block as the leaf is opened, and is wedged tighter and tighter against the wall block as external loads tend to rotate the leaf downstream. Width of girders depends on the length of the leaf and the water head on the gate. To simplify framing, all girders are made the same width. The face of the girders with the leaf in the recessed or open position are located a sufficient distance back of the face of the lock wall to allow for protective fenders. These fenders were formerly made of wood, but with the increased cost of timber, and of labor for replacing fenders, steel is now coming into general use. Fenders are provided on the downstream face of the gate leaves, so as to protect this exposed face when the gates are recessed. Upper gate leaves are protected from the top girder to the minimum upper pool level, while lower gate leaves are protected from about 10 ft above the maximum lower pool to the minimum lower pool. Between these limits, fenders are mounted on each horizontal girder, and between girders where necessary to obtain a spacing of not more than about 3 ft 6 in. on centers. Each gate recess should be provided with two bumper blocks, one near the top of the gate leaf, and one near the bottom, fastened to the face of concrete. They should be of proper thickness so the leaf will bear against them when in the open position, so that any barge impact will be taken into the lock walls and not be transferred to the operating machinery.

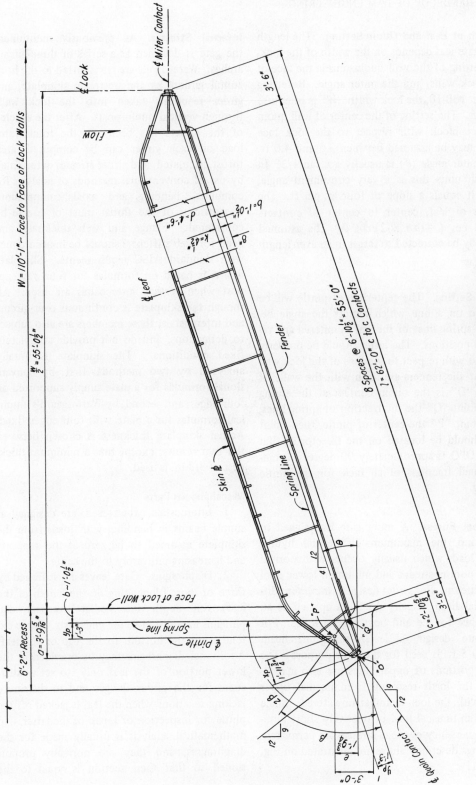

Figure 10B-10. Typical horizontally-framed miter gate leaf.

Length of Leaf and Quoin Setting. The length of a gate leaf depends on the width of the lock, the setting of the wall quoins inside the face of the lock walls, and the miter angle. Referring to Fig. 10B-10, the lock width "W" is of course known. The setting of the center of wall quoin contact block with respect to the lock face ("a") may be assumed between 3.5 and 4.0 ft. The miter angle (θ) is usually arc tan 1/3. In English units this is a very convenient angle, since it equals a slope of four in. per ft. The length of leaf, center to center of contacts would be: $L = (a + W/2)/\cos \theta$. The assumed "$a$" may be corrected to result in an even length of leaf.

Pintle Setting. The centerline of pintle will be located on a line which bisects the angle between spring lines of the leaf in mitered and recessed positions. The latter should be properly located with respect to the face of the lock wall so that the fenders are flush with the wall. If point "O" is the quoin contact at the spring line, point "Q" the intersection of spring lines, and point "P" the center of pintle, then point "P" should be located on the bisector so that angle OPQ is approximately 90 degrees, avoiding small fractions of an inch for the pintle setting.

External Forces. A miter gate is designed to withstand the maximum differential hydrostatic load, which usually occurs with normal upper pool upstream and minimum lower pool downstream. For girder design, a minimum differential head, usually 15 ft, is applied as an allowance for ice and boat impact loads. For skinplate design, a lesser minimum head, usually 5 ft, is used for the same purpose. The lower portions of upper gates are also investigated for loads resulting from normal upper pool with the lock chamber unwatered. Since this is an unusual load of relatively short duration, emergency overload stresses are permitted. Loadings described above are illustrated on Fig. 10B-11.

Internal Stresses. As previously mentioned, the gate is designed as a series of three-hinged arches. Water loads are transmitted to the horizontal girders by the upstream skinplate, and girder reactions taken into the lock walls through vertical quoin posts. After the spacing of the girders has been set, the total water load on each girder can be computed, arch thrust calculated, and girder stresses determined by use of conventional methods of analysis for combined bending and axial compression. Eccentricity of the thrust must of course be considered. Flange and web thickness, and need for web stiffeners should be in accordance with standard AISC requirements. Skinplate design is based on formulas for four-way support when vertical intercostals are used. Although the skinplate is continuous over girders and intercostals, these members are also subject to deflections, and do not provide completely fixed conditions. The skinplate is usually analyzed by two methods; first, by conventional formulas for a plate simply supported on four edges, and second, by Westergaard's empirical formulas for a plate with four edges fixed. Actual skinplate thickness is chosen between the two results, except that a minimum thickness of 3/8 in. is used.

Miscellaneous Parts

1. Intercostals, when used, are designed as simple beams in bending, with loads from the skinplate assumed to be zero at the supports and increasing uniformly to the center.

2. Diaphragms. Gate leaves are stiffened by three or more (depending on the length of the leaf) full depth intermediate vertical diaphragms in addition to the end posts. These extend from the bottom to the top of the leaf. Additional diaphragms may be provided in the lower portion of the leaf only, to act as stiffeners for the girder flanges and to distribute jacking reactions when the leaf is jacked off the pintle for inspection or repair of the latter. No mathematical analysis is usually made for the diaphragms, and they are normally proportioned so that their section is equal to the

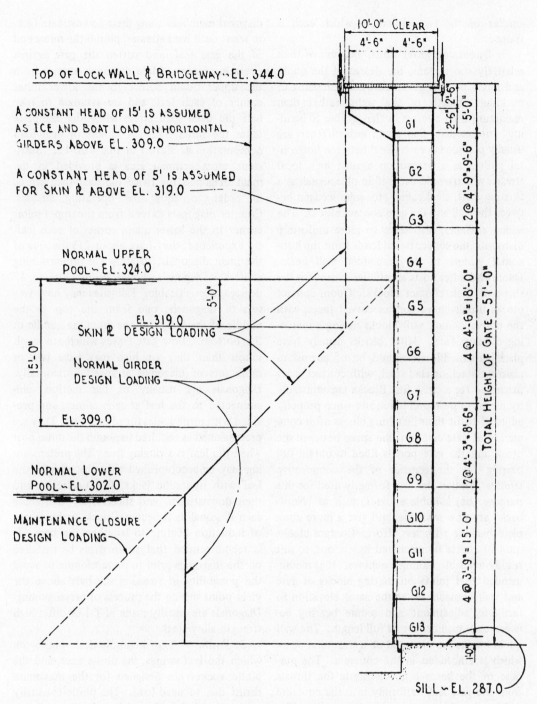

Figure 10B-11. Water loads on miter gate leaf.

smaller of the two girders to which each is framed.

3. Quoin and Miter Ends. Because of their relatively short spans, the design of the quoin and miter posts is governed by the shearing or buckling strength of the webs, rather than maximum fiber stresses in flanges due to bending. Intermediate flange and web stiffeners are usually provided in each panel between horizontal girders as a precaution against high local stresses which would occur if an object such as a floating drift, steel cable, etc. were wedged between the wall and the gate contact blocks. The added stiffeners also serve to more uniformly distribute the concentrated loads from the horizontal girders to the embedded wall beam. Quoin and miter posts are similar except for the shape of their contact blocks. Quoin contact blocks usually have convex curved faces, with the corresponding wall blocks having a matching concave face. Miter blocks usually have plane faces. Blocks should be of corrosion-resistant steel, or clad steel, with contact faces machined for a close fit. Blocks are adjustable by means of push-pull bolts, and when properly adjusted to fit their matching blocks after completion of gate erection, the space between the block and the gate post is filled to obtain full bearing for the transfer of the compressive stresses. Molten zinc was formerly used for this purpose, but suitable epoxies, such as "Nordback" are now available, and give a more complete backing with less effort. Contact blocks must of course be machined in sections, to suit readily available milling machines. It is recommended that joints on mating blocks of gate and wall contacts be at the same elevation to facilitate adjustment and secure bearing between each member for its full length. The wall quoin blocks are mounted on a structural beam which is embedded in the concrete. The purpose of the beam is to distribute the thrusts from the gate leaf uniformly into the concrete within allowable concrete bearing stresses. The wall blocks may be either fixed or adjustable, but should be removable for replacement if damaged or worn.

4. Diagonals. Miter gates are equipped with diagonal members along their downstream face, to resist dead load stresses, plumb the miter end of the gate leaf, and stiffen the gate against operating forces. Main diagonals extend from the upper quoin corner to the lower miter corner of each leaf, and are assumed to take half the dead load stress while the skinplate takes the other half. A sample of dead-load computation is shown in Fig. 10B-12. Sufficient cross-sectional area is provided in the main diagonals for twice the dead-load stress, in order to allow for operating stresses. Counter diagonals extend from the upper miter corner to the lower quoin corner of each leaf. By experience, they have about 2/3 the area of the main diagonals, to allow for pretensioning stresses. Diagonals should be as close to 45 degrees as practicable. Tall gates may have two sets of diagonals, one from the top to the middle of the leaf, and one from the middle to the bottom. Low gate leaves which are much longer than they are high may have two or more sets of diagonals arranged horizontally. Diagonals are usually of bar section, pin-connected to the leaf at each corner, and provided with turnbuckles for adjustment. They are pretensioned as required to plumb the miter post when the leaf is swinging free. The pretensioning may be accomplished either by warping the leaf with hydraulic jacks, first upstream, and then downstream, and successively tightening each diagonal as it becomes slack, or by means of induction heating to lengthen the bars. It is recommended that strain gages be installed on the diagonals prior to pretensioning to avoid the possibility of stressing the bars above the yield point during the process of pretensioning. Diagonals are usually made of T-1 or other high strength alloy steel.

5. Pintle. The hemispherical pintle on which the leaf swings, the pintle base, and the pintle socket are designed for the maximum thrust due to dead load. The pintle is usually made of nickel steel, and its spherical radius is governed by the allowable unit bearing stress on the pintle socket bushing. This bushing is usually aluminum-bronze, with an allowable stress of 3000 psi. Pintles and mating bushings

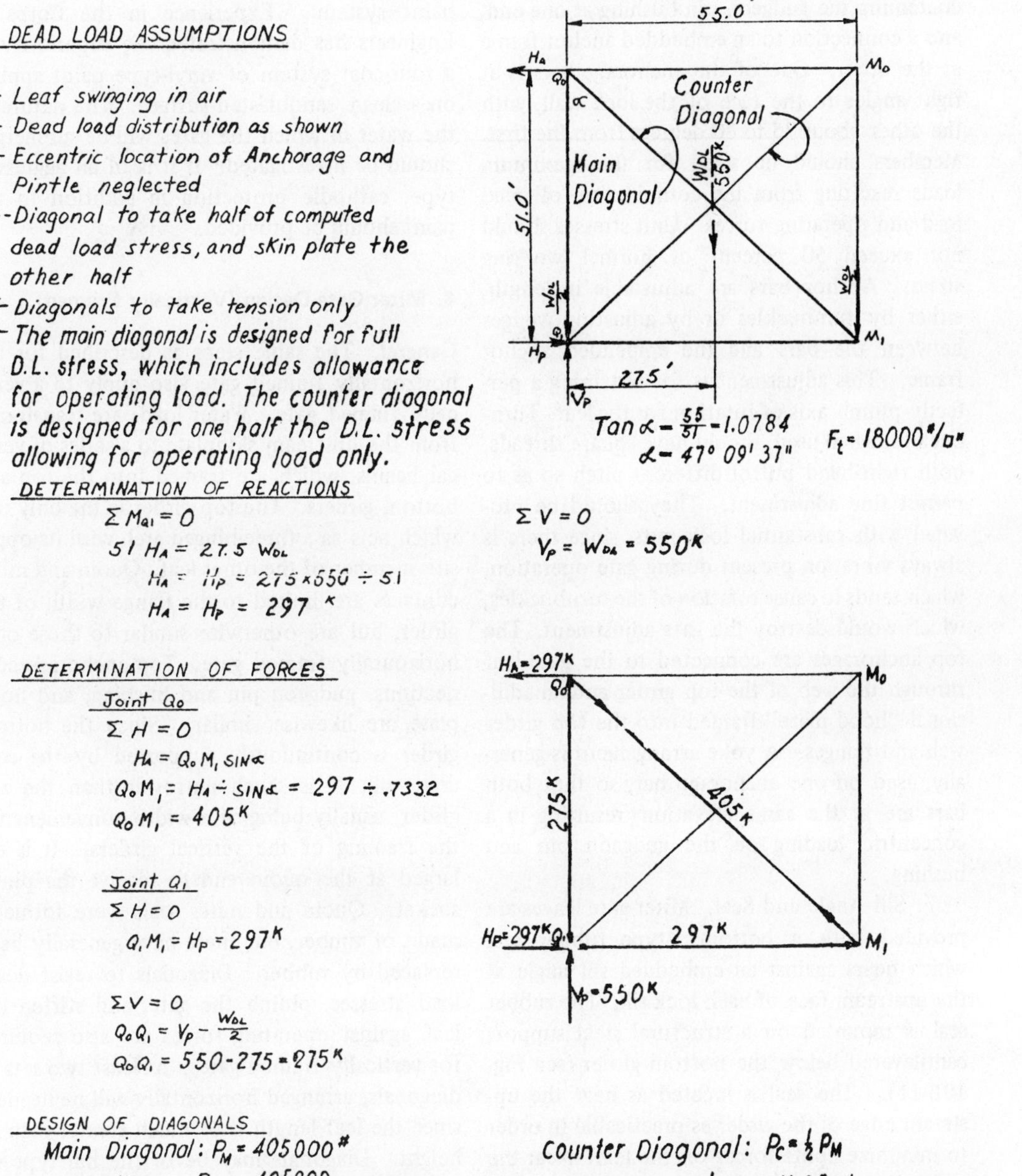
DEAD LOAD ASSUMPTIONS
1- Leaf swinging in air
2- Dead load distribution as shown
3- Eccentric location of Anchorage and Pintle neglected
4-Diagonal to take half of computed dead load stress, and skin plate the other half
5-Diagonals to take tension only
6-The main diagonal is designed for full D.L. stress, which includes allowance for operating load. The counter diagonal is designed for one half the D.L. stress allowing for operating load only.
55.0'
H_A
Q_0
M_0
α
Counter Diagonal
Main Diagonal
W_{DL} 550K
51.0'
W_{DL}
W_{DL}
Q_1
H_P
M_1
27.5'
V_P
$Tan\,\alpha = \frac{55}{51} = 1.0784$
$\alpha = 47°\ 09'\ 37''$
$F_t = 18000\#/\square''$
DETERMINATION OF REACTIONS
$\Sigma M_{a1} = 0$
$51\ H_A = 27.5\ W_{DL}$
$H_A = H_P = 27.5 \times 550 \div 51$
$H_A = H_P = 297^K$
$\Sigma V = 0$
$V_P = W_{DL} = 550^K$
DETERMINATION OF FORCES
Joint Q_0
$\Sigma H = 0$
$H_A = Q_0 M_1\ SIN\alpha$
$Q_0 M_1 = H_A \div SIN\alpha = 297 \div .7332$
$Q_0 M_1 = 405^K$
Joint Q_1
$\Sigma H = 0$
$Q_1 M_1 = H_P = 297^K$
$\Sigma V = 0$
$Q_0 Q_1 = V_P - \frac{W_{DL}}{2}$
$Q_0 Q_1 = 550 - 275 = 275^K$
$H_A = 297^K$
Q_0
M_0
275^K
405^K
297^K
$H_P = 297^K$ Q_1
M_1
$V_P = 550^K$
DESIGN OF DIAGONALS
Main Diagonal: $P_M = 405000\#$
$Req'd\ A_M = \frac{P_M}{F_t} = \frac{405000}{18000} = 22.5\square''$
Counter Diagonal: $P_C = \frac{1}{2} P_M$
$Req'd\ A_C = \frac{1}{2} A_M = 11.25\square''$

Figure 10B-12. Dead-load diagram–miter gate leaf.

should be machined and scraped for maximum contact area, and then match-marked for installation together in the field. Bushings are provided with grease grooves, and lubrication lines are extended to the top of the gate for ready access. The pintle is set on a cast iron embedded base, and affixed to it to prevent rotation of the pintle on the base, and to assure that the movement is always between pintle and bushing.

6. Top Anchorage. The top of the gate leaf is held in position by a pair of anchorage bars,

containing the gudgeon pin bushing at one end, and a connection to an embedded anchor frame at the other. One of the anchorage bars is at right angles to the face of the lock wall, with the other about 75 to 80 degrees from the first. Members should be sized for the maximum loads resulting from the combination of dead load and operating forces. Unit stresses should not exceed 50 percent of normal working stress. Anchor bars are adjustable in length, either by turnbuckles or by adjusting wedges between the bars and the embedded anchor frame. This adjustment is for obtaining a perfectly plumb axis of rotation for the leaf. Turnbuckles, when used, should have square threads, both right-hand but of different pitch so as to permit fine adjustment. They should be provided with substantial lock nuts, since there is always vibration present during gate operation, which tends to cause rotation of the turnbuckles, which would destroy the gate adjustment. The top anchorages are connected to the gate leaf through the web of the top girder and an additional "hood plate" framed into the top girder web and flanges. A yoke arrangement is generally used on one anchorage bar, so that both bars are at the same elevation, resulting in a concentric loading on the gudgeon pin and bushing.

7. Sill Angle and Seal. Miter gate leaves are provided with a bottom J-type rubber seal which bears against an embedded sill angle at the upstream face of each lock sill. The rubber seal is mounted on a structural steel support cantilevered below the bottom girder (see Fig. 10B-11). The seal is located as near the upstream edge of the girder as practicable in order to minimize uplift forces on the leaf. Near the quoin end of the leaf the seal curves around the pintle in a compound reverse curve, and ends at the quoin contact block. All points on the reverse curves are eccentric with respect to gate rotation, so that a wedging action for a tight seal is produced as the leaf miters, and a quick breakaway of contact occurs when the leaf is opened, to minimize seal wear.

8. Corrosion Mitigation. Miter gates must be protected against corrosion by an adequate paint system. Experience in the Corps of Engineers has demonstrated the superiority of a four-coat system of vinyl-type paint applied on a clean, sandblasted surface. The nature of the water in which the gates will be submerged should be investigated. If it is of an aggressive type, cathodic protection in addition to the paint should be provided.

4. Miter Gate Design (Vertically Framed).

General. The same steps as described for the horizontally framed gate also apply to a vertically framed gate. Water loads are transferred from the upstream skinplate to a series of vertical beams, which are framed into the top and bottom girders. The top girder is the only one which acts as a three-hinged arch with its opposite member of the other leaf. Quoin and miter contacts are limited to the flange width of the girder, but are otherwise similar to those on a horizontally framed gate. Top anchorage connections, gudgeon pin and bushing, and hood plate are likewise similar. Since the bottom girder is continuously supported by the concrete sill, it is much narrower than the top girder, usually being of a width convenient for the framing of the vertical girders. It is enlarged at the quoin end to receive the pintle socket. Quoin and miter seals were formerly made of timber, but these have generally been replaced by rubber. Diagonals to resist dead-load stresses, plumb the gate, and stiffen the leaf against operating forces are also required for vertically framed gates. At least two sets of diagonals, arranged horizontally will be needed, since the leaf length is so much greater than its height. Diagonals may be of the bar type, as used on horizontally framed gates. However, very good results have been obtained from rolled beam sections framed solidly into top and bottom girders.

5. Culvert Valves

General. As previously mentioned, many types of culvert valves have been used in locks. However, for modern, medium to high lift locks, the reversed tainter type valve is used almost exclusively. This will be the only type discussed here. Even among reversed tainter valves, there have many shapes of body. Double skinplate, with both plates convex, double skinplates with concentric plates, and single skinplates have all been used. Side frame arms are usually streamlined by circular edge plates and full or partial plating between the arms. Hydraulic models of various shapes of valves have been tested at the Corps of Engineers Waterways Experiment Station for dynamic downpull, uplift, and stability. It was found that best results are obtained with a single-skinplate, vertically framed body. Accordingly, this type of valve is recommended for all but very low lifts.

External Loads. Culvert valves will normally be subjected to an unbalanced head equal to the height from the normal upper pool to the minimum tail water, plus variable dynamic forces during lock filling and emptying. Filling valves may also be subject to hydrostatic loads from the normal upper pool to the floor of the culvert during an unwatered lock chamber. In view of the dynamic forces, valves should be designed for a hydrostatic head from the top of the lock wall to the floor of the culvert at a maximum unit stress of 0.45 times yield strength, and be investigated for a hydrostatic load of two times the normal lift at an allowable unit stress of 0.67 times yield strength.

Design of Component Parts
1. Skinplate may be analyzed as a continuous member supported on the vertical ribs, with loads as described in the previous paragraph.
2. Curved Ribs. Since the radius of curvature is quite large compared with the rib depth, the slight theoretical increase in stress due to curvature is usually neglected, and they are designed by formulas applicable to straight members. Ribs are continuous over the two supporting horizontal beams, and are formed of structural tees with outstanding legs welded to the skinplate. The space between the skinplate and horizontal beams is left open for free flow of water, so as to minimize dynamic forces on the valve.
3. Rigid Frame. The rigid attachment of side arms to each horizontal girder warrants the analysis of the assembly as a rigid frame with hinged ends at the trunnions. Loads on each girder are the sum of the reactions of the vertical ribs supported by it.
4. Seals. The bottom seal is usually metal-to-metal, with the lower edge of the skinplate bearing on an embedded beam in the culvert floor. Side seals may be of J-type molded rubber, fastened to each side of the skinplate by adjustable curved supporting angles and clamping bars, and sealing against corrosion-resistant embedded rubbing plates in the culvert walls. These plates are extended above the top of culvert walls for the full travel of the valve, to provide lateral support of the valve at all times. This is very desirable in view of the extreme turbulence in the valve pit during filling operations. Rubber J-type seals have been used for the top seal, but they have given a great deal of trouble unless fully supported on both sides. A caisson-type rubber seal, which can be clamped on both edges, has given very good results. The top seal bears on a curved embedded plate, which extends out from the face of the valve pit and provides a sloping face for the rubber seal to contact. The bottom, side, and top embedded plates are located in recesses in the lock concrete, are provided with threaded anchor bolts, and are grouted into place after being adjusted to the valve seals for proper contact.
5. Miscellaneous Details. The trunnion girder may be designed as part of the valve, or it may be a separate girder carrying the trunnion clevises, into which the valve arms with their trunnion bushings fit. For the first type, the trunnion bearings are set on shelves at either side of the culvert, so as to transmit the main valve reactions into the concrete in direct compression. Bearings are of the split type to permit installation or removal of the valve as a complete unit, and are adjustable on their embedded supports to insure proper adjustment and rotation of the valve. The second type is

also adjustable, and since the trunnion and bushing on each arm can be inserted between the members of each clevis, and the pins inserted, the bearings need not be split.

6. Culvert Bulkheads

Each valve has bulkhead recesses extending from the top of the lock wall to the bottom of the culvert located both upstream and downstream of the valve. Structural steel (or aluminum) bulkheads are provided so that any valve can be isolated, unwatered, and repaired without the necessity of shutting down the entire lock. The number of bulkheads required, and the material needed for construction depend on the facilities available for transporting the bulkheads from storage to each valve location, for lowering and raising them in the recesses, and on the number of valves to be unwatered simultaneously. Two valves in one wall may be taken out of service concurrently, if desired, while the lock is operated with the valves in the opposite wall. Bulkheads may be constructed of structural steel, high-strength alloy steel, or aluminum, and may be one piece or sectional. They should be designed for hydrostatic head from the normal upper pool to the culvert floor, at a basic stress not to exceed 0.67 times the yield strength of the material used. Some means must be provided for bypassing the bulkheads and filling the valve pit so that the bulkheads may be removed under balanced head. Recesses downstream from the filling valves are often equipped with horizontal sealing diaphragms just above normal lower pool, to prevent air from being sucked into the culvert during the lock filling operations. The diaphragms must be removable to allow for placement of the bulkhead when desired.

E. MITER GATE AND CULVERT VALVE MACHINERY

1. General

Operating machinery for the miter gates and culvert valves may be either electrically driven or hydraulically driven. The machinery is usually located in recesses below the top of lock walls. Electric drive is limited to those projects where the lock walls are high enough so that the machinery is not flooded by the maximum upper pool. Hydraulic machinery does not have this limitation, and is always used where the lock walls are periodically overtopped, since the hydraulic pumps and their driving motors may be located in a building above maximum high water. Since the development of high-pressure pumps, control valves, and other equipment, use of hydraulic machinery is becoming more popular, even where threat of flooding does not exist. One reason is that all parts of electrically driven machinery must be designed for a motor stalling torque of about 2.8 times normal full-load value, while hydraulic machinery can be controlled so that stalling torque is about 1.25 times the normal full load. Also, hydraulic machinery can be easily varied in speed by use of flow-control valves, or variable-delivery pumps, while electrical machinery is limited to not more than two speeds by use of two-speed motors.

2. Miter Gate Machinery

Arrangement. Each gate leaf is provided with its operating machinery which, for the electrically-operated type, consists of a motor, brake, limit switch, primary and secondary speed reducers, sector gear and arm, and a connecting strut arm to the gate leaf. Fig. 10B-13b shows a typical arrangement of electrical machinery. In hydraulically-operated machinery, the motor, brake, and speed reducers are replaced by a horizontal cylinder with piston and rod, and a toothed rack attached thereto, which meshes with the sector gear. Tangential movement of the rack produces rotation of the sector gear and opening or closing of the gate leaf. A typical arrangement of hydraulic machinery is shown in Fig. 10B-13a.

Kinematics. The sector gear rotates at a constant angular velocity, proportional to the motor rpm or the cylinder piston speed. The relationship between the sector gear and gate leaf is such, however, that the leaf movement is slowest at each end of its travel and fastest at the mid-point. There are two types of linkage between sector gear and gate; the Panama type, and the Ohio River type. In the Panama linkage, the connecting strut is on "dead center" with the sector arm and gear when the leaf is recessed, and is in a direct line with the sector arm when the leaf is mitered. Thus the leaf theoretically has zero velocity at each end of its travel. The two positions are illustrated on Fig. 10B-13b. The connecting strut must, of course, be mounted above the sector arm and gear to reach the "dead center" recessed position, and its connection to the strut arm must be cantilevered. In the Ohio River linkage, the sector arm and connecting strut form an angle of about 15 degrees in the recessed position, and fail to reach a straight line at the mitered position by about the same angle (see Fig. 10B-13a). Strut and sector arm are in the same plane, and their connection has no eccentricity. In both types of linkage, the connecting strut is provided with a spring assembly to absorb any shock loads during gate operation. These springs may be nests of coil springs, or an assembly of ring springs. They are arranged to compress when the strut is subjected to either tension or compression loads, and are preloaded to a spring compression of about 30,000 lb. Length of strut is adjusted so there will be an additional spring compression of about ½ in. at each end of the gate travel, in order to hold the gate firmly in the recess against its bumpers, and to miter it firmly until a water load is applied.

Time of Operation. Miter gate machinery is usually designed to open or close a gate leaf in about 90 sec., regardless of the length of the leaf. Electrically operated machinery will have equal opening and closing times. Hydraulic machinery, especially if driven by constant delivery oil pumps, will open a leaf faster than it will close one, since the rod end of the cylinder has a smaller net area than the head end, and will therefore move further with an equal volume of oil.

Operating Forces. The principal load on the miter gate machinery is due to the resistance to moving the gate through the water. This resistance varies with the length of the leaf, its submergence, the total angle of travel, and the time for completing a closing or opening cycle. Measurement of torque around the gate pintle has been measured by model tests. In 1941 and 1942 the Special Engineering Division, Panama Canal ran model tests in connection with the design of the Third Locks Project.[7] About 1950, a number of generalized model tests were conducted at the Waterways Experiment Station, Vicksburg, Miss. on operating forces on miter-type lock gates, using the Panama linkage, Ohio River linkage, and a modified Ohio River linkage.[8] The model was on a 1:20 scale, and the results contained in the report can be converted to any size prototype desired by use of the following factors of hydraulic similitude:

Dimension	Ratio	Scale Relation
Length, width, or gate submergence	L_r	1:20
Time	$t_r = L_r^{1/2}$	1:4.472
Force	$F_r = L_r^3$	1:8,000
Torque	$T_r = L_r^4$	1:160,000

From these ratios a gate leaf of any length, submergence, and time of operation may have its torque computed from the model test results. This report also compares its results with the Panama tests previously mentioned.

Design of Parts

General. Forces on the various parts of the machinery may be determined graphically by constructing force polygons at a series of 10 or more intervals of an opening or closing cycle. For each interval, the direction of the gate axis, connecting strut, and sector gear arm can be drawn. From these lines a force polygon may be constructed as shown in Fig. 10B-14, for one position of an opening cycle.

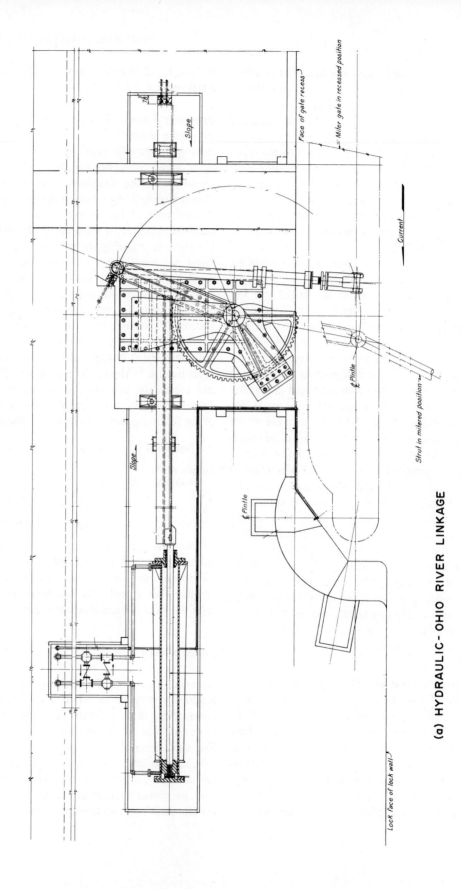

(a) HYDRAULIC-OHIO RIVER LINKAGE

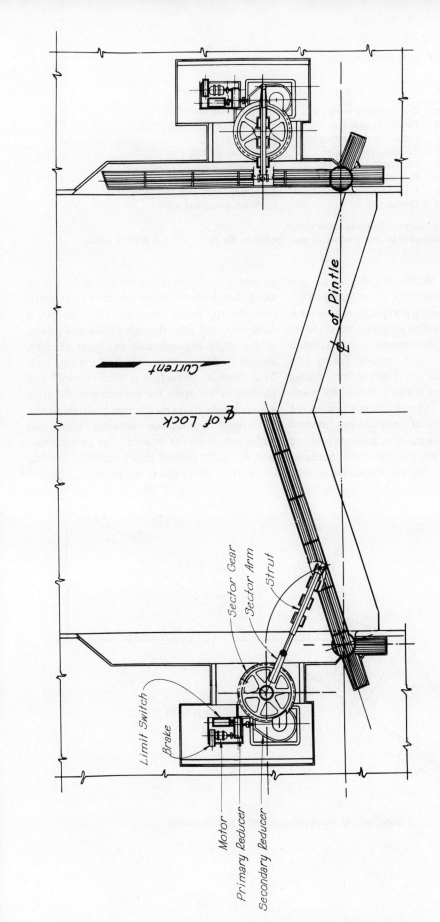

(b) ELECTRIC-PANAMA LINKAGE

Figure 10B-13. Typical miter gate machinery: (a) hydraulic (b) electric.

Line	Direction	Length
O-A	Perpendicular to gate axis	Given opening force (1)
A-B	Parallel to gate axis	Force in gate leaf
O-B	Parallel to connecting strut	Force in strut
B-C	Parallel to sector arm	Force in sector arm
O-C	Perpendicular to sector arm	Force perpendicular to sector arm
B-D	Parallel to piston rod	Piston rod force (2)
D-E	Perpendicular to piston rod	Separating force, rack and sector gear
E-O	Closing line	Force on sector gear pin

(1) From model test, converted to gate under design
(2) Stress C-O times radius to center of sector gear divided by the sector gear tooth pitch radius.

Connecting Strut. Maximum tension and compression are picked off the force polygons described in the preceding paragraph. The strut is usually made from a pipe section, with diameter and wall thickness determined from maximum forces at allowable unit stresses, taking into account L/R ratios for compression. Spring capacity in the strut is also governed by maximum strut forces. The strut connection to the gate leaf is by means of a vertical pin fastened to the gate leaf. Since the strut is normally above the top girder, the pin may be of the cantilever type, fastened to the top and the second girders, or it may be supported by the top girder and an extra "hood plate" above the strut and framed into the top girder. In either case, the pin is fastened to the strut through a yoke, and rotates as the angle between gate and strut changes. Bronze bushings are provided on the gate leaf. This flexible connection is used to avoid any binding in the event the pin axis and the strut are not perfectly perpendicular throughout the leaf travel. Connection between strut and sector arm is bronze bushed. The gate connection is usually located about one-third the leaf length from the center of the pintle.

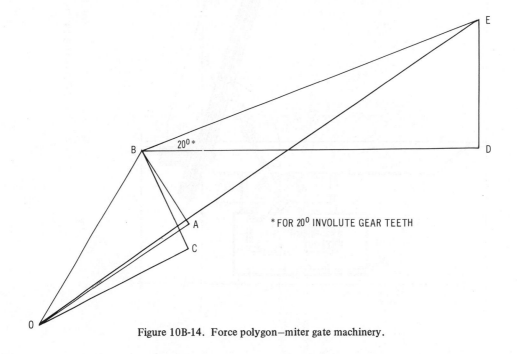

Figure 10B-14. Force polygon—miter gate machinery.

Sector Arm. This piece is of structural steel weldment or cast steel, and is bolted onto the sector gear. It is designed to resist the maximum force transmitted to it by the connecting strut as a cantilever beam.

Sector Gear. For a Panama linkage, the sector gear is made a full 360 degrees, since the connecting strut is mounted above the gear. For the Ohio River linkage, however, the sector gear must be a partial circle to allow the connecting strut to move as near the hub of the gear as possible in the recessed, or open position. The diameter of the sector gear is determined by the general magnitude of the tangential force desired, and its width depends on the allowable unit stress in the gear teeth.

Cylinder and Rack (Hydraulic Machinery). Length of cylinder is set by the stroke required to move the sector gear through its required rotation. Length of stroke then sets the unsupported length of the piston rod, and its diameter for allowable L/R ratio. The cylinder diameter is determined from the allowable working pressure in the hydraulic system and the net cylinder area (gross area minus rod area) needed to develop the computed cylinder rod force. Working pressures have tended to increase from 300 psi on old locks to 2000 psi or more on modern locks. However, the higher the working pressure, the greater the difference in available force between the head end and the rod end of the cylinder, and the higher the cost of the piping system to withstand the increased pressure. Thus a balance must be struck between working pressure, size of cylinder, required pump capacity, and cost of oil lines.

Reducers, Brakes and Motors (Electrical Machinery). In the force polygon for hydraulic machinery, Fig. 10B-14, the line B–D representing piston rod force also is the force at the pitch radius of the secondary reducer for an electrically driven hoist. Using this force, the required output torque of the reducer can be determined. Knowing the desired overall operating time for the gate leaf, the total reduction from an 1800 rpm motor is calculated. This is divided between the primary and secondary reducers so as to provide units having overall speed ratios readily available from gear reducer manufacturers. Ratings of reducers selected may be based on intermittent rather than continuous service. When the motor horsepower requirement has been determined, taking into account the efficiency of each unit in the assembly, a standard size motor must of course be selected. A 15 percent overload may be accepted in sizing the motor, in view of the intermittent type service and the ability of most motors to take this amount of overload without harm. Once the motor size is determined, the speed reducers, sector gear and arm, and strut must be checked against the stalling torque of the motor, which should not exceed 280 percent of full-load torque. Stresses in any part of the machinery should not exceed 75 percent of the yield strength of the material used. Reducers should have oil-lubricated, sealed anti-friction bearings, and at least the secondary reducer should be equipped with an externally-mounted oil pump for positive lubrication of each bearing. If the project is located where winter temperatures drop below freezing, thermostatically controlled oil heaters should be provided to keep the oil flowing freely. A spring operated, solenoid-release brake is provided on the extended shaft of the motor opposite the primary reducer. A rugged positive type limit switch is needed to stop the machinery at either end of its travel, and for shifting the motor from one speed to the other when a two-speed motor is used. A safety limit is also mounted on the strut, which will stop the machinery if the spring is compressed more than the normal amount. All electrical equipment should be suitable for the damp conditions normally prevailing at a lock.

3. Culvert Valve Machinery

General. The culvert valve operating machinery may also be either electrically or hydraulically

604 HANDBOOK OF DAM ENGINEERING

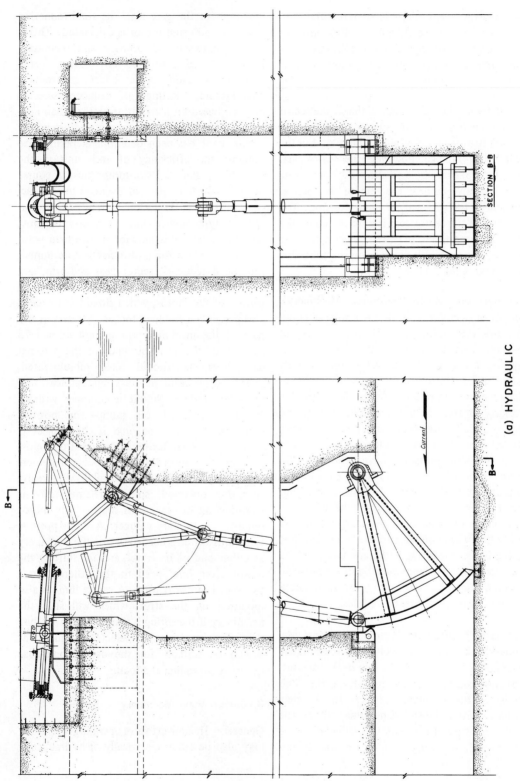

Figure 10B-15. Typical valve machinery: (a) hydraulic (b) electric.

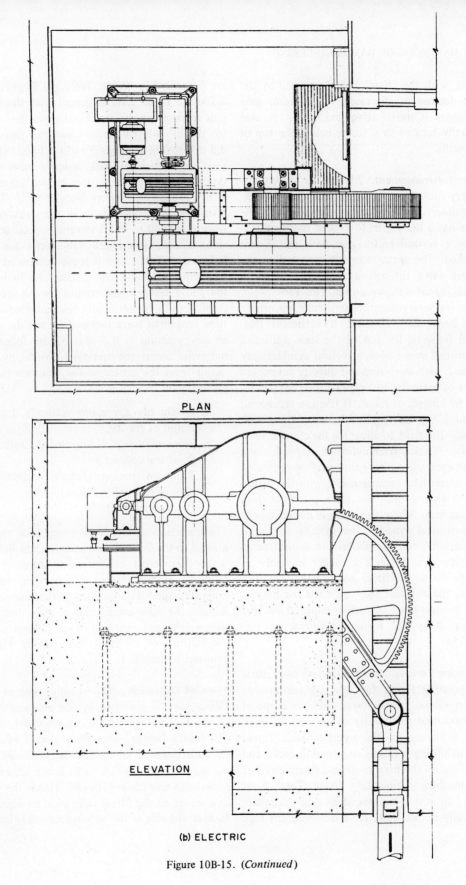

PLAN

ELEVATION

(b) ELECTRIC

Figure 10B-15. (*Continued*)

driven, with the choice being governed by the same factors as discussed under miter gate machinery. Culvert valve machinery is also generally located in a recess below the top of lock walls.

General Arrangement. Electrically driven machinery includes a motor, brake, limit switch, and primary and secondary speed reducers. The latter has a horizontal low-speed pinion which drives a vertical sector gear with sector arm attached. The sector arm is connected to the culvert valve through a rigid pipe strut. A typical layout is shown on Fig. 10B-15b.

Several arrangements of hydraulic machinery have been used. If there is sufficient freeboard between the top of the lock walls and the normal upper pool, a vertical cylinder may be used, with the piston rod directly connected to the culvert valve by a pipe strut. The cylinder may be hinged, or fixed. If fixed, a cross-head assembly will be required between the piston rod and the strut to allow for the circular valve motion. Vertical fixed cylinders may be located in the open valve pit, or they may be protected by a watertight compartment, with the piston rod extending into the valve pit through a stuffing box. Where freeboard is not sufficient for a vertical cylinder, the cylinder is placed horizontally, and the piston rod connected to the pipe strut through a rocker assembly to change the motion from horizontal to vertical. Again, the cylinder may be fixed, or hinged, with a cross-head assembly required for the former. A typical layout is shown on Fig. 10B-15a.

Operating Forces. Model tests have been made of operating forces for reversed tainter-type culvert valves. It was found that the shape of the valve, and particularly of the bottom, has a great influence on the hydrodynamic forces present during a raising or lowering cycle, and therefore on the magnitude and fluctuations of the machinery operating loads. Total hoist loads on the recommended single-skinplate vertically framed valve for a moderate lift lock are contained in W.E.S. Technical Report No. 2-561.[9] These tests indicated that the strut pull throughout the entire valve was less than for the dry weight of the valve, and therefore did not govern the capacity of the hoist. Other tests on a high-lift lock, using a valve with double concentric skinplates are contained in W.E.S. Technical Report No. 2-552.[10] These tests indicated that forces in the strut varied from downpull to uplift at various positions of valve travel. Generalized valve tests are not believed available, and it is recommended that if a proposed valve falls outside the limits of the above tests, model studies for the specific conditions to be met should be conducted. The total tangential force required to raise the valve at any position is the sum of the following moments about the trunnion divided by the radius from the center of the trunnion to the pickup point:

a. Weight of valve acting vertically through center of gravity.
b. Side seal friction. (Assumed at 300 lb./lin. ft. of seal contact.)
c. Trunnion friction. (Usually neglected in view of its small moment)
d. Hydrodynamic forces.

The direction of this total force is of course normal to the line between trunnion and pickup. Since the weight of the valve decreases the required closing force, the opening forces determine the maximum load on the machinery. In fact, the valve often tends to close under its own weight, and must be held back by restricting the outflow of oil from the cylinder with a counterbalance valve.

Time of Operation. The opening time of the filling valve is governed by the characteristics of the lock filling system. As discussed under "Hydraulic Design," the rate of rise of water in the lock chamber vs. cushion depth should not exceed a value which will avoid excessive turbulence and hawser stresses. Hence the rate of opening of the filling valve must be adjusted to keep the rate of rise within acceptable limits.

This may result in a total opening time between 2 and 8 min. The rate of closing for the filling valves, or for opening or closing the emptying valves may be at any desired speed, and is usually between 1½ and 2 min. One advantage of hydraulically operated valve machinery is the ease of varying valve operating time to almost any value desired.

Design of Parts. Forces on the various parts of the valve operating machinery may be determined graphically in a manner similar to that described for miter gate operating machinery. For a number of valve positions, the total force required to open the valve, calculated as described on p. 606 above, is drawn to scale, normal to the line from trunnion to valve pickup. Components parallel to the strut and to the trunnion determine these forces. The strut force is then transferred through the rocker frame to the piston rod, and from the piston rod force at each position, the maximum force and cylinder diameter can be determined. In electrical machinery, the strut force at each position is resolved graphically into a sector arm force, sector gear force, and required torque through the secondary and primary reducers to the motor. As in the gate operating machinery, the various parts of the valve operating machinery must be checked against the stalling torque of the motor selected. In hydraulic machinery, relief valves are provided, set to open at about 15 to 25 percent above normal required pressure, so that stalling torque analysis is not necessary if parts are designed for the relief pressure. The strut between the valve and rocker or sector gear is usually a pipe section, with diameter such as not to exceed allowable L/R ratio. If the lock wall is quite high, resulting in a long strut, the strut may be provided with intermediate supports in the form of A-frame rockers hinged to the concrete wall of the valve pit. All connections are made with corrosion-resisting pins and aluminum-bronze bushings grooved for pressure lubrication, and provided with extended lube lines for convenient access. The strut is also provided with a coil spring assembly near its top to take care of shock loads. Cylinders are sized for the same working pressures as are used for the miter gate machinery.

Hydraulic System

General. Hydraulically operated machinery requires a set of pumps together with a distributing system and control valves to direct the hydraulic oil under pressure to the desired location. There are two principal types of hydraulic systems; the open, or circulating system, and the closed no-flow system. In the first type, the four-way valves which direct the oil to the miter gate and culvert valve cylinders are of the "X" center type, which, in the neutral position connect both pressure ports and block both cylinder ports. They are installed in series along the pressure line. When a pump is started, oil flows through each four-way valve in turn, and thence back to the oil storage tank through the return line. When any four-way valve is operated, oil flow is diverted to one end of the cylinder which it controls, and oil flows out of the opposite end of that cylinder to the return line and back to the storage tank. In the second, or "closed" type, the four-way valves are on individual branch lines off the main pressure line. They are of the "closed center" type, and when in the neutral position block both pressure and cylinder ports. A separate small pressure pump is used to build up pressure in the system and open the main pump unloading valves. Thus, when a main pump starts, it is not under any load. Operating a four-way valve causes line pressure to drop, the unloading valve to close, and the main pump delivers oil to the cylinder controlled by the valve which was operated. Oil from the opposite end of the cylinder flows to the return line and back to the storage tank. The circulating system is the simpler, and a typical schematic is shown in Fig. 10B-16. Usual practice is to provide one pump for each wall, and one spare that can be valved to either wall. The closed system is used on locks having heavy

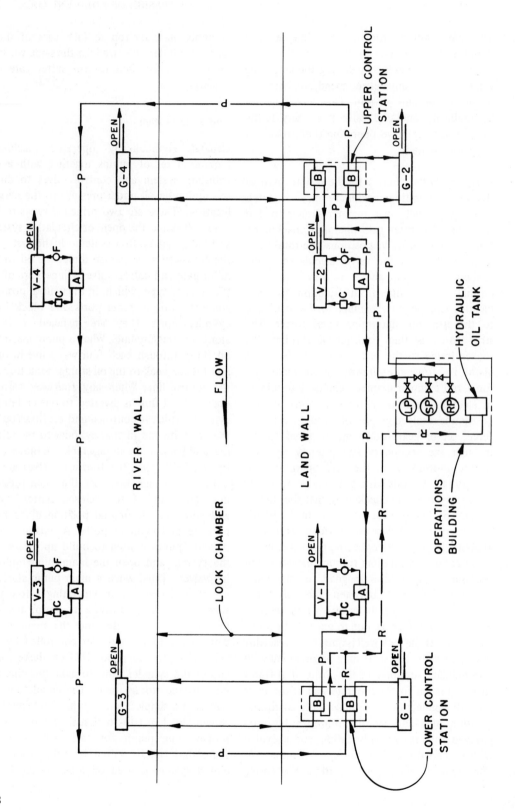

LEGEND:

— P —— PRESSURE LINE
– – R – – RETURN LINE
(LP) LAND WALL PUMP
(SP) STAND-BY PUMP
(RP) RIVER WALL PUMP
[A] 4-WAY CONTROL VALVE, PILOT OPERATED, SPRING CENTERED, TANDEM SPOOL, DOUBLE SOLENOID, EXTERNAL PILOT CHOKE TYPE WITH PROVISION FOR MANUAL OPERATION AND 40 MICRON EXTERNAL FILTER.
[B] 4-WAY CONTROL VALVE, MANUALLY OPERATED, SPRING CENTERED, TANDEM CENTER.
⌀F FLOW CONTROL AND OVERLOAD RELIEF VALVE, PRESSURE COMPENSATED.
⌀C COUNTERBALANCE VALVE
⋈ PLUG VALVE

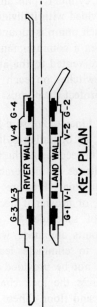

Figure 10B-16. Schematic—hydraulic system.

609

traffic, such as the modern Ohio River 1200 ft locks.

Pumps. The pumps are usually located inside the lock operation building, above maximum high water. They may be of the constant delivery or variable delivery type, depending on the scheme of operation. They should be designed specifically for hydraulic oil, and have sufficient capacity to operate the gates and the valves in the desired time, at a pressure which will provide proper operating pressure at the farthest cylinder, plus line losses. Pumps may be provided with individual oil reservoirs, on which each pump is mounted, or they may draw oil from a common tank. In either case the tanks are vented to the atmosphere. Low-oil cutoff switches on each tank are recommended to protect the pumps.

Distribution Lines. Pressure lines connect the pumps with all operating cylinders. The lines may be run in recesses in the top of lock walls, but pipe galleries are preferable from a maintenance standpoint. Pressure lines should be schedule 80, or heavier if necessary, while return lines of schedule 40 pipe are usually adequate. Joints are made with socket welding fittings to eliminate leaks. If crossover galleries can not be provided in the lock sills, lines may cross the lock chamber in recesses in the walls and floor. These lines are usually made of stainless steel to increase their life and reduce maintenance.

Valves. Pressure relief valves are provided in each pump discharge line, near each end of the miter gate machinery cylinders, and at the four-way control valves for the culvert valve machinery. These valves prevent excessive pressures in the system. Plug valves are provided at various points, so that any pump may be removed, or any culvert valve or gate cylinder or their control valves can be taken out of service without interfering with the remainder of the system. Check valves are used to prevent reverse flow in lines when such flow would result in improper operation. Counterbalance valves are provided in the oil line from the head end of each culvert valve hoist cylinder. This type of valve puts a back pressure on the cylinder and prevents the culvert valve from closing from its own weight at a rate faster than oil is introduced into the opposite end of the cylinder.

Other Equipment. A bucket type strainer is located in the pump suction line, and a filter is provided in the return line before it empties into the oil reservoir. These serve to keep any foreign matter out of the circulating oil. Pressure indicating gages are located at each pump, and at each operating stand to give a visual check on operating pressures.

F. ELECTRICAL SYSTEMS

1. General

A navigation lock will require electric power for operating the gates and valves, for lighting service, and for maintenance equipment, such as unwatering pumps, welding machines, etc. It should be kept in mind that the majority of the electrical installation will be in areas of high humidity due to the proximity of the reservoir. All applicable provisions of the National Electric Code, and other standards should be adhered to for the safety of operating personnel.

2. Sources of Power

If the project includes a hydroelectric plant, power for the lock would probably be obtained from the station service board in the powerhouse. If not, it may be necessary to utilize a commercial source for power. In the latter case, it may also be desirable to provide a standby generating unit at the lock, unless the incoming power is very reliable. Since the principal motors and other power equipment for the lock normally operate at 480 volts, it will usually be necessary to provide a transformer substation at the lock.

3. Electrical Requirements

Switchboard. A main switchboard usually serves as the supply and distribution point for all 480-volt circuits on the lock. It is ordinarily located in the lock operation building, above maximum high water. A switchboard of the control-center type is recommended, with cubicles for incoming power, standby unit (when present), for all 480-volt motor starters, and for distribution circuit breakers for all other power circuits. Necessary meters and switches for current and voltage indication, and ground detector lights should also be provided. It is well to have space for spare circuits which may be added in the future. The general arrangement of a typical switchboard and one-line diagram are shown on Fig. 10B-17.

Power Circuits. The power circuits from the switchboard will include some or all of the following items, as applicable for the particular site conditions:

1. Circuits from motor starters to the main hydraulic pumps, or to gate and valve machinery; to sump pumps, air compressor, raw water and fire pumps, and tow-haulage unit.
2. Electric space heaters for the operation building.
3. Power receptacles at gate and valve machinery recesses for unwatering pumps, or for maintenance equipment such as welding machines, etc.
4. Lighting transformers.
5. Adjacent spillway gate feeders.
6. Control circuits. If variable-delivery pumps are used for the hydraulic system, or two-speed motors for an electrical operation, relays for controlling the operation of the miter gates and culvert valves will be located in the switchboard, and circuits will run to the control stands on the lock walls. If a traffic-signal system is proposed, relays for operation of this system will also be located on the switchboard, and circuits provided to the traffic lights. All control circuits should be of the 120-volt type.

Lighting Circuits. The lighting circuits are fed from the lighting transformers, and originate from distribution circuit-breakers located in panelboards in the lock operation building. Major circuits are 120/208 volts, with loads equally divided to reduce voltage drop. Lighting circuits may include as many of the following as are applicable to the lock under design:

1. Lock wall and chamber lighting. If the lock is to be operated during hours of darkness, a general lighting system will be required to permit operating personnel to perform their duties safely. A common type of system consists of 400-watt mercury lights mounted on poles about 30 ft high, and spaced at about 125 ft centers along the main lock walls. The poles are set 10 to 12 ft back of the face of the lock wall, with the lights set 1 to 2 ft back of the face to minimize dark shadows in the lock pit. Spacing of lights on approach walls may be increased to 180 ft spacing. Luminaires should have I.E.S. type III light distribution, to direct the light principally along the walls. They should also be equipped with lowering devices for safety in maintenance.

2. Miter gate recess lights. A floodlight is located at the upstream end of each gate recess for additional illumination of the gate area during operation.

3. Machinery area lights. Each miter gate machinery and culvert valve machinery recess should have additional illumination, especially if these recesses are covered with grating. Convenience outlets should also be provided.

4. Floating mooring bitts. Each floating mooring bitt recess may have a floodlight mounted at its top, and pointing straight down to illuminate the recess and facilitate its use during hours of darkness.

5. Discharge area lighting. Where culverts discharge into the river outside the lower lock approach, it may be desirable to furnish one or more floodlights to light the general area, and also to provide an illuminated warning sign. This is particularly applicable if there is fishing from small boats in the area.

6. Navigation lights. U.S. Coast Guard regulations specify the location and type of navigation lights to be displayed at locks in various

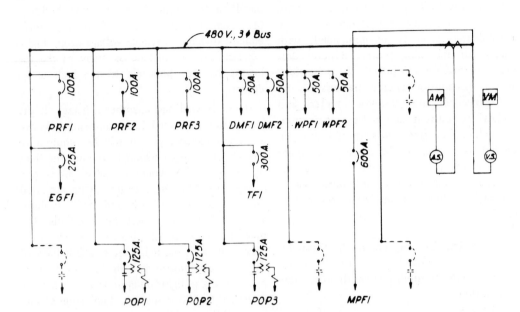

ONE LINE DIAGRAM
NO SCALE

FEEDER REFERENCE IN ONE LINE DIAGRAM	NAME PLATE SCHEDULE			
	NO.	REQ	FIRST LINE	SECOND LINE
PRF1	1	1	Power Outlets Feeder	Downstream-PRF1
EGF1	2	1	Emergency Gate Feeder	
PRF2	3	1	Power Outlets Feeder	Upstream-PRF2
——	4	1	Visual	Traffic Signals
——	5	1	Horn	
POP1	6	2	Land Wall	Oil Pump
PRF3	7	1	Power Outlets Feeder	Shop-PRF3
POP2	8	2	Standby	Oil Pump
POP3	9	2	River Wall	Oil Pump
DMF1	10	1	Davit Motor Feeder	Downstream-DMF1
DMF2	11	1	Davit Motor Feeder	Upstream-DMF2
TF1	12	1	Lighting and Heating	Transformer
WPF1	13	1	Raw Water Pump Feeder	Land Wall
WPF2	14	1	Raw Water Pump Feeder	River Wall
MPF1	15	1	Main Power Feeder	Switchboard Bus

DESIGN OF ADJACENT LOCKS 613

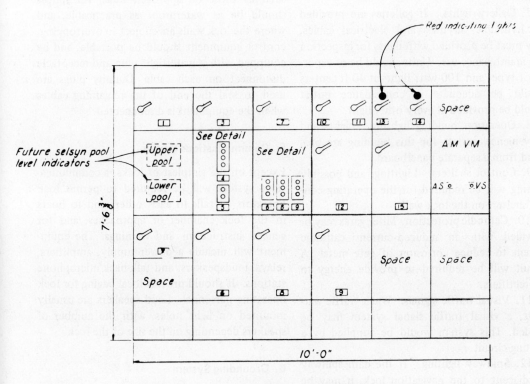

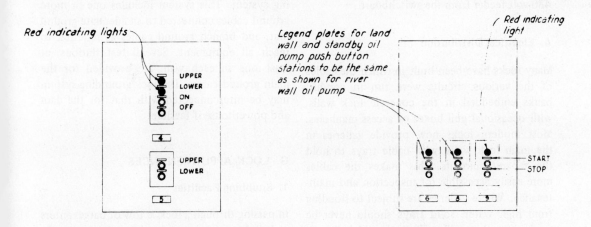

Figure 10B-17. One-line diagram—electrical system.

614 HANDBOOK OF DAM ENGINEERING

areas. These regulations should be complied with.

7. Gallery lights. If galleries are provided for hydraulic oil lines and electrical cables, they must be provided with lights for inspection and maintenance use. Units should be of vaporproof type, and 100-watt lights at 40 ft centers should be adequate. A convenience outlet should be provided at every other light.

8. Operation building lighting. Lights and convenience outlets for this building are provided from a separate panelboard.

9. Control shelters. Lighting, and possibly heating will be required for the operating control shelters on the lock walls.

10. Cathodic protection. Miter gates may be provided with an induced-current cathodic system to reduce corrosion of gate metal. A circuit will be required to provide energy to the rectifiers.

11. Visual traffic signals. When traffic warrants, a visual traffic signal system may be needed. This system would be supplied by a lighting circuit.

12. Spillway lighting. If the dam spillway is adjacent to the navigation lock, it may be desirable to supply the spillway lighting from the lock, either from a lock lighting circuit, or if the voltage drop would be excessive, by a 480-volt feeder from the switchboard.

4. Electrical Distribution

Many locks have been built in which the cables of the various circuits were run in conduit banks embedded in the concrete lock walls, with occasional pull boxes or access manholes. Most modern locks now provide galleries in the main lock walls, with cable trays to hold the separate circuits. This makes the cables more easily accessible for inspection and maintenance. Where galleries are subject to flooding from high water, solid trays should never be used, since they will collect silt which is very tedious and time-consuming to clean up. Open ladder-type trays are preferred, even where flooding is not a problem. Conduits will of course still be used for access to equipment from the nearest gallery, and for remote circuits, such as those on approach walls. All splices should be as waterproof as practicable, and where the lock walls are subject to overtopping, control equipment should be portable, and be equipped with a watertight plug and receptacle disconnect on each cable. Dummy plugs are used to seal the end of the remaining cables when the equipment is disconnected.

5. Communications System

Except on the simplest of locks, a communications system will be required to permit lock operators to talk to each other, and to boats in the lock chamber or approaches, and for general instructions and warnings. The equipment will include a power supply, amplifiers, relays, loudspeakers, and talk-back microphone stations. It should provide local paging for lock operating personnel. Loud speakers are usually mounted on light poles, with the number of speakers depending on the size of the lock.

6. Grounding System

All electrical equipment, including lighting poles, switchboard, panelboards, and switch boxes should be connected to the main grounding system. This system includes one or more ground cables connected to an adequate ground mat, and branch ground cables as required to reach all equipment. Several test stations, at least one in each wall, are provided for the main ground cable. The lock grounding system may be interconnected with that for the dam and powerhouse if feasible.

G. LOCK APPURTENANCES

1. Snubbing Facilities

In passing through a lock, a tow of barges enters and leaves slowly under its own power. In order to control movement and bring the tow to a positive stop without ramming the closed lock gates, check posts are provided along the top of approach walls and chamber walls. These should be solidly anchored to the concrete, and withstand the breaking strength of the heaviest hawser likely to be used to tie up a barge. Check posts are set about 2 ft from face of wall, and should be spaced at about 100-ft centers (or less, on small locks) on approach walls, and for at least 2 similar spaces at either end of the lock chamber. Line hooks, set into the face of the approach and chamber walls, are provided at each check post location. These line hooks are spaced at 5-ft vertical intervals, from about 5 ft below top of walls to just above minimum upper or lower pool as applicable. Check posts and line hooks are also used to tie up a tow to the approach wall while waiting to enter the lock, or to secure a portion of a tow which is too long to be locked through at one time. Originally, tows were snubbed to line hooks during a lockage, with lines being changed from one hook to the next as the water level in the chamber rose or fell. In order to eliminate this process, the floating mooring bitt was developed. A floating tank, equipped with guide rollers and a mooring bitt, operates in a recess provided in the face of the lock wall. When a tow is snubbed to a pair of floating mooring bitts, it is held in position throughout an entire filling or emptying process. Four to eight floating bitts are usually provided in each chamber wall, depending on the length of the lock, with a variable spacing to fit tows of different size barges. Floating bitts should be ruggedly designed, with corrosion-resistant roller faces and positive lubrication of all bearings. Tanks should be provided with an access hatch for inspection and maintenance, and an automatic means for flooding to prevent sudden release if the bitt is temporarily jammed during a filling operation.

2. Esplanade

All locks will require a working area for storage, maintenance, parking, and other miscellaneous uses. Features to be considered include access to the lock, parking, water supply and sewage disposal, protected storage for maintenance equipment and spare parts, and security fencing.

3. Wall Face Protection

Concrete lock walls receive constant impacts and abrasion from tows entering or leaving the lock, and to a lesser extent while it is being raised or lowered in the lock chamber. Protection of vertical wall faces is provided by horizontal rows of rolled structural steel T-shaped sections having a slightly convex surface which protrudes about ½ in. from the face of the wall. This type of armor is adapted to inland waterway tows of barges having vertical sides. The rows are spaced at 20-in. intervals, extending through the following limits:

Lower approach wall, from slightly below minimum lower pool to ordinary high water, plus the height of an empty barge (10 ft on a major waterway).

Lock chamber, from just below minimum pool to maximum upper pool, plus the height of an empty barge. This often results in armor to the top of lock walls.

Upper approach wall, from just below minimum upper pool to maximum upper pool, plus the height of an empty barge.

Wall armor is usually continuous for the full length of the approach walls, and for about 50 ft from the gate recesses at either end of the lock chamber. For the remainder of the lock chamber, wall armor is normally limited to about 6 ft on either side of all vertical recesses, such as ladders and floating mooring bitts. In addition to the rolled wall armor, the edges of all vertical recesses, and the horizontal corner at the top of the chamber face of all walls should be protected with a bent steel plate. A typical protection plate has 8-in. legs, 6-inch radius, and is made from ¾-in. plate. All armor is provided with bent anchor straps to hold it securely in place. Horizontal wall armor and corner protection will of course not extend across monolith joints.

4. Recess Covers

Recesses in the top of lock walls are provided with covers, usually of steel grating, to permit safe movement of operating personnel. If the walls are to be used by a mobile crane, the grating must be designed for resulting wheel loads, or in the case of large openings, such as miter gate recesses or valve pits, these may be left open and protected by guard rails.

5. Guard Rail

Guard rails should be provided on both sides of gate or bridge walkways, and for the entire length of lock and approach walls, except where these are backfilled to within a few feet of their top. Any openings in top of lock walls or elsewhere, not equipped with covers, should also be protected with rail. Collapsible railing consisting of posts and chain may be used in cold climates or where the lock walls are overtopped by high water. Guard rails adjacent to the lock face of walls should be set back a sufficient distance so that overhanging barge rakes do not wipe them off.

6. Ladders

Vertical ladders in the face of lock walls are needed for access between vessels and walls by crew members during locking operations, and for use of anyone accidentally falling overboard during a lockage. At least 3 ladders should be provided on the main approach walls; one near the end, one near the center, and one near the gate sill. A least 3 ladders should also be provided in each chamber wall. Maximum spacing of ladders should not exceed 400 ft. Ladders on approach walls should extend from top of wall to minimum pool; those in the lock chamber should extend to minimum lower pool. Former practice was to provide extra ladders to each sill, and to extend at least one ladder to the lock floor for use in maintenance operations. It has been found, however, that usefulness of these ladders is limited to a height of about 40 ft, and they are seldom used for maintenance access on high-lift locks.

7. Water Level Indicators and Recorders

Tile gauges are provided in recesses in the lock walls to indicate the depth of water over the upper and lower miter sills. The zero of each gauge is at the top of the adjacent sill, so that available depth is read directly in feet and tenths. Numbers should extend a few feet below water and above high water. Gauges are located a short distance upstream and downstream from each gate recess, on the wall opposite the normal operating wall so as to be readily visible from it. These gauges also give a lock operator a ready indication of when the pools are equalized, and the gate may be opened. Recording gauges may be desired in each pool to provide a continuous record of water levels, unless such facilities are provided in the adjacent dam or powerplant. These gauges can be arranged to record and/or indicate pool levels in the lock operation building for convenience.

8. Air Bubblers

Each gate recess is often provided with a vertical air pipe discharging near the gate pintle a few feet above the bottom of the recess. This bubbler is very useful in moving floating drift out of the quoin area during gate operation, so as to prevent its being caught between gate and wall contact blocks. Bubblers may be manually operated, or automatic, as desired.

9. Water and Compressed Air Systems

A potable water supply is needed for use of lock personnel. A municipal or water district supply is ideal, but if not available, water must be treated at the site. It is often possible to combine lock requirements with those for the powerplant and serve both facilities with one system. Sanitary disposal facilities, complying with State and Federal requirements are also needed. A raw water system is generally provided on the lock chamber walls for washing down walls and galleries. Outlets for 1-in. hose are spaced at 150 to 200 ft centers on the main chamber walls and in the pipe galleries. A supply of 50 to 100 gpm at a pressure of 100 psi should be adequate for this purpose. Where flammable cargoes are handled, it may be desirable to provide a fire protection system for the lock. This system primarily protects the miter gates from heat damage by providing a continuous water spray on the skinplate face of each gate leaf from a series of spray nozzles mounted at the top of the leaf. In addition, 1½ in. outlets with 100 to 150 ft of hose for each outlet may be located near each gate bay. The 1½ in. size is considered the largest that can be handled by one man. Hose nozzles should be of the combination water-fog type. Fire pump or pumps for a single lock should have a capacity of about 800 gpm at a pressure of 100 psi. The pump should be remote-controlled, so that it can be started from any outlet. Gate sprays should likewise be remote-controlled. One piping system can serve both washing down and fire outlets, if properly sized. They are run in the galleries wherever possible, and are normally drained when not in use to prevent feeezing in cold weather. Compressed air at 100 psi is distributed to the same outlet recesses provided for raw water. About 100 cfm should be sufficient for normal maintenance. Air supply for major repairs can be supplied by portable equipment mobilized for the specific job.

10. Navigation Aids

A number of aids may be provided for the help of tows approaching or using the lock. Among them are the following:

Wall Stripes. The ends of approach walls may be painted with diagonal stripes above minimum pool level. These stripes are 12 in. wide, alternately black and white or aluminum, and are sloped downwards toward the lock chamber face.

Distance Markers. Numbers indicating the distance to the lock sill may be painted at the top of the lock approach and chamber walls. Figures should be 12 in. high for visibility from the pilot house of the incoming vessel, and are normally spaced at 50 ft intervals. Markers on the lower approach wall show the distance to the downstream end of the lower gate recess; those on the upper approach wall show the distance to the upstream end of the upper gate recess; markers in the lock chamber show the distance to the downstream face of the upper miter sill, which is also indicated by a vertical yellow stripe on the face of each lock wall. The distance markers may be painted directly on the concrete, or they may be painted on aluminum plates, mounted in shallow recesses in the lock wall, or supported on the guard railing. Figures are generally white on a green background, and may be of reflective type paint for easier reading at night.

Lifebouys. Ring type lifebuoys with 100 ft of $3/8$-in. nylon line attached, may be provided at 150 to 200 ft intervals along the approach and chamber walls for use in the event a crew member falls overboard.

Signs. Warning signs in the area of the lock discharges, portable "no smoking" signs (used during lockage of hazardous cargoes), and other directional signs may be provided as conditions indicate.

Tow Haulage Unit. When a tow is longer than the lock chamber, it must be split and locked through in two sections, with the towboat remaining with the second section. Some means must be provided to pull the first section out of the chamber, so that the chamber can be prepared to lock the second section through. A number of types of haulage units have been used, and the type selected for a particular project will depend on the amount of use expected. The simplest installation is a pair of single drum hoists, either electric or air driven. One hoist is located on the top of the lock wall upstream from the upper gate bay, and the other similarly mounted downstream from the lower gate bay. The free end of the line is paid off the drum and fastened to a bitt on the back barge of the first section. The barges are then pulled out of the chamber and allowed to drift until they are clear of the lock gate, and snubbed to check posts until the second section is locked through. A line capacity of about 5000 to 7000 lb with a speed of 50 to 60 fps is recommended. Electric drive must have a variable-speed control, as the barges will accelerate very slowly. Another type unit consists of a single reversible hoist located at about the center of the lock, and an endless cable running along the face of the lock wall and around sheaves near the gate recesses. This cable is provided with a line long enough to reach pool level, and is fastened to the back barge either directly or with an intermediate hawser. Barges are pulled out as described above. The hoist drum is designed so that as the cable is paid out at one end, it returns on to the drum at the other end. The most sophisticated system uses a reversible hoist and endless cable which pulls a wheeled towing bitt on the top of the wall, between the gate bays. The bitt may travel in a recess provided for it, or it may be mounted on a rail fastened to the concrete. The latter is used for locks already constructed, or in cold climates where snow and ice would clog the recessed type. A hawser furnished by the tow is slipped over the traveling bitt and fastened to the back barge as before. A maximum line pull of 10000 lb is recommended, so as to not break the normal hawsers used on barges. A speed of 80 to 100 fpm may be used. Speed and line pull control on recent units is through an eddy-current coupling controlled by a stepless auto transformer unit. It is very important that the lock operator be able to see the barges he is pulling during the entire operation.

REFERENCES

1. EM 1110-2-2601 "Navigation Locks"—Department of Army, Office of Chief of Engineers, Washington, D.C.
2. EM 1110-2-1604 "Hydraulic Design of Navigation Locks"—Department of Army, Office of Chief of Engineers, Washington, D.C.
3. "Filling System for Lower Granite Lock" by G.C. Richardson: Paper #6718, Journal of Waterways and Harbor Division, Proceedings ASCE, Vol 95, No. WW3, August 1969.
4. "U.S. Development of Hydraulic and Structural Designs for Locks" by J. P. Davis, M. E. Nelson and R. E. Patton. XXIst International Navigation Congress, PIANC Section 1, Subject 2, page 191.
5. "Hydraulic Design of Columbia River Navigation Locks" by G. C. Richardson and M. J. Webster. Transactions ASCE Vol. 125, Part 1, pp. 345–364.
6. EM 1110-2-2602 "Planning and Design of Navigation Lock Walls and Appurtenances" Dept. of Army, Office of Chief of Engineers, Washington, D.C.
7. "Hydraulic Model Investigation of Miter Gate Operation" by Maurice N. Amster: Proceedings of ASCE, Vol. 70, No. 3, March 1944, Page 303.
8. "Operating Forces on Miter Type Gates"—Technical Report No. 2-651, U.S. Army Engineer Waterways Experiment Station, Vicksburg, Miss.
9. "Filling and Emptying System—New Poe Lock, St. Mary's River, Sault Ste. Marie, Michigan"—Technical Report No. 2-561, U.S. Army Engineer Waterways Experiment Station, Vicksburg, Miss.
10. "Hydraulic Prototype Tests of Tainter Valve, McNary Lock, Columbia River", Technical Report No. 2-552, U.S. Army Engineer Waterways Experiment Station, Vicksburg, Miss.
11. "Filling and Emptying System, Cannelton Main Lock, Ohio River, and Generalized Tests of Sidewall Port Systems for 110 X 1200-ft. Locks", Technical Report No. 2-713, U.S. Army Engineer Waterways Experiment Station, Vicksburg, Miss.

11
Design of Reservoirs

HENRY L. HANSEN, P.E.
Environmental Control Officer
U.S. Bureau of Reclamation
Auburn, California

11-1 GENERAL CONSIDERATIONS

This chapter is written with the intent to stimulate and direct the reader's thinking rather than to present complete, in depth discussion of the subject matter. With the variety of technological disciplines that bear on the subject, it would be difficult to make any other claim.

A reservoir, as considered in this chapter, is defined so as to be consistent with the subject matter of the entire Handbook. Reservoirs that would contain anything other than water are not discussed. One definition which has been given, but which is too broad, is: "A reservoir is a body of water on which man exerts major control over the storage and use of the water." In this text, however, underground reservoirs, bodies of salt water, and very small covered reservoirs that are really a form of storage tank are excluded. This chapter considers a reservoir to be a man-made lake, which has been defined in the literature as: "A freshwater body created or enlarged by the building of dams, barriers, or excavations."

A reservoir can be considered as being designed only in a broad sense. It is really selected, with such improvements or remedial work being devised as may be considered necessary to assure safe and satisfactory performance of its intended function. While the association of "design" with "reservoir" is commonly found in the literature, the hydrologist thinks of "reservoir design" in a very different way from a structural engineer.

The investment of effort properly required for the design study of a reservoir depends on each individual case. For a number of years now the multipurpose concept for projects has been ascendant. The trend toward ever larger developments appears to relate to the multipurpose concept. Larger reservoirs are usually designed for long-period storage carry-over from wet to dry years. Yet there are many cases where a single purpose concept is appropriate. And, of course, projects have been and will continue to be built in a great variety of settings. These factors—size, location, and purpose—do influence the extensiveness of investigation warranted. Seepage, for instance, is relatively unimportant at a flood detention basin. On the other hand, there has been conjecture that failure of a dam (and loss of the water) may have been responsible for the fall of an ancient civilization.

This chapter contains five sections in addition to this first, introductory section. They are discussions of structural integrity and adequacy, relocations and right-of-way, associated structural work, reservoir clearing, and environmental considerations. Detailed discussions of design procedures are not generally included, because these are available elsewhere. Design of a saddle dam, for example, is not different than the design procedure for dams generally. It is recognized that individual topics may also be discussed in other chapters; nevertheless, their discussion in this chapter is deemed appropriate. A lengthy bibliography is included, although no attempt was made to compile a complete one. Rather, a representative bibliography was attempted. Many of the writings listed were used liberally for reference in preparing the chapter.

In searching the literature it was soon concluded that no comparable work on reservoirs existed. It is sincerely hoped that this initial focus on reservoirs will elicit constructive comments leading to improvement. More emphasis on reservoirs does appear to be mandated by the growing criticism of water development projects, as it is mostly the reservoirs that are objected to rather than the dams which create them.

In contemplating his works, the water resources engineer will do well to be aware that all lakes, natural or artificial, are only temporary obstructions to the process of water flowing downhill. Therefore, natural forces are constantly at work to remove such obstructions.

11-2 STRUCTURAL INTEGRITY AND ADEQUACY

One key consideration in developing a project is the structural integrity and adequacy of the reservoir. The ability of the reservoir to safely contain the projected volumes of water for use when needed must be assured, at least throughout the planned project lifetime. Not only must the reservoir be safe but the water must be available. Principal factors in this consideration are rim stability, water holding capability, seismicity, bank storage, evaporation, and sedimentation. Off-stream sites are generally more difficult to select and design because a location outside a valley makes loss by leakage more likely. The engineer must depend on the geologist for much of the information on which he bases his evaluation of a reservoir site. Geologic investigations also are important for other considerations relating to reservoirs, including potential sources of water pollution and the siting of associated structures.

Examination of existing geologic maps and other geologic records is one of the first steps during the initial stage of investigation. Field investigation is required for a reconnaissance grade study. It includes surface exploration and data from existing developments such as wells and mines. Subsurface investigations are rarely omitted, and only if there are no indications of foundation problems. Since investigation of potential dam sites will also be underway, some information from drill holes is likely to be available.

A feasibility grade study requires more comprehensive information. At this point, one site will have been tentatively selected, thus permitting a concentration of data-gathering efforts. Although aerial photography is valuable, other forms of remote sensing can provide additional information on ground water conditions and vegetation. Geologic investigations of both reservoir and dam site should proceed concurrently. If a major defect is uncovered in one or the other, investigations can be reoriented or stopped before a disproportionately large exploration investment accrues on a site which must be abandoned.

Rim Stability

Rim stability and water holding capability can be closely related. Seepage of water through a segment of reservoir rim could be the first step in a sequence leading to failure. Rim failure obviously could reduce the water holding capability of a reservoir. Major slides into a reservoir could significantly reduce its capacity. A failure that lowers the rim profile below the maximum water surface elevation reduces or prevents water storage. The mechanism of rim failure may be either sliding or erosion.

Disclosures of potential sliding may or may not be serious. A potential slide may be evaluated as small relative to reservoir size, slow moving, and located so as to not threaten any other developed facilities. Or it may be classified as large, capable of sudden movement in a critically important location, and compromising the structural integrity of the reservoir. It may be somewhere in between these extremes. The potential slide could be capable of stabilization by preventive treatment, or such efforts may be determined to be impractical.

Knowledge of potential slides can affect right-of-way acquisition—the establishment of the "take line." For example, the nose of a peninsula jutting out into a proposed reservoir may be stable, yet potential slide conditions along the neck of the peninsula could indicate future creation of an island. In order to avoid future legal complications the whole peninsula would have to be included in the right-of-way acquisition program. All potential slide areas must be within the limits of the project take line.

A potential slide could in the future isolate one part of the reservoir from the remainder. If this may not be tolerated and if corrective measures are feasible, appropriate remedies should be undertaken beforehand. Should an anticipated slide be large relative to the size of the reservoir, the likely reduction in reservoir capacity should be considered. Reduced minimum rim elevation due to slides may also limit reservoir capacity. Knowledge of such potential enables timely evaluation of a site.

Knowledge of potential slides can be essential in planning for the relocation of existing improvements as well as in planning for new reservoir related structures. Some potential slides are peculiar to colder climates. Snow avalanches and masses of ice falling from hanging glaciers can be serious safety problems. The need to avoid an anticipated slide area is obvious. Perhaps less obvious might be the need to consider wave action due to a rapidly moving slide.

The shape of a reservoir is a major influence on wave action. A sinuously-shaped reservoir might thoroughly dissipate waves generated by a slide, but a straight, narrowing canyon of a funnel-shaped reservoir could actually amplify such waves. A sufficiently massive movement could cause overtopping of a dam, either by wave action, or simply by raising the water surface faster than the spillway is discharging.

If an anticipated slide cannot be safely ignored, and the site is not to be abandoned, some form of accommodation is required. It may be that incorporation of restraints into the operating criteria for the reservoir will be a sufficient control. Limiting the rates of filling and drawdown may suffice. Imposing a maximum allowable water surface at some level less than the maximum normal water surface until slide activity has been completed could be adequate. For a reservoir where recreational use is planned, temporary restrictions may be necessary.

Should simply allowing for a slide by including it as a parameter in determining reservoir operating criteria be insufficient, then movement must be more directly prevented or the potential slide must be removed. Measures to prevent saturation or increased water pressure along the slip surface could be adequate. If surface runoff is a contributor to the problem, surface drainage should be directed away from critical areas. Installation of drains to relieve water pressure along slip surfaces is sometimes effective. Some form of impervious lining may be practical. A rock bolting program including tendons may be needed to pin the unstable mass to its parent formation. An alternative may be some form of shear wall.

Stabilization may be achieved by strengthening or replacing weak material. Injecting grout is probably the most familiar remedy. The reader is cautioned at this point that grouting is a subject which itself could require a textbook.

Grouting, however, can easily be a self-defeating measure. Although strengthening weak material, grouting would tend to increase the driving load if injected into the rock mass. Grouting would reduce the drainage capabilities, and, even with low injection pressures, could cause jacking along the sliding plane during injection. Removal and replacement of incompetent materials, if very extensive, is likely to be very costly.

Simply placing a sufficient mass as a buttress against the toe of the potential slide could prove to be the most feasible solution, especially if this location can be utilized as a waste area for structural excavation. Unloading of the slide may be required, possibly to the point of complete removal. This can be more feasible if material to be removed can be used as borrow for structural fill. Combinations of some or all of the above may be required. For example, a filter may be needed between buttress material and the questionable mass.

Planning to ameliorate effects of a potential slide after it occurs, if it does occur, may be satisfactory. In such a case, of course, assurance of safety must be convincing. Possible disruption of normal reservoir operations should also be considered in contemplating such a decision.

The general public should be advised through the medium of the environmental impact reports and public hearings of any special elements of reservoir designs.

The slide shown in Fig. 11-2-1 occurred many years after project completion. Under normal operation, the reservoir filled yearly. It became necessary, however, to draw the reservoir down considerably below any level that had been an-

Figure 11-2-1. A large slide in a reservoir after unusually great drawdown. Note the slope of the intact portion of the reservoir bank and the high water line. (*Courtesy U.S. Bureau of Reclamation*)

ticipated. When refilling of the reservoir began, the slide took place.

Water Holding Capability

Reservoir water losses occur either to the air or to the ground. These losses can be controlling factors in site selection and must be studied and evaluated in the planning process. The technical and economic feasibility of remedial measures must be considered. Discussions of water holding capability usually pertain to seepage and leakage; accordingly, evaporation and bank storage are covered in subsequent parts of this section.

Loss of reservoir water is not always a negative consideration. For a single purpose development such as a flood control project, ground water loss may only be of concern if it relates to safety. If ground water recharge is a planned benefit, seepage is desirable. Small reservoirs may be designed with filter bottoms to encourage seepage.

Loss of reservoir water to the ground can be either by seepage or leakage. The term "seepage" as used here designates minor water losses. "Leakage" designates water losses which may either compromise safety or represent a distinct economic loss.

Care should be taken to guard against "tunnel visions" in considering reservoir seepage. Losses permissible as seen from the standpoints of safety and economics may be intolerable for improvements outside and below the reservoir. Farmland subjected to reservoir seepage could become too saturated to farm. Of less importance would be project maintenance costs due to required control of phreatophyte growth encouraged by seepage. From another viewpoint, much of the reservoir seepage may be recoverable from diversions of downstream flows and pumping from ground water supplies, and so should not be thought of as complete loss to beneficial use.

Because a dam site is usually at the lowest point of a reservoir it becomes the zone of reservoir past which most seepage tends to flow. This is fortunate because it is at a dam site that the most intense geological exploration effort is undertaken. Here, too, foundation treatment necessary for stability and safety of the dam is carried out. Beneficiation of the foundation to control and reduce seepage is considered by designers virtually automatically with dam design requirements.

Certain rocks and soils are generally poorer than others in terms of water holding capability. Their presence does not automatically mean a reservoir site is unsatisfactory, but it is an indication that greater investigation is required in order to evaluate the site. Loess, other aeolian deposits, and sand are more permeable soils. The worst rocks for leakage are the soluble ones such as limestone and gypsum, basalts and other volcanics, and gravelly deposits of glacial drift and alluvium. Leakage in limestone formations can occur through caverns, channels or tubular cavities along joints, bedding planes, and faults. Volcanics are generally permeable because of well developed joints, fracturing, or structure. Unconsolidated sediments leak for lack of fines.

It is important to trace the path of leakage or unacceptable seepage. Various techniques using dye or other visibly identifiable substances, or of using radioactive tracers can be employed. Keeping records of the ground water table (from well observations) and streamflow data can be very helpful in tracing these paths, as well as in estimating seepage losses. It is necessary in the planning stage to estimate the seepage losses which would occur during operation of the project.

Remedial or preventive measures can be costly and/or of limited effectiveness. When a potential leakage or seepage condition is localized rather than general, preventive measures may be feasible even for a large reservoir site. This may be the case, for example, where a permeable fault or shear zone, an ancient buried channel, or a rim feature resulting from glacial, fluvial, or aeolian deposition exists.

If unacceptable water losses are predicted, but a reservoir is small enough (and alternatives are limited) it may be feasible to provide a lining. A variety of materials, including concrete, butyl rubber, compacted earth, and asphaltic compounds have been used. When linings are to be used, underdrain systems must be considered. This is especially true where reservoir drawdown will be rapid, as is typical at many pumped storage projects.

For large reservoirs where extensive lining is out of the question, it may be feasible to line critical areas. A pervious zone through the rim of a reservoir site may require lining or such an area may need to be removed and replaced by a saddle dam. Rather than remove an entire unsatisfactory area, it may be sufficient to provide a cutoff trench through the area, with compacted backfill of impervious material.

Occasionally, in connection with a dam, a blanket of impervious material extending from the upstream toe of the structure is required. In at least one instance the natural impervious mantle on a valley floor was used as a blanket. Holes cut through the mantle by stream erosion were refilled, and surface compaction was carried out over the entire area adjacent to the dam. It was also necessary to install drainage wells just below the downstream toe of the earthfill dam to relieve water pressures.

Grouting a questionable area to prevent leakage and to prevent or minimize seepage is sometimes feasible. Although it is typically thought of in terms of creating an impervious "curtain" created from a line of holes, grouting can also be considered as creating a three-dimensional impervious zone, especially if more than one row of holes is used. Grouting may be prescribed alone or in combination with other measures to minimize water losses. It may be performed from the bottom of a cutoff trench or from a tunnel. To meet various requirements, the ingredients, mix proportions, pattern and depth of holes, application pressure, rate of pumping, sequence of holes grouted, and sequence of grouting selected zones of depth in holes, can all be varied.

Soft materials in open discontinuities are susceptible to leakage and washing out. Solution cavities in limestones and related carbonates are often filled or partially filled with soft materials. Highly fractured zones containing both soluble and insoluble rocks can become spongeworks capable of passing large flows. When treatment is required it is often necessary to remove the soft materials before attempting to seal off the escape passages. At the dam site it would, of course, be necessary to excavate to a depth below all cavities and seams sufficiently large, numerous, or open to affect bearing strength. Below that depth and in the abutments larger openings can be mined out and refilled with concrete. Smaller openings can often be cleaned out with a mixture of air and water under pressure using the different holes of a group of drill holes. For larger areas along a reservoir rim, however, this can become very expensive. When washing close to the surface, care must be taken to prevent upheaval.

In an attempt to minimize costs, percussion drilling is sometimes substituted for diamond drilling. Percussion drilling for this work is usually unsatisfactory because it creates difficulties in determining just where smaller openings and undesirable materials are. There is no recovery of core to log. It is more difficult for the driller to accurately equate changes in rates of advance of his drill with depth of hole and to note loss of water in the hole. With percussion drilling it is more difficult to maintain a desired hole alignment. Also, the wall of a percussion drilled hole is rougher and more disturbed from the original condition; this makes it more difficult to identify discontinuities and changes in strata using the various types of instruments available (these include mechanical and optical devices as well as the television bore hole camera).

Seismicity

The relationship between reservoir impoundment and seismic events, although not fully understood, is receiving much attention. At a number of large reservoirs an increase in local seismic activity is thought to have occurred in association with impoundment of water. Other large reservoirs have been formed with no increased seismic activity. Because of the general failure to instrument in advance and because most reservoirs are in sparsely populated areas, a lack of prior records causes some uncertainty. Attempts to correlate changes in reservoir water level with seismic activity have been made.

A number of mechanisms have been suggested by which reservoirs could cause earthquakes. The direct effect of additional load causing crustal adjustment is now almost discounted. Changes in the normal effective stresses in the underlying rock because of increased pore pressures has been suggested as a cause. It has been suggested that transmission of hydrostatic pressure through discontinuities in the underlying rock can have a triggering effect where a critical state of stresses already exists. Other explanations as well as variations and elaborations of those just mentioned have been given.

As we build ever higher dams and impound ever larger reservoirs the need for understanding this aspect of the behavior of reservoir foundations becomes more important. Mitigation or prevention require first an understanding of the impoundment-earthquake relationship. Every large reservoir site should be subjected to geologic, geodetic, and seismic studies. One empirical definition of a "large" reservoir is one with a storage capacity of 12×10^8 m^3 (1,000,000-acre-ft) or more, usually impounded behind a dam 90 m. (300 ft.) or greater in height.

At the time of the feasibility study, records of seismic activity in the area should be evaluated. Since this is also necessary in designing the dam, little extra effort is required. Should frequent seismic events or substantial intensity of events be noted, further investigation becomes prudent. It would make necessary a more detailed stage of geologic mapping of the reservoir. Discovery of an active fault within the reservoir site would, of course, raise questions about the various project structures as well as the reservoir itself.

Since one condition of possibly major significance is the state of stress existing in the rock which will underlie the reservoir, investigation of stress conditions in that rock may be warranted for large reservoir sites, particularly if the site geology also raises concern about seismicity. The existence of a generally horizontal stress condition of substantial magnitude would be an indicator that further study should be undertaken. In-situ stresses can be determined by stress relief testing. Usually, however, any such testing has only been performed during preconstruction investigations of dam sites and at depths insufficient to provide an indication of the representative condition of the underlying rock mass.

Geodetic studies before and after reservoir filling can also provide useful information on the rock masses containing and supporting a reservoir. Although this information is sometimes obtained at dam sites, it has seldom been obtained for reservoirs. Measurement of crustal strain by various types of instruments including quartz strain gauges has also been suggested. Seismically induced reservoir subsidence as well as deformation under direct loading could be recorded by these studies.

Seismic studies should be carried out for every large reservoir. Monitoring systems of from 3 to 10 high gain short period stations have been stated as necessary to accurately determine the locations of microearthquakes of local origin. A network of 3 stations should be installed 5 years before any storage begins. If increased activity occurs as the reservoir fills, the network should be expanded. One of the seismograph stations should be considered a permanent installation and the others temporary. There has recently been considerable discussion among experts, without general agreement, of whether or not it is necessary for seismometers to be attached to rock foundations. For design of a permanent installation, see Section 11-4, Associated Structural Works.

With the accumulation of data, understanding of any impoundment-earthquake relationship will surely grow. In the meantime, engineers will be faced with decisions to build or not to build. Whether increased seismic activity may be expected only during a period of adjustment or whether it will be dependent on reservoir water surface fluctuation could be a decision-influencing factor. Surely a large reservoir close to a metropolitan area may require a decision differing from that of a reservoir located in an isolated area.

Bank Storage

The term "bank storage" is used to describe water from a river or lake which spreads out from that body, filling interstices in the surrounding earth and rock mass. Such water is assumed to remain in the surrounding mass, in contrast to seepage or leakage which would continue to move or to ultimately enter the ground water supply or surface flows. The water is theoretically recoverable. If the body of water were to be removed, the bank storage would return in time to the site of the river or lake. In one sense bank storage is considered a loss because it may be water that is not available on demand. For very large reservoirs with very slow drawdown rates bank storage may be substantially available on demand, whereas the rapid drawdown rate at a small reservoir may be much greater than return flows from bank storage.

The location and slope of the existing ground water table will be a major influence on bank storage at a site. If the relationship between an existing stream and the water table is such that ground water feeds the stream, bank storage will be minimized. Where an existing stream feeds the ground water, bank storage will be a problem. Bank storage will also be influenced by the permeability and porosity of the confining rock or soil. Discontinuities are not significant contributors to bank storage. Bank storage is less likely to be a problem in metamorphosed rock, especially where considerable overburden has built up.

Bank storage is not a condition which can be prevented or mitigated. It is a condition which must be estimated for feasibility studies and measured during reservoir operation because it provides additional regulation of the annual streamflow. Estimating the quantity of bank storage to be anticipated is a responsibility usually assigned to ground water geologists.

Evaporation

In addition to loss of reservoir water to the ground, losses occur to the atmosphere. This takes place by direct evaporation and indirectly by transpiration from plants. Evapo-transpiration is influenced by the types and amount of vegetations. Evaporation, of course, depends on the climate. But these losses are also affected by the shape of the reservoir. Wind conditions, humidity, and temperatures should all be assessed. A reservoir site having a small surface area to volume ratio will suffer lower evaporation losses than a saucer-shaped reservoir of equal capacity. There is at least one case of a reservoir at high elevation where wind-induced evaporation influenced the determination of optimum water surface.

The motivation for research into ways of reducing evaporation can be easily understood when the order of magnitude of the losses is considered. In the western United States alone, loss per year due to direct evaporation is 31,000 Mm^3 (25 million acre-ft). Evaporation retardant chemicals are available for use in reservoir operations. They reduce evaporation by increasing the surface tension of water through the formation of a mono-molecular film. Although the best of these are not without other undesirable qualities, their susceptibility to wind drift and their cost are the principal limiting factors. One of the most difficult aspects of evaporation reduction research has been quantifying results to permit evaluation.

Certain types of plants known as phreatophytes are particularly high consumers of water. They grow in areas having a high water table, are rapid-spreading, and have roots that can go down 23 m. (75 ft). Phreatophytes use from 50 to 100 percent more water than most agricultural crops. It is estimated that they consume 31×10^9 m^3 (25 million acre-ft) of water in the western United States. Except for site selection, this is an operation and maintenance problem. Eradication by physical or chemical methods followed by replacement with plant having lower water requirements, and continuing with a vegetation management program seems to be the only effective solution.

Sedimentation

As with bank storage and evaporation, sedimentation is basically a problem which can only be accounted for; man's efforts can at best result in some control of it rather than elimination of the problem. Ultimately, sedimentation is the determining factor in the life expectancy of a reservoir. As a reservoir silts up, its effectiveness diminishes. It is a standard practice to allocate a certain portion of the reservoir space for sediment storage. Prediction of a reservoir's life expectancy (by hydrologists) is based on the amount of sediment delivered to it, the reservoir size, and its ability to retain the sediment. Sediment surveys conducted at intervals during the lifetime of a reservoir will update the capacity data necessary for efficient operation.

The sediment transport capacity of the stream is influenced by velocity of stream flow. When the stream enters a reservoir the velocity diminishes and sediment deposition begins. A delta forms at the inlet and will gradually extend toward the outlet. This is perhaps an oversimplification of a complex process; nevertheless, the space nearest the outlets from the reservoir tends to be the last to silt up. Although examples of completely silted-up reservoirs do exist, it is not a foregone conclusion that a given reservoir will ultimately become completely silted-up.

A number of measures have been tried in attempts to minimize sediment deposition in reservoirs. The most effective and most costly is to provide watershed protection through a vegetative management program attempting to prevent soil erosion. Silt detention basins at inlets of smaller reservoirs may have some success. Low level outlets in dams to provide a degree of flushing action have had a limited success at some smaller reservoirs. Sediment removal can be economical for small reservoirs. Not all the effects of sedimentation are necessarily negative. With time, deposition of sediments in reservoir waters can have the effect of a natural blanket, resulting in lessened seepage losses. Mitigation measures can be taken to transform accumulations near inlets into such things as fish breeding ponds, wild fowl refuges, and recreation areas.

Growth of a delta at an inlet also affects the stream channel for some distance upstream. Because of the backwater effects on sediment transport the stream channel will probably aggrade. Channel gradients become flatter, channel cross sections are reduced, and flooding occurs more frequently. If the natural topography in the inlet area is a relatively level basin, drainage could be hindered. It could become necessary to dredge the main inlet channel, a potentially difficult construction problem due to lack of floatation for large machines and a tendency for bank slippage and caving.

Other problems, of a less general occurrence or of lesser importance, sometimes develop. Shoreline retreat is a problem to a certain extent at practically all reservoirs. The primary factor causing it is the magnitude and direction of the prevailing wind over the reservoir. If the reservoir has long reaches of its area which coincide with the prevailing direction of the wind, a bank erosion problem can be anticipated. Severity of the problem will also be influenced by wave height and the erodibility of the shoreline material. The redistribution of these surficial materials will also result in sediment deposition elsewhere in the reservoir. Some form of bank protection could be feasible if the problem is serious enough.

Wind can have other associations with sedimentation of reservoirs. The occurrence of high winds at times when a reservoir is drawn down exposing large, flat areas can cause serious erosion problems. Protective measures would be the same as those considered good farming practices. Also, rarely, wind deposited sediments in desert areas could be significant.

The location and planning of suitable marinas on reservoirs must receive careful consideration. The marina needs to be sheltered from the wind and must not be subjected to siltation from any tributary drainage areas. For further discussion of marinas, see Section 11-4, Associated Structural Works.

In colder climates, the buildup of deltas with their flat channel gradients, divided channels, islands, and submerged bars is an ideal environment for the formation of river ice jams. The problem created is not the ice jam itself but the flooding that may be caused by the temporary impoundment above the water surface of the permanent reservoir.

11-3 RELOCATIONS AND RIGHTS-OF-WAY

Relocations and rights-of-way are two related subjects which are seldom even mentioned in the formal education of civil engineers. In order to build virtually any project, however, engineers must deal with such matters. Relocation and right-of-way considerations must be included in any realistic programming of a schedule and budget. Provision for adequate lead time is particularly important in the acquisition of rights-of way.

Before beginning a project the right of the project owner to use the land as well as the water must be assured. The right to use the land is assured by obtaining title to it. The removal of people and the cessation of the use of improvements must then be accomplished. Acquisition of real property and relocation assistance may both be involved in the removal of people. Under the U. S. Uniform Relocation Assistance and Land Acquisition Policies Act of 1970 (most states have similar acts), people cannot be removed until the latter of: (1) 90 days have elapsed since the occupant has been notified to move; or (2) the owner has received payment for the property. Cessation of the use of improvements within a reservoir area is achieved by physically transporting them from the right-of-way, razing them, or simply abandoning them. Within the concept of just compensation the project owner obtains control from the owner of the improvement by either relocating, replacing, or purchasing the improvement.

Rights-of-Way

Right-of-way is usually thought of in the context of land acquisition as a strip of land over which a road, railroad, or some utility line passes. As used here, the term is broadened to include the lands to be taken for a project, including the reservoir. The limits of taking of right-of-way around a dam and reservoir can be called the project "take line." If the owner of the project is to be a governmental agency or a public authority, acquisition of privately owned lands may be assured by the right of eminent domain. Webster's New International Dictionary defines this right as "superior dominion exerted by a sovereign state over all property within its boundaries that authorizes it to appropriate all or any part thereof to a necessary public use, reasonable compensation being made." Privately owned utility companies also usually have this right. In those instances where the buyer and the landowner are unable to negotiate an agreement, the buyer with eminent domain rights can condemn the land in question and a court then decides the compensation to be paid the owner.

When the use of privately owned lands is required for a project, the right to use it may be obtained in two different ways; the land may be purchased outright, the buyer acquiring fee simple title, or the right to use the land may be purchased by an easement. Easement rights can be permanent or temporary, and include flowage easements, drainage easements, borrow pit and spoil area easements, construction and maintenance easements, and pipeline or other utility line easements. They are appropriate when land is needed only for temporary or intermittent use, and is otherwise available for at least limited use to the landowner.

In connection with the construction or operations of projects it usually becomes necessary to construct project structures across various existing facilities. These include railroads, highways, transmission lines, communication lines, oil and gas lines, canals, and other facilities. An agreement with each owning entity is required, granting the project owner the right to cross the facility in question. The engineering provisions of a crossing agreement should include: the need for the construction; description of the facility; rights-of-way or easements; responsibility for temporary measures; standards of work; work to be accomplished by each party; maintenance obligations; furnishing of work or materials records; future rights and obligations; and liability insurance and indemnity provisions. Crossing agreements which provide for the project owner to bear a share of the cost of reconstructing or altering the facilities of the other party would also be subject to the same provisions as relocation agreements, discussed later in this section.

Another term unfamiliar to many civil engineering students is "patent" as applied to land. In the United States and at least some other countries a patent is the title deed by which a government, either state or federal, conveys its lands. The term appears frequently in areas where there has been considerable mining activity. Government lands available for development may be prospected for minerals. When a claim is filed on such land without purchasing the land, it is known as an unpatented mining claim. Title to valid unpatented claims within the right-of-way must be acquired separately and distinctly from title to the land itself.

Right-of-way acquisition includes a number of distinct phases or steps. The land must be described and evidence of title must be obtained. Appraisal must be made of the land, improvements to the land, crops, and other resources. Negotiations are required. Acquisition itself must take place, either through direct purchase or condemnation proceedings. Finally, relocation assistance may be required.

Coordination between engineering and right-of-way work is required throughout the development of a project. During the planning and design stages, the right-of-way group must initially be a party to the programming operations. Engineering and right-of-way should jointly establish the tentative take line. Estimated costs and time for acquisitions should be furnished. Timely discovery of unusual conditions sometimes occurs in this way. The existence of old mine shafts and tunnels, for example, may be discovered from right-of-way investigations.

Some right-of-way work may be required to enable preconstruction investigations at the project site. Initially, right of access may be secured by obtaining permits from landowners. Before actual construction can begin, of course, control of the work sites must be achieved by the project owner. In those cases where the organization retained for engineering is not responsible for right-of-way acquisition, definite procedures or work routines insuring coordination should be established.

One of the earliest activities at any construction site is to establish survey control points tied into the regional or state coordinate system. Work can then begin on laying out in the field the location of the various structures to be constructed. (Sometimes a structure is sited first and its location then determined.) Once control has been established the surveying of property boundaries and the tentative project take line must begin. This data is needed in preparing land descriptions for right-of-way acquisition. At the same time title evidence is assembled. Real property appraisals are then made. During this period, various existing or potential problems will be brought to the attention of the engineering staff. The appraisers will sometimes seek clarification and engineering opinions. They should make suggestions and point out the effects of various physical conditions on property appraisal values.

A negotiator quite often finds it necessary to consult with the engineering staff as questions, offers, and counter-offers arise during more involved negotiations. Minor adjustments to the established take line may be made. Matters such as the removal and building of fences, time allowances for the removal of improvements, and possible modifications of entrances typically arise.

An orderly acquisition program will usually continue throughout the construction of the project. Initiation of reservoir impoundment is the controlling event in the acquisition program. All lands to be inundated must be acquired early enough that completion of the necessary timber and brush clearing operations can be assured ahead of the rising waters.

Right-of-way acquisition work always enters a legal phase for a government agency when legal staff check every proposed acquisition for sufficiency of title and compliance with regulations. Acquisition work also enters a legal phase when condemnation proceedings are contemplated. Condemnation is initiated when negotiations reach an impasse. It is also frequently used, however, as a means of establishing clear title to property. Such cases may be known as "friendly condemnations." During this legal phase the engineer is often asked for various estimates, exhibits, and even valuations. He is sometimes called to appear at a trial as an expert witness.

The final step in the right-of-way acquisition process is the relocation of improvements and people. Assistance by the project developer is becoming increasingly common. Although this has long been the practice for improvements such as roads and utilities at reservoir sites, its application to people individually is relatively new. This assistance includes reimbursement for the seller's monetary expenses and help in locating suitable replacement housing. Removal of a residence may not take place until resettlement is available. Relocation assistance is required by any U. S. agency obtaining right-of-way and at any project where federal money is being used. In the U. S., most states also have similar laws which apply to all acquisitions by the state, counties, and cities where any public funds are involved.

Relocations

Relocations to an engineer will at first thought mean utilities and transportation facilities, rather than people as discussed above. But there are sometimes other relocations required. These may include buildings and other structures (with associated improvements such as fencing), whole communities, cemeteries, schools, and parks. Examination of archeologic and historic sites and the effects of relocation on people are also involved. The need, cost, and scheduling for relocations should be given sufficient attention in the planning and design process.

In a large organization, right-of-way staff will usually prepare agreements necessary for relocation of utilities and other existing facilities. In a small organization an engineer or technician may prepare the documents. Either the project owner or the engineering consultant may be responsible. Regardless of how relocations are handled, the project engineering staff must be very much involved.

Once the need for a relocation is decided, the next and the most important step is to determine the new location and alignment of the improvement. Many other factors must be considered, such as whether the relocation will be a betterment, when the facility's owner is to resume responsibility, imminent improvements already prepared by the owner, operation and maintenance (O&M) costs during construction, and changes in relocation plans.

A relocation agreement (usually for a road, railroad, or utility line) with the facility's owner may include provision for responsibility for abnormal maintenance during a seasoning period after construction is finished. Items such as settlement of fills, consolidation of base for railroad trackage, erosion or sluffing of unconsolidated and unstable areas, excessive ravelling of cut sections, or other problem areas which would be over and above routine maintenance are usually considered abnormal. For a railroad relocation, responsibility for lateral movement of the track should be covered.

Consequential damages, such as the following, are usually considered to be costs beyond the obligation of the project owner to pay:

1. Increased taxes or damage to the owner's business resulting from the relocation of the facility.
2. Increased cost of operation and maintenance because of location in rougher terrain and longer facilities.
3. The potential of an existing facility to be improved in the future at minimum cost in order to benefit or increase the operating capacity.

To help in establishing the project owner's limit of obligation for the relocation of a facility, an inventory of the property and comparative studies should be made of the existing works versus the proposed new works. The inventory should include quantities, sizes, types, lengths, or other key facets of the existing works to provide a basis for comparison for the new relocated facilities.

The relocation agreement itself could contain many provisions. Possible paragraphs are merely identified here as a checklist. No attempt is made to prepare the reader for actually writing the document. The relocation agreement should include the identity of the parties; the need for the relocation; scope of the contract; definitions; rights-of-way; standards of work; betterments; participation of third parties; plans and specifications; procurement of necessary permits and licenses; what work is to be performed by each party; and commencement, prosecution, and completion of the work. The agreement could also cover: abnormal maintenance; inspection; transfer of operations (for a railroad); salvage; estimated costs; payments; contracting for the work; insurance; release of obligations; ownership and conduct of the work; maintenance of completed work; hold harmless clause; indemnification of the improvement owner; interference; interruption; termination; disputes; notices; covenant against contingent fees; waiver; right to cross relocated facility; and successors and assigns. If the project owner is a governmental agency the document could also contain clauses about: landscape preservation and natural beauty; freight rates on materials transported to the site; records and audits; facility owner not to act as agent for the government; liability of the government contingent on appropriations; work hours legislation; equal employment opportunity; certification of non-segregated facilities; convict labor; government policy provisions; and officials not to benefit. Items of a more unusual nature could also be encountered in negotiations, peculiar to the project involved.

Utilities Relocations

The various utilities include electricity, gas, sewers, telephone, water, and other pipelines. Many are subject to the review and control of regulatory agencies. Water may be conveyed by canal or ditch as well as pipeline. A utility may be a privately-owned corporation or may be publicly owned (such as a municipal water department).

It is good practice to build a close liaison at the working level with members of a utility's engineering staff. To avoid confusion later, one man in each organization should be assigned the overall responsibility and authority for coordination. Note that relocations may not be the only business transacted with a utility, because utility services will also be needed by the project. As soon as a project construction office is staffed, contact should be made with the engineering staff of each local utility likely to be affected by construction. Because of the specialized nature of much of the utility engineering work it is the general practice for the utilities themselves to be responsible for much of the design and construction required for relocations. Handling their own relocations, they can better arrange outages and are more certain to be satisfied with the work.

Relocations often include a need for additional right-of-way. It could be decided that each entity actually doing the relocating should secure its own right-of-way. Landowners in the project area may then be exposed to the personnel and policies of several different right-of-way departments; in fact it would be possible that a given owner could have to deal with two or more entities. This could result in inconsistencies, confusion, and bitterness on the part of the land-owning public affected. In order to minimize this possibility, it is recommended that all right-of-way acquisition associated with a project be handled by one organization.

In the United States it would be very unusual to have a large project that did not require the relocation of any utilities. From the beginning of field work at the site, the project surveyors and other staff members should be alert to note the existence of each utility facility observed. Maps of all facilities in the project area should be obtained through early contacts with the utility engineering staffs.

As soon as the construction plans of each

project component or facility have been tentatively established the utilities should be notified. A map showing the location and alignment of the feature and the suspected locations of utility facilities should be sent to all utilities, with a request that they list and indicate all of their facilities in the vicinity.

When a utility has had time to prepare preliminary relocation designs, an early jobsite conference should be held. This is an important meeting and it should be documented. Engineering decisions are agreed upon and the utility obtains the information necessary to formulate its proposal. Special requirements, reimbursable and nonreimbursable items, and other pertinent information are discussed.

The utility then prepares a cost estimate for the work. After review by engineering representatives of the project owner, a formal agreement is negotiated. The agreement should include the mutually acceptable cost estimate *as an estimate*. If it is obvious that a number of different relocations will have to be negotiated with a particular utility, a master agreement is sometimes negotiated. As each relocation is engineered, a simple amendment is added to the master agreement.

The next step is for the utility to be formally notified to proceed with construction. Continued close liaison and coordination with proper inspection will result in a mutually satisfactory relocation with a minimum of disruption. It is usually helpful if the same inspector is assigned to all the relocation work for a given utility, thus encouraging a more uniform interpretation and treatment of requirements. An inspector is only needed, of course, where a utility relocation alignment crosses a project facility alignment. When the work has been completed, the utility should present a billing itemized to correlate to the cost estimate included in the agreement. The billing should be accompanied by "as built" drawings of the relocated facilities.

Transportation Facilities Relocations

Probably the major relocations costs at a project are incurred in relocating transportation facilities. These include railroads, highways, roads, and trails. River traffic would be only rarely a problem. The need to deal with political entities, often involving public hearings, is also time consuming. Careful planning and good public relations can be a wise investment, leading to overall time saving.

In the U.S. a railroad relocation may involve a State Public Utilities Commission. The relocation work itself is not necessarily carried out by the railroad. Preparation of the new track alignment to subgrade and construction of required structures including bridges and tunnels is usually the responsibility of the project. The railroad will sometimes place the ballast, ties, rail, and signal equipment. The actual connecting of the relocated segment of line and disconnecting of the old portion is almost always done by the railroad's own work crews.

Inventory data on the existing line serve as a good beginning for the collection of information required to design the relocated segment of line. For further discussion of required data see Section 11-4.

Public highways and roads are owned by state, county, or city governments. Although some highways are included in various federal assistance programs, they are the responsibility of state highway departments. If relocation of an interstate highway is necessary, the state will have to obtain the approval of the Federal Government.

A typical procedure for relocating a state highway is for the highway department's staff, working closely with the project staff, to prepare several alternate preliminary plans. Some initial road relocation planning during the planning phase of the project itself is strongly recommended. This initial planning will determine what roads must be relocated, their classifications, if major relocations are required, and if difficult or unusual construction conditions can be expected. Once construction status for a project is reached, however, more formal preliminary plans are developed. For state highways, preliminary plans are publicized in the news media, in talks to civic groups, and in communication with other units of government affected. A public meeting is then held by the concerned State Agency. The public's input is considered and the alternates are restudied and reworked. Then a public hearing is held by the appropriate decision making body, such as the State Highway Commission. The State Highway Department proposes its choice of routes. At each of these meetings the project owner and the consultant should speak. Finally, the decision is made by the Commissioners. Similar procedures are usually followed by county and city governments.

Quite often the state will want to do the design work and supervise construction of the relocations. Whether or not they do, reimbursement for incurred engineering costs will be required, as provided in the relocation agreement. It is desirable to have the state prepare a report of their preliminary studies which can be used in preparing the project environmental impact statement. This, of course, is something which should be done in the project planning stage.

Two questions in particular that should be addressed in the relocation agreement are the design standards to which the road will be built and the traffic capacity that is to be accommodated. Most likely the road being replaced will be an older road, built to design standards considerably lower than those currently existing. The present traffic burden of the road may exceed its designed capacity or projected increased demand will be project generated traffic. This could mean that a present two-lane road should be replaced by a four-lane road. From an engineering viewpoint, design to current standards to accommodate anticipated traffic is most logical, but the extent of the project's financial obligation should be clearly stated in the relocation agreement.

The "current" highway design standard may be based on the average daily traffic count as of the date of the relocation contract, projected according to criteria of the specific highway department. It should be established that the standards have been and are being maintained generally on rebuilding or new construction of comparable roads within the area. Projected traffic volumes, normally 20 years in the future, provide for reasonable traffic growth as indicated by the anticipated future normal economic growth of the area.

It is reasonable to expect the cost of a betterment to be borne by the owning highway authority. Construction to standards higher than current design standards could be required by a highway department. This would be a betterment. Perhaps the present daily traffic count indicates a two-lane road is presently sufficient, but volume projections in accordance with current design standards indicate the need for a four-lane road, a higher classification. This may or may not be considered a betterment. In cases like these, the relocation agreement should clearly indicate the obligation of each party.

The phrase "replacement in kind to present day standards" may be used in relocation agreements. Its meaning would appear to be very clear and sufficient for use in an uncomplicated document. Yet consider what can happen if sufficient thought and planning has not been given in preparing the agreement:

The time lapse between initiation of final design for a road and the beginning of construction is considerably longer than had been expected. Perhaps this is due to unanticipated design difficulties, or to delays in obtaining funding, or to public disagreement over alignments. With the design work nearly complete, major revisions are made in the official "present day standards," which would affect both alignment and cost. Which present day standards are to be required?

A one-lane dirt road with hairpin turns winds down a canyon, crosses a one-lane bridge, and winds back up the other side of the canyon. The canyon is to be partly filled by the project's reservoir. When design work for relocation is initiated, it is realized there are no present day standards for one-lane roads. The owner of the road argues that a two-lane 70-km/hr (43 mph.) paved relocated road is required. The project owner argues that no replacement at all is needed. The consultant realizes that a one-lane bridge with the required span would be impractical from an engineering viewpoint. What is the project obligation?

The two examples recited above illustrate the need for adequate planning for relocation agreements. In the following example, a relocation agreement was satisfactory to the signatory parties but the relocation accomplished still may not have been the best solution:

A local secondary road is to be inundated by a reservoir. This road is not heavily used and its severance, although inconveniencing some people, will not cause any hardships because other routes and access are available. The topography is such that a replacement road located completely out of the reservoir would be very expensive. If the road is relocated to a minimum elevation where it would be above the reservoir water surface for all but the 25-year and greater floods, the cost would be quite modest. Both parties agree to the less expensive solution. Two years after completion of the project the 25-year storm occurs and irate citizens begin pounding on politicians' desks and contacting the news media. The headlines—"Engineers Goof Again!"

Other considerations about roads usually arise at a large project. Although they are not actually relocation associated, these discussions occur with the same entities and often at the same time as relocation discussions. Their timely discussion promotes good engineering solutions. Construction access to the work site may require new roads. Permanent access for operations and maintenance and for recreation may also be necessary. It may be that state or county units are planning roads in the area (perhaps to meet project-generated development). Construction traffic may be predicted to ruin existing roads not built to meet such demands; yet construction traffic cannot legally be excluded from most public roads. This could be a consideration in road relocation agreement negotiations.

With the increasing attention being given to recreation, the number of trails and trail systems is growing. There are bicycle trails, equestrian trails, off-road-vehicle trails, and pedestrian trails. Relocations can be handled fairly easily when recreation facilities are incorporated into a project; otherwise, special measures may be necessary, just as for other relocations. It is important to note that the various trail uses are usually incompatible with each other, so they should seldom be combined.

Canals occasionally exist at reservoir sites. Canal flows can sometimes be routed through reservoirs, in which case water quality should be mentioned in any relocation agreement. If pumping requirements are part of the relocation solution this must be covered in the agreement. In the U. S., relocations or crossings of power canals require the approval of the Federal Power Commission.

Other Relocations

It is sometimes decided that certain buildings or other structures will be relocated rather than simply razed or salvaged. Most often this will be due to a present owner's wishes, but occasionally a project owner may deem relocation more reasonable than replacement or simple purchase. This kind of relocation, however, cannot be forced on the present owner. The only right the project owner can enforce is the right of eminent domain. This requires the payment of just compensation which must be in cash. For example, if a railroad refuses to enter into a relocation agreement and to relocate, the project owner's only relief is to condemn the existing line and to pay cash therefor. It would be impossible to list all the types of structures possible but they could include radio or television towers, highly specialized industrial plants, unique buildings, and even gold mining dredges. As a general rule it is best to have the owners of such facilities be responsible for their removal.

Relocation of a cemetery requires special treatment. In addition to the normal considerations, efforts must be made to contact descendants of all persons buried there. This must be done before each body can be disinterred. Relocations of this type should be accomplished by contract with entities specializing in this work. Although the whole cemetery may be relocated as a unit, relatives may direct that individual bodies be moved to other existing cemeteries. The layout and siting of individual graves in the relocated cemetery is a matter best left to the cemetery administrators. Existing graves not in a cemetery are a variation of this type of relocation. Most state or regional governments have statutes providing the procedure which must be followed in removing graves.

The relocation of churches, parks, schools, and other public use facilities also requires special consideration. Obviously, an important factor in selecting new locations is the relationship of proposed sites to the users of the facilities. It would be unreasonable to relocate a neighborhood church completely away from the parishioners. Betterment of public use facilities may be mandatory under existing state laws. A public school, for example, may have to meet more modern seismic design criteria. This should be covered in the relocation agreement. If recreation facilities are to be included in the project, their locations can be a desirable influence in selecting a new site for a relocated park. It should be noted that government entities usually have no obligation to provide replacements.

Almost inevitably, some monument, historical landmark, or place of archeologic, historic, scenic, or even religious interest will be subject to reservoir inundation. These may be of national or merely local significance. Attempting to evaluate the importance of such sites easily becomes an emotional exercise. Yet this must be done in order to decide the extent of the effort to be made and cost incurred in salvage and/or relocation measures.

Examples of this type of relocation problem abound. Perhaps the most famous were the salvage and relocation efforts undertaken with the building of Egypt's high Aswan Dam. In that case, Egyptian government funds were augmented internationally by private and public donations. Salvage efforts were so extensive that even an ancient temple carved into a hillside was relocated out of the reservoir.

A number of laws have been enacted in the U. S. to protect the cultural heritage. Their significance in relation to relocation requirements is that they shift the responsibility for evaluation from the project owner to state or federal entities. The process can be lengthy, so adequate time must be allotted for associated investigations and paper work, to avoid delaying construction. It is the responsibility of the project owner to make archeological and historical searches, surveys, and investigations. Local historical societies can sometimes be enlisted to help in finding sites and relocating artifacts.

It is sometimes necessary to relocate an entire community. Not only does this require the moving of individual families, but community facilities must also be relocated. Representatives of the community (incorporated or unincorporated) should participate heavily in site selection and other relocation planning. It is not unusual that total acreage required for a new community will exceed the total occupied by the existing community. Because the existing community is likely to have developed over a period of many years, the physical facilities and the development plan itself will probably constitute a betterment.

One way of accomplishing relocation of a community is for the project owner to provide the land for and completely develop the new site. Individual residents would simply trade old for new. This approach has been used in less-developed societies. The U. S. Government does not relocate communities as such. It purchases all houses and businesses from the owners and will purchase platted streets and alleys from incorporated communities. Many more moderate possible solutions with less social impact exist.

One approach used was to purchase individual lands and improvements outright. As payment for inundated streets and utilities, the project owner furnished land for the new community. The community then was able to use income from lot sales to pay for the installation of new utilities and community improvements.

In some cases a community to be inundated

is not re-established. This is more likely with very small unincorporated hamlets. Their acquisition and elimination is not always viewed in a negative way. In at least one instance everyone welcomed the opportunity to move away.

In relocating facilities, care should be exercised to preserve the natural beauty of the area. In addition, when the facility has been relocated and the old structure abandoned, the area should be cleaned up. Those portions of the old location that will not be inundated by the new reservoir should be restored to conform to their natural appearance. Costs associated with preserving or restoring the natural beauty of the work sites are properly considered part of the cost of relocating the facility.

11-4 ASSOCIATED STRUCTURAL WORKS

When developing a reservoir there are often structures associated with the reservoir itself in addition to the dam, its appurtenances, and ancillary works. These associated structures are discussed in this section. It is not intended here to present discussions in depth which will enable design of the structures themselves, where that information is already commonly available. Road, bridge, and building design textbooks and manuals exist in abundance. Instead what is intended here is to point out considerations relevant to their association with reservoirs. For those features for which design information is not so readily available, somewhat more discussion is presented.

Reservoir-associated structural works include recreation and visitor facilities, those resulting from required relocations, and miscellaneous service features. Measures required to assure reservoir integrity and adequacy were discussed in Section 11-2. Other project features can include power and pumping plants, canals, pipelines and tunnels for water conveyance, switchyards and powerlines to transmit electrical energy, fishery facilities, downstream channel improvements, and access roads. Many of these are major highly-specialized facilities which themselves justify a handbook. It would be inappropriate to comment about them in this text.

Obviously, continued awareness of all features and functions is required for a well-executed project.

Recreation Facilities

Although recreation experts may be expected to prepare more efficient and more desirable plans and designs than civil engineers, many of the components of a total recreation plan are properly civil engineering works. Recreation facilities are more and more frequently being included in water development schemes. If recreation is to be a project benefit, the engineer will do well to be familiar with basic recreation design considerations as the planning and design of the project and its features progress. Recreation facilities may include swimming beaches, boat launching facilities, marinas, navigation buoys, picnic areas, campgrounds, meeting areas, and trails. Archeologic, historic, and natural features included may require development and/or protection. Where intensive development of recreation potential is planned, recreation experts should definitely be consulted. In any case, vehicle access and parking, sanitary facilities, solid waste collection and disposal, drinking water supply, and police and fire protection must be considered.

The importance of signs at recreation areas cannot be overstated. It would be a mistake to merely incidentally consider them as their need becomes apparent. They should be unobtrusive yet easily visible, their messages should be unmistakable, sufficient numbers of them should be used, and their siting should leave no room for confusion. Reasonable uniformity is desirable and simplicity is essential.

Sites for swimming beaches must have suitable terrain, including consideration of soil, gradient, composition, and stability. Proximity to picnic and campgrounds, availability of sand, crosscurrents favorable to washing away flotages, water quality (including temperature), wind and wave action, and reservoir drawdown are important. Swimming beach slopes below water should be between 5 and 15 percent, with 5 to 8 percent desirable. Above water, 25 percent is the maximum feasible slope.

It can be assumed that 55 percent of the visitors at adjacent camp or picnic grounds will use the swimming and sunbathing facilities. Beaches, including turfed sunbathing areas, should be at least partially screened from parking or day use areas other than picnic facilities by trees or shrubs. Designed capacity should assume a desirable minimum beach area of 9 sq m (100 sq ft)/person. One of 10 beach users will normally be swimming at one time.

Sanitary facilities should be provided for a maximum of 70 people/toilet fixture. If outdoor showers are to be provided, a maximum of 200 people per shower fixture is recommended. Swimming beaches are often located adjacent to campgrounds or picnic areas, enabling the siting of sanitation facilities to serve both areas. Facilities should be located not more than 120 m (400 ft) from maximum normal water surface at the beach.

Water supply requirements will be from 10 to 60 liters (2.5 to 15 gal)/person/day, depending on the type of toilets and whether or not shower facilities are provided. Parking should be provided within walking distance, 150 m (500 ft) from the beach. Safety provisions should include lifeguard towers every 90 m (300 ft) and buoys or floating markers to enclose the outer limits of the swimming area. Stumps and trees should be removed and stump holes filled in the underwater areas and on the beach itself.

Boat launching facilities may include a maneuvering area for a vehicle and trailer, a boat ramp, a fairway for maneuvering boats in the water, loading docks, and a channel linking the launching facility with the main body of water. Adequate parking areas, sanitary facilities, and public service areas (gasoline, oil, emergency repair items, and berthing facilities) should be provided. Because it is desirable to have access to the water even at minimum pool elevation, boat launching facilities should be under construction prior to permanent inundation of the site.

A minimum radius of 11 m (35 ft) is required for a maneuvering area. A fairway should have a minimum water depth of 1.2 m (4 ft), should be the width of the launching ramp plus the width of the shoulders or loading dock, and should have a minimum length of 15 m (50 ft) free of navigation hazards at the lowest operating water level. A channel should have a minimum water depth of 1.2 m (4 ft) and a minimum bottom width of 23 m (75 ft), and should be properly marked and buoyed.

Each lane of a boat launching ramp will handle 40 launchings and 40 retrievings/day under average conditions. For a multilane ramp, each lane should be 3.7 m (12 ft) wide. A single lane ramp should be 5.5 m (18 ft) wide. If a ramp is less than 18.5 m (60 ft) wide, a vehicle maneuvering area should be provided for every 60 m (200 ft) of ramp length. The ramp should extend a minimum of 1.2 m (4 ft) below the lowest water level at which it is designed to operate. A slope of 12 to 15 percent is required. One other important design requirement is the protection of the ramp pavement edges to prevent undercutting due to wave action. A ramp should be finished with a non-skid surface. Ramp pavement designs should be checked for uplift stability, particularly where reservoir drawndown rates and water surface fluctuations are substantial.

Loading docks, floating structures that provide pedestrian access to and from boats, may be part of the boat launching facilities or, with public services, may be part of a separate marina. Minimum width of a loading dock should be 2 m (6 ft). Docks made of wood should be pressure-treated with a preservative. Metal barrels should not be used to provide flotation; polyurethane foam works well. A gangway, 1.2 m (4 ft) wide, will probably be needed to make the loading dock accessible from the shore; it should have a maximum slope of 30 percent and a live load capacity of 200 kgs/sq m (40 lbs/sq ft). A reasonable live load capacity for the floating dock itself is 100 kgs/sq m (20 lbs/sq ft). Freeboard under any loading condition should not exceed 50 cm (20 in.) or be less than 20 cm (8 in.). The lateral stability of loading docks should be sufficient to withstand wind, wave, current, and impact loading that may be expected during the lifetime of the structure. At reservoirs where seasonal or longer term water surface fluctuation is great, loading docks would have to be readily capable of relocation to adjust for substantial changes in shoreline location.

At any reservoir where boating is permitted there will be some need for navigation buoys. There are a number of more or less standard types of buoys, floats which must stay where placed. Government, either national or provincial, usually has safety regulations for inland waterways. These regulations specify the type of buoy required for the various kinds of navigation hazards. At reservoirs, buoys usually must be relocated with major changes in water surface elevation.

All picnic areas should be planned so as to take advantage of the topography and native vegetation with minimal manipulation or damage. Picnic areas should be located, when possible, in combination with other types of day use areas at the reservoir. Areas should be designed in clusters of not less than 10 nor more than 100 units. Distance between units should be not less than 10 m (35 ft), with a maximum density of 50 units/hectare (20 units per acre). A 60 m (200 ft) buffer strip should be maintained to provide separation from all developments other than day use recreation areas. Provision of a grill, stove, or small fireplace with each unit is desirable. Maximum slope for a picnic area is 20 percent.

Picnic areas must have a water supply and toilets. Nonthreaded self-closing faucets are recommended on waterline outlets. Drinking fountains may also be included. They should be located throughout the picnic area no more than 30 m (100 ft) away from each unit. Toilet facilities are required on the basis of not less than one fixture per sex for each 15 units. Water supply requirements are 60 liters (15 gals)/person per day with flush toilets or 10 liters (2.5 gals)/person per day without flush toilets.

Picnic areas set aside for use by groups of people are sometimes also desirable. These should be located apart from general picnic units but within day use areas. Group areas of 25 units and two cooking areas with multiple cooking surfaces work well.

Refuse collection receptacles are required. For smaller developments, one anchored container should be provided for each 2 units. At large day use areas, bin type receptacles designed for mechanical handling by collection trucks may be appropriate.

For larger day use facilities play areas for various group activities should be considered. Parking space for vehicles is necessary at all picnic areas. If a picnic area adjoins the shore of a reservoir on which boating is permitted, a portion of the picnic area shoreline should be made available for boat parking.

A campground is an area designed and developed for overnight occupancy usually for a period not to exceed 14 days. One type of campground is the remote, primitive camp, accessible only by foot, horse, or boat. Most common, however, is the campground readily accessible by various types of recreational vehicles (motor homes, pickup truck campers, travel trailers, tent trailers). At larger recreational developments, campgrounds for organized group use should also be provided.

In selecting and designing a campground a number of factors should be considered. The location must be compatible with the overall recreation plan and the development in general. Additionally, however, the site must be suitable for its intended purpose. It should be sheltered from prevailing winds and have some shade from the sun. Surface drainage should not interfere with the use of the site. For best results, the maximum slope should be less than 20 percent and it is desirable to have some level areas. Minimizing alterations to the landscape and using native species for necessary plantings will encourage a more successful campground.

Each campsite should include an off-road parking spur, table, stove or barbecue pit, and a 5 × 5 m (15 × 15 ft) tent area. Units should be spaced 25 to 30 m (75 to 100 ft) apart along either side of the campground road, depending on the topography and cover. Concentration of units should not be more than 12 units/gross hectare (5 units/gross acre). At least half the units at a campground should have sufficient parking space for recreational vehicles and may include individual electrical and water outlets and sewage disposal hookups.

Circulation roads for vehicular traffic should have a minimum 5.5 m (18 ft) width and an all-weather, dust free, paved surface with maximum 0.6 m (2 ft) shoulders. The minimum sight distance at intersections should be 20 m at 15 km/hr (75 ft at 10 mph). The maximum grade recommended is 12 percent. Loops and one-way circuits may be single lane, 4 m (12 ft) wide. All campground roads must be capable of accommodating the maintenance vehicles which will be required to service the area.

Sanitary facilities should be provided on the basis of one comfort station for each 25 campsite units. A comfort station is a permanent restroom with flush-type plumbing fixtures. Comfort stations should be located no more than 90 m (300 ft) to 120 m (400 ft) from the farthest campsite and no closer than 15 m (50 ft) from the nearest unit. The size of comfort station considered here is one which contains three toilets and two lavatories for females, and two toilets, one urinal, and two lavatories for males.

For large campsites, if greater convenience is desired, a combination station could be substituted for every third comfort station. In addition to the fixtures provided with a comfort station, a combination station as considered here has two individual shower stalls per sex and at least one laundry tub. The comfort stations and combination stations as discussed above would be sewered. For smaller campsites where a central sewer system is not feasible, chemical toilets are a less desirable but sometimes necessary alternative. Chemical toilets should be pumped weekly during the peak recreation season. Generally, septic tanks should not be permitted at recreation areas near reservoirs.

Instead of individual sewer hookups for recreation vehicles, a central sanitation station may be made available to receive the discharge from their holding tanks. It is suggested that one station per 100 camping sites be provided.

Facilities for wash water disposal should be provided, desirably within 30 m (100 ft) but in case more than 45 m (150 ft) from the farthest camp unit. A 1 m (3 ft) length of 50 cm (18 in) concrete pipe embedded in the soil and filled with rock can make an effective facility. The disposal point should be at least 8 m (25 ft) from campsites and water points.

For water supply, one nonthreaded self-closing faucet should be provided for each eight camping spaces. They should be located desirably within 30 m (100 ft) but in no case more than 45 m (150 ft) from the farthest camp unit. Hose bibbs with threaded faucets may be provided for those units designed with a spur for recreational camping vehicles. Campground water requirements will range from 20 to 150 liters (5 to 40 gals.)/person/day.

Permanent comfort stations should be lighted, and electrical outlets may be provided at those units designed with spurs for recreational camping vehicles. One anchored refuse disposal receptacle should be provided for each two camp units, within 30 m (100 ft) of each campsite.

Provision of facilities to accommodate camping by organized groups is desirable. Where the degree of development permits, an organized group camp should consist of several smaller (25-person capacity) camps within the overall group camp area. This can accommodate several small groups separately, or can accommodate larger groups by utilizing several individual smaller areas as one camp. Each of the small group areas should be separated by a buffer zone.

A centralized headquarters with camp administration, cooking and dining facilities, and first-aid station should be provided at an organized group campground. Individual parking spaces at each campsite should not be provided. Instead, centralized parking should be provided near the headquarters area, with 25 parking spaces for each 100 persons of the camp's capacity. No vehicles should be permitted in the camp area except service vehicles.

A special combination building for sanitation should be provided to serve a maximum of 150 persons. It would contain hot water and two laundry tubs, with four shower fixtures, four lavatories, and four toilet fixtures for each sex. The building should be sited not more than 185 m (600 ft) from the farthest sleeping area. Either a sewage treatment facility or connection to a central sewer system is required to service the headquarters and wash houses. Bin-type refuse disposal receptacles designed for mechanical handling by collection trucks can be very efficient at centralized locations such as a group campground headquarters area.

Each of the small group areas should have a potable water supply outlet with a nonthreaded self-closing faucet. A supplemental vault-type toilet for each sex and a waste water drain (dry well) should also be provided. Each of these areas should be furnished with a sufficient number of anchored refuse disposal receptacles.

Remote primitive camps are particularly attractive to those vacationers looking for relative isolation from other groups of people. Accordingly, not more than 8 campsite units should be developed at each primitive camp. Buffer spaces between units are important. Each unit should consist of one table, one graded tent space approximately 5 × 5 m (15 × 15 ft), and a fireplace.

A water supply would be provided only where a source is readily available and may be developed at little cost. One vault-type toilet per sex should be provided. Pit-type toilets (unlined) should not be permitted at recreation areas near reservoirs. For each 4 campsite units there should be one wash water drain (dry well). One anchored, animal-proof refuse disposal receptacle should be provided for each two camp sites. If a camp is located along an equestrian trail, a watering trough (with water) and a hitching rack should be provided.

Remote primitive camps are usually reachable by trails, although some sites may be selected that would only be accessible by boat. They, of course, must all be serviced either by boat or by service trails negotiable by 4-wheel drive vehicles. A centralized parking lot with one space for each campsite unit should be provided at the roadhead where trails to primitive camp areas begin. Where a camp is located close to the lake, the shoreline should be cleared as necessary to provide an inconspicuous shore boat landing.

It is appropriate to provide meeting areas at group and family campgrounds and sometimes at group picnic areas. Seating around some focal point can be accomplished using benches made from fallen logs. For a meeting area that can accommodate larger groups, addition sanitary facilities are desirable.

Sewage collection, treatment, and disposal is particularly important because of the potential influence that recreation development can exert on the quality of reservoir waters. A remote reservoir with minimal recreation development and use may require only seasonal pumping of vault toilets into a tank vehicle for removal. Connection to existing (or expanded) municipal facilities may be most desirable for a development within or close to a populated area. Consideration of the overall long range project impact on the general area could be a key factor in sanitation planning decisions. For a large recreation development the best solution could well be the creation of one or more centralized sewage collection, treatment, and disposal systems.

In reaching sanitation planning decisions, a number of considerations peculiar to recreation facilities should be weighed. The service area is likely to consist of a number of separated locations. The demand is likely to be seasonal, with high peak periods such as weekends and holidays. The facilities required could be rather small scale. Finally, construction access could be difficult, particularly for the collection system, and environmental restraints could be substantial.

A waste treatment and disposal system commonly utilized in recreation areas for multiple units is a combination of a waste stabilization pond and spray field. More sophisticated systems providing a higher degree of treatment may be warranted at very large, intensively-developed recreation areas. By using holding tanks to contain peak flows, treatment facilities can sometimes be more efficiently and economically sized. Factors used in selecting the location for treatment and disposal facilities include distance from habitation and recreation points, prevailing winds, surface runoff, water pollution potential, and terrain.

Water supply designs should meet peak demand conditions and provide firefighting capability in the recreation area. Emergency storage can be provided by estimating at double the anticipated daily use for the capacity of the water storage reservoir or tank. If the lake is the only permissible source for potable water, planning should consider the relationship between recreation activity areas and the water intake. Planning and design should also note any special requirements necessary to provide safe drinking water.

Trails are used to enhance public enjoyment of the environment and utilization of fish and wildlife resources. It is essential that trail systems be closely coordinated with the other recreational facilities being developed. There are hiking and bicycle trails, horse trails, and off-road vehicle trails. In general, each is incompatible with the others. Important factors to consider in establishing a trail alignment are terrain, cover, aesthetic value, points of interest, road crossings or other potential dangers, and final destination. Where there is to be an extensive trail system, connections to other trails and roads will be another significant factor.

A hiking trail can be anything from an undeveloped deer track to a park-like walkway. The anticipated traffic and environmental scene should be the basis for selecting appropriate design criteria. In a rugged remote area a hiking trail could be only 0.6 m (2 ft) wide, with occasional short gradients as steep as 60 percent. A short trail within a heavily used recreation area may require a 3 m (10 ft) width and a maximum grade limit of 6 percent. As a general rule a minimum width of 1.2 m (4 ft) and a maximum gradient of 15 percent are desirable.

Overhead clearance for a hiking trail should be a minimum of 2.5 m (8 ft). The desirable minimum clearing width is 0.6 m (2 ft) beyond each edge of trail tread, with all brush removed and stumps flush cut. Drainage should be developed where necessary to prevent erosion or ground saturation. Where sidehill construction is necessary, backslopes may be as steep as the material can stand without serious erosion.

When it is determined that a trail or a segment of trail should be surfaced, natural materials should be used wherever feasible. Bituminous material may be used on trails in heavily traveled areas. A natural material that can be very effective, but is perhaps not immediately thought of by engineers, is wood in the form of chips, shaving, or sawdust. If chipping is used in clearing the reservoir (see Section 11-5), a plentiful supply could be available. They provide protection for roots and soil and make good trail beds.

Gradient is more important for bicycle trails than for any other kind of trail. Bicycle trails should be horizontal wherever possible. Except for very short distances where a maximum of 10 percent can be permitted, the gradient should not exceed 6 percent. When long grades are unavoidable, frequent level areas should be provided, where the cyclist can dismount without difficulty and rest. Another consideration more important for bicycle trails is turning radius. Care should be taken when laying out trails to avoid sharp angles and short radius curves. The minimum turning radius may vary from 2 to 6 m (7 to 20 ft), depending on the speeds likely to be attained along the given reach of trail.

Bicycle trails should be 2 m (6.6 ft) wide. A technique sometimes employed for variety is to divide a trail when bypassing features such as boulders, trees, and descriptive markers. The same clearance requirements apply for bicycle trails as for hiking trails.

Bicycle trails require surfacing. Although the recommended material is some form of bituminous paving, other acceptable choices are concrete and soil cement. Bases and subbases should be adequately prepared to support the surfacing material. Highly organic topsoil should be removed. Base material may consist of crushed stone or properly compacted selected fill material. The paved bicycle trail is very similar to a road and has the same requirements and solutions for drainage.

Equestrian or horse trails, like hiking trails, can vary in acceptable width. The recommended minimum tread width is 1 m (3 ft). Where terrain permits and traffic justifies it, horse trails could be as much as 4 m (14 ft) wide. Allowable gradient should depend largely on the type and condition of the soil (erosion potential and safeness of footing) and the length of the grade.

Overhead clearance for horse trails should be at least 3 m (10 ft). A clearing width of 1 m (3 ft) beyond the tread of the trail is recommended. Selected fill material as well as wood chips, shavings, and sawdust may be used on horse trails.

Another feature recommended for horse trails is the "staging area." One such area should be located at the trail head and at each intersection of a trail and a road where people are likely to be starting or ending their riding. A staging area should contain off-road parking for vehicles, a small corral, a hitching rail, and water trough for horses, and a drinking foundation and comfort station for riders.

Off-road vehicles (ORVs) are very popular in the United States as recreation vehicles, but public recreation authorities have been slow to provide the trails to accommodate them. The two types of ORVs (off-road vehicles) are short wheel base 4-wheel drive vehicles (such as Jeeps) and certain types of motorcycles (trail bikes).

The Jeep type of vehicle trail must have a minimum radius of curvature of 5.5 m (18 ft). If they can obtain the necessary traction these vehicles can climb very steep grades. A recommended practical limit is 60 percent but for very short distances steeper grades can be permitted, depending on the traction available. On these grades firm soil or rock is desirable, not only for vehicle traction but for erosion protection as well. An occasional fording of a stream is permissible provided good traction can be maintained. Required trail width is 1.9 m (75 in). Vertical clearance required is 2.0 m (77 in) and ground clearance of these vehicles is 16 cm (6.4 in). Trail improvement needed aside from clearing should be minimal.

Maximum permissible grade for trail bikes is 55 percent where trails are located on firm soil or rock. This type of terrain is desirable throughout trail bike trails. Loose rock or fragile soils are generally less suitable. Chronically wet areas must be avoided for trail bikes. Trails should have a minimum radius of curvature of 1.8 m (6 ft), although occasional turns of shorter radius are acceptable. The minimum tread of trail is 0.6 m (2 ft), with a desirable clearing width of 0.6 m (2ft) on each side of the tread. Vertical clearance of 2.0 m (77 in) is required.

Visitor Facilities

Before considering the type and extent of visitor facilities to be built at a project, another question must first be answered. Why have any? The reason is that the owner senses either a benefit from public relations or an obligation to the public. A private entity would naturally wish to inform the public of the need for and value of the project as well as the effort required to build it. A public agency, in addition to the foregoing, would feel an obligation to the public for whose best interests it functions.

Visitor facilities could vary from a simple roadside turnout with a sign to an elaborate visitor center complex and guided tours. The extent of the visitor facilities appropriate at a project depends on a number of factors in addition to the effort the owner wishes to make. These are the public's interest in the project, the convenience of access as it relates to the magnitude of a population, and the size of the project.

The facilities could provide a view of a project's major features, the project purposes, a description of the project, the story of how the project was or is being built, tours of major features, a description of the area, and the area's history. For something a little more substantial than a simple sign, there could be a sunshade containing a map display and a little descriptive material. A comfort station could be included. An information center with rest rooms and an observation deck would be a larger facility. A really elaborate visitor complex may be justified for a project that is monumental. Such a complex could include an exhibit building with information desk, an auditorium, food and refreshment facilities, an observation deck, and guided tours of the dam, powerplant, pumping plant, switchyard, spillways and outlet works, fish passage facilities, fish hatchery, or any other appropriate features.

The process of planning and designing visitor facilities requires a certain sequence of steps. An estimate must be made of the expected number of visitor days/year. The best possible vista point commensurate with the investment to be made must be selected. Then it must be decided what information will be presented and how it will be presented. Finally, the site and the physical plant must be designed. There is one pitfall of which the designer of an elaborate visitor complex should be aware. He can easily lose sight of the fact that seeing the project features is the primary reason for a visit.

Estimating the future number of visitor days/year is best done by a group effort. Examination of influencing factors in comparison with other projects and their visitor counts appears to be a good approach. Influencing factors include the beauty of the area after completion of the project, impact of the project on the public during construction, proximity of population centers and major highways, future population growth, extent of recreation development, and publicity given to the project (during the construction of a very large project there can be public requests for hundreds of speeches). It is typical for visitor days/year to be greater during construction and for the first year or two after completion than the normal volume in later years. If temporary facilities are constructed initially, visitation records will be valuable in projecting future use.

Within the basic audio and visual methods of communicating information a great variety of possibilities exist. Signs, brochures, and maps are the standard techniques. Others include photos (during both construction and operation), viewing telescopes, geologic samples, exhibits of construction equipment and materials, historic artifacts displaced during construction, automated slide presentations, and relief maps or topographic models. More elaborate techniques could include motion pictures, working and other types of three-dimensional models, and regularly scheduled lectures. Tours of facilities themselves could be conducted by tour guides or could be self-guided.

Considerable latitude is often given in developing the basic architectural presentation of a visitor center, yet it must be compatible with the setting even if it is merely a simple sunshade. Temporary structures should be located so that they will not interfere with future permanent facilities. Once anticipated visitation, site selection, and displays and interpretative data have been established, design of the structure is straightforward. It should consider a drainage plan, parking, utilities, traffic flow patterns, pathways, trails, observation points, and landscaping.

Relocations

Major project costs may be incurred in relocating existing developments. The process required to accomplish relocations was discussed in Section 11-3 but some planning and design considerations are noted here. While not actually relocation work, modification and protection of existing developments is also discussed here. Most often the relocations involve roads or railroads, which must be realigned to avoid the new reservoir or to cross it.

Although a reservoir may be crossed by ferry service or by tunnel, the common solution is a bridge. Ferry service is impractical where water surface fluctuations are significant, and tunnels are very costly. As can be seen in Fig. 11-4-1, some of the bridges required can be major structures indeed. It may be necessary to design

Figure 11-4-1. A large bridge required for a road relocation. The maximum normal water surface of the future reservoir will be close to the top of the concrete piers. In the lower foreground is the old road and bridge. (*Courtesy U.S. Bureau of Reclamation*)

bridge piers and abutments for static and dynamic stability under various reservoir conditions.

It has been a common practice in the past to combine a necessary crossing with the dam itself by locating the roadway on the crest of the dam. As a general rule this is a poor solution. Although initial savings in cost may be substantial, future operation and maintenance difficulties due to the presence of a public road may also be substantial. This solution can be especially undesirable when the dam is a structure visited by large numbers of tourists and the road is a regional or national highway.

These difficulties can be avoided if it is possible to somehow separate through traffic from tourists and operation and maintenance activities. For an earth or rockfill dam it may be feasible to design the road as an elevated structure whose piers rest on the sloping downstream face of the dam. The solution for a concrete dam may be to elevate the road above the crest of the dam as a viaduct. It would be necessary to design these elevated structures for seismic loads transmitted through the dam. It could also be necessary to study possible changes in prevailing wind currents caused by the dam and to alleviate traffic crosswinds on the new road.

In relocating roads or railroads out of a reservoir it is good practice to keep the toes of fill slopes above the maximum reservoir water surface. In those cases where this is not practical, fills should be constructed as impervious embankments or designed for stability under saturated conditions. Protection against wave action must be provided.

The scheduling of road and rail relocations during project construction can be important. When timing is not critical, such work can be scheduled to promote the orderly and efficient prosecution of the project work in terms of manpower and funding requirements. Sometimes, however, the timing can be critical. A relocated road may be needed early during construction because it will provide construction access, or because the abandoned segment of road may be needed for the exclusive use of construction traffic. A road or railroad may run right through the dam site. Occasionally, an existing road eventually to be relocated may be subject to potential inundation due to temporary impoundment behind a construction cofferdam during high river flows. If uninterrupted use of the road is essential, then the relocation must be completed before temporary impoundment can occur.

Relocations of utilities are also often necessary. Viable alternatives may include burial within the future reservoir site or routing across the dam, as well as circumventing or bridging the reservoir. Uplift stability may need to be checked for pipelines buried within a reservoir site. In a more unusual case, manmade islands were constructed within a reservoir site and transmission line towers were erected on them. Although this was economically the most feasible solution for the required transmission line relocation, it was aesthetically undesirable.

There may be existing facilities that will be within or adjacent to the reservoir which are to remain in use, for which modification or protection will be required. Embankment slopes may need to be riprapped to provide shoreline protection. Guniting or shotcreting may provide the needed protection. Retaining walls may be necessary. Jetties are usually impractical due to the relatively steep slopes of reservoirs. Vegetation is usually ineffective protection against shoreline erosion.

An existing tunnel for water conveyance may need to continue in service after completion of the project. Such a structure demands careful reanalysis for the new loading conditions to which it will be subjected. If the tunnel is a gravity flow structure, at least a portion of it nearest the reservoir will have to be redesigned. A new intake structure and a gate or valve control may be required. It may be necessary to do pressure grouting.

It is sometimes necessary to protect some natural landmark or historic site from inundation. Occasionally, such protection may also be decided upon for a community, rather than carrying out a relocation. The resulting dike will be a barrier not only to reservoir waters but also to waters collecting on the other side. Pumping facilities will be required to remove reservoir seepage, natural flows, and community effluents.

Miscellaneous Service Features

During construction of a project, some form of office space and support facilities will be needed by the owner or his representative. The extent and nature of these facilities can vary greatly with the circumstances. Principal factors are the size of the project and the remoteness of the site. The scheduled rate of construction, number of contractors, site conditions, area encompassed by the facilities being constructed, climate, and responsibilities of the supervising entity are other influencing factors.

For a small project with only one contractor, close to an urban area, a typical construction trailer or larger camping trailer may suffice. At a very large, remotely located project the facilities required may include several buildings for offices, laboratories, storage, equipment servicing, and housing. The contractors would be facing many of the same requirements, and with these circumstances, contract provisions would usually require the construction of a camp containing all the needs of a small community.

Where permanent operation and maintenance facilities will be required, construction of at least some of these facilities initially will minimize the cost of temporary facilities. Perhaps permanent site preparation and installation of utilities can be accomplished but only temporary buildings can be erected. Sometimes a site can be prepared for some future permanent use, but can be utilized in the interim by temporary structures. Occasionally, facilities required for use during construction but not permanently needed, can be so located and designed that their later acquisition and use by other interests can be considered.

At a very large project the owner or his consultant may need facilities for some or all of the following:

A right-of-way staff to make several hundred property acquisitions and relocate displaced persons.

Contract administration personnel to administer as many as 100 contracts; staff to prepare engineering data, do supplemental design work, and develop construction planning and coordination.

Field personnel, including geologists, inspectors, and surveyors.

A materials control laboratory and staff.

Administrative services, including office services, property and supply, garage, photography, and guards. If at a camp a manager, maintenance personnel, and a staff to provide community services may be needed.

Besides the Project Construction Engineer and assistants to handle programming, safety, and environmental control, there could be drill crews and archaeologists.

Modern dams and power plants are often operated from control centers at some other location, especially if the development is one unit in a system of dams and power plants. At isolated places, however, it may be necessary to have a permanent staff. When a permanent community will be required, some of the facilities and the services at construction camps should be permanent construction.

At some reservoir sites it may be necessary to do some channelization or grading to prevent the creation of backwater ponds when the reservoir is drawn down. This is desirable in order to discourage the growth of water borne vectors such as mosquitoes. It also would be beneficial in preventing fish from becoming stranded.

Fencing is an unglamorous subject that is not usually thought of in terms of design. The fact is that a great deal of fencing is used at every dam-reservoir project. Proper selection of the type and dimensions depends on its purpose. Fencing is used for safety (of the public and workmen), for land and wildlife management, to assure unimpeded operations of facilities, and as a security measure.

One standard measure taken to deter floating debris from interfering with the operation of spillways, penstocks, and outlet works at dams is the emplacement of a log boom just upstream from the dam or intake. Log booms derive their name from the fact that they have traditionally been made from timber available as the result of reservoir clearing operations. They can, however, be constructed of other materials. In fact, the apparently inexpensive solution of using logs may not be so inexpensive when operation and maintenance costs are considered. As the logs become waterlogged, they tend to submerge and become ineffective. Periodic removal from the water for inspections, drying out, replacement, and other maintenance can be cumbersome.

One alternative to wood in "log" booms uses tunnel ventilation pipe. Each length of pipe is filled with styrofoam and sealed at each end. Effective centered swivel connections are easy to make. The booms are relatively easy to emplace or remove from the water and do not become waterlogged. They are spray painted (with highly visible colors) and can remain in service for longer intervals of time than the wooden type of log boom.

There are a few key considerations in designing log booms. The tendency of individual segments to roll as the result of wave action may make the use of some type of swivel connection desirable. Anchorages must be provided which will satisfactorily resist the forces that may be transmitted to them by the boom; these are primarily wind and wave drag, although other loadings are conceivable. To avoid the possibility of the boom becoming hung up and allowing flotsam to slip past it when the reservoir is being drawn down, shorter lengths of segments should be used near shore.

The need for recording the seismic history of a reservoir site and the capability required of the equipment was discussed in Section 11-2, Structural Integrity and Adequacy. The environment demanded by the equipment, its size and weight, and the utilities required must be known in order to design the necessary housing. For this information the engineer must depend on the seismologist.

Although the location generally should be determined in relation to other stations in the network and the overall topography of the reservoir area, the site geology, availability of power and telephone utilities, and accessibility of the site are also important. If the seismologist determines that the recording instruments should be founded on rock attached to the underlying rock mass, then a site with a satisfactory rock formation at or nearer to the surface is most desirable. An independent power source (batteries) and radio transmission of recorded motion, both costly, are the only alternatives if the required utilities are not available. The site must be accessible but should not be so easily accessible that it attracts unwanted visitors.

The design of one seismograph installation is discussed as an example. The station has a 4-component system with 3 sensitive seismometers (gains of 55,000 to 77,000) and one strong motion seismometer (gain of 2,200). Because the selected site is 610 m (2000 ft) from the nearest telephone and powerlines, it was decided to separate the transmitting facilities (which require the utilities) from the seismometers which are the actual pick-up units. The transmitter is located close the the utility lines and is connected to the seismometers by a buried cable. This was less costly than extending the utilities out to the seismometer site.

Since the area will be developed for recreation, the seismometers and the transmitter are in underground vaults. The transmitter is on a hillside and has a side entrance as shown in Fig. 11-4-2. Its interior dimensions are 2.44 × 2.44 × 2.44 m (8.0 × 8.0 × 8.0 ft). The seismometer vault is on the spur of a ridge and can be entered through a hatch. Interior dimensions of this vault are 2.44 × 2.74 × 2.44 m (8.0 × 9.0 × 8.0 ft). Each vault is oriented N-S, E-W.

The seismometers are set on a 1.22 × 1.22 m (4.0 × 4.0 ft) pier in the vault floor with 61 cm

Figure 11-4-2. Entrance to a seismograph vault located on a hillside in a rural area. The doorway is out of sight along the wall containing the utility meter. Note the safety fencing. (*Courtesy U.S. Bureau of Reclamation*)

(24 in) clearance on three sides, projecting above the floor by 30 cm (12 in). Waterproof mastic separates the pier from the vault floor. The pier is supported on four 20 cm (8 in) diameter columns extending 12.2 m. (40 ft) to competent rock. Both the pier and its columns are of unreinforced concrete.

Other miscellaneous service features include stream gaging stations, theodolite piers for the precise measurements of the dam required in the monitoring of structural behavior, and weather stations. Stream flow and weather data are useful in project operations, as well as before and during construction. Although the watershed presumably would have a number of stream gaging stations in operation for many years prior to project construction, the station referred to here would be constructed immediately below the dam site.

Other Project Features

Power plants and pumping plants can be located upstream from, on, or downstream from a reservoir. Canals, pipelines, and tunnels for water conveyance can lead to or from a reservoir. The discharge or intake points for these structures must be well marked and separated from recreationers both in the water and on land for safety reasons. They may need to be protected for security reasons. Wildlife safety should also be considered. Intakes to such structures will probably require protection from floating debris in the reservoir. In very steep canyons, buildings and valve houses, etc. may be subjected to falling rocks or snowslides.

Electric substations, switchyards, and transmission lines must be separated from the public and from wildlife. Options sometimes are available in the siting of these facilities, which can be exercised to help achieve the needed separation. Environmental considerations, discussed in Section 11-6, also are compatible with separation from the public.

In at least one case a series of wells was required at a reservoir. They were installed to lower a saline water table existing at the site, in order to prevent contamination of reservoir waters.

For mitigation of adverse effects or for enhancement, some kind of fishery facilities may be part of a project. The purpose could be to build up a reservoir fishery for commerce or sport; or, it could be to deal with spawning beds lost to fish by construction of a dam. Facilities could consist of various kinds of screens, fish passageways such as "ladders," or hatcheries. Simply trapping migrating fish and transporting them around the dam in a tank truck could be the most practical solution. Other solutions can easily become rather exotic and quite expensive. The design of fishery facilities is not a routine problem. For the greatest chance of success, the fisheries biologist and the engineer together must develop the needed solution.

Virtually every dam and reservoir project requires the construction of access roads. For a project at a remote, rugged site, access can be a very significant factor in terms of both money and time. In order to create the most satisfactory design, the engineer must consider both the initial and subsequent use of an access road. Will it be used only during construction? Will it be limited to project use only? Will the public travel on it? Will it become part of the local, regional, or national road system?

If an access road will be heavily used by large off-highway construction equipment initially, but have a later continuing purpose, it may be advisable not to lay the base course and surfacing until later. Another solution could be to place a temporary layer of material over the finished road on a short segment subjected to such use.

Maximum loads must be anticipated. Wheel loadings not only are a design factor, but allowable wheel loads should be mentioned in other specifications for construction in which the road may be used. In addition to the road section, maximum loads may influence the minimum radius of curvature, maximum grade, and design of intersections. Where overpasses and/or tunnels are involved, minimum clearance requirements become important.

Planning ahead, the designer may provide for future intersections. An access road through a future recreation area may be laid out for eventual use as an interior recreation road. If the design criteria for future use are incompatible with present temporary requirements, at least sufficient right-of-way may be acquired and utilities relocation cognizant of future use may be carried out.

Before designing an access road, it could be important to know if the road will have to be passable under any weather condition. If a road will only be needed during construction seasons, a river ford may be adequate, but if the road will be needed during a potentially major flood, it may be necessary to build a major bridge. A permanent road that will be the sole means of access to a dam must be designed and aligned so it would not be flooded out if the spillway operates at its designed capacity.

11-5 RESERVOIR CLEARING

This may be that part of the project with the greatest impact on the general public during construction. People dislike seeing trees destroyed. Yet there are mitigating possibilities. For those sites in forested areas, conservation and utilization measures of the clearing operations can be emphasized. Although these measures have in the past usually meant added expense, this should not automatically be assumed to be the case for a new project. Whether the motivation is public relations or genuine conservation, such measures warrant consideration.

Aerial photography can be a most useful tool in planning for clearing work. Newer developments in remote sensing can enable identification of the types and condition of vegetation, as well as ground conditions. Reservoir clearing, however, is seldom carefully considered by the designer. All too often the only early thoughts on the subject are for cost estimating purposes. One major reason for this lack of consideration is that clearing operations usually are scheduled late in the construction sequence, particularly for larger projects. When the subject is finally given detailed attention it may be competing with the imminent possibility of project cost overruns and a pressing need to economize.

Factors Influencing Clearing

The first decision which must be made is whether or not to clear at all. A divergence of views among interests exists on the subject. This becomes more significant when a project is multipurpose rather than single purpose. If a reservoir is to be solely for municipal water supply, for example, water quality may be the overriding consideration. In contrast, power, flood control, irrigation, fish and wildlife, navigation, silt detention, and recreation purposes each may lend themselves to different viewpoints. At one extreme there may be a requirement to not only clear a reservoir but to completely grub and rake it as well. At the other extreme it may be desirable to not clear anything at all.

Power generation needs a substantial minimum pool elevation for minimum head, and requires adequate protection and unimpeded flow at the penstock inlets. Flood control and irrigation functions result in large fluctuations of the reservoir water surface elevation. Navigation and water-oriented recreation uses including boating, swimming, and water skiing require removal of all obstructions. A large dead storage volume may be provided for silt detention. Effect on future operation and maintenance costs must be an important consideration in all clearing decisions.

Efforts to encourage fish and wildlife populations at large reservoirs usually include the retention of vegetation. The retained vegetation provides protection or cover for fish against their predators. During those times when a reservoir is not full, exposed vegetation will give covered access to wildlife wanting to reach the water. One vegetation retention measure is to leave selected areas of a reservoir uncleared. This can be quite unsightly and can dampen the attractiveness of boat fishing. A compromise is sometimes achieved by clearing only the trees but leaving the brush. For this purpose a practical definition of a tree could be: "Any vegetation 5 cm (2 in) or more in diameter at a point 15 cm (6 in) above the gound on the uphill side, and 2 m (6 ft) or greater in height." In the case of at least one large reservoir felled trees were actually anchored in place by chaining them to tree stumps. A variation of this is to anchor down piles of cleared brush for fish and bird cover. Obviously these measures are somewhat experimental and can be very expensive.

A number of other factors in addition to project purpose influence reservoir clearing. These can be grouped under scale of project and accessibility, nature of cover, configuration of work zone, environmental restraints, and disposal of materials.

For a large project, reservoir clearing work is usually accomplished separately from the prime construction contract. Any clearing work required at construction sites themselves may be the responsibility of the prime contractor.

He may then have a subcontractor handle the clearing. The more difficult and extensive the clearing work is, and the larger the prime contractor, the greater seems to be the likelihood that a subcontractor will be engaged. If the site for a major structure requires extensive investigation for design data, it may be necessary to award a special contract to clear the site. This can also be very useful for accurately determining original ground surfaces by aerial photography, for making surveying work easier, and for enabling more accurate cost estimates of the future clearing work in the reservoir. Of course the work may incur disadvantages associated with having clearing done too early, such as erosion and regrowth.

Most clearing contractors are relatively small businesses in comparison with large construction companies. A case where a reservoir to be cleared is located within a forested stand of commercial timber could, however, attract a large forest products firm. Nevertheless, clearing work for larger reservoirs usually lends itself to division into more than one contract. Economy can often result by issuing one bid invitation and specifications with multiple bid schedules. Each schedule would be based on a geographical segment of the total area to be cleared. In this way, a contractor would be free to bid on as little as one schedule or on the entire invitation. Award of a single contract or a combination of contracts can then be based on lowest resulting combined cost to the owner. Dividing the work into separate schedules based on elevation will seldom lead to efficient clearing operations. Only in the case of a reservoir which is expected to fill gradually or in steps over a long period of time would this be a recommended procedure.

At larger reservoirs, zones of clearing may be established rather than planning to clear the entire site. Such zones refer to significant levels of water surface elevation associated with the future operation of the facilities. In general, the zone to be cleared is defined by the limits of anticipated reservoir fluctuation. The upper limit may be determined by the maximum water surface elevation anticipated during normal or routine operations, or by the limits which would be reached by the occurrence of the most severe flood expected (excluding extremely rare occurrences). It would be unusual to clear a reservoir all the way up to a level determined by the spillway design flood (the routing of which may include surcharge space). A lower limit for zoned clearing may be related to minimum power pool (inactive storage) elevation or to dead storage elevation. Allowance for backwater effect toward the inlet end of the reservoir is sometimes considered. It is good practice to make allowance for wave action and shoreline erosion by adding some increment to the water surface elevation being used to determine the upper limit. For example, if the maximum normal anticipated water surface elevation is to be 30.5 m (98.4 ft) a clearing limit of 32.0 m (105 ft) may be chosen. In establishing the lower limit of a clearing zone, an extension should be made. If, for example, it is decided that no vegetation should be allowed to remain above elevation 8.5 m (27.9 ft), then clearing down to elevation 5.0 m (16.4 ft) could be required. This allows for the height of the brush. In addition it may be necessary to provide in the specifications a requirement for the topping of all trees extending above the lower limit.

For smaller reservoirs, full clearing is likely to be practical as well as desirable. Reservoirs with extreme drawdown (such as certain pumped storage schemes) will require full clearing. Clearing is an obvious necessity for those offstream sites where all sides and the floor are to be shaped. Where a reservoir is to have a lining, stripping of unsuitable material will probably be required as well as clearing.

Any work schedule meant to be realistic must consider prevailing climatic conditions. The climate is an influence on reservoir clearing. It is a determinant of the nature of vegetation. Working conditions and productivity are affected by it. The climate can control access to the work areas due to rain, snow, frozen rivers, etc. If cleared vegetation is to be disposed of by burning, atmospheric conditions such as wind, temperature, and humidity may control when the work will be performed.

One aspect of accessibility is the degree of remoteness of the site from any towns or cities. If a reservoir site is so remote that a construction camp is required this will be a cost increasing factor. On the other hand, remoteness from populated areas will tend to decrease the need for and the cost of air and noise pollution control measures. Site remoteness will generally be more detracting to smaller clearing contractors.

Accessibility may be difficult or restricted because of topography or, infrequently, right-of-way limitations. This difficulty can be affected by the selection of schedules; in other words, the divisions of the clearing work. Topography can be the deciding influence in the selection of the clearing method to be used. The more rugged and steep-sloped the site is, the less desirable will be the use of heavy equipment. It may dictate the use of some type of barge or floating operation. The use of helicopters is conceivable. A shallow, basin-shaped reservoir site may be perfect for a completely mechanized clearing operation.

Reservoir sites are possible virtually anywhere throughout nature's fully diverse geography, from arctic tundra to tropical jungle. The vegetation and the soil cover existing at the site are important factors in developing clearing plans.

It is easy enough to envision arid or alpine sites where no clearing at all would be required. Sites have been located in dense forests or jungles, where clearing is a major expense. The character of the vegetation is, then, a most significant factor. In using the term "character," we include species occurring, distribution, density, proportions of the mix, and even the vigor of its growth. There could be sufficient, accessible vegetation of commercial value to warrant considering sales. Whether trees are hardwood or softwood could influence the selection of the cutting process. The means of disposal selected could be influenced by the character of the vegetation. Certain species of vegetation impart undesirable qualities to water if left to rot. If significant quantities of such vegetative types were present, they would have to be removed, including stumps and roots. The distribution and density of vegetation may influence the method of clearing chosen. Some types of brush, for example, will simply lay over for a bulldozer blade and then spring back up afterward. Decisions on when clearing is to be done may be affected by regrowth capability.

Soil cover could be substantial or very thin, sandy, clayey, or rocky. A soil that is highly susceptible to erosion could, in concert with climatic and topographic conditions, require a method of clearing and disposal that minimizes disturbance of the soil. Clearing operations may need to be scheduled just prior to reservoir filling in order to prevent erosion of the bared ground during rainy seasons. If conditions are such that hand clearing could make the difference between soil stability and instability, then that method should be required. Where depth of cover is thin, slopes are steep, and the soil will obviously erode or slough off, a method which uproots trees will be desirable. It then will be unnecessary to dispose of a lot of stumps later on when soil erosion and decay frees them.

Configuration of the work zone is determined by topography and by the division of the work. This configuration obviously influences clearing costs. A discontinuous work zone may in effect impose two separate operations involving either more men and equipment or additional moving and setting up. Depending on the degree of barrier involved, a river coursing through a clearing work zone could make the zone discontinuous. Where a large reservoir site contains two or more distinctly different parts, they should be separated in the clearing schedule. An example of this could be where a dam site and a portion of the reservoir are in a river canyon and the remainder of the reservoir will be in a wide basin upstream from the canyon.

Another factor influencing reservoir clearing

is environmental considerations and restraints. Air pollution, aesthetics, fire safety, noise pollution, population, regional and local economy, and water quality are some of these. Various local, regional, and national regulations may require compliance, including permits or granting of variances.

Air pollution resulting from clearing work consists of dust and smoke. It has been contended that wood smoke is not a pollutant since fire is a naturally occurring phenomenon in the forest cycle. Nevertheless, most people would consider smoke from a clearing operation to be a pollutant.

Dust is no stranger to the heavy construction scene. It is associated with the use of heavy equipment. It is controlled by more or less standard measures such as treatment of haul road and work area surfaces. Unfortunately, this is not very practical for clearing work, where the work site itself typically covers such a widespread area. Dust, however, is only objectionable in proximity to populated areas.

As a product of burning, smoke is obviously not a problem where the method of clearing and disposal does not involve burning. However, smoke need not be a problem even where burning is involved. Dirt-clogged stumps and roots are smoke producers. If clearing of stumps and roots is required for specific reasons, they can be disposed of by other means such as burial and/or removal. Shaking and restacking burn piles two or more times helps to knock dirt off. Subjecting the burnable material to water can help. This can be accomplished by letting the material lay through a rainy season before burning, or by floating it in the collection process (this can have other drawbacks).

A common practice employed in burning is to use old rubber tires in each burn pile. Once it has started burning, rubber makes a relatively hot fire. However, this is not a recommended practice. Burning rubber gives off large quantities of smoke, and this is a sign of incomplete combustion. Other means exist to create hot fires in which combustion will be more nearly complete. The essential ingredient in the combustion process is an abundant supply of oxygen. By providing plenty of air and some containment of heat, a virtually smokeless fire can be attained.

One of the aesthetically detractive appearances at reservoirs, which people frequently mention, is the "bathtub ring" effect created when the reservoir water surface is below the upper limit line of clearing. This is most pronounced where nearby vegetation is plentiful and reservoir drawdown is substantial. Not much can be done about this. The sharpness of the line demarking the upper limit of clearing can sometimes be made less pronounced by making it somewhat irregular. Clearing for recreational development can help accomplish this. Establishing a zone of selective clearing may also help. Species of brush or trees known to be tolerant of occasional inundation may be left growing in the upper portions of cleared zones. These efforts will be fruitless where slopes are steep, soil cover is thin, and wave-induced erosion occurs. The author's viewpoint, admittedly subjective, is that a reservoir with a lot of exposed dead timber and brush standing in it looks much worse than a completely cleared ring. If recreation use of the reservoir is planned, it will be desirable to integrate recreational planning with access routes for the clearing contractors. Otherwise, access roads for clearing should be restricted to the zones to be cleared insofar as practical.

Any major construction project should have a fire safety program. Where vegetative clearing is required and burning may be the means of disposal, fire safety consideration is imperative. Local fire districts may, in fact, insist on some kind of formal agreement that includes a thoroughly detailed program. Firebreaks are usually necessary. There could be requirements controlling the location, shape, and size of piles of the material to be burned. Perhaps burning will be permitted only at certain times of the year, and when the material has aged a specified minimum length of time. Such requirements must, of course, be spelled out in the contract specifications. An adequate fire safety program could involve more equipment, and even more manpower. Local or area governments may require the contractor to obtain burn permits.

Noise pollution usually becomes a problem only with an adjacent populated area. Sources are heavy equipment (including mechanical chippers) and chain saws. If sale of commercial timber is involved or if disposal is to be off-site, truck traffic may also be a noise contributor. Whatever noise abatement devices are available for today's equipment could be used and work could be scheduled only during the least disturbing hours.

The influence of the presence of a sizable local population on reservoir clearing has already been mentioned. But a populated area can influence matters other than air and noise pollution control. In some parts of the world it is the practice to lease or rent reservoir lands subject to periodic inundation for farming, when compatible with other requirements. In such circumstances some clearing could be left up to the individual farmer. In concert with regional and local economic conditions a nearby population could be the source of abundant and cheap manpower for hand clearing methods. Such a decision would also be encouraged by limited availability of capital with which to purchase expensive equipment. A local population may also be a potential market for firewood.

Concern for water quality will also be an environmental restraint influencing reservoir clearing decisions. Water quality as it relates to clearing can be affected by floating debris, erosion, and rotting vegetation.

Uncleared inundated lands can be a source of debris, floating or fixed, for many years. The problems of debris with recreation use are obvious. Even if the reservoir is not to be used directly for recreation, floating debris can be a major continuing operating and maintenance expense. Trashracks at the intakes of outlet works, power plant penstocks, and diversion works are usually installed to protect gates, passageways, turbines, and valves against the hazards of floating material. Periodically submerged brush and trees gradually deteriorate and break free. Vegetation within the zone of fluctuating water surface becomes freed by soil erosion due to wave wash. Floating debris also results when disposal of cleared vegetation is incomplete or improper. Even a well planned and executed clearing program will contribute some floating debris with the first inundation of the cleared areas.

Unnecessary erosion can result from poorly done clearing. Ideally, clearing work should be accomplished just prior to filling of the reservoir. This minimizes water turbidity due to rainfall runoff from denuded slopes. The rising water level eliminates this problem. If reservoir lands are highly susceptible to erosion, as mentioned in discussing soil cover earlier, proper clearing can minimize the problem. Temporary plantings of grasses can sometimes prevent the unwanted water pollution.

As discussed when talking of the influence of vegetation, certain species affect water quality when decomposing. For example, decaying vegetation rich in tannins, such as live oak, cause strong taste and odor problems. Where a municipal water supply reservoir has substantial growths of such species, complete removal by grubbing and raking even the roots may be necessary. It has been recommended for certain reservoirs that grass and other herbage be mowed and removed just prior to inundation. Stripping all vegetation, however, releases bottom sediments and nutrients into the waters.

Yet another factor influencing reservoir clearing is the means of disposing of the material to be cleared. From a conservation standpoint it would be desirable to make some use of all cleared materials. This, however, is simply impractical in most cases. Disposal possibilities are basically to bury, burn, remove, or just "let it lay."

Combustible materials, particularly stumps, are sometimes buried within a reservoir area. This is usually practical only when there is a very limited quantity of materials to be disposed of in this way, and a large fill (perhaps a waste area) is to be placed within the reservoir. Possibly the need to contour a borrow area would provide another burial site. In any case, the principal concern is to be sure that buried materials stay buried and do not work their way free to float and become operation and maintenance hazards. Typical requirements are for a minimum 5 ft of cover.

By far the most commonly used means of disposal is to burn. In planning a burn operation, attention must be given to fire safety and air pollution. These matters have already been discussed. The simplest approach would be just to create suitable firebreaks and burn the material in place as it stands. Such a procedure would most likely be unsatisfactory both with regard to air polluting smoke and incomplete combustion of the burnable materials. Felling the materials before burning (and perhaps even crushing them or chopping them up) would be some improvement but would still be a smokey operation. It would tend to leave considerable quantities of slash after burning. Use of this means of disposal would be subject to the restraints associated with highly mechanized clearing methods. In a rain forest or jungle setting this could be a very economical operation. A further improvement would be to collect and deck the materials before burning. With the materials thoroughly dried out and with the site topography providing for some heat retention and a chimney effect, this can be a satisfactory operation. A still further improvement would be to introduce forced air into the burn piles. Use of a furnace appears to be the final step, with the best fire safety, virtually no smoke (except at start-up), and possibly the highest cost.

The third means of disposal mentioned is removal of the materials from the area to be cleared. This, of course, means there must be some other site or uses for the materials. Quite often this results in removal of only some of the materials, leaving the remainder to be buried or burned. Selling marketable timber is a long recognized possibility. Income from such sales is sometimes greater than the total clearing costs. Selling or even giving away firewood is often possible. Having the general public enter a project area is usually undesirable from a liability standpoint. It could require additional personnel to guide and control their activities. One alternative would be to bring the firewood to a suitable location at the edge of the construction area and offer it there. Another alternative would be to contract with local woodcutters, who in turn would sell and deliver to customers. Obviously, using cleared materials for firewood is only reasonable if there are enough nearby users. Other unusual and more specialized marketing possibilities can occasionally occur. These are likely to depend on the species of vegetation. At one project, for example, the stumps and roots of black walnut trees were sold.

Chipping or shredding is usually included in discussing disposal, although it is actually a form of processing. The processed material must still be removed from the area to be inundated. A chipping operation is shown in Fig. 11-5-1. Note the type of vegetation in the general area. The chips could be used as a mulch spread within the project area surrounding the reservoir. It would be important to study the possible effect on water quality before proceeding with this action. There are important commercial uses for wood chips. Such applications may require selective chipping. Debarking the material may be necessary before chipping. The wood chips desired may only be from certain species of trees. Wood chips are used in the paper industry. In an experimental application wood chips were used in a mix with garbage as a fuel for a thermal powerplant. Finally, with respect to chipping, piled chips can become susceptible to spontaneous combustion; therefore, they require special attention.

Clearing Methods

Clearing operations include four basic steps. These are felling, collection, disposal, and cleanup. Normally the fourth step, cleanup, is per-

Figure 11-5-1. A chipping operation in clearing a reservoir. Note the relatively level terrain and the sparseness of trees. (*Courtesy U.S. Bureau of Reclamation*)

formed by operations and maintenance forces after the project has changed over from construction status. Clearing by hand is the least disturbing method but is also probably the most expensive. Using heavy equipment will usually get the work done faster. Various methods and adaptations of methods have been developed, sometimes with surprising originality. In this discussion of methods no claim of all-inclusiveness is made.

Basically, tree felling can be either by cutting or by uprooting. In modern thinking, "hand clearing" has come to include tree felling using hand-held gasoline-powered chain saws. The use of heavy equipment for tree felling usually means specially rigged bulldozers.

For the cutting method a bulldozer may have a blade with a cutting edge rigged at a horizontal plane angle with the front of the bulldozer. Usually the blade will be "V" shaped. The cutting edge may be serrated. Other techniques have been developed to assist in the cutting process. Varieties of "stingers" (a heavy steel spear attached to the front of the bulldozer in place of the blade) have been used to spear trees, thus splitting them and making them easier to cut. Another technique is to mount some type of bar arrangement as high up on the front of the bulldozer as practical. The bar must be a little forward of the cutting blade. The tree surface exposed to the blade is thus in as much tension as possible, making it easier to cut. The cutting method works best with larger trees, although it works well with any tree sufficiently rooted that the stump will stay in place. Trees should be cut so the stumps remaining are no more than 15 cm (6 in) in height on the uphill side.

Trees may be uprooted either by pushing against them or by pulling them down. A conventionally rigged bulldozer can, of course, uproot a tree, but special rigging which exerts the dozer's force high up on a tree will do the same work much faster. Sometimes a root cutting attachment is used for greater effectiveness. Generally, this method of felling is most efficient for medium-sized trees.

Trees and some brush can be pulled down by dragging a heavy anchor chain between two tractors. Swivel links used with the chain control twisting, and one or more heavy balls near the center will control the tendency of the chain to climb. This method is very efficient where the land is relatively flat and sandy, with few rock outcrops. Trees must be shallow rooted and vegetation not too dense. Substituting a cable for the chain provides the flexibility and convenience of using winches on the tractors. This may result in higher maintenance costs and greater tendency of the cable to climb over brush and small trees.

Another more unusual variation of the chain method was employed at one project. It works well on sloping forested land and is particularly suited for terrain with rock outcrops and recently logged land with many well-anchored stumps. Anchor chains were connected to an axle through a 2.4 m (8 ft) diameter steel ball. Two large tractors with heavy duty winches holding up to 275 m (900 ft) of cable were used for pulling the anchor chains and ball. The chains and cables were thus held 1.2 m (4 ft) off the ground and able to pass over most outcrops and stumps. Fig. 11-5-2 shows the equipment in use. In light going, the tractors would move parallel to each other pulling the cable behind them. In heavy stands of fairly large trees or in soft ground the tractors would pay out the cable, anchor to large trees, and then winch in the cable. The distance between tractors varied with the terrain and the size of the trees. This type of clearing operation is inherently safer than other methods because, when using their winches, tractor operators can always stay well clear of the area in which trees are falling.

A variety of heavy duty rakes and plows have been developed for clearing brush. The effort is to minimize soil disturbance while efficiently clearing. When land must be grubbed, however, heavy duty rakes or several small ripper "teeth"

Figure 11-5-2. Clearing a reservoir site using a large ball and cable pulled by two large tractors. The apparatus is moving toward the right. Note the size of the trees and the density of the forest. (*Courtesy U.S. Bureau of Reclamation*)

are mounted at the rear of the bulldozer and the ground is ripped. The soil is then usually windrowed and the roots raked up.

Collection of the downed vegetation is necessary unless it is to be left or burned in place. If it is to remain in place it is sometimes crushed or chopped to facilitate decomposition or burning. Methods of collection are either land based or water based. When materials are to be sold the downed trees must be trimmed, sawed to acceptable size, and collected. Even when material is to be disposed of by chipping, it must first be collected to points convenient for the chipper.

Land based collection can be done completely by hand. This requires more work in reducing the material to sizes which can be manhandled. Burn piles must be smaller. A

DESIGN OF RESERVOIRS   657

Figure 11-5-3. Collecting downed timber with disposal by burning. A specially designed grappling device is being used by the crane. Note the blower forcing air to the fire beyond the left of the photo.

completely hand operation should be the cleanest collection method (perhaps the most expensive also). Felled timber and brush can be decked or piled by bulldozers. With special rigging they can stack materials higher. Roots and smaller slash can be collected by rear-mounted rakes.

Water based collection depends on the floatation of the downed vegetation. These methods are attractive at sites where access is difficult. If a reservoir is large and will not fill in one year, however, the work becomes complicated. Collection of downed vegetation must then be done in several different years. At a reservoir which will fill in a relatively short time, floatation should work best. For a slowly filling reservoir, however, problems can develop. One part of a downed tree can become waterlogged before enough of the tree mass is immersed to cause floatation. It may be that care in not allowing trees to fall downhill would solve the problem. Water based collection requires some way of concentrating the floating materials. Tugboats or other craft are used to put the material in coves or maneuver it around in barges. Note the boat shown in Fig. 11-5-3. The material is then either picked up for disposal or held in the coves by log booms until the water level recedes.

Disposal by burial and removal has already

been mentioned. They require only standard equipment with the exception of chipping. Mechanical chippers with their own power units are often mounted on "low boy" semitrailers. The chippers can usually be adjusted to produce chips of any of several sizes. They can discharge directly into the hauling truck, as shown in Fig. 11-5-1. Disposal by chipping and removal does require a haul road system unless a water based operation is used. In that case, a practical scheme would need large barges for conveyance of the chipped material. Chipping is a slow process. For the large volumes of materials at many sites, large numbers of mechanical chippers would be needed.

In response to concern with air pollution, considerable attention is being given to improvement of the burn method of disposal. As discussed previously, the key to complete combustion is a sufficient supply of oxygen. All improvement efforts embody this concept. Portable blowers are available to introduce forced air to burning materials. A blower can be seen in Fig. 11-5-3.

One method of burning involves the excavation of an open pit, which serves as a furnace. Heat is retained by the pit walls. A continuous curtain of air is forced across the top of the pit, deflecting downward at the far wall. Rising ash particles are carried back down into the fire. Properly done, combustion can be virtually complete and particulate emissions very low. The method requires some type of crane for continuous feeding of the fire, because the heat created prevents the use of a bulldozer.

Another method uses a self-contained air curtain furnace unit mounted on a "low boy" semitrailer. With this method air is introduced to the fire at additional places in the furnace, as well as at the top. This furnace can be fed by a rake equipped bulldozer, although a crane could be more efficient. This method can also be adapted for use on a barge.

Periodic cleanup of floating debris is a more or less routine operation and maintenance function for on-stream reservoirs. With the first filling of a reservoir, however, cleanup can be expected to be something more than routine. Provision should be made for a water surface raking operation as a construction item during that initial filling. Water raking may need to be repeated for a large reservoir which will take more than one year to fill. Sometimes debris can be collected in or restricted to arms or coves of a reservoir. Tugboats and log booms are needed for this work. It is then often possible to take advantage of later reservoir drawdown to accomplish disposal by land based methods.

It was pointed out earlier in this section that reservoir clearing seldom receives the consideration it should because the field work is scheduled late in the construction sequence. By way of summary, this section has discussed the elements of work and complex planning and study involved in reservoir clearing. It points to the urgency for an early start on all the preliminary work.

11-6 ENVIRONMENTAL CONSIDERATIONS

With the rapid growth in awareness and concern for man's natural and cultural environment, engineers and the engineering profession became interested and responsive to the criticism of their activities by newly enlightened "environmentalists." The civil engineering profession has not been as sufficiently concerned in the past with the effects of civil works upon the environment as their critics thought they should. This section discusses the impact of environmental problems on the dams and reservoirs.

A key obstruction in the past to the clear and broad environmental awareness by the civil engineer working with water resources is an insufficient appreciation of the work that water accomplishes (or the purposes it serves) in nature without man's "help." Water flowing unregulated down a stream is doing work. We should not think of it as being wasted or unused. Engineers have traditionally considered only man-regulated or managed water as water being put to use. The question becomes: "What are the wisest uses in the best interests of mankind?"

Development of a water resource will have environmental impacts in more than one geographical area. The reservoir and work site areas will be impacted. There will be impacts downstream from the reservoir due to the changes made in the regimen of the river. Service areas receiving either water or electric power will certainly have impacts. Other portions of the nation or province, depending on the magnitude of the project, may be affected to some extent. The various impacts may be permanent or temporary, occurring during construction or during operation of the completed project.

Consideration of environmental impacts and their mitigation during the design phase of a project presumes earlier consideration and evaluation involved in the decision to build the project. One such basic environmental consideration would be the minimum flows and flow patterns to be maintained in the river below a project. Another would be the timing of project development as it relates to the exploitation of mineral resources within the area of a future reservoir. Even environmental evaluation of return flows from project irrigated lands should be weighed in making the decision.

Field data should be obtained and interpreted during the planning of a project. They will indicate conditions existing before project development begins, establishing a "base line" against which to measure project impacts. In the United States these data are used in the preparation of an Environmental Impact Statement (EIS), which is a companion document accompanying the Feasibility Report. Both documents are used in the decision-making process. The contents and preparation of an EIS (Environmental Impact Statement) are discussed in Chapter 1, "Planning." A good EIS will not only identify all significant environmental impacts, whether negative or positive, but describe them in such a way that their relative significance can be evaluated.

When discussing environmental effects it can be helpful to group under a few basic headings the many subjects of environmental concern involved in water resource development. One division of groups could be in terms of the natural environment, the cultural environment, and the water resource. Under the natural environment one could discuss physiography, climate, geology, seismicity, soils, vegetation, fish and wildlife, rare or endangered species, vectors and health organisms, ecosystems and communities, water quality, air quality, timber and mineral resources, fire, and tidal influence. Discussion under cultural environment could include local government and institutions, economy and improvements, traffic patterns, sound levels, archeology and history, scenic attractions and recreation, lighting, and safety hazards. Under water resources could be discussed ground water supplies, existing water development, present water supplies, predictions of water demand, present flood protection, need for additional flood protection, other functions of existing water development, and predictions of future needs for other functions. It can be seen that not all of these listed may be applicable to the geographical area being considered. Other subjects may appropriately be included.

Compilation of field data must continue throughout the design and construction phases of the project. Some of these data, such as stream water temperatures and flow conditions, traffic counts, and weather observations, have regularly been obtained for engineering purposes. But other types of data including water quality, fish and wildlife census, air quality (including vehicle emissions), and sound levels are now required for environmental reasons. A typical sound level monitoring and recording apparatus is shown in Fig. 11-6-1. Portability is usually a requisite for sampling, monitoring, and recording equipment. Infrared imaging by remote sensing can be a very useful tool. Infrared scanning systems provide a pictorial display of the infrared energy emitted by terrain features, such as saturated areas and different types of vegetation.

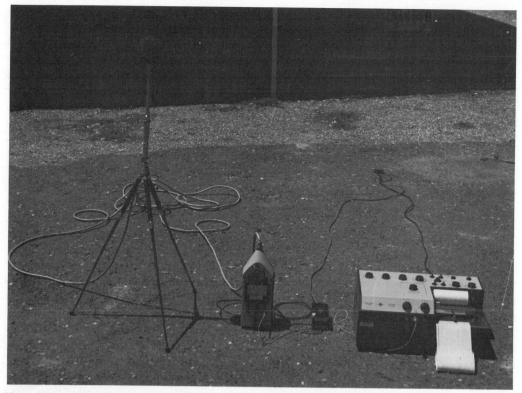

Figure 11-6-1. Typical sound level monitoring equipment will include a microphone (covered by windscreen in this photo), meter, and recorder in addition to a power source. (*Courtesy U.S. Bureau of Reclamation*)

At this point it must be strongly stated that reconnaissance and feasibility level field investigations for a project should be carefully planned so as to minimize altering the landscape. At that stage of development the starting date for construction has not been established, site selection is undetermined, and project construction is uncertain.

Some of the negative environmental impacts of a project cannot be mitigated. An inundated canyon is such an example. Its replacement by a reservoir is not mitigation no matter what positive values the reservoir may have. Yet a surprising number of positive measures of substantial value can be taken during planning, design, and construction. Establishment of the project take line can be environmentally most significant. Determining the take line should involve relating it to the lake, the terrain, and the purpose of the project. The degree of control of the lands around the reservoir will affect water quality. Control of these lands to assure a "greenbelt" around the reservoir will enable wildlife mitigation and recreation benefits. Whatever development is most suitable around the reservoir can best be assured when the peripheral lands are controlled by the project owner.

Selecting the type of dam to be built can also include environmental considerations. Use of the reservoir site as a source for construction materials is environmentally desirable where the ensuing scars will be permanently inundated. It is sound resources stewardship as well to utilize materials present within a reservoir site (high quality natural aggregates, for example) that would later become unavailable to man's use. Where a dam will be viewed by large numbers of people, and estimated costs of different dam types are close to equal, aesthetic considerations could be the deciding factor.

Project planners in particular should be aware of the environmental sensitivity of the physical location of a contemplated project. Local variations in sensitivity can be very significant but can only be considered individually. On an area-wide basis, however, some generalizations can be offered. An environment marked by the presence of man will be less sensitive to encroachment. Tropical lakes differ from temperate lakes in that events move faster, are very complex, and tend to be seasonal. Tropical reservoirs appear to have a maximum impact on flora and fauna, whereas reservoirs in arid zones create the least impact.

Forming a lake can cause an effect on the climate. Beyond this we as yet know very little. More to the point is the fact that there appears to be no way of controlling any impacts on the climate short of not forming the lake. Engineers have long been aware of evaporation of water from a water surface, but most evaporation research has been directed toward accurate prediction and estimates of reservoir evaporation losses. The efforts to control evaporation are discussed in Section 11-2. Changes in occurrence of fog and clouds, in the absolute humidity, in air temperatues, in wind patterns, and in precipitation are all potential factors. A possible additional factor is the impact on the climate in service areas associated with a project.

By altering the environment we do affect vegetation. These effects are due to changes in the presence of moisture and the intrusion of man and animals. The impact on service area vegetation due to the application of more water and changes in farming practices are readily apparent. Uniform downstream releases, being usually a change in existing pre-project conditions, are undesirable as they relate to vegetation (and fish and wildlife). Downstream from a dam and reservoir, riparian vegetation will gradually constrict a stream channel when flood flows are eliminated. If a project is operated so as to provide a periodic flushing action, encroachment can be largely prevented. Since a new ecosystem is created by the impounding of waters in a reservoir, lake level stability is biologically desirable. Timely fluctuations in a lake level can help to control the growth of undesirable plants and organisms. Benefits (or at least changes) in vegetation often will result from efforts primarily for other purposes. Examples of this are erosion prevention measures and reforestation for watershed management. Given a suitably adaptable species and proper soil type, a lowlands area reservoir will exert a significant influence on development of area forests because of raising the ground water table.

In contrast to climatic impacts, much can be done to mitigate and even enhance fish and wildlife impacts. True, riverine ecosystems are replaced by lake ecosystems. But management of watershed lands, the funds for which are usually not available without the project, can greatly increase the carrying capacity of watershed lands set aside for wildlife mitigation. Such measures can include creation of water supplies by small impoundments on tributary streams and simply protecting wildlife from the predations of man. Use of fire to control vegetation (often jointly for fire protection) can encourage desirable vegetation for browse and cover. Direct planting of vegetation can be employed both to increase browse and to create "edge effects" at the periphery of the protected area. Much literature is available on fisheries, lake and river. Controlled releases of reservoir waters can particularly enhance downstream fisheries. With regard to anadromous fish it is often necessary to either incorporate some kind of facility to enable fish to surmount the obstruction caused by damming their migration path, or to build and operate a fish hatchery. Except for low dams where fish ladders can be successful, fish hatcheries are more effective solutions. A somewhat novel solution to the anadromous fish mitigation problem has been the combining of fish spawning beds with a diversion canal. The key to this solution is the required periodic cleaning of the gravel spawning beds.

Much also has been written about water quality with respect to reservoirs. The tendency of lakes to stratify and how this occurs is the subject of considerable study. Attempts to ameliorate this condition where its occurrence is objectionable are of two kinds. For smaller reservoirs, vertical mixing of waters (and sometimes sediments) is sometimes attempted by reaeration with air released from pipe systems underwater. More effective control can be accomplished by releasing water from various levels of the reservoir at times compatible with downstream requirements. Multilevel outlets provide for flexibility of reservoir temperature releases and power plant operation which enables the desired control.

Dams, reservoirs, and their associated structures cause major impacts on the cultural environment. The resettlement of large numbers of people is sometimes necessary. The least disruptive way of accomplishing resettlement will depend entirely on the type of society in which the affected people live. Group or individual resettlement are the two basic choices. Vocational retraining is often necessary and often unsuccessful. Even in more urbanized societies, relocation can mean a change of occupation for people. Those remaining nearby will have changed incomes. If recreation is a significant project purpose, a local economy stimulated by construction itself will have the best chance of maintaining the higher level. New industry such as commercial lake fisheries and development of forest products can be an enhancement.

Local government will see a changed income. Project construction is likely to increase sales tax revenues. Higher property taxes due to increasing property valuations may be more than offset by the removal of lands from the tax rolls. With the development of a project will come increased demands for the services provided by local government. Transportation in the area adjacent to a project is often greatly improved, with new roads and bridges replacing structures built to out of date standards.

During construction some mitigation of noise (unwanted sound), air pollution, and light pollution can be accomplished. Mitigation measures would be unnecessary in a remote area but major effort is properly required in heavily populated areas. Noise can often be affected by the selection of types and quantitites of equipment. An air compressor operating at maximum or greater than rated capacity will create more objectionable noise than a larger compressor, or even two units operating at a lower percentage of capacity. By making electric power convenient to use, the project owner can influence the types of construction equipment selected. Note the electric powered air compressor station in Fig. 11-6-2. Since sound waves travel in straight lines, judicious siting of stationary and semiportable equipment can minimize noise. Stationary equipment also lends itself to insulating enclosures. Air pollution can be mitigated by dust prevention measures, particularly on haul road and contractor use area surfaces. Frequent watering suffices for temporary surfaces but some type of treated or even paved surface may be desirable for more permanent locations. Lighting can be controlled by careful placement and adjustment of the direction in which a light is aimed, and by selection of directional fixtures.

Many steps can be taken to effect mitigation of aesthetic impacts, yet this can also be most difficult. What is pleasing to one person's eye may not be so to another's. As an illustration of this, consider engineering structures themselves. To an engineer a structure may look pleasing but to someone else it may be a blight on the landscape. The location and alignment of structures out of view from most of the public is usually the most desirable measure. Electric power switchyards and transmission lines can sometimes be so sited. Although technological progress is being made, it is still usually too costly to place transmission lines underground except for short distances or in urban areas. When a structure cannot be placed out of the line of most viewers' sight it sometimes can be camouflaged by plantings of vegetation, by configuration of its silhouette, or by

Figure 11-6-2. An electric powered air compressor station located in a confining area. (*Courtesy U.S. Bureau of Reclamation*)

painting. Good architectural treatment may dictate blending a structure into its environment or may dictate that it stand out in bold contrast to its surroundings.

Temporary construction facilities should be completely removed if they have not been designed for conversion to some permanent use after completion of a project. Waste areas can often be located within the reservoir area below the future water surface. Sometimes much of the excavated materials can be used as structural materials for the dam or as structural fills. Boat ramps can sometimes accommodate large quantities of fill. It is desirable to limit haul road locations to elevations below the normal water surface when they are upstream from the dam. The sites of removed temporary facilities should be obliterated and restored after construction. Although some revegetation planting may be necessary, the natural revegetation process will often be sufficient.

A need to search for archeologic and historic sites in the path of a project and to conduct salvage of "worthwhile" or "significant" finds is readily understood by most engineers. Yet all too often archeologist-historian groups and engineers find themselves in disagreement. Differences over what is "worthwhile" or "significant" are understandable, but other factors causing disagreement can be avoided. During the programming of project development, the input of a qualified archeologist and historian with knowledge of the area should be sought in order to provide adequate lead time and cost estimates for required work. If it can be demonstrated that project development is likely to be the only vehicle by which thorough archeologic and historic investigations and reports will be made, enthusiastic cooperation is more likely. Any definition of what is historic must be relative to the recorded history of the given culture. What is considered old in one society may be thought of as relatively recent by another. Engineers should be aware of a number of factors. An archeologist tends to be an archeologist first, and only secondly

Figure 11-6-3. Looking across a reservoir just before sunset. (*Courtesy U.S. Bureau of Reclamation*)

an historian. In published reports, archeologists and historians will only indicate general locations of sites, so that the sites will remain intact. The project engineer must therefore be sure to obtain an unpublished listing of sufficiently precise locations. To an archeologist, salvage is definitely a second choice, a mitigation rather than an enhancement. His first choice will always be preservation. In all honesty he must admit, however, to the possibility that continuous inundation can be a better medium for preservation than air.

Many measures can be reasonably undertaken to mitigate any undesirable impacts of a water resources project such as a dam. A water resources project including a dam and reservoir can be used to arrest or correct environmental deterioration, and where necessary, force improvement of an area environment. A man-made lake can be beautiful, as demonstrated in Fig. 11-6-3.

BIBLIOGRAPHY

Kiersch, George A. "Vaiont Reservoir Disaster," *Civil Engineering*, March 1964.

Ringheim, A. S. "Design and Performance of Provisions for Foundation Seepage Control at Gardiner Dam," *Eng. J.*, Can. **53**, (No. 4), pp. 17-25, APC 1970.

Lewandowski, E. R. and Zlaten, A. "Foundation Problems of Ruedi Dam and Reservoir," Paper, Water Resource Eng. Conf., Amer. Soc. Civ. Eng., New Orleans, La. Feb. 1969.

Wiegel, R. L., et al. "Water Waves Generated by Landslides in Reservoirs," ASCE Proceedings, *J. Waterways and Harbors Div.*, **96**, (No. WW2), pp. 307-333, May 1970.

Lehnert, J. and Robertson, F. P. "Bituminous Blanket for Dike at Ludington Pumped Storage Project," *Civ. Eng.* **42**, (No. 12), pp. 54-57, Dec. 1972.

Redfield, R. C. "Brantley Reservoir Site–an Investigation of Evaporite and Carbonate Facies," *Eng. Geol.*, **4**, (No. 2), pp. 14-30, July 1967.

Jones, J. C. and Kleiner, D. E. "The Upper Reservoir for the Seneca Pumped Storage Plant," Paper, Conf. Recent Develop. Design Constr., Earth and Rockfill Dams, U. of Cal., Berkeley, Mar. 1968.

Pecherkin, I. A. "Forecasting the Reworking of Karstic Shores of Reservoirs Composed of Sulfate Rocks." *Hydrotech. Constr.* (No. 12), pp. 1060-1063, Dec. 1968.

Jaeger, C. "The Stability of Partly Immerged Fissured Rock Masses, and the Vajont Rock Slide," Civ. Eng. Public Works Rev., Vol. 64, No. 761, pp. 1204-1207, Dec. 1969.

Casagrande, A. Wilson, S. D., and Schwantes, E. D., Jr. "The Baldwin Hills Reservoir Failure in Retrospect," Proc. ASCE Spec. Conf. Performance Earth-Supported Struct., Vol. 1, Part 1, Purdue Univ., Lafayette, Ind., pp. 551-588, June 1972.

Bulletin of the Association of Engineering Geologists, "Reservoir Leakage and Groundwater Control," Vol. 6, No. 1, Spring 1969.

Castle, R. O. Yerkes, R. F., and Youd, T. L. "Ground Rupture in the Baldwin Hills–an Alternative Explanation," Bull. Assn. *Eng. Geol.*, Vol. 10, No. 1, pp. 21-46, Winter 1973.

Resources Agency of California, Department of Water Resources. "Investigation of Failure–Baldwin Hills Reservoir," April 1964.

Rollins, M. B. and Dylla, A. S. "Bentonite Sealing Methods Compared in the Field," ASCE Proceedings, *Journal of the Irrigation and Drainage Division*, **96**, (No. IR2), Paper 7379, p. 193, June 1970.

Hight, H. W. "Cabin Creek Pumped Storage Hydroelectric Project," ASCE Proceedings, *Journal of the Power Division*, **97**, (No. PO 1), Paper 7826, pp. 135-149, Jan. 1971.

Burden, W. W. "Design of Carters Pumped Storage Project," ASCE Proceedings, *Journal of the Power Division*, **96**, (No. PO3), Paper 7353, pp. 383-400, June 1970.

Chuck, R. T. "Largest Butyl Rubber Lined Reservoir," *Civil Engineering*, **40**, (No. 5), pp. 44-47, May 1970.

Wermenlinger, D., Thomson, J.G.S., and Gardiner, J. M. "Churchill Falls Power Facilities," ASCE Proceedings, *Journal of the Power Division*, **97**, (No. PO2), Paper 9774, pp. 515-537, March 1971.

Gunwaldsen, R. W., and Ferreira, A. "Northfield Mountain Pumped Storage Project," *Civil Engineering*, **41**, (No. 5), pp. 53-57, May 1971.

The Committee on Hydro Power Planning and Design. "Pumped Storage: State-of-the-Art," ASCE Proceedings, *Journal of the Power Division*. **97**, (No. PO3), Paper 8262, pp. 675-695, July 1971.

Binger, W. V. "Tarbella Dam Project, Pakistan," ASCE Proceedings, *Journal of the Power Division*, **98**, (No. PO2), Paper 9265, pp. 221-245, Oct. 1972.

Hunt, B. W. "Seepage From Shallow Reservoir," ASCE Proceedings, *Journal of the Hydraulics Division*, **99**, (No. HY1), Paper 9476, Jan. 1973.

Comminallis, E. "Ludington Pumped Storage Project," ASCE Proceedings, *Journal of the Power Division*, **99**, (No. PO1), Paper 9696, pp. 69-88, May 1973.

Whitehead, C. F. and Ruotolo, D. "Ludington Pumped Storage Project Wins 1973 Outstanding C.E. Achievement Award," *Civil Engineering*, **43**, (No. 6), pp. 64-68, June 1973.

Chan, H. W. "Ludington Pumped Storage Project," *U.S.C.O.L.D. Newsletter*, No. 43, p. 6, March 1974.

"Geology and Foundation Treatment, Tennessee Valley Authority Projects," Tennessee Valley Authority, Technical Report No. 22, 1949.

Carter, R. K., Lovell, C. W., Jr., and Harr, M. E. "Computer Oriented Stability Analysis of Reservoir Slopes," Purdue University Water Resources Center, Technical Report No. 17, January 1971.

DaCosta-Nunes, A. J. "Stabilization of Taluses of Residual Soils of Granito-Gneisic Origin," Paper, Nat. Eng. School, Rio de Janeiro, Brazil, 1969. Trans. from French, USBR No. 600.

Zajicek, V. "Hydrological Documentation for the Exploitation of Reservoirs in Permeable Solid Rocks for the Purpose of Artificial Infiltration," Symp. Int. Union Geod. Geophys.–Int. Ass. Sci. Hydrol., Haifa, Israel, pp. 124-131b, Mar. 1967.

Leech, T. D. J. "Diffusion Blasting and its Potential for the Development of Australia's Inland Surface Water Resources," *Journal of the Institution of Engineers*, Australia, **41**, (No. 10-11), pp. 165-173, Oct.-Nov., 1969.

Retief, J. V. and Kruger, P. "The Use of Nuclear Explosives for Water Resources Development in Arid Regions," Dept. of Civil Engineering, Stanford Univ., June 1971.

Anon. "Peru Carves a Reservoir in a Mountain," Eng. New-Rec., Vol. 184, No. 13, pp. 26 and 31, March 1970.

Anon. "Hawaiian Butyl Rubber Lined Reservoir is World's Largest," *World Irrigation*, **20**, (No. 1), pp. 15-18 & 20, Jan.-Feb. 1970.

Molinari, J., Guizerix, J., and Chambard, R. "A New Method of Detecting Leaks in Reservoirs or Canals Using Labeled Bitumen Emulsions," Int'l. Atomic Energy Agency, Isotope Hydrology 1970, Vienna,

Series IAEA-SM-129/4, pp. 743-760 (translated from French for Natl. Science Foundation and U. S. Dept. of Interior, Bureau of Reclamation).

U. S. Bureau of Reclamation Geology Manual, Technical Instructions No. 2–Reservoir Site Geologic Studies (Tentative Edition).

Muller, Leopold. "New Considerations on the Vaiont Slide," *Rock Mechanics and Engineering Geology*, VI/1-2, 1968.

Tolobre, I. A. "The Behavior of Rock Foundations of Large Dams and Associated Reservoirs," *Construction*, 22, No. 7 (1967).

Hast, N. "The State of Stresses in the Upper Part of the Earth's Crust," *Engineering Geology*, 2(1), 1967 pp. 5-17.

Mickey, Wendell V. "Seismic Effects of Reservoirs," *The Military Engineer*, July-August 1972.

Housner, G. W. "Seismic Events at Koyna Dam," Proceedings, Eleventh Symposium of Rock Mechanics, Berkeley, California, June 16-19, 1969.

The Joint Panel on Problems Concerning Seismology and Rock Mechanics, "Earthquakes Related to Reservoir Filling," Division of Earth Sciences, National Research Council, Washington, D.C., January 1972.

Rothe, J. P. "Fill a Lake, Start an Earthquate," *New Sci.*, 39, (No. 605), pp. 75-78, July 1968.

Kall, R. and Charalambakis, S. "Impounding of Manicougan 5 Reservoir as Possible Trigger Cause of Local Earthquakes," 10th Int. Congr. Large Dams, Montreal, Can., Vol. 3, Quest. No. 38, pp. 795-814, June 1970.

Raleigh, C. B. "Investigation of Seismic Activity Related to Reservoirs," Bull. Assn. Eng. Geol., Vol. 9, No. 3, pp. 177-183, Summer 1972.

Baker, R. G. "A Basic Discussion of Seismology," California Dept. of Water Resources, Office Manual, Aug. 1970.

Rouse, G. C. "Seismic Activity in the Vicinity of San Luis Reservoir," Bureau of Reclamation, REC-OCE-70-52, Nov. 1970.

Rechard, P. A. "Determining Bank Storage for Use in Reservoir Operation," unpublished report, Upper Colorado River Commission. (circa 1964).

U. S. Dept. of Interior, Bureau of Reclamation–Region 4, "Bank Storage–Lake Powell," September 1969.

Library, U. S. Bureau of Reclamation, Denver, CO. "Evaporation from Open Water Surfaces," Bibliography No. 51, March 1940.

Harbeck, C. E., Jr., and Meyers, J. S. "Present Day Evaporation Measurement Techniques," ASCE Proceedings, *J. Hydraul. Div.* 96, (No. HY 7), pp. 1381-1390, July 1970.

Horton, J. S. "Evapotranspiration and Watershed Research as Related to Riparian and Phreatophyte Management," an Abstract Bibliography, Forest Service–U.S. Dept of Agriculture, Misc. Publication No. 1234, Jan. 1973.

Turner, P. M. "Annual Report of Phreatophyte Activities–1968," Bureau of Reclamation, REC-OCE-70-27, July 1970.

University of Arizona–Agricultural Experiment Station, Institute of Water Utilization, "Final Report on Evaporation Reduction Investigation Relating to Small Reservoirs," prepared for the U.S. Bureau of Reclamation, Technical Bulletin 177, October 1966.

Borland, W. M. "Reservoir Sedimentation," *River Mechanics*, II, Shen, H. W., ed. Ch. 29, 1971. LOC 70-168730.

Szechowych, R. W. and Qureshi, M. M. "Sedimentation in Mangla Reservoir," ASCE Proceedings, *Journal of the Hydraulics Division*, 99, (No. HY9), Paper 10033, Sept. 1973.

Flaxman, Elliott M. "Some Variables which Influence Rates of Reservoir Sedimentation in Western United States," International Association of Scientific Hydrology, Publication No. 71, pp. 824-838, 1966.

Roehl, John W. "Sediment Design Requirement for Small Reservoirs," International Association of Scientific Hydrology, Publication No. 71, pp. 795-803, 1966.

Chee, S. P. and Sweetman, A. P. "Sedimentation Characteristics of Gorge-Type Reservoirs," Water Resources Bulletin, Vol. 8, No. 5, pp. 881-886, Oct. 1972.

"The Appraisal of Real Estate," American Institute of Real Estate Appraisers, Chicago, Ill., LOC #59-15823.

"Uniform Appraisal Standards for Federal Land Acquisitions," Interagency Land Acquisition Conference, Washington, D. C., 1972, U. S. Govt. Printing Office.

"Acquisition for Right-of-Way," American Association of State Highway Officials, Washington, D. C., 1962.

"A Procedural Guide for the Acquisition of Real Property by Governmental Agencies," Department of Justice, Land and Natural Resources Division, 1972, U. S. Govt. Printing Office, Washington, D.C.

Brown, C. M. *Boundary Control and Legal Principles*, John Wiley & Sons, Inc., New York, 1957, LOC 57-10804.

Brown, C. M. and Eldridge, W. H. *Evidence and Procedures for Boundary Location*, John Wiley & Sons, Inc., New York, 1962, LOC 62-18988.

Rutledge, A. J., *Anatomy of a Park*, McGraw-Hill Book Co., 1971, LOC 78-141925.

Pederson, Erling F. "Construction Roads and Appurtenances for Dillon Dam," ASCE Proceedings, *Journal of the Construction Division*, 93, (No. CO2), pp. 53-72, Sept. 1967.

Anon. "What's New in Land Clearing?" *World Construction*, 20, (No. 2), pp. 22-25, Feb. 1967.

Sowers, G. F. "Remote Sensing for Water Resources," *Civil Engineering*, **43**, (No. 2). pp. 35-39, Feb. 1973.

Hickey, Ben F. "Fast Clearing Methods for Big Reservoir Area," *World Construction*, pp. 49-51, May 1963.

Taylor, J. I. and Stingelin, R. W. "Infrared Imaging for Water Resources Studies," ASCE Proceedings, *Journal of the Hydraulics Division*, **95**, (No. HY1), Jan. 1969.

LeSchack, L. A. and Brown, R. G. "Estimating Land-Clearing Costs Using Aerial Photography," *Civil Engineering*, **39**, (No. 2), pp. 66-68, Feb. 1969.

"Consequences on the Environment of Building Dams," Question 40, Transactions of the Eleventh Congress on Large Dams, Madrid, Spain, Vol. I, Vol. V, R. G. Q. 40, 11-15–June 1973.

"Man-Made Lakes: Their Problems and Environmental Effects," Geophysical Monograph 17, Ackerman, W. C. White, G. F., and Worthington, E. B., editors, *American Geophysical Union*, Washington, D.C., 1973.

"Man-Made Lakes, Planning and Development," Karl F. Lagler, editor, Food and Agriculture Organization of the United Nations, Rome, 1969.

Van Hylckama, T. E. A. "Water Resources," in *Environment*, Murdoch, W. W., ed., Sinauer Assoc. Inc., Stamford, Conn., 1971.

Howarth, David. *The Shadow of the Dam*, Collins, London, 1961.

Clements, Frank. *Kariba–The Struggle with the River God*, Methuen, London, 1959.

Smith, Charles R. "Anticipations of Change: A Socio-Economic Description of a Kentucky County Before Reservoir Construction," Univ. of Kentucky Water Resources Institute, Lexington, Ky. Research Report No. 28, 1970.

Bishop, A. B. "An Approach to Evaluating Environmental Social and Economic Factors in Water Resources Planning," Water Res. Bull. Vol. 8, No. 4, pp. 724-734, Aug. 1972.

Smith, C. L. and Hogg, T. C. "Cultural Aspects of Water Resources Development–Past, Present, and Future," Water Resources Bulletin, Vol. 7, No. 4, pp. 652-660, August 1971.

Henning, D. H. "Environmental Policy and Politics: Value and Power Context," *Natural Resources Journal* (UNM), **11**, (No. 3), pp. 447-454, July 1971.

Hagan, R. M. and Roberts, E. B. "Ecological Impacts of Water Projects in California," ASCE Proceedings, *Journal, Irrig. Div.*, **98**, (No. III), pp. 25-48, March 1972.

Schad, T. M. "Water Resources and the Environment," Water Resources Bulletin, Vol. 8, No. 2, pp. 404-409, April 1972.

Wright, J., "Nature and Man and Water," *Water Spectrum*, **4**, (No. 1), pp. 11-15, 1972.

Waller, G. H. "MOP for EQ," paper, ASCE Irrigation and Drainage, Special Conference, Spokane–September 1972.

Mann, D. "Ethical and Social Responsibility in the Planning and Design of Engineering Projects," ASCE, *Journal of Professional Activities*, **98**, (No. PP1), January 1972.

Haugen, A. O. and Lenning, R. E. "Pre-Impoundment Recreational Use Pattern and Waterfowl Occurrence in the Saylorsville Reservoir Area," Iowa State University Water Resource Research Inst. Completion Report, June 30, 1970.

Kwiatkowski, R. W. and Pierce, L. D. "Environmental Design of Bear Swamp Project," ASCE Proceedings, *Journal of the Power Division*, **99**, (No. PO1), Paper 9738, pp. 205-215, May 1973.

Dudley, E. F., "Environmental Aspects of Site Selection," ASCE Proceedings, *Journal of the Power Division*, **98**, (No. PO1), Paper 8971, pp. 21-28, June 1972.

Sterling, C. "Superdams: The Perils of Progress," *Atlantic*, **229**, (No. 6), pp. 35-41, June 1972.

Cain, S. A. "Ecological Impacts on Water Resources Development," Water Resources Bulletin, Vol. 4, No. 1, pp. 57-74, March 1968.

Wisely, W. H. "People, Ecology, and the Aswan High Dam," *Civil Engineering*, **42**, (No. 2), pp. 37-39, Feb. 1972.

Kemp, L. E., McKee, G. D., Raabe, E. W., and Warner, R. W. "Water Quality Effects of Leaching from Submerged Soils," *Journal of American Water Works Association*, **62**, (No. 6), pp. 391-396, June 1970.

ASCE, "Register of Selective Withdrawal Works in United States," ASCE Proceedings, *Journal of the Hydraulics Division*, **96**, (No. HY9), paper 7533, pp. 1841-1872, September 1970.

Reservoir Water Quality Management Task Force of Interior Committee on Water Resources Research. "Report on Reservoir Water Quality Management Problems and Related Research Needs," U. S. Dept. of Interior, unpublished report, Aug. 1, 1966.

Sylvester, R. O. and Seabloom, R. W. "Influence of Site Characteristics on Quality of Impounded Water," *Journal of American Water Works Assn.* **57**, (No. 12), pp. 1528-1546, Dec. 1965.

Chen, C. W. and Orlob, G. T. "Predicting Quality Effects of Pumped Storage," ASCE Proceedings, *Journal of the Power Division*, **98**, (No. PO1), Paper 8984, pp. 65-75, June 1972.

"Sinamwenda, The Nuffield Lake Kariba Research Station, Report 1962-1968," University College of Rhodesia and University of the Witwatersrand, University College of Rhodesia, Salisbury, 1969.

Fraser, J. C. "An Annotated Bibliography on the Establishment of Acceptable Flows for Fish Life in Controlled Streams," Food and Agriculture,

Organization of the United Nations, E1 FAC 70/SC III-4, April 13, 1970.

Jenkins, Robert M. "Reservoir Fishery Research Strategy and Tactics," U. S. Dept. of Interior, Bureau of Sport Fisheries and Wildlife, Circular 196, July 1964.

"The Storage Lakes of the U.S.S.R. and Their Importance for Fishery," edited by P. V. Tyurin, U. S. Dept. of Commerce, Springfield, VA, translated from Russian, 1966.

"Reservoir Fishery Resources Symposium, April 5-7, 1967," American Fisheries Society, Washington, D.C.

"Reservoir Fisheries and Limnology," edited by Gordon E. Hall, Special Publication No. 8, 1971, Washington, D.C. LOC #70-168771.

Jacobson, C. B. "Benefits to Environment, Glen Canyon to Hoover Dams," ASCE Proceedings, *Journal of the Power Division*, 99, (No. PO2), Paper 10146, pp. 395-404, Nov. 1973.

Sciandrone, J. C. "Environmental Protection at California Dams," *Civil Engineering*, 44, (No. 3), pp. 80-83, March 1974.

"Design of Small Dams," Bureau of Reclamation, U. S. Dept. of Interior, 2nd Edition, Revised Reprint, U. S. Govt. Printing Office, Washington, D.C., 1974.

McCullough, C. A., and Nicklen, R. R. "Control of Water Polution During Dam Construction," ASCE Proceedings, Sanit. Eng. Div., Vol. 97, No. SA1, pp. 81-89, Feb. 1971.

Orlob, G. T., and Selna, L. G. "Temperature Variations in Deep Reservoirs," ASCE Proceedings, *J. Hydraul. Div.*, 96, (No. HY2), pp. 391-410, Feb. 1970.

Sefchovich, E. "Condenser Cooling and Pumped Storage Reservoirs," ASCE Proceedings, *Journal of the Power Division*, 97, (No. PO3), Paper 8247, pp. 611-621, July 1971.

Huber, W. C. Harleman, D. R. F. and Ryan, P. J. "Temperatures Prediction in Stratified Reservoirs," ASCE Proceedings, *Journal of the Hydraulics Division*, 98, (No. HY4), Paper 8839, April 1972.

Benton, W. B., Wall, W. J. and McKeever, S. R. "Select Reservoir Withdrawal by Multilevel Intakes," ASCE Proceedings, *Journal of the Power Division*, 96, (No. PO1), Paper 7029, pp. 109-115, Jan. 1970.

Riesbol, H. S., Anderson, J. B., Wend, F. H., and Holmes, H. T. "Thermal-Hydraulic Study: Arkansas Cooling Reservoir," ASCE Proceedings, *Journal of the Power Division*, 97, (No. PO1), Paper 7811, pp. 93-113, Jan. 1971.

"Water Quality Behavior in Reservoirs," a Compilation of Published Research Papers, compiled by James M. Symons, U. S. Dept. H.E.W., 1969, Washington, D. C.

Posey, F. H., Jr., and DeWitt, J. W. "Effects of Reservoir Impoundment on Water Quality," ASCE Proceedings, *Journal of the Power Division*, 96, (No. PO1), pp. 173-185, Jan. 1970.

Eff, Kenneth S. "Water Quality Management Data Associated with Reservoir Operations," Proceedings of the National Symposium on Data and Instrumentation for Water Quality Management, Conference of State Sanitary Engineers and Wisconsin University, July 21-23, 1970, Madison, Wis., pp. 384-394, 1970.

12
Specifications for Dam Construction

JAMES T. MARKLE
Staff Counsel Three
State Water Resources Control Board
Sacramento, California

12-1. INTRODUCTION

A. Functions of the Specifications

In its most basic form, the term "specifications" as it relates to dam construction means a detailed written description of the owner's design concepts. This written description, together with the project drawings, has as its primary function the communication of these design concepts to the dam constructor.

Specifications perform an additional function of great importance from an engineering point of view. Whether the project is to be constructed through a formal competitive bidding process (as is typically the case where the owner is a public agency), or by an informal but nevertheless competitive solicitation of offers (as is more typical of nonpublic owners), or by negotiated procurement of the construction service (rare in either the public or private sector), the specifications have an administrative-legal function. The provisions of the specifications become conditions, in a contract law sense, affecting the legal rights and duties of the owner and the constructor. This is so because all construction procurement processes contemplate creation of a contractual relationship between the owner and the constructor, having as its objective the construction of the project as contemplated by the owner for the price offered by the constructor. The specifications are the terms and conditions of this contract. The Appendix of this chapter includes the Table of Contents of a complete set of specifications for construction of a dam.

B. Categorization of the Specification Provisions

For purposes of clarity in communicating to the bidder-constructor both the owner's technical requirements and the contractual framework within which the project is to be constructed and the price paid, specification provisions are typically divided into categories. These divisions are usually bound in a single document and the provisions are collectively referred to as "the specifications." The number and title of the divisions of the specifications vary among owners.

1. The Technical Requirements. There is a need for one division comprising those provisions which describe the owner's technical requirements in achieving his construction objective. Included in this category, for example, may be provisions which describe the constituent materials of, and the process to be used in producing concrete, together with a description of the process to be employed in its placement. Depending on the owner's preference, this division may be entitled "Technical Provisions," "Technical Specifications," or "Technical Conditions."

2. The General Requirements. There is also a need for a division to describe the administrative and legal conditions under which the work is to be performed. These provisions normally do not deal with technical matters. When the owner is a public agency, this division of the specifications will contain provisions reflecting public policy requirements imposed by governmental authority; an example of which is a requirement that the constructor not compensate project workmen at a rate less than that determined to be prevailing in the locality of the project.

Depending on owner preference, this division may be entitled "General Provisions" or "General Conditions." Many owners are involved in procuring construction of a number of engineering or architectural projects. These owners often find it useful to develop general provisions for standardized use in specifications for all of their projects. For this reason one often finds that the specifications division containing the general provisions is entitled "Standard Provisions." In addition to administrative convenience, use of standardized general provisions by multi-project owners has the advantage of familiarizing the construction industry with the administrative-legal rules and procedures of the particular owner, which also contributes to efficient management of the individual projects.

3. Special Requirements. A final division of the specifications consists of a group of provisions which are chiefly non-technical in nature but which are nevertheless tailored to the particular project and therefore cannot conveniently be placed with the "general" or "standard" provisions.

An example of such a provision is a requirement that the project be completed by a certain date or within a specified number of days. Such a completion date or completion period is based on a number of factors, such as the magnitude of the construction effort and the needs of the owner.

4. Miscellaneous Requirements. Additional provisions and documents, which are not in the nature of specifications, may nevertheless be part of the construction contract between the owner and contractor. Foremost among such additional provisions, and of particular importance to competitive procurement of construction service, are bidding provisions. These provisions communicate to the interested bidder-constructors in the industry the rules for preparation and submission of bids. They may also describe the process by which the owner will award the contract (that is, accept an offer). In a public project, failure of a bidder-constructor to follow these rules will usually make his bid "non-responsive" and the owner will be legally prohibited from awarding the contract to him, although his bid price may appear to be the lowest. In the private sector, failure of an offeror-constructor to follow the bidding rules may mean that no contract can be created, even though the owner may wish to accept the offer, for the reason that the offeror-constructor may not legally be bound by such an acceptance where the terms of his offer do not substantially conform to the conditions of the owner's solicitation for offers.

Another contract document included in the Appendix of this chapter is the "Notice to Contractors." This document specifies the time and place of reception of bids, contains a general description of the project's principal features, and an itemized listing, together with estimates of quantities, of the work, materials and equipment required. It also informs prospective bidders where complete sets of the bid documents may be obtained. In the case of formal competitive procurement of construction services, this notice will typically be published in one or more newspapers of general circulation and in one or more trade papers circulated in the area covering the construction site. It may also be mailed to constructors known to be interested in performing the type of work described.

C. An Outline of a Model Contract for Dam Construction

The Appendix to this chapter is taken from the specifications used by the State of California's Department of Water Resources to construct Perris Dam and Lake, a unit of California's State Water Project. The State Water Project consists of a complex system of large dams and reservoirs, large pumping plants, and hundreds of miles of canal, pipeline and tunnel. Project purposes include water conservation and distribution, hydroelectric power generation, flood protection, fish and wildlife enhancement, water quality enhancement, and all forms of water-oriented recreation. The Perris Dam specifications reflect more than a decade of dam specification development by the owner agency.

Additional sources of model specifications are the dam construction contracts of United States governmental agencies long experienced in the design and construction of dams: the Bureau of Reclamation of the U.S. Department of the Interior, and the Corps of Engineers of the U.S. Department of the Army. Many large municipalities, such as the Department of Water and Power of the City of Los Angeles, have developed comprehensive dam construction specifications.

California's Department of Water Resources is, of course, a public owner, and the many units of the State Water Project were constructed by privately owned engineering contractors. Under state law, construction services are procured through formal competitive bidding. Preconstruction engineering, including design and specifications development, and contract administration during construction were performed by engineers and technicians employed by the owner.

Reference to the Table of Contents of the Specifications (Appendix pp. 685-690) show that the specifications for Perris Dam are divided into three groups of provisions: Standard Provisions (Sections 1 through 9), Special Provisions (Sections 10 through 13), and Technical Provisions (Sections 14 through 24). Sections 1 through 13 only appear in the Appendix. Sections 14 through 24, the Technical Provisions, are not included in the Appendix because of their volume and because they are specific to the particular project. Example of Technical Provisions are available from the sources noted above.

Since the users of this handbook will have a major interest in the technical requirements of the specifications, this group is treated first in the discussion which follows.

12-2. THE TECHNICAL PROVISIONS

A. Theoretical View

The requirements of the technical provisions may be thought of as being of two general types.

1. Performance Specifications. This type of specification describes the work in terms of the owner's objectives or needs. An often-used example of a performance specification is that found in a contract between the United States Department of the Army and the Wright brothers in 1907. Although not related to dam construction, this specification illustrates communication of requirements in terms of performance characteristics desired by the owner. The specification provided in part is as follows:

> "2. It is desirable that the flying machine should be designed so that it may be quickly and easily assembled and taken apart and packed for transportation in Army wagons. It should be capable of being assembled and put in operating condition in about one hour.
> "3. The flying machine must be designed to carry two persons having a combined weight of about 350 pounds, also sufficient fuel for a flight of 125 miles.
> "4. The flying machine should be designed to have a speed of at least forty miles per hour in still air..."

The distinguishing characteristic of performance specifications is that the constructor is assigned responsibility for engineering design, and as a consequence will have great discretion as to design details. Before employing the performance specification technique for any portion of the work, the owner must determine that this transfer of discretion will not result in a completed project whose design details are unacceptable to him. In dam design it is used often for major hardware, such as gates, valves and pumps.

2. Design Specifications. This type of specification describes the work in terms of precise measurements, tolerances, materials, and, often, in-process tests. Here, the owner is assigned— as between the parties to the contract—responsibility for design, and, as a consequence, for omissions, errors, and deficiencies in the specifications and plans. The great advantage to the owner in employing the design specification technique is that the owner—and therefore the owner's managing engineer—retains maximum control over the work and the possibility that the completed project will have unacceptable design details is avoided.

An example of a design specification, taken from Section 17 of the project outlined in the Appendix, related to dam embankment materials, follows:

> "(1) General.—The embankment and riprap materials shall conform to the requirements of this article and shall be derived from sources specified herein. Pursuant to Section 16, Article (b), the Contractor shall make maximum use of materials derived from mandatory excavations in Zone 2 embankment, blanket, backfill, dam embankment facing, and site development embankments, when such materials can be processed or selected to meet specification requirements, before resorting to the use of borrow areas. Zone 1 embankment material shall be obtained exclusively from the Clay Borrow Area, Zone 3 and Zone 4 material, and riprap shall be obtained exclusively from the rock source.
>
> "(2) Zone 1 Dam Embankment.—Material for Zone 1 dam embankment shall be clay material obtained from an area northeast of the lake and designated Clay Borrow Area. Zone 1 embankment material as compacted in embankment shall be at uniform moisture content that is within ±1 percent of that designated by the Engineer. The moisture content required will be approximately 1 percent to 2 percent above optimum moisture content determined by the soil tests specified in Article (m). The designated moisture content will be furnished by the Engineer to the Contractor at daily intervals or at changes in the composition of material being placed.
>
> "(3) Zone 2 Dam Embankment.—Zone 2 dam embankment shall be composed of material derived from the lake borrow area and designated material from mandatory excavations. Zone 2 embankment material as compacted in the embankment shall be at a uniform moisture content that is within one percent of that designated by the Engineer. The moisture content required will be approximately one percent above optimum moisture content determined by the soil tests specified in Article (m). The designated moisture content will be furnished by the Engineer to the Contractor at

daily intervals or at changes in the composition of material being placed.

"(4) Dam Embankment Facing.—Material for dam embankment facing shall be obtained from required stripping or excavation for dam foundation which is free from excessive organic matter as determined by the Engineer. Moisture content and compaction requirements shall be as specified for Zone 2 dam embankment in Subarticle (3).

"(5) Zone 3 Dam Embankment.—Material for Zone 3 dam embankment shall be obtained from excavating, crushing, and processing material from rock source, and shall conform to the following gradation requirements after compaction in the dam embankment:

Size	Percent by Weight Passing
$1\frac{1}{2}$ inch U.S. Standard Sieve	100
$\frac{3}{4}$ inch U.S. Standard Sieve	75-100
$\frac{3}{8}$ inch U.S. Standard Sieve	50-85
No. 16 U.S. Standard Sieve	10-55
No. 50 U.S. Standard Sieve	0-25
No. 200 U.S. Standard Sieve	0-5

(with an average not exceeding 3 percent for any 5 consecutive working days in which Zone 3 is being placed)

"Material used in the embankment shall be free of weathering and objectionable matter as determined by the Engineer.

"(6) Zone 4 Dam Embankment.—Material for Zone 4 dam embankment shall be obtained from excavating, crushing and processing material from the rock source conforming to the following gradation after compaction in the dam embankment:

Size	Percent by Weight Passing
6 inch U.S. Standard Sieve	100
3 inch U.S. Standard Sieve	25-100
$1\frac{1}{2}$ inch U.S. Standard Sieve	10-40
No. 4 U.S. Standard Sieve	0-5

"Material used shall be free of weathering and objectionable matter as determined by the Engineer.

"(7) Blanket—Material for blanket shall be designated material from the sources specified in Subarticle (3). Moisture content shall be as specified for Zone 2 dam embankment in Subarticle (3). Stripping from Lake Borrow Area and foundation excavation free of excessive vegetation and other deleterious material, as determined by the Engineer, may be used in the lower lift of the blanket.

"Excavation for removal of buried water pipe shall be filled with compacted material as specified in this article to the level of the surrounding foundation and the blanket constructed to the lines shown."

3. Combined performance and design specifications.

It is not uncommon to find that the technical specifications, taken as a whole, contain elements both of performance and design specifications. For example, the illustrative specification quoted above, provides further as follows:

"The distribution and gradation of materials throughout the dam embankment shall be such that the dam embankment will be free from lenses, pockets, streaks, and layers of materials differing substantially in gradation from the surrounding material within each zone.

"Material shall be placed and spread in successive and approximately horizontal layers of the thickness hereinafter specified. The placing of material shall be done so as to obtain a layer of uniform thickness without spaces between successively deposited loads. Placing and spreading shall be done in such a manner as to prevent segregation of the material. Should the embankment surface become rutted or uneven subsequent to compaction, it shall be releveled before placing the next layer of material."

This requirement may be thought of as specifying performance characteristics of the completed embankment since it does not specifically describe the construction method of achieving embankment that is free from lenses, pockets, streaks, and layers of material of substantially differing gradation.

Selection of the specification approach to be used in developing a given set of technical specifications depends upon several considerations. Chief among these are the engineering design and construction supervision resources available to the owner. If the owner has adequate engineering resources, he can accept the responsibility inherent in developing design specifications so that he may maximize his control over the design features of the completed project. Use of performance specifications is best reserved for those portions of the work where detailing construction methods would result in an undue infringement upon the constructor's management discretion in selecting construction techniques, or where the objectives sought approach the state-of-the-art and it is desirable that the constructor have maximum engineering design discretion in developing methods to achieve the objective.

B. Principles

Application of the following general principles to development of the technical provisions will help assure that communication of the owner's requirements to the constructor is precise, accurate, complete, and free from ambiguity.

1. Clarity. Every effort should be made to develop technical provisions which can be understood by the average constructor who is reasonably familiar with the type of construction contemplated. Laws governing procurement of construction by public owners often include a requirement to the effect that the specifications be such as can be understood by the "average competent mechanic."

The purpose of this principle is twofold. First, in competitive procurement it is essential that all of the bidders bid on a commonly understood project. To the extent bidders have a dissimilar understanding of project requirements from that contemplated by the owner, the competitive procurement procedure tends toward invalidity.

Second, lack of clarity in the technical requirements means that differing reasonable interpretations of the requirements during performance of the work may be made by the constructor and the owner. This in turn can lead to formal disputes between the parties, which are often expensive to resolve and unproductive in terms of project construction.

2. Completeness. Performance of the requirements described by the technical provisions should lead to proper completion of the whole project as contemplated by the owner. This means development of specifications which reflect a careful thinking-through of the entire project, prior to selection of the constructor, by the owner and his design staff.

Although dam construction specifications typically include—in the General Conditions— a provision permitting the owner to make changes and additions to the work (the "Changes Clause"), every effort should be made to include requirements covering every facet of the work necessary to assure proper completion of the project. Frequent use of the "Changes Clause" to add work which should have been included in the original technical specifications means that the cost of the added work will not reflect the pricing advantage of competitive or negotiated bidding.

A technique to be avoided in the preparation of technical provisions is the attempt to describe the method of performance or extent of a particular portion of the work by use of the phrase "as directed by the Owner" or "as directed by the Engineer." When the bidder-constructor, in developing his bid, encounters this type of description, he will have difficulty pricing that portion of the work since he cannot be certain of what the Engineer will direct. If he prices the portion high, because the owner's direction may result in a contingency with a relatively costly performance, the bidder-constructor's competitive position on the bid may be damaged. If he prices the portion low, or does not price the contingency at all, he risks incurring costs during performance for which he has made inadequate provision. The owner may then be faced with a claim upon the theory that the directed work is a change for which the constructor is entitled to additional compensation.

Overuse of the "as directed" phrase in the technical provisions generally indicates a failure on the designer's part to think through fully the requirements for complete performance of the whole work, prior to the start of construction. Use of the "as directed" phrase is justified only when its use is limited to authorizing the owner to coordinate sequential performance of portions of the work, where cost implications are very minor, and the specifications should so state.

The model Outline of Technical Provisions in the Appendix (Sections 14 through 24) emphasizes project completeness from control of water and unwatering through instrumentation of the dam foundation and embankment.

3. Consistency: Internal and External. Of very great importance to proper completion of the project is the principle that the Technical Provisions not include conflicting requirements and that the Technical Provisions do not conflict with the requirements of the Standard Provisions or the Special Provisions (see Section 2 of the Standard Provisions in the Appendix of this chapter).

While this principle is easy to state, its successful application depends upon a thorough knowledge of the requirements of the Technical Provisions as a whole, as well as a thorough knowledge of the Standard Provisions and the Special Provisions.

C. Review of Technical Provisions

It is highly desirable that the Technical Provisions and the Special Provisions as a whole be comprehensively reviewed by a person or persons, on behalf of the owner, other than those primarily charged with the responsibility of developing them.

The advantage of such an independent, comprehensive engineering review, particularly of the specifications for a relatively sophisticated engineering project, is that any deficiencies in application of the principles of clarity, completeness, and consistency are more apparent to the persons who review the entire package of technical requirements as a whole. Such review is especially important where the specifications contain design requirements related to more than one engineering discipline—such as civil, mechanical, and electrical—developed by different designers. The owner's interests will be best served if such review is intensive; that is, if any *possible* ambiguity or absence of clarity is identified and brought to the attention of the responsible managing engineer.

If legal counsel is available to the owner, a legal review of the specifications as a whole is desirable. For any major dam project legal review is essential before the specifications are released to the bidder-contractors. Legal review help assure that any specific legal requirements imposed by the particular jurisdiction are met—an objective that is of special importance in the case of public contracts. In addition, while counsel is normally not expected to review engineering design concepts, his legal training is particularly suited to identifying instances of lack of clarity, ambiguity, and inconsistency in the specifications, and bringing such instances to the attention of the managing engineer. The impact of recurring legislation and of the interpretations by the courts of prior legislation on the proposed project should be determined by counsel.

12-3. THE STANDARD PROVISIONS

The following discussion of standard, or general provisions is based on the Standard Provisions which are set out in sections 1 through 9 of the Appendix of this chapter. Each section is generally discussed, and those articles of general application are given special attention.

Section 1. Definitions

This section defines certain words, phrases, and abbreviations for the purpose of usage in the contract. This is useful for two reasons: (1)

repetition of descriptive language in the specification is avoided since many frequently used concepts can be expressed by a single word or phrase, and (2) disputes over the meaning of these frequently used terms are avoided since the defined words, phrases, and definitions carry the same meaning throughout the specifications.

It is highly important that specification writers and reviewers be aware of the definition in the general provisions, and of the defined meanings, so that these words, phrases, and abbreviations are not inadvertently used when another meaning is intended.

Section 2. Interpretation of Contract

This Section creates a "law of the contract" with respect to interpretation. It contains a provision which sets forth a method of correcting "errors"—that is, conflicts—that may creep into the specifications or plans, despite the owner's best efforts. Of special importance is the following rule set forth in Section 2:

> "The drawings will govern over the Standard Provisions; the Special Provisions will govern over both the Standard Provisions and the drawings, the Technical Provisions will govern over the Special Provisions, the Standard Provisions and the drawings.
>
> "Detail drawings will govern over general drawings; figures or dimensions written on drawings will govern over scaled distances."

Section 3. Applicable Laws and Regulations

This Section collects public policy requirements, usually expressed in statutes, regulations, or local ordinances, that are applicable to performance of the work.

Many of the Articles within this section are required to be included in these illustrative specifications because the owner is a public agency and the law of the jurisdiction requires their inclusion.

However, private owners, in the interest of clarity, should include many of these requirements in their specifications, even though not required by law to do so. For example, inclusion of provisions similar to Articles (1) Health and Sanitation, (m) Accident Prevention, and (n) Fish Protection, together with Article (p) Payment, will make completely clear that, as between the parties, it is the constructor's responsibility to insure that performance of the work will comply with laws of general application relating to health and safety of workmen and protection of water quality and fisheries, and, moreover, that no separate compensation will be paid by the owner for meeting these responsibilities.

Section 4. Contractual Relations of Parties

This Section fixes, as between the parties, additional non-technical responsibilities. Although the Articles in this and succeeding Sections may not be required by the law of the jurisdiction to be included in specifications for dam construction, experience has shown that the subjects covered in this Section should be dealt with in the specifications issued by public agencies.

With few exceptions, the private owner should also cover the matters set forth in this Section in his general provisions. The private owner might wish to select appropriate Articles from Section 3 and combine them with the Articles of Section 4 into one section entitled "Contractual Relations of the Parties."

Section 5. Prosecution of the Work

The provisions of this Section are generally concerned with time constraints upon the constructor in his performance of the work.

Article (a) identifies the event—that is, the giving of notice to begin the work, by the engineer—the occurrence of which authorizes the constructor to commence operations. This event also becomes the beginning date of the period of time within which the constructor is contractually obligated to complete the whole work, and, in certain cases the beginning date of various periods of time within which the constructor is obligated to complete specified portions of the work (see Section 11, Article (a), for specification of the completion periods under the model contract).

Article (b) imposes upon the constructor the duty to perform the work diligently and authorizes the owner to order the constructor to take specified steps if performance lags. Should the constructor fail to respond to an order pursuant to this Article, the owner may terminate the constructor's control over the work—a default termination—under the provisions of Article (h).

Article (c) makes time of performance an essential condition of the contract, and provides the necessary legal basis for a later provision which quantifies an amount of money—"liquidated damages"—which normally will be deducted from payments to the constructor for each day he is late in completing the whole work or any specified portion (see also Section 11, Article (a)).

Article (d) sets forth the occurrences which will legally entitle the constructor to an extension of the completion deadline (or deadlines) specified in Section 11. The effect of such an entitlement is to excuse all or a portion of a constructor's late performance. This in turn means that, to the extent of his entitlement, the constructor will be relieved of his obligation to suffer liquidated damages which have been assessed or for which he is liable for assessment. A typical example of the operation of this Article occurs where the owner finds it necessary to order additional work under the specifications' changes clause (see Section 7). It is not reasonable to expect the constructor to perform a significant amount of extra work within the time constraints calculated for performance of the originally specified work. Therefore, where the constructor can demonstrate that the owner's order for a change in the work has caused him delay in meeting a completion deadline, he is entitled to an extension of the specified time (or times) for completion and, correspondingly, relief from assessment of liquidated damages.

Article (f) specifies the grounds upon which the owner may suspend—that is, order the stoppage of—all or a portion of the constructor's performance of the work. Needless to say, where a work suspension is not due to any fault of the constructor, an order to suspend the work will entitle him to an extension of the specified completion date. Furthermore, where an order to suspend the work is issued for the owner's convenience and benefit, the constructor is entitled both to an extension of the specified completion date and to additional payment ("adjustment in compensation").

Article (h) identifies those conditions under which the constructor's performance of the work may legally be terminated by the owner. The specifications permit the owner to take this course of action in those situations where the constructor has refused direct orders of the owner to continue work, has demonstrated inability or unwillingness to complete the work in a timely manner, or to perform it in accordance with the requirements of the plans and specifications. This is a very strong remedy, and the prudent owner will attempt to utilize all other management tools available before invoking his right to terminate. The constructor will usually not agree that he is in default in performance and, since he normally stands to lose financially or otherwise if his performance is terminated, the constructor will often seek to vindicate his position in a court of law if termination is ordered by the owner.

Section 6. Control of the Work

Article (a) contains a general grant of contract management authority to the owner's representative, the Engineer. Its most significant provision—and one of the most significant provisions to be found anywhere in the specifications—is that provision which authorizes the Engineer "to decide all questions as to interpretation and fulfillment of contract requirements, including, without limitation, all questions as to the prosecution, progress, quality and acceptability of the work." This grant of authority to one of the parties to the contract carries with it a very considerable responsibility, essential to conduct of the work in accordance with the specifications. The applicable law of jurisdictions generally implies a condition upon the exercise of this authority; that is, that it not be fraudulent, capricious, arbitrary, or so grossly erroneous as necessarily to imply bad faith. A decision by the Engineer as to interpretation and fulfillment of contract requirements may ultimately be reviewed in a court of law, should the constructor believe that the decision violates this standard.

Article (b) identifies the general scope of supervision over the work to be exercised by the field representative of the constructor. This Article requires the constructor to designate such a representative and specifies the scope of authority that must be delegated to him to act on behalf of the constructor.

Several articles in this Section specify processes which are intended to aid the Engineer in exercising his contract management authority. These include Article (c), which provides for the constructor's submission and the Engineer's review of working drawings and data; Article (k), which deals with testing of materials incorporated into the work; Articles (1) and (o), providing for the Engineer's inspection and ultimate acceptance of the work; and Article (m), which provides the Engineer with specific authority to reject defective work and order its removal and replacement with work which meets specification requirements.

Section 7. Changes and Unforeseen Site Conditions

a. The "Changes Clause." This section contains those articles which, taken together, are popularly called the "Changes Clause." (Articles (b), (c), (d), (e), and (f)). They include provisions for requiring the constructor to make changes in the work, under certain circumstances, from that originally required by the specifications and plans, including performance of additional work. They are considered absolutely essential in specifications for major construction projects, including especially dam construction. The rationale underlying inclusion of the Changes Clause is the likelihood that during construction circumstances and conditions will arise (1) that could not have been foreseen by the most competent designer, and (2) which, if not met by changes to the original plans and specifications, would result in construction of an inadequate or incomplete project.

The Changes Clause is therefore a most important part of the specifications for a major construction project in that it enables the owner, through the Engineer, to react to unforeseen circumstances encountered during performance by ordering changes in the work to the end that an adequate, complete project is achieved. The Changes Clause is also significant because the Engineer's exercise of his authority thereunder is generally the source of a majority of disputes between the owner and the constructor that will arise in a given project.

Several important concepts should be kept in mind in considering the development, inclusion, and exercise of a Changes Clause.

First, the Engineer's authority to order changes in the work is limited. Article (b) (1) contains the typical limitation: "The Engineer may order such changes in the work *as are required for its proper completion . . .* " The Changes Clause does not authorize material changes in the project which alter the scope of the work as originally contemplated, nor does it authorize the addition of "frills" not required for proper completion. A change to the specifications, for example, which would convert an earthfill dam project to a concrete arch dam project would likely be beyond the authority given the owner by the Changes Clause. However, an order by the Engineer for a change in the axis of the dam, or a change in its dimensions, may well be a proper exercise of change authority, depending upon the circumstances creating the need for such a change.

Second, an order for a proper change in the work—that is, a change required for proper completion of the work—is not an amendment to the contract. This concept is important because under contract law, an amendment requires agreement of both contracting parties. An order for a proper change in the work does not require the constructor's consent prior to his being bound to perform the change work. In entering into the contract, the constructor agrees to all of the contract's terms and conditions, including the authority of the owner to make proper changes, and the constructor therefore has agreed to prosecute proper change work with the same diligence as the originally specified work (see Article (d) (1)). Refusal to do so gives the owner the right to terminate the constructor's control over the work under the provisions of Section 5, Article (h).

Third, an order for a change in the work which results in an increase in the constructor's cost of performance creates a right in the constructor to an upward adjustment in the compensation due him under the contract. Article (e) specifies in detail adjustments in compensation by reason of changes in the work, as well as for unforeseen site conditions (discussed below) and for certain types of owner-caused delay to the work. It should be noted that a proper change in the work may also be ordered which results in a decrease in the constructor's cost of performance. In such a case, the owner has a right to a downward adjustment in compensation.

b. Unforeseen Site Conditions. The provisions of Article (g) entitle the constructor to an adjustment in compensation if during the performance of the work, he encounters either (1) subsurface or latent physical conditions at the work site which differ materially from those conditions as represented in the bid documents, or (2) unknown and unusual physical conditions at the work site which differ materially from conditions generally recognized as inherent in the particular type of work being performed.

In the absence of inclusion in the specifications of the provisions of clause (1) above, a constructor who, having reasonably relied upon the owner's representation of conditions at the site in developing his bid, encounters differing conditions which increase his cost of performing the work, is usually required to bring an action for misrepresentation against the owner in a court of law in order to recover such increased costs. These provisions permit the Engineer to adjust the constructor's compensation, without the necessity of a lawsuit, pursuant to the contract method specified in Article (d) (3).

In the absence of inclusion of the provisions of clause (2) above, the constructor who encounters such unknown and unusual conditions has no remedy at all and is required to bear any increased costs of performance without an adjustment in compensation. Prudent bidder-constructors, therefore, are presumed to include in their bids a contingency cost to cover this risk. Inclusion of these provisions permits payment for increased costs of performance arising from the constructor's encountering the specified unique conditions pursuant to the Article (d) (3) method. There is, therefore, no necessity for the bidder-constructors to inflate their bids by including contingency costs in their bids.

Section 8. Claims

This section provides a procedure for the constructor's submission of claims resulting from disputes arising during performance of the work, and for decision of such claims by the Engineer. There are a multitude of sources of disputes between the constructor and owner which may lead to the constructor's invocation of the claims procedure. Typical sources include disputes arising from the pricing of adjustments in compensation for changes in the work ordered by the Engineer. Another fertile source of disputes leading to claims is the constructor's disagreement with an interpretation of the specification requirements made by the Engineer under Section 6, Article (a).

It is important to note that the provisions of Article (c) place claims-deciding authority in the Engineer, the representative of the owner, who is one of the parties to the contract. This is consistent with the general contract management authority given the Engineer by Section 6, Article (a). The contractual grant of claims-deciding authority carries with it the responsibility to decide claims fairly; that is, in a manner that is not fraudulent, capricious, arbitrary, or so grossly erroneous as necessarily to imply bad faith (see Section 6, Article (a)). The Engineer's claim decisions may be reviewed in a court of law, should the constructor believe his claim has been unfairly considered against this standard.

Section 9. Measurement and Payment

Article (c) of this section contains detailed specifications for the pricing of change work, work resulting from encountering unforeseen site conditions, delay to the work, or waiver of defects in the work, by means of "force account" computation. Force account payment is essentially a pricing method involving identification of actual necessary costs of labor, materials, equipment and construction equipment used in performance, plus an allowance on such costs for superintendence, general expenses and profit; it is a cost-plus-percentage-of-cost pricing method.

Force account pricing is not the exclusive method of pricing an adjustment in compensation. Other methods are negotiation of the parties and unilateral pricing by the Engineer (see Section 7, Article (e) (3)). These methods, however, are also cost-related, and the detailed force account pricing provisions contained in this section provide a useful standard against which to measure the reasonableness of the prices determined by these other methods.

Article (d) specifies the method of computation and payment of compensation to the constructor during the progress of the work. Normally, the constructor is paid, in the form of monthly progress payments, the value of the work performed to date, computed in accordance with his bid prices, less a percentage of such value retained for final payment when the work is completed and accepted (see Section 6, Article (o), and Article (e) of this section).

12-4. THE SPECIAL PROVISIONS

As previously noted, the special provisions portion of the specifications contains non-technical requirements of special application to the particular project.

Section 10 is the first section of the Special Provisions. It generally describes the principal features of the particular project and its location.

Several articles of Section 11, Special Conditions, merit comment. Article (a) specifies the time periods within which the work must be completed, pursuant to the provisions of Section 5, Article (c). In this particular project, the constructor is required to meet interim completion dates with respect to two groups of portions of work, and an overall completion date. This article also specifies the amounts of liquidated damages that will be assessed should the constructor fail to meet these time constraints.

Article (b) consists of a complete listing of the contract drawings and requires that the work shall conform to them.

In Article (j), the owner agrees to furnish the constructor all rights-of-way for the permanent works to be constructed, for borrow and waste areas, and for construction roads.

Article (k) implements the provisions of Section 4, Article (g), by specifically identifying portions of the work of which the owner is entitled to take possession and use prior to acceptance of the whole work.

Article (t) imposes detailed requirements for the form and submittal of construction schedules. Articles (u), (v), and (w) specify other types of documents required to be submitted by the constructor, including specific working drawings and data, final drawings, and operation and maintenance instructions.

Section 12 implements Section 6, Article (e), by specifying in detail the work layout and survey responsibilities as between the owner and the constructor.

Section 13 specifies the requirements for a "construction facility" to be used by the Engineer and his staff in managing the contract, including a headquarters office building, a materials testing laboratory building, a project construction sign, and a visitors site overlook and information booth. These structures are intended to be permanent facilities, and the requirements for their construction may have been included in the Technical Provisions. The reason for their inclusion in the Special Provisions is that they are administrative in function and are not a part of the dam and reservoir construction proper.

12-5. SUMMARY

This chapter has discussed the functions of specifications for dam construction and has presented a conceptual format for specification development: technical provisions, general or standard provisions, and special provisions.

The chapter has suggested certain general principles applicable to development and review of the technical provisions. It has presented an outline of a model of a dam contract and commented upon the more significant standard provisions in an effort to establish a theoretical foundation for contract management. It has presented and commented upon certain important special provisions to give a flavor of the kinds of requirements that are properly included in that category.

Implicit in this discussion is the author's view that the completed specifications for a major construction project, including but not limited to a dam project, should comprise an integrated whole, capable of being understood and executed by the "average competent mechanic."

In the final analysis, a set of construction specifications is the vital link between the owner's perception of his project needs and the realization of those needs in the form of a completed dam and reservoir built by the constructor. If the specification provisions are inadequate, the owner cannot expect to realize an adequate dam and reservoir at the agreed upon price; if the provisions are adequate, the owner can expect it.

CHAPTER 12
APPENDIX

	Page
Table of Contents of Specifications	685
Notice to Contractors	691
Standard Provisions	694
Special Provisions	730

TABLE OF CONTENTS OF SPECIFICATIONS

NOTICE TO CONTRACTORS

STANDARD PROVISIONS

SECTION 1. DEFINITIONS

SECTION 2. INTERPRETATION OF CONTRACT

SECTION 3. APPLICABLE LAWS AND REGULATIONS
- (a) General
- (b) Permits and Licenses
- (c) Fair Employment Practices
- (d) Labor Code Requirements—Application
- (e) Prevailing Wages
- (f) Hours of Labor
- (g) Travel and Subsistence Payments
- (h) Alien Labor
- (i) Apprentices
- (j) Workmen's Compensation
- (k) Subletting and Subcontracting Fair Practices Act
- (l) Buy American Act
- (m) Health and Sanitation
- (n) Accident Prevention
- (o) Fish Protection
- (p) Suspension of the Work
- (q) Payment

SECTION 4. CONTRACTUAL RELATIONS OF PARTIES
- (a) Contractor's Responsibility for Subcontractors and Suppliers
- (b) Liability Insurance
- (c) Preservation of Property
- (d) Maintenance and Protection of the Work
- (e) Injury or Death of Persons
- (f) Indemnification of Other Governmental Authorities
- (g) Cooperation With Other Contractors and Forces
- (h) Patents
- (i) Payment of Taxes
- (j) Property Rights in Materials and Equipment
- (k) Use of Property at the Site
- (l) Assignment
- (m) Stop Notices, Tax Demands and Claims of State Agencies
- (n) Personal Liability
- (o) Conditions on Indemnification
- (p) Waiver of Rights
- (q) Notices
- (r) Suspension of the Work
- (s) Payment

SECTION 5. PROSECUTION OF THE WORK
- (a) Commencement of the Work
- (b) Prosecution of the Work
- (c) Time of Completion and Liquidated Damages
- (d) Time Extensions

686 HANDBOOK OF DAM ENGINEERING

    (e) Acceleration in Lieu of Time Extension
    (f) Suspension of the Work
    (g) Payment for Delay
    (h) Termination of Control Over the Work

SECTION 6. CONTROL OF THE WORK
    (a) Authority of the Engineer
    (b) Contractor's Supervision of the Work
    (c) Working Drawings, Data on Materials and Equipment, and Architectural Samples
    (d) Supplemental and Revised Drawings
    (e) Surveys
    (f) Construction Equipment and Plant
    (g) Use of Construction Equipment Different From That Specified
    (h) Materials and Equipment
    (i) Trade Name Materials and Equipment and Alternatives
    (j) Preparation of Equipment for Shipment
    (k) Testing of Materials
    (l) Inspection of the Work
    (m) Defects in the Work and Unauthorized Work
    (n) Cleanup
    (o) Final Inspection and Acceptance of the Work
    (p) Payment

SECTION 7. CHANGES AND UNFORESEEN SITE CONDITIONS
    (a) Variations in Quantities
    (b) Authority for Changes
    (c) Ordering of Changes
    (d) Prosecution of Changes in the Work
    (e) Adjustments in Compensation
    (f) Acceptance and Protest of Changes
    (g) Unforeseen Site Conditions
    (h) Materials and Data on Conditions at Site

SECTION 8. CLAIMS
    (a) Notice of Potential Claim
    (b) Submission and Documentation of Claims
    (c) Decision of Claims
    (d) Release of Undecided Claims

SECTION 9. MEASUREMENT AND PAYMENT
    (a) Measurement and Computation of Quantities
    (b) Payment at Contract Prices
    (c) Force Account Payment
    (d) Progress Payments
    (e) Final Payment
    (f) Interest
    (g) Corrections in Final Payment

SPECIAL PROVISIONS

SECTION 10. GENERAL DESCRIPTION AND LOCATION OF THE WORK

SECTION 11. SPECIAL CONDITIONS
    (a) Time of Completion and Liquidated Damages
    (b) Drawings
    (e) Definitions

SPECIFICATIONS FOR DAM CONSTRUCTION 687

    (d) Fair Employment Practices
    (e) Alien Labor
    (f) Materials and Equipment from Foreign Countries
    (g) Determination of Rights
    (h) Payment of Withheld Funds
    (i) Data Available to Bidders and the Contractor
    (j) Right-of-Way
    (k) Advance Possession and Use of the Work by the Engineer
    (l) Cooperation with Other Contractors and Forces
    (m) Cooperation with Riverside County
    (n) Materials and Equipment Furnished by the Department
    (o) Contractor's Schedules
    (p) Contractor's Roads and Construction Plants
    (q) Protection of Existing Facilities
    (r) Temporary Gates and Fences
    (s) Water
    (t) Dust Control
    (u) Specific Submittal Requirements for Working Drawings and Data
    (v) Final Drawings
    (w) Operations and Maintenance Instructions
    (x) Indemnification of Third Parties
    (y) Guarantee
    (z) First Aid Services and Facilities
  (aa) Safety Engineer
  (bb) Measurement of Materials Paid For by the Ton
  (cc) Funds Available for Payment
  (dd) Prevailing Wages
  (ee) Payment
  (ff) Award of Contract

SECTION 12. LAYOUT OF THE WORK AND SURVEYS
    (a) Description
    (b) Layout of the Work
    (c) Tolerances in Setting Survey Stakes
    (d) Payment

SECTION 13. CONSTRUCTION FACILITY
    (a) Description
    (b) Sanitary Sewer System
    (c) Headquarters Office Building and Materials Testing Laboratory Building
    (d) Project Construction Signs
    (e) Visitors Information Booth
    (f) Payment

TECHNICAL PROVISIONS

SECTION 14. CONTROL OF WATER AND UNWATERING
    (a) Description
    (b) Protection Against Floods During Construction
    (c) Unwatering Foundations and Borrow Areas
    (d) Payment

SECTION 15. CLEARING AND GRUBBING, AND EROSION CONTROL
    (a) Description
    (b) Clearing

- (c) Grubbing
- (d) Removal of Buried Water Pipe
- (e) Disposal of Debris
- (f) Erosion Control
- (g) Measurement
- (h) Payment

SECTION 16. EXCAVATION
- (a) Description
- (b) General Excavation Requirements
- (c) Blasting
- (d) Excavating Beyond Established Lines
- (e) Excavation for Dam Foundation
- (f) Excavation for Grout Caps
- (g) Excavation for Inlet Works and Spillways
- (h) Excavation for Outlet Works Intake
- (i) Structure Excavation
- (j) Production of Materials from Borrow Areas
- (k) Excavation in Rock Source
- (l) Measurement
- (m) Payment

SECTION 17. DRILLING HOLES AND GROUTING
- (a) Description
- (b) Definitions
- (c) Procedure
- (d) Mobilization for Grouting Operations
- (e) Construction Equipment for Grouting Operations
- (f) Materials
- (g) Drilling Grout Holes
- (h) Drilling Exploratory Holes
- (i) Installing Pipe and Fittings
- (j) Washing and Pressure Testing
- (k) Grouting
- (l) Repair and Cleanup
- (m) Records
- (n) Measurement
- (o) Payment

SECTION 18. EMBANKMENTS, BACKFILL, AND SLOPE PROTECTION
- (a) Description
- (b) General Requirements
- (c) Materials
- (d) Preparation of Foundation in Rock for Dam Embankment
- (e) Preparation of Foundation in Alluvium for Dam Embankment
- (f) Placement and Compaction of Embankment
- (g) Compactors
- (h) Sprinklers and Disk
- (i) Riprap
- (j) Spillway Backfill
- (k) Impervious Backfill
- (l) Soil Tests
- (m) Measurement
- (n) Payment

SECTION 19. ROADWAYS AND SITE DEVELOPMENT
- (a) Description
- (b) Roadway Excavation
- (c) Site Development
- (d) Ditch and Channel Excavation
- (e) Embankment Construction
- (f) Compacting Original Ground
- (g) Structure Backfill
- (h) Corrugated Metal Products
- (i) Subgrade Construction Equipment
- (j) Subgrade Preparation
- (k) Aggregate Base
- (l) Liquid Asphalt Prime Coat
- (m) Asphalt Concrete
- (n) Pavement Striping
- (o) Precast Concrete Bumpers
- (p) Stone Slope Protection
- (q) Fences
- (r) Padlocks
- (s) Measurement
- (t) Payment

SECTION 20. CONCRETE PRODUCTION AND PLACEMENT
- (a) Description
- (b) Composition
- (c) Cement and Pozzolan
- (d) Admixtures
- (e) Water
- (f) Sand
- (g) Coarse Aggregate
- (h) Production of Sand and Coarse Aggregate
- (i) Batching
- (j) Mixing
- (k) Maximum Temperature of Concrete
- (l) Forms
- (m) Preparation for Placing
- (n) Placing
- (o) Repair of Concrete
- (p) Finishes and Finishing
- (q) Protection Against Freezing
- (r) Curing
- (s) Tolerances for Concrete Construction
- (t) Payment

SECTION 21. CONCRETE CONSTRUCTION
- (a) Description
- (b) Reinforcing Steel
- (c) Joints
- (d) Waterstop
- (e) Concrete in Backfill and Grout Cap
- (f) Structure Concrete
- (g) Measurement
- (h) Payment

SECTION 22. INSTRUMENTATION
- (a) Description
- (b) General
- (c) Materials and Equipment to be Furnished by the Department
- (d) Instrumentation Trenches and Risers
- (e) Piezometers
- (f) Accelerometers
- (g) Cable Identification Markers
- (h) Crest Settlement Monuments
- (i) Measurement
- (j) Payment

SECTION 23. INSTRUMENTATION TERMINALS
- (a) Description
- (b) Materials
- (c) Graded Gravel Fill
- (d) Masonry Construction
- (e) Steel Roof Decking
- (f) Built-up Roof
- (g) Metal Doors and Frames and Locks
- (h) Aluminum Thresholds
- (i) Application of Epoxy Coating
- (j) Miscellaneous Items of Work
- (k) Painting
- (l) Mechanical Work
- (m) Electrical Work
- (n) Payment

SECTION 24. MISCELLANEOUS FACILITIES
- (a) Description
- (b) Drain Pipe
- (c) Drain Pipe Installation
- (d) Precast Concrete Pipe Riser
- (e) Measurement
- (f) Payment

BID FORM
- BIDDING REQUIREMENTS AND CONDITIONS
- BIDDER'S AGREEMENT
- BID SCHEDULE
- SUBCONTRACTORS
- BIDDER'S SIGNATURE PAGE
- FAIR EMPLOYMENT PRACTICES CERTIFICATION
- WORKMEN'S COMPENSATION INSURANCE CERTIFICATION
- BIDDER'S BOND FORM

CONTRACT FORM

APPENDIX I SCHEDULE OF FORCE ACCOUNT EQUIPMENT RATES

APPENDIX II WELDING PROCEDURE SPECIFICATION FORM
      FILLET WELD PROCEDURE TEST REPORT FORM
      GROOVE WELD PROCEDURE TEST REPORT FORM

STATE OF CALIFORNIA
THE RESOURCES AGENCY
DEPARTMENT OF WATER RESOURCES

NOTICE TO CONTRACTORS

Sealed bids for

CONSTRUCTION OF
PERRIS DAM
AND LAKE
STATE WATER FACILITIES
CALIFORNIA AQUEDUCT
SANTA ANA DIVISION
RIVERSIDE COUNTY
CALIFORNIA
SPECIFICATION NO. 70-25

will be received by the Department of Water Resources of the State of California at the office of the Director of Water Resources, Room 1202, 849 South Broadway, Los Angeles, California, until 10:00 a.m., on **THURSDAY, OCTOBER 1, 1970**, at which time they will be publicly opened and read at an announced location in the vicinity of such office.

Bids will be considered only if submitted for all of the work required for the above project. The work is defined in Section 1 of the Standard Provisions of the Specifications and includes the following principal features:

A zoned embankment type dam approximately 11,600 feet long at the crest and 120 feet in height; open channel spillway, clearing of lake area; excavation and blanketing for outlet works intake; materials haul road.

THE SPECIFICATIONS PROVIDE FOR OBTAINING ONE TYPE OF BORROW MATERIAL FROM TWO ALTERNATIVE SOURCES UNDER SCHEDULE "A" AND SCHEDULE "B". BID SHALL INCLUDE BID PRICES FOR BOTH SCHEDULE "A" AND SCHEDULE "B". (SEE SECTION 10, AND SECTION 11, ARTICLE (ff)).

Quantities of work, materials and equipment required for completion of the work as estimated to be as follows:

Item 1	Control of water and unwatering
Item 2	Clearing and grubbing
Item 3	85 Acres erosion control
Item 4	1,750,000 Cubic yards excavation for dam embankment foundation to Station 122+00
Item 5	80,000 Cubic yards excavation for dam embankment foundation beyond Station 122+00
Item 6	280 Cubic yards excavation for grout caps
Item 7	70,000 Cubic yards excavation for spillway
Item 8	300,000 Cubic yards excavation for outlet works intake
Item 9	10,900 Cubic yards structure excavation
Item 10	8,670,000 Cubic yards excavation in clay borrow area
Item 11	Deleted
Item 12	Mobilization for grouting operations
Item 13	3,400 Linear feet drilling blanket and curtain grout holes between depths of 0 and 25 feet
Item 14	1,900 Linear feet drilling curtain grout holes between depths of 25 and 50 feet
Item 15	1,900 Linear feet drilling curtain grout holes between depths of 50 and 100 feet

Item 16	800 Linear feet drilling exploratory holes
Item 17	1,600 Pounds pipe for grouting
Item 18	288 Each connection to grout holes
Item 19	8,600 Cubic feet grouting
Item 20	1,300 Cubic feet slush grout
Item 21	6,360,000 Cubic yards compacting Zone 1 dam embankment
Item 22	8,230,000 Cubic yards compacting Zone 2 dam embankment
Item 23	390,000 Cubic yards compacting dam embankment facing
Item 24	1,230,000 Cubic yards Zone 3 dam embankment
Item 25	1,150,000 Cubic yards Zone 4 dam embankment
Item 26	1,160,000 Cubic yards compacting blanket
Item 27	320,000 Cubic yards riprap
Item 28	810 Cubic yards spillway backfill
Item 29	75 Cubic yards impervious backfill
Item 30	18,600 Cubic yards roadway excavation
Item 31	3,400 Cubic yards site development embankment
Item 32	430 Cubic yards ditch and channel excavation
Item 33	20,000 Square yards compacting original ground
Item 34	145 Cubic yards structure backfill
Item 35	194 Linear feet 18-inch, 14-gage, corrugated metal pipe, galvanized
Item 36	66 Linear feet 54-inch, 12-gage, corrugated metal pipe, galvanized
Item 37	116 Linear feet 43-inch by 27-inch, 14-gage, corrugated metal pipe arch, galvanized
Item 38	2 Each 54-inch, metal flared end section, galvanized
Item 39	3,410 Tons aggregate base
Item 40	20 Tons liquid asphalt prime coat
Item 41	940 Tons asphalt concrete
Item 42	2,570 Linear feet pavement striping
Item 43	81 Each precast concrete bumper
Item 44	6 Cubic yards stone slope protection
Item 45	672 Linear feet 4-foot high chain link fence
Item 46	744 Linear feet 6-foot high chain link fence
Item 47	14,000 Pounds reinforcing bars
Item 48	2,970 Barrels cement
Item 49	22 Linear feet waterstop
Item 50	300 Cubic yards concrete in backfill and grout cap
Item 51	85 Cubic yards structure concrete
Item 52	230 Linear feet drilling and backfilling holes for foundation piezometers
Item 53	24 Each furnishing and installing piezometers
Item 54	Furnishing and installing terminal facilities for piezometers
Item 55	5 Each installing accelerometer
Item 56	27 Each furnishing and installing crest settlement monuments
Item 57	Instrumentation terminal No. 1
Item 58	Instrumentation terminal No. 2
Item 59	11,000 Linear feet 12-inch diameter perforated drain pipe
Item 60	24 Linear feet 36-inch diameter precast concrete pipe riser
Item 61A	9,320,000 Cubic yards excavation in lake borrow area—Schedule "A"
Item 61B	9,770,000 Cubic yards excavation in lake borrow area—Schedule "B"

WAGE RATES

Pursuant to Sections 1770 and 1773 of the California Labor Code, the Department has determined the general prevailing rates of wages, including employer payments for health and welfare, pensions, vacations and similar purposes as provided in Section 1773.1 of the Labor Code, for the crafts, classifications, or types of workmen required for the work, in the locality of the work. Copies of prevailing wage rates are on file in the Resources Building, Room 406-2, 1416 9th Street, Sacramento, California, as provided in Sections 1773 and 1773.4 of the California Labor Code as amended.

The general prevailing rate of wages for overtime, Sundays, and holidays for each craft, classification, or type of workmen required for the work, in the locality of the work, is determined to be not less than one and one-half ($1\frac{1}{2}$) times the basic hourly rate for the craft, classification, or type of workmen, as determined above, plus employer payments for that craft, classification, or type of workmen, as determined.

The general prevailing rate of wages for apprentices is determined to be the standard wage paid to apprentices under the regulations of the trade at which he is employed.

BIDDING

Forms of bid and contract, and drawings and specifications for the project, hereinafter called bid documents, may be obtained only at the office of the Department of Water Resources, Room 406-2, Resources Building, 1416 Ninth Street, Sacramento, California, or by mail upon written request to the Department of Water Resources, P.O. Box 398, Sacramento, California 95802. Bid documents may be seen at the above location or at the offices of the Department at Glen Drive, Oroville, California; 909 South Broadway, Los Angeles, California; 601 California Avenue, Bakersfield, California; 406 East Palmdale Boulevard, Palmdale, California; and 31849 No. Lake Hughes Road, Castaico, California.

A CHARGE OF $5.00, WHICH IS NOT REFUNDABLE, WILL BE MADE FOR EACH SET OF SPECIFICATIONS AND DRAWINGS.

Bid documents furnished to bidders not meeting prequalification or joint venture bidding requirements, or to prospective subcontractors, suppliers, or other parties not interested in bidding on the work, shall not be used for bidding purposes and will be so stamped.

All bidders shall be prequalified by the Department, in accordance with the State Contract Act and bid form requirements. Bid documents for use in bidding will be furnished only to prequalified bidders, and will be furnished to joint venture bidders only if they meet the joint venture bidding requirements set forth in the bid form.

All bidders shall be licensed for the work when and as required by the provisions of Chapter 9 of Division 3 of the California Business and Professions Code.

Questions relating to bidding may be directed to the Office Engineer of the Department of Water Resources in Sacramento, at the location of address given above, or at Telephone (916) 445-5018. A conducted tour of the site of the work will be scheduled for all interested prospective bidders.

The Director of Water Resources may reject any or all bids.

Director of Water Resources

Dated:

694   HANDBOOK OF DAM ENGINEERING

1-15-71 EDITION
Supersedes 7-1-69
SECTION 1

STANDARD PROVISIONS

SECTION 1. DEFINITIONS

Whenever in this contract the following abbreviations and terms or pronouns in place of them are used, their meanings shall be as follows:

Abbreviations.—

AAMA	Architectural Aluminum Manufacturers Association
AASHO	American Association of State Highway Officials
AGMA	American Gear Manufacturers Association
AISI	American Iron and Steel Institute
AISC	American Institute of Steel Construction
ANS	American National Standard
ANSI	American National Standards Institute, Inc.
ASCE	American Society of Civil Engineers
ASHRAE	American Society of Heating, Refrigerating and Air Conditioning Engineers
ASME	American Society of Mechanical Engineers
ASTM	American Society for Testing and Materials
AWS	American Welding Society
AWPA	American Wood-Preservers' Association
AWWA	American Water Works Association
ESO	Electrical Safety Orders of State of California Department of Industrial Relations, Division of Industrial Safety
IEEE	Institute of Electrical and Electronic Engineers
IPCEA	Insulated Power Cable Engineers Association
NEC	National Electrical Code
NEMA	National Electrical Manufacturers Association
NFPA	National Fire Protection Association
SAE	Society of Automotive Engineers
UL	Underwriters' Laboratories, Inc.

Acceptance.—The formal written acceptance of the work by the Director or his authorized representative.

Approved, directed, ordered, or their derivatives.—Approved, directed, or ordered by the Engineer, unless otherwise expressly indicated.

As shown.—As shown on the drawings, unless otherwise expressly indicated.

Bid or proposal.—The offer of a bidder to perform the work.

Bidder.—Any person, firm, partnership, corporation or combination thereof submitting a bid, either directly or through a duly authorized representative.

Bid documents.—The Notice to Contractors, forms of bid and contract, and drawings and specifications for the project furnished to bidders by the Department.

Change.—Any change in the contract ordered by the Engineer under authority of the specifications.

Change in the work.—A change which modifies the physical nature or extent of the work, as distinguished from a change which adjusts compensation or extends time for performance of the work.

Change order.—A formal written order of the Engineer for a change or changes designated "change order". The only form of order for changes which may effect an adjustment in compensation or extension of time for performance of the work.

Construction equipment.—Equipment used in the performance of the work but not incorporated therein, exclusive of manufacturing equipment at plants removed from the site of the work.

Contract.—The written agreement for the performance of and payment for the work which includes by reference the Notice to Contractors, Contractor's Bid, drawings, specifications, contract bonds, certificate of liability insurance, and all addenda and changes to the foregoing documents.

Contractor.—The person, firm, partnership, corporation, or combination thereof which has entered into the contract with the Department, or the legal representatives of such party. The Contractor is referred to throughout the contract as if of the singular number and masculine gender.

Contract prices.—The prices for the work set forth in the Contractor's Bid.

Days.—Calendar days, unless otherwise expressly indicated.

Department.—The Department of Water Resources of the State of California.

Deputy Director.—The Department's Deputy Director, State Water Project.

Director.—The Director of Water Resources of the State of California.

Drawings.—All drawings listed in the Special Provisions and any addenda thereto, all supplemental drawings, and revised drawings furnished by the Engineer, and exact reproductions of any of the foregoing.

Engineer.—The Deputy Director, acting either directly or through authorized representatives, each representative acting within the scope of his delegated authority.

Equipment.—Equipment incorporated or to be incorporated in the work, unless otherwise expressly indicated.

Highways Specifications.—The Standard Specifications of the State of California Division of Highways, Department of Public Works in effect on the date of the Notice to Contractors.

Materials.—Materials incorporated or to be incorporated in the work, unless otherwise expressly indicated.

Project.—The improvement to State property, named in the bid documents, to be accomplished under the contract.

Site of the work.—The area to be occupied by the project and all nearby areas occupied or used by the Contractor or his subcontractors during performance of the work, including areas for the production, procurement, storage and disposal of earthwork, concrete and paving materials, and similar materials.

Specifications.—These Standard Provisions and the Special Provisions and Technical Provisions for the project, and all appendices, addenda and changes to such documents.

State.—The State of California.

State Contract Act.—Chapter 3, Part 5, Division 3, Title 2, of the California Government Code, commencing at Section 14250.

Subcontractor.—Any person, firm, partnership, corporation, or combination thereof which contracts with the Contractor to perform any portion of the work, excluding those whose function is limited to the furnishing of materials, equipment, or personal services.

The work.—(1) All the facilities specified, indicated, shown, or contemplated by the contract as comprising and necessary for completion of the project, including any portions of such facilities furnished to the Contractor by the Department, and (2), except as otherwise expressly provided in the specifications, the provision and furnishing by the Contractor of all materials, equipment, labor, methods, processes, construction and manufacturing materials and equipment, tools, plants, supplies, power, water, transportation and other things necessary to complete the foregoing facilities in accordance with the contract.

Work, used in conjunction with or as opposed to materials and equipment.—Those activities of the Contractor, completed or in process, which are exclusive of the furnishing of materials and equipment, unless otherwise expressly indicated.

SECTION 2. INTERPRETATION OF CONTRACT

The Standard Provisions, Special Provisions, Technical Provisions, drawings, changes, and all other parts of the contract are essential thereto, and a requirement occurring in one shall be as binding as though occurring in all. The parts of the contract are complementary and describe and provide for completion of the entire project.

References in the contract to specifications, codes, or test methods published by governmental or private authorities shall be to the specification, code, or test method in effect on the date of the Notice to Contractors, unless otherwise expressly indicated.

The Contractor shall, either before commencing the work or during its performance, promptly report to the Engineer in writing, all apparent ambiguities, conflicts, discrepancies, omissions, or other errors in the contract. On receipt of such notice, the Engineer will promptly investigate the matter and give appropriate orders or instructions to the Contractor. If the Contractor, before receiving orders or instructions from the Engineer, performs any portion of the work affected by such apparent error, such performance shall be at his own risk and he shall not be entitled to additional compensation or time by reason of the error or its later correction.

Subject to the foregoing provisions and the provisions of Article (e) in the Bidding Requirements and Conditions, errors in the contract will be corrected in accordance with the following rules of interpretation:

The drawings will govern over the Standard Provisions; the Special Provisions will govern over both the Standard Provisions and the drawings; the Technical Provisions will govern over the Special Provisions, the Standard Provisions and the drawings.

Detail drawings will govern over general drawings; figures or dimensions written on drawings will govern over scaled distances.

SECTION 3. APPLICABLE LAWS AND REGULATIONS

(a) General.—The Contractor shall keep informed of and comply with all federal, state, county and municipal laws, ordinances and regulations applicable to the work, or to those engaged or employed in the work, and all orders and decrees of bodies or tribunals having any jurisdiction or authority over the work which are so applicable. The Contractor shall indemnify and save harmless the State and all officers and employees of the State connected with the work from all claims, suits, or actions of any nature brought for, or on account of, the violation of any such law, ordinance, regulation, order, or decree by the Contractor or his subcontractors, suppliers, agents, or employees.

The Contractor shall, either before commencing the work or during its performance, promptly report to the Engineer in writing, all apparent conflicts or discrepancies between the drawings, specifications, or other part of the contract and any such law, ordinance, regulation, order or decree. On receipt of such notice, the Engineer will promptly investigate the matter and give appropriate orders or instructions to the Contractor. If the Contractor, before receiving orders or instructions from the Engineer, performs any portion of the work affected by such apparent conflict or discrepancy, such performance shall be at his own risk and he shall not be entitled to additional compensation or time by reason of the conflict or discrepancy or its later correction.

(b) Permits and Licenses.—Except as otherwise provided in the Special Provisions, the Contractor shall procure all permits and licenses, pay all charges and fees, and give all notices necessary and incident to the due and lawful prosecution of the work.

(c) Fair Employment Practices.—In connection with the performance of the work within the State of California, the Contractor agrees as follows:

(1) The Contractor will not willfully discriminate against any employee or applicant for employment because of race, color, religion, ancestry, or national origin. The Contractor will take affirmative action to ensure that applicants are employed and that employees are treated during employment without regard to their race, color, religion, ancestry, or national origin. Such action shall include, but not be limited to, the following: employment, upgrading, demotion or transfer;

recruitment or recruitment advertising; lay-off or termination rates of pay or other forms of compensation; and selection for training, including apprenticeship. The Contractor agrees to post in conspicuous places, available to employees and applicants for employment, notices, to be provided by the awarding authority setting forth the provisions of this Fair Employment Practices article.

(2) The Contractor will send to each labor union or representative of workers with which he has a collective bargaining agreement or other contract or understanding, a notice, to be provided by the awarding authority, advising the said labor union or workers' representative of the Contractor's commitments under this article, and shall post copies of the notice in conspicuous places available to employees and applicants for employment.

(3) The Contractor will permit access to his records of employment, employment advertisements, application forms, and other pertinent data and records by the Fair Employment Practices Commission, the awarding authority or any other appropriate agency of the State designated by the awarding authority for the purposes of investigation to ascertain compliance with the Fair Employment Practices article of this contract.

(4) A finding of willful violation of the Fair Employment Practices article of this contract or of the Fair Employment Practices Act shall be regarded by the awarding authority as a basis for determining the Contractor to be not a "responsible bidder" as to future contracts for which the Contractor may submit bids, for revoking the Contractor's prequalification rating, if any, and for refusing to establish, re-establish or renew a prequalification rating for the Contractor.

The awarding authority shall deem a finding of willful violation of the Fair Employment Practices Act to have occurred upon receipt of written notice from the Fair Employment Practices Commission that it has investigated and determined that the Contractor has violated the Fair Employment Practices Act and has issued an order under Labor Code Section 1426 or obtained an injunction under Labor Code Section 1429.

Upon receipt of any such written notice, the awarding authority shall notify the Contractor that unless he demonstrates to the satisfaction of the awarding authority within a stated period that the violation has been corrected, his prequalification rating will be revoked at the expiration of such period.

(5) The Contractor agrees, that should the awarding authority determine that the Contractor has not complied with the Fair Employment Practices article of this contract, then pursuant to Labor Code Sections 1735 and 1775, the Contractor shall, as a penalty to the awarding authority, forfeit, for each calendar day, or portion thereof, for each person who was denied employment as a result of such non-compliance, the penalties provided in the Labor Code for violation of prevailing wage rates. Such monies may be recovered from the Contractor. The awarding authority may deduct any such damages from any monies due the Contractor from the State.

(6) a. Nothing contained in this Fair Employment Practices article shall be construed in any manner or fashion so as to prevent the awarding authority of the State from pursuing any other remedies that may be available at law.

b. Nothing contained in this Fair Employment Practices article shall be construed in any manner or fashion so as to require or permit the hiring of an employee not permitted by the National Labor Relations Act.

(7) In his bid, the Contractor shall certify to the awarding authority that he has or will meet the following standards for affirmative compliance, which shall be evaluated in each case by the awarding authority:

a. The Contractor shall provide evidence, as required by the awarding authority, that he has notified all supervisors, foremen and other personnel officers in writing of the content of the anti-discrimination clause and their responsibilities under it.

b. The Contractor shall provide evidence, as required by the awarding authority, that he has notified all sources of employee referrals (including unions, employment agencies, advertisements, Department of Employment) of the content of the anti-discrimination clause.

c. The Contractor shall file a basic compliance report, as required by the awarding authority.

Willfully false statements made in such reports shall be punishable as provided by law. The compliance report shall also spell out the sources of the work force and who has the responsibility for determining whom to hire, or whether or not to hire.

d. Personally, or through his representatives, the Contractor shall, through negotiations with the unions with whom he has agreements, attempt to develop an agreement which will:

1. Spell out responsibilities for nondiscrimination in hiring, referral, upgrading and training.

2. Otherwise implement an affirmative anti-discrimination program in terms of the unions' specific areas of skill and geography, to the end that qualified minority workers will be available and given an equal opportunity for employment.

e. The Contractor shall notify the contracting agency of opposition to the anti-discrimination clause by individuals, firms or organizations during the period of its prequalification.

(8) The Contractor will include the provisions of the foregoing paragraphs (1) through (7) in all subcontracts and in any supply contract to be performed within the State of California, so that such provisions will be binding upon each subcontractor and, to the extent provided, each supplier.

(d) Labor Code Requirements—Application.—The requirements of the California Labor Code set forth in Articles (e) through (h) below shall apply to all workmen employed on the work by the Contractor or any subcontractor or supplier to the full extent provided in said code. The following classes of workmen will be considered exempt from such requirements, subject to any contrary interpretation of the code provisions by judicial or administrative authority:

1. Workmen engaged in installing, assembling, repairing, or reconditioning contruction equipment and tools used in the work, if employed by a repair shop, garage, blacksmith shop, or machine shop not owned or operated by the Contractor which was established and operating on a bona fide commercial basis for a period of at least two months prior to award of the contract;

2. Workmen engaged in the construction or operation of production, proportioning or mixing plants from which concrete, aggregate, or similar materils are supplied to the work, if employed by a plant not owned or operated by the Contractor which was established and operating on a bona fide commercial basis for a period of at least two months prior to award of the contract; and

3. Workmen engaged in manufacturing work at plants removed from the site of the work.

(e) Prevailing Wages.—In accordance with Section 1775 of the California Labor Code: (1) the Contractor shall forfeit, as a penalty to the State, and the Department may deduct, from money due or to become due the Contractor under the contract, $25 for each workman employed on the work by the Contractor or any subcontractor for each day or portion thereof during which such workman is paid less than the stipulated prevailing wage rates set forth in the contract, for the craft or classification or type of work in which such workman is employed, in violation of Labor Code Section 1770 through 1780; and (2) the Contractor shall pay to each such workman the difference between the stipulated prevailing wage rates and the amount paid to him for each day or portion thereof for which he was paid less than such rates.

In accordance with Section 1773 of the Labor Code, the Contractor shall post at each jobsite a copy of the prevailing rates of per diem wages in the locality in which the work is to be performed for each craft, classification, or type of workman needed to execute the contract.

(f) Hours of Labor.—Eight hours of labor constitutes a legal day's work. The Contractor shall forfeit, as a penalty to the State, and the Department may deduct, from money due or to become due the Contractor under the contract, $25 for each workman employed on the work by the Contractor or any subcontractor for each day during which such workman is required or permitted to work more than 8 hours in any one day and 40 hours in any one calendar week in violation of the provisions of the California Labor Code, Sections 1810 through 1815; provided, that a workman may be required or permitted to work in excess of 8 hours per day and 40 hours in any one calendar week without penalty if he is paid for all hours in excess of 8 hours per day at not less one and one-half times the basic rate of pay, as provided in Section 1815.

(g) **Travel and Subsistence Payments.**—The Contractor shall make travel and subsistence payments to each workman needed to execute the work in accordance with the requirements of Section 1773.8 of the California Labor Code.

(h) **Apprentices.**—The Contractor shall comply with the provisions of Section 1777.5 and 1777.6 of the California Labor Code concerning the employment of apprentices by the Contractor or any subcontractor.

Information relative to apprenticeship standards, wage schedules, and other requirements may be obtained from the Director of Industrial Relations, ex officio the Administrator of Apprenticeship, San Francisco, California, or from the Division of Apprenticeship Standards and its branch offices.

(i) **Workmen's Compensation.**—The Contractor shall secure the payment of workmen's compensation to his employees in accordance with the provisions of Section 3700 of the California Labor Code and, when applicable, the provisions of Section 3600.5(b) of that code.

(j) **Subletting and Subcontracting Fair Practices Act.**—The Contractor shall comply with the "Subletting and Subcontracting Fair Practices Act" in the California Government Code, commencing at Section 4100. Violation of any of the provisions of that act by the Contractor shall be considered a breach of contract for which the Department may, in its direction, cancel the contract, assess the Contractor a penalty in an amount not more than ten percent of the subcontract involved, or both. The Department may deduct any such penalty, in whole or in part, from money due or to become due the Contractor under the contract.

(k) **Materials and Equipment from Foreign Countries.**—If the Contractor elects to supply materials or equipment produced or manufactured in a country or countries other than the United States of America, the following provisions shall apply:

1. The taxes which the Contractor is required to pay pursuant to Section 4, Article (i), shall include all charges of any kind imposed by any foreign governmental entities, and shall include all charges imposed by the Federal Government as a result of the importation of materials or equipment produced or manufactured in a country or countries other than the United States.

2. All documents pertaining to the contract, including but not limited to, correspondence, bid documents, working drawings and data shall be written in the English language, and all numerical data shall be in the English system of units as used in the United States.

3. Notwithstanding the provisions of Section 5, Article (d) (1)h., the acts of a foreign governmental entity will not be a cause for time extension.

4. Unless otherwise specified, the design, construction, dimensions, performance and tests of all materials and equipment shall conform to the standards of applicable United States authorities listed under abbreviaions in Section1.

(l) **Health and Sanitation.**—The Contractor shall comply with all State and local laws and regulations pertaining to health and sanitation in the performance of the work, and to the establishment and maintenance of dwellings and camps for housing and feeding of employees.

(m) **Accident Prevention.**—

(1) *General.*—The Contractor shall be responsible for accident prevention and safety in the performance of all of the work, and shall be governed by the requirements of this article in all operations at the site of the work.

The Contractor shall comply with all applicable safety regulations and orders of the State of California, Department of Industrial Relations, Division of Industrial Safety and all applicable provisions of the California Health and Safety Code, and shall take or cause to be taken such additional measures as may be necessary for the prevention of accidents. He shall impose the foregoing requirements on all subcontractors and enforce compliance therewith.

The Contractor shall plan and revise his operations as necessary to meet the foregoing requirements; provided, that if safety requirements conflict with requirements of the drawings or specifications which are both express and specific in nature, and such conflict cannot reasonably be avoided by revision of the Contractor's operations, then, subject to the provisions of Article (a), an appropriate order for changes will be issued pursuant to Section 7.

(2) *Guarding the Work.*—The Contractor shall furnish, erect and maintain such fences, barriers, lights and signs, and provide such flagmen and guards as are necessary to give warning to the public or others of any dangerous condition to be encountered as a result of the work and to prevent injury or damage to persons or property therefrom.

(3) *Accident Prevention Program.*—Within 30 days after notice to begin the work, the Contractor shall prepare, submit to the Engineer, and disseminate among his employees and subcontractors a written program for accident prevention. The Engineer will inspect the program for safety of Department employees and the public, and the Contractor shall promptly revise the program to correct any defects noted.

Within such 30 day period and periodically thereafter the Contractor shall meet with the Engineer to review accident prevention practices on the work, as they affect Department employees and the public. Promptly after each meeting, necessary revisions to the accident prevention program shall be prepared, submitted and disseminated, in the same manner as the original program.

The Engineer's inspection of the program and of revisions thereto shall in no way waive any of the requirements of this article or excuse the Contractor of any of his obligations under Section 4.

(4) *Accidents Reports.*—The Contractor shall report to the Engineer on forms furnished by the Department, within the time and in the manner specified therein, all accidents incident to the work which result in death or injury to persons, or in damage to property, and all cases of occupational disease incident to the work. Such reports will be considered confidential and will be used solely to develop information for use in prevention of future accidents and cases of occupational disease.

(5) *Board of Inquiry.*—The Contractor shall notify the Engineer immediately of any fatal or serious injury to any person engaged in the work, and shall convene a board of inquiry to investigate such injury within 72 hours after its occurrence. The Engineer may attend all meetings of the board, and the Contractor shall notify the Engineer 24 hours in advance of the date, time and place of each meeting. The Contractor shall furnish to the board all witnesses, records, documents, and equipment necessary to determine the cause of the injury. No transcript will be made of the board's meetings, unless desired by the Contractor, and all records, documents, written statements of witnesses and copies thereof will be returned to the Contractor upon completion of the investigation. Attendance at board meetings will be limited to the Contractor's representatives and employees, witnesses, and the Engineer and his representatives. The board's sole purpose will be to develop information for use in prevention of future accidents, and it will not determine fault or liability for the injury.

(6) *Violation of Safety Requirements.*—If the Engineer discovers a violation of the requirements of this article which endangers any person or property and notifies the Contractor thereof in writing, the Contractor shall take corrective action within the time specified in the notice. In case of imminent danger, the Engineer may dispense with such notice and order immediate corrective action. Should the Contractor fail to comply with a notice or order so given, the Engineer may suspend the work pursuant to Section 5, Article (f)2. for failure to carry out orders given and to perform provisions of the contract.

At any time, the Engineer may refer violations of safety requirements to the Division of Industrial Safety or other proper authority, or take other appropriate actions with respect to such violations.

(n) **Fish Protection.**—The Contractor shall comply with Section 5650 of the California Fish and Game Code, and with all other provisions of that code and of the Special Provisions relating to water pollution. Before using explosives in State waters inhabited by fish, the Contractor shall obtain a permit from the Fish and Game Commission of the State of California, in accordance with Section 5500 of the California Fish and Game Code.

(o) **Suspension of the Work.**—A suspension of the work for any cause shall not relieve the Contractor of his responsibilities under this section.

(p) **Payment.**—The contract prices shall include full compensation for all costs incurred under this section, except insofar as direct payment is provided in the Special or Technical Provisions for work, materials, or equipment employed in meeting the requirements of this section. Such direct payment provisions shall in no way excuse the Contractor from employing other work, materials and equipment as necessary in meeting the requirements of this section.

SECTION 4. CONTRACTUAL RELATIONS OF PARTIES

(a) **Contractor's Responsibility for Subcontractors and Suppliers.**—The Contractor shall be responsible under the contract for the acts and omissions of his subcontractors, suppliers, and persons either directly or indirectly employed by them, as fully as he is for the acts and omissions of his own employees. Nothing in the contract shall create any contractual relation between any subcontractor or supplier and the Department, or any obligation on the part of the Department to pay or cause to be paid any money to any subcontractor or supplier.

(b) **Liability Insurance.**—Until notified in writing of the acceptance of the work, the Contractor shall maintain in full force and effect a policy or policies of comprehensive bodily injury and property damage liability insurance meeting all the requirements of Article (1) of the Bidding Requirements and Conditions. If, prior to acceptance of the work, any policy of insurance obtained by the Contractor pursuant to said article expires, or is canceled, or the total coverage as provided by all policies is reduced below the minimum limits of coverage specified in said Article, the Contractor shall submit to the Director, at least 15 days prior to the date of such expiration, cancellation, or reduction, a revised Certificate of Insurance on a form furnished by the Department, certifying new or renewed insurance coverage meeting all the requirements of said Article (1). The Contractor's failure to maintain insurance coverage in compliance herewith shall be considered a material breach of contract for which the Director may terminate the Contractor's control over the work pursuant to Section 5, Article (h), in addition to pursuing other available remedies against the Contractor.

(c) **Preservation of Property.**—

(1) *General.*—The Contractor shall exercise due care in the performance of the work to avoid injury, damage, or loss to existing improvements, utility facilities, and other property of any nature on or near the site of the work. The Contractor shall be responsible as follows for any injury, damage, or loss to property arising out of the work, irrespective of fault or negligence, excepting only such injury, damage, or loss as is caused solely by the negligence or willful misconduct of the State or its officers or employees.

a. The Contractor shall repair, replace, or otherwise restore the property to a condition as good as it was in when he commenced the work, or as good as required by the specifications, if applicable. In addition, the Contractor shall perform such temporary repairs of any facility, necessary to restore it to service, as are ordered by the Engineer. At the request of the property owner, or upon the failure of the Contractor to perform promptly, the Engineer may cause such repair, replacement, restoration, or temporary repair to be performed by the owner or his designee, in which case the costs shall be borne by the Contractor and may be deducted, in whole or in part, from any money due or to become due him under the contract.

b. The Contractor shall idemnify and save harmless the State and all officers and employees of the State connected with the work from all claims, suits, or actions of any nature brought for or on account of any such injury, damage, or loss. Claims, suits, or actions for recovery of fire damages or fire suppression expenses under California Health and Safety Code Sections 13007-13009 are included within this indemnification obligation.

(2) *Underground Improvements.*—The Contractor shall exercise due care to locate any underground improvements or facilities which may be affected by his operations, and shall be responsible for injury, damage, or loss thereto as provided in Subarticle (1), whether or not such improvements or facilities are shown on the drawings. If improvements or facilities not shown on the drawings are encountered within the Department furnished right-of-way, any removal, replacement, alteration, or relocation thereof necessary for performance of the work and to be performed by the Contractor will be ordered by the Engineer pursuant to Section 7, except to the extent that such work is necessitated by injury, damage, or loss for which the Contractor is responsible under Subarticle (1).

(3) *Compliance With Entry Permits.*—Except as otherwise expressly provided in the Special or

Technical Provisions, the Contractor shall comply with all terms and conditions of any instrument granting the State, the Department, or the Contractor the right to enter upon property which pertain to the time, method, or manner of performing the work, including, but not limited to, all terms and conditions pertaining to the prevention and suppression of fires. The Contractor shall indemnify and save harmless the State and all officers and employees of the State connected with the work from all claims, suits or actions of any nature brought for or on account of the violation of any such terms and conditions by the Contractor or his subcontractors, suppliers, agents, or employees.

(d) Maintenance and Protection of the Work.—
(1) *General.*—Until notified in writing of acceptance of the work, and subject to the provisions of Subarticle (2) below, the Contractor shall be responsible for:

    a. Maintenance and protection of the work, including, but not limited to, the storage of materials and equipment, erection of temporary structures and provision of drainage as necessary to protect the work from injury, damage, or loss.

    b. Any injury, damage, or loss to the work resulting from the action of the elements or any other cause, irrespective of fault or negligence, excepting only such injury, damage, or loss as is caused solely by the negligence or willfull misconduct of the State or its officers or employees.

Any portion of the work suffering injury, damage, or loss for which the Contractor is responsible under b. above will be considered defective and shall be corrected or replaced pursuant to Section 6, Article (m); except that if the injury, damage, or loss is to Department-furnished equipment, the Engineer may cause correction or replacement to be performed in the first instance by the equipment manufacturer, and all costs thereof shall be borne by the Contractor and may be deducted, in whole or in part, from any money due or to become due him under the contract.

The Contractor shall indemnify and save harmless the State and all officers and employees of the State connected with the work from all claims, suits, or actions of any nature brought for or on account of any injury, damage, or loss for which he is responsible under b. above.

(2) *Advance Possession and Use of the Work by the Engineer.*—Prior to acceptance of the work, the Engineer, by written notice to the Contractor, may take possession of and use or open to use portions of the work designated in the Special Provisions, or any other portion of the work if he determines that its possession and use has become essential in meeting early project operational requirements. Upon receipt of such notice, the Contractor will be relieved of his responsibilities under this article for maintenance and protection of the portion of the work described therein, and for any subsequent injury, damage, or loss to such portion of the work, except that which results from the Contractor's own operations or negligence. The Contractor shall remain responsible for completion of such portion of the work in accordance with the drawings and specifications, if it is incomplete at the time of possession and use by the Engineer; for defects in the work and unauthorized work under Section 6, Article (m); and for timely performance and liquidated damages under Section 5, subject to the provisions of Article (d) (1) of that section.

(e) Injury or Death of Persons.—The Contractor shall be responsible for the injury or death of any workman, member of the public, or other person which arises out of the work, irrespective of fault or negligence, excepting only such injuries or deaths as are caused solely by the negligence or willful misconduct of the State or its officers or employees. The Contractor shall indemnify and save harmless the State and all officers and employees of the State connected with the work from all claims, suits or actions of any nature brought for or on account of any such injury or death.

(f) Indemnification of Other Governmental Authorities.—When so provided in the Special Provisions, the Contractor shall indemnify and save harmless any county, city, public district, or agency of the federal government designated therein whose limits or jurisdictions any portion of the work is to be performed, and its officers and employees connected with the work, all in the same manner and to the same extent as he is required to indemnify and save harmless the State and its officers and employees under Articles (c), (d), and (e).

(g) Cooperation With Other Contractors and Forces.—The Department may at any time perform or cause the performance of other work of any nature at or near the site of the work, including use of materials sources, storage areas and disposal areas in use under this contract. It may also perform or cause the performance of other work at locations removed from the site of the work which is related to the project or which otherwise affects the work under this contract.

The Contractor shall cooperate with the contractors and forces employed on such other work in all phases of contracting activity, both at and away from the site of the work. He shall conduct his operations in such manner as not to unnecessarily delay or hinder their work, and shall adjust and coordinate his work with theirs so as to permit proper and timely completion of all projects in the area. If the Contractor is responsible for furnishing and installing major items of specialized equipment, he and the contractors for related equipment shall exchange working drawings, templates, gages and other pertinent materials as necessary to insure proper connection and integration of the Contractor's equipment with such related equipment.

The Contractor shall cooperate on the foregoing terms with contractors and forces employed on any nearby or related projects of other entities, including but not limited to those designated in the Special Provisions.

The Contractor shall be responsible for any injury, damage, or loss to other contractors or forces resulting from his breach of the foregoing duties, and he shall indemnify and save harmless the State and all officers and employees of the State connected with the work from all claims, suits, or actions of any nature brought for or on account of any such injury, damage, or loss.

If the Contractor discovers any defect in work being performed by other contractors or forces such as to jeopardize proper and timely completion of the work under this contract, he shall promptly notify the Engineer thereof in writing.

(h) **Patents.**—Except as otherwise provided herein, the Contractor shall bear all costs arising from the use on or incorporation in the work of patented materials, equipment, devices, or processes furnished by him under the contract, and shall indemnify and save harmless the State and all officers and employees of the State connected with the work from all claims, suits, or actions of any nature brought for or on account of the use on or incorporation in the work of any such patented items.

The foregoing requirements will not apply to any patented item for which the Department has supplied detailed design in addition to design criteria and operating requirements, or which is patented only in combination with items furnished by the Department, provided: that the item is not a stock item available on the open market or an item specified by trade names or manufacturers' names; that neither the Contractor nor any of his subcontractors or suppliers owns or has an interest in the patent in question; and that before ordering or using the item, the Contractor notifies the Engineer in writing of the patent thereon.

(i) **Payment of Taxes.**—Except as otherwise provided in the Special Provisions, the contract prices shall include full compensation for all taxes which the Contractor is required to pay, whether imposed by federal, state, or local government, and no tax exemption certificate or any other document designed to exempt the Contractor from payment of tax will be furnished to the Contractor by the Department.

(j) **Property Rights in Materials and Equipment.**—Nothing in the contract shall be construed as vesting in the Contractor any right of property in materials and equipment furnished by him after they have been attached or affixed to the work or the ground, or after payment has been made of 90 percent of their value under Section 9, Article (d) following their delivery on the ground or storage subject to or under the control of the State, whether or not they have been so attached or affixed, and all such materials and equipment shall become the property of the State upon being so attached or affixed, or upon such payment. Property rights in materials and equipment furnished by the Department shall remain in the Department at all times.

The transfer to the State of property rights in materials and equipment shall in no way relieve the Contractor of his responsibilities with respect to such materials and equipment under Article (d), "Maintenance and Protection of the Work", Section 6, Article (m), "Defects in the Work and Unauthorized Work", and other provisions of the contract.

(k) **Use of Property at the Site.**—The Contractor may make arrangements with third parties for the occupancy or use of land or improvements at the site of the work, with or without compensation, if done for the purposes of the contract and not in conflict with any agreement between the State and any owner, former owner, or tenant of the land or improvement.

(l) **Assignment.**—The Contractor shall not assign performance of the work except upon the written consent of the Director, and in no event shall be assign performance of part of the work. An assignment shall not relieve the Contractor or his sureties of their responsibilities under the contract.

The Contractor may assign money due or to become due under the contract and such assignment will be recognized by the Department, if given proper notice thereof, to the extent permitted by law, subject to all proper set-offs in favor of the Department and too all deductions and retentions of money provided in the contract.

(m) **Stop Notices, Tax Demands and Claims of State Agencies.**—The State, by and through the Department or other appropriate State office or officers, may retain out of any money due or that may become due the Contractor under the contract, amounts sufficient to cover: claims filed in accordance with Section 3179 et seq. of the California Civil Code, plus the reasonable costs of litigation of such claims; tax demands filed in accordance with Section 12419.4 of the California Government Code; and claims of state agencies offset under Section 12419.5 of the California Government Code. Such retentions may be in addition to amounts retained under Section 9, Article (d) (3).

(n) **Personal Liability.**—No officer or employee of the State connected with the work shall be personally responsible for any liability arising under or in connection with the contract.

(o) **Conditions on Indemnification.**—The Department, promptly after receiving notice thereof, will notify the Contractor in writing of the commencement of any claim, suit, or action against the State or its officers or employees for which he must provide indemnification under this contract. To the extent permitted by law, the Department will authorize the Contractor or his insurer to defend such claim, suit, or action and will provide him or his insurer, at his expense, information and assistance for such defense. Failure of the Department to give such notice, authorization, information, or assistance shall not relieve the Contractor of his indemnification obligations.

The State may retain so much of the money due or that may become due the Contractor under the contract as the Department considers necessary to protect the State's interests until disposition has been made of any such claim, suit, or action, in addition to pursuing other available remedies for enforcement of the Contractor's indemnification obligations. Such retention may be in addition to amounts retained under Section 9, Article (d) (3).

The provisions of this article shall not apply to claims, suits, or actions against other governmental authorities for which the Contractor must provide indemnification under Article (f) and the Special Provisions.

(p) **Waiver of Rights.**—Any waiver at any time by either party of rights with respect to a default or other matter arising under the contract shall not be considered a waiver of rights with respect to any other default or matter.

(q) **Notices.**—Except as otherwise expressly provided in the specifications, any notice, order, instruction, claim, drawing or data submittal, or other written communication required or permitted to be given under this contract shall be deemed to have been delivered or received:

1. Upon personal delivery to the Contractor or his authorized representative, or to the Engineer, as the case may be; or

2. Upon expiration of the normal period for delivery of first-class mail between the place of mailing and the place of address, not to exceed five days following its deposit in the United States mail in a sealed envelope addressed to the Contractor or the Engineer, as the case may be, with first-class or airmail postage paid, as appropriate; or

3. Upon expiration of 12 hours following its submission to a telegraph company, addressed to the Contractor or the Engineer, as the case may be.

For purposes hereof, the address of the Contractor shall be the business address given in his bid, and the address of the Engineer shall be as designated in the notice to begin the work. Either party may change his address at any time by written notice to the other.

(r) **Suspension of the Work.**—A suspension of the work for any cause shall not relieve the Contractor of his responsibilities under this section.

(s) **Payment.**—The contract prices shall include full compensation for all costs incurred under this section, except insofar as direct payment is provided in the Special or Technical Provisions for work, materials, or equipment employed in meeting the requirements of this section. Such direct payment provisions shall in no way excuse the Contractor from employing other work, materials and equipment as necessary in meeting the requirements of this section.

SECTION 5. PROSECUTION OF THE WORK

(a) **Commencement of the Work.**—None of the work shall be performed before approval of the contract by the Attorney General of the State.

The Contractor shall begin the work within 30 days after receiving notice to begin the work from the Engineer. The Contractor may begin the work before receiving such notice if he gives the Engineer at least 24 hours advance written notice of the date on which he will commence operations and the portions of the work to be performed, and receives the Engineer's prior written approval thereof. The Contractor shall not be entitled to additional compensation or time for any delay or hindrance caused by or attributable to his commencement of the work prior to receipt of notice to begin the work.

(b) **Prosecution of the Work.**—The Contractor shall diligently prosecute the work and all portions thereof to completion within the times specified therefor or any extensions thereof. The capacity of the Contractor's equipment and plants, his sequence and methods of operation, and his forces employed, including management and supervisory personal, shall be such as to insure completion of the work and all portions thereof within such times.

If, at any time, the Engineer determines that the Contractor's progress is not sufficient to insure completion of the work or any portion thereof within the time specified therefor or any extension thereof, he may so notify the Contractor and order him in writing to do any or all of the following within the time limits specified in such order:

1. Submit his plan or schedule for improving progress;
2. Take such steps as may be necessary to improve progress and advise the Engineer thereof in writing;
3. Increase his labor force, employ overtime operations, increase the number of shifts of work per day, increase the number of days of work per week, increase the capacity of his equipment and plants, change his sequence of operation, change his methods of operation, or take other specific steps to improve progress.

Such action by the Engineer shall not be considered a notice of default within the meaning of Section 14394 of the California Government Code, but may be taken preliminary to such notice.

In determining the Contractor's progress for purposes of this article, the Engineer may consider, among other factors, failure of the Contractor's last submitted schedule to provide for timely completion of the work or of any portion thereof, or failure of the Contractor to meet such schedule in any material respect, and the Engineer may consider such factors whether or not he has approved such schedule.

The contract prices shall include full compensation for all costs incurred under this article.

(c) **Time of Completion and Liquidated Damages.**—Time is of the essence of the contract. The Contractor shall complete all of the work within the time specified in the Special Provisions, or any extension thereof, and, if so provided in the Special Provisions, shall complete designated portions of the work within the times specified therefor, or any extensions thereof.

If all of the work or any portion thereof is not completed within the time specified therefor in the Special Provisions or any extension thereof, damage will be sustained by the State. It is and will be extremely difficult and impracticable to determine the actual damage which the State will sustain by reason of such delay. Therefore, the Contractor shall pay to the State, as liquidated damages, the amount specified in the Special Provisions for each day's delay in completing the work or any portion thereof beyond the time specified therefor or any extension thereof. The Department may deduct such liquidated damages, in whole or in part, from any money due or that may become due the Contractor under the contract. If the Department is responsible for any part of a delay in completing the work or any portion thereof beyond the time specified therefor, an extension of time and adjustment in compensation will be made to the extent provided in this section and Section 7, and the Contractor shall pay liquidated damages to the State for the balance of the delay in accordance with the foregoing provisions.

(d) Time Extensions.—

(1) Causes for Time Extension.—Subject to the provisions of Subarticles (2) and (3) below, the Contractor will be entitled to an extension of the time for completion of the work, or any designated portion thereof, for that part of any delay in completion beyond such time, including delay of subcontractors or suppliers, which is due to any cause specified in a. through h. below, provided that such cause is unforeseeable, beyond the control and without the fault or negligence of the Contractor and his subcontractors or suppliers, and, if there are other causes of delay, that it is the controlling cause of delay, as determined by the Engineer.

a. Any change in the work ordered pursuant to Section 7, Article (b) (1).

b. Failure of the Department or its other contractors to furnish within the time specified, or within a reasonable time if none is specified, access to the work, right-of-way, working areas, utility relocations, completed facilities of related projects, drawings, materials, equipment, or services for which the Department is responsible under the contract. The Engineer's advance possession and use of an incomplete portion of the work pursuant to Section 4, Article (d) (2) may be considered a failure to furnish working areas for purposes hereof.

c. Survey error by the Department.

d. Use of a portion of the work by the Department following discovery of defects therein, such as to defer its correction or replacement under Section 6, Article (m).

e. Unforeseen site conditions, as defined in Section 7, Article (g).

f. An increase in the quantity of any unit price item of work over that quantity estimated in the bid schedule.

g. Earthquake, fire, flood, cloudburst, cyclone, or other natural phenomenon of a severe and unusual nature, or a suspension of the work pursuant to Article (f) 1. due to any such phenomenon.

Inclement weather will not be considered severe and unusual, in and of itself, unless it results in precipitation which, either in amount, frequency, or duration, is not equalled or exceeded at the location and during the time of year in question on an average of more than once in ten years, as determined from National Weather Service records.

h. Act of the public enemy, act of another governmental entity, epidemic, quarantine restriction, freight embargo, strike or labor dispute.

Priority rating of orders or materials by the Federal Government for national defense purposes will be considered on the same basis and subject to the same conditions, including unforeseeability, as any other governmental act, unless the Engineer is specifically directed otherwise by judicial or federal administrative authority. A priority rating on work being performed by the Contractor or a subcontractor or supplier will be considered unforeseeable only if the rating was placed on the work following award of this contract, unless the Engineer is specifically directed otherwise by judicial or federal administrative authority.

(2) Processing of Time Extensions for Changes in the Work.—Upon receipt of an order for a change in the work, the Contractor shall submit written evidence to the Engineer of any delay in completion of the work or any portion thereof beyond the time specified therefor caused by such change, unless the question of delay was previously considered and the order either grants or disallows a time extension for the changes. As soon as practicable following receipt of such evidence, the Engineer will determine the extent of such delay, if any, and will issue a change order extending time, subject to the provisions of Section 7, Article (c), or will make other adjustment pursuant to Article (e) of this section.

(3) Processing of Time Extensions for Other Specified Causes.—

a. Notice of Potential Delay.—The Contractor shall notify the Engineer in writing of any occurrence specified in Subarticle (1), paragraphs b. through h. which may cause delay to the work as soon as such occurrence comes to his attention and in no event later than 30 days following its commencement, provided, that if the occurrence is so far removed from the Contractor's immediate operations that it cannot reasonable be identified as a potential cause of delay within the above time limit, notice thereof shall be given as soon as it can be so identified, and provided further, that only one notice need be given of a continuing occurrence. Notice of unforeseen site conditions, in addition to meeting the above requirements, shall be given before such conditions are disturbed, as provided in Section 7, Article (g). Failure to give notice of any such occurrence in the manner and within the time above stated shall constitute a waiver of all claims in connection therewith, whether direct or consequential in nature, irrespective of the time of commencement of any delay resulting from the occurrence.

b. Evidence of Delay.—If the Contractor determines that an occurrence noticed in accordance with a. above will delay completion of the work or any portion thereof beyond the time specified therefor, he shall submit written evidence to the Engineer of such occurrence and delay. As soon as practicable following receipt of such evidence, the Engineer will determine the cause and extent of the delay and will, if he finds that a time extension is justified, issue a change order extending time or make other adjustment pursuant to Article (e) of this section. If the Engineer determines that a time extension if not justified, he will so advise the Contractor in writing. Should the Contractor disagree with such determination, he may submit a notice of potential claim to the Engineer as provided in Section 8, Article (a).

(e) **Acceleration in Lieu of Time Extension.**—In lieu of granting any time extension, in whole or in part, which may be due under Article (d), the Engineer may, pursuant to Section 7, Article (b) (1), order the Contractor to accelerate the work or any portion thereof to overcome the delay for which such extension or part thereof is due.

Unless otherwise approved by the Engineer, additional compensation provided for such acceleration shall be subject to reduction on a pro-rata basis for each day's delay in completion of the work or portion thereof beyond the accelerated time therefor. Such reduction in compensation shall in no way excuse the Contractor's obligation to pay liquidated damages for such delay.

(f) **Suspension of the Work.**—By written order to the Contractor, the Engineer may suspend the work wholly or in part, for such period as he may deem necessary, for any of the following reasons:

1. Weather conditions or other conditions which are unfavorable for the proper prosecution of the work;
2. Failure of the Contractor to carry out orders given or to perform any provisions of the contract; or
3. The convenience and benefit of the Department, in which case the order will constitute an order for a change in the work pursuant to Section 7, Article (b) (1).

The Contractor shall immediately comply with such written order, and shall resume the suspended work only upon the Engineer's written order to do so.

A suspension ordered under authority of 1. above will be considered a cause for time extension only to the extent provided in Article (d) (1) g. No additional compensation will be paid for any such suspension, or for the failure or refusal of the Engineer to order any such suspension.

A suspension ordered under authority 2. above shall not be cause for a time extension or additional compensation.

A suspension ordered under authority of 3. above will be considered a cause for time extension and adjustment in compensation to the same extent as other changes in the work.

(g) **Payment of Delay.**—The contract prices shall include full compensation for all costs incurred by reason of any delay to the work, except as otherwise provided in Section 7, Article (e).

(h) **Termination of Control Over the Work.**—If the Contractor fails to comply with the requirements of Article b). "Prosecution of the Work," or with any orders of the Engineer given thereunder, fails to perform work or furnish materials or equipment of the quality required by the contract, fails to prosecute orders for changes in the work in accordance with Section 7, Article (d), fails to maintain insurance coverage as provided in Section 4, Article (b), or fails in any other material respect to fulfill the requirements of the contract, the Director may terminate the Contractor's control over the work and take over its completion as provided in the State Contract Act.

In the event of such termination, the Contractor shall:

1. Preserve all construction materials, equipment and plant at the site of the work until notified in writing of those items which the Director, acting pursuant to the State Contract Act, will take over and use in completing the work;

2. Upon receipt of the foregoing notice, remove from the site of the work all construction materials, equipment and plant not designated for use by the Director in such notice;

3. Assist the Engineer in making an inventory of all materials and equipment in storage at the site of the work, en route to the site of the work in storage or manufacture at other locations, and on order from suppliers; and

4. Assign to the Department or the Department's completion contractor, as directed by the Engineer, subcontracts and supply contracts designated by the Engineer, to the extent such contracts are legally assignable.

The Contractor's liability to the State upon such termination shall be as provided in the State Contract Act and shall specifically include, as part of those damages sustained or to be sustained by the State, liquidated damages for delay through the actual time of completion of the work.

SECTION 6. CONTROL OF THE WORK

(a) **Authority of the Engineer.**—In exercising the specific authority granted him under other provisions of the contract and in any case not covered by such specific authority, the Engineer shall have authority to decide all questions as to interpretation and fulfillment of contract requirements, including, without limitation, all questions as to the prosecution, progress, quality and acceptability of the work. He may implement and enforce his decisions by orders, instructions, notices, and other appropriate means.

The Contractor will be advised in the notice to begin the work of delegations of authority by the Deputy Director to his representatives, and will be notified in writing of any changes in such delegations.

Any oral decision, order, instruction, or notice of the Engineer will be confirmed in writing on the Contractor's written request. Such request shall state the specific subject of the decision, order, instruction, or notice and its date, time, place, author, and recipient.

(b) **Contractor's Supervision of the Work.**—

(1) *General.*—The Contractor shall provide competent, efficient supervision of the work.

All of the work shall be performed in a skillful, workmanlike and orderly manner, and the Contractor and his supervisory personnel shall enforce this requirement at all times.

(2) *Contractor's Representative.*—Before beginning the work, the Contractor shall designate for the Engineer in writing one person within his organization, satisfactory to the Engineer, who shall have complete authority to supervise the work, to receive orders from the Engineer, and, subject to the provisions of Subarticle (3), to represent and act for the Contractor in all matters arising under the contract. The Contractor shall not remove his representative without first designating in writing a new representative meeting all of the foregoing requirements.

The Contractor's representative shall normally be present at or about the site of the work while the work is in progress. Before leaving the site of the work for any extended period, whether or not the work is in progress, the Contractor's representative shall designate for the Engineer in writing an assistant, satisfactory to the Engineer, with full authority to act for the representative in his absence, or shall make substitute arrangements satisfactory to the Engineer. When neither the Contractor, his representative, nor the representative's authorized assistant is present on a part of the work, the Engineer may give necessary orders relative thereto, and such orders shall be received and obeyed by the superintendent, foreman, or other employee of the Contractor in charge of that part of the work.

(3) Execution of Documents.—In designating his representative under Subarticle (2), the Contractor shall expressly state whether, and to what extent, that representative is delegated authority to execute and bind the Contractor to change orders, claim releases and similar documents, and shall furnish to the Engineer legally sufficient evidence of any such delegation. A delegation of such authority to any other person will not be recognized until supported by similar written notice and evidence. If the Contractor fails to state or limit the authority of his representative, any documents executed by the representative shall be fully binding on the Contractor, notwithstanding any lack of actual authority in the representative with respect to such documents.

(4) Removal of Personnel.—If so ordered by the Engineer, the Contractor shall immediately remove any employee, subcontractor, or supplier, or any employee of a subcontractor or supplier, who fails or refuses to carry out orders properly given, or who is, in the judgment of the Engineer, insubordinate, disorderly, incompetent, or lacking in requisite skill, and such person shall not again be employed on the work.

(5) Notice of Unlisted Subcontracts.—Immediately upon letting any subcontract which is exempt from the Subletting and Subcontracting Fair Practices Act, is not listed in the bid, and is in an amount in excess of one-half of one percent of the total contract price, the Contractor shall notify the Engineer in writing of the portion of the work subcontracted and the name and location of the place of business of the subcontractor. Immediately upon the substitution of any other subcontractor on such portion of the work, the Contractor shall notify the Engineer in writing of the original subcontract, the portion thereof to be performed by the new subcontractor, and the name and location of the place of business of the new subcontractor.

(c) **Working Drawings, Data on Materials and Equipment, and Architectural Samples.**—

(1) General.—The Contractor shall furnish to the Engineer such working drawings, data on materials and equipment (hereinafter in this article called data), and architectural samples (hereinafter in this article called samples) as are required for the proper control of the work, including but not limited to those working drawings, data and samples specifically required elsewhere in the specifications and in the drawings.

All working drawings, data and samples shall be subject to inspection by the Engineer for conformity with the drawings and specifications.

(2) Working Drawings and Data Defined.—Working drawings include, without limitation, shop detail drawings, fabrication drawings, falsework and formwork drawings, pipe layouts and similar classes of drawings. They shall contain all required details and information in reasonable scale, and enough views to clearly show work to be performed or items to be furnished.

Data on materials and equipment include, without limitation, materials and equipment lists, catalog data sheets, cuts, performance curves, diagrams and similar descriptive material. Materials and equipment lists give, for each item thereon, the name and location of the supplier or manufacturer, trade name, catalog reference, size, finish and all other pertinent data.

(3) Form of Submission.—Working drawings, data and samples shall be plainly identified by contract name, specification number, and the description and location of the subject portion of the work.

Two legible transparent prints and two contact prints of each drawing shall be furnished. Eight copies of data shall be furnished.

The Contractor shall properly check and correct all working drawings and data before their submission, whether they are prepared within his own organization or by a subcontractor or supplier.

(4) Time of Submission.—

a. General.—Subject to such specific time requirements as may be stated in the Special and Technical Provisions, the Contractor shall furnish working drawings, data and samples in such manner and sequence that they may be inspected in an orderly manner before the subject portions of the work are performed, and that all related information necessary for such inspection is available to the Engineer when the drawing, data, or sample is received. In addition, working drawings and materials and equipment lists shall be furnished in accordance with the schedule specified in b. below.

b. **Working Drawings and Materials and Equipment Lists.**—Within 60 days following notice to begin the work, the Contractor shall furnish to the Engineer a schedule for submission of all working drawings and materials and equipment lists. The schedule will be inspected by the Engineer and the Contractor shall correct any defects noted therein. The schedule shall at all times present a complete plan for orderly submission of such drawings and lists, and shall be revised as necessary to meet this requirement. The Contractor shall promply notify the Engineer of any occurrence requiring substantial revision of the schedule and shall furnish a revised schedule within 15 days of such occurrence. Revised schedules will be inspected and corrected in the same amount as the original schedule. Inspection of the original and revised schedules shall in no way waive the requirements of a. above.

(5) **Inspection and Revision.**—The Engineer will inspect and return working drawings, data and samples as provided below within 30 days after receipt thereof, or within 30 days after receipt of all related information necessary for such inspection, whichever is later.

One print of each drawing and one copy of data will be returned, marked "NO APPARENT DEFECTS", "DEFECTS NOTED", or "REJECTED". Defects discovered on inspection will be indicated on the drawing or data, or otherwise communicated to the Contractor in writing on return of the drawing or data.

Samples to be incorporated in the work will be returned, together with a written notice designating the sample "NO APPARENT DEFECTS", "DEFECTS NOTED", or "REJECTED" and indicating defects discovered on inspection. Other samples will not be returned, but the same notice will be given with respect thereto, and such notice shall be considered a return of the sample for purposes of this subarticle.

The Contractor shall revise any working drawing or data marked "DEFECTS NOTED" or "REJECTED" and shall correct any sample so designated, unless, in the case of a drawing, data, or a sample marked or designated "DEFECTS NOTED", revision or correction is expressly waived by the Engineer in writing. Revised drawings and data and corrected samples shall be furnished to the Engineer and will be inspected and returned by him in the same manner as original drawings, data and samples, and within 15 days after receipt thereof or within 15 days after receipt of all related information necessary for such inspection, whichever is later. Any revised drawing or data marked "DEFECTS NOTED" or "REJECTED" and any corrected sample so designated shall be further revised for corrected in accordance with the foregoing procedure.

The Contractor may proceed with any of the work covered by a working drawing, data, or a sample marked or designated "NO APPARENT DEFECTS" upon its return to him. He may also proceed with the nondefective portions of the work covered by a working drawing, data or sample marked or designated "DEFECTS NOTED"; and, if resubmitted is expressly waived in writing, may proceed with all work covered by such working drawing, data or sample, provided that he proceeds in accordance with the Engineer's notes and comments.

The Contractor shall not begin any of the work covered by a working drawing, data or sample marked or designed "REJECTED", or any portion of the work noted as defective on a drawing, data or sample marked or designated "DEFECTS NOTED" if resubmittal is not expressly waiyed in writing, until a revision or correction thereof has been inspected and returned to him.

A drawing, data, or sample marked or designated "REJECTED" or "DEFECTS NOTED" and requiring resubmittal shall be revised or corrected and resubmitted to the Engineer within 20 days after its return to the Contractor.

Neither the inspection nor lack of inspection of any working drawing, data, or sample shall waive any of the requirements of the drawings or specifications, or relieve the Contractor of any obligations thereunder, and defective work, materials, and equipment may be rejected notwithstanding conformance with drawings, data, or samples inspected by the Engineer.

(d) **Supplemental and Revised Drawings.**—During the progress of the work, the Department may issue drawings supplemental to those listed in the Special Provisions, showing additional details required for the performance of the work, and may issue revised drawings pursuant to Section 7, "Changes and Unforeseen Site Conditions".

(e) Surveys.—

(1) General.—The Engineer will perform all surveys required for measurement of quantities of the work for payment. Performance of surveys required for layout and performance of the work will be allocated between the Contractor and the Engineer as provided in the Special Provisions. Any surveys required for layout and performance of the work not so allocated will be performed by the Engineer.

(2) Availability of Engineer's Survey Services.—The Engineer's survey personnel will not be present on the work at all times, and when present, will generally be available during only one shift per day. The Contractor shall give the Engineer advance notice, of not less than two working days, of his requirements for survey services. He shall give the Engineer such assistance and provide such drill holes, forms, ladders, spikes, nails and light as may be required by the Engineer in performing survey services, and shall adjust his operations as necessary for such purposes.

(3) Removal of Engineer's Survey Marks.—Where construction operations require removal of the Engineer's principal stakes or other key survey marks, the Engineer will reference and remove them, or, if the Engineer's survey personnel are not available, the Contractor shall reference and remove them in an approved manner. Such stakes and marks will be replaced by the Engineer as necessary.

All of the Engineer's principal stakes and key marks shall be carefully preserved by the Contractor until he is authorized to remove them. If the Contractor damages, destroys, or removes such stakes or marks without authorization, he shall be responsible under Article (m) for any resultant defects in the work or unauthorized work, and shall bear the costs of restoration or replacement of the stakes or marks. Such costs may be deducted, in whole or in part, from any money due or that may become due him under the contract. Such costs will include a reasonable charge for use of Department materials and equipment, and for overhead.

(4) Contractor's Surveys.—

a. Equipment and Personnel.—The Contractor's instruments and other survey equipment shall be accurate, suitable for the surveys required, and in proper condition and adjustment at all times. All surveys shall be performed under the direct supervision of a qualified surveyor.

b. Field Notes and Records.—The Contractor shall properly record his surveys in duplicate page field notebooks. The original pages of such records shall be furnished to the Engineer at intervals directed by him. Each field notebook shall be furnished to the Engineer when filled or completed.

c. Use by Engineer.—The Engineer may at any time use line and grade points and markers established by the Contractor.

d. Checking by Engineer.—The Contractor's surveys are a part of the work and may be checked at any time by the Engineer pursuant to Article (1). The Contractor shall be responsible under Article (m) for any lines, grades, or measurements which do not comply with specified or proper tolerances, or which are otherwise defective, and for any resultant defects in the work or unauthorized work.

e. Spacing and Tolerances in Setting Survey Stakes.—Spacing and tolerances generally applicable in setting survey stakes shall be as set forth in the Special Provisions. Such spacing and tolerances shall not supersede stricter spacing or tolerances required by the drawings and specifications, and shall not otherwise relieve the Contractor of responsibility for measurements in compliance with the drawings and specifications.

(f) Construction Equipment and Plant.—

(1) General.—The Contractor shall provide and use on the work only such construction equipment and plant as are capable of producing the quality and quantity of work and materials required by the contract within the time or times specified. Upon the written order of the Engineer, the Contractor shall remove unsatisfactory construction equipment from the work and discontinue the operation of unsatisfactory plants.

(2) Equipment Identification.—Before using any unit of construction equipment of the work larger than a hand tool, the Contractor shall plainly stencil or stamp an identifying number thereon

at a conspicuous location, and shall furnish to the Engineer, in triplicate, a description of the unit and its identifying number. The Contractor shall similarly stencil or stamp on each unit of compacting equipment, before its use, the make, model number and empty gross weight of the unit. The empty gross weight shall be either the manufacturer's rated weight or the scale weight.

(g) **Use of Construction Equipment Different from That Specified.**—Specifications providing that construction equipment of a particular size or type is to be used in performing a designated portion of the work will not be construed to discourage the development and use of new or improved equipment, and the Contractor may request, in writing, permission from the Engineer to use equipment different from that specified. Such request shall be submitted in sufficient time to permit action thereon without delaying the work.

Before considering such request, the Engineer may require the Contractor to furnish satisfactory evidence that the equipment proposed for use is capable of producing work equal to or better than that which can be produced by the equipment specified. Permission to use such equipment will be granted only in writing, and will be granted only if the equipment is new or improved and its use is deemed to be in furtherance of the purposes of the contract. Such permission will not be considered a change in the equipment specification; however, if use of the different equipment requires changes in other specifications or in the drawings, permission to use it will be granted, if at all, under and subject to the provisions of Section 7, except that the Contractor shall not be entitled to additional compensation or time for any such changes.

If permission to use different construction equipment is granted by the Engineer, it shall be granted for the purpose of testing the quality of work actually produced by such equipment and shall be subject to continuous attainment of results which, in the judgment of the Engineer, are equal to or better than those which can be obtained with the equipment specified. The Engineer may withdraw such permission at any time that he determines that such equipment is not producing work that is equal in all respects to that which can be produced by the equipment specified. Upon withdrawal of such permission by the Engineer, the Contractor shall use the equipment specified and shall correct or remove and replace any defective work produced with the different equipment, subject to the provisions of Article (m).

The Contractor shall have no claim against the Department for withholding or withdrawal of permission to use different construction equipment. Permission to use particular equipment under this contract shall in no way be construed as permission to use such equipment under any other contract.

Nothing in this article shall relieve the Contractor of his responsibility for performing work and furnishing materials of the quality required by the drawings and specifications.

(h) **Materials and Equipment.—**

(1) Furnished by the Contractor.—The Contractor shall furnish all materials and equipment to be incorporated in the work, except those items designated in the specifications to be furnished by the Department and those furnished by the Department pursuant to any order for extra work under Section 7.

Only materials and equipment conforming to the requirements of the drawings and specifications shall be incorporated in the work. Except as otherwise specified or approved in specific instances, all such materials and equipment shall be new and unused and of the highest quality available. Materials and equipment for which no specification requirements are given in the drawings or specifications shall be those best suited for the specified use, considering function, strength, durability and resistance to corrosion. Manufactured materials and equipment shall be obtained from sources which are currently manufacturing such materials or equipment, except as otherwise approved in specific instances.

If so ordered by the Engineer, sources of materials shall be approved by him before delivery from those sources is commenced. Approval of a source of materials may be withdrawn by the Engineer at any time that the materials delivered from the source are found to be defective, and the Contractor shall thereupon cease all deliveries from that source.

Manufacturers' warranties, guarantees, manuals, instruction sheets and parts lists provided with materials and equipment shall be furnished to the Engineer before final payment is made.

(2) *Furnished by the Department.*—Materials and equipment to be furnished by the Department will be available to the Contractor at locations designated in the specifications. The Contractor shall remove them from the carrier or storage facility at which they are made available, load and haul them to the site of the work, and there unload them. The Contractor shall be responsible for such materials and equipment from the time of their removal from the carrier or storage facility in accordance with Section 4, Article (d), "Maintenance and Protection of the Work". He shall bear all demurrage and storage charges arising from his failure to promptly remove them from the carrier or storage facility, and any amount of such charges paid by the Department may be deducted, in whole or in part, from any money due or that may become due him under the contract.

If the Contractor discovers any defects in Department-furnished materials or equipment or finds that they are not in suitable condition for their intended use, he shall promptly notify the Engineer thereof in writing. Pursuant to Section 7, the Engineer may order the Contractor to return defective or unsuitable materials or equipment, to perform corrective work thereon, or to take other appropriate action with respect thereto.

(i) Trade Name Materials and Equipment and Alternatives.—Certain materials and equipment to be incorporated in the work may be specified by trade names or by names of manufacturers and catalog information, followed by the words "or equal" or "or equivalent". Upon the Contractor's written request, and subject to the following requirements, the Engineer will approve the use of alternative materials or equipment, if he is satisfied that the alternative is at least equal in quality to the name items specified and of the required characteristics for the purpose intended. Such approval will be given only in writing.

The burden of proof as to the comparative quality and utility of alternative materials and equipment shall be upon the Contractor. He shall furnish complete data on any requested alternative pursuant to Article (e), and such other pertinent information as may be requested by the Engineer. The Contractor's request and supporting data shall be submitted in sufficient time to permit approval action without delaying the work, but need not be submitted earlier than 35 days after award of the contract. If such request is disapproved, the Contractor will not be permitted to use the same alternative materials or equipment in modified form.

(j) Preparation of Equipment for Shipment.—The Contractor shall prepare equipment for shipment in such manner as to protect it from damage or loss in transit.

Heavy parts shall be mounted on skids or crated. If slings will be used in handling, the parts shall be prepared for ready attachment of slings. If slings cannot be safely attached to a crate, the crated parts shall be made to project through the crate for attachment of slings.

Small parts shall be boxed or wired in bundles if they might otherwise be lost.

Each crate, box, bundle or other packing unit and each major part shall be plainly marked for identification. Such identification shall include the name of the manufacturer, name and location of the manufacturer's plant, manufacturer's shop number, name of the part and unit number if applicable, name of the contract and specification number.

(k) Testing of Materials.—Materials incorporated or to be incorporated in the work shall be tested during the progress of the work as requested by the Engineer, and samples of such materials shall be furnished by the Contractor for testing when requested by the Engineer.

Tests shall be performed by the Contractor, except as otherwise expressly provided in the drawings and specifications. All tests performed by the Contractor shall be witnessed by the Engineer unless the requirement therefor is waived in writing. The Contractor shall give the Engineer reasonable advance notice of all such tests. The Engineer may perform additional tests of materials tested by the Contractor, and the Contractor shall furnish samples for this purpose as requested.

Whenever a specification or test method of the ASTM, AWWA, or other authority is made applicable to materials, the Contractor shall test the materials in accordance therewith, shall submit to the Engineer two certified copies of test results, and shall not use such materials until test results have been approved by the Engineer; provided, that the requirement for test results may be waived in writing by the Engineer.

(l) Inspection of the Work.—

(1) *General.*—All of the work shall be subject to inspection by the Engineer for conformity

with the drawings and specifications. Working drawings, data on materials and equipment, and architectural samples will be inspected under Article (c). Inspection of the balance of the work will be in accordance with this article, unless otherwise expressly indicated. Materials tests conducted pursuant to Article (k) and all other specified tests will be considered part of the inspection process and shall be subject to all the provisions of this article.

(2) Engineers' Access to the Work.—The Engineer shall have access to, and may inspect the work at all times and places at the site of the work. He shall have access to, and may inspect materials and equipment to be incorporated in the work at all times at the place of production or manufacture and at the shipping point, as well as at the site of the work.

The Engineer will designate for the Contractor materials and equipment to be inspected at the place of production or manufacture. The Contractor shall give the Engineer 14 days advance written notice of the start of the manufacture or production of materials and equipment so designated. The Engineer's failure to so designate materials and equipment shall in no way limit his right to inspect them at the place of production or manufacture.

The Contractor's materials and equipment contracts shall include a notice to the supplier or subcontractor of the inspection requirements of this article.

(3) Cooperation and Safety.—The Engineer will perform inspections in such manner as not to delay the work unnecessarily, and the Contractor shall perform the work in such manner as not to delay inspection unnecessarily. The Contractor shall give the Engineer reasonable advance notice of operations requiring special inspections or tests, and he may request inspection of a portion of the work at any time by reasonable advance notice to the Engineer.

The Contractor shall bear any additional inspection costs resulting from his failure to have a portion of the work ready for inspection at the time requested by him for its inspection, or from reinspection of any previously rejected portion of the work where the defects requiring such rejection were due to the Contractor's fault or negligence. Such costs may be deducted, in whole or in part, from any money due or that may become due the Contractor under the contract.

The Contractor shall furnish the Engineer all reasonable facilities for his safety and convenience in inspecting the work, at all times and at all places where inspection may take place. If the Engineer finds that conditions are unsafe for inspection at a particular location, he may, upon notice to the Contractor, refuse to inspect in that location until such conditions are corrected. The Contractor shall bear any additional costs resulting from such action, including any costs incurred to permit subsequent inspection of any portion of the work covered or completed at that location before correction of the conditions, whether or not such portion of the work is found to meet contract requirements. The Engineer may take such action in addition to or in lieu of any actions authorized under Section 3, Article (m), "Accident Prevention".

(4) Inspection of Covered or Completed Portions of the Work.—If so ordered in writing by the Engineer, the Contractor shall uncover, remove, tear out, or disassemble, in whole or in part, any covered or completed portion of the work to permit its inspection. If that portion of the work is found to be defective or unauthorized, the Contractor shall bear all costs of uncovering, removal, tearing out, or disassembly, and the provisions of Article (m) shall apply. If such portion of the work is found to conform with the drawings and specifications, it shall be recovered, replaced, reassembled, or otherwise restored by the Contractor to its original condition and, except as stated below, all work required in connection with the inspection will be considered extra work under Section 7, Article (b) (1). If such portion of the work was covered or completed without the approval of the Engineer, where such approval was required by the specifications or required in advance by the Engineer, or if the provisions of Subarticle (3) apply, the Contractor shall bear all costs involved in the inspection, notwithstanding conformance of such portion of the work with the drawings and specifications.

(5) Inspection Not a Waiver or Acceptance.—Neither the inspection nor lack of inspection of any portion of the work, nor the presence or absence of the Engineer during performance of any of the work shall waive any of the requirements of the drawings and specifications, or relieve the Contractor of any obligations thereunder, and defective work, materials, and equipment may be

rejected notwithstanding their prior inspection or lack of inspection by the Engineer and notwithstanding their conformance with the drawings and specification at the time of any prior inspection.

(m) **Defects in the Work and Unauthorized Work.**—Any work, materials, or equipment not conforming with the drawings and specifications will be considered defective and, except as hereinafter provided, will be rejected by the Engineer, whether in place or not.

Minor, inconsequential defects may be waived in writing by the Engineer, but the Engineer's failure or refusal to exercise such authority shall not be subject to claim by the Contractor. If a waiver will result in an appreciable saving of costs to the Contractor, exclusive of potential costs of rejection under this article, it will be made only upon an equivalent adjustment in compensation pursuant to Section 7, Article (e).

Other defects may be waived only as expressly authorized by Special or Technical Provisions which establish liquidated damages for the defects or make other provision for relief to the Department upon waiver of the defects.

Upon rejection, the Contractor shall correct defective work, materials, or equipment, or shall remove and replace them with work, materials, or equipment conforming with the requirements of the drawings and specifications, as ordered by the Engineer. Corrected materials and equipment shall be used in the work only upon the written approval of the Engineer.

Any work done beyond the lines and grades shown on the drawings or established by the Engineer, or any extra work done without order of the Engineer pursuant to Section 7, Article (c) will be considered unauthorized and will not be paid for. Such work shall be removed, removed and replaced with authorized work, or otherwise corrected, as ordered by the Engineer.

The Contractor shall remove, tear out, or disassemble any portion of the work when necessary to give access to defects in the work or unauthorized work, and shall replace, reassemble, or otherwise restore it to its original condition following correction, replacement, or removal of the defective or unauthorized portion of the work. If so ordered in writing by the Engineer, the Contractor shall remove, tear out, or disassemble any portion of the work of other contractors when necessary for the above purpose, and shall replace, reassemble, or otherwise restore it to its original condition, all as extra work under Section 7, Article (b) (1).

If the Engineer has taken possession of a portion of the work pursuant to Section 4, Article (d), the Department may use that portion of the work, notwithstanding the discovery of defects therein, until it can be taken out of use without injury to the Department; provided, that the Contractor will not be responsible for aggravation of the defects caused by negligent use or maintenance of that portion of the work by the Department.

If the Contractor fails to comply promptly with any order of the Engineer given under this article, the Engineer may cause the defective work, materials, or equipment or unauthorized work covered by the order to be corrected, removed and replaced, or removed, as the case may be, and the costs thereof shall be borne by the Contractor and may be deducted, in whole or in part, from any money due or that may become due him under the contract.

(n) **Cleanup.**—The Contractor shall at all times maintain the work and the site of the work in as clean a condition as practicable, free from hazardous or unsightly accumulations of excess materials or rubbish.

The Contractor shall leave the work and the site of the work in a neat and presentable condition, removing all rubbish, falsework, temporary structures, other construction materials and equipment, and excess materials, excepting only such temporary structures or other items as have become the property of the State under the terms of the contract or other agreement. Such final cleanup work shall be performed within the time specified for completion of all of the work, with such exceptions as may be approved in writing by the Engineer. Unless otherwise provided in the specifications, the Contractor shall clean any portion of the work for which a separate time for completion is specified and the site thereof to the above standards within that specified time, with such exceptions as may be approved in writing by the Engineer.

(o) **Final Inspection and Acceptance of the Work.**—When the Engineer has made a final inspection and has determined that the work, including final cleanup, has been completed in all respects in accordance with the drawings and specifications (without consideration of timeliness of completion), he will recommend that it be accepted by the Director or the Director's authorized representative. The Contractor will be promptly notified in writing of acceptance of the work, and upon receipt of such notice, will be relieved of his responsibilities under Section 4, Article (d) for care and maintenance of the work, and for subsequent injury, damage, or loss thereto. Acceptance of the work will conclusively establish its conformity with the drawings and specifications, except as regards delays in completion, latent defects, fraud, or such gross errors as amount to fraud, and subject to any guarantee or warranty, express or implied, provided by the Contractor under the contract.

(p) **Payment.**—The contract prices shall include full compensation for all costs incurred under this section.

SECTION 7. CHANGES AND UNFORESEEN SITE CONDITIONS

(a) **Variations in Quantities.**—The quantities of work, materials and equipment given in the bid schedule are estimated only, being given as a basis for the comparison of bids, and the Department does not, expressly or by implication, represent that the quantities of work, materials, and equipment actually required will correspond therewith. A variation between the estimated and actual quantities of a unit price item of the work will in no event be considered a change in the work, in and of itself, but may be a cause for a time extension as provided in Section 5, Article (d) (1)f., or for an adjustment in compensation as provided in Article (e) (2).

(b) **Authority for Changes.**—

(1) Changes in the Work.—The Engineer may order such changes in the work as are required for its proper completion, including, but not limited to: design modifications which increase or decrease the quantities of unit price items of the work; extra work; changes in specified methods or sequences of performing the work or establishment of methods or sequences not otherwise authorized; changes in the availability or use of the site of the work or of Department-furnished materials, equipment, facilities, or services; acceleration of the work or any portion thereof in lieu of a time extension therefor; suspension of the work, wholly or in part, for the convenience and benefit of the Department; and elimination of any portion of the work no longer required for its completion. He may also order the elimination of a portion of the work, even though required for proper completion, if, due to unforeseen causes, the Contractor would be unduly delayed in performing that portion of the work or his performance thereof would otherwise be adverse to the Department's interests, and if its elimination will not materially change the nature and extent of the work.

Extra work consists of that portion of any new and unforeseen work, materials, or equipment required for proper completion of the work which is determined by the Engineer not to be covered by items of the work for which there are contract prices, and of any work expressly designated as extra work in the drawings and specifications.

A specification provision stating that certain materials, equipment, or services will be furnished by the Department or by others will not preclude an order for changes directing the Contractor to furnish such materials, equipment, or services.

(2) Adjustments in Compensation.—The Engineer may order adjustments in compensation for changes in the work, unforeseen site conditions, delays to the work, waivers of defects in the work, and variations between estimated and actual pay quantities of unit price items of work (hereinafter referred to as variations in quantities), all to the extent provided by Article (e).

(3) Time Extensions.—The Engineer may order extensions of the time or times for completion of the work for delays due to changes in the work and other specified causes, all to the extent provided by Section 5, Article (d).

(c) **Ordering of Changes.**—All orders for changes will be given in writing. An order which provides for an adjustment in compensation or an extension of time will be given only by change order.

An order for a change in the work may be given in advance of any necessary adjustment in compensation or extension of time, and pending issuance of a covering change order, the Contractor shall proceed with the change as provided in Article (d). A change order covering a change in the work will specify all additions to and reduction in the work and other alterations therein by reason of the change, the adjustment in compensation, if any, by reason of the change; and the time extension, if any, granted for the change or, if it is impracticable to determine the time extension justified by the change prior to issuance of the change order, a reservation of such determination until such time as it is practicable. If the Engineer determines that no adjustment in compensation or extension of time is due for an ordered change in the work, he will so advise the Contractor in writing.

The Contractor shall not be entitled to additional compensation or time for any costs or delay incurred in performance of a change in the work before receipt of a written order for the change.

(d) Prosecution of Changes in the Work.—

(1) General.—Upon receipt of an order for a change in the work the Contractor shall comply therewith and prosecute all portions of the work ordered or affected thereby with the same diligence and in the same manner as if such change was originally included in the contract, except as otherwise provided in the order. If, in the judgment of the Contractor, an oral order of the Engineer effects a change in the work, the Contractor may obtain confirmation of the order in writing in accordance with Section 6, Article (a) before complying therewith.

(2) Maintenance of Cost Records.—If the Contractor is ordered to proceed with a change in the work in advance of any necessary adjustment in compensation therefor, then, pending issuance of a covering change order, he shall maintain records of and report all costs of any portions of the work performed pursuant to or materially affected by such change in accordance with Section 9, Article (c) (2), as though such portions of the work were to be paid for on a force account basis.

(3) Disposition of Surplus Materials and Equipment.—When an order for change eliminates an item of the work in whole or in part, and prior to such order the Contractor has ordered acceptable materials or equipment for the eliminated portion of the work, the Contractor shall:

    a. Cancel his order for such materials or equipment; or

    b. If such order cannot be cancelled, and the materials or equipment cannot be returned to the vendor, either retain such materials or equipment as his property or deliver them to the Department as the property of the State, as ordered by the Engineer; or

    c. If such order cannot be cancelled, but the materials or equipment can be returned to the vendor, then, at his option, either return such materials or equipment to the vendor, retain them as his property, or, if approved by the Engineer, deliver them to the Department as the property of the State.

(e) Adjustments in Compensation.—

(1) General.—As used in this article, an adjustment in compensation shall mean a change in contract prices or the establishment of a new price or prices in consideration of a change in the work, unforeseen site condition, delay to the work, waiver of defects in the work, or variation in quantities.

(2) Bases for Adjustments.—

    a. Changes in the Work.—A change in the work will be a basis for adjustment in compensation if it increases or decreases the quantity of any unit price of the work in such manner as to change materially the unit cost of that item, as determined by the Engineer; if it requires the performance of extra work; or if, for any other reason, it cannot in the judgment of the Engineer be fairly and reasonably paid for at contract prices.

    b. Unforeseen Site Conditions.—An unforeseen site condition will be a basis for adjustment in compensation to the extent provided in Article (g).

    c. Delay to the Work.—A delay to the work will be a basis for adjustment in compensation if if is caused by failure of the Department or its other contractors to furnish within the time specified, or within a reasonable time if none is specified, access to the work, right-of-way, working areas, utility relocations, completed facilities of related projects, drawings, materials, equipment, or services for which the Department is responsible under the contract; provided that:

    1. Such cause of delay is unforeseeable, beyond the control and without the fault or negligence of the Contractor and his subcontractors and suppliers, and, if there are other causes of delay, that it is the controlling cause as determined by the Engineer;

    2. Notice of potential delay is given in accordance with Section 5, Article (d) (3); and

    3. The delay is unreasonable, considering all the circumstances, and materially increases the costs of the work, as determined by the Engineer.

The Engineer's advance possession and use of an incomplete portion of the work pursuant to Section 4, Article (d) (2) may be considered a failure to furnish working areas for purposes of this subarticle.

If the Contractor determines that a delay to the work noticed in accordance with Section 5, Article (d) (3) provides a basis for adjustment in compensation hereunder, he shall submit written evidence to the Engineer of the cause and extent of the delay (if not submitted pursuant to said subarticle) and of the effect of the delay on the costs of the work. As soon as practicable following receipt of such evidence, the Engineer will determine the cause, extent and effect of the delay and will issue a change order, subject to the provisions of Subarticle (3) of this article, for any adjustment in compensation found due. If the Engineer determines that no adjustment is due, he will so advise the Contractor in writing. Should the Contractor disagree with such determination, he may submit a notice of potential claim to the Engineer as provided in Section 8, Article (a).

    d. *Waiver of Defects.*—A waiver of defects in the work will be a basis for adjustment in compensation to the extent provided in Section 6, Article (m).

    e. *Variations in Quantities.*—A variation between the estimated and actual pay quantity of a unit price item will be a basis for adjustment in compensation as follows:

Where the total pay quantity of a unit price item is more than 125 percent (overrun), or less than 75 percent (underrun) of the estimated quantity an adjustment will be made in the contract price upon written demand of either party. Such written demand by either the Contractor or the Engineer shall be made within 60 days following the date when it becomes apparent to such party that there will be such an overrun or underrun in an estimated quantity. For purposes of this subarticle, the estimated quantity is the quantity shown in the bid schedule as modified by addenda, and by change orders.

For overruns, the adjustment shall be limited to the number of units by which the actual pay quantity exceeds 125 percent of the estimated quantity.

For underruns, final payment for the item will be computed by applying the unit price bid in the schedule to the actual pay quantity, and then adding to the result an amount obtained by applying to the number of units of underrun below 75 percent of the estimated quantity, a reasonable allowance per unit for the Contractor's mobilization and other fixed costs relating thereto, as determined by the Engineer. The payment for the total pay quantity of an underrun item of work will in no case exceed the payment which would be made for the performance of 75 percent of the estimated quantity of such item at the original contract price.

    (3) *Extent and Method of Adjustment.*—Subject to the provisions of this subarticle, an adjustment in compensation will account for those increases or decreases in the costs of the work which are the necessary foreseeable consequence of a change in the work, unforeseen site condition, delay to the work, waiver of defects in the work, or variations in quantities. Where contract prices cover the affected work, materials, or equipment, increases or decreases in costs will be added to or subtracted from those prices, without repricing of base costs, to the maximum extent possible.

An adjustment in compensation based on a change in the work, unforeseen site condition, or variations in quantities will not account for delay costs, even though a necessary, foreseeable consequence of the change, condition, or variations, unless the Engineer determines that the delay involved is unreasonable, considering all the circumstances.

Except as hereinafter provided, an adjustment in compensation will include a reasonable allowance for profit on increases in costs and a reasonable reduction in profit for decreases in costs accounted for therein. No profit will be allowed on delay costs, irrespective of the basis for adjustment under Subarticle (2). No allowance will be made for loss of anticipated profit on any portion of the work not performed by reason of a change in the work, or variations in quantities.

Adjustments in compensation will be determined as follows:

 a. By negotiation of the parties on the basis of estimates of those increases or decreases in costs to be accounted for in the adjustment, provided, that upon the Engineer's written request, the Contractor submits his detailed estimate of such increases or decreases in costs, together with cost breakdowns and other appropriate data supporting such estimate, within the time specified in such request; or

 b. By the Engineer, on the basis of his estimate of those increases or decreases in the costs to be accounted for in the adjustment, if the Contractor fails to submit his estimate of such increases or decreases in costs in the manner and within the time provided in a. above, or if the parties fail to negotiate an adjustment within a reasonable time as determined by the Engineer; or

 c. At the option of the Engineer, whether or not negotiations are initiated under a. above, by force account payment for those portions of the work required or materially affected by the change in the work, unforeseen site condition, delay to the work, waiver of defects in the work, or variations in quantities, as provided in Section 9, Article (c).

(f) **Acceptance and Protest of Changes.**—The Contractor's written acceptance of a change order or other order for changes shall constitute his final and binding agreement to the provisions thereof and a waiver of all claims in connection therewith, whether direct or consequential in nature.

Should the Contractor disagree with any order for changes, he may submit a notice of potential claim to the Engineer in accordance with Section 8, Article (a) at such time as the order is set forth in a change order or, in the case of changes in the work, at such time as he receives written notice from the Engineer, pursuant to Article (c), that no adjustment in compensation or extension of time is due.

Disagreement with an order for a change in the work shall in no way excuse the Contractor from complying with the order and prosecuting the change in accordance with Article (d).

(g) **Unforeseen Site Conditions.**—The Contractor shall notify the Engineer in writing of the following conditions, called unforeseen site conditions, promptly upon their discovery and before they are disturbed:

 1. Subsurface or latent physical conditions at the site of the work differing materially from those represented in the bid documents and the materials and data referenced therein; or

 2. Unknown physical conditions at the site of the work of an unusual nature differing materially from those ordinarily encountered and generally recognized as inherent in work of the character provided for in this contract.

The Engineer will promptly investigate any condition of which he is so notified or any condition discovered by him which appears to fall within the above categories, and will, as soon as practicable, issue appropriate orders or instructions. If he determines that the condition is an unforseen site condition, that it will materially increase or decrease the costs of any portion of the work, and that such change in costs is not or will not be accounted for in any order for a change in the work, he will issue a change order adjusting the compensation for such portion of the work in accordance with Article (e).

If the Engineer determines that a condition of which he has been notified by the Contractor is not an unforeseen site condition, or that as an unforeseen site condition, it does not justify an adjustment in compensation, he will so advise the Contractor in writing. Should the Contractor disagree with such determination, he may submit a notice of potential claim to the Engineer as provided in Section 8, Article (a).

The discovery of an unforeseen site condition, actual or alleged, shall in no way excuse the

Contractor from prosecuting the work in accordance with Section 5, Article (b) pending orders or instruction from the Engineer, except for such stoppages as may be necessary to permit investigation of the conditions. Should the Engineer suspend the work of any portion thereof by reason of a condition determined to be an unforeseen site condition, such suspension will be for the convenience and benefit of the Department under Section 5, Article (f). A suspension of the work by reason of a condition alleged by the Contractor to be an unforeseen site condition, but determined not to constitute such a condition, will be for conditions unfavorable for the proper prosecution of the work under said article.

The Contractor's failure to give written notice of an unforeseen site condition promptly upon its discovery and before it is disturbed shall constitute a waiver of all claims for such condition, whether direct or consequential in nature.

(h) **Materials and Data on Conditions at Site.**—The Department's investigations of conditions at the site of the work are made primarily for design purposes. Samples, drill cores, logs of test borings and similar materials and data obtained in such investigations, included or referenced in the bid documents and made available to the Contractor prior to submission of his bid, will indicate the character of conditions only at the points from which the samples, cores, or other materials or data are taken, and shall not, in themselves, be considered representations within the meaning of Article (g)1. that similar conditions exist throughout the site of the work, or any part of it, or that other conditions may not be encountered. Opinions, conclusions, interpretations, or deductions concerning conditions at the site which may be expressed or implied in any such materials and data, or any matters contained therein of which the Contractor has, or reasonably should have, independent knowledge, shall not be considered representations of site conditions within the meaning of Article (g)1.

In no event shall the Contractor be relieved of his responsibilities under Article (g), "Examination of the Work" in the Bidding Requirements and Conditions, or under other provisions of the contract, by the Department's failure to investigate site conditions, by its inadvertent failure to include or reference in the bid documents or make available to the Contractor materials and data obtained in site investigations, or by oral representations of Department officers and employees concerning site investigations or conditions prior to submission of the Contractor's bid.

SECTION 8. CLAIMS

(a) **Notice of Potential Claim.**—Should the Contractor disagree with any decision, order, instruction, notice, other action, or omission of the Engineer, he may, within 30 days after receipt or occurrence of the same, subject to such express limitations as are provided elsewhere in these Standard Provisions, submit to the Engineer a written notice of potential claim stating clearly and in detail his objections thereto; provided, that if the objection is to an oral decision, order, instruction, or notice, the Contractor shall have first requested that it be confirmed in writing in accordance with Section 6, Article (a). Such notice shall indicate, insofar as possible, the basis of the stated objections and the nature and amount of any additional compensation or extension of time which the Contractor believes will or may be due. If the Contractor has the same objection to a series of decisions, orders, instructions, notices, or other actions concerning the same specific subject matter, he may submit a continuing objection in his notice of potential claim with respect to the first such action and need not submit further notices of such objection.

Any decision, order, instruction, notice, other action, or omission of the Engineer shall be final and conclusive on the Contractor if he fails to submit a notice of potential claim with respect thereto in the manner and within the time above stated, and such failure shall constitute a waiver of all claims in connection with the action or omission, whether direct or consequential in nature.

(b) **Submission and Documentation of Claims.**—The Contractor may submit a claim to the Engineer concerning any matter noticed in accordance with Article (a) within 60 days following the submission of such notice, unless, due to the nature of the claim or the incomplete state of the work, it is impracticable to determine the amount or extent of the claim within such period, in which case the claim may be submitted at the earliest time thereafter that such determination can be made, but in no event later than execution of the release of claims provided for in Section 9, Article (e).

A claim shall set forth clearly and in detail, for each item of additional compensation or extension of time claimed, the reasons for the claim, references to applicable provisions of the specifications, the nature and amount of the costs or time involved, the computations used in determining such costs or time, and all other pertinent factual data. The Contractor shall furnish such clarification and further available information as may be requested in writing by the Engineer within the time specified in such request.

Where not otherwise required by these Standard Provisions, the Contractor shall maintain complete and accurate daily records of the costs of any portion of the work for which additional compensation is claimed, and shall give the Engineer access thereto or certified copies thereof, as requested. All such records shall be retained by the Contractor and opened to inspection and audit by the Department and its authorized representatives on the same terms as are set forth in Section 9, Article (c) (2) for retention, inspection and audit of the cost records of force account work.

The Contractor shall have no claim for loss of anticipated profit on portions of the work not performed.

Except as provided below, any decision, order, instruction, notice, other action, or omission of the Engineer noticed by the Contractor under Article (a) shall be final and conclusive on the Contractor if he fails to submit and document a claim with respect thereto in the manner and within the time above stated, and such failure shall constitute a waiver of all claims in connection with the action, or omission, whether direct or consequential in nature.

If a claim is submitted later than the time above stated, but is submitted in properly documented form and under timely notice in accordance with Article (a), the Contractor may include in the claim a statement of the reason for delay in submittal and a request that the Engineer consider and decide the merits of the claim notwithstanding such delay. The Engineer may grant such request if good cause is shown for the delay, the factual and cost basis of the claim may be ascertained and verified notwithstanding the delay, and the Department's interests are not otherwise prejudiced; however, the Engineer's failure or refusal to exercise such authority shall not be subject to claim by the Contractor.

(c) Decision of Claims.—The Engineer shall decide all claims of the Contractor arising under and by virtue of the contract, and his decision, whether on the merits of the claim or on its timeliness, shall be final and conclusive unless it is fraudulent, capricious, arbitrary, or so grossly erroneous as necessarily to imply bad faith.

The foregoing authority of the Engineer will be administered as follows, unless another procedure, uniform in application, is adopted for Department contracts and the Contractor is notified thereof in writing.

Claims will be decided by a designated representative of the Deputy director, who will furnish his decisions to the Contractor in writing. The representative's decision on a claim shall be final and conclusive, as provided above, unless within 30 days after receipt thereof the Contractor makes written request for review and decision of the claim by the Deputy Director. Following such request, the Deputy Director will consider the entire claim, as submitted to his representative, and issue his decision to the Contractor in writiing, which shall be final and conclusive as provided above.

Details of the foregoing procedure and supplemental procedures for the consideration and decision of claims will be furnished to the Contractor in writing by the Engineer upon receipt of a notice of potential claim. The Engineer may alter such detailed and supplemental procedures from time to time.

Each claim will be processed and decided as soon as practicable following its submission and the submission or availability of any additional information and data necessary to its decision.

(d) **Release of Undecided Claims.**—Undecided claims submitted to the Engineer before execution of the release provided for in Section 9, Article (e) and not specifically excepted therefrom shall be deemed released by the Contractor and will not be further considered by the Engineer.

(e) **Determination of Rights.**—Claims totaling in the aggregate twenty-five thousand dollars ($25,000) or less arising from this contract may, at the option of the Contractor or the Department, be subject to determination of rights under this contract in accordance with Sections 14378–14380 of the California Government Code, the rules and regulations adopted in implementation thereof, and the provisions of this article.

All claims arising from this contract shall be submitted to the Engineer for his decision, pursuant to Articles (b) and (c), and a final Engineer's decision shall be rendered thereon prior to giving notice of claim pursuant to Section 14379(b) of the California Government Code, except that the Contractor may give such notice of claim on or after the 30th day following final payment to the Contractor pursuant to Section 9, Article (e), if the Engineer's final decision on such claim has not been rendered on or before such date.

In no event shall a notice of claim be given pursuant to Section 14379(b) of the California Government Code later than six months after final payment pursuant to Section 9, Article (c).

The Contractor may apply to the Department for consent to appointment of a hearing officer prior to completion of this contract. Such application shall be in writing. Consent to such early appointment shall rest in the sole discretion of the Department, but in no event shall such consent be given if the total of the amount claimed and of amounts claimed on all prior claims, if any, on which consent to such earlier appointment has been given exceeds twenty-five thousand dollars ($25,000).

A decision by a hearing officer pursuant to a determination of rights proceeding shall be final if supported by law and by substantial evidence.

Nothing in this article shall be construed as waiving any right of the Department under this contract to receive any notice or other document required thereby within any period of time stated herein including, but not limited to, the Notice of Potential Claim required by Article (a).

SECTION 9. MEASUREMENT AND PAYMENT

(a) **Measurement and Computation of Quantities.**—All items of the work to be paid for at a contract price per unit of measurement will be measured by the Engineer in accordance with United States Standard Measures. A ton shall mean 2,000 pounds, avoirdupois. Except as otherwise expressly provided in the specifications, the methods of measurement and computation of quantities of such items will be determined by the Engineer, taking into account the price of the item relative to its quantity and the costs of measurement.

The weights of metalwork, pipe, and other metal parts to be paid for by weight will be determined by the Engineer on the basis of handbook weights, scale weights, or manufacturer's catalog weights, or in the absence of any of the foregoing, on the basis of estimated weights; provided, that weights of nonmetallic coatings will be excluded.

(b) **Payment at Contract Prices.**—The contract price for an item of the work shall include full compensation for all costs of that item, including the costs of any work, materials and equipment incidental to the item but not specifically shown or described in the drawings and specifications, subject only to such express limitations as may be stated in the specifications defining the item or prescribing payment therefor.

The contract prices shall include full compensation for all costs of any work, materials, and equipment required by the drawings and specifications at the time of contract award, but not covered by a contract price or otherwise expressly made the subject of direct payment.

(c) **Force Account Payment.**—

(1) *General.*—Force account payment will be made only for those portions of the work required or materially affected by a change in the work, unforeseen site condition, delay to the work, or waiver of the defects in the work, as provided in Section 7, Article (e). Such payment will be

determined separately for each such change, condition, delay, or waiver. Such payment will also be determined separately for any delay costs payable in connection with a change in the work or unforeseen site condition and for other costs of such change or condition. Force account work, as used in this article, shall mean any portion of the work for which separate force account payment is being determined hereunder.

Force account payment will consist of the actual necessary costs of labor, materials, equipment and construction equipment used in the force account work, as determined by the Engineer in accordance with Subarticles (3), (4) and (5) respectively, plus an allowance on such costs for superintendence, general expense and profit* determined in accordance with the following schedule, which allowance shall constitute full compensation for all costs of such work not expressly included as actual necessary costs under Subarticles (3), (4) and (5).

Total Increment of Actual Necessary Costs of Labor, Materials, Equipment and Construction Equipment	On Costs of Labor	Allowance for Superintendence, General Expense and Profit* —In Percent On Costs of Materials, Equipment and Construction Equipment
$0 to $25,000	20%	17%
$25,000 to $50,000	19%	16%
$50,000 to $100,000	18%	15%
$100,000 to $150,000	17%	14%
$150,000 to $200,000	16%	13%
$200,000 to $300,000	15%	12%
$300,000 to $400,000	14%	11%
Over $400,000	13%	10%

The allowance will be computed for each increment of actual necessary costs by applying to the respective components of that increment the respective percentages for that increment and totaling the results obtained. The percentage rates of allowance for any cost increment shall not be affected by the amount of subsequent cost increments. Thus, if the actual necessary costs of the force account work** total $120,000, the total allowance thereon will be computed as follows:

20% of the labor component of the first increment ($25,000); plus
19% of the labor component of the second increment ($25,000); plus
18% of the labor component of the third increment ($50,000); plus
17% of the labor component of the fourth increment ($20,000); plus
17% of the M,E&CE*** component of the first increment; plus
16% of the M,E&CE component of the second increment; plus
15% of the M,E&CE component of the third increment; plus
14% of the M,E&CE component of the fourth increment.

No additional payment will be made by reason of the performance of force account work by a subcontractor or other force not within the Contractor's organization.

(2) Cost Records.—The Contractor shall maintain complete and accurate daily records of the costs of force account work, clearly distinguishing such costs from the costs of other portions of the work, and shall give the Engineer access thereto or certified copies thereof as requested.

The Contractor shall submit to the Engineer, at the end of each shift of force account work, a written report of the costs of all labor, materials, equipment and construction equipment used in

---
*Where actual necessary costs are delay costs, then irrespective of the basis for force account payment under Section 7, Article (e) (2), the stated percentage rates of allowance will be reduced by the Engineer to eliminate allowance for profit in accordance with the provisions of Subarticle (3) of said article.
**Exclusive of delay costs.
***Materials, equipment and construction equipment.

performing such work during that shift, whether furnished by the Contractor, a subcontractor, or other force. The report shall itemize: names or identifications and classifications of workmen and their hourly rates of pay, hours worked, and total pay; materials and equipment used and unit and total prices therefor; and the size, type and identification number of each unit of construction equipment, the hours it was operated, and the unit and total charges therefor; provided, that rates of pay, unit prices and charges, and extensions of amounts may be omitted from the report and submitted as soon as practicable thereafter. Before submission to the Engineer, the report shall be verified and signed by the representative of the Engineer in charge of inspection of the force account work and by the representative of the Contractor in charge of such work. The representative of the Engineer will state on the report any exceptions thereto.

The Engineer will compare the reports submitted by the Contractor for each day of force account work with his own records, make any necessary adjustments, and, on the basis of such adjusted reports, compile the costs of such work in a daily force account work report. Such report, when submitted to the Contractor, shall be the basis of payment for the force account work unless the Contractor, within 15 days after receipt thereof, submits a written statement of his objections thereto, together with valid copies of payrolls, vendors' invoices and other documents substantiating such objections. Upon receipt of such statement and substantiating information, the Engineer will make any revisions to the report which he finds justified and resubmit it to the Contractor, whereupon it shall become the basis of payment for the force account work. Should the Contractor disagree with the resubmitted report, he may submit a notice of potential claim to the Engineer as provided in Section 8, Article (a). If the Contractor has the same objection to a series of reports, he may submit a continuing objection in his written statement and notice of potential claim with respect to the first such report, and need not submit further statements or notices of that objection.

The Contractor shall retain all records of the costs of force account work and other records bearing on such costs during the period of contract and for three years after acceptance of the work, or, if later, until the final resolution of all claims, whether by administrative or judicial action. During such period the Contractor shall, upon reasonable notice, open such records to inspection and audit by the Department and its authorized representatives. Where payment for force account work is based on costs to a subcontractor or other force not within the Contractor's organization, the Contractor shall require that party to retain his records of such costs and open them to inspection and audit by the Department and its authorized representatives on the foregoing terms.

All force account payments shall be subject to adjustment on the basis of audit. Each party shall pay to the other any amount found due the other as a result of such audit, provided that if the total amount found due is less than $100, no payment need be made.

(3) Labor.—The actual necessary cost of labor used in performing force account work, whether employed by the Contractor, a subcontractor, or other force, will be the sum of the following:

    a. Actual wages paid to workmen, which will include any payments to or on behalf of workmen for health and welfare, pensions, vacations and similar fringe benefits.

    b. Payments to or on behalf of workmen, other than actual wages as defined above, which are required by State and Federal laws, including, but not limited to, compensation insurance, social security and unemployment insurance payments.

    c. Subsistence and travel allowances paid to workmen under collective bargaining agreements, or as a regular practice of the employer.

As used in this subarticle, the term "workmen" will include persons classified as working foremen by the Engineer.

(4) Materials and Equipment.—The actual necessary cost of materials and equipment used in performing force account work will be the price paid by the purchaser, whether the Contractor, a subcontractor, or other force, to the supplier thereof, subject to the following conditions:

    a. The cost of Department furnished materials and equipment will be excluded from the actual necessary cost of materials and equipment.

b. Available cash or trade discounts will be deducted from the price paid for materials or equipment, if not taken by the purchaser.

c. If the purchaser procures materials or equipment by any method other than a direct purchase from the original supplier, the actual necessary cost thereof will be the price paid to the original supplier by the secondary supplier, as determined by the Engineer, with no markup except for the actual necessary cost of handling such materials or equipment.

d. The actual necessary cost of materials or equipment procured from any source owned wholly or in part by the purchaser will not exceed the price paid currently by the purchaser for similar materials or equipment procured from that source for contract items of the work, or the lowest current wholesale price for such materials or equipment in the quantity required, in the same area, delivered to the site of the work or other place of use, whichever price is lower.

e. If the price paid for materials or equipment is determined to be excessive by the Engineer, or is not substantiated by the Contractor as required in Subarticle (2), the actual necessary cost thereof will be the lowest current wholesale price for such materials or equipment in the quantity required, delivered to the site of the work or other place of use, less any deductions for cash or trade discounts required to be made under b. above.

f. In the case of disposal of surplus materials and equipment pursuant to Section 7, Article (d) (3):

1. The actual necessary cost of materials or equipment delivered to the Department will include the actual necessary cost of any transportation and handling of the materials or equipment ordered by the Engineer.

2. The actual necessary cost of non-returnable materials or equipment retained by the Contractor will be reduced by the amount of the salvage value of such materials or equipment, as determined by the Engineer.

3. The actual necessary cost of materials or equipment returned to the supplier will be the charges made by the supplier for such return and the actual necessary cost of handling the materials or equipment.

4. Returnable materials or equipment retained by the Contractor will not be included in actual necessary costs.

(5) Construction Equipment.—

a. General.—Construction equipment, for purposes of this article, will not include any tools or other items of equipment having a unit replacement value of $50 or less, whether or not consumed by use, unless the item is listed in the Schedule of Force Account Equipment Rates attached to the specifications as Appendix I. The costs of all such unlisted items are included in the allowances provided in Subarticle (1).

All construction equipment used in performing force account work shall be approved by the Engineer. Equipment requiring separate tractive or power units shall be matched to such units according to the manufacturer's minimum recommended ratings. The Contractor shall furnish the Engineer a written description of each item of construction equipment, listing its identification number and, as applicable, its make, model, size, capacity, mounting, horsepower, type of power, and fuel, which description will be used to identify and classify the equipment for purposes of this subarticle. The foregoing requirements are in addition to the requirements of Section 6, Article (f).

The actual necessary cost of construction equipment used in performing force account work will be the sum of charges for operation, idle time and moving of equipment furnished by the Contractor and by subcontractors, and for equipment rented from third parties, all as determined in accordance with b. and c. below.

b. Construction Equipment Furnished by Contractor and Subcontractors.—

1. Operation and Idle Time of Construction Equipment Furnished Under Schedule.—The charges for operation and idle time of any construction equipment furnished by the Contractor or a subcontractor which is listed in the Schedule of Force Account Equipment Rates, attached to the specifications as Appendix I, will be determined from the hourly rates listed

in that schedule, as provided in i., ii., and iii., below. Such rates include the costs of depreciation, interest, taxes, insurance, storage, fuel, oil, lubricants and lubrication, supplies, tires, repairs, necessary attachments, maintenance, tools used for maintenance, and all incidentals. In determining the charge for any such item of equipment for fractional parts of an hour, less than 30 minutes of operation or idle time will be considered one-half hour thereof, and 30 to 60 minutes of operation or idle time will be considered one hour thereof. Up to 30 minutes of idle time required during any operating shift for regular servicing, minor repair, or maintenance of equipment will be considered operating time. A shift will be considered eight hours.

i. First Shift Operation Charges.—The charge for operation of any item of the above equipment for the first shift of its operation on any day will be the product of the hourly rate for that equipment listed in the schedule and the period of time the equipment was operated during that shift.

ii. Extra Shift Operation Charges.—The change for operation of any item of the above equipment for the second or third shift of its operation any day will be the product of the hourly rate for that equipment listed in the schedule, multiplied by the applicable extra shift factor listed in the schedule, and the period of time the equipment was operated during that shift.

iii. Idle Time Charges.—The charge for idle time of any item of the above equipment on any day will be the product of the hourly rate for that equipment listed in the schedule, multiplied by the applicable idle time factor listed in the schedule, and the period of time the equipment was idle during that day, provided that such period shall not exceed eight hours, or a period of time which, when added to the time the equipment was operated during that day, equals eight hours, whichever is less. Such charge will be made only if the Engineer determines that the idle equipment could be used to advantage on other portions of the work, but that such use is precluded by its assignment to the force account work. No charge will be made for idle time of equipment on holidays observed on the work, Sundays, or other nonworking days; for periods of idle time required during operating shifts for regular servicing, minor repair, or maintenance; for idle time caused by the existence of weather conditions which are unfavorable for the proper prosecution of the work; or for idle time caused by any suspension of the work pursuant to Section 5, Article (f), other than a suspension for the convenience and benefit of the Department.

2. Operation and Idle Time of Construction Equipment Not Furnished Under Schedule.—Charges will be allowed for operation and idle time of any construction equipment furnished by the Contractor or a subcontractor which is not listed in the Schedule of Force Account Equipment Rates only if the Engineer determines that such equipment is necessary for proper prosecution of the force account work. Such charges will be based on hourly rates, extra shift factors, and idle time factors established by the Engineer, but otherwise will be determined in the same manner as the charges for operation and idle time of equipment listed in the schedule, and subject to all of the provisions of 1. above. The Contractor shall furnish to the Engineer an accurate statement of the ownership expense and operating cost of each item of the above equipment, and any other cost data which will assist the Engineer in establishing rates therefor.

3. Moving of Construction Equipment.—Charges will be allowed for moving of any construction equipment furnished by the Contractor or a subcontractor to and from the force account work, to the extent that such moving is required by use of the equipment on such work. Such charges will consist of those costs of loading, unloading and transporting such equipment or of moving it under its own power which are not included in the charges provided for in 1. and 2. above, subject to the following conditions:

i. No charge will be allowed for moving of equipment not subject to charge under 1. or 2. above.

ii. The location from which the equipment is to be moved to the force account work shall be approved by the Engineer prior to the move.

iii. The cost of transportation on trailers will be determined at rates not exceeding applicable minimum rates established by the California Public Utilities Commission, or in the absence of such established rates, not exceeding those of established trucking concerns in the area of the work.

iv. If equipment is moved from the force account work to a location other than that approved by the Engineer under ii, above, the charge therefor will not exceed the costs, as specified above, of moving it to the approved location.

v. If the equipment is used on other portions of the work, the moving charges therefor will be limited to the moving costs, as specified above, resulting solely from its use on the force account work.

c. Construction Equipment Rented from Third Parties.—Charges will be allowed for construction equipment rented from third parties only if the Engineer determines that such equipment is necessary for the proper prosecution of the force account work, and that neither the Contractor nor any of his subcontractors has such equipment on the work. Such charges will consist of: the rental paid for the equipment, up to the current prevailing rental rate for such equipment of equipment rental dealers in the area of the work; the charges for moving the equipment to and from the force account work, determined in the same manner as the charges for moving equipment furnished by the Contractor or a subcontractor and subject to all of the provisions of b. 3. above; and payments made for equipment operators, if the equipment must be rented with operators and the cost thereof is not included in the rental rate.

(d) **Progress Payments.—**

(1) Progress Estimates.—Subject to the provisions of Subarticle (4), the Department, once each month, will prepare a written estimate of the total amount and value of:

a. Work done to the time of such estimate, and

b. Unused materials and equipment furnished by the Contractor, to which he has acquired full title as against suppliers or other third parties, either delivered on the ground or stored subject to or under the control of the Department to the time of such estimate, hereinafter in this article called materials and equipment furnished;

provided, that such work done and materials and equipment furnished have reached a sufficient stage of completion or of incorporation in the work that their value may be estimated as hereinafter provided.

The value of work done and materials and equipment furnished will be estimated on the basis of contract and change order prices, except that where a single price covers various phases of work, or various types or units of materials or equipment, or any combination of work, materials and equipment, the estimate may be based upon a reasonable allocation of that price by the Engineer among any separable units of the total item covered, as determined by the Engineer. At the Engineer's request, the Contractor shall submit work control plans and other information on given items of the work to assist the Engineer in price allocation. No price allocation will be made to preliminary or incidental activities such as design, mobilization, site preparation, form work and cleanup. No price allocation will be made to production or manufacture of materials and equipment, except that allocations may be made to production work on Department-furnished earth materials, and to manufacturing work on specialized mechanical and electrical equipment where an extended manufacturing period and substantial expenditures by the Contractor are required.

The amounts and values of materials and equipment furnished to be considered in each estimate shall be reported to the Engineer by the Contractor on a Department furnished form properly completed and signed. At the Engineer's request, the Contractor shall furnish with such report valid copies of paid vendors' invoices, evidence that materials and equipment not delivered on the ground are stored subject to or under the control of the Department, and other documentation of the report. The estimated value of materials and equipment furnished for an item of the work will in no event exceed the total price of the item less that portion of such price, if any, allocated to installation or other incorporation of such materials and equipment in the work.

(2) Progress Payments.—Upon completion of each monthly estimate of work done and materials and equipment furnished, the Department, subject to the provisions of Subarticles (3) and (4), will pay to the Contractor the estimated value of such work, materials and equipment less the amount of all prior payments and all liquidated damages and other amounts to be deducted or retained under the contract.

(3) Retention from Progress Payments.—The Department will retain from each progress payment 10 percent of the estimated value of work done and materials and equipment furnished, except as follows. In any progress payment made after the work is 50 percent complete in estimated value, taking into account all prior change orders, the Department may retain less than 10 percent of, or may pay the full amount of the increment in the estimated value of work done since the preceding progress estimate, if the Engineer determines that the Contractor is making sufficient progress as of the time of such payment to insure completion of the work and all portions thereof within the times specified therefor or any extensions thereof. Amounts retained from progress payments under this subarticle will not be paid to the Contractor until the time of final payment and release of claims as provided in Article (e), notwithstanding subsequent improvement in the progress of the work.

(4) General Conditions.—No progress estimate or payment need be made when, in the judgment of the Engineer, there may be cause for termination of control over the work under Section 5, Article (h), or when, in his judgment, the increment in the estimated value of work done and materials and equipment furnished since the preceding estimate is less than $300.

The Engineer may withhold all or any part of a progress payment otherwise payable upon the Contractor's failure to submit any required schedule for the work, whether preliminary or detailed in nature, in the manner and within the time specified in the Special Provisions. Progress payments or portions thereof so withheld may be paid to the Contractor, at the discretion of the Engineer, following submission of the delinquent schedule in proper form.

No progress estimate or payment shall be considered an approval or acceptance of any work, materials, or equipment. Estimated amounts and values of work done and materials and equipment furnished will be conformed with actual amounts and values, as they become available, in subsequent progress estimates and the final estimate. All estimates and payments will be subject to correction in subsequent progress estimates and the final estimate.

(e) **Final Payment.**—Within 45 days following acceptance of the work, or within such longer period of time as may be required due to causes within the control or due to the fault or negligence of the Contractor, the Engineer will prepare in writing and furnish to the Contractor a final estimate of money due under the contract, itemizing the actual quantities of the work, all amounts earned, all prior payments, all amounts retained under Article (d) (3), all amounts deducted and retained under other provisions of the contract, and the amount due. Within 30 days after issuance of the final estimate, or, if later, upon the Contractor's submission of all operation and maintenance data, guarantees, and similar materials required by the specifications to be submitted before final payment, the Department will pay to the Contractor the amount determined to be due under the contract; provided, that payment will be made of the total amount retained under Article (d) (3) only after the Contractor has furnished the Department a release of all claims against the State arising under and by virtue of the contract, and provided further, that payment will be made of amounts retained under other provisions of the contract only to the extent and at the times permitted thereby. In executing his release of claims, the Contractor may state as an exception thereto any claim in a definite amount submitted to the Engineer in accordance with Section 8, Article (b), or noticed in accordance with Article (a) of that section and submitted with the release.

Within 45 days after all claims excepted from the Contractor's release of claims have been decided by the representative of the Deputy Director, or within 45 days after any such claims reviewed by the Deputy Director have been decided by him, the Engineer will prepare in writing and furnish to the Contractor a supplement to the final estimate stating the amount, if any, found due on such claims, and within 30 days after issuance of such supplement, the Department will pay the stated amount to the Contractor.

(f) **Interest.**—The Contractor shall not be entitled to interest on any amount paid on a claim prior to execution of his release of claims. The Contractor's entitlement to interest, if any, on any claim payment made after such release, whether by administrative or judicial action, shall not commence before the following times:

   1. 30 days after issuance of a supplement to the final estimate stating an amount due on the claim; or
   2. Absent the issuance of any such supplement, 30 days after receipt of a decision of the claim by the representative of the Deputy Director, unless the claim is reviewed by the Deputy Director, in which case, the time of receipt of a decision of the claim by the Deputy Director.

If final payment is not made within the time provided in Article (e), entitlement to interest, if any, will commence upon the expiration of such time.

The Contractor shall not be entitled to interest on any other amount paid under the contract by reason of its retention: pursuant to a provision of the contract, because of error in a progress estimate or the final estimate, or pending the determination of other amounts due the Contractor to be included in the same payment.

(g) **Corrections in Final Payment.**—Nothing in the provisions of Section 8, Article (c), or in Article (e) of this section shall preclude adjustment of the final estimate, any supplement thereto, and payments made thereunder as provided in Article (c) (2) of this section or as hereinafter provided in this article.

For a period of three years after acceptance of the work, or, if later, until the final resolution of all claims, whether by administrative or judicial action, the foregoing estimates and payments shall be subject to correction for clerical errors in the calculations involved in the determination of actual quantities of the work and payments therefor. Each party shall pay to the other any amount found due the other as a result of such corrections, provided that if the total found due is less than $100, no payment need be made.

(h) **Payment of Withheld Funds.**—Attention is directed to Article (d), and in particular to the retention provisions of said article.

Upon the Contractor's request the Department will make payment of funds with held from progress payments pursuant to the requirements of Government Code Section 14402 if the Contractor deposits in escrow with the State Treasurer securities eligible for the investment of state funds under Government Code Section 16430 or bank certificates of deposit, upon the following conditions:

   1. The Contractor shall bear the expenses of the Department and the State Treasurer in connection with the escrow deposit made;
   2. Securities or certificates of deposit to be placed in escrow shall be subject to approval of the Department and, unless otherwise permitted by the escrow agreement, shall be of a value at least 110 percent of the amount of the retention to be paid to the Contractor pursuant to this section;
   3. The Contractor shall enter into an escrow agreement satisfactory to the Department, which agreement shall include provisions governing inter alia:
      a. the amount of securities to be deposited;
      b. the providing of powers of attorney or other documents necessary for the transfer of the securities to be deposited;
      c. conversion to cash to provide funds to meet defaults by the Contractor including, but not limited to, termination of the Contractor's control over the work, stop notices filed pursuant to law, assessment of liquidated damages or other amounts to be kept or retained under the provisions of the contract;
      d. decrease in value of securities on deposit;
      e. the termination of the escrow upon completion of the contract;
   4. The Contractor shall obtain the written consent of the surety to such agreement.

SPECIAL PROVISIONS

SECTION 10. GENERAL DESCRIPTION AND LOCATION OF THE WORK

The work is defined in Section 1 and includes the following principal features:

A zoned embankment type dam approximately 11,600 feet long at the crest and 120 feet in height.

Open channel spillway.

Clearing of lake area.

Excavation and blanketing for outlet works intake.

Materials Haul Road.

The specifications provide for obtaining one type of the borrow material from two alternative sources called Excavation in Lake Borrow Area Schedule "A" and Schedule "B". Schedule "A" provides for construction of the dam utilizing borrow material from the lake area closest to the dam embankment. Schedule "B" provides for obtaining borrow material by grading a proposed recreational development area along the north shore and eastern end of the reservoir. Schedule "B" is the preferred borrow source and subject to the provisions of Section 11, Article (ff), will be utilized if the low bid received for that schedule is within the financing capability of the State.

The site of the work is located in Riverside County, approximately 13 miles southeast of the City of Riverside.

SECTION 11. SPECIAL CONDITIONS

(a) **Time of Completion and Liquidated Damages.**—Pursuant to the provisions of Section 5, Article (c), the Contractor shall:

(1) Complete the construction signs; Construction Facility and Visitor's Viewsite complete with grading, paving, fencing, Headquarters Office Building and Materials Testing Laboratory Building; and Instrumentation Terminals, before the expiration of

90 DAYS

(2) Complete the excavation for outlet works intake and construction of the blanket within the limits reserved for others as shown, before the expiration of

210 DAYS

and

(3) Complete the balance of the work before the expiration of

940 DAYS

Liquidated Damages for failure to complete the portion of the work described in (1) above within the time specified shall be $200 per day.

Liquidated Damages for failure to complete any portion of the work described in (2) above within the time specified shall be $500 per day.

Liquidated damages for failure to complete the balance of the work within the time specified shall be $1,200 per day.

(b) **Drawings.**—The work shall conform to the following drawings:

STATE OF CALIFORNIA
THE RESOURCES AGENCY
DEPARTMENT OF WATER RESOURCES
DIVISION OF DESIGN AND CONSTRUCTION
STATE WATER FACILITIES

CALIFORNIA AQUEDUCT
SANTA ANA DIVISION
PERRIS DAM AND LAKE

Sheet No.	Drawing No.	Title

GENERAL

1	V-4K1-1	Location and Vicinity Maps
2	V-4K2-1	List of Drawings (Revision No. 1 7/24/70)
3	V-4K3-1	Project Map (Revision No. 1 7/24/70)
4	V-4K3-2	General Plan (Revision No. 1 7/24/70)

EMBANKMENT

5	V-4D1-1	Sections (Revision No. 1 7/24/70)
6	V-4D1-2	Sections and Details (Revision No. 1 7/24/70)
7	V-4D2-1	Drainage Details (Revision No. 1 7/24/70)
8	V-4D3-1	Instrumentation; Plan and Sections
9	V-4D3-2	Instrumentation; Piezometer Installation Details
10	V-4D3-3	Instrumentation: Miscellaneous Details
11	V-4S1-1	Instrumentation Terminal No. 1; Plan, Elevations, Sections and Details
12	V-4S1-2	Instrumentation Terminal No. 1; Details and Notes
13	V-4T1-1	Instrumentation Terminal No. 1; Mechanical and Electrical Installation
14	V-4S2-1	Instrumentation Terminal No. 2; Plan, Elevations, Sections and Details

SPILLWAY

15	V-4D4-1	Plan, Profile and Sections (Revision No. 1 7/24/70)
16	V-4D4-2	Chute; Plan, Sections and Details (Revision No. 1 7/24/70)

OUTLET WORKS INTAKE EXCAVATION

17	V-4D5-1	Plan, Profile, and Sections

MATERIALS HAUL ROAD

18	V-4M1-1	Typical Sections
19	V-4M2-1	Plan and Profile; Station 8+25 to Station 37+00
20	V-4M2-2	Plan and Profile; Station 37+00 to Station 68+73.30
21	V-4M3-1	Culvert List
22	V-4M3-2	Drainage Details
23	V-4M3-3	Miscellaneous Details

CONSTRUCTION FACILITY

24	V-4M4-1	Parking Lot and Access Road

VISITORS VIEWSITE

25	V-4K4-1	Plan, Profile, Section and Detail
26	V-4J1-1	Information Booth
27	V-4J2-1	Project Construction Sign; 8 Feet by 16 Feet
28	V-4J2-2	Project Construction Sign Details; 8 Feet by 16 Feet

MISCELLANEOUS

29	V-4Z1-1	Lake Borrow Area; Plan—Schedule A
30	V-4Z1-2	Lake Borrow Area; Sections—Schedule A
31	V-4Z1-3	Lake Borrow Area; Plan—Schedule B
32	V-4Z1-4	Lake Borrow Area; Plan—Schedule B
33	V-4Z2-1	Clay Borrow Area; Plan and Section
34	V-4X1-1	Exploration Location Map (Revision No. 1 7/24/70)
35	V-4X1-2	Exploration Location Map (Revision No. 1 7/24/70)
36	V-4X1-3	Exploration Location Map

Sheet No.	Drawing No.	Title
37	V-4X1-4	Exploration Location Map
38	V-4X1-5	Exploration Location Map
39	T-5A4-1	Standard Fencing Details; 4' Chain Link
40	T-5A3-1	Standard Fencing Details; 6' Chain Link
41	T-0E1-2	Format for CPM Type Contractor's Schedule
42	T-0E1-4	Format for Precedence Type Contractor's Schedule

(c) **Definitions.**—In the list of abbreviations in Section 1, delete the following at the end of the listing:

USAS United States of America Standard Institute

and insert the following after the abbreviation AISC:

ANS American National Standard
ANSI American National Standards Institute, Inc.

(d) **Fair Employment Practices.**—The requirements of Section 3, Article (c) (6) b. are replaced by the following:

b. Nothing contained in this Fair Employment Practices article shall be construed in any manner of fashion so as to require or permit the hiring of an employee not permitted by the National Labor Relations Act.

(e) **Alien Labor.**—The requirements of Section 3, Article (h), are deleted.

(f) **Materials and Equipment from Foreign Countries.**—Delete Article (1) of Section 3, and replace with the following:

(1) Materials and Equipment from Foreign Countries.—If the Contractor elects to supply materials or equipment produced or manufactured in a country or countries other than the United States of America, the following provisions shall apply:

1. The taxes which the Contractor is required to pay pursuant to Section 4, Article (i), shall include all charges of any kind imposed by any foreign governmental entities, and shall include all charges imposed by the Federal Government as a result of the importation of materials or equipment produced or manufactured in a country or countries other than the United States.

2. All documents pertaining to the contract, including but not limited to, correspondence, bid documents, working drawings and data shall be written in the English language, and all numerical data shall use the foot-pound-second system of units of measurement.

3. Notwithstanding the provisions of Section 5, Article (d) (1) g., the acts of a foreign governmental entity will not be a cause for time extension.

(g) **Determination of Rights.**—After Article (d) of Section 8 add the following:

(e) **Determination of Rights.**—

(1) Claims totaling in the aggregate twenty-five thousand dollars ($25,000) or less arising from this contract may, at the option of the Contractor or the Department, be subject to determination of rights under this contract in accordance with Sections 14378-14380 of the California Government Code, the rules and regulations adopted in implementation thereof, and the provisions of this article.

(2) All claims arising from this contract shall be submitted to the Engineer for his decision, pursuant to Articles (b) and (c), and a final Engineer's decision shall be rendered thereon prior to giving notice of claim pursuant to Section 14379 (b) of the California Government Code, except that the Contractor may give such notice of claim on or after the 30th day following final payment to the Contractor pursuant to Section 9, Article (e), if the Engineer's final decision on such claim has not been rendered on or before such date.

(3) In no event shall a notice of claim be given pursuant to Section 14379 (b) of the California Government Code later than six months after final payment pursuant to Section 9, Article (e).

(4) The Contractor may apply to the Department for consent to appointment of a hearing officer prior to completion of this contract. Such application shall be in writing. Consent to

such early appointment shall rest in the sole discretion of the Department, but in no event shall such consent be given if the total of the amount claimed and of amounts claimed on all prior claims, if any, on which consent to such earlier appointment has been given exceeds twenty-five thousand dollars ($25,000).

(5) A decision by a hearing officer pursuant to a determination of rights proceeding shall be final if supported by law and by substantial evidence.

(6) Nothing in this article shall be construed as waiving any right of the Department under this contract to receive any notice or other document required thereby within any period of time stated herein including, but not limited to, the Notice of Potential Claim required by Article (a).

(h) **Payment of Withheld Funds.**—After Article (g) of Section 9 add the following:

(h) **Payment of Withheld Funds.**—Attention is directed to Article (d), and in particular to the retention provisions of said article.

Upon the Contractor's request the Department will make payment of funds withheld from progress payments pursuant to the requirements of Government Code Section 14402 if the Contractor deposits in escrow with the State Treasurer securities eligible for the investment of state funds under Government Code Section 16430 or bank certificates of deposit, upon the following conditions:

(1) The Contractor shall bear the expenses of the Department and the State Treasurer in connection with the escrow deposit made;

(2) Securities or certificates of deposit to be placed in escrow shall be subject to approval of the Department and, unless otherwise permitted by the escrow agreement, shall be of a value at least 110 percent of the amount of the retention to be paid to the Contractor pursuant to this section;

(3) The Contractor shall enter into an escrow agreement satisfactory to the Department, which agreement shall include provisions governing inter alia:

    a. the amount of securities to be deposited;

    b. the providing of powers of attorney or other documents necessary for the transfer of the securities to be deposited;

    c. conversion to cash to provide funds to meet defaults by the Contractor including, but not limited to, termination of the Contractor's control over the work, stop notices filed pursuant to law, assessment of liquidated damages or other amounts to be kept or retained under the provisions of the contract;

    d. decrease in value of securities on deposit;

    e. the termination of the escrow upon completion of the contract;

(4) The Contractor shall obtain the written consent of the surety to such agreement.

(i) **Data Available to Bidders and the Contractor.**—The following data related to the work will be furnished upon written request to the office of the Department, P.O. Box 388, Sacramento, California, 95802, and may be inspected during regular office hours at the office of the Department, Room 406-2, 1416 Ninth Street, Sacramento, California:

Testing procedures contained in the Department of Water Resources "Manual of Testing Procedures for Soils."

Attention is directed to Section 7, Article (h). Project Geology Report No. D-121, titled "Geology and Construction Materials Data, Perris Dam and Lake, Riverside County, California", will be furnished upon written request and may be inspected as provided above.

Drill cores and soil samples taken from the site of the work may be inspected during regular office hours upon request to the office of the Department, 406 East Palmdale Boulevard, Palmdale, California.

Additional data related to the work may be inspected during regular office hours at the office of the Department, Room 406-2, 1416 Ninth Street, Sacramento, California.

(j) **Right-of-Way.**—The Department will furnish all right-of-way for the permanent works and for lands shown on the drawings for use as borrow areas, waste areas, stockpiles, and roads. The Contractors will be permitted to use for roads and for construction plant purposes all right-of-way shown, except for reserved areas shown on the drawings or right-of-way required by other contractors pursuant to Article (1). Limits of right-of-way will be flagged for the Contractor's guidance during construction operations.

Should the Contractor require additional lands outside of the project area for work space, housing or other purposes, he shall make his own arrangements and bear all costs for such additional lands or easements.

(k) Advance Possession and Use of the Work by the Engineer.—Pursuant to the provisions of Section 4, Article (d), the Engineer will take possession of and use or open to use the following portions of the work prior to acceptance of the work, subject to the terms and conditions of that article.

1. The portion of the work specified in Articles (a) (1) and (a) (2).

After completion of excavation and blanketing for outlet works intake specified in Article (a) (2), the Contractor shall permit others engaged in performing work for the Department to use the area shown on the drawings as directed by the Engineer.

Attention is directed to the provisions of Section 13, Article (a), as to the Contractor's responsibility for injury, damage or loss to the portion of the work specified in Article (a) (1).

(1) Cooperation with Other Contractors and Forces.—Provisions relating to cooperation with other contractors and forces engaged in work in this area are contained in Section 4, Article (g).

Contracts under construction or scheduled for award, during the period of this contract include but are not limited to:

	Estimated Award Date	Estimated Completion Date
Santa Ana Pipeline	June 1971	April 1973
Inlet Works and Outlet Works	April 1971	May 1973
Recreation Development	July 1972	July 1973
Metropolitan Water District Bernasconi Tunnel No. 2	September 1970	April 1972
Metropolitan Water District Perris Control Facility to 2nd San Diego Aqueduct Canal	January 1971	June 1973
Metropolitan Water District Perris Control Facility and Connection to DWR	January 1972	January 1973
Relocation of Southern California Edison Transmission Lines	Under Construction	December 1970

(m) Cooperation with Riverside County.—Prior to start of the work, the Contractor shall contact the Director of Public Works, Riverside County, California, obtain all permits and licenses required, and obtain all other authorization required in the performance of his work. The cost of the permits and licenses shall be borne by the Contractor.

The Contractor shall bear all costs involved in curtailing and disposing of tumbleweeds, and for conformance with any other ordinances of Riverside County.

(n) Materials and Equipment Furnished by the Department.—Materials and equipment to be furnished by the Department pursuant to Section 6, Article (h), are specified in Section 22, and will be delivered to the Contractor's storage facilities at the job site. Testing equipment to be installed in the Materials Testing Laboratory Building as specified in Section 13, Article (c) (1) b., will be delivered to the Construction Facility.

(o) Contractor's Schedules.—

(1) Preliminary Schedule.—Within 30 days after receiving notice to begin the work, the Contractor shall furnish to the Engineer a preliminary schedule for the work showing his general plan

for orderly completion of the work and showing in detail his planned mobilization of plant and equipment, sequence of early operations, and timing of procurement of materials and equipment. The Contractor shall assist the Engineer in reviewing and evaluating such schedule.

(2) Detailed Schedule.—Within 60 days after receiving notice to begin the work, the Contractor shall furnish to the Engineer a detailed schedule for orderly completion of the work, showing his planned sequences of operations, and the dates for commencement and completion of all salient features of the work.

At the time the detailed schedule is furnished, the Contractor in addition shall furnish his estimate of earnings by months.

The schedule shall be comprehensive, covering both activities at the site of the work and off-site activities such as design, procurement, and fabrication. The schedule shall be orderly and realistic, and shall be revised as necessary to meet this requirement. The Contractor shall promptly advise the Engineer of any occurrence requiring substantial revision of the schedule and shall furnish a revised schedule within 15 days of such occurrence.

The detailed schedule and each revision thereof shall be subject to approval by the Engineer for conformity with the requirements of this article. The Contractor shall assist the Engineer in reviewing and evaluating each schedule furnished. Disapproved schedules will be returned to the Contractor, shall be revised by him to correct the defects noted, and shall be resubmitted to the Engineer within 15 days after receipt. Approval of a schedule shall in no way waive any of the requirements of Section 5; Section 9. Article (d); and other applicable provisions of contract, or excuse the Contractor of any of his obligations thereunder.

(3) Form of Schedules.—The Contractor shall submit two transparent prints of each schedule and revised schedule furnished.

The preliminary schedule shall be of the bar chart, critical path method, or precedence method type, at the Contractor's option.

The detailed schedule shall be of the critical path method type or precedence method type at the Contractor's option. Schedules shall be in the form of a network diagram and activity listing for critical path type schedules and a network diagram, activity listing, and input listing for precedence type schedules. Network diagrams and listings shall be generally in the form shown on the drawing entitled, "Format for CPM Type Contractor's Schedule" or "Format for Precedence Type Contractor's Schedule."

The network diagram shall show in detail and in orderly sequence all activities, their descriptions, durations, and dependencies or precedences, necessary to the completion of the work. The activity listing shall show the following information for each activity on the network diagram:

1. Identification by activity number and description
2. Duration
3. Earliest start and finish dates
4. Latest start and finish dates
5. Total float time

If the precedence method is selected, the input listing shall show the following information for each activity on the network diagram:

1. Identification numeric activity label
2. Duration
3. Activity description
4. Following activity number or numbers
5. Delay factor

Node numbers for critical path method type schedules shall consist of no more than four digits. Activity numbers for precedence method type schedules shall be numeric and consist of no more than three digits. Listing of activities for activity and input listings shall be by order of beginning activity number, lowest to highest.

The criteria used in determining which days are considered as nonworking days, for the purpose of preparing each schedule, shall be clearly shown.

(p) **Contractor's Roads and Construction Plants.**—Access to the work from existing transportation facilities and construction roads within the site of the work shall be provided by the Contractor.

The Contractor shall furnish working drawings pursuant to Article (u), of all roads and trails, and all borrow pits, spoil disposal areas, and drainage facilities related to his roads and plants.

When optimum use of construction plant sites or access roads requires joint use of roads in such sites by the Contractor and others engaged in work for the Department, the Contractor shall negotiate joint use and maintenance of such roads. Terms and conditions of agreements resulting from negotiations shall be subject to approval of the Engineer.

(q) **Protection of Existing Facilities.**—

(1) General.—The provisions of Section 4, Articles (c) and (g) shall apply to all utility facilities in the project area until they are removed from service or relocated by others.

Pursuant to Section 4, Article (c), the Contractor shall be responsible for the protection of existing triangulation surveying stations, bench marks, and all public land survey monuments, and private property corner markers within the project area.

The following facilities are existing and shall be protected pursuant to this article:

Seismic Facility and Power Line to Facility located at Martin Street and Bradley Rd.

Drill Holes so designated.

Southern California Edison transmission lines.

Pacific Lighting Service gas lines.

The above installations will be flagged by the Engineer.

The Contractor shall notify the Department 7 days prior to commencement of any excavation or embankment construction for the Materials Haul Road within 50 feet of the Pacific Lighting Service Company's gas lines crossing.

(2) Maintaining Public Traffic.—The Contractor shall make all necessary provisions for maintenance of public traffic and shall conduct his operations for the work so as to offer a minimum of obstruction and inconvenience to public traffic.

Pursuant to Section 3, Article (n), the Contractor shall provide flagmen as necessary to avoid accidents and prevent damage and injury to passing traffic. Barricades and obstructions shall be illuminated at night; lights shall be kept burning from sunset until sunrise.

(3) Country Roads.—The Contractor shall repair any damage to the existing paving and base on the county roads resulting from his operations. Repairs shall conform as nearly as practicable to its original condition before being damaged.

(r) **Temporary Gates and Fences.**—The Contractor shall install temporary gates wherever it is necessary to operate construction equipment through existing fences. The temporary gates shall be constructed of materials and to standards at least equal to those of the existing fence. Before cutting the fence, the Contractor shall install braces and additional posts on each side of the opening and shall anchor the fence as necessary to maintain tension on the wires. Gates shall be kept closed except when construction equipment is passing through. Prior to acceptance of the work, the fence shall be restored as nearly as practicable to its original condition. Deviation from the above requirements will be premitted only where the Contractor furnishes to the Engineer advance written approval from the property owner of a different method of operation.

Temporary fences shall be provided by the Contractor wherever it is necessary to remove or alter portions of existing fences for performance of the work. Prior to acceptance of the work fences shall be rebuilt in original position unless removal or relocation is required by the contract. Temporary fences shall prevent livestock from straying from or onto adjacent lands.

(s) **Water.**—The Contractor shall develop his source or sources of water supply at his expense and shall furnish all equipment necessary to collect, load, transport, and apply water necessary for performance of the work.

All temporary facilities for furnishing and distributing water shall be removed from the project area by the Contractor prior to acceptance of the work.

Watering equipment used for the application of water shall be equipped with positive means of shutoff and shall have pressure type distributors equipped with spray systems that will ensure uniform application of water.

(t) Dust Control.—Pursuant to the provisions of Section 4, the Contractor shall assume all responsibility for dust control and shall carry out proper and efficient measures wherever and as often as necessary to prevent his operations from producing dust in amounts harmful to persons, damaging to property, hazardous to public traffic, or causing a nuisance to persons living nearby or occupying buildings in the vicinity of the work.

(u) Specific Submittal Requirements for Working Drawings and Data.—

(1) General.—The Contractor shall furnish to the Engineer working drawings and data on materials and equipment in accordance with the provisions of Section 6, Article (c), and the specific submittal requirements of this article.

(2) Contractor's Construction Roads.—The Contractor shall furnish working drawings for construction roads provided by him within the site of work pursuant to Article (p). The drawings shall show widths of roads, direction of traffic, curves, grades, and related information in sufficient detail for inspection of the drawings for compliance with the specifications and drawings.

(3) Control of Water and Unwatering Plan.—The Contractor shall furnish detailed plans of the facilities for control of water and unwatering as specified in Section 14. The plan shall be submitted prior to September 1 of each year or at least 30 days prior to the time of construction of such facilities, whichever is earlier.

(4) Clearing Plan.—The Contractor shall furnish a detailed plan for clearing work specified in Section 15. The plan shall subdivide the areas of clearing for orderly progression of the clearing work and shall give the dates for commencement and completion of the clearing.

(5) Earthwork.—The Contractor shall furnish his proposed methods of excavating and earth moving and of moisture conditioning embankment materials obtained from borrow and stockpiles. He shall also furnish, for each zone of embankment, the number of pieces of excavating, transporting, and compacting equipment which he intends to use together with the capacity of each such piece of equipment.

(6) Contractor's Concrete Plant.—The Contractor shall furnish working drawings and data for his concrete plant and construction equipment, including proposed procedures for the following: processing, handling, transporting, storing, and proportioning materials for concrete; and mixing, transporting, forming, placing, and curing concrete.

(7) Instrumentation.—The Contractor shall furnish working drawings for the riser form and cable carrier, and any other drawings necessary for this work.

(8) Rock Crushing and Screening Plants.—The Contractor shall furnish working drawings and data for his rock crushing and screening plants including his proposed procedures for crushing, screening, sorting and stockpiling materials; and details of the crusher.

(9) Rock Source.—The Contractor shall furnish plans showing the layout, slopes and benches of the excavation in the rock source.

(10) Headquarters Office Building and Materials Testing Laboratory Building.—The Contractor shall furnish working drawings for the Headquarters Office Building and Materials Testing Laboratory Building.

(v) Final Drawings.—The Contractor shall furnish two complete sets of full size black and white prints or blueprints, and one complete set of full size reproducible prints of all drawings and revisions for equipment furnished under Section 22 made up to the time the equipment is complete.

(w) Operation and Maintenance Instructions.—

(1) General.—The Contractor shall furnish complete operation and mainteance (O&M) instructions, for the equipment under Section 22. The instructions shall incorporate the requirements specified in Subarticle (2). The Contractor shall integrate instructions from subcontractors with his submittal.

The Contractor shall submit to the Engineer for inspection, 6 sets of O&M instructions for each category of equipment, specified in Section 22, referenced above, not later than 135 days before scheduled shipment of the first complete unit of each such catagory.

The Engineer will inspect and return O&M instructions as provided below within 45 days after receipt thereof, or within 45 days after receipt of all related information necessary for such inspection, whichever is later.

One set of O&M instructions will be returned, marked "NO APPARENT DEFECTS", "DEFECTS NOTED", or "REJECTED". Defects discovered on inspection will be indicated on the O&M instructions or otherwise communicated to the Contractor in writing on return of the O&M instructions.

Within 30 days after receipt of O&M instructions marked "REJECTED", the Contractor shall revise the instructions in accordance with the directions for revision and shall resubmit 6 sets of the revised instructions for inspection. The Engineer will inspect and return the resubmitted O&M instructions in the same manner and time as specified above for the original submittals.

Within 30 days after receipt of O&M instructions marked "NO APPARENT DEFECTS" or "DEFECTS NOTED", the Contractor shall revise the instructions in accordance with revisions noted, if any, and shall furnish 15 sets of O&M instructions in final form.

The 6-set submittals are not to be considered included in the fifteen sets of instructions in final form.

(2) Content.—As specified in the following paragraphs, the instructions shall consist of title page, frontispiece, and information covering description, installation, operation, preventive maintenance, corrective maintenance, overhaul, parts list and list of recommended spare parts, and an appendix.

The title page shall include the name and function of the equipment, manufacturer's identification number, and the Department's specifications number and title.

The contents shall list all sections and subsection titles of the instructions with reference to the page on which each starts and a list of included drawings.

The frontispiece shall be a recognition illustration of the equipment described in the instructions.

The descriptive information shall consist of drawings and diagrams, and a physical and a functional description of the equipment including major assemblies and subassemblies.

The installation information shall cover preinstallation inspection, installation, calibration, and preparation for operation, both for initial installation and for installation after overhaul.

The operation information shall include step-by-step procedures for starting, restarting, operating, shutdown, and emergency requirements. The information shall also include performance specifications and operating limitations.

The maintenance information shall include step-by-step procedures for inspection, operation checks, cleaning, lubrication, adjustments, repair, overhaul, disassembly, and reassembly of the equipment for proper operation of the equipment. A list of special tools which are required for maintenance shall be included with the maintenance information.

The complete parts list and a list of recommended spare parts shall provide all necessary information, including part numbers and catalog item numbers if applicable, for identifying parts. Parts or assemblies obtained from another manufacturer shall be identified by the name of that manufacturer and his identifying part number. The size, capacity, or other characteristics of the part shall be supplied if required for identification.

The appendix shall include safety precautions, a glossary and, if available at time of submittal, copies of test reports, and other relative material not specified to be submitted otherwise.

(x) Indemnification of Third Parties.—The Contractor shall indemnify and save harmless the Federal Government, its officers and employees connected with the work, all in the same manner and to the same extent as he is required to indemnify and save harmless the State and its officers and employees under Section 4, Articles (c), (d), and (e).

(y) Guarantee.—The Contractor guarantees that all materials and equipment furnished will conform to the requirements of this contract and be fit for use as specified. The Contractor guarantees all material and equipment against defects in material and workmanship. To the extent that the Contractor is responsible for design of the work in compliance with the design criteria and operating requirements of the specifications and drawings, the Contractor guarantees that the materials and equipment furnished will perform in accordance with such criteria and requirements, except for failures as are caused solely by the negligent use or maintenance of the materials and equipment by the Department.

The period of this guarantee shall commence upon acceptance of the work, and shall terminate one year after acceptance, provided that such period shall be extended from the time of discovery of any defect or failure under this guarantee for one year following its correction, as to the corrected portion of the work.

The Contractor shall correct all defects or failures within the terms of this guarantee which are discovered within the guarantee period. The Department will give the Contractor prompt written notice of such defects or failures following their discovery. The Contractor shall commence corrective work within 10 days following notification by the Department of the defect or failure and shall diligently prosecute such work to completion. The aforementioned corrective work shall include redelivery of the equipment, and in no event shall such corrective work be completed later than 6 months after the date of notification to correct. The Department may, however, defer commencement of corrective work until the materials or equipment can be taken out of service without injury to the Department, but in no event will corrective work be so deferred longer than six months from the time of discovery of the defect or failure requiring correction. Replacement parts and repairs shall be subject to approval of the Department. The Contractor shall bear all costs of corrective work, which shall include necessary disassembly, transportation, reassembly and retesting, as well as repair or replacement of the defective materials or equipment, and any necessary disassembly and reassembly of adjacent work, provided that the Department will disassemble and reassemble at its expense adjacent materials or equipment not furnished by the Contractor, where necessary to give access to the defective materials or equipment.

If the Contractor fails to perform corrective work in the manner and within the time stated, the Department may proceed to have such performed at the Contractor's expense and he shall honor and pay the costs thereof upon demand, and his sureties shall be liable therefor. The Department shall be entitled to all costs and expenses, including reasonable attorneys' fees, necessarily incurred as a result of the Contractor's refusal to honor and pay such costs.

The Contractor's performance bond and labor materials bond shall continue in full force and effect during the period of this guarantee, but during such period shall apply only to corrective work required to be performed hereunder.

The rights and remedies of the Department under this article are not intended to be exclusive, and do not preclude the exercise of any other rights or remedies provided by this contract or by law with respect to unsatisfactory work performed by the Contractor.

The guarentee under this article is in addition to any manufacturer's guarantee or warranty. Manufacturer's guarantees or warranties that extend over a period of time greater than the guarantee period specified herein shall not be modified or voided by any requirement of this article.

(z) First Aid Services and Facilities.—

(1) General.—The Contractor shall provide and maintain the first aid services and facilities specified herein. Such services and facilities shall be for the use of Contractor and Department personnel assigned to the work and shall be available throughout the contract period. First aid facilities, equipment and supplies shall conform to the requirements of the Construction Safety Orders of the California Division of Industrial Safety, Title 8, Chapter 4, Subchapter 4, Article 3, Section 1509 et seq., and these specifications.

(2) Personnel.—A licensed ambulance driver, and qualified first aid attendant shall be on duty at the infirmary specified in Subarticle (3), each working shift. The ambulance driver may, at the option of the Contractor be assigned to other work in the immediate vicinity of the infirmary.

(3) Infirmary.—An infirmary shall be provided near the dam. It shall be located near existing access roads and such that dust and other unsanitary conditions are minimal. The infirmary shall have an unobstructed floor area of at least 300 square feet after installation of the required furnishings. Interior walls and ceilings shall be painted with a prime coat and two coats of white

enamel. Windows and doors shall be screened. The infirmary shall be identified by a sign reading "Infirmary" with letters not less than 12 inches high, painted in red over a white background.

The infirmary shall be equipped with a surgical sink with hot and cold water, water closet and enclosure, plumbing, waste disposal system, and air conditioning. A standby electrical energy source shall be provided, rated at a minimum of 50 amperes 110 volts.

The infirmary shall be equipped with first aid supplies and furnishings as necessary to adequately care for injured employees. It shall be maintained to standards of sanitation satisfactory to the Engineer.

(4) Ambulance.—An ambulance shall be stationed at the infirmary. The ambulance shall be equipped with red lights and siren as required by the Administrative Procedures of the California Highway Patrol. The ambulance and its operation shall conform to the requirements of the California Department of Motor Vehicles Regulations as provided in the California Administrative Code, Title 13, Chapter 2, Subchapter 5, Article 1, and laws relating to the operation of ambulances in the State of California.

(5) Hospital Accommodations.—The Contractor shall furnish evidence in writing to the Engineer of a continuing arrangement with an approved hospital for the emergency admittance, hospitalization, and medical care of injured employees.

(6) Coordination.—The Contractor may, at his option, provide the first aid facilities and services specified herein jointly with other contractors engaged in work for the Department at the site of the work.

(aa) Safety Engineer.—The Contractor shall employ a competent and experienced safety engineer who shall devote his full time during working hours toward accident prevention during construction. The individual shall be fully qualified, possessing safety engineering experience in heavy construction. The qualifications of the safety engineer shall be subject to approval of the Engineer prior to his assignment to the work.

(bb) Measurement of Materials Paid for by the Ton.—Pursuant to the provisions of Section 9, Article (a), materials to be paid for by the ton shall be weighed on scales furnished by the Contractor or on other sealed scales regularly inspected by the State Bureau of Weights and Measures, at the option of the Contractor. All scales shall be suitable for the purpose intended and shall conform to the tolerances and specifications of the State Bureau of Weights and Measures. The Contractor shall have all scales inspected and scaled by the State Bureau of Weights and Measures as often as the Engineer may determine necessary to ascertain the accuracy of such scales.

All platform scales shall be of sufficient size to permit the entire vehicle or combination of vehicles to rest on the scale platform while being weighed. Combination vehicles may be weighed as separate units provided they are disconnected while being weighed.

If materials are weighed on scales furnished or owned by the Contractor, a weigher furnished by the Department, shall do all weighing.

If materials are weighed on scales not furnished or owned by the Contractor, a representative of the Department, at the discretion of the Engineer, may be present to witness the weighing and to verify and compile the daily record of scale weights. If the weighing is not witnessed by a representative of the Department, the Contractor shall furnish a public or private weighmaster's certificate, or a certified daily summary weigh sheet. A duplicate weigh slip or a load slip shall be furnished to each vehicle weighed and the slip shall be delivered to the Engineer at the point of delivery of the material. The form of weighmaster certificate shall be as prescribed by Section 12704, Chapter 7, California Business and Professions Code.

Vehicles used for hauling materials shall be weighed empty daily at such time as may be directed, and each vehicle shall bear a plainly legible identification mark.

If material is shipped by rail, the car weights will be accepted, provided that the actual weight of material only will be paid for and not minimum car weights used for assessing freight tariff. Car weights will not be acceptable for material passed through mixing plants.

Quantities of sand, aggregate, or similar materials to be paid for will be determined by deducting from scale weights the weight of water in the material in excess of 6 percent of the dry weight of the material.

The weight of liquid asphalt or asphalt emulsion will be determined from volumetric measurements in the manner prescribed and employing conversion factors contained in Section 93-1.04 of the Highways Specifications.

(cc) Funds Available for Payment.—

(1) General.—The Project is a part of the State Water Facilities as defined in the California Water Resources Development Bond Act (Chapter 8 of Part 6 of Division 6 of the California Water Code). Other contracts for construction of such Facilities, and contracts for equipment to be incorporated into such Facilities, are now in progress or will be placed subsequent to award of this contract, or both. The Department believes that sufficient funds are available or will be made available to complete such Facilities, but the sources of funds available to the Department at any time for such Facilities are limited.

Prior to July 31 of any fiscal year in which this contract is executory, the Department will furnish the Contractor a general statement of the funding status of the State Water Facilities. The statement will include a forecast of the availability of funds for contract payments and other costs under this contract during such fiscal year. The statement will be for general information purposes only and shall not be construed as warranting the availability of funds for contract payments under this contract during such fiscal year.

At any time the Director may determine that contract payments and other costs due and to become due for performance of this contract and the above specified contracts exceed the funds available, or to become available by law. After making such determination the Director may allocate funds available and funds so to become available among this contract and such other contracts. Such allocation will be made in accordance with the Director's determination of the relative necessity of this contract and such other contracts to the completion of the State Water Facilities.

(2) Notice of Insufficient Funds.—Should the Director's allocation of funds to this contract result, in his judgment, in insufficient funds being available to cover all contract payments and other costs due and to become due under this contract, the Director will notify the Contractor in writing of the following:

    a. The insufficiency of funds.

    b. The amount of funds allocated to this contract.

    c. The date, determined by the Director on the basis of the Contractor's approved progress schedule, upon which funds allocated to this contract will be insufficient to cover further contract payments, which date shall be no less than 30 days following the date of the Notice. Such date is hereinafter called the date of insufficiency of funds.

(3) Contractor's Options.—Within 10 days of receipt of a Notice of Insufficient Funds, the Contractor shall select one of the following two options and shall notify the Director in writing of his selection:

    a. The Contractor may continue to perform the work after the date of insufficiency of funds, so long as funds are available to the Department for inspection and superintendence; provided, that he shall have only the contingent rights to payment for such performance provided by this subarticle and Subarticle (4), and that he otherwise assumes the risk of nonpayment therefor.

In the event the Contractor selects this option, the Department will continue to prepare progress estimates as specified in Section 9, Article (d). If the Director allocates additional funds to this contract, the Department will pay to the Contractor, to the extent of such allocation, the total of progress payments which would have been made but for the insufficiency of funds, computed as specified in Section 9, Article (d) (2), plus interest computed as follows:

    1. Per annum interest will be computed and will accrue on each progress payment separately.

    2. The interest rate will be determined by reference to the table entitled "Bank Rates on Short-Term Business Loans" of the Federal Reserve Bulletin. The sub-table entitled "Weighted Average Rates" will be used. The "Size of Loan" column which corresponds to the amount of the progress payment, for the month in which such progress payment would have been

made, will be entered, and the interest rate selected by reference to the line for the West Coast centers.

3. Should the manner of presenting bank rates on short-term business loans be changed by the Federal Reserve Bulletin, the new presentation corresponding to the intent of Paragraph 2. above will be used to determine the interest rate.

Should the Engineer determine, at any time after the Contractor has selected the foregoing option, that the Contractor is failing to perform the work diligently in accordance with Section 5, Article (b), the remedies specified in that article will not apply and in lieu thereof, the Engineer may suspend the work for failure of the Contractor to carry out orders given or to perform any provisions of the contract, within the meaning of Section 5, Article (f).

Should the Engineer determine at any time after the Contractor has selected the foregoing option, that funds are no longer available to the Department for inspection and superintendence of the work, he will proceed to terminate the contract pursuant to Subarticle (4).

b. The Contractor may treat the Notice of Insufficient Funds as an order to suspend the work commencing upon the date of insufficiency of funds. The Contractor's election of this option will be considered a suspension for the convenience and benefit of the Department within the meaning of Section 5, Article (f). The Contractor shall resume the suspended work as provided in Section 5, Article (f).

(4) Termination.—Subject to the provisions of Subarticle (8), the Director, at any time after the date of insufficiency of funds, and the Contractor, at any time after the expiration of 90 days from such date, may elect to terminate the contract. The terminating party shall notify the other in writing of his election and of the effective date of termination, which shall not be less than 5 days from the date of receipt of such notice. The party issuing a termination notice may rescind it at any time prior to the effective date of termination by written notice to the other party.

Should the contract be terminated pursuant to this article, the Contractor shall:

a. Stop performance of the work on the effective date of such termination.

b. Remove from the site of the work all construction materials, equipment and plant.

c. Place no further subcontracts or supply contracts.

d. Terminate all executory subcontracts or supply contracts.

e. Settle all outstanding liabilities and all claims arising out of termination of such executory subcontracts or supply contracts, with the approval or ratification of the Engineer to the extent required by the Engineer.

f. Assist the Engineer in making an inventory of all materials and equipment in storage at the site of the work, en route to the site of the work, in storage or manufacture at other locations, and on order from suppliers.

g. Transfer title (if title has not passed pursuant to Section 4, Article (j)) and deliver to the Department in the manner and at the times directed by the Engineer all materials and equipment, completed or partially completed. He shall also furnish completed or partially completed working drawings, data, operation and maintenance instructions, and other property being held or prepared for submittal to the Department under the contract.

h. Take such action as may be necessary, or as the Engineer may direct, for the maintenance and protection of the work, including the materials and equipment specified in Paragraph f.

(5) Liability for Termination.—Termination pursuant to Subarticle (4) shall be without liability to the State, except for the following entitlements in the Contractor. The entitlements described in a. and b. below are expressly conditioned upon availability of funds at the time of termination.

a. If the Contractor has elected option a. under Subarticle (3) and continued to perform the work following the date of insufficiency of funds, then, upon termination and to the extent that funds are then available therefor, he will be entitled to payment of the total of progress payments which would have been made, but for the insufficiency of funds, for work done and materials and equipment furnished to the effective date of termination, plus interest on each such progress payment to such date. Such progress payments and interest payments will be determined as provided in Subarticle (3) a.

b. If the Contractor has elected option b. under Subarticle (3) and treated the Notice of Insufficiency of Funds as an order suspending the work for the convenience and benefit of the Department, then, upon termination and to the extent that funds are then available therefor, he will be entitled to an adjustment in compensation for the suspension determined in accordance with Section 7, Article (e).

c. Upon termination, the Contractor will be entitled to payment for the following, to the extent not covered by payments under a. or b. above or by payments made prior to the date of insufficiency of funds:

1. The value of completed or partially completed materials and equipment, working drawings, data, operation and maintenance instructions, and other property delivered to the Department pursuant to Subarticle (4) g.

2. The cost of storage and delivery of materials and equipment under Subarticle (4) g.

3. The costs of such maintenance and protection of the work under Subarticle (4) h. as determined by the Engineer to be in excess of the Contractor's obligations under Section 4, Article (d).

Payment for the foregoing items will be determined as an adjustment in compensation under Section 7, Article (e).

d. Upon termination, the Contractor will be entitled to payment of retentions from progress payments and other amounts retained under the contract as part of the final payment made under Subarticle (6) and Section 9, Article (e), subject to all of the conditions stated therein.

(6) *Final Payment and Release of Claims.*—After termination pursuant to this article, final payment for performance of the work will be made in accordance with the provisions of Section 9, Article (e), except as otherwise provided herein. For the purpose of payment only, the work will be deemed accepted, within the meaning of Section 9, Article (e), on the effective date of termination. The final estimate of money due under this contract, as that term is used in Section 9, Article (e), shall include the value of the materials and equipment specified in Subarticle (4) g.

The provisions of Section 9, Article (e), respecting release of claims and exceptions thereto are hereby expressly made applicable to final payment pursuant to this subarticle. The Contractor may submit with his release of claims any claim based upon disagreement with the amount of any payment included pursuant to Subarticle (5).

(7) *Liquidated Damages and Guarantee.*—This article shall not be construed as waiving the provisions of Section 5, Article (c), nor as waiving the Department's rights, as to complete work, arising from any guarantee provision contained in the contract; provided, that, for the purpose of computing the period of guarantee under any such provision, the work shall be deemed accepted on the effective date of any termination pursuant to this article.

(8) *Allocation of Additional Funds.*—At any time after the Director makes a determination and allocation as provided in Subarticle (1), and prior to the effective date of any termination pursuant to Subarticle (4), he may determine that additional funds have become available, or that additional funds will become available by law, or both, and may allocate such funds among this contract and other than executory contracts for the State Water Facilities. Should the Director's allocation of additional funds to this contract result, in his judgment, in sufficient funds being available to cover all payments and other costs due and to become due under the contract, including any payments due under Subarticle (3) of this article, he will so notify the Contractor in writing. Such notice shall operate, upon receipt, to cancel the Notice of Insufficient Funds and there shall be no further exercise of rights under this article until such time as another Notice of Insufficient Funds may be given.

(9) *Director's Determinations.*—Determinations by the Director pursuant to this article will be final and conclusive.

(dd) Prevailing Wages.—In accordance with Sections 1770 and 1773 of the California Labor Code, the Department has determined the prevailing rates of per diem wages, including employer payments for health and welfare, pensions, vacations and similar purposes in the locality in which the work is to be performed for each craft, classification, or type of workman required for the work. Such determination is filed in the office of the Department of Water Resources, Room 406-2, Resources Building, 1416 Ninth Street, Sacramento, California, and is hereby made a part of the contract. Copies are available as provided in the Notice to Contractors. The Contractor shall post at the site of the work a copy of the Department's determination of prevailing wages.

(ee) Payment.—The contract prices shall include full compensation for all costs incurred under this section, except payment made pursuant to Article (cc) (5) in the event of termination of contract due to insufficient funds.

(ff) Award of Contract.—Award of contract will be made pursuant to Article (j), "Award of Contract" of the Bidding Requirements and Conditions and in accordance with the provisions specified herein.

Bids will be considered only if submitted on the complete work described in Section 10. Attention is directed to Section 16, Article (j) (3), which specifies two alternative borrow areas; and to Bid Schedule Item No. 61A, Excavation in Lake Borrow Area Schedule A; and to Bid Schedule Item No. 61B, Excavation in Lake Borrow Area Schedule B. Bids shall include bid prices for both alternative borrow areas and any bid which omits a bid price on either alternative borrow area item will be rejected. All bids will be evaluated and compared on the basis of the total price bid using each alternative borrow area price and award of the contract, if it be awarded, will be to the lowest responsible bidder on one or the other of the alternatives, at the option of the Director.

SECTION 12. LAYOUT OF THE WORK AND SURVEYS

(a) Description.—Pursuant to Section 6, Article (e), this section covers survey services to be provided by the Department and by the Contractor for layout and performance of the work, and details the minimum acceptable standards for Contractor performed surveys.

(b) Layout of the Work.—The responsibility for layout of the work and surveys required therefor shall be divided between the Department and the Contractor as follows:

(1) Right-of-Way, Contractor's Construction Facilities, Borrow and Waste Areas.—The Engineer will stake out the right-of-way lines for the work, where necessary, and the limits of construction areas, borrow areas, borrow pits, and spoil or waste areas within the right-of-way and will establish a bench mark with or adjacent to each such area.

The Contractor shall perform all other surveying and related services required for properly laying out and performing the construction of construction plant and other facilities, including the Contractor's roads, required by the Contractor, and the spoil or waste and borrow.

(2) Clearing and Grubbing.—The Engineer will provide four bench marks near the project clearing lines and shall mark on the ground by stakes the project clearing lines at intervals of 500 feet or less. The Contractor shall perform all other surveying and related services required for proper layout and control of clearing and grubbing.

(3) Earthwork for Roads.—For roadway excavation and embankment, the Engineer will establish the centerlines by setting and referencing the PI's (if accessible), BC's, EC's, points along tangents at intervals of not more than 500 feet and points along curves at intervals of not more than 300 feet; set all necessary slope stakes at the top of cuts and bottom of fills; and set bench marks on or adjacent to the right-of-way at convenient locations generally not more than 1,000 feet apart. The Engineer will reset the centerline points only upon completion of the excavation, as provided in Subarticle (4), below.

The Contractor shall perform all other surveying and related services required for proper layout and construction of this earthwork, including, but not restricted to setting grade stakes on slopes and berms and stakes for rough grading of the subgrade.

(4) Drainage Facilities.—The Engineer will stake the centerlines or reference lines for the road and dam embankment drainage, and will set grade reference points or stakes for the invert grades of these facilities as required.

The Contractor shall perform all other surveying and related services required for proper layout and construction of these portions of the work.

(5) Earthwork and Grouting for Dam.—The Engineer will provide survey services for layout of earthwork for the dam and grouting for the dam as follows:

a. Set approximately a total of 20 bench marks at convenient locations adjacent to and outside the toes of the dam.

b. Set and reference horizontal control points as required.

c. Set and reference slope stakes at the tops of cuts outlining the dam foundations and slope stakes outlining zones. Such slope stakes will be set on cross section lines spaced generally 50 feet apart on the abutments and 100 feet apart elsewhere, and where desirable and practicable, will be referenced.

d. Stake the centerline of the grout curtain as required.

The Contractor shall perform all other surveying and related services required for proper layout and construction of this earthwork.

(6) Excavation for Outlet Works Intake.—The Engineer will provide survey services for layout of the outlet works intake as follows:

a. Set approximately 3 bench marks at convenient locations adjacent to and outside the excavation for the outlet works intake.

b. Set and reference control points as necessary to establish a centerline through, or baselines adjacent to, the excavation.

c. Set and reference slope stakes on natural ground at the tops of cut slopes at intervals of approximately 50 feet measured in a direction parallel to the structure centerline.

The Contractor shall perform all other surveying and related services required for proper layout and performance of this portion of the work.

(7) Instrumentation.—The Engineer will perform all surveying required for the instrumentation.

(8) Earthwork and Concrete for Spillway.—The Engineer will provide survey services for layout and inspection of the earthwork and concrete for spillway as follows:

a. Set bench marks at convenient locations adjacent to the spillway.

b. Establish the structure centerline by setting and referencing the PI's (if accessible), BC's, EC's, BVC's, EVC's, and points along tangents at intervals of not more than 1,000 feet and each end of the structure; and re-establish such stakes upon completion of the excavation.

c. Set and reference slope stakes on natural ground at the tops of the cut slopes at intervals of approximately 50 feet measured in a direction parallel to the structure centerline.

d. Set grade stakes as required for checking and correcting finish grading of the structure foundation.

e. Set a stake for line and grade at each end of the structure and at each transverse joint therein.

The Contractor shall perform all other surveying and related services required for proper layout and performance of the earthwork and concrete work for the spillway.

(9) Riprap, Bedding and Backfill.—The Engineer will perform all surveying related services required for the riprap, bedding, and backfill.

(10) Other Parts of the Work.—The Department will provide the survey services reasonably required for proper layout and performance of such parts of the work, if any, as are not covered above in this section.

(c) Tolerances in Setting Survey Stakes.—Tolerances in setting survey stakes shall be as stated below:

Kind of Survey Stake or Mark	Tolerance on Error in Line		
	Distance	Tangent	Curve
Markers on hubs and monuments on centerlines and offset centerlines	1:5,000	0.02'	¼ min.
Intermediate stakes or marks on centerlines and offset centerlines for:			

Kind of Survey Stake or Mark	Tolerance on Error in Line		
	Distance	Tangent	Curve
Rough excavation and embankment for roads and other work not otherwise provided	1 : 1,000	0.10'	1 min.
Trimming of excavation and embankment, unless otherwise provided	1 : 1,000	0.10'	1 min.
Trimming or preparation of earth subgrade for roadways, concrete pipe, and other concrete structures	1 : 2,000	0.05'	½ min.
Roadway subbase and base, steel and asbestos cement pipe, and other work not otherwise provided for	1 : 2,000	0.05'	½ min.
Roadway surfacing, steel reinforcement, concrete pipe and other formed concrete	1 : 5,000	0.02'	½ min.

Grade Stakes or Marks for: — Elevation

Rough excavation and embankment for roads and other work not otherwise provided for -------- 0.10'

Trimming of excavation and embankment, unless otherwise provided -------- 0.10'

Trimming or preparation of earth subgrade for roadways, concrete pipe, and other concrete structures -------- 0.05'

Roadway subbase, and base, steel and asbestos cement pipe, and other work not otherwise provided for -------- 0.05'

Roadway surfacing, steel reinforcement, concrete pipe and other formed concrete -------- 0.02'

(d) Payment.—The contract prices shall include full compensation for all costs incurred under this section.

SECTION 13. CONSTRUCTION FACILITY

(a) Description.—This section covers the portion of the Construction Facility for which bid items are not provided and other project facilities and services to be provided by the Contractor for the Department, as follows:

1. Constructing a Headquarters Office Building.

2. Constructing a Materials Testing Laboratory Building.

3. Constructing one 8-foot by 16-foot project construction sign and removing and reinstalling an existing 8-foot by 16-foot project construction sign.

4. Constructing a visitor's information booth.

Concrete shall be of commercial plant quality and shall contain not less than 376 pounds of cement per cubic yard.

This section does not cover grading and paving the Construction Facility and the access road thereto, or fencing the Construction Facility, which portions of the work are provided for elsewhere in these specifications.

The Contractor shall be responsible for any injury, damage, or loss to the Headquarters Office Building, Materials Testing Laboratory Building, project signs, and visitor's information booth as specified in Section 4, Article (d)(1) b., until acceptance of the work.

Upon acceptance of the work, the portion of the Construction Facility to be constructed under this section will become the property of the Department and shall be left in place.

(b) Sanitary Sewer System.—The Contractor shall construct a sanitary sewer system for the Construction Facility where shown. The system shall consist of sanitary sewer lines, a precast concrete septic tank, a distribution box, effluent line, and an absorptive field.

(c) Headquarters Office Building and Materials Testing Laboratory Building.—

(1) General.—The Contractor shall construct a Headquarters Office Building and a Materials

Testing Laboratory Building for the exclusive use by the Department at the locations shown on the drawings.

The Contractor shall provide utilities, power, drinking water, telephone service, lighting, heating, cooling air conditioning, maintenance and janitor services for the duration of this contract.

    a. Headquarters Office Building.—The Headquarters Office Building shall be a wood frame structure 28 feet wide by 60 feet long with partitions to enclose a 12-foot by 28-foot room at one end of the building; a 4-foot wide hallway located at the centerline of the building for the remaining 48 feet; 12-foot by 12-foot rooms on each side of the hallway; and two rest rooms. The rest rooms shall each accommodate two water closets complete with doors and partitions, and two lavatories with hot and cold water and sewer services connected.

    b. Materials Testing Laboratory Building.—The Materials Testing Laboratory Building shall be a wood frame structure 16 feet wide by 56 feet long constructed on a concrete slab 28 feet by 56 feet. The slab shall extend 12 feet beyond the building in the long direction, with a roof overhang supported by posts for the full length of the concrete slab. Partitions to enclose a rest room and a utility room shall be provided. The rest room shall accommodate a water closet complete with door and partition and a lavatory with hot and cold water and sewer services connected.

The Contractor shall install the following Department furnished testing equipment as directed:

    1—Gilson Shaker, 120V, 6 amps.
    1—Specimen Extruder, pneumatic
    2—Compacting Machines, 120V, 5 amps.
    1—Mixer, density test, 120V, 7 amps.
    1—Concrete Specimen Compressive Machine, 120V, 1 amp.
    1—Cylcap Pot, 120V, 12.8 amps.
    2—Floor Ovens, 3 phase, 10 kw, 240V, 50 amps.
    3—Bench Ovens, 1 phase, 120V, 15 amps.
    2—Nutone Fan Heaters, 120V, 15 amps.
    1—Ro-Tap, 120V, 6 amps.
    1—Vibrating Table, 240V, 40 amps.
    1—Air Compressor, 1 h.p. motor, 120V, 13 amps.

The following equipment shall be furnished and installed by the Contractor:

    2—Sink with deck mounted, extended 12-inch goose neck faucet and 2-inch straight drain to outside (no trap needed)
    1—Water Heater, 30 gal.
    2—Heating and Cooling Unit, 240V, 50 amps.
    2—Flood Lamps.
   16—Fluorescent Lighting in building, under porch roof and rest room.
    1—Finish Counter Top, 18-gage galvanized sheet metal, 6 inches up back and end of counter, and one inch over front edge.
    1—Wash Basin (in rest room) with mixer faucet, drain, and trap.
    1—Toilet (in rest room).
    1—Mirror (in rest room).
    1—Powdered Soap Dispenser (in rest room).

(2) Construction.—The buildings shall have a concrete slab on grade, typical wood plate, sill and stud partitions and walls with wood siding, wood roof rafters, plywood roof sheathing, roll roofing, wood doors, metal windows, gypsum wallboard ceilings and wall surfaces, mineral wool insulation in exterior wall and on the top of ceiling, adequate electric lighting, painting inside and outside. All interior woodwork shall comply with the requirements of 1962 "Manual of Millwork" Woodwork Institute of California, WIC, for "Economy" grade work. The site shall be graded as shown.

    a. Floor Slab.—The floor slab shall be 4-inches thick. It shall be reinforced with 10 x 10-6 inch x 6-inch welded wire mesh and a 12-inch wide by at least 12-inch deep perimeter beam reinforced in the bottom with two continuous No. 3 reinforcing bars. The slab shall be placed on 6 inches of gravel fill. At entrances, provide a 4-foot by 4-foot concrete slab on grade or stoop and stairs as necessitated by the grade. Finish slabs shall be made smooth and level by steel trowelling. Concrete shall be damp cured for at least 72 hours.

b. Wood Framing.—Framing lumber shall be Douglas Fir, S4S, "Construction Grade," Paragraph 122-b. All connections shall have at least 2 nails with 16d for 2-inch material and 8d for 1-inch material. All framing shall be securely connected and braced. Studs shall be placed 16 inches on centers. Sheathing shall be Douglas Fir plywood, exterior type, "B-C" grade with "B" side exposed at eaves and rakes. Rafters shall be placed at 24 inches on centers maximum. Board and batten siding shall have solid blocking at third points between studs. Joists shall be 16 inches on centers with crown side up.

c. Roofing.—Material shall be heavy, mineral surfaced asphalt roll roofing, applied with cemented laps and nailed as recommended by the manufacturer.

d. Siding.—Siding shall be $1'' \times 8''$ lap joint or 12-inch board and batten, Douglas Fir or Redwood. All shall be applied over 30-pound asphalt felt.

e. Doors.—Doors shall be paint grade Birch or vertical grain Douglas Fir solid core to comply with WIC. Exterior doors shall be glazed upper panel, other doors shall be solid flush veneer. Thickness shall be $1^3/_4$ inches for all doors.

f. Windows.—Windows shall be aluminum projected with extruded sections not less than 0.125-inch thickness and at least $1^1/_8$-inch in depth. The area of the windows shall be approximately $1/_8$ of the area of the building. Vents shall be screened. Hardware shall be the manufacturer's standard heavy duty type having sufficient strength and rigidity for the intended service.

g. Wallboard.—Gypsum wallboard shall be $1/_2$-inch thick to comply with the requirements of ASTM Designation: C 36. All joints shall be taped and nail heads finished flush and smooth.

h. Insulation.—Insulation shall be mineral wool with vapor seal and shall be full-thick (nominal 4 inch) to solidly cover the spaces between exterior studs and ceiling joists.

i. Painting.—All materials shall comply with the requirements of State of California "Standards and Specifications" dated June 10, 1960, Department of Finance, Purchasing Division. Colors will be selected from the manufacturers standard colors. Primer and finishing coats shall be fully compatible.

Exterior wood trim and doors shall be painted one prime coat and one coat of sash and trim enamel. Sheathing—one coat LTZ exterior wood paint over primer. Metal shall be treated, primed and finished with one coat of metal enamel.

Interior wallboard shall be painted two coats with best grade interior pigmented synthetic resin emulsion paint. Wood trim and doors—one coat of semigloss enamel over primer. Paint rest rooms same as wood trim.

j. Finish Hardware.—Hinges shall be prime paint finish, steel, five knuckle, nonrising button pin, size 4×4 minimum. Furnish $1^1/_2$ pair of butts per door. Exterior doors shall have keyed lockset except for storage closets and as directed. All locks shall be keyed to a master keying system specified in Section 23, Article (g). Finish for locksets shall be dull chrome. Backset shall be $2^3/_4$ inches.

k. Glazing.—Glass shall be clear double strength of $3/_{16}$-inch sheet glass. Water closet shall have obscure glass on window.

l. Sheet Metal.—The material shall be 24-gage minimum galvanized sheet steel. Flashing shall be provided to weatherproof the building.

m. Rest Room Accessories.—Furnish and install one chrome plated or stainless steel powdered soap dispenser at each lavatory, Boraxo, Powdurn, Luron or equivalent; also one bottom dispensing, paper towel holder of white enamel metal, towel size $10^1/_4 \times 10^3/_4$ inches, capacity 150 folded towels. Toilet paper dispensers, one to each compartment, shall be white enamel metal, slant top, single fold double fold, bottom dispensing, capacity 800 sheets. Mirrors shall be metal framed, 16×22 inches in size, No. 1 quality, $1/_4$-inch mirror glazing quality plate glass, place one over each lavatory or as directed.

(3) Mechanical Work.—The buildings shall each be provided with an automatic thermostatically

controlled heating and refrigerated air conditioning system of sufficient capacity to maintain even temperature air conditioning in all parts of each building.

Drinking water shall be cooled with an electric water cooler of proper type to accommodate the type of water supplied, bottle or tap water.

Each rest room in the office building shall have two water closets and two lavatories with bright chrome trim. The rest room in the laboratory building shall have one water closet and one lavatory with bright chrome trim. Water shall be heated with an electric water heater of 30-gallon storage capacity.

A sump with sand trap and $2'$-$0''$ by $2'$-$0''$ steel drainage grating shall be installed in the Material Testing Laboratory Building, and shall be Bufnel M-3-225, Blaw-Knox 4-P-21, or equivalent.

All mechanical items shall be connected to services as necessary and maintained in good repair until acceptance of the work.

(4) Electrical Work.—Lighting shall provide uniformly distributed, shielded 80-foot candles light in all working spaces. All room systems shall be provided with wall switches. Convenience outlets shall be located as directed at not more than 16 feet apart; receptacles shall be double flush type. All lighting fixtures shall be fully lamped.

Electrical installation shall provide at least 1-foot candle of illumination 5 feet above the ground throughout the fenced area.

(5) Telephone Service.—The Contractor shall provide telephone service to each building and in the concrete placement area. The telephone service shall provide communication between each location and also with Contractor's telephone system and to the outside. Long distance toll charges from telephone service provided by the Contractor will be paid for separately by the Department.

(d) Project Construction Signs.—

(1) Description.—This article covers the construction and maintenance of a 8-foot by 16-foot wooden signboard, and removing and reinstalling an existing 8-foot by 16-foot project construction sign as directed by the Engineer. The existing sign shall be repainted as required prior to erection at new location. Maintenance of the signs as herein specified shall apply to both the new and to existing relocated signs. The new signboard shall be of plywood, glued and nailed to each side of wood framing members to form a "stressed-skin" structural panel. The panels shall be secured to wooden posts, set in concrete, and braced, as shown.

(2) Erection and Maintenance.—Immediately upon commencement of the work the Contractor shall remove and reinstall the existing sign, and erect and complete the 8-foot by 16-foot information sign in the locations designated by the Engineer.

The Contractor shall maintain the signs in a clean and undamaged condition until acceptance of the work.

(3) Materials.—Materials shall conform to the following requirements:

Concrete.—Concrete shall be of commercial quality and shall contain not less than 376 pounds of cement per cubic yard.

Metalwork.—Structural shapes shall conform to the requirements of ASTM Designation: A 36. Bolts, lag screws, and other fastenings shall be high grade commercial quality. Sheet metal shall be galvanized. Galvanizing shall conform to the requirements of ASTM Designation: A 525, Coating Class 1.25 Commercial. All metalwork and fastenings shall be galvanized. Structural shapes shall be galvanized after fabrication.

Wood and Wood Products.—All wood and wood products shall be Douglas fir and shall be of clean, kiln-dried, bright, stock. Lumber grades are those of the West Coast Lumber Inspection Bureau (WCLIB), utilizing Grading and Dressing Rules No. 15, plywood grades are those of the American Plywood Association, referenced by DFPA. No wood to receive glue shall contain more than 12 percent moisture at the time of fabrication.

Framing members of 1 inch thick shall conform to the requirements of Paragraph 151-b, of (WCLIB) "B&BTR" Industrial, shall be full 1 inch thick and dressed on edges only to standard width. Framing members 3 inches thick shall conform to the requirements of Paragraph 152-b of (WCLIB) "B&BTR" Industrial. Framing members 2 inches thick and 4 inches thick shall conform to the requirements of Paragraph 153-c of (WCLIB) "1200-f Industrial" and shall be dressed four sides to standard sizes. Posts shall conform to the requirements of Paragraph 124-b of (WCLIB). Braces shall conform to Paragraph 123-b of (WCLIB). Posts and braces shall be dressed four sides to standard sizes.

Plywood shall be Exterior-type Medium Density Overlaid, designated EXT-DFPA-MD Overlay. Adhesive for the fabrication of the signboard shall be a resorcinol-phenol-formaldehyde, waterproof cold-setting type.

Primer.—Primer for wood and metal shall conform to the requirements of Federal Specification TT-P-25. Vinyl type wash coat shall conform to the requirements of Military Specification MIL-C-15328A. Thinner shall conform to the requirements of Federal Specification TT-E-489, Class A. Colors shall be pigment in oil conforming to the requirements of Federal Specification TT-P-381.

(3) Fabrication of Panels.—The sign panels shall be fabricated in a dry, closed area with temperature above 60 F. Framing members shall be positioned and secured in a common plane before and during the application of the plywood sheets. All contact surfaces of framing members shall first be spread with adhesive to provide a film of adhesive between all contact surfaces between the framing and the plywood. The plywood sheets shall be well fitted and nailed in place to provide "stressed-skin" construction. All nails shall be set for puttying. Fabricated panels shall not be subjected to any bending stresses before the adhesive has set.

(4) Field Erection.—Posts shall be set accurately and braced in vertical position in prepared holes and embedded in concrete as shown on the drawings. Concrete shall be sloped to drain away from the embedded post and trowelled to a smooth hard finish. The sign panel members shall be installed level and plumb and shall be bolted securely to the supporting members. Steel washers shall be used under all bolt heads and nuts bearing on wood surfaces. All field joints in panel members shall be glued and nailed as specified for the fabrication of the panels.

After erection of the sign panel, the top and end surfaces shall be covered with a sheet metal cap, fitted and nailed in place using annular grooved nails. The cap shall be formed with edges folded and flanged inward $1/2$ inch.

(5) Painting.—Prior to painting, all exposed nails in wood shall be set and after priming shall be puttied flush. All metal surfaces shall first be treated with vinyl wash primer metal pretreatment. All exposed wood and metal surfaces shall receive a coat of primer and two coats of machinery enamel of color selected by the Department. The paint shall be applied uniformly without sags, runs, or other defects. Each coat shall be thoroughly dry before the succeeding coat is applied. Painting shall not be done in inclement weather.

Lettering of the type and size shown shall be applied using machinery enamel of colors selected.

An emblem decal, to be furnished by the Department, shall be installed. Two coats of clear lacquer shall be applied to the surface of the decal.

(e) Visitors Information Booth.—

(1) Description.—This article covers the construction of a visitors information booth. The information booth shall be located in the visitors viewsite.

(2) Materials.—Materials shall conform to the following requirements. Each piece of lumber shall bear the official grade mark of the appropriate inspection bureau or association. Redwood shall be graded under the rules of the Redwood Inspection Service referenced herein by RIS. All other soft woods shall be graded under the rules of the West Coast Lumber Inspection Bureau, referenced herein by WCLIB; or under the rules of the Western Pine Association, referenced herein by WPA. Types and grades of plywood specified are those of the American Plywood Association, as referenced herein by DFPA.

Framing Lumber.—All framing shall be of select heart redwood having an average moisture content of not more than 15 percent and shall be surfaced on four sides with edges eased. All wood members placed on concrete shall be "Foundation Grade" redwood, surfaced four sides.

Finish Lumber.—All exterior finish lumber shall be vertical grain redwood. Exterior tongue and groove siding shall be "Select Heart," Paragraph 105, RIS, with best face saw textured in accordance with the Standards of the California Redwood Association; all other finish lumber shall be "Clear All Heart," Paragraph 103, RIS, manufactured in accordance with the details shown and specified.

Plywood.—Plywood shall be Douglas fir and shall be Grade A for exposed surfaces and Grade C for all other surfaces. Surfaces to be painted or similarly treated shall be overlaid with a resin impregnated fiber, medium density grade. All plywood shall be furnished in largest commercial sizes obtainable for the required use.

Doors.—Doors to shelf areas shall be constructed of 100 percent acrylic plastic of the type and thickness shown on the drawings.

Metal Fastenings.—Fastenings of plate steel thicker than 16 gage shall be as specified in ASTM Designation: A 36.

Exposed nails in redwood shall be of aluminum with annular ringed shanks; screws in exposed work shall be of stainless steel or aluminum. Siding nails shall be used for all exterior finish carpentry, casing nails shall be used for all interior finished carpentry of redwood; finish nails shall be used in Douglas fir trim. All other nails for woodwork shall be of steel wire with common heads. Unless shown otherwise on the drawings, nails for 1-inch material shall be 8d, for 2-inch material 16d, and for 3-inch material 20d; however, nails for exterior 1-inch board placed over reversed battens shall be 10d. Nails for sheathing paper shall be $1/2$ inch long with annular ringed shanks and with integral disc head 1-inch in diameter. Sizes of nails for other uses shall be as approved.

Bolts shall conform to the requirements of ANSI Specification No. B18.2. All bolts and nuts shall be galvanized and where bearing on wood shall be provided with galvanized cut washers.

Sealants.—Sealants shall be a nonstaining, nondrying, nonshrinking permanently plastic, paintable sealer and shall be equivalent to "Vulcatex" as manufactured by A. C. Horn Co., or Sealant EC-895 as manufactured by the Minnesota Mining and Manufacturing Company.

Wood Preservative.—Wood preservative shall be an oil-borne pentachlorophenol paintable type conforming to the requirements of Federal Specification TT-W-571.

Hardware.—All hardware shall be of a type and size suited to the particular application. All finish hardware, not specified to be primed for painting shall have U.S. 26 D finish and shall be complete with installation accessories of similar material and finish. All exposed screws in finish hardware shall be "Phillips" type.

Hinges shall be spring loaded and shall comply with the requirements of Federal Specification FF-H-116.

Roof Decking.—Metal roof deck shall be aluminum V beam as shown on the drawings. Decking shall be secured with cadmium plated lag bolts at alternate high rib with cadmium plated steel washers over neoprene washers.

Display Cases.—Aluminum display cases shall be Poblocki No. C 612, Gotham No. 350, or equivalent, complete with glass, cork board, and locks. Corkboard shall be $1/4$ inch bulletin board grade self-sealing standard tan permanently bonded to $1/4$ inch tempered hardboard backing.

(3) Framing.—All framing members shall be cut, fitted, and securely fastened together to produce work that is rigid, plumb and level or sloping, as shown, and to form true planes free of warp, crook, and twist. Individual framing members shall not be spliced between bearings. Faces of all framing members shall be joined with correct alignment and shall match all members at points of joining. Nails including toed nails shall be fully driven with heads embedded into the wood surface. Toed nails shall be driven so that approximately $2/3$ of the nail is in the supporting members. Nailing shall be in accordance with the requirements of the Uniform Building Code unless shown otherwise on the drawings.

(4) Exterior Carpentry.—All redwood shall be left free of all stains and other discoloration. Members defining lines and intersection of planes shall be selected for straightness and shall be wedged, blocked and securely nailed to produce the required lines. All exposed nails shall be evenly and symmetrically spaced and neatly driven leaving the wood free of hammer marks.

Sheathing paper shall be applied over all framing before the wall finish is applied. The paper shall be applied vertically with continuous sheets and with side joints made over studs; all edges shall be neatly fitted at openings and the paper nailed in place, free of wrinkles and tears. Exterior wall finish shall be cut to fit tight against all frames, trim and other intersecting members. Members shall be continuous without horizontal joints and shall be securely nailed to the framing; the spacing of nails and method of nailing of redwood shall be as recommended by the California Redwood Association unless otherwise shown on the drawings.

(f) Payment.—The contract prices shall include full compensation for all costs incurred under this section.

13
Construction Procedures and Equipment

CHARLES F. PALMETIER
Assistant Chief Construction Engineer (retired)*
U.S. Bureau of Reclamation
Denver, Colorado

*Deceased

I. INTRODUCTION

The topic of construction, even when limited to dams, is so broad and complex that this chapter is confined to a brief discussion of some of the more common concepts and problems in dam construction. Those interested in further study of construction are referred to the *Handbook of Construction Management and Organization* by John B. Bonny and Joseph P. Frein, published by Van Nostrand Reinhold Company.

As most dams are a part of a governmental project, their construction is usually obtained by awarding a firm fixed price contract to the lowest responsible bidder. In this type of construction the bidders must determine, with a reasonable degree of accuracy, the total requirement of the project in men, materials and equipment and they must do this in a time period usually not exceeding 60 days. The low bidder, having been awarded the contract, has a somewhat different objective than the owner's engineer. His objective is to win profits through effecting every economy that is possible under the contract.

In the private sector current practice is for the owner and his engineer to select a constructor as a third member of the team to assist in the final preparation of the designs and to supervise construction. In this type of construction administration the constructor works for a fee as does the engineer to the end that the owner receives the desired final product at the minimum cost.

The skill and art of estimating construction cost is a specialty that is adequately covered by numerous texts and is not within the realm of this chapter. Therefore the remainder of this chapter will assume that construction is to be performed under the administrative structure briefly outlined in the preceding paragraph.

When the constructor is brought into the team the elements of the project requiring his expertise are:

1. Preparation of a reasonably accurate estimate of money required by months (or quarters) for the duration of the con-

struction so that the owner may plan his financing and so minimize the costs of interest during construction.
2. Preparation of a construction schedule for the entire project. If prepared by well-informed and experienced construction engineers, the critical path method of scheduling is well suited to dam construction.
3. Preparation of a construction organization chart and the selection of qualified personnel to fill the required positions.
4. Preparation of a list of all equipment required for the construction together with a schedule indicating when each piece of equipment must be delivered and when it may be disposed of.
5. Layout of the construction plant, roads, utilities, fire protection and the like as well as the determination of what permits and rights-of-way will be required for construction purposes.
6. The environmental impact statement must include the effects during construction and the constructor must work with the owner and the engineer for this phase of the statement.
7. Labor relations. It is probable that the constructor will enter into all contracts with labor unions (if a union job) and that neither the owner nor the engineer will be a party to these contracts. The owner and the engineer usually have their own labor problems and should not be concerned with the problems involved with construction labor.
8. Purchasing. The administrative team must decide what purchases will be made by the constructor in his name and those that will be made by the owner in his name. Generally the owner will buy all items of permanently installed major equipment with the specs prepared by the engineer.

II. CONSTRUCTION SCHEDULING

The scheduling of the proposed construction forms the basis of most actions relative thereto. The Owner must know when the facility is to be available for operation and when he will be required to have funds available to finance the designs and construction. The Owner's Engineer must schedule the design production, the issuance of construction specifications and the purchasing of permanently installed machinery and equipment. The constructor must schedule his work within the framework of the over-all project schedule to insure that the work will be staffed, equipped and supplied in a timely and efficient manner. In the initial phase of the project, the major scheduling responsibility falls on the Engineer whether he is retained on a fee basis by the Owner or represents an in-house capability of the Owner (as is usually the case in government construction of civil works). For the purposes of this chapter it will be assumed that the Engineer has full responsibility for over-all project scheduling and that the Owner has only supplied the desired completion date and the money.

The Initial Construction Schedule

In the scheduling of the construction of most dams, the completion date is established first often without adequate information or study and all subsequent schedules are required to meet this promised completion date. The reasons for this are numerous and complex but can be over-simplified by pointing out that in most fields of human effort procrastination will defeat progress unless deadlines are established and maintained. Initially, in the case of construction by contract, the Engineer must proceed with the construction program without the expert advice of the experienced constructor. The usual method is to delineate the work that can be efficiently done under a single contract and establish the contract performance period and the contract award date. When more than one on-site contract is required, then the scheduling of the work under each contract and the timing must be coordinated to avoid on-site interference of one contractor with another. Having established contract award dates, the schedule must be analyzed to insure that there is sufficient time allowed for the preparation of final designs and specifications. It may well be that it will be necessary to divide the work into several contracts simply because designs and specifications cannot be prepared for the overall structure in the time allowed. For example, where it is a comparatively simple matter to prepare the designs and specifications for a diversion tunnel and cofferdam, the work for the entire structure including foundation treatment, concrete details and gates will take possibly more time than actually required to construct the diversion tunnel. The initial schedule must consider the status of rights-of-way and relocations. Contracts should not be awarded until all of the rights-of-way required for the work have been obtained. The schedule for diversion of the river and reservoir filling is restrained by status of relocations as well as land taking. No roadway or utility can be flooded, even for short intervals, until their replacements are substantially completed and in service.

The Critical Path

This method of scheduling is an effective tool; however, it is more elaborate than necessary for the simpler jobs. Its use is advisable for complicated jobs involving the coordination of several contracts with each other and with the delivery of machinery and equipment. The greatest danger in the use of CPM is erroneous assumptions as to the time required for the various operations being scheduled. Unless these assumptions are reasonably accurate, the resulting schedule will be worse than useless in that it may lead to erroneous administrative decisions. The critical path method if used should not be confined to construction but should include design progress as no design engineer enjoys being identified as a schedule obstruction.

Working Construction Schedules

This is concerned with the coordination of details and of resources. The schedule will form the time basis for the purchasing of all materials, supplies, machinery and equipment required for the construction. The ideal situation is that purchased items are used in the construction within a few days after they arrive on the job as large inventories represent the sterile investment of interest-bearing capital. Conversely, if construction is delayed because of the lack of required items, the ongoing costs of overhead and lost labor production can be a disaster. For this reason, most experienced constructors prefer to have all purchased items on-site about 30 days before the acutal requirement. The cost of interest on this investment is considered cheap insurance against an unwanted shutdown.

The working construction schedule must also provide for manpower leveling. When the construction forces are assembled for a phase of construction, they must be kept busy continuously—weather permitting. Intermittent employment will reduce not only the on-site productivity, but will also reduce the quality of labor available. This is to say that the truly skilled journeyman will not stand for part time employment but will go elsewhere. Thus the construction administration must not only concentrate on the work on the critical path but must also be prepared to work on non-critical phases of the project to achieve continuity of employment for the construction forces.

Manpower leveling must also be coordinated with the optimum utilization of the construction plant. The schedule must attempt to attain at least 80% of plant production capacity on a sustained basis throughout each major phase of the project. For example, if a concrete plant has a capacity of 150 cu yd/hr and is planned to operate 15 hr a day, then the schedule should provide for the placing of at least 1800 cu yd of concrete each day. This example brings up another point. On large concrete dams, it is customary to schedule the major effort on form building during the day shift and the concrete placements during the evening and night shifts. Some constructors of earth dams prefer to schedule the fill placing mainly during daylight hours even if this means two 9-hour shifts with the start of work at 4 o'clock in the morning. Where there is major investment in equipment it may be economical to schedule overtime as in the above example remembering that while the operator receives overtime premium pay, the machine over-all costs go down as utilization is increased. For details of scheduling techniques, see Chapter 17 of the *Handbook of Construction Management and Organization*.

III. CONSTRUCTION ORGANIZATION

Construction Supervision—The Contractor

The successful management of a construction project depends on three principal elements— equipment, schedules and job personnel. Generally speaking, because of the numerous social and reporting requirements written into Federal construction law, a Federal construction contract will require more overhead personnel than those in the private sector. Construction organizations vary; however, most field organization will have the following positions with responsibilities approximately as described (see Fig. 13-1.).

Project Manager

The project manager is the contractor's representative on-site and usually has authority to commit his employer in all matters relating to the project. Customarily, he is provided with a power-of-attorney by the corporate board of directors. However, he is usually limited in his authority to accept contract modifications in excess of a determined amount—say $50,000. Authority to purchase major items of construction equipment is usually retained by the corporate headquarters office. Final decisions on labor contracts are also beyond his authority. He does have a major say in staffing the job, but once initially staffed, he usually has full authority for hiring needed additional personnel and for terminations. He must approve all purchase orders issued by the project. On small jobs all of the functions of this position may be retained by the Contractor's home office and the position eliminated.

General Superintendent

The general superintendent is responsible for the coordination of all day to day construction operations. He must requisition those materials, supplies and tools necessary for these operations

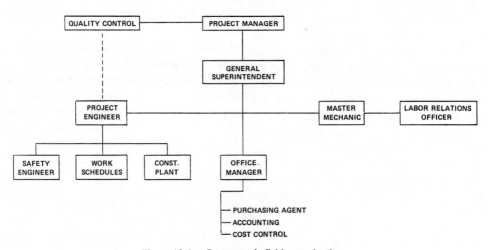

Figure 13-1. Contractor's field organization.

or approve the requisitions of his craft superintendents. He must furnish the Master Mechanic with priorities on equipment repair and maintenance. Note that some contractors prefer to have the Master Mechanic report directly to the Project Manager. He works with the Project Engineer on the design of the construction plant. He also consults with the Labor Relations officer in the assignment of work to the various crafts. His direction of the work must be in accordance with the project construction schedule and he is not free to deviate from this schedule without the approval of the project manager. The background of the general superintendent is that of a craftsman and he has achieved this position through his ability and drive. He is usually the second highest paid man on the project.

Project Engineer

The project engineer has the greatest variety of responsibilities of any man on the project manager's staff. If the contractor is required to establish the lines and grades required by construction operations, the staffing and workmanship are his responsibility. The details of the construction plant design are for his determination although the over-all elements and capacities have usually been established when the bid was submitted. He reviews and establishes prices on all contract modifications and either negotiates or assists in the negotiation of these prices with the owner's engineer. In the event of disputed work, the preparation of the claim is his job. The coding of the purchase orders and foremens' daily labor and equipment for cost keeping records is done under his supervision. He prepares requisitions for all materials and equipment which the specifications require the contractor to furnish as well as the requisition for materials and equipment required by his designs. Administration of all subcontracts, except the coordination of their daily operations, is under the project engineer. While the over-all construction schedule was prepared during the bidding period, after award of contract he must prepare a detailed schedule, keep it up to date, and coordinate daily with the general superintendent to insure that they are both in agreement regarding the schedule. On small jobs, the project engineer may have no staff, so most of the paper work described above becomes a function of the "home office."

Safety Engineer

The safety engineer is responsible to the project manager for all facets of job safety. He provides material for and organizes "tool box" safety meetings and conducts, at least monthly, a safety meeting of supervisors and assistant supervisors. He requisitions all safety equipment and supplies as well as first aid supplies. He supervises first aid personnel and arranges with local doctors and hospitals for emergency care of injuries. He prepares all reports on safety and health required by governmental agencies and insurance companies. He assists in the development of a workmens' compensation and liability insurance program: requires compliance by subcontractors with the project safety program as well as with existing laws. Finally, he has full authority to shut down any operation he considers unsafe.

Office Manager

The office manager directs and supervises all activities involving office administration, accounting, financing, timekeeping and payroll, banking, accumulation of cost data and preparation of financial and accounting reports. He also exercises accountability and procedural authority in areas of purchasing and warehousing, and, in general, is the guardian of his employers funds and property.

Purchasing Agent

The purchasing agent is responsible for proper purchasing and expediting functions. His efforts are directed to achieving the above objectives—also the coordinating of all efforts with the procurement requirements of the general superintendent and the project engineer. The purchasing agent is directed by the Office Manager in procurement policy in areas of accounting and internal control matters. He and the buyers obtain price quotations and negotiate purchase prices. It is not unusual for the "home office" to retain the purchasing function on all but the very large projects.

Labor Relations

The labor relations officer must be thoroughly familiar with all labor agreements with each craft likely to be employed on the job, whether by the prime contractor or by a subcontractor. If there is a type of work to be done not clearly covered by these agreements, then he must make a survey of any ongoing or recently completed jobs in the area that had a similar type of work and determine what assignments were made at these jobs. Only with this back-up data is he in a position to recommend an assignment that will avoid a jurisdictional dispute and a possible work stoppage. From the labor agreements he prepares a schedule of pay rates and fringe benefits for the payroll clerks. He, and only he, requisitions labor from the appropriate unions, or if an open shop, locates and signs-on qualified craftsmen. He conducts all exit interviews. He attempts to mediate disputes between workers and supervisors at the start so as to avoid negotiations at higher levels. He attends all labor negotiations (or is immediately available to the negotiators) the results of which will affect the project. Finally, he is responsible for seeing that supervisory personnel are familiar with the terms and conditions of the labor agreements that apply to the work supervised. This function is a responsibility of the home office on small jobs.

Quality Control

Some owners place the responsibility for inspection with the contractor. If such is required, an inspection unit must be added to the organization. While the head of such an organization nominally reports to the Project Manager, he is (or should be) unique in that he can only be discharged by an officer of the corporation holding the contract.

Construction Supervision—The Owner

Those agencies which construct dams have a variety of organizational structures. There are numerous variations but the on-site organization is usually divided into the following stated elements (see Fig. 13-2).

The Resident Engineer

He is the owner's representative on the job although, particularly on Federal projects, his

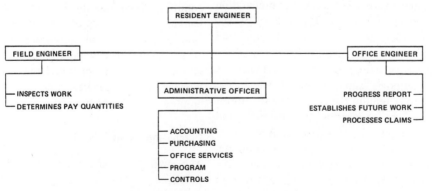

Figure 13-2. Owner's field organization.

authority to make decisions may be severely limited. As a general rule, he does sign all correspondence addressed to the contractor even though the letters, other than routine, may be prepared in detail at a higher office.

The Field Engineer

He and his staff are responsible for all the field inspection and inspection of materials and permanent equipment delivered to the job. He is responsible for determining the quantities of work done for payment purposes unless this work is required to be performed by the contractor, in which case, he checks their quantities. Again, he is responsible for all surveys unless the contract places this responsibility with the contractor. His inspection forces prepare daily reports on the work to provide documentary evidence that the work was done in accordance with the specifications and to document any circumstances relating to a dispute or possible dispute which may lead to a claim.

The Office Engineer

He and his staff are responsible for computation of all quantity calculations, for preparing or verifying all progress payment estimates, for preparing estimates of future fund requirements and for preparing and updating the over-all project construction schedule. This office is also responsible for compiling all data relative to claims or possible claims and for preparing cost estimates and data required for negotiations leading to equitable settlement of claims. Routine correspondence relative to contract matters is usually prepared by the Office Engineer.

The Administrative Officer

The Administrative Officer with his staff is responsible for all housekeeping functions, including accounting, purchasing, warehousing, office services and budgetary controls. He usually has only limited contact with the contractor.

IV. CONSTRUCTION EQUIPMENT

No discussion of construction equipment would be complete without considering the function of the Manufacturer's Representative or Sales Engineer. Nowadays the MR or Sales Engineer is usually a graduate engineer who has specialized in the capabilities of equipment produced by one manufacturer. It is a mistake to regard these specialists as just salesmen or "peddlers."

Normally the sales engineer will have become familiar with any construction project where his employer's equipment could be used. He will have analyzed the work and is prepared to recommend to the construction engineer the type, capacity and number of units required. Thanks to the competition in the industry, there will probably be available recommendations from each of two or more representatives of competing manufactures.

The construction engineer is advised to take these recommendations seriously, but he must, on the basis of his independent studies, make the final decision. He must also recognize that the primary interest of the Sales Engineer is to sell his particular brand of equipment whereas his responsibility as construction engineer is to efficiently equip the job with the least capital investment.

The following discussions of equipment assume that for a specific project, the construction engineer will contact the appropriate manufacturers' representatives to obtain complete details on the capabilities of any equipment he proposes to purchase.

It is sometimes economical to equip a project, at least in part, with used equipment. This has the initial advantage of reducing the capital investment required, but usually results in higher operation and maintenance costs. Associated with the higher O&M costs are the delay costs when an essential piece of equipment becomes unserviceable. Used equipment should be purchased only after a careful examination by a master mechanic.

Equipment for Excavations

Construction considerations. Whether the required excavation is common or rock or a combination, the layout of haul roads and disposal areas will have a considerable effect on the economy of the over-all excavation operation. For example, a 45-ton capacity dump truck can travel loaded at 20 mph up a 2 percent grade with this speed being reduced to 10 mph for a 6 percent grade and to 5 mph for a 12 percent grade. Also, the steeper grades will increase maintenance and repair costs. The importance of the haul roads and disposal (or storage) area becomes evident when the load-haul-dump-return cycle for a unit of hauling equipment is considered. For example:

Load	4 min	
Haul	12 min	2 mi at 10 mph
Dump	2 min	
Return	6 min	2 mi at 20 mph
	24 min at 80% efficiency = 30 min cycle	

Assume the unit has a 25 cu yd capacity and the desired production rate is 200 cu yd/hr then four hauling units will be required. As haul units are as a rule only 80 percent serviceable, the job would actually require 5 units. (One unit should be counted as being in the shop for repairs or service all the time). Now if the grades can be reduced or the haul shortened so that the cycle time can be halved, the need for a least two haul units would be eliminated. With haul units costing in excess of $100,000 each and operating costs generally exceeding $10.00/hr, the need for careful engineering of haul roads is apparent.

Excavating Equipment

When considering all types of dams, the requirements for excavating equipment will be so varied that selection of equipment will depend on the individual job. The requirements for concrete dams will differ from the requirements for earth and rockfill dams. In the case of a concrete dam, after the excavations are completed, the requirement for most of this equipment is completed and the equipment can be salvaged. Conversely, in the case of an earth or rockfill dam, much of the equipment used in excavating will also be used in placing the fill. In fact, the selection of equipment may sacrifice efficiency in the excavating operation so that optimum efficiency may be obtained in the placing of the fills.

Loading Equipment

Loading equipment covers a vast array of machinery and new devices are continuously being developed by the Construction Machinery Industry. As an example, up to the early 1960s the most popular loading machine for rock was the power shovel. At that time, development of the front-end loader accelerated and today the front-end loader has displaced the power shovel in most applications. The basic power shovel machine has two other applications—the dragline and the backhoe. Both of these applications are primarily useful where the machine must rest at a higher elevation than the material to be excavated. Thus a careful analysis of the over-all requirements would be necessary to make a final selection. The fact that the dragline can be effectively converted into a crane for subsequent concrete operation, is also a factor to be considered.

Continuous Loaders

For excavations for dam foundations and borrow pits the continuous belt loader is widely used—especially where the borrow pits are at such a distance from the structure that the use of scrapers to load and transport the material is uneconomical. The use of the continuous loader dictates the use of trucks on all but the largest jobs such as the Oroville Dam constructed by the California Department of Water Resources.

Scrapers

Scrapers come in a variety of types and sizes and also underwent considerable improvement during the 1960s. They are chiefly valuable where large volumes of common material have to be excavated and moved comparatively short distances. It is usually necessary to have a bulldozer push the scraper during loading to reduce the time of this operation.

Belt Loaders

The belt loader is essentially a hopper over a conveyor belt which discharges the material into haul units. The hopper is usually set well below the material to be excavated and loading is accomplished by pushing the material into the hopper with one or more bull dozers. This is an extremely efficient system where physical conditions and job requirements permit its use. A modification of this system is frequently used in earth dam construction by discharging the material over vibrating screens so as to classify the raw material into the gradations required by the various zones in the embankment.

Transportation of Materials

The transportation of excavated materials and materials required for the structure represents an equipment requirement of considerable cost magnitude. The construction engineer must carefully consider the total job requirement including types and character of materials to be moved, distances to be transported, climatic conditions in the job area, and job schedule. He also must consider the possible low salvage value if the equipment selected is inordinately specialized.

Figure 13-3. Earth moving equipment at work on core, Oroville Dam, California.

Rear Dump Trucks

Rear dump trucks (off highway) are commonly used for the transportation of excavated rock. Their bodies are constructed to withstand the impact of loading. They may be obtained in capacities from as little as 7 tons to well in excess of 60 tons. They are not normally equipped with tail gates so that the haul road grades should be kept low enough—say 6 percent—so that loaded material will not fall out. If properly maintained, this type of equipment has a comparatively low rate of depreciation as they are extensively used in all types of construction.

Bottom Dump Trucks

Bottom dump trucks (off highway) are used in transporting common material, raw and graded aggregates, graded materials for embankments, in fact any material that will pass through the gates and will not arch over and so jamb in the body. They come in practically all capacities up to 100 tons and more. In dam construction their most common use is conveying material to the fill for earth dams and transporting raw aggregates for large concrete dams see Fig. 13-3.

Conveyor Belts

These are chiefly used in the processing of aggregates and in transporting wet concrete to the placement. Except for some of the concrete placing units, conveyor belts usually involve high first cost and low salvage. They have the advantage of continuous operation, and, in some instances, can be used to transport materials where the cost of haul roads would be prohibitive. With the emphasis on environmental consideration, conveyor belts may come into greater use because of the comparative freedom from noise and dust.

V. FOUNDATION EXCAVATIONS

General

As has been covered in previous chapters, the foundation of a dam is one of the most important, if not the most important, considerations in the design of the over-all structure. It is usually necessary to restrict the size of the explosive charges as final foundation lines and grades are approached to avoid shattering the rock that is to form the structure foundation. To accomplish this end, objective current practice is to require that the final lift of rock to be removed be no more than 5 feet. Also, where there are extensive vertical or near vertical cuts in rock, it is customary to require line drilling or presplitting to insure the integrity of the rock remaining in place.

Rock Drilling

The net effect of these requirements is that it is seldom economical to use the quarry type rock drill (6 in. bit and larger) in dam excavation. The customary tracked drill using bits from 3 inches to $4\frac{1}{2}$ inches in diameter is the most widely used tool.* In determining the size and number of units required for a specific excavation project, the most important information required is the rate of penetration that can be anticipated with a pneumatic rock drill. This information is seldom available from the foundation investigations as the core drilling for geologic studies is done with diamond bits with rotary drills and there is no definite correlation between the rates of penetration for the two types of drills. Where test drilling has been done, however, an experienced construction engineer can usually make a reasonably accurate estimate of the penetration rate from an examination of the drill cores obtained during the investigations. It should be noted that, for competitively bid work, the bidders seldom have the time to do test drilling and must rely on the judgment of their construction personnel.

The selection of the most effective type of drill bit is a specialized study depending largely on the type of rock to be drilled. In general, manufacturer's recommendations can be relied on.

---

*It should be noted here that jackhammers are seldom used because of low productivity as well as the physical fatigue effect on the operator.

Having established the rate of drill penetration, the number of drills required for the excavation can be determined. The over-all rate of excavation desired is the prime factor involved not only in the determination of the number of drills, but in the determination of all required excavation equipment. Other factors include the number of hours to be worked each day, the amount of line drilling and drilling for presplitting to be done, and the frequency of blasting. Note that the loading of holes with explosive is usually done during daylight hours as a safety precaution. Detonation usually follows the end of the day shift to avoid lost time on that shift.

Compressed Air Station

With the number of rock drills required determined, the volume of compressed air required is, in effect, also determined. On small jobs where only three or four rock drills are required, it may prove economical to have a portable compressor for each or each pair of rock drills. With a greater number of rock drills required, consideration must be given to the installation of a central air compressor station. If such a station proves economical, then the type of power must be determined. In general, if electric power is available in sufficient quantity, electric powered stationary compressors are preferred as they usually have lower operating costs, higher serviceability and lower depreciation than do diesel powered units. In making the determination of the number and size of units to be installed, consideration must be given to the compressed air requirements after excavation is substantially complete. Air will be required for cleanup of rock and concrete prior to placing of concrete, for concrete vibrators and for other pneumatic tools. While the excavation operation will usually demand the greatest amount of compressed air, the selection of the unit size should be such that units may be shut down when maximum capacity is not required. Most constructors feel that a central air compressor station should have at least three units for flexibility of operation and scheduling of maintenance.

With a central air station, it is necessary to layout pipe line to the areas of use. It should be noted that in addition to the air receiver at the central station, air receivers at the ends of long lines will do much to maintain even pressures at the operating equipment and to avoid repeated stopping and starting of the compressors due to load fluctuations. As with the other elements of the plant, in the design of the over-all air station, full use should be made of the expertise available from manufacturers' representatives.

VI. CONSTRUCTION PLANT—INSTALLATION

In the preceding discussion of rock drilling, the compressed air station was discussed; however, this is only one part of the fixed plant required by most dam construction projects. The size of the dam, the duration of construction, the topography surrounding the site and the location of existing highways and utilities will have a major impact on the requirements for the fixed construction plant. There follows a discussion of the several elements of the plant which should be used more as a check list.

Administrative Structures

Administrative structures, including the job office access gates, guard structures, first aid facilities, employee parking and the like, will depend on the anticipated maximum level of employment. The job office should be located so that persons, other than employees, do not have to enter any construction area to get to the project manager or members of his administrative staff. If the job requires guards and identification of employees, parking for visitors should be provided outside of the guarded gate. First aid facilities may be contiguous with the administrative office or may be contained in a separate structure nearer the site of the work. Wherever they are, ambulance parking must be provided. A common mistake in providing employee parking, is to ignore the fact that the on-coming shift must park in the space not occupied by the shift still at work. Further, employee parking should be within walking distance of the reporting point or points to avoid the expense of bussing. For the office, serious consideration should be given to a trailer or a complex of trailers because of their early availability and comparatively high salvage value.

Shops and Warehouses

The siting of the warehouse with proper consideration given to the flow of items in and out as well as to security for outside storage areas, can do much to reduce losses due to pilfering. Some constructors prefer to have their equipment repair shop adjacent to the warehouse so that the warehouse man can also tend the tool room and spare parts supply. On jobs having considerable form work, it is usual to have a carpenter shop. The accepted rule here is to arrange the lumber storage with relation to the shop so that the lumber is moved into one end of the shop, fabricated into forms which are then moved out the other end to the place of use or temporary storage. Storage of forms in the same area as the lumber is stored does not as a rule contribute to job efficiency.

On most dam jobs, it is necessary to have refueling and lubricating facilities. While it would appear convenient to have these facilities handy to the repair shop and warehouse, fire danger dictates that petroleum products be reasonably isolated.

Craft Buildings and Change Houses

Union agreements covering heavy construction generally require that a building—usually 8 ft by 16 ft—be provided for each craft as the assembly point at the start of the shift, for storage of personal tools and lunch boxes and for eating lunch. On jobs with high employment, space for these "shacks" can become a problem. Change houses are usually only required by miners working underground but when required, must be equipped with showers, sanitary facilities and lockers.

Sanitary Facilities

Chemical or portable toilets are used on most jobs because the changing character of the work makes more permanent installations uneconomical. Such toilets are furnished and serviced by specialty contractors with their own equipment for collecting and legally disposing of the sewage. Where shops, offices, warehouses and the like have sufficient occupancy, it may be that a sewerage system is indicated. If so, a check of local laws covering sewage disposal is indicated.

Water and Air Systems

Water of a quality suitable for drinking is seldom provided by the job water system. Drinking water is usually supplied to the workmen in 16-gal capacity portable coolers with paper cups. The job water supply can come from the river, a well or other source. The capacity of the system will depend on the amounts required by fire protection, watering borrow pits and fill, dust supression, washing foundations and concrete surfaces and for curing concrete.

As a part of the system, the removal of excess water and its treatment in accordance with State and/or local regulations before release into the water course is required. The delivery system should contemplate the placing of hydrants convenient to points of use and at points required for fire protection, as for example, the warehouse and lumber yard. The compressed air system (previously described) and the water system are usually installed at the same time and in some instances in the same trench as they are both installed by pipefitters. It should be noted that for concrete construction, air is probably required wherever job water is required.

Electric Power and Lighting

The proximity of electric utility lines with sufficient capacity to meet job requirements will probably dictate the character of much of the job equipment. If commercial power is only available at some distance, then the cost of transmission line construction (with a comparatively low kilowatt hour cost) must be compared with the cost of diesel-electric generating equipment (with its comparatively high operating cost). If commercial power is not available or not economical, then each operation such as pumping, aggregate processing, screening plants, batching plant, compressor station and the like, must be examined to determine if each should have its own generating station or if there should be a central generating station. Job lighting for excavation and fill operations can usually best be provided by semiportable light towers each with its small gasoline-electric generator so that they can be moved without extensive rewiring. Lighting for concrete operations is normally supplied electricity from job load centers. The requirements for lighting of construction are covered by Federal Regulations under the National Health and Safety Act.

VII. PROBLEMS IN DAM CONSTRUCTION

Experience has indicated that some problems in dam construction reoccur more frequently than others. The following paragraphs will attempt to outline these problems and means of avoiding them, or at least recognizing their existence. Earth dams and concrete dams are considered, but buttressed dams have not because of limited experience in recent years.

Earth Dams

From the constructors point of view, the elements of earth dam construction are: (a) diversion and care of the river during construction; (b) excavation for the foundation, (c) grouting, (d) placing the main embankment, and (e) concreting of appurtenant structures. Careful scheduling of these elements with relation to each other is essential for an efficient job. Experience has shown some of the trouble spots in constructing earth dams and the following paragraphs will cover these.

Diversion and Care of River

The usual earth dam design provides for an outlet works that will eventually be used to control discharges from the reservoir but during construction is used to carry the flow of the river so as to permit the unwatering of the dam foundation. It is essential that this outlet works (usually a tunnel through one of the abutments) be completed sufficiently for river diversion before the cofferdams can be closed and work in the river bed started. A common failing in scheduling this operation is to consider it as a comparatively short tunnel requiring lining. Some place in the tunnel will be a gate chamber with a shaft extending up to the top of the dam. There is no way to construct this shaft (except for concreting the upper portions) after the tunnel is in use for diversion. The tunnel itself is heavily reinforced and some portions may be steel lined. (Note that timely delivery of the steel lining is a scheduled item). The gate chamber is also heavily reinforced and requires intricate form work. For diversion, all concrete in the tunnel, the tunnel intake, the gate chamber (except for the concrete for embedding the gates to be installed at a later date) and downstream stilling basin must be complete.

With the completion of the first stage of the outlet works, the cofferdams can be closed and the river diverted. Detail runoff studies were made in connection with the design and these studies should be used in determining the height of the cofferdams. The hydraulic capacity of the tunnel will be more than enough to carry normal flows but the engineer must determine what storage capacity and pond above the upstream cofferdam he is to provide to contain flood peaks. It is seldom economical to provide for the containment of floods greater than the estimated 1-year flood. In making this determination, however, a careful analysis of the damage to the partially completed structure and the cost of cleanup and repair in the event of cofferdam overtopping must be made. It may be that such an analysis was made during the design stage but this study should be carefully checked with the advent of the actual start of construction.

Diversion of the river and the subsequent unwatering of the foundations will increase stream turbidity to unacceptable levels unless facilities are provided to remove sediment. The discharge from unwatering pumps, at least, should pass through a settling basin, and in some cases, even more sophisticated devices may be required. If the construction operation alters the Ph factor of the effluent objectionably, chemical treatment may be indicated. The actual closure of the cofferdam will, for a brief period, increase the stream turbidity. This can be minimized by making the closure during a period of low flow and by making the initial closure with clean rock.

Excavation for the Foundation

During the construction of the first stage outlet works and cofferdams, required excavations above river surface can be accomplished. Earth dam designs are based on the utilization of practically all material including that from required foundation excavation. Therefore a detail study is necessary of possible storage and/or processing areas to minimize rehandling and hauling. This study should be carried out in conjunction with the preparation of a flow diagram and layout haul routes for the selected borrow areas. In scheduling excavations, a requirement often overlooked is the required excavation for the "grout cap." This is usually a trench cut into the foundation rock from 5 to 8 ft deep and about 6 ft wide. Its excavation is slow, requiring line drilling and/or presplitting. If local conditions permit, this trench should be excavated as the over-all excavation proceeds down the abutments so that when the river bottom is finally excavated and the trench completed, concreting of the cap can immediately follow and grouting can be started.

Foundation Grouting

The grouting of dam foundations is more of an art than a science and even the experts in this field can seldom accurately predict either the time required or the "grout take" to adequately seal a foundation. Thus the construction engineer is faced with a fully equipped job ready to start placing embankment yet the start of this operation must await release of the foundation by completion of the grouting operation. The scheduling of the time required for grouting is frequently overlooked by both those who plan the over-all project schedule and those who plan the detailed construction schedule.

The actual drilling and grouting of foundations is usually done by construction firms specializing in this work. They have the specialized diamond drilling equipment as well as the mixing and pumping equipment. Most difficulties and disputes arise in this phase of the work when an inexperienced firm is employed for drilling and grouting.

Embankment Construction

An essential part of embankment construction is the conditioning of the borrow material. In most instances, the moisture content of the soils from Zones 1 and 2 material is less than optimum, so that water must be added. Analysis of the in-situ material should be a continuing process as in most climates it is much more difficult to remove water than to add it. Climate should be studied, as it is costly to bring material to optimum water content only to have seasonal rains follow and soak the material.

The gradation of material for filter blankets frequently requires screening and there are numerous ingenious schemes for accomplishing this operation in concert with the over-all construction of the embankment. As local conditions vary so widely, no recommendation is being made other than to secure the advice of equipment manufacturers representatives.

The actual work on the embankment is more a traffic control and dispatching problem than anything else. The various zones have to be brought up concurrently including the rock blankets on the faces. (An exception is the design that calls for a soil-cement facing.) The special compaction at the abutments must

Figure 13-4. Fill compacting equipment, Oroville Dam, California.

be accomplished on schedule so that the over-all embankment placement is not slowed. Testing of the compaction and moisture content in the fill is a continuing operation. When properly equipped and organized, the actual fill placement is one of the simplest parts of the project (see Fig. 13-4).

Concreting of Appurtenant Structures

When the construction of earth dams goes awry, the problem is usually with the concrete structures. Constructors focus their attention on the excavations and embankment construction and neglect the concrete requirements in their planning. The difficulties with the first-stage outlet works are pointed out under the heading "Diversion and Care of River" which follows. The scheduling of the spillway with its stilling basin, gate structure (if any) and possibly a bridge is frequently cursory. The time required for drilling and installing anchor bars needed to hold the completed structure on a steep slope is often neglected in construction planning. The under drainage systems are required for the spillway and the stilling basins so as to avoid destructive hydrostatic uplift forces. They are not complicated but do not lend themselves to rapid construction usually requiring considerable hand labor. The formwork and the reinforcement designs seldom permit pre-assembly of the rebar or extensive re-use of forms.

The transportation of mixed concrete to the forms requires special attention because of the length of the average spillway, and the great difference in elevation between the stilling basin at the lower and upper end intake or gate structure.

Finally, because of the location on a hillside, access for men, materials and equipment is usually difficult.

The concrete structures required are as varied

as the dam sites themselves and no hard rules covering their construction can be set forth, but the construction engineer is cautioned to spend at least as much time studying the scheduling of these structures as he does on the main dam.

Concrete Dams

The construction of a concrete dam generally requires more individualized equipment than does an earth dam. Thus one of the elements in the construction of the former is the selection of plant arrangement whereas for the earth structure the problem is primarily one of selection of type and number of units readily available on the equipment market. Elements of concrete dam construction are: (a) diversion and care of the river; (b) selection of the construction plant; (c) foundation excavation and treatment; (d) concreting; and (e) installations of gates and mechanical items. If a hydroelectric power plant is involved, concreting and installations become a much more difficult problem.

Diversion and Care of the River

Concrete dam designers and constructors usually have more options in ways and means of passing river flows during construction than with earth dams. Basically, this is because of the ability of hardened concrete to withstand flowing water. Thus it is possible to pass water over portions of the partially completed structure as was done at the Grand Coulee Dam on the Columbia River in the 1930s or to build half the structure, including most of the spillway, behind a cofferdam from one abutment and then complete the structure from the other abutment behind cofferdams diverting through the partially completed spillway. Variations of these two basic schemes are widely used. Also some concrete dams, particularly those in narrow canyons, divert the river flows through tunnels similar to the scheme used at the world-famous Hoover Dam on the Colorado River. For concrete dams it is more common for the design engineer to specify the details of river diversion during construction, particularly if the dam is being constructed in connection with a navigation lock, and water borne traffic must be maintained during construction. However, for storage reservoirs for flood control or water conservation purposes, the construction engineer usually has some latitude in planning for river care.

The problems are basically the same as for earth dams. That is, what is the economical height for the cofferdams, considering probable damage to the work in the event of overtopping. The partially completed concrete dam is not likely to be damaged by flooding; however, construction progress would be interrupted and cleanup after the flood recedes would be costly.

Selection of the Construction Plant

The priority item in planning the construction of a concrete dam is the selection of an efficient system for the delivery of the freshly mixed concrete to the forms. The delivery system can be by a cableway or cableways spanning the canyon, by cranes operating from one or more trestles, or by other means such as conveyor belts or concrete pumps. In the case of conveyors or pumps, some sort of hoist equipment must be provided for moving materials, forms, and the conveyors or concrete lines.

The biggest advantage of cableway systems is that their installation can be started long before the final foundation is uncovered so that concrete can be placed as soon as the rock is ready. A further advantage is that the cost of the equipment and its installation is usually less than for other systems. Operating costs of cableway systems are high in relation to other systems. For wide canyons the required sag of 10 percent of the span (when loaded) may require head and tail tower heights that are either impractical or uneconomical or both.

If a cableway system is selected, then the location of the concrete mixing facilities can be determined. Since during a concrete placement, it is too time consuming to hook and unhook concrete buckets to the load block of the cableway, a system is required that will permit filling the buckets while they are attached to the cableway. (The last major job where buckets were changed was Hoover Dam in the early 1930s.) Customarily, the concrete mixing plant is located so as to discharge mixed concrete into a transfer car (or cars) which operate on a railroad above a loading dock upon which the concrete buckets are landed by the cableway for refilling. These transfer facilities are usually located at an elevation so that approximately 75 percent of the concrete mass will be lowered to the forms.

When a trestle system is used, as for example at the Grand Coulee Dam, it is usually necessary to embed some portion of the trestle in the permanent structure. Therefore, final, acceptable foundation rock must have been reached before trestle erection can be started. Operating on the trestle will be one or more cranes (whirleys or hammerheads are usual). It is customary to locate the concrete mixing plant at one end of the trestle with concrete being discharged directly into the concrete buckets. The concrete buckets are moved on the trestle by rail to a location where they can be hooked by the crane. With the recent development of automatic hooks, it is possible for the crane operator to set the empty bucket on the transfer car and unhook and hook a full bucket without the use of a hooktender. The automatic hook has not been used with cableways because the cableway operator does not have the accurate control of the hook that is available to the crane operator. The advantages of the trestle system are low operating costs and flexible capacity. With the trestle system additional cranes can be added if production increases are needed, but once installed, the capacity of the cableway system is fixed.

The other systems find their application in buttress type dams and thin arch dams where the volume of concrete to be placed is comparatively low. The Bartlett Dam, a multiple arch buttressed structure on the Salt River Project in Arizona, was concreted by pumping the concrete. Hoisting equipment is still required for movement of materials, forms, belts or pipes and the like, but its capacity can be limited as it is not required to lift 12 or 25 ton loaded concrete buckets.

Foundation Excavation and Treatment

Excavation operations will be required to be conducted so that the final foundation rock will not be damaged by blasting. It is also customary to require any seam or localized areas of poor rock to be excavated several feet below final foundation grades and backfilled with concrete. If it is anticipated that it will be required to clean the foundation for inspection to determine the suitability of the rock, the construction schedule must allow time for these inspections as well as for the treatment of seams. Foundation grouting for concrete dam foundations prior to concreting is usually limited to shallow holes and hence does not delay over-all construction to the extent that foundation treatment delays earth dam construction. Deep drilling of both grout and drain holes is done from the foundation galleries in the dam after concreting is substantially complete and this operation is seldom a schedule factor in over-all dam completion.

Concreting

The scheduled rate of concreting will determine the size of the aggregate processing plant, the concrete mixing plant, the refrigeration plant (if cooling is required) and the delivery and placing systems. All of these interrelated plants can operate at peak efficiency if there is a place for the concrete. Which brings us to the subject of the set-strip-move-set cycle for the concrete forms. (For complete information on the designing of forms, the constructor should have a copy of the book "Formwork for Concrete," American Concrete Institute publication SP-4). Forms must be designed so that they can be conveniently set, filled with concrete, stripped, cleaned and moved to the next placement and reset. They must also be designed to withstand the pressures generated when the fresh concrete is vibrated. For dam construction, where multiple use of forms is possible, shop fabricated steel forms are considered preferable for economic reasons. The first cost of steel forms is high in relation to job built wooden forms, but whereas the wood form can only be used 4 to 6 times without extensive reconditioning, the comparable steel form can be used dozens of times and still have considerable salvage value. In general, steel forms have a faster set-strip-move-set cycle and maintain their shape better than do wooden forms. The importance of form movement is exemplified by the contractor's development of self-raising steel forms for the construction of the Dworshak Dam—a 693 ft gravity structure located on the North Fork Clearwater River in Idaho.

Another factor in the concreting is the preparation and cleanup of concrete surfaces prior to placing concrete thereon. Specification requirements vary—some owners' engineers require all surfaces be wet sandblasted, some permit use of high pressure wet jets in lieu of wet sandblast and others allow green-cutting with a simple air and water jet cleanup prior to concreting. The wet sandblast requirement is the most time consuming, involving as it does the move of blasting apparatus into then out of the area to be blasted and then the subsequent cleanup of the sand. The green-cut and clean method is the quickest, but only on a job operating on an around the clock basis as the green-cutting must be done within about 3 hours of the completion of the placement or the laitance will have hardened to the extent that the air water jet will not remove it.

The schedule for concreting must also consider the amount of embedded material to be installed. Reinforcement is not usually required in gravity and arch dams except around openings such as galleries, wherever it is used there will be a delay to the normal set-strip-move-set cycle. When embedded piping and/or conduit are required there will be further delays. Miscellaneous metal and anchor bolts generally do not involve material delays as these items are set by the same craft that sets the forms.

Installations of Gates and Mechanical Items

The most frequent construction difficulty encountered is the lack of timely delivery. In the case of sophisticated machinery and equipment, it is advisable to have the manufacturer's erection engineer on site for the assembly and installation both to insure proper installation and to validate any guarantees.

A source of increased costs and time consuming delays is the improper installation of anchor bolts during concreting, or the damage to these bolts by subsequent operations. Some contractors and designers have found it economical to use weld plates anchored in the concrete with the anchor bolts (stud bolts) welded to these plates at the time of equipment installation. Weld plates have a further construction advantage in that they do not require that the sheathing of the concrete form be pierced thus increasing the form re-use and reducing the time required for setting and stripping.

14
Public Safety Controls for Dams and Reservoirs

ALFRED R. GOLZÉ
Consulting Civil Engineer
Sacramento, California

INTRODUCTION

Any dam constructed any place must be so designed and built that it is a safe structure, that is, the areas below the dam are protected from the consequences of a failure or untimely release of its reservoir contents. Design and construction of a dam and its appurtenant facilities requires the application of superior professional engineering in the public interest. Such is the theme of this handbook.

The foundation of the dam must be stable under all conditions and capable of carrying the weight of the structure. The dam must impound its reservoir without undue strain and fully resist the application of external forces such as those resulting from earthquakes. The reservoir area must be water-retentive and free of the possibilities of dangerous slides. The dams and their appurtenant facilities must be maintained in excellent condition throughout their life. Operation and surveillance through the years must be conducted in such a manner that any physical change in the structure of the dam, including its foundation, can be detected promptly and corrections made. If abandoned, at any time, the dam must be removed or breeched to eliminate any hazard to downstream areas.

Unfortunately, this has not always been the case. In the United States there have been some notable failures of major dams. Abroad we find a similar situation. A few examples will illustrate the point involved.

In the United States, on March 12, 1928, the St. Francis dam, located on San Francisquito Creek near Los Angeles, failed. It was a curved concrete gravity structure, 184 ft (56 m) in height. The dam's foundation failed under full reservoir conditions, due to inherent weakness in the underlying material.

In 1963, the Baldwin Hills reservoir in Los Angeles, California, an offstream earth structure 262 ft (80 m) in height, failed due to gradual deterioration of the foundation during the life of the structure (see Fig. 1). In the same year the Vajont reservoir in Italy was destroyed when a catastrophic slide squeezed the water

772 HANDBOOK OF DAM ENGINEERING

Figure 14-1. Failure—Baldwin Hills Dam and Reservoir, Los Angeles, California, 1963.

content of the reservoir almost instantaneously over the top of the 870 ft (265.5 m) high concrete arch dam. The dam was but slightly damaged.

In 1959, the Malpassant dam in France, a thin arch 216 ft (66 m) in height, suffered a foundation failure, and was swept away. 1959 also saw the Vega de Tera dam in Spain, a buttressed masonry structure 111 ft (34 m) high, fail from use of different types of materials in the dam and an incorrectly computed modulus of elasticity for the masonry.

1976 witnessed the failure of the Teton Dam, a 305 ft. (93 m) earthfill dam in Idaho, caused by leakage in the right abutment.

These are but illustrations and it will be noted that geology has played an important part in the failure of the cited structures. There are a number of other structures which over the years have reacted similarly. In other words, it does not matter how good the design of a structure may be from the technical viewpoint, it can be worthless if the structure is built on a foundation incapable of bearing the load or remaining intact under saturated conditions.

With examples of recent failures such as

those cited above before us, and the ever expanding development of civilization in the valleys and flood plains below dams, extreme care in the design and construction, operation and maintenance of all dams and their appurtenant works—past, present and future—must be exercised. The state of the art of design progresses constantly. The chapters on dam design included in this volume are examples. The guards against foundation failures are given extensive treatment in Chapter 4 of this volume. The operation of the completed facility includes the dam, immediately adjacent power plants, locks, spillways and recreational appurtenances. Such operations geared to maximize multiple purpose project functions must always be within the factors of safety of the individual facility's units.

Perhaps the most important of all, maintenance to a condition approaching initial quality must be maintained throughout the life of the facilities. Deterioration of a dam, any part thereof, or any of its key appurtenances, could ultimately lead to disaster. To avoid such disaster extensive surveillance become a basic requirement. While the owners of most dams provide for systematic reading of instruments and gauges, for inspection of sensitive areas, and annual overhauls of basic machinery and equipment, this is not always enough. The charts and diagrams derived for instrument readings need to be studied and compared with previous data. Unscheduled examination of dams, particularly during periods of high water, can provide additional measures of safety.

The use of qualified personnel is another basic element. The design must not only be by competent engineers but should be supervised by registered or licensed engineers for the State in which the structure is to be located. In likemanner, construction is a professional undertaking which, again, for maximum performance and safety, should be supervised and directed by graduate registered or licensed engineers. The operation of completed facilities should be done by men skilled by years of experience working under the supervision of one or more licensed engineers qualified in this particular field.

In connection with the use of qualified personnel which may either be persons in the employ of the owner or persons employed by firms of private engineers, independent consultants should be available at all stages of design and construction of the facility to advise and counsel the engineers in charge. In connection with the periodic and special inspections of the structure, independent consultants should likewise be employed. Their advice to the dam owners concerning the physical condition and the efficiency of his facility brings to bear a professional judgement supported by years of experience.

FEDERAL CONTROLS

In the United States, controls to achieve the full objective of the desirable approach to design, construction, operation and maintenance, is accomplished in part at the Federal level and in part at the State level. Considering, first, the Federal level; the Federal Power Commission makes periodic inspections of licensed projects with its own staff throughout the United States. This includes over 400 licensed hydro-power projects, or about 80 percent of all the total kilowatts for non-Federal hydro capacity. Included in these projects are more than 600 dams and reservoirs ranging in age from those constructed before 1900 to those being currently built. These dams range from 35 ft (10.7 m) in height, to that of Oroville dam, the highest in the United States at 770 ft (235 m). It covers dams made of every conceivable type of material and all types of ownership other than the Federal Government.

The authority for the Federal Power Commission inspection is its Order No. 315.[1] FPC inspections are made at individual dams at least every 5 years, including their appurtenances but excluding transmission lines and generating equipment. To quote the Order No. 315; the Federal inspectors look for:

deficiencies or potential deficiencies in the

---

[1] U.S. Federal Power Commission, Order No. 315, Promulgating New Part 12 of Regulations Under the Federal Power Act (18 CFR Part 12).

condition of project structures, the quality and adequacy of maintenance or methods of operation which might endanger public safety. Such inspections shall provide pertinent data with respect, but not limited, to such matters as the settlement, movement, erosion, seepage, leakage, cracking, examination of internal conditions of stress and hydrostatic pressures in structures, their foundations and abutments, functioning of foundation drains and relief wells, and stability of critical sections of the reservoir, shorelines, and back slope above structures.

All inspections are made under the direction of a qualified independent consultant although the owner's personnel may participate. A report is made under the direction of the consultant which must have their approval and is filed with the FPC upon completion. Owners are required to advise the Commission within 30 days of its plan for action to implement corrective measures that may be required as a result of the inspection.

Other Federal agencies concerned with dam design, construction, operation and maintenance, principally the Bureau of Reclamation, Army Corps of Engineers and the Tennessee Valley Authority, have their own programs to insure complete safety of the Federal Government-owned water retention structures in their custody. The experience of these agencies has been reflected in much of the material included in this volume.

STATE CONTROLS

At the State level: Many States have enacted laws which to some degree exercise control over the design and construction, maintenance and operation of dams and their appurtenances and reservoirs. California leads the States in its legislation which maintains a very tight control over the complete cycle of the design and construction, operation and maintenance of dams. It was enacted in 1929 following the failure of the St. Francis dam. Emphasis has been placed on the use of competent and qualified personnel and the employment of registered licensed engineers. Controls exercised by other States vary from California's strict regulations to none at all in certain States.

The United States Committee on Large Dams of the International Commission on Large Dams, known as "USCOLD" is a professional society devoted to the improvement in all phases of activity related to large dams. A "large" dam is defined as one 25 ft (7.63 m) or more in height.

In 1965, concerned over the recent series of dam failures, USCOLD undertook a questionnaire survey of the 50 states, inquiring as to the exercise of their police powers in supervision of dams extending from the planning stage, through design and construction, into operation and maintenance.

The replies received from the States were summarized by USCOLD:[2]

"There is a great difference among the States in the manner in which each State is carrying out its responsibilities to the public for the safety of dams built within each State's jurisdiction, as the following will indicate.

"Thirty-three States specifically require that a permit or license be obtained prior to the commencement of construction of a dam. Seventeen States provide on-site inspection by State personnel during construction. Twenty-nine of the States review the plans and designs in preliminary form and thirty-two require that contract plans and specifications be reviewed by the State.

"Nine States reported that they are actively engaged in considering modifications of existing regulations and fifteen States acknowledged the need for improvements in regulations to meet the present needs of the State.

"However, five States stated that no per-

---

[2] "Supervision of Dams by State Authorities," published by the United States Committee on Large Dams, New York, N.Y., July 1966.

mit or license was required prior to the commencement of construction of a dam. Thirteen States replied that they had no printed regulations or instructions for filing applications relating to the construction of dams. Nine States said that no supervision of design of a dam was being exercised. Eleven States answered that supervision or inspection of the construction of a dam by the State was not required. Insofar as existing dams are concerned, seventeen States indicated no current authority over the manner in which the dams are operated and fifteen States did not exercise control over the maintenance of dams once they have been constructed. Ten States do not require that the design of dams must be supervised or approved by a registered professional engineer.

"The answers also revealed that in many of the States, the annual budget of the office and staff directly related to dam and reservoir supervision was limited and in some States, with hundreds of dams subject to supervision and with a requirement that they be inspected annually, inadequate funds were provided to carry out this responsibility."

MODEL LAW

USCOLD, as a public service with particular attention to the needs of the several States, drafted a "Model Law for State Supervision of Safety of Dams and Reservoirs."[3] It recognizes that a dam and reservoir cannot be half safe— it must be 100 percent safe against failure. The model law emphasizes the great responsibilities that rest with professional engineers and public officials charged with the design, construction, operation and maintenance of dams and reservoirs.

The USCOLD Model Law follows: (Table of Contents on page 776).

---

[3] Published by the United States Committee on Large Dams, New York, N.Y. 1970.

MODEL LAW FOR STATE SUPERVISION OF SAFETY OF DAMS AND RESERVOIRS

Chapter 1. Definitions

1000. Unless the context otherwise requires, the definitions in this chapter govern the construction of this Act.

1001. "Agency" means that Agency, Department, Office, or other unit of State Government designated by State Law to be responsible for implementation or direction of this Act. (This section to be replaced in enactment of the law by a reference to the State unit created or selected to implement and direct the Act which may be regular State employees or specialists and consultants, including consulting engineering firms or organizations, for any or all of the provisions of this Act.)

1002. Jurisdiction applies to any artificial barrier, herein called a "dam," including appurtenant works, which does or will impound or divert water, and which (a) is or will be 25 feet or more in height from the natural bed of the stream or watercourse measured at the downstream toe of the dam, or from the lowest elevation of the outside limit of the dam, if it is not across a stream channel or watercourse, to the maximum water storage elevation or (b) has or will have an impounding capacity at maximum water storage elevation of 50 acre-feet or more.

1003. No obstruction in a canal used to raise or lower water therein shall be considered a dam. A fill or structure for highway or railroad use or for any other purpose, which does or may impound water, shall be subject to review by the Agency and shall be considered a dam if the criteria of Section 1002 are found applicable.

1004. "Reservoir" means any basin which contains or will contain impounded water.

1005. "Owner" includes any of the following who own, control, operate, maintain, manage, or propose to construct a dam or reservoir:

(a) The State and its Departments, institutions, agencies, and political subdivisions.

MODEL LAW FOR STATE SUPERVISION OF SAFETY OF DAMS AND RESERVOIRS
TABLE OF CONTENTS

	Sections
SUPERVISION OF SAFETY OF DAMS AND RESERVOIRS	1000–1210
Chapter 1. Definitions	1000–1010
Chapter 2. General Provisions	1025–1033
Chapter 3. Administrative Provisions	1050–1052
Chapter 4. Powers of the Agency	1075–1085
Article 1. Powers in General	1075–1080
2. Investigations and Studies	1081–1082
3. Action and Procedure to Restrain Violations	1083–1085
Chapter 5. Applications	1100–1118
Article 1. New Dams and Reservoirs or Enlargements of Dams and Reservoirs	1100–1105
2. Repairs, Alterations, or Removals	1106–1111
3. Approval of Applications	1112–1118
Chapter 6. Fees	1125–1133
Chapter 7. Inspection and Approval	1150–1173
Article 1. New or Enlarged Dams and Reservoirs	1150–1152
2. Certificates of Approval	1153–1155
3. Repaired or Altered Dams and Reservoirs	1156–1158
4. Removal of Dams and Reservoirs	1159–1161
5. Complaints as to Unsafe Conditions	1162–1165
6. Inspection During Progress of Work	1166–1173
Chapter 8. Maintenance, Operation and Emergency Work	1174–1180
Article 1. Maintenance and Operation	1174–1176
2. Emergency Work	1177–1180
Chapter 9. Offenses and Punishment	1185–1187
Chapter 10. Dams and Reservoirs Existing Prior to the Effective Date of this Law	1200–1210
Article 1. Dams and Reservoirs Completed Prior to Effective Date of this Law	1200–1206
2. Dams and Reservoirs Under Construction Before Effective Date of this Law	1207–1209
3. Fees for Dams or Reservoirs Under Construction Before Effective Date of this Law	1210

(b) Every municipal or quasi-municipal corporation.
(c) Every public utility.
(d) Every district.
(e) Every person.
(f) The duly authorized agents, lessees, or trustees of any of the foregoing.
(g) Receivers or trustees appointed by any court for any of the foregoing.

"Owner" does not include any agency of the United States Government, including those who operate and maintain dams owned by the United States.

"Person" means any person, firm, association, organization, partnership, business trust, corporation, or company.

1006. "Alterations," "repairs," or either of them, mean only such alterations or repairs as may directly affect the safety of the dam or reservoir, as determined by the Agency.

1007. "Enlargement" means any change in or addition to an existing dam or reservoir, which raises or may raise the water storage elevation of the water impounded by the dam.

1008. "Water storage elevation" means the maximum elevation of water surface which can be obtained by the dam or reservoir without encroaching on the approved freeboard at maximum design flood.

1009. "Days" used in establishing dealines, means calendar days, including Sundays and holidays.

1010. "Appurtenant works" include, but are not limited to, such structures as spillways, either in the dam or separate therefrom; the reservoir and its rim; low level outlet works; and water conuits such as tunnels, pipelines or penstocks, either through the dam or its abutments.

Chapter 2. General Provisions

1025. It is the intent of the Legislature by this Act to provide the regulation and supervision of all dams and reservoirs exclusively by the State to the extent required for the protection of public safety.

1026. No city or county has authority, by ordinance enacted by the legislative body thereof or adopted by the people under the initiative power, or otherwise, to regulate, supervise, or provide for the regulation or supervision of any dams or reservoirs in this State, or the construction, maintenance, operation, or removal or abandoment thereof, nor to limit the size of dam or reservoir or the amount of water which may be stored therein, where such authority would conflict with the powers and authority vested in the Agency by this Act. This Act shall not prevent a city or country from adopting ordinances regulating, supervising, or providing for the regulation or supervision of dams and reservoirs that (a) are not within the State's jurisdiction, (b) are not subject to regulation by another public agency or body, or apply only to appurtenances such as roads and fences not germane to the safety of the structure.

1027. All plans and specifications for initial construction, enlargement, alteration, repair or removal of dams and supervision of contruction shall be in charge of a civil engineer, licensed by this State, experienced in dam design and construction, assisted by qualified engineering geologists and other specialists when necessary.

1028. No action shall be brought against the State or the Agency or its agents or employees for the recovery of damages caused by the partial or total failure of any dam or reservoir or through the operation of any dam or reservoir upon the ground that such defendant is liable by virtue of any of the following:

(a) The approval of the dam or reservoir, or approval of flood handling plans during construction.

(b) The issuance or enforcement of orders relative to maintenance or operation of the dam or reservoir.

(c) Control and regulation of the dam or reservoir.

(d) Measures taken to protect against failure during an emergency.

1029. Nothing in this Act shall be construed to relieve an owner or operator of a dam or reservoir of the legal duties, obligations, or liabilities incident to the ownership or operation of the dam or reservoir.

1030. The findings and orders of the Agency and the certificate of approval of any dam or reservoir issued by the State are final and conclusive and binding upon all owners, and State agencies, regulatory or otherwise, as to the safety of design, construction, maintenance, and operation of any dam or reservoir.

1031. Nothing in this Act shall be construed to deprive any owner of such recourse to the courts as he may be entitled to under the laws of this State.

1032. All records of official actions of the Agency and its correspondence pertaining to the supervision of dams and reservoirs are public documents.

1033. All owners shall notify the Agency of any change in ownership of any dam or reservoir subject to this Act at the time the transfer of ownership occurs.

Chapter 3. Administrative Provisions

1050. The Agency shall be administered and directed by a civil engineer, licensed by this State, experience in the design and construction of dams and reservoirs, and it shall employ such clerical, engineering, and other assistants as are necessary for carrying on the work of dam and reservoir supervision in accordance with this Act.

1051. When the safety considerations pertaining to a certificate of approval, dam, reservoir, or plans and specifications require it, or when requested in writing to do so by the owner, the Agency may appoint a consulting board of two or more consultants not previously associated with the structure, to report to the Agency on its proposed action with respect to these considerations.

1052. The cost and expense of a consulting board if appointed on the request of an owner shall be paid by the owner.

Chapter 4. Powers of the Agency

Article 1. Powers in General

1075. The Agency, under the police power of the State, shall review and approve the design, construction, enlargement, alteration, repair, maintenance, operation, and removal of dams and reservoirs for the protection of life and property as provided in this Act.

1076. All dams and reservoirs in the State are under the jurisdiction of the Agency, except those dams which are Federally owned.

1077. It is unlawful to construct, enlarge, repair, alter, remove, maintain, operate or abandon any dam or reservoir coming within the purview of this Act except upon approval of the Agency, provided that this section shall not be deemed to apply to routine maintenance and operation not affecting the safety of the structure.

1078. The Agency shall adopt and revise from time to time such rules and regulations and issue such general orders as may be necessary for carrying out, but not inconsistent with, the provisions of this Act.

1079. In making any investigation or inspection necessary to enforce or implement this Act, the Agency or its representatives may enter upon such private property of the dam owner as may be necessary.

1080. In determining whether a dam or reservoir or proposed dam or reservoir constitutes or would constitute a danger to life or property, the Agency shall take into consideration the following conditions, not necessarily all inclusive: the possibility that the dam or reservoir might be endangered by overtopping, seepage, settlement, erosion, cracking, earth movement, earthquakes, failure of bulkheads, flashboard, gates and conduits, which exist or which might occur in any area in the vicinity of the dam or reservoir. Whenever the Agency deems that any conditions endanger a dam or reservoir, it shall order the owner to take such action as necessary to the satisfaction of the Agency to remove the resultant danger to life and property.

Article 2. Investigations and Studies

1081. For the purpose of enabling it to make decisions as compatible with public safety and economy as possible, the Agency shall make or cause to be made such investigations and shall gather or cause to be gathered such data including advances made in safety practices elsewhere, as may be needed for a proper review and study of the various features of the design, construction, repair and enlargement of dams, reservoir, and appurtenances.

1082. The Agency shall also make or cause to be made from time to time such watershed investigations and studies as may be necessary to keep abreast of developments affecting stream run-off and as required to facilitate its decisions.

Article 3. Action and Procedure to Restrain Violations

1083. The Agency may take any legal action proper and necessary for the enforcement of this Act.

1084. An action or proceeding under this article may be commenced whenever any owner or any person acting as a director, officer, agent, or employee of any owner, or any contractor or agent or employee of such contractor is:

(a) Failing or omitting or about to fail or omit to do anything required of him by this Act or by any approval, order, rule, regulation, or requirement of the Agency under the authority of this Act, or

(b) Doing or permitting anything or about to do or permit anything to be done in violation of or contrary to this Act or any approval, order, rule, regulation, or requirement of the Agency under this Act.

1085. Any action or proceeding under this article shall be commenced in a court of appropriate jurisdiction in which (a) the cause or some part thereof arose, (b) the owner or person complained of has its principal place of business, or (c) the person complained of resides.

Chapter 5. Applications

Article 1. New Dams and Reservoirs or Englargements of Dams and Reservoirs

1100. Construction of any new dam or reservoir or the enlargement of any dam or reservoir shall not be commenced until the owner has applied for and obtained from the Agency written approval of plans and specifications.

1101. A separate application for each reservoir and its dams shall be filed with the Agency upon forms to be provided by it.

1102. The applications shall give the following information:

(a) The name and address of the owner.

(b) The location, type, size, and height of the proposed dam and reservoir and appurtenant works.

(c) The storage capacity and reservoir surface areas for normal pool and maximum high water.

(d) Plans for proposed permanent instrument installations in the dam.

(e) As accurately as may be readily obtained, the area of the drainage basin, rainfall and streamflow records and flood-flow records and estimates.

(f) Maps and general design drawings showing plans, elevations, and sections of all principal structures and appurtenant works or other features of the project in sufficient detail, including design analyses, to determine safety, adequacy and suitability of design.

(g) Such other pertinent information as the Agency requires, such as proposed time for commencement and completion of contruction.

1103. The Agency shall, when in its judgment it is necessary, also require the following:

(a) Data concerning subsoil and rock foundation conditions and the materials entering into construction of the dam or reservoir.

(b) Investigations of, and reports on, subsurface conditions, involving such matters as exploratory pits, trenches and adits, drilling, coring, geophysical surveys, tests to determine leakage rates, and physical tests to measure in place and in the laboratory the properties and behavior of foundation materials at the dam or reservoir site.

(c) Investigations of, and reports on, the geology of the dam or reservoir site and its vicinity, possible geologic hazards, including seismic activity, faults, weak seams and joints, availability and quality of construction materials, and other pertinent features.

(d) Such other appropriate information as may be necessary in a given instance.

1104. In instances wherein the physical conditions involved and the size of the dam or reservoir are such as to render the above requirements as to drainage areas, rainfall, streamflow, flood flow, and drilling or prospecting of site unnecessary, the Agency may waive the requirements.

1105. The application shall set forth the purpose or purposes for which the impounded or diverted water is to be used.

Article 2. Repairs, Alterations, or Removals

1106. Before commencing the repair, alteration or removal of a dam or reservoir, including the alteration or removal of a dam or reservoir so that it no longer constitutes a dam or reservoir as defined in Sections 1002 and 1004 of this Act, the owner shall file an application and secure the written approval of the Agency, except as provided in this article. Repairs shall not be deemed to apply to routine maintenance and operation not affecting the safety of the structure.

1107. The application shall give such pertinent information or data concerning the dam or reservoir, or both, as may be required by the Agency and such information as to other matters appropriate to a thorough consideration of the safety of such a change as may be required by the Agency.

1108. The application shall state the proposed time of commencement and of completion of remedial construction.

1109. The application shall give the name and address of applicant, shall adequately detail, with appropriate references to the existing dam or reservoir, the changes which it is proposed to effect, and shall be accompanied by maps and plans and specifications which shall be a part of the application and which shall be of such character and size and set forth such pertinent details and dimensions as the Agency may require. The Agency may waive any of the requirements of this section if found by it unnecessary.

1110. In case of an emergency where the Agency declares repairs or breaching of the dam are immediately necessary to safeguard life and property repairs or breaching shall be started immediately by the owner, or by the Agency at the owner's expense, if he fails to do so. The Agency shall be notified at once of proposed emergency repairs or breaching and of work under way when instituted by the owner.

1111. The proposed repairs, breaching and work shall be made to conform to such orders as the Agency issues.

Article 3. Approval of Applications

1112. Upon receipt of an application the Agency shall give its consideration thereto and shall approve or disapprove the same within the time provided in Section 1114.

1113. If an application is defective, it shall be returned to the applicant for such action as necessary to correct the defects, endorsed so that in order to retain its validity, it must be corrected and returned to the Agency within 30 days or such further time as may be given by the Agency. If the application is not returned, it shall be rejected.

1114. No applications shall be approved or disapproved in less than 30 days after the receipt of the fee required by Section 1125, but all applications shall be approved or disapproved as soon as practicable thereafter. At the discretion of the Agency hearings may be held on each application.

1115. Approvals shall be granted under terms, conditions, and limitations necessary to safeguard life and property.

1116. Actual construction shall be commenced within one year after date of approval; otherwise the approval becomes void.

1117. The Agency may, upon written application and for good cause shown, extend the time for commencing construction.

1118. Notice shall be given to the Agency at least ten days before construction is to be commenced and such other notices shall be given to the Agency as it may require.

Chapter 6. Fees

1125. The application for a new dam and reservoir or enlargement shall set forth the estimated net cost, as defined in this chapter, of the dam and reservoir or enlargement and shall be accompanied by a filing fee based upon the estimated cost and according to the following schedule: (Schedule below will of necessity vary in each State.)

(a) For the first one hundred thousand dollars ($100,000) a fee of 2 percent of the estimated costs.

(b) For the next four hundred thousand dollars ($400,000) a fee of 1½ percent.

(c) For the next five hundred thousand dollars ($500,000) a fee of 1 percent.

(d) For all costs in excess of one million dollars ($1,000,000) a fee of one-half of 1 percent.

In no case, however, shall the fee be less than one hundred dollars ($100) or more than fifty thousand dollars ($50,000).

1126. One fee only shall be collected for an enlargement to be affected by flashboards, sandbags, earthen levees, gates, or other works, devices, or obstructions which are, from time to time, to be removed and replaced or opened and shut and thereby operated so as to vary the surface elevation of the impounded water.

1127. For the purpose of this Act, the estimated net cost of the dam and reservoir or enlargement involved shall include the following:

(a) The cost of all labor and materials entering into the construction of the dam and appurtenant works or reservoir, including right of way.

(b) The cost of preliminary investigations and surveys.

(c) The cost of the construction plant properly chargeable to the cost of the dam or reservoir.

(d) Any and all other items entering directly into the cost of the dam or reservoir.

1128. Excluded from the costs listed in Section 1127 shall be the costs of:

(a) Right of way for other than the dam and reservoir.

(b) Detached or underground powerplants, including switchyards and substations.

(c) Electrical generating, or pump-generating machinery.

(d) Roads, railroads, helioports and landing strips affording access to the dam or reservoir.

1129. An application shall not be considered by the Agency until the filing fee is received. All or part of the filing fee may be returned to the applicant only if he withdraws or cancels the application any time prior to the start of construction. The amount of the refund will be determined by the Agency with due regard to funds actually expended by the Agency in consideration of the application.

1130. As soon as possible after giving the notice of completion required in Section 1150, the owner shall file an affidavit with the Agency stating the actual cost of the dam and reservoir or enlargement thereof in such detail as the Agency requires to determine whether a further fee is due. In the event the owner of a new or enlarged dam or reservoir, because of loss of records, recent change of ownership, or other causes beyond his control, is unable to report the actual cost of construction or enlargement, he shall file an affidavit to this effect, stating the reasons therefor, within thirty days after receiving a written request therefor from the Agency. The Agency shall then make its own appraisal of the cost of construction or enlargement and determine what further fee, if any, is required.

1131. In the event the actual cost exceeds the estimated net cost by more than 15 percent, a further fee shall be required by the Agency computed under the schedule set forth in Section 1125 upon the actual cost, plus a penalty of 15 percent of the actual cost. No further fee shall be required, however, if such fee is to be computed at less than twenty dollars ($20). Upon making a determination that a further fee is required, the Agency shall notify the owner that he may appear within sixty days thereafter before an authorized representative of the Agency to protest the amount of the fee, in whole or in part, determined by the Agency to be required, and the sufficiency of the appraisal upon which such determination was based.

1132. All filing fees and other charges collected under the provisions of this Act shall be paid into a special fund in the State Treasury immediately after the Agency has certified as to the correctness of the amounts received, to be available to the Agency for expenditure for the purposes authorized by this Act.

1133. The fees provided for in this article shall be required of all enumerated in the definition of owner in Section 1005 of this Act.

Chapter 7. Inspection and Approval

Article 1. New or Enlarged Dams and Reservoirs

1150. Immediately upon completion of a new dam and reservoir or enlargement of a dam and reservoir the owner shall give a notice of completion to the Agency, and as soon thereafter as possible shall file with the Agency a certificate signed by the responsible engineer supervising construction for the owner, certifying that the project was constructed in conformance with approved plans and specifications, accompanied by supplementary drawings or descriptive matter showing or describing the dam or reservoir as actually constructed, which shall include but not be limited to the following:

(a) A record of all geological boreholes and grout holes and grouting.

(b) A record of permanent location points, benchmarks and instruments embedded in the structure.

(c) A record of tests of concrete or other material used in the construction of the dam and reservoir.

(d) A record of seepage flows and embedded instrument readings.

1151. In connection with the enlargement of a dam and reservoir, the supplementary drawings and descriptive matter need apply only to the new work.

1152. A certificate of approval shall be issued by the Agency upon a finding by the Agency that the dam and reservoir are safe to impound water within the limitations prescribed in the certificate. No water shall be impounded by the structure prior to issuance of the certificate.

Article 2. Certificates of Approval

1153. Each certificate of approval issued by the Agency under this Act may contain such terms and conditions as the Agency may prescribe.

1154. The Agency may revoke or suspend any certificate of approval whenever it determines that the dam or reservoir constitutes a danger to life and property. Whenever it deems such action necessary to safeguard life and property, the Agency may also amend the terms and conditions of any such certificate by issuing a new certificate containing the revised terms and conditions.

1155. Before any certificate of approval is revoked by the Agency, the Agency shall hold a hearing. Written notice of the time and place of the hearing shall be mailed, at least twenty days prior to the date set for the hearing, to the holder of the certificate. Any interested persons may appear at the hearing and present their views and objections to the proposed action. Any petition to a court of appropriate jurisdiction to inquire into the validity of action of the Agency revoking a certificate of approval shall be commenced within thirty days after service of notice of the revocation on the holder of the certificate.

Article 3. Repaired or Altered Dams and Reservoirs

1156. Immediately upon completion of the repair or alteration of any dam or reservoir, the owner shall give notice of completion to the Agency and as soon thereafter as possible shall file with the Agency a certificate signed by the responsible engineer supervising the work for the owner that the repairs or alterations were completed in accordance with the approved plans and specifications, accompanied by supplementary drawings or descriptive matter showing or describing the dam or reservoir as actually repaired or altered together with such maps, data, records, and information pertaining to the dam or reservoir as repaired or altered as the Agency requires.

1157. A certificate of approval shall be issued by the Agency upon a finding by the Agency that the dam and reservoir are safe to impound water within the limitations prescribed in the certificate. Pending issuance of a new certificate of approval, the owner of the dam or reservoir shall not, through action or inaction, cause the dam or reservoir to impound water beyond the limitations prescribed in the existing certificate.

1158. The certificate of approval shall supersede any previous certificate of approval issued for the dam or reservoir so repaired or altered.

Article 4. Removal of Dams and Reservoirs

1159. Upon completion of the removal of a dam or complete drawdown of a reservoir such evidence as to the manner in which the work was performed and as to the conditions obtaining after the removal as the Agency requires shall be filed with the Agency.

1160. This evidence shall show that a sufficient portion of the dam has been removed to permit passage of floods down the watercourse, across the site where the dam was located, and on downstream, the flow to be within flooding criteria required by the Agency.

1161. Before final approval of the removal of a dam or reservoir is issued, the Agency shall inspect the site of the work and determine that it fully conforms with its flood criteria.

Article 5. Complaints as to Unsafe Conditions

1162. Upon receipt of a written complaint alleging that the person or property of the complainant is endangered by the construction, enlargement, repairs, alterations, maintenance, or operation of any dam or reservoir the Agency shall cause an inspection to be made unless the data, records, and inspection reports on file with it are found adequate to make a determination whether the complaint is valid.

1163. If the Agency authorizes an inspection the complainant shall deposit with the Agency a sum estimated by it to be sufficient to cover costs of the inspection. The Agency may utilize independent consultants of its selection to make the inspection and a report to the Agency.

1164. If it is found that an unsafe condition exists, the Agency shall notify the owner to take such action as is necessary to render or cause the condition to be rendered safe, including breaching or removal of any dam found beyond repair, and any money deposited to secure an inspection shall be returned.

1165. If, after an inspection is made on account of a complaint, the complaint is found by the Agency to have been without merit, the cost therefor shall be payable into the Special Fund in State Treasury from the money deposited, with any excess returned to the complainant. The Complainant will be provided with a copy of the official report of the inspection.

Article 6. Inspection During Progress of Work

1166. During the construction, enlargement, repair, alteration, or removal of any dam or reservoir the Agency shall make either with its own engineers or by consulting engineers or engineering organizations, periodic inspections at State expense for the purpose of ascertaining compliance with the approved plans and specifications. The Agency shall require the owner to perform at his expense such work or tests as necessary, provide adequate supervision during construction by a civil engineer registered or licensed by the laws of this State, and to disclose information sufficient to enable the Agency to determine that conformity with the approved plans and specifications is being secured.

1167. If, after any inspections, investigations, or examinations, or at any time as the work progresses, or at any time prior to issuance of a certificate of approval it is found by the Agency that amendments, modifications, or changes are necessary to ensure safety, the Agency may order the owner to revise his plans and specifications, provided, however, the owner may, pursuant to Section 1051, request an independent consulting board to review the order of the Agency.

1168. If conditions are revealed which will not permit the construction of a safe dam or reservoir the Agency's approval shall be revoked.

1169. In the event that conditions imposed may be waived or made less burdensome in its judgment without sacrificing safety, the Agency may authorize an owner to revise the plans and specifications accordingly.

1170. If at any time during construction, enlargement, repair, or alterations of any dam or reservoir the Agency finds that the work is not being done in accordance with the provisions of the original approved plans and specifications or in accordance with the approved revised plans and specifications, it shall give a written notice thereof and order compliance by registered or certified mail or by personal service to the owner.

1171. The notice and order shall state the

particulars in which the original approved plans and specifications or the approved revised plans and specifications are not being or have not been complied with and shall order the immediate compliance with the original approved plans and specifications or with the approved revised plans and specifications as the case may be.

1172. The Agency may order that no further work be done until such compliance has been effected and approved by the Agency.

1173. A failure to comply with the approval and approved plans and specifications shall render the approval subject to revocation by the Agency, if compliance is not made in accordance therewith after notice and order from the Agency as provided in this article. If compliance is not forthcoming in a reasonable time, the Agency may order the incomplete structure removed sufficiently to eliminate any safety hazard to life or property.

Chapter 8. Maintenance, Operation and Emergency Work

Article 1. Maintenance and Operation

1174. Supervision over the maintenance and operation of dams and reservoirs in this State, other than those owned by the Federal Government, insofar as necessary to safeguard life and property from injury by reason of the failure thereof is vested in the Agency.

1175. The Agency shall require owners or their agents to keep available and in good order records of original and any modification construction and to report annually with respect to maintenance, operation and engineering including piezometric data and geologic investigations. The Agency shall issue such rules and regulations and orders as necessary to secure adequate maintenance, operation and inspection by owners or their agents and shall require engineering and geologic investigations by owners or their agents which will safeguard life and property. In addition, the owner of a dam or reservoir or his agent shall fully and promptly advise the Agency of any sudden or unprecedented flood or unusual or alarming circumstances or occurrence existing or anticipated which may affect the dam or reservoir.

1176. The Agency, from time to time, but not less often than once every five years, either with its own engineers, or by consulting engineers or engineering organizations, shall make inspections of dams and reservoirs at State expense for the purpose of determining their safety but shall require owners to perform at their expense such work as may reasonably be required to disclose information sufficient to enable the Agency to determine conditions of dams and reservoirs in regard to their safety and to perform at their expense other work which may reasonably be required, including installation of instruments necessary to secure maintenance and operation which will safeguard life and property.

Article 2. Emergency Work

1177. The Agency shall be responsible for determining that an emergency exists and through normal disaster communication channels shall warn the public, immediately employing any remedial means necessary to protect life and property, if in its judgment either:

(a) The condition of any dam or reservoir is so dangerous to the safety of life or property as not to permit of time for the issuance and enforcement of an order relative to maintenance or operation.

(b) Passing or imminent floods or any other condition which threaten the safety of any dam or reservoir.

1178. In applying the remedial means provided for in this article, the Agency may in emergency with its own forces, or by other means at its disposal, do any of the following:

(a) Take full charge and control of any dam or reservoir.

(b) Lower the water level by releasing water from the reservoir.

(c) Completely empty the reservoir.

(d) Perform any necessary remedial or protective work at the site.

(e) Take such other steps as may be essential to safeguard life and property.

1179. The Agency shall continue in full

charge of such dam or reservoir, or both, and its appurtenances until they are rendered safe or the energency occasioning the action has ceased and the owner is able to take back such operations. The Agency's take over will not operate to relieve the owner of a dam or reservoir of liability for any negligent acts of the owner or his agents.

1180. The cost and expense of the remedial means provided in this article, including cost of any work done to render a dam or reservoir or its appurtenances safe, shall be collected by presentation of bills to owners in the same manner as other debts to the State are recoverable, provided that if such bills are not promptly paid by the owners in the same manner as other debts to the State are recoverable, provided that if such bills are not promptly paid by the owners the cost shall be recovered by the State from the owner by action brought by the Agency in a court of appropriate jurisdiction.

Chapter 9. Offenses and Punishment

1185. Every person who violates any of the provisions of this Act or of any approval, order, rule, regulation, or requirement of the Agency is guilty of a misdemeanor and punishable by a fine of not more than _____ ($___) or by imprisonment in _____. In the event of a continuing violation each day that the violation continues constitutes a separate and distinct offense.

1186. Any person who wilfully obstructs, hinders, or prevents the Agency or its agents or employees from performing the duties imposed by this Act or who wilfully resists the exercise of the control and supervision conferred by this Act upon the Agency of its agents or employees is guilty of a misdemeanor and punishable as provided in this article.

1187. Any owner or any person acting as a director, officer, agent, or employee of an owner, or any contractor or agent or employee of a contractor who engages in the construction, enlargement, repair, alteration, maintenance, or removal of any dam or reservoir, who knowingly does work or permits work to be executed on the dam or reservoir without an approval or in violation of or contrary to any approval as provided for in this Act, or any inspector, agent, or employee of the Agency who has knowledge of such work being done and who fails to immediately notify the Agency thereof is guilty of a misdemeanor and punishable as provided in this article.

Chapter 10. Dams and Reservoirs Existing Prior to the Effective Date of this Law

Article 1. Dams and Reservoirs Completed Prior to Effective Date of this Law

1200. Every owner of a dam or reservoir that falls within the definition of a dam or reservoir in Sections 1002 and 1004 of this Act that was completed prior to the effective date of this Law shall immediately file an application with the Agency for the approval of such dam or reservoir.

1201. A spearate application for each reservoir and its dams shall be filed with the Agency upon forms to be supplied by it and shall include or be accompanied by such appropriate information concerning the dams or reservoirs as the Agency requires.

1202. The Agency shall give notice to file an application to owners of such dams or reservoirs who have failed to do so as required by this article, and a failure to file within thirty days after such notice shall be punishable as provided in this Act.

1203. The notice provided for in this article shall be given by certified mail to the owner at his last address of record in the office of the county assessor of the county in which the dam is located and such mailing shall constitute service.

1204. The Agency shall make inspections of such dams or reservoirs at State expense.

1205. The Agency shall require owners of such dams or reservoirs to perform at their expense such work or tests as may reasonably be required to disclose information sufficient to enable the Agency to determine whether to issue certificates of approval or to issue orders directing further work at the owner's expense necessary to safeguard life and property. For this purpose, the Agency may require an owner to lower the water level of, or to empty, the reservoir.

1206. If, upon inspection or upon completion to the satisfaction of the Agency of all work that may be ordered, the Agency finds that the dam and reservoir are safe to impound water, a certificate of approval shall be issued. The owner of the dam or reservoir shall not, through action or inaction, cause the dam or reservoir to impound water following receipt by the owner of a written notice from the Agency that a certificate will not be issued because the dam or reservoir will not safely impound water. Before such notice is given by the Agency, the Agency shall hold a hearing. Written notice of the time and place of the hearing shall be mailed, at least twenty days prior to the date set for the hearing, to the owner of the dam or reservoir. Any interested persons may appear at the hearing and present their views and objections to the proposed action.

Article 2. Dams and Reservoirs Under Construction Before Effective Date of this Law

1207. Any dam or reservoir that falls within the definition of a dam or reservoir in Sections 1002 and 1004 of this Act and which the Agency finds was under construction and based on its findings not 90 percent constructed on the effective date of this Law shall, except as provided in Section 1208, be subject to the same provisions in this Act as a dam or reservoir commenced after that date. Every owner of such a dam or reservoir shall file an application with the Agency for the Agency's written approval of the plans and specifications of the dam or reservoir.

1208. Construction work on such a dam or reservoir may proceed, provided an application for approval of the plans and specifications therefor is filed, until a certificate of approval is received by the owner from the Agency approving the dam and reservoir or an order is received by the owner from the Agency specifying how the construction must be performed to render the dam or reservoir safe. After receipt of an order specifying how construction of the dam or reservoir must be performed, work thereafter must be in accordance with the order.

1209. Such dams or reservoirs as are based on Agency findings 90 percent or more constructed on the effective date of this Law shall be subject to the same supervision as dams or reservoirs which were completed prior thereto.

Article 3. Fees for Dams or Reservoirs Under Construction Before Effective Date of this Law

1210. The owners of dams or reservoirs that, based on Agency findings, are 90 percent or more constructed on the effective date of this Act and that are subject to the provisions of this Act shall be be required to pay a fee but shall submit an application for approval and issuance of a State certificate as provided in Section 1209. Applications for the approval of dams and reservoirs that are made subject to this Act that are found by the Agency to have been less than 90 percent constructed on the effective date of this Law shall be accompanied by fees as much less than provided for dams and reservoirs commenced after that date as the percentage of construction found by the Agency to have been completed on that date.

Index

Index

Index

Act of Congress (Environmental Policy) PL 91-190, 91-94
aerial photographs—construction materials, 150
aggregates, concrete, 173, 196
alternatives, environmental impact, 80
analysis—storms of record, 121-124
    flood runoff, 124, 125
appendix—chapter 12—specificiations for construction, 685-752
appurtenant structures—earth dams, 313-314
arch dams—concrete, 393
    see also concrete dam design
arch dam stress analysis system (ADSAS), 408

basic data collection—planning, 46
benefit-cost analysis, 33
benefits, project, 8
bibliography (see also references)
    chapter 1   96-97
    chapter 4.02   202-204
    chapter 9   525-526, 548
    chapter 11   664-668
buttress dams, 445

cement, Portland, types (Table 8-7-1), 473
commitments of resources, 90
concrete aggregates, 173, 196
concrete dam analysis
    arch dam stress analysis system (ADSAS), 408, 409-413
    dynamic analysis, 423-431
    finite element method, 408, 413-415
    foundation analysis, 431-436
    temperature, 415-423
    trial-load method, 408-409
concrete dam design
    definitions, 385-386
    general, 385
    selection of site and type of dam, 388
concrete dam design—arch dams
    description, 393
    fundamental principles, 408
    layouts, 394
    level of design, 394
    required data, 394
concrete dam design—buttress dams
    buttresses, 457-458

    deck and transition section, 449-457
    dynamic analysis, 458
    general, 445-446
    layout, 446-449
    miscellaneous design features, 458-460
concrete dam design criteria
    cracking, 391, 392
    earthquake, 391
    factors of safety, 392
    load combinations, 391
    loads, 390
    material properties, 389
concrete dam design—gravity dams
    gravity method, 439-445
    layout, 437-439
    methods of analysis, 439
concrete dam foundations
    effective deformation modulus (EDM), 462
    geologic investigation, 235-248, 460-461
    stability, 463
    testing, 461-462
    treatment, 463-469
concrete dam instrumentation
    embedded instruments, 487-489
    general, 486-487
    other instruments, 494-496
    precise survey methods, 489-493
    uplift pressure measurements, 493
concrete dam openings
    galleries, 481-485
    joints, 475-481
    openings, 481-485
    waterways, 485
concrete dam references
    chapter 8, 497-498
concrete dam temperature control
    control, 469-475
    cooling, 421-423, 470-474
    Portland cement types, (Table 8-7-1), 473
construction materials
    classification of rocks, 166
    concrete aggregates, 173
    environmental restraints, 185
    for embankments, 178
    general, 149
    investigations, 149, 195
    sampling and logging, 167

788   INDEX

construction materials (*continued*)
　soil classification, 157
　sources, 195
　subsurface exploration, 155
　surface exploration, 151
construction procedures and equipment
　construction equipment, 759-762
　　conveyor belts, 762
　　excavators, 760
　　dump trucks, 762
　　loaders, 760, 761
　　scrapers, 760-761
　　transportation, 761
　construction organization, 756-758
　　administrative officer, 759
　　field engineer, 759
　　general superintendent, 756-757
　　labor relations, 758
　　office engineer, 759
　　office manager, 757
　　project engineer, 757
　　purchasing agent, 757-758
　　quality control, 758
　　resident engineer, 758-759
　　safety engineer, 757
　　supervision—the contractor, 756
　　supervision—the owner, 758-759
　construction plant—installation, 763-765
　　administrative structures, 763
　　craft buildings and change houses, 764
　　electric power and lighting, 764-765
　　shops and warehouses, 764
　　water and air systems, 764
　construction scheduling, 754-756
　　critical path, 755
　　initial construction schedule, 754-755
　　working construction schedule, 755-756
　foundation excavations, 762-763
　　compressed air stations, 662, 663, 763
　　general, 762
　　rock drilling, 762-763
　introduction, 753-754
　problems in dam construction, 765-770
　　earth dams, 765
　　　diversion and care of river, 765-766
　　　embankment construction, 766
　　　foundation excavation, 766
　　　foundation grouting, 766
　　concrete dams, 768
　　　concreting, 769
　　　diversion and care of river, 768
　　　foundation treatment, 769
　　　installation of mechanical items, 760
　　　selection of construction plant, 768-769
costs
　allocation method, 36
　environmental improvements, 40
　environmental protection, 40
　estimates, 28
　limited analysis, 36
　principles equitable cost allocation, 41
　project, 13
critical path, 755

dam failures, 187-188, 248, 771-772
dam sites and sizes
　site investigation, 204
　site selection, 1, 2, 6, 15, 44, 90, 267-289
　size, 1, 2, 8, 90
　see also selection of dams
dam types
　composite, 287-289
　concrete—arch, 275-280
　concrete—buttress, 284-287
　concrete—gravity, 280-283
　embankment, 270-273, 291-318
　rockfill, 273-275
design criteria—concrete dams, 388-393
design criteria—project functions, 21
　aeration of water, 27
　cost estimates, 28
　earthquake effects, 261-265
　fishery requirements, 23
　flood control, 22
　hydroelectric power, 23
　river navigation, 23
　water conservation, 22
design floods, 117-120
　diversion, 118
　general, 117
　reservoir, 119
　spillway, 119-120
development of design floods, 131-134
　addition of base flow, 133-134
　application unit hydrograph, 133
　determination of unit hydrograph, 132-133
　losses, 131-132
　procedures, 131
development of design storms, 125-131
　adjustments in storm transposition, 126
　general, 125
　statistical storm maximization, 128-131
　transposition of depth-area-duration relations, 125
　transposition of isohyetal patterns, 125
　transposition of maximized depth-area relations, 126
diversion and care of river—concrete, 768
diversion and care of river—earth, 765-766
dynamic analysis—arch dams, 423-431
dynamic analysis—buttress dams, 458

earth dam design
　appurtenant structures, 313-314

INDEX 789

compaction requirements, 309-311
construction and maintenance, 314-316
criteria, 291-292
embankment, 302-313
embankment cracking, 295-296
embankment dams, 303-308
field and laboratory investigations, 296-297
foundations and abutments, 298-302
general considerations, 292-296
instrumentation, 316-318
sections, 304
seismic problems, 295
selection of type, 292
slope protection, 311-313
stage construction, 294
test fills, 297
earthfill materials, 198
earthquake, nature of
concrete dam, 391
generation, 251-253
ground motions, 256-261
hazard, 248-266
magnitude and intensity, 253-255
acceleration table, 259
intensity scale, 254
magnitude table, 253
reservoir loadings, 249-251
economic evaluations, 32, 62, 66, 67, 70
embankment materials, 178
environmental categories and classifications, 63
Battelle-Columbus labs, 64
Geological Survey, 63
Water Resource Council, 63, 89
Wild and Scenic River classification, 64
environmental effects, 72
adverse, 72-78
adverse—summary table, 79
beneficial, 72
beneficial and adverse, 72
environmental factors
cost allocations, 41
cost of improvements, 40
cost of protection, 40
plan formulation, 17
environmental impact alternatives, 80-84
environmental uses—relationship short-term and long-term productivity, 84
environmental investigations, 61
concern with dam site and sizes, 62
purpose, objective, definitions, 61
Environmental Policy Act of 1969 (PL 91-190), 91-94
Environmental Quality Improvement Act of 1970 (PL 91-224), 94-96
equipment, construction, 759-762
evaporation map, U.S. water surface, 114-115

explorations
concrete dam sites, 235-248
drilling in bedrock, 222-226
earth dam sites, 230-235
geophysical, 207-214
powerhouse site, 228-230
soil, 214-221
sources of construction materials, 195-204

factors of safety, concrete dams, 392
finite element method of analysis, 408-409
fishery requirements, 23
flood control, 22, 29
design, 117
runoff, analysis, 124, 125
flood, development of design, 131-134
flood discharges, estimates past floods, 102
foundation analysis, concrete arch dams, 431-436
foundation excavations, 762-763, 766-769
foundation investigations, 187, 230-248
foundation treatment, rockfill dams, 357
foundations and abutments, earth dams, 298-302

galleries in concrete dams, 481-485
geologic investigations, 187-189, 235-248, 460-461
geologic mapping, 204-207, 213, 620
geophysical exploration, 136, 207
gravity dams—concrete, 437-445

hydroelectric power, 23, 549-562
hydrograph, 104, 132, 133
hydrologic events, 108
frequency analysis, 108-113
hydrologic studies, 99, 108

ice pressure, 391
investigations
construction materials, 149, 195
environmental, 61
feasibility grade, 2, 20
foundation, 230-234
geologic (concrete dams), 460-461
geologic, foundation and seismicity, 187
geophysical, 207-214
preconstruction, 2
programming, 59
reconnaissance grade, 2, 18
instrumentation
concrete dams, 486-496
earth dams, 316-318
rockfill dams, 369

joints, concrete dams, 475-481

laboratory testing—rockfill dams, 361
locks adjacent to dams

locks adjacent to dams (*continued*)
  design—electrical systems, 610-614
    communication system, 614
    electrical distribution, 614
    electrical requirements, 610-614
    grounding system, 614
    sources of power, 610
  design—general, 562-567
    filling and emptying systems, 566
    lock types, 565-566
    scope, 563-564
    table—standard lock sizes, 564
  design—hydraulic, 567-579
    bottom longitudinals, 575
    discharge manifolds, 571
    general design discussion, 567-569
    hydraulic characteristics of culvert valves, 575
    intake manifolds, 569-571
    lock chamber laterals, 574
    lock chamber wall ports, 571-574
    multiple wall ports, 574-575
  design—lock appurtenances, 614-618
    air bubblers, 616
    esplanade, 615
    guard rail, 616
    ladders, 616
    navigation aids, 617-618
    recess covers, 615-616
    snubbing facilities, 614-615
    wall face protection, 615
    water and compressed air systems, 616-617
    water level indicators, 616

locks
  design—lock gates and culvert valves, 588-598
    lock gate types, 589
    miter gates, 589, 590-598
    rolling gates, 589
    tainter gates, 589-590
    vertical lift gates, 590
  design—lock walls and sills, 580-588
    approach walls, 581-582
    culvert intake and discharge walls, 581
    design criteria, 583-585
    loadings, 585-587
    lock chamber walls, 580
    lock sills, 582-583
    reinforcing of gravity walls, 587-588
    upper and lower gate bay walls, 580-581
  design—miter gate and culvert valve machinery, 598-610
    culvert valve machinery, 603-610
    miter gate machinery, 598-603

maintenance—earth dams, 314-316
maps—evaporation water surface, 114-115
  construction materials, 149-150
  geologic, 204-207

medium head power plants, 554-558
model law—safety of dams, 775-786

National Environmental Policy Act of 1969 (PL 91-190), 91-94
navigation, 23, 617-618

operation and routing studies, 113-117
organization, construction, 756-758
Oroville Dam, California, 343-345, 505, 557
outlet works
  arrangement, 530-531
  bibliography, 548
  conditions determining layout, 529-530
  controls, 521-535, 537-539
  determination capacities, 526-527
  entrance and outlet channels, 542
  functions, 526
  intake structures, 539-541
  relation to reservoir storage levels, 527-529
  structural design details
    cut-and-cover conduit details, 545-548
    general, 542-543
    tunnel details, 543-545
  terminal structures and dissipating devices, 541-542
  waterways, 535-537

plan formulation, 4, 13
  analysis, 14
  environmental factors, 17
planning
  basic data collection, 46
  preconstruction, 2, 46
  process, 3
  sedimentation, 48, 49
  water rights, 50
planning and environmental studies, 1-97
power
  generation, 16-17
  hydroelectric, 23, 549-562
power plants adjacent to dams
  basic design principles, 549
  low head arrangements, 553, 561
  medium head arrangements
    Aswan, 556
    Boundary, 554
    Bull Shoals, 555
    Castaic, 558
    Churchill Falls, 558
    Estreito, 555
    Glen Canyon, 555
    Grand Coulee, 556
    Guri, 556
    Kemano, 558
    Krashoyarsk, 556
    Ludington, 556
    Mica, 557
    Mossyrock, 553

INDEX 791

Northfield Mountain, 557
Nurek, 551, 557
Oroville, 557
Plyavinsk, 554
nonembedded penstocks, 561
powerhouses, 551, 561
pumped storage arrangements, 559
river basin developments
    Arkansas, 558
    Coosa, 559
    James Bay, 559
    Kama-Volga, 558
    Missouri, 558
    Savannah, 559
    Tennessee, 558
surface plants, 561-562
surge tanks, 551, 561
underground plants, 562
water passages, 550, 560
problems in dam construction, 765-770
    concrete dams, 768
    earth dams, 765
project
    benefits, 9
    costs, 13
    effects, 2
    staged development, 16
project functions, basic data, 57-59

quality control—construction, 758

references (*see also* bibliography)
    chapter 2,   146, 147
    chapter 3,   186
    chapter 4.01,   194-195
    chapter 4.031,   207
    chapter 4.032,   214
    chapter 4.034,   226
    chapter 4.11,   234-235
    chapter 4.12,   247-248
    chapter 4.2,   265-266
    chapter 5,   289
    chapter 6,   318
    chapter 7,   378-384
    chapter 8,   497-498
    chapter 10A,   553, 560, 562
    chapter 10B,   618
repayment of costs
    analysis, 34
    rates, 34
reservoir
    geologic data, 51
    geologic investigations, 189
    landslides, 190-191
    leakage, 190
    measurement and computation, 56
    seismicity, 191-192
    sites, 7

sizing, 17
subsidence, 192-194
reservoir design—associated structural works
    boat landing facilities, 637-638
    campgrounds, 638
    circulation roads, 639
    headquarters, 639-640
    operation and maintenance facilities, 645
    picnic areas, 638
    recreation facilities, 636
    remote camps, 640
    sanitary facilities, 637, 640
    swimming beaches, 636-637
    trails, 641-642
    visitor facilities, 642, 643
    water supply, 637, 639, 641
reservoir design—clearing
    air pollution, 652
    burn disposal, 654, 657-658
    factors of influence, 649
    fire safety, 652
    large reservoirs, 650
    noise pollution, 653
    small reservoirs, 650
    tree collection, 656-657
    tree felling, 655-656
reservoir design—environmental considerations
    cultural impact, 662
    effect on climate, 661
    field data, 659-661
    fish and wildlife, 661
    water quality, 662
reservoir design, general
    bank storage, 626
    consequential damages, 630
    considerations, 619-620
    definition, 619
    earthquakes caused by, 625
    evaporation, 626
    geodetic studies, 625
    geologic maps, 620
    grouting, 624
    miscellaneous relocations, 634
    procedure—relocating state highway, 632
    potential slides, 621-622
    relocations, 630-636, 643
    relocation agreement, 631
    reservoir lining, 624
    rights-of-way, 628-630
    rim stability, 620
    sedimentation, 627
    seismic studies, 625
    seismicity, 624-626
    structural integrity, 620
    transportation, 632
    utilities, 631
    water holding capability, 623
reservoir design, relocations

reservoir design, relocations (*continued*)
   churches, parks, schools, and other public facilities, 635
   entire community, 635
   general, 632, 634, 635
river basin developments, 552, 558-559
rock classification, construction materials, 166
rockfill dam design
   asphaltic concrete deck, 332-336
   concrete facings, 327-328
   cutoff wall, 327
   definition and history, 319-320
   deformation and cracking of dams, 373
   early dams with impervious facings, 320
   earth core dams, 339-354
   excavation and surface preparation, 357-359
   factors influencing alignment and cross section, 355-357
   field tests, 364
   foundation treatment, 357
   grouting, 359
   impervious facings, 338
   instrumentation, 369
   laboratory testing, 361
   materials, 320-322, 360-361
   modern concrete face, 328-332
   placed rock, 326-327
   placement of dumped rockfill, 323-325
   settlement of dumped rockfill, 325-326
   stability analysis, 365
   steel deck, 337-338
   timber deck, 336
rockfill materials, 198, 320-322
routing of spillway design flood, 134
runoff records
   analysis, 104
   extending period, 102
   flow-duration curves, 105
   hydrograph, 104
   mass curve, 106
   storage draft curve, 108

safety factors—concrete dams, 392
safety of dams
   introduction, 771-773
   model law for states, 775-786
   safety controls—Federal, 773-774
   safety controls—State, 774-775
   USCOLD, 774
sedimentation, general, 8, 48, 49
sedimentation in reservoirs, 142-146, 627
sedimentation—table, rates of reservoir sedimentation, 146
seismic hazards, 248-266
seismicity investigations, 187
seismic problems—earth dams, 295
seismic zoning map of the United States, 250

selection of dams
   concrete dams—arch, 275-280
   concrete dams—buttress, 284-287
   concrete dams—composite, 287-289
   concrete dams—gravity, 280, 283
   embankment dams, 270-273
   general, 267-270
   process, 1, 44
   rockfill dams, 273-275
   site, 1, 2, 6, 15, 44, 90, 267-289
   size, 1, 2, 8, 44, 90
selection of type—earth dams, 292
selection of type—concrete dams, 388
sites for appurtenant structures, 226
sizes—standard—for shiplocks, 564
soil classification—construction materials, 157, 220
soil classification system—unified, 160, 221
soil triangle, 151
specifications for dam construction
   appendix for chapter 12, 685-752
     notice to contractors, 691-693
     special provisions, 730-752
     standard provisions, 694-729
     table of contents, 685-690
   categorization, 669-671
   functions, 669
   model contract for dam construction, 671
   special provisions, 680-681
     priority of special provisions, 676, 696
   standard provisions, 675-680
     applicable laws and regulations, 676
     changes and unforeseen site conditions, 678-679
     claims, 679-680
     contractual relation of parties, 676
     control of the work, 677-678
     definitions, 675-676
     interpretation of contract, 676
     measurement and payment, 680
     priority of standard provisions, 676, 696
     prosecution of the work, 676-677
   technical provisions, 671-675
     clarity, 674
     combined performance and design, 673-674
     completeness, 674-675
     consistency, 675
     design, 672-673
     performance, 671-672
     principles, 674
     priority of technical provisions, 676, 696
     review, 675
     theoretical view, 671
spillways, design of
   controlled crests
     flashboards and stoplogs, 517-519
     general, 517
     radial gates, 519
     rectangular lift gates, 519

emergency spillways, 504-505
flood routing 134-137, 501-502
function of spillways, 499
relation surcharge storage to spillway capacity, 500-501
selection of inflow design flood, 500
selection of size and type, 502-504
service spillways
  components, 506-510
  selection of layout, 505-506
spillway bibliography, 525-526
spillway types
  baffle apron drop, 515
  chute (open channel), 512-513
  conduit and tunnel, 513-514
  culvert, 515-516
  drop inlet (Morning Glory), 514-515
  general, 510
  ogee (overflow), 511
  side channel, 511-512
  siphon, 516-517
  straight drop, 510
structural details
  crest structure and walls, 520-521
  cutoffs, 524
  general, 519-520
  open channel linings, 521-524
  riprap, 524-525
standard sizes shiplocks, table, 564
storage selection—final, 113

storms, development of design, 125-131
storms of record, analysis, 121-124
stream gauging, 6, 100, 101

temperature control—concrete dams, 469-475
topographic maps—location construction materials, 149
Toronto power plant, 550
trial-load analysis method, 408-409
trial-load twist analysis method, 439
tunnels, alignment studies, 227
tunnels, outlet works, 543-545
tunnels, spillways, 513-514

unified soil classification system, 160
USCOLD, 774

visitor facilities, 642, 643, 750-752

water
  aeration, 27
  conservation, 22
  rights, 50
water resource development
  dams—key feature, 3
  evaluation of records, 99
  purpose, 2
  revision of existing records, 100
water supply, available for development, 5
wave action—dam freeboard, 137-147
wild and scenic river class, 64